P9-DCM-678

About Wine

Join us on the web at

culinary.delmar.com

About Wine

J. Patrick Henderson

Dellie Rex

THOMSON
DELMAR LEARNING™

Australia Canada Mexico Singapore Spain United Kingdom United States

About Wine
by J. Patrick Henderson and Dellie Rex

Vice President, Career Education Strategic Business Unit:
Dawn Gerrain

Acquisitions Editor:
Matthew Hart

Managing Editor:
Robert Serenka, Jr.

Product Manager:
Patricia M. Osborn

Editorial Assistant:
Patrick B. Horn

Director of Production:
Wendy A. Troeger

Content Project Manager:
Matthew J. Williams

Project Editor:
Maureen M.E. Grealish

Technology Project Manager:
Sandy Charette

Director of Marketing:
Wendy E. Mapstone

Marketing Channel Manager:
Kristin B. McNary

Marketing Coordinator:
Scott A. Chrysler

Cover and Text Design:
Potter Publishing Studio

Cover Photography:
© James Scherzi Photography, Inc. 2005.

Printed in the United States of America
1 2 3 4 5 XXX 10 09 08 07 06

For more information contact Delmar Learning,
5 Maxwell Drive, Clifton Park, NY 12065-2919.

Or find us on the World Wide Web at
www.delmarlearning.com. or
www.culinary.delmar.com

Library of Congress Cataloging-in-Publication Data

Henderson, J. Patrick.
About wine/by J. Patrick Henderson and Dellie Rex.—1st ed.
p. cm.
ISBN 1-4018-3711-5 (alk. paper)
1. Wine and wine making. I. Rex, Dellie. II. Title.
TP548.H395 2006
641.2'2--dc22 2006016239

NOTICE TO THE READER

DEDICATION

This book is dedicated to the memory of William S. Gray.

William S. Gray, known as "Scotty" to his friends, spent more than 30 years in the field of hotel management. In addition to his work in the business of hospitality, William Gray also taught courses in the subject of Hotel and Resort Management at New York University, Purdue, and Southern Vermont College. He used the knowledge he acquired in his professional and academic career as author of *Hospitality Accounting* and as coauthor of *Hotel & Motel Management and Operations.* In 2000 his lifelong interest in wine inspired him to begin work on his third book, the text that was ultimately to become *About Wine.* Before he could complete his project, William Gray passed away in 2002. He is survived by his wife of 37 years, Diane Gray. The authors wish to gratefully acknowledge William Gray's vision for the book and his contribution toward its publication.

LETTER FROM THE EDITOR

Thomson Delmar Learning is excited to announce the *About* series, the first installment in a robust line of culinary arts textbooks from a leader in educational publishing. You'll soon discover why it's all *About* baking, garde manger, and wine! These essential textbooks for culinary arts students present the tools and techniques necessary to ensure success as a culinary professional in a highly visual, accessible, and motivating format. It is truly the first culinary arts series written for today's culinary arts students.

The following principles represent the vision behind the *About* series:

- A highly visual, accessible, and motivating format
- A comprehensive instructor support package that provides the tools necessary to make your life easier and your in-class time as effective and stimulating as possible
- A thorough and complete review by a team of academic and industry professionals ensuring that the *About* series is the most up-to-date and accessible culinary arts series ever published

About Wine **by J. Patrick Henderson and Dellie Rex.** This introductory wine textbook presents culinary arts and hospitality students with practical and detailed knowledge necessary to manage wine and wine sales. The five distinct sections of the text cover the basics of wine, the wine regions of the world, types of wine, and the business of wine. Special features of *About Wine* include detailed color diagrams, maps, and photographs throughout to keep the text interesting and engaging. Useful appendices designed for use as a quick reference or as a basis for more research are also included, making this text a valuable resource even after formal training has ended.

About Professional Baking **by Gail Sokol.** With over 700 full-color photographs demonstrating best practices and key techniques, students will be motivated and prepared for each and every baking laboratory exercise. Features include profiles of professional bakers; an entire chapter on *mise en place;* 125 fully kitchen-tested recipes written in an easy-to-comprehend format; clearly stated objectives and key terms presented at the beginning of each chapter; and hundreds of detailed step-by-step procedural photographs.

Modern Garde Manger **by Robert Garlough and Angus Campbell.** This innovative and comprehensive text is designed to meet the educational needs of both culinary arts students and experienced culinary professionals. Carefully researched content and fully tested recipes span the broad international spectrum of the modern garde manger station. Seventeen chapters are divided between five areas of instruction, each focusing on a different aspect of the garde manger chef's required knowledge and responsibilities. With nearly 600 color photographs, more than 250 recipes, and 75 beautifully illustrated graphs and charts, *Modern Garde Manger* is the most comprehensive text of its kind available for today's culinary arts student and the professional chef.

We look forward to providing you with the highest quality educational products available. Please contact us at (800) 477-3692 to preorder your desk copies of this exciting new series.

All the Best!

Matt Hart
Culinary Arts Acquisitions Editor
matthew.hart@thomson.com

Instructor Resources

INSTRUCTOR'S MANUAL

The *Instructor's Manual* provides chapter outlines, answers to end-of-chapter review questions, and additional assessment questions and answers. The *Instructor's Manual* is available at no charge to adopters of the text.

ELECTRONIC CLASSROOM MANAGER (ECM)

The *Electronic Classroom Manager* is a CD-ROM designed as a complete teaching tool for *About Wine*. It assists instructors in creating lectures, developing presentations, constructing quizzes and tests, and offers additional lesson plans. This valuable resource simplifies the planning and implementation of the instructional program. This complimentary resource package is available upon adoption of the text and consists of the following components:

Electronic Support Slides are designed as a visually-appealing way to extract key points from the textbook to enhance class lectures.

Computerized Test Bank, which consists of a variety of test questions including multiple choice and true/false.

Lesson Plans and **Activities** are available as additional resources to aid the instructor in preparing lessons.

Online Resources

A Companion web site to accompany *About Wine* is available to instructors and students featuring **electronic support** slides. They are on a chapter-by-chapter basis designed as a visually appealing way to extract key points of the book to enhance class lectures.

Key Features

LEARNING OBJECTIVES: Answering the question "What am I about to learn?" will best describe this chapter-opening feature. These learning objectives are used as a structure in helping students understand that by the end of the chapter they will have a working knowledge of the material presented.

After reading this chapter, you should be able to

- display an understanding of how both red and white table wines are produced.
- describe how sparkling wine is produced.
- explain how dessert wines are produced.

KEY TERMS: Key terms are listed in the beginning of each chapter enforcing the importance of new terminology presented in each chapter.

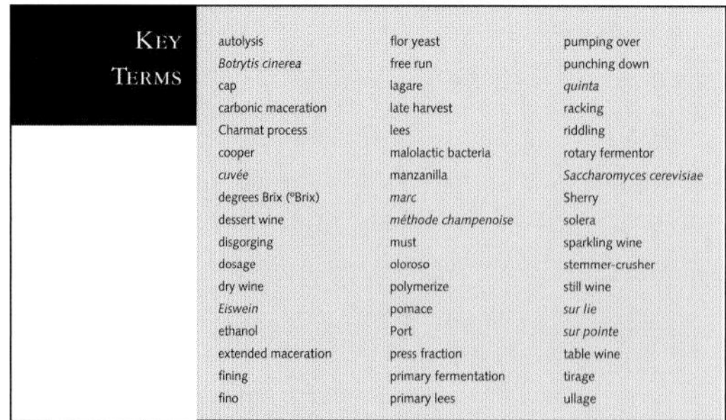

KEY TERMS

autolysis	flor yeast	pumping over
Botrytis cinerea	free run	punching down
cap	lagare	*quinta*
carbonic maceration	late harvest	racking
Charmat process	lees	riddling
cooper	malolactic bacteria	rotary fermentor
cuvée	manzanilla	*Saccharomyces cerevisiae*
degrees Brix (°Brix)	*marc*	Sherry
dessert wine	*méthode champenoise*	solera
disgorging	must	sparkling wine
dosage	oloroso	stemmer-crusher
dry wine	polymerize	still wine
Eiswein	pomace	*sur lie*
ethanol	Port	*sur pointe*
extended maceration	press fraction	table wine
fining	primary fermentation	tirage
fino	primary lees	ullage

Key terms are bolded at first use within the chapters for easy identification.

METHODS OF CAP MANAGEMENT

Punching down is the oldest, simplest, and most gentle method of mixing the cap of skins and the juice. A punch-down device is used to press down the cap into the juice. Done by hand, it

Special Features

REVIEW QUESTIONS: Each chapter ends with a series of assessment questions for the student for further consideration of content to encourage application and synthesis of material presented.

GLOSSARY: Key terms with phonetic spelling and definitions are provided at the back of the book.

PHOTOGRAPHS: This beautifully illustrated textbook contains over 300 full-color photos.

MAPS OF WINE REGIONS: Presented throughout the chapters, maps highlight specific regions being discussed and feature some top wine producers of that region.

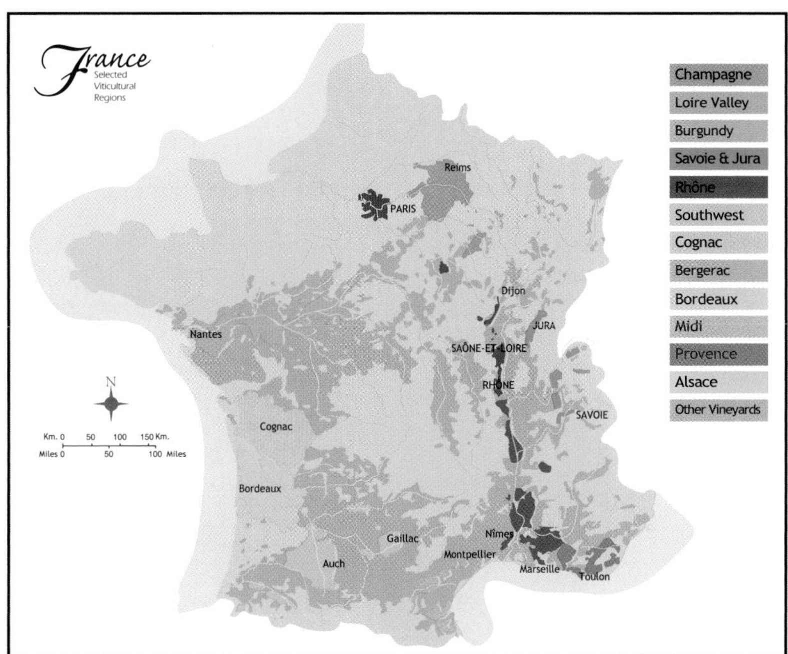

WINE LABELS: A special feature on how to read a wine label is integrated throughout the chapters. This allows students to understand the sophisticated nature of the world's wine labels and interpret the abundance of available information.

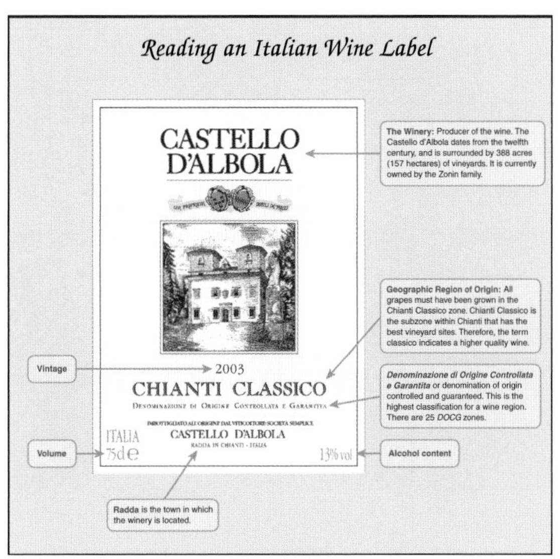

DETAILED ILLUSTRATIONS: Colorful line art is displayed throughout the text emphasizing and simplifying different processes with ease.

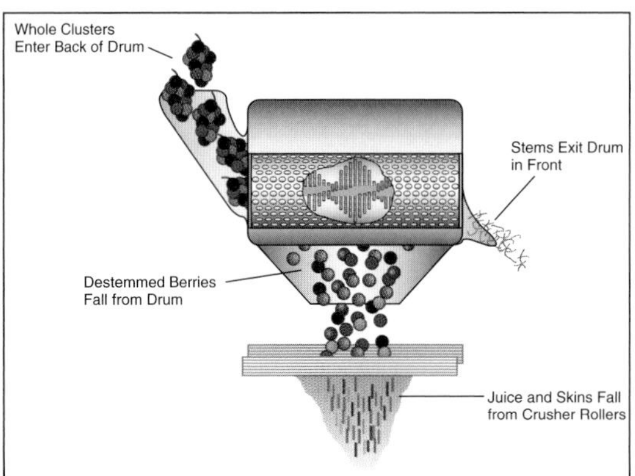

TABLES AND CHARTS offer visual representation of important information for easy comprehension.

Table 3-2 Types of Sherry

TYPE	WINEMAKING METHOD	WINE STYLE
Fino (Manzanilla)	Aged in partially full barrels under a surface layer of flor yeast growth.	Pale yellow in color with a distinctive nutty aroma. Dry and served slightly chilled. Alcohol 15.5 to 18%
Amontillado	A mature Fino sherry that is allowed to oxidize as it ages in partially filled barrels after the flor yeast is removed.	Darker than Fino sherries with a richer flavor and nutty aroma, dry with alcohol from 16 to 22%
Oloroso	Fortified before aging to prevent flor yeast growth, aged in partially filled barrels.	Dark amber in color and dry to slightly sweet (1 to 3% sugar), alcohol from 17 to 22%
Cream Sherry	Fortified before aging to prevent flor yeast growth, aged in partially filled barrels. Sweetened before bottling.	Dark amber in color and sweet (7 to 10% sugar), alcohol from 17 to 22%
Baked (California Style)	Flor yeast is grown throughout the wine (submerged culture) by bubbling oxygen through the tank. After fermentation the wine is heated to 130°F (55°C) for 1-3 months.	dark amber in color and usually sweet (7 to 10% sugar), alcohol from 17 to 22%

BRIEF CONTENTS

SECTION I The Basics

1 What Is Wine?

2 The Vineyard—From Soil to Harvest

3 The Winery—From Grapes to Bottle

4 Tasting Wines

SECTION II Wine Regions of Europe

5 France

6 Italy

7 Spain and Portugal

8 Germany

9 Other European Regions and the Mediterranean

SECTION III Wine Regions of North America

10 California

11 Washington and Oregon

12 New York, Canada, and Other North American Regions

SECTION IV Wine Regions of the Southern Hemisphere

13 Australia and New Zealand

14 Chile and Argentina

15 South Africa

SECTION V The Business of Wine

16 Selling and Serving Wines

17 Developing and Managing a Wine List

18 Buying and Cellaring Wines

APPENDIX A Wine Law in the United States

APPENDIX B American Viticultural Areas

APPENDIX C The 1855 Official Classification of the Médoc

APPENDIX D The 1855 Official Classification of Sauternes and Barsac

APPENDIX E The 1959 Official Classification of Graves

APPENDIX F The Revised Official Classification of Saint-Émilion, 1996

APPENDIX G Pomerol (Not Classified)

APPENDIX H Germany Communes

APPENDIX I Wine Organizations and Publications

APPENDIX J Wine Tasting Sheet

GLOSSARY

REFERENCES

INDEX

CONTENTS

SECTION I

The Basics, 1

CHAPTER 1

What Is Wine?, 2

- The History of Wine, 5
  - Egypt and Greece, 5
  - Roman Era, 6
  - The Middle Ages, 8
  - Twelfth Centry to Modern Times, 10
  - Golden Age of Wine, 10
  - Wine Today, 12
- Economic Cycles in the Wine Business, 13
- Summary, 13

CHAPTER 2

The Vineyard—From Soil to Harvest, 16

- Grapes Used for Winemaking, 19
- Soil and Site, 20
- Climate, 21
- *Terroir, 22*
- Techniques of Grape Growing, 24
- The Growing Season, 26
  - Budbreak, 26
  - Bloom, 28
  - Véraison, 29
  - Harvest, 30
  - Dormancy, 31
- Major Grape Varieties, 32
  - Barbera, 32
  - Cabernet Franc, 32
  - *Organic Viticulture, 33*
  - Cabernet Sauvignon, 34
  - Chardonnay, 35
  - Chenin Blanc, 36
  - Gewürztraminer, 36
  - Grenache, 37
  - Merlot, 37
  - Muscat Blanc, 37
  - Petite Sirah, 38
  - Pinot Blanc, 38
  - Pinot Gris/Pinot Grigio, 38
  - Pinot Noir, 39
  - Riesling, 40
  - Sangiovese, 40
  - Sauvignon Blanc, 40
  - Syrah, 41
  - Tempranillo, 41
  - Viognier, 41
  - Zinfandel, 42
- Summary, 43

CHAPTER 3

The Winery—From Grapes to Bottle, 44

- The Process of Fermentation, 47
  - Red Wine Crush and Fermentation, 49
- *Carbonic Maceration and Extended Maceration, 53*
  - White Wine Crush and Fermentation, 55
- Barrels and Aging, 58
- *Malolactic Fermentation, 59*
- Finishing a Wine, 61
  - Bottling, 63
- Sparkling Wine, 64
  - Other Methods of Sparkling Wine Production, 67
- Dessert Wines, 68
  - Late-Harvest Wines, 68
  - Port-Style Wines, 69
  - Sherry, 70
- The Attributes of Wine, 71
- Summary, 72

CHAPTER 4

Tasting Wines, 74

- Sensory Evaluation: How the Senses Respond to Wine, 76
  - The Sense of Sight—Appearance, 77
  - The Sense of Smell—Aroma, 77
  - The Sense of Taste—Flavor, 79
  - The Sense of Touch—Texture, 79
- Organizing a Tasting, 80
  - The Proper Setting for Tasting, 81
  - Presenting the Wines, 82
  - Other Considerations, 84
- *Proper Tasting Techniques, 85*
  - Evaluation by Sight, 86
  - Evaluation of Aroma, 86
  - Evaluation by Mouth, 88
  - Group Discussion of the Wines, 89
- Discussing and Evaluating Wine, 89
  - Difficulties in Evaluating Wine, 90
  - Understanding Wine Descriptors, 91
- *The Wine Aroma Wheel, 92*
- Wine and Food, 94

A Simple Experiment in Pairing Wine and Food, 96
Wine and Health, 97
Negative Effects from Excessive Alcohol Consumption, 97
Positive Effects from Moderate Wine Consumption, 98
Special Considerations for Women, 98
Summary, 99

SECTION II

Wine Regions of Europe

CHAPTER 5

France, 101

French Wine—Historical Perspective, 104
Appellations Controlée Laws, 106
Weaknesses of the System, 110
Wine Regions of France, 110
Bordeaux, 111
Burgundy, 127
The Classification System of Burgundy, 133
Côtes-du-Rhône, 148
The Southern Côtes du Rhône, 155
Champagne, 157
Classification of Champagne Styles, 161
Alsace, 163
The House of Léon Beyer, 165
The Loire Valley, 167
The South of France, 173
Summary, 177

CHAPTER 6

Italy, 178

Italian Wine—Historical Perspective, 180
The Denominazione d'Origine Controllata (DOC) Laws, 184
Quality Designations, 184
Naming of Italian Wines, 187
Wine Regions of Italy, 189
Piedmont, 190
Tuscany, 195
Tre Venezia, 204
Southern Italy, 213
Summary, 216

CHAPTER 7

Spain and Portugal, 218

History of Wine Production, 221
Spanish Wine—Historical Perspective, 221
Portuguese Wine—Historical Perspective, 223
Government Involvement, 224
Wine Laws , 226
Wine Regions, 229
The Wine Regions of Spain, 230
The Wine Regions of Portugal, 239
Summary, 246

CHAPTER 8

Germany, 248

German Wine—Historical Perspective, 251
Wine Laws, 255
Wine Categories, 255
Reading the Labels, 259
Revision to the Wine Laws, 260
The Wine Regions, 263
Climate, 263
Grape Varietals, 264
The Mosel-Saar-Ruwer, 266
The Rheingau, 270
Nik Weis of St. Urbans-Hof, 271
Rheinhessen, 273
The Pfalz, 273
Other Regions, 274
Summary, 276

CHAPTER 9

Other European Regions and the Mediterranean, 278

Central Europe, 280
Austria, 280
Austrian Wine—Historical Perspective, 281
Switzerland, 282
Summary—Central Europe, 284
Eastern Europe, 284
Hungary, 285
Eastern Europe: Other Countries, 287
Summary—Eastern Europe, 287

Eastern Mediterranean Countries, 287
Greece, 288
Israel, 290
Lebanon, 291
Summary—Eastern Mediterranean, 291

SECTION III

Wine Regions of North America

CHAPTER 10

California, 293

California Wine—Historical Perspective, 297
Commercialization, 298
Prohibition, 300
The Wine Revolution, 302
American Viticultural Areas, 303
The Wine Regions of California, 304
The Napa Valley, 306
Sonoma County, 314
Lake and Mendocino Counties, 319
The Central Coast, 322
The Central Valley, 329
Other Grape Growing Regions of California, 330
Summary, 331

CHAPTER 11

Washington and Oregon, 332

Washington State, 335
Washington State Wine—Historical Perspective, 336
Prohibition and Rebirth, 337
The Wine Regions of Washington, 339
The Columbia Valley, 341
The Puget Sound Region, 343
The Columbia Gorge Appellation, 345
Oregon State, 345
Oregon State Wine—Historical Perspective, 346
The Beginning of an Industry, 346
The Wine Regions of Oregon, 348
The Willamette Valley, 348
The Umpqua, Rogue, and Applegate Valleys, 350
The Columbia and Walla Walla Valleys, 351
Summary, 351

CHAPTER 12

New York, Canada, and Other North American Regions, 352

New York State, 354
New York State Wine—Historical Perspective, 355
Wine Regions of New York, 356
Other Wine Regions in the Eastern United States, 358
Virginia, 359
The Western United States, 360
Texas, 360
New Mexico, 360
Colorado, 361
Canada, 361
Canadian Wine—Historical Perspective, 362
Wine Regions, 364
Summary, 367

SECTION IV

Wine Regions of the Southern Hemisphere

CHAPTER 13

Australia and New Zealand, 369

Australia, 373
Australian Wine—Historical Perspective, 375
The Wine Regions of Australia, 377
New South Wales, 380
Victoria, 381
South Australia, 383
Western Australia, 386
Tasmania, 387
New Zealand, 387
New Zealand—Historical Perspective, 388
The Wine Regions of New Zealand, 389
Gisborne, 389
Hawkes Bay, 393
Marlborough, 393
Other New Zealand Wine Regions, 394
Summary, 394

CHAPTER 14

Chile and Argentina, 396

Chile, 399
Chilean Wine—Historical Perspective, 399
The Wine Regions of Chile, 401
The Atacama and Coquimbo Region, 403
Pisco, 405
The Aconcagua Region, 404
The Central Valley Region, 405
The Southern Region, 409
Argentina, 408
Argentine Wine—Historical Perspective, 408
The Wine Regions of Argentina, 410
The Mendoza and San Juan Regions, 410
The La Rioja and Salta Regions, 412
The Rio Negro Region, 413
Summary, 413

CHAPTER 15

South Africa, 416

South African Wine—Historical Perspective, 421
The Wine Regions of South Africa, 424
The Olifants River Region, 429
Klein Karoo (Little Karoo) Region, 429
Breede River Valley, 430
Chenin Blanc and Pinotage, 431
Coastal Region, 433
Summary, 439

SECTION V

The Business of Wine

CHAPTER 16

Selling and Serving Wines, 441

Wine Service and the Role of the Sommelier, 444
Start Early, 445
Promoting Successful Wine Sales, 446
Driving On-Premise Sales, 447
Featured Wines, 447
Setup for Sales, 448
Server Suggestions, 449
Menu Suggestions, 450
Visible Storage and Special Seating, 450
Offer Options, 451
Offer Flights, 451
Bottle Sizes, 452
Bringing Their Own, 454
Tableside Wine Service, 455
Glasses First, 455
Proper Serving Temperatures, 458
Opening the Wine, 458
Wine Keys, 459
Corks and Cork Taint, 464
Taste Test, 465
Sending Back the Wine, 466
Sparkling Wine, 469
Decanting the Wine, 469
Staff Training, 472
Training During Staff Meetings, 472
Focus on the USP, 474
Summary, 475

CHAPTER 17

Developing and Managing a Wine List, 476

Choosing Which Wines to Sell, 478
Special Pricing, 480
Trade Tastings, 480
Selling Wines by the Glass, 481
Choosing Wines to Sell by the Glass, 482
What to Have, 484
Preserving Open Bottles of Wine, 485
Storage and Inventory Levels, 488
Pricing the Wine List, 489
Sliding Scale Pricing, 490
Incentives, 491
Filling Out the Wine List, 491
Bottle Formats, 492
Organizing the List, 493
By Region of Origin, 493
By Varietal, 494
By Style, 495
By Price Point, 496
To Describe or Not to Describe?, 496
Staff Favorites Section, 496
List Formats, 497
Summary, 497

CHAPTER 18

Buying and Cellaring Wines, 498

What Happens as Wines Age?, 501
Color, 501
Aromatics, 501
Aging and Tannins, 502
Acids, 502
Effects of Temperature on Aging, 502
Sediment, 503
Bottle Size, 503
The Effect of Vintage on Ageability, 503
Styles of Wine to Age, 504
Which Wines Should *Not* Be Aged?, 504
Ideal Cellar Conditions, 505
Temperature and Humidity Control, 505
Passive Cellars, 506
Renting Storage Space, 507
Alternatives to Cellars, 507
Purchasing Wine, 508
Racking for the Wine, 508
Buying Futures, 509
Buying Wine at Auctions, 510
Older Wine and Restaurants, 510
Accessing Older Wines, 511
Private Cellars, 511
Serving Older Wines in the Restaurant, 511
Summary, 512

APPENDIX A

Wine Law in the United States, 513

APPENDIX B

American Viticultural Areas, 518

APPENDIX C

The 1855 Official Classification of the Médoc, 523

APPENDIX D

The 1855 Official Classification of Sauternes and Barsac, 525

APPENDIX E

The 1959 Official Classification of Graves, 526

APPENDIX F

The Revised Official Classification of Saint-Émilion, 1996, 527

APPENDIX G

Pomerol (Not Classified), 529

APPENDIX H

Germany Communes, 530

APPENDIX I

Wine Organizations and Publications, 531

APPENDIX J

Wine Tasting Sheet, 532

GLOSSARY, 533

REFERENCES, 542

INDEX, 545

Wine has been a part of civilization for thousands of years and over time has become closely associated with food and fine dining. Dining out in a fine restaurant is a very popular leisure activity. The selection, presentation, and ultimate enjoyment of a wine accompanying a meal is an integral part of the dining experience. Beyond the importance of wine to those who consume it, it is also an important source of a restaurant's income. There is a saying in the food industry that a restaurant breaks even with the kitchen and makes a profit off the wine list. This may or may not be true, but nonetheless wine is a vital attraction to customers deciding whether or not to frequent a particular restaurant. Therefore, it is essential for restaurant managers and chefs to have knowledge of wine, how to taste and evaluate it, an understanding of how wine is made and where it comes from. This knowledge coupled with the ability to manage wine inventory allows restaurateurs to fully meet their diners' needs and maximize the returns on their investment in wine.

About Wine provides the basic information that serves as the foundation for both a successful career in wine and increased enjoyment of wine. Chefs, restaurant managers, and retailers are just a few of the professionals in the culinary and hospitality industries who sell and serve wine on a daily basis. These professionals require a fundamental understanding of wine and how to both manage and sell it. ***About Wine*** is a valuable resource for culinary students and professionals, as well as amateur wine enthusiasts.

The Purpose of This Textbook

Wine is a diverse field of study that can be intimidating to the novice learner, yet extensive knowledge about wine is not necessary to enjoy its qualities. The purpose of this book is to provide a solid base in the subject that can be built upon with further practice. Once the foundation in wine knowledge is established, new experiences with wine will be easier to comprehend and to incorporate into your work. This book is a reference that can aid in the understanding of new wines you encounter long after the coursework has been finished. To those who already have a familiarity with wine, the book will offer a deeper grasp of areas in wine studies and winemaking regions you may be less acquainted with. ***About Wine*** interprets the comprehensive subject of wine in a way that can be easily understood and applied.

Organization of Text

The chapters are organized in six sections that deal with specific areas of wine studies.

Section I:	The Basics
Section II:	Wine Regions of Europe
Section III:	Wine Regions of North America
Section IV:	Wine Regions of the Southern Hemisphere
Section V:	The Business of Wine
	Appendices, Glossary, References, and Index

SECTION I

The Basics includes four chapters and covers the history of wine and its place in society over the centuries. Chapter 2 discusses grape growing and the importance of the vineyard in winemaking. Chapter 3 explains how table wines, dessert wines, and sparkling wines are produced. Section I closes with an extensive chapter on tasting and evaluating wines, which is valuable to anyone who deals with wine in a professional or amateur basis.

SECTION II

Wine Regions of Europe are covered in the next five chapters: France; Italy; Spain and Portugal; Germany; and Other European Regions and the Mediterranean, which may be less familiar to readers. Knowledge of these regions is important to all students of wine because Europe is the birthplace of modern winemaking and most of the styles of wine and grape varieties grown throughout the world have their origins on this continent. Each chapter covers the individual regions within the country, their history of winemaking, and the wines that they produce. Full-color, detailed maps of the wine-producing areas help readers familiarize themselves with each region.

Section III

Wine Regions of North America are covered in three chapters: California; Washington and Oregon; and New York, Canada, and other North American Regions. Domestic wines come in as many forms as their European counterparts and make up the majority of wine consumed in the United States. The wine regions of North America are covered in much the same way as the European wine regions,with sections on local grape growing and winemaking techniques as well as detailed maps of grape-growing appellations.

Section IV

Section IV consists of three chapters discussing the **Wine Regions of the Southern Hemisphere.** This is one of the fastest growing and innovative regions of winemaking in the world today. Chapters devoted to Australia and New Zealand; Chile and Argentina; and South Africa include their winemaking history, grape growing and wine production methods, and maps of their wine regions.

Section V

The Business of Wine and how it is presented and managed in food service is covered in depth in the three chapters of this section. Topics include buying and cellaring wines, developing and maintaining a wine list, and selling and serving wine. The chapters clearly illustrate how to develop a successful and profitable wine program. Important subjects include staff training and working with distributors. ***About Wine*** differentiates itself from many other wine books with this detailed section containing practical knowledge for those in the restaurant and hospitality trade.

Appendices

The book concludes with 10 appendices: Appendix A, Wine Law in the United States; Appendix B, American Viticultural Areas; Appendix C, The 1855 Official Classification of the Médoc; Appendix D, The 1855 Official Classification of Sauternes and Barsac; Appendix E, The 1959 Official Classification of Graves; Appendix F, The Revised Official Classification of Saint-Émilion, 1996; Appendix G, Pomerol (Not Classified); Appendix H, Germany Communes; Appendix I, Wine Organizations and Publications; and Appendix J, Wine Tasting Sheet. After formal training in the study of wine is complete, this section will be a valuable resource for learning about new wines and wine regions.

Supplements

Instructor's Manual

The *Instructor's Manual* provides chapter outlines, answers to end-of-chapter review questions, and additional assessment questions and answers. The *Instructor's Manual* is available at no charge to adopters of the book.

Electronic Classroom Manager (ECM)

The *Electronic Classroom Manager* is a CD-ROM containing all of the instructor resources available to accompany the *About Wine* text. This valuable resource simplifies the planning and implementation of the instructional program. Included in the ECM are chapter-specific PowerPoint presentations, a computerized test bank, additional lesson plans, and activities.

Online Resource

A Companion Web site to accompany *About Wine* is available to instructors and students featuring electronic support slides.

ACKNOWLEDGMENTS

I would like to thank my friends, family, and coworkers at Kenwood Vineyards and Santa Rosa Junior College for their assistance and cooperation during the writing of this book. Most of all, I give special thanks to my wife, Stephanie, whose daily support and encouragement and whose skill in editing helped me to write a much better book than I could have on my own.

PAT HENDERSON

First, I want to thank my son, Dan Rex, for his unquestioning support over the years—both of me and of my work. Also, many thanks to my three sisters for their patience. (Now I am free to get back to work on our book!) Special thanks to my wine-tasting buddies, Ralph Protsik, Liz Weiner, Bob Farrar, and Ernie and Robin Krieger who, over the 22 years we have tasted together, have kept my mind open to all types of wine and my palate attuned to all levels of quality. I also want to acknowledge my students, at Boston University's School of Hospitality Administration and now at New England Culinary Institute; they are a constant source of inspiration.

DELLIE REX

The authors wish to thank the following individuals and organizations for their many contributions to the writing of *About Wine.*

Marc Beyer, Jeff Brooks, Jennifer Burns, Jerry Comfort, Adam Dial, Wilma Dull, F. Korbel & Brothers, Ed Flaherty, German Wine Information Bureau, Joel Green, Italian Trade Commission, Kenwood Vineyards, Kobrand Corporation, Deidre Magnello, New England Culinary Institute, New York Wine and Grape Foundation, Ann C. Noble, Oregon Wine Board, Mel Sanchietti, Steve Schukler, Rhonda Smith, SOPEXA-USA, Mark Stupich, Valley of the Moon Winery, Washington Wine Commission, Nik Weis, Wine Institute, Wines from Spain Program, and Zonin Corporation, among many others listed in the photography credits.

The authors also wish to thank all of the people at Thomson Delmar Learning, especially Patricia Osborn for her constant encouragement and hard work, Matthew Hart, Pat Gillivan, Matt Williams, Maureen Grealish, and Kristin McNary. Their skill and patience combined with an incredible enthusiasm for the project made this book possible.

SPECIAL THANKS

Delmar Learning and the authors would like to give a special thanks to the contributing authors of *About Wine,* David Garaventa and Angela Lloyd, for their dedication to this project.

REVIEWERS

The authors and Thomson Delmar Learning would like to thank the following reviewers for their invaluable feedback and suggestions for improvement:

DONNA ALBANO
Asssitant Professor, Hospitality Management
Atlantic Cape Community College
Mays Landing, New Jersey

MARIAN BALDY
Instructor
California State University
Chico, California

TIMOTHY DODD
Texas Wine Marketing Research Institute
Texas Tech University
Lubbock, Texas

JILL DOEDERLEIN
Travel & Tourism Instructor
Lansing Community College
Lansing, Michigan

BARRY GOLDMAN, FMP
Chef Instructor
Florida Culinary Institute
West Palm Beach, Florida

CAROL GUNTER CEC, MBA
Chef Instructor
Hospitality, Travel & Tourism Management
Purdue University
West Lafayette, Indiana

NINA JARRETT
Chef Instructor
Kapi`Olani Community College
Kapi`Olani, Hawaii

GEOFF LABITZKE
Wine Educator
Young's Market Company
Sonoma, California

JACK LARKIN
Vice President
Pacific Southern Wine Company
Atlanta, Georgia

ABBY NASH
Lecturer
School of Hotel Administration
Cornell University
Ithaca, New York

GEORGE STAIKOS
Professor
School of Hotel, Restaurant & Tourism Management
Fairleigh Dickinson University
Madison, New Jersey

ANTHONY STRIANESE
Certified Culinary Educator by the American Culinary Federation
Schenectady County Community College
Schenectady, New York

PETER SZENDE
Assistant Professor School of Hospitality Administration
Boston University
Boston, Massachusetts

JAMES TAYLOR
Assistant Professor, Hospitality Management
Columbus State Community College
Columbus, Ohio

MICHAEL WRAY
Associate Professor
Director, Restaurant and Culinary Administration
Metropolitan State College of Denver
Denver, Colorado

PHOTOGRAPHY CREDITS

CHAPTER 01

Chapter Opener, Alison Woollard/Shutterstock
Top Inset, Diane N. Ennis/Shutterstock
Bottom Inset, April Turner/Shutterstock
Figure 1–1, Bettmann/Corbis
Figure 1–2, Gianni Dagli Orti/Corbis
Figure 1–3, Archivo Iconografico, S.A./Corbis
Figure 1–4, Bacchus/Michelangelo Buonarroti/The Bridgeman Art Library/Getty Images
Figure 1–5, Bettmann/Corbis
Figure 1–6, Wine Institute

CHAPTER 02

Chapter Opener, Falk Kienas/Shutterstock
Top Inset, Jennifer Burns
Bottom Inset, Charles O'Rear/Corbis
Figure 2–1, Oregon Wine Board, Patrick Prothe Photography 2004
Figure 2–2, Jeffrey Granett
Figure 2–3, Oregon Wine Board, Frank Barnett Photography 2004
Figure 2–4, K. Hackenberg/Corbis
Figure 2–5, Owen Franken/Corbis
Figure 2–10, Jennifer Burns
Figure 2–23, Oregon Wine Board, Frank Barnett Photography 2004
Figure 2–24, Steve Shukler
Figure 2–31, Oregon Wine Board, Frank Barnett Photography 2004
Figure 2–37, Charles O'Rear/Corbis

CHAPTER 03

Chapter Opener, Romilly Lockyer/Getty
Figure 3–2, ETS Laboratories
Figure 3–18, Washington Wine Commission

CHAPTER 04

Chapter Opener, Tom Schmucker/Shutterstock
Bottom Inset, New England Culinary Institute
Figure 4–5, Charles O'Rear/Corbis
Figure 4–7, Riedel Glassware
Figure 4–9, New England Culinary Institute
Figure 4–10, New England Culinary Institute
Figure 4–11, New England Culinary Institute
Figure 4–12, New England Culinary Institute
Figure 4–14, Copyright 1990 A C Noble. Colored laminated plastic wine aroma wheels may be obtained from www.winearomawheel.com
Figure 4–15, Royalty–Free/Corbis
Figure 4–17, A. Inden/Corbis

CHAPTER 05

Chapter Opener, Agence Images/Alamy
Top Inset, Cephas Picture Library/Alamy
Bottom Inset, Cephas Picture Library/Alamy
SF 5–1, Kobrand Corporation
UNF 5–1, Kobrand Corporation
UNF 5–2, Kobrand Corporation
UNF 5–3, Kobrand Corporation
UNF 5–4, Kobrand Corporation
UNF 5–5, Kobrand Corporation
UNF 5–6, Kobrand Corporation
UNF 5–7, Kobrand Corporation
UNF 5–8, Kobrand Corporation
UNF 5–9, Kobrand Corporation
UNF 5–10, Kobrand Corporation
Figure 5–1, Eitan Simanar/Alamy
Figure 5–2, Agence Images/Alamy
Figure 5–3, Per Karlsson–BKWine.com/Alamy
Figure 5–4, Sébastian Baussais/Alamy
Figure 5–5, Per Karlsson–BKWine.com/Alamy
Figure 5–6, Cephas Picture Library/Alamy
Figure 5–7, Cephas Picture Library/Alamy
Figure 5–8, Cephas Picture Library/Alamy
Figure 5–9, Cephas Picture Library/Alamy
Figure 5–10, Cephas Picture Library/Alamy
Figure 5–11, Cephas Picture Library/Alamy
Figure 5–12, Alex Griffiths/Alamy
Figure 5–13, Per Karlsson–BKWine.com/Alamy
Figure 5–14, Peter Horee/Alamy
Figure 5–16, Vineyard Brands
Figure 5–17, Vineyard Brands
Figure 5–18, Kobrand Corporation
Figure 5–19, Vineyard Brands
Figure 5–20, Cephas Picture Library/Alamy
Figure 5–21, Ray Roberts/Alamy
Figure 5–22, Stockfolio/Alamy
Figure 5–23, Cephas Picture Library/Alamy
Figure 5–24, Pixfolio/Alamy
Figure 5–25, Charles O'Rear/Corbis
Figure 5–26, Kobrand Corporation
Figure 5–27, Jon Arnold Images/Alamy
Figure 5–28, Cephas Picture Library/Alamy
Figure 5–29, Cephas Picture Library/Alamy
Figure 5–30, PCL/Alamy
Figure 5–31, Per Karlsson–BKWine.com/Alamy
Figure 5–32, Cephas Picture Library/Alamy
Figure 5–33, Cephas Picture Library/Alamy
Figure 5–34, Kobrand Corporation
Figure 5–35, Kobrand Corporation
Figure 5–36, Mr. J.L Delpal
Figure 5–37, Léon Beyer
Figure 5–38, Cephas Picture Library/Alamy
Figure 5–39, Per Karlsson–BKWine.com/Alamy
Figure 5–40, Per Kerlsson–BKWine.com/Alamy
Figure 5–41, Philippe Roy/Alamy
Figure 5–42, Peter Horree/Alamy

CHAPTER 06

Chapter Opener, Sergio Pitamitz/zefa/Corbis
Top Inset, Charles O'Rear/Corbis
Bottom Inset, Cephas Picture Library/Alamy

SF 6–1, Zonin USA Inc.
UNF 6–1, Kobrand Corporation
UNF 6–2, Kobrand Corporation
UNF 6–3, Kobrand Corporation
UNF 6–4, Kobrand Corporation
Figure 6–1, North Carolina Museum of Art/Corbis
Figure 6–2, Bettmann/Corbis
Figure 6–3, Copyright RS/RG Associates
Figure 6–4, Zonin USA Inc.
Figure 6–5, Zonin USA Inc.
Figure 6–6, Martyn Goddard/Corbis
Figure 6–7, Zonin USA Inc.
Figure 6–8, Fabio Muzzi/Corbis Sygma
Figure 6–9, David Lees/Corbis
Figure 6–10, Sergio Pitamitz/zefa/Corbis
Figure 6–11, Charles O'Rear/Corbis
Figure 6–12, Castello Banfi
Figure 6–13, Castello Banfi
Figure 6–14, Fattoria Il Palagio
Figure 6–15, Cephas Picture Library/Alamy
Figure 6–16, Chantal Comte Diffusion
Figure 6–17, Zonin USA Inc.
Figure 6–18, Zonin USA Inc.
Figure 6–19, G. Rossenbach/zefa/Corbis
Figure 6–20, Jonathan Blair/Corbis
Figure 6–21, Cephas Picture Library/Alamy

CHAPTER 07

Chapter Opener, Grupo Osborne, permission granted by Janet Kafka and Associates
Top Inset, Jaime Gonzalez/ShutterStock
Bottom Inset, Upperhall Ltd/Getty
UNF 7–1, Shutterstock
UNF 7–2, Grupo Osborne, permission granted by Janet Kafka and Associates
UNF 7–3, Kobrand Corporation
UNF 7–4, Spanish Institute of Foreign Trade (ICEX)
UNF 7–5, Kobrand Corporation
Figure 7–1, Visual Arts Library (London)/Alamy
Figure 7–2, Grupo Osborne, permission granted by Janet Kafka and Associates
Figure 7–3, Grupo Osborne, permission granted by Janet Kafka and Associates
Figure 7–4, Jaime Gonzalez/ShutterStock
Figure 7–5, Jaime Gonzalez/ShutterStock
Figure 7–6, Richard Langs/ShutterStock
Figure 7–7, Grupo Osborne, permission granted by Janet Kafka and Associates
Figure 7–8, Grupo Osborne, permission granted by Janet Kafka and Associates
Figure 7–9, Mila Petkova/ShutterStock
Figure 7–10, Grupo Osborne, permission granted by Janet Kafka and Associates
Figure 7–11, Grupo Osborne, permission granted by Janet Kafka and Associates
Figure 7–12, Joe Gough/Shutterstock
Figure 7–13, Alan Smillie/Shutterstock
Figure 7–14, Grupo Osborne, permission granted by Janet Kafka and Associates
Figure 7–15, Photo Kobrand Corporation
Figure 7–16, Grupo Osborne, permission granted by Janet Kafka and Associates
Figure 7–17, Upperhall Ltd/Getty

CHAPTER 08

Chapter Opener, G. Rosenbach/zefa/Corbis
Top Inset, Bildagentur Franz Waldhaeusl/Alamy
Bottom Inset, Cephas Picture Library/Alamy
UNF 8–1, Holt Studios International Ltd/Alamy
UNF 8–2, Kobrand Corporation
UNF 8–3, St. Urbans–Hof
UNF 8–4, St. Urbans–Hof
Figure 8–1, Bildarchiv Monheim GmbH/Alamy
Figure 8–2, RF|Binder Partners
Figure 8–3, St. Urbans–Hof
Figure 8–4, Von Othegraven, permission granted by Classical Wines
Figure 8–5, St. Urbans–Hof
Figure 8–6, Von Othegraven, permission granted by Classical Wines
Figure 8–7, Cephas Picture Library/Alamy
Figure 8–8, Georg Breuer, permission granted by Classical Wines
Figure 8–9, Bildagentur Franz Waldhaeusl/Alamy
Figure 8–10, Frithjof Hirdes/zefa/Corbis
Figure 8–11, Michael Pole/Corbis
Figure 8–12, Georg Breuer, permission granted by Classical Wines
Figure 8–13, Cephas Picture Library/Alamy
Figure 8–14a, allOver photography/Alamy
Figure 8–14b, allOver photography/Alamy
Figure 8–15, G. Rosenbach/zefa/Corbis
Figure 8–16, Holt Studios International Ltd/Alamy
Figure 8–17, Cephas Picture Library/Alamy
Figure 8–18, Cephas Picture Library/Alamy
Figure 8–19, German Wine Institute
Figure 8–20, Cephas Picture Library/Alamy
Figure 8–21, Cephas Picture Library/Alamy
Figure 8–22, Cephas Picture Library/Alamy
Figure 8–23, f1online/Alamy
Figure 8–25, Cephas Picture Library/Alamy
Figure 8–26, Chad Ehlers/Getty

CHAPTER 9

Chapter Opener, Villiger/ zefa/Corbis
Top Inset, lavigne hervÃ/Shutterstock
Bottom Inset, Hermann Danzmayr/Shutterstock
Figure 9–1, Adam Woolfitt/Corbis
Figure 9–2, A. Villiger/zefa/Corbis
Figure 9–3, lavigne hervÃ/Shutterstock
Figure 9–4, Per Karlsson-BKWine.com/Alamy
Figure 9–5, Alan King/Alamy

CHAPTER 10

Chapter Opener, Aaron/Shutterstock
Top Inset, Stephen Beaumont/Shutterstock
UNF 10–1, Kobrand Corporation
UNF 10–2, Kobrand Corporation
UNF 10–3, Kobrand Corporation
UNF 10–4, Kobrand Corporation

UNF 10–5a, Kobrand Corporation
UNF 10–5b, Kobrand Corporation
Figure 10–2, Bettmann/Corbis
Figure 10–3, Corbis
Figure 10–4, Wine Institute
Figure 10–5, Bettmann/Corbis
Figure 10–6, Wine Institute
Figure 10–8, Jennifer Burns
Figure 10–9, Mark Stupich
Figure 10–13, Mark Stupich

CHAPTER 11

Chapter Opener, Chuck Pefley/Alamy
Top Inset, Charles O'Rear/Corbis
Bottom Inset, Wolfgang Kaehler/Corbis
UNF 11–1, Kobrand Corporation
Figure 11–1, Washington Wine Commission
Figure 11–2, Oregon Wine Board, Patrick Prothe Photography 2004
Figure 11–3, Washington Wine Commission
Figure 11–4, Charles O'Rear/Corbis
Figure 11–5, Washington Wine Commission
Figure 11–6, Washington Wine Commission
Figure 11–7, Cephas Picture Library/Alamy
Figure 11–8, Cephas Picture Library/Alamy
Figure 11–9, Washington Wine Commission
Figure 11–10, Wolfgang Kaehler/Corbis
Figure 11–11, Oregon Wine Board, Patrick Prothe Photography 2004
Figure 11–12, Oregon Wine Board, Patrick Prothe Photography 2004
Figure 11–13, Oregon Wine Board, Patrick Prothe Photography 2004
Figure 11–14, Cephas Picture Library/Alamy

CHAPTER 12

Chapter Opener, Randall Tagg Photography, provided by the NY Wine & Grape Foundation
Top Inset, Randall Tagg Photography, provided by the NY Wine & Grape Foundation
Bottom Inset, Jessie Eldora Robertson/Shutterstock
UNF 12–1, Kobrand Corporation
UNF 12–2, Kobrand Corporation
UNF 12–3, Kobrand Corporation
Figure 12–1, Randall Tagg Photography, provided by the New York Wine & Grape Foundation
Figure 12–2, Randall Tagg Photography, provided by the New York Wine & Grape Foundation
Figure 12–3, New York Wine & Grape Foundation
Figure 12–4, Bill Russell
Figure 12–5, Westport Rivers Vineyard & Winery, Bill Russell and Julie Fox Communications
Figure 12–6, Randall Tagg Photography, provided by the New York Wine & Grape Foundation
Figure 12–7, Lakeview Cellars Winery

CHAPTER 13

Chapter Opener, Ben Goode/Shutterstock
Top Inset, Royalty–Free/Corbis
Bottom Inset, Charles O'Rear/Corbis
UNF 13–1, Kobrand Corporation
UNF 13–2, Kobrand Corporation
UNF 13–3, Kobrand Corporation
UNF 13–4, Kobrand Corporation
UNF 13–5, Kobrand Corporation
Figure 13–1, Royalty–Free/Corbis
Figure 13–2, Cephas Picture Library/Alamy
Figure 13–3, Charles O'Rear/Corbis
Figure 13–4, Royalty–Free/Corbis
Figure 13–5, Oliver Strewe/Getty
Figure 13–6, Charles O'Rear/ Corbis
Figure 13–7, Charles O'Rear/ Corbis
Figure 13–8, Charles O'Rear/ Corbis
Figure 13–9, Charles O'Rear/ Corbis
Figure 13–10, Clay McLachlan/Getty
Figure 13–11, Cephas Picture Library/Alamy

CHAPTER 14

Chapter Opener, Cephas Picture Library/Alamy
Top Inset, Charles O'Rear/Corbis
Bottom Inset, Iberimage.com/Heinz Hebeisen
UNF 14–1, Kobrand Corporation
UNF 14–2, Kobrand Corporation
UNF 14–3, Via Wines
UNF 14–4, Via Wines
Figure 14–1, Charles O'Rear/Corbis
Figure 14–2, Eduardo Longoni/Corbis
Figure 14–3, Charles O'Rear/Corbis
Figure 14–4, Richard T. Nowitz/Corbis
Figure 14–5, Danita Delimont/Alamy
Figure 14–6, Gary Cook/Alamy
Figure 14–7, Viña Errázuriz
Figure 14–8, Viña Errázuriz
Figure 14–9, Charles O'Rear/Corbis
Figure 14–10, Viña Errázuriz
Figure 14–11, Cephas Picture Library/Alamy
Figure 14–12, Cephas Picture Library/Alamy
Figure 14–13, Cephas Picture Library/Alamy
Figure 14–14, Iberimage.com/Heinz Hebeisen
Figure 14–15, Carlos Barria/Reuters/Corbis

CHAPTER 15

Chapter Opener, Stuart Taylor/Shutterstock
Top Inset, Stuart Abraham/Alamy
Bottom Inset, Luminous/Alamy
SF 15–1, Kumkani, Stellenbosch, South Africa, Omnia Wines
UNF 15–1, Kobrand Corporation
UNF 15–2, Wines of South Africa
Figure 15–1, Martin Harvey/Alamy
Figure 15–2, Sean Nel/Shutterstock
Figure 15–3, Stuart Abraham/Alamy
Figure 15–4, AFP/Getty Images
Figure 15–5, Cephas Picture Library/Alamy
Figure 15–6, Peter Titmuss/Alamy
Figure 15–7, Stuart Taylor/Shutterstock
Figure 15–8, Cephas Picture Library/Alamy
Figure 15–9, luminous/Alamy
Figure 15–10, Peter Titmuss/Alamy

CHAPTER 16

Chapter Opener, Peter Doomen/ Shutterstock
Top Inset, New England Culinary Institute
Bottom Inset, New England Culinary Institute
UNF 16–1, New England Culinary Institute
Figure 16–1, allOver photography/ Alamy
Figure 16–2, New England Culinary Institute
Figure 16–4, Royalty–Free/Corbis
Figure 16–5, New England Culinary Institute
Figure 16–7, New England Culinary Institute
Figure 16–8a, New England Culinary Institute
Figure 16–8b, New England Culinary Institute
Figure 16–9a, New England Culinary Institute
Figure 16–10, New England Culinary Institute
Figure 16–11, New England Culinary Institute
Figure 16–12, New England Culinary Institute
Figure 16–13, New England Culinary Institute
Figure 16–14, New England Culinary Institute
Figure 16–15, New England Culinary Institute
Figure 16–16, New England Culinary Institute
Figure 16–17, Peter Dooman/ Shutterstock
Figure 16–18a, Royalty–Free/Corbis
Figure 16–18b, Royalty–Free/Corbis
Figure 16–19, Bruce Shippee/ Shutterstock
Figure 16–20a, New England Culinary Institute
Figure 16–20b, New England Culinary Institute

CHAPTER 17

Chapter Opener, Anita/Shutterstock
Top Inset, Feng Yu/Shutterstock
Bottom Inset, A.L. Spangler/Shutterstock
Figure 17–2, WR Publishing/Alamy
Figure 17–5, New England Culinary Institute
Figure 17–6, Sasha Davas/Shutterstock

CHAPTER 18

Chapter Opener, Michael Mattox/ Shutterstock
Top Inset, Andre Nantel/Shutterstock
Bottom Inset, Richard Langs/ Shutterstock
Figure 18–1, Royalty Free/Corbis
Figure 18–2, Royalty Free/Corbis
Figure 18–3, Lola/Shutterstock
Figure 18–4, Steve Lovegrove/ Shutterstock
Figure 18–5, Royalty Free/Corbis

Pat Henderson and Dellie Rex come to the world of wine by way of diverse professional backgrounds, yet they have in common a passion for the topic as well as a keen interest in sharing their knowledge with others. They each have more than 20 years experience in their fields making and selling wine, and both have also taught classes on wine at the college level. Each has a practical understanding of the business of wine yet neither has lost his appreciation for the romance of wine that makes it such an enjoyable and fascinating subject of study.

J. Patrick Henderson

Senior Winemaker, Kenwood Vineyards; Instructor, Santa Rosa Junior College

Pat Henderson began his career in the winemaking program at the University of California at Davis where he received a degree in Fermentation Science. While at school learning the technical aspects of winemaking, Pat interned at a number of wineries doing harvest and part-time work to supplement his formal education. The winemakers at the six different wineries where he worked gave him broad exposure to many different styles of wine and how to make them. After graduating, he returned to the site of one of his internships and was hired as enologist for Kenwood Vineyards in California's Sonoma Valley and worked there for nine vintages. In 1995 Pat became winemaker at Hedges Cellars in Washington State where he established a new winery at the Estate Vineyard in the Columbia Valley. Within one year, wines he produced won both best of show and ratings of more than 90 points. After two vintages in the Northwest, Pat returned to the Sonoma Valley to become winemaker/general manager at Valley of the Moon Winery. Here he worked to rebuild the historic winery and introduce a new line of wines featuring less common varietals such as Syrah, Sangiovese, and Pinot Blanc. The transformed winery quickly gained a reputation for producing wines of excellent quality and good value. After seven vintages as winemaker at Valley of the Moon, he returned to Kenwood Vineyards as senior winemaker, 20 years after his first vintage at Kenwood. Pat strives to make unique wines that are multidimensional in character, and his philosophy on winemaking is one that relies heavily on the quality of a wine's flavors and balance. Beyond his role as a winemaker, Pat has shared his interest in wine with others as an instructor in a course in winemaking at Santa Rosa Junior College and gives educational seminars throughout the country. Since 1991 Pat has taught hundreds of students the fine points of wine and the craft of winemaking. He has been a member of the American Society of Enologists and Viticulturists for 25 years. Pat currently resides with his wife, Stephanie, in Napa, California.

Dellie Rex

Instructor of Wines, New England Culinary Institute

Dellie Rex has been in the business of wine for over 25 years, with experience in retail, wholesale and marketing, and as an educator. For 13 of those years, she worked independently as president of Rex Associates, a regional marketing company through which she coordinated all sales and marketing efforts for her eight client wineries and importers.

Dellie also acted as marketing consultant for a variety of clients, including the Trade Commission of Portugal, the Commercial Office of Spain, and the Oregon Wine Marketing Coalition.

Although she concentrated on marketing, Dellie has worked throughout her career as a wine educator. For over 12 years, she ran a wine education service for consumers while also serving as Adjunct Professor at Boston University's School of Hospitality Administration. In the summer of 2005, Dellie became a wine instructor at NECI (New England Culinary Institute). She teaches in all three programs: the Bachelors in Hotel and Restaurant Management; Culinary AOS; and Baking and Pastry Arts.

In its December 1999 issue, *Boston Magazine* listed Dellie Rex as one of the 50 most intriguing women in New England, dubbing her "the wine expert's wine expert." In an article in August 2000, the *Boston Globe* featured Dellie as one of Massachusetts' five leading wine educators, describing her "greatest pedagogical resource" as a "combination of plain speech and genuine enthusiasm that makes you believe in your potential to learn."

Dellie Rex is a member of the Society of Wine Educators and of La Commanderie de Bordeaux, an honorary society for wine connoisseurs. She serves on the Board of the Elizabeth Bishop Wine Resource Center in Boston. She graduated from the University of Colorado and, in 1978, was granted a MBA with honors from Babson College where she concentrated on entrepreneurial studies. In 1983, Dellie earned a certificate in French regional wines from L'Académie du Vin in Paris. She resides in Essex Junction, Vermont.

Contributors

DAVID J. GARAVENTA
Academic and Wine Instructor
New England Culinary Institute
Essex Junction, Vermont
Chapters 16, 17, and 18: The Business of Wine

ANGELA LLOYD
Wine Writer
Cape Town, South Africa
Chapter 15: South Africa

SECTION I

The Basics

The Basics include four chapters covering the history of wine and its place in society over the centuries. This section discusses grape growing, the importance of the vineyard in winemaking, how table wines, dessert wines, and sparkling wines are produced, and tasting and evaluating wines.

CHAPTER 1

What Is Wine?

After reading this chapter, you should be able to

- discuss the historical origins of winemaking.
- describe the influence of the Greek and Roman civilizations on winemaking.
- explain the role of the church in winemaking during the Middle Ages.
- describe the economic cycles of grape growing and winemaking.

KEY TERMS

amphorae
Bacchus
boom and bust
Champagne
coopers
Dionysus
Dom Perignon
enology
fighting varietals
phylloxera
Prohibition
trellising
vintage
viticulture
Volstead Act

Wine has been an integral part of the human experience for nearly 70 centuries (Figure 1–1). Though its origins are somewhat disputed, it has been made and consumed wherever grapes or, for that matter, a variety of fruits are grown. Though people still make wine from tree fruits, berries, grains such as rice, and even flowers, for the purpose of our study, we shall hold to a narrower definition. Wine, as is commonly understood worldwide today, is the result of processing and subsequent fermentation of juice from grapes. Fermentation is a natural process that acts to stabilize grape juice and allow it to be stored as wine for later consumption. The alcohol in wine that is produced by fermentation also prevents the growth of pathogenic microorganisms. This means that wine was always safe to drink even when the local water supply was contaminated. Over the centuries differ-

FIGURE 1–1 The Wine Tasters. Tasting and enjoying wine has been a part of civilization for many centuries.

ent cultures have had contrasting opinions on wine. While many cultures have regarded it as an essential and healthful beverage often incorporated into religious rites, others have shunned its use and considered it sinful. These competing ideas have influenced the development of wine during the course of history, and continue to affect its place in society today.

The History of Wine

Because civilization itself began in the Middle East, it is not surprising that the origins of wine are in the same region. Popular belief is that wine was first consumed in the areas of Persia (modern day Iran) around 5000 to 6000 BC. Although the exact nature of the wine is uncertain, it was probably made from dates or other tree fruits native to the region rather than grapes and was undoubtedly rudimentary in nature.

Egypt and Greece

Around 3000 BC winemaking from grapes began. During this period both the Egyptians and the Phoenicians produced wines from grapevines that were specifically cultivated for that purpose. Ancient Egyptian artwork and sculptures provide a great deal of information about the winemaking practices of the time. Paintings and reliefs on the walls of tombs document how grapes were grown, picked, crushed, and fermented and how the resulting wine was stored (Figure 1–2). It is clear from these paintings that wine production had evolved into an elaborate procedure. The paintings also show that wine was an essential part of meals and celebrations of the Egyptian aristocracy. Containers of wine have been found in royal burial chambers for the dead to

FIGURE 1-2 Egyptian wall painting of winemaking showing vines of ripe grapes and crushing and amphorae for storage, circa 1400 BC.

enjoy in the afterlife. Much like the practice today, the vessels were marked with information on the origin of the grapes that produced the wine and the year, or **vintage,** they were harvested. It is also believed that wines were made in China during the same period, although it is not clear whether they were made from grapes or rice (Johnson, 1989).

By 2000 BC, wine had become an important part of Greek culture, its praises lavishly sung by the poets of the day. Beginning around 1000 BC, the expansion of the Greek empire brought vineyards and winemaking to regions throughout the Mediterranean basin, including parts of North Africa, southern Spain, southwest France, Sicily, and much of the Italian mainland. Apart from the fact that it was made from fermented grape juice, the wine of ancient Greece bore little resemblance to that of today. It was most likely made from dried grapes or raisins, as are some wines still made today. The result would have been a heavy, sweet, almost syrupy liquid, perhaps even concentrated by cooking. Barrels and bottles were not yet invented so the Greeks stored their wines in containers called **amphorae.** These were cylindrical jars with narrow necks and two handles similar to those used by the Egyptians. There was little understanding of the microbiology of winemaking so contamination and early spoilage was undoubtedly a frequent problem. Wine was often served with jugs of warm water (sometimes seawater) to dilute it. Like the Egyptians before them, wine occupied a large place in Greek society. The Greeks created a deity, **Dionysus,** in honor of wine, and no festival or banquet was complete without it.

Roman Era

The Romans took viticulture and winemaking to a new height. Though the growing of wine grapes in Italy predated the rise of the Roman Empire by many centuries, it was the Romans who began the practice of **trellising** vines off the ground by training them to grow up trees, a practice that is still followed in parts of Italy and Portugal.

As was the practice in other fields, the Romans' technological advances in viticulture (grape growing) and enology (the study of winemaking) were thoroughly documented in literature and

FIGURE 1-3 Men pulling a barge full of wine barrels. The Roman civilization brought much advancement to the practice of winemaking and extended vineyards throughout their empire.

art (Figure 1–3). Even the great poet Virgil offered advice to grape farmers ("Vines love an open hill . . ."). Unlike the Greeks, the Romans were first-rate barrel makers or **coopers,** and storage in wooden barrels, not unlike the barrels used today, as well as in clay amphorae, was common.

Although no one can say for certain what the wines of the Roman era tasted like, we do know that some wines from the best vintages were stored and drunk for up to a century or longer. In its most common form, Roman wine was probably similar to the inexpensive table wines of today that one finds throughout the Mediterranean growing region—young, light, and somewhat rough. Its best examples were probably somewhat more robust and flavorful and could age for many years.

As it had with the Greeks, Roman viticulture and winemaking followed closely on the heels of the Roman legions as they pushed the boundaries of their empire north and westward. The Romans grew grapes throughout Italy, expanded the vineyards of Spain north to the Pyrenees, and planted vines throughout what is now modern Portugal. Around the first century, they also began a steady expansion north from Provence through a wild and savage territory that would later become France and into parts of what is now modern Germany. By AD 500 the Romans were growing grapes in Languedoc, Auvergne, the Rhône and Loire Valleys, Burgundy, Bordeaux, Trier, Paris, Champagne, and along the Rhine and Moselle Rivers.

This great expansion of vineyards laid the foundation for modern viticulture, as it included all the principal regions that would come to make the wine world of modern Europe. Not to be outdone by their Greek predecessors, the Romans adopted their own god of wine, **Bacchus** (Figure 1–4). Moreover, like the Greeks, no holiday was complete without its Bacchanalia, or drunken feast.

In addition to the Greeks and Romans, wine was also significant to Jewish and early Christian cultures. Wine is mentioned more than 150 times in the Old Testament and is an important part of Jewish religious celebrations such as weddings and Passover. In fact, during the Exodus, several Israelis expressed their regret at leaving the Egyptian vineyards behind. As Christianity grew in popularity, the religious significance of wine also grew, spreading throughout the Roman Empire. The first miracle that Christ performed was the conversion of water to

FIGURE 1-4 Renaissance statue of Bacchus, the Roman deity of wine. Called Dionysus by the Greeks, the deity was also portrayed as a child.

wine, and tradition has it that the wine Christ consumed at the Last Supper was from the northwest Italian region of Lombardia. Christians also consumed wine as a sacrament during Mass in a reenactment of the Last Supper, referring to the consecrated wine as "the blood of Christ."

The medicinal qualities of wine were also highly regarded by the Romans. The physician Galen, doctor to Emperor Marcus Aurelius, freely prescribed wine in moderated daily doses as a cure for most illness. Galen, who had free access to the imperial cellars, believed that the older the wine, provided it had not soured in storage, the better was the cure.

By the first century AD, the Roman understanding of viticulture and winemaking had reached new heights. Though they lacked the knowledge and technology to perform even rudimentary chemical analysis, they were keen observers of the agricultural process. Several books, dating from as early as the third century BC, describe grape growing and winemaking in considerable detail. Columella, a second-century naturalist, dedicated his life's work to the study and improvement of wine. Another naturalist, Pliny the elder, writing in the first century AD, provided instructions on planting and tending more than 90 varieties of wine grapes. The Roman devotion to the physical, intellectual, and spiritual attributes of wine remains unique in history. Their knowledge of matching grape varieties to soils and climates, trellising, and other growing techniques forms the basis for many contemporary practices.

The death of Marcus Aurelius in AD 180 marked the end of the Roman Empire. By then, Rome had been fighting a war of attrition with the Gothic tribes of northern Europe. Having battled them for nearly two centuries, Rome finally proved unable and unwilling to raise the armies necessary to keep them subdued.

The Middle Ages

The fall of Rome ushered in a long period of great strife throughout the civilized world. Competing groups battled for control of territory and commerce. France and Spain, both of which eventually would evolve into great nation-states, provided the landscape for much of the turmoil. Wars between the Franks, Teutons, and Goths brought widespread destruction. In the seventh century AD, the Moors of North Africa crossed the Straits of Gibraltar and invaded Spain. Their occupation, reaching as far north as the Pyrenees and lasting until 1492, would eventually unite Christian Europe against them. Although the Moors' Islamic faith prevented them from consuming alcohol, they were tolerant of others living in the region, allowing viticulture and winemaking to continue in Spain, albeit at a more limited scale than during the Roman period.

During the Middle Ages, the 1,000-year period between the fall of Rome and the beginning of the Renaissance, the practice of agricultural activity on any meaningful scale fell to the Catholic Church. Monasteries were established by the various religious orders throughout Europe. The goal was both to expand Christian teaching and beliefs and to broaden the political power of the church. Some of the monasteries became great centers of study and knowledge, the forerunners of Europe's finest universities, while others fostered the learning of trades and still others became important commercial outposts.

No matter what their primary focus, the monasteries of Europe engaged in all aspects of agriculture, from the growing of food and feed crops to raising animals for milk and meat to growing grapes and winemaking (Figure 1–5). In fact, in exchange for accepting the teaching of the church, the peasants of Europe soon came to expect the much-needed assistance by the monastic orders in meeting their daily needs.

Given the by-then lengthy history of wine in Europe, it is little wonder that the church took over stewardship of the continent's vineyards. In the early seventh century, Pope Gregory the Great instructed the monastic orders to expand wine production, and the planting of wine grapes again began to spread. As was true with much of its activities, the church kept strict

FIGURE 1-5 A sixteenth century French tapestry depicting grapes being first crushed, then pressed into barrels.

control of winemaking. All grapes were required to be pressed in monasteries, for which a "donation" of 10 percent of production was taken. The church also controlled the commerce in wine, ensuring that the monasteries' stocks were fully sold before others were allowed to market their wines.

The wine wealth allowed the monks to continue **viticulture** (grape growing) and the study of winemaking **(enology)** begun by their Roman predecessors. Matching grape varieties to soil conditions and climate, propagation and planting, trellising, crushing, fermenting, fining, and storage were all meticulously studied and improved, resulting in great leaps in the quality of their wines. Meanwhile, the peasants were forced to make inferior wines from lesser grapes for their own consumption.

Despite the inherent inequities, the relationship between peasants and monks was a fruitful one for both sides. As the church grew in wealth and power, the countryside prospered with the flourishing of trades and crafts, not the least of which were farming and winemaking. Villages formed around monasteries, many growing into bustling towns and some into cities of 20,000 or more. Farmers began using the skills learned from the monks to improve and expand their own vineyards, and viticulture quickly became a major form of agricultural activity.

During the reign of Charlemagne (771–814), medieval viticulture and enology reached a peak. A great scholar as well as king, Charlemagne nurtured the monasteries, bestowing great gifts of land and other wealth to the church. In exchange, he required that vines be planted and well maintained throughout Europe. Consequently, viticulture and winemaking reached new heights.

Twelfth Century to Modern Times

By the twelfth century, the political landscape of Europe had undergone great changes. The Crusades, or holy wars against the Moslem world, were well underway. With the cooperation of a grateful church, the European monarchies were rapidly consolidating their power. City-states were developing into nation-states. The wine world of the time profited greatly from the benevolent urges of a rising aristocracy. Like Charlemagne, the nobles of the day bequeathed large tracks of land to the church, much already planted to vineyards. The monastic orders made good use of their increasing endowment. Some of the finest vineyards in France, Germany, Italy, and Spain trace their origins to monastery plantings of the period.

A major event during this period for the wine world took place in 1152, when Henry II, King of England and Duke of Normandy, married Eleanor of Aquitaine, the divorced wife of Louis VII of France. The Norman Conquest in 1066 had created a situation where English kings were also French nobility with the right to own lands on both sides of the English Channel. Eleanor's dowry included the entire winemaking region of what is now Bordeaux in southwest France. These vast vineyards were combined with those of Henry's Anjou estates in the Loire Valley, and represented perhaps the single largest holding of vineyards of its time.

During the reigns of Henry II and his son, Richard the Lion Hearted, the English developed an enormous thirst for the wines of France. A huge fleet, the forerunner of the British Navy, developed to accommodate the demand on shipping. Wines from Languedoc, Loire, and Bordeaux poured across the channel in a seemingly never-ending flow.

It took nearly three centuries, including a continuing series of wars between 1337 and 1453 (known collectively as the Hundred Years War) for the French to dislodge the English from their precious vineyards. In 1429, led by Joan of Arc, the French drove the English out of the Loire Valley, and in 1453, they succeeded in expelling them from Bordeaux. With the loss of their French territories, the British turned to other regions such as Germany, Italy, Portugal, and Spain to satisfy their thirst for wine.

By this time, the end of the fifteenth century, the great European Renaissance was well underway. Literally a "rebirth" in creative thinking, the Renaissance was to have a profound and lasting effect on religion, philosophy, science, and art. The church, though a leader in many areas of the Renaissance, soon found its authority shaken by such free thinkers as Martin Luther. The monastic orders became easy targets for religious reformers, and it wasn't long before their economic and political hold on the populace began to fade. By the end of the seventeenth century, much of the church's vineyard holdings throughout Europe had been broken up and passed back into private hands.

The early eighteenth century saw the wide spread use of cork as a bottle stopper and the development of sparkling wine or **Champagne.** This is more than a coincidence because Champagne bottles require a good seal. Although the Benedictine monk, **Dom Perignon,** is often credited with the discovery of Champagne, he probably produced them by accident and others developed the techniques of production (Kolpan, Smith, & Weiss, 2002). Champagne soon became fashionable, not only in Europe, but also throughout the then-extensive French colonies and settlements in Africa, Asia, and the New World. Cork, with its unique sealing properties, revolutionized the storing and aging of wines, making it possible to age wines for long periods and to ship them in bottles to distant markets for sale and consumption.

Golden Age of Wine

It was in the nineteenth century, though, that wine enjoyed its greatest advances and suffered one of its most devastating blows. The advent of modern studies of chemistry and microbiology brought a deeper understanding of the winemaking process, and the laboratory soon began playing a major role in winemaking. This better understanding of technology, combined with the knowledge gained by centuries of trial and error in European vineyards, resulted in huge advances in the quality of wine. As the science and caliber of wine took a leap forward, wine appreciation in the modern sense was born. Attracted to the glamour of winemaking, the wealthy soon began buying up vineyards throughout Europe. These new wine "lords" showed off their holdings by labeling their products with both their family and estate names (bottles were typically unlabeled before). The French established a system to classify their vineyards. Great vintage followed great vintage throughout the first half of the century, creating what some have called a "Golden Age of Wine."

In the second half of the century, disaster struck in the form of a microscopic root louse, **phylloxera,** that is native to the eastern United States. The pest was brought to France on a merchant ship carrying grapevines that were native to North American. By 1868, phylloxera had been identified in the vineyards of southern France. Within 20 years, it spread throughout the country, destroying most of the vineyards. By 1874, it had also infected Germany. Some French producers migrated to Spain, taking their grape varieties with them. However, eventually the pest followed, and soon all the vineyards of Europe were infected. It was not until growers began replanting their vineyards with rootstocks from North America that Europe's winemaking industry was revived. These rootstocks, being native to the region phylloxera was from, had evolved to be resistant to the pest.

The phylloxera epidemic, coupled with economic and political turmoil, sent many wine makers, both wealthy and of modest means, in search of new vineyard land. It was during the late nineteenth and early twentieth centuries that the wine industry in the New World became commercially important. Vineyards in North and South America, Australia, and South Africa flourished with the influx of emigrants from Europe. Though some areas, especially California, had an already established wine industry, many of today's new-world wine regions trace their start to the years between 1880 and 1910. The New World producers took their cue from their European predecessors, in many cases borrowing grape varieties, techniques, and technologies (Figure 1–6). However, when winemakers from different growing regions of Europe came to the New World they adapted Old World methods to the particular conditions in their new homes. This combination of winemaking techniques from around Europe helped to create many innovations.

The first half of the twentieth century, with its two devastating world wars, again saw setbacks in winemaking worldwide. In America, this was compounded by **Prohibition,** which outlawed the sale or consumption of alcoholic beverages from 1919 until 1933. The temperance movement had been gaining ground in the United States for 100 years and at the end of World War I, the **Volstead Act** was passed implementing Prohibition as the 18th Amendment. During this time most wineries went out of business save for a few that were allowed to make sacramental wine or medicinal "wine tonics." Prohibition did little to control alcohol consumption, and Americans continued to drink "bootleg" alcohol obtained illegally or primitive wines made at home. Organized crime flourished distributing alcohol, and the government lost the alcohol sales tax revenues it had received before Prohibition. The 18th Amendment was repealed in 1933 when it became obvious that it was not working. During Prohibition, Americans' tastes changed because wine drinkers became used to drinking substandard homemade wines that were often sweetened and fortified with distilled alcohol to cover up the flaws. As part of repeal, each state was allowed to make its own regulations concerning the commerce of alcohol that led to a confusing patchwork of state laws that survive to this day.

Following the Second World War, both the Old and New world wine industries saw a resurgence as reconstruction monies flowed to Europe and returning U.S. servicemen came home with a newly acquired interest in wine. By the 1950s, wine, as a beverage and as a business, was again on the rise. Throughout the 1960s and 1970s, wine production and consumption grew at an in-

FIGURE 1-6 Pressing grapes and pumping juice into fermentation tanks at a California winery in 1911.

creasing pace. In America, some producers took the revolutionary step of naming their wines after the grape varieties they were made of (i.e., Cabernet Sauvignon or Chardonnay) instead of following the common practice of using European geographic names, such as Burgundy or Chablis to identify their wines.

Wine Today

Since the early 1970s, the wine world has been undergoing another huge transformation. Where before, in both the Old and New Worlds, there were only a handful of producers making high quality, premium wines, today there are thousands of producers throughout the world making excellent wines (Figure 1–7). Behind this explosion of quality producers lies a greater consumer interest in fine wines and the broader availability of state-of-the-art technology and winemaking expertise. There are now excellent viticulture and enology schools in Europe, North and South America, Australia, and South Africa. New technology in both the vineyard and winery is widespread.

In Europe, the lesser known regions of southern France, Italy, Spain, Greece, Hungary, and even the former countries of the Eastern Bloc are now making wines that are on par with those of some of the best traditional growing regions. In the United States, New York, Washington, Oregon, Virginia, and Texas are now recognized wine producers. In fact, North America's reputation as a winemaking region today rivals that of Europe, and many consumers are just beginning to discover the moderate price and excellent value of the wines of Australia, New Zealand, Chile, Argentina, Hungary, and South Africa. This global competition from new wine regions has put pressure on producers in traditional regions, such as California and Europe, to keep their prices competitive, and it has given consumers access to a multitude of quality wines at attractive prices.

FIGURE 1-7 A young vineyard in California.

Over the past 25 years, the world's wine industry has become vastly more consistent in product. The worldwide focus on quality improvement has led to the increasing standardization of taste and styles. This is particularly evident in the growing vineyard acreage dedicated to the so-called **fighting varietals** (Chardonnay, Cabernet Sauvignon, and Merlot). In France, Italy, Spain, and even California, growers are removing more traditional vines to compete in the fighting varietal market. There has also been a movement toward making softer, less tannic wines that require little bottle aging before they are consumed. This global viewpoint and increased competition has resulted in consolidation of many winemaking companies. Frequently smaller family-owned wineries are purchased by large multinational corporations that import and export wine as well as produce it. While this globalization of the industry has led to better prices and more consistently good product, some wine enthusiasts feel it has also made wines from around the world more homogenous and uninteresting.

Economic Cycles in the Wine Business

Wine, because it is made from grapes, is considered an agricultural product, and like many other agricultural products, exhibits a **boom and bust** economic pattern. Because of the time it takes to establish a new vineyard, get a crop from its vines, and then produce a wine from its grapes, it is very difficult for growers and vintners to respond to changing market conditions. For example: the popularity of a certain variety of wine will lead to scarcity of the grapes used to produce it, resulting in high prices. At this point, many growers will plant the variety to take advantage of the higher prices resulting in overproduction and ultimately lower prices for their crop. Another factor influencing the economics of winemaking is the fact that many wine consumers consider wine a "luxury item" and not a food. In difficult economic times, fine wines are one of the things consumers give up to save money.

Summary

Over the next 20 years, grape growing and winemaking will likely continue the course it has been on since the 1970s. Technological advancement, including sophisticated software for use in both the vineyard and the winery, should continue at its current or at an even greater pace. There is now a movement to blend new technology with traditional grapes and winemaking techniques, especially in Spain, Italy, and parts of France. In the New World, countries like Chile and Australia have invested heavily in new vineyards and wineries to take advantage of the export market. This reflects the continued growth in consumer wine knowledge and interest.

While total wine consumption worldwide is falling, demand for premium wines in Europe, North America, and the Far East continues to grow. This has been matched by increased plantings throughout the world to meet the anticipated need. In the United States demand for California wines has been tempered by inexpensive imports from South America and Australia, putting pressure on California vintners to keep their prices low and quality high. The wine business climate at the turn of the new century is similar to one that existed during the early 1980s. During that time excess production of grapes and wine led to reasonably priced, high quality wines that brought in many new consumers. These new wine drinkers in turn helped to fuel the growth in the wine industry that was to come in the 1990s. Despite the uncertain, cyclical nature of the wine business, it will undoubtedly remain an integral part of fine dining, as it has for thousands of years.

Review Questions

1. When did winemaking begin and what fruits were used to make the first wines?
2. What were the differences in winemaking between the ancient Greeks and Romans?
3. When did winemaking come to Europe, and what culture brought it?
4. What was the role of the Church in relation to wine in the Middle Ages?
5. What effect did the introduction of phylloxera have on the vineyards of Europe?
6. What were the causes of Prohibition, and what factors influenced its repeal?
7. How do economic cycles affect grape growing?

CHAPTER 2

The Vineyard—From Soil to Harvest

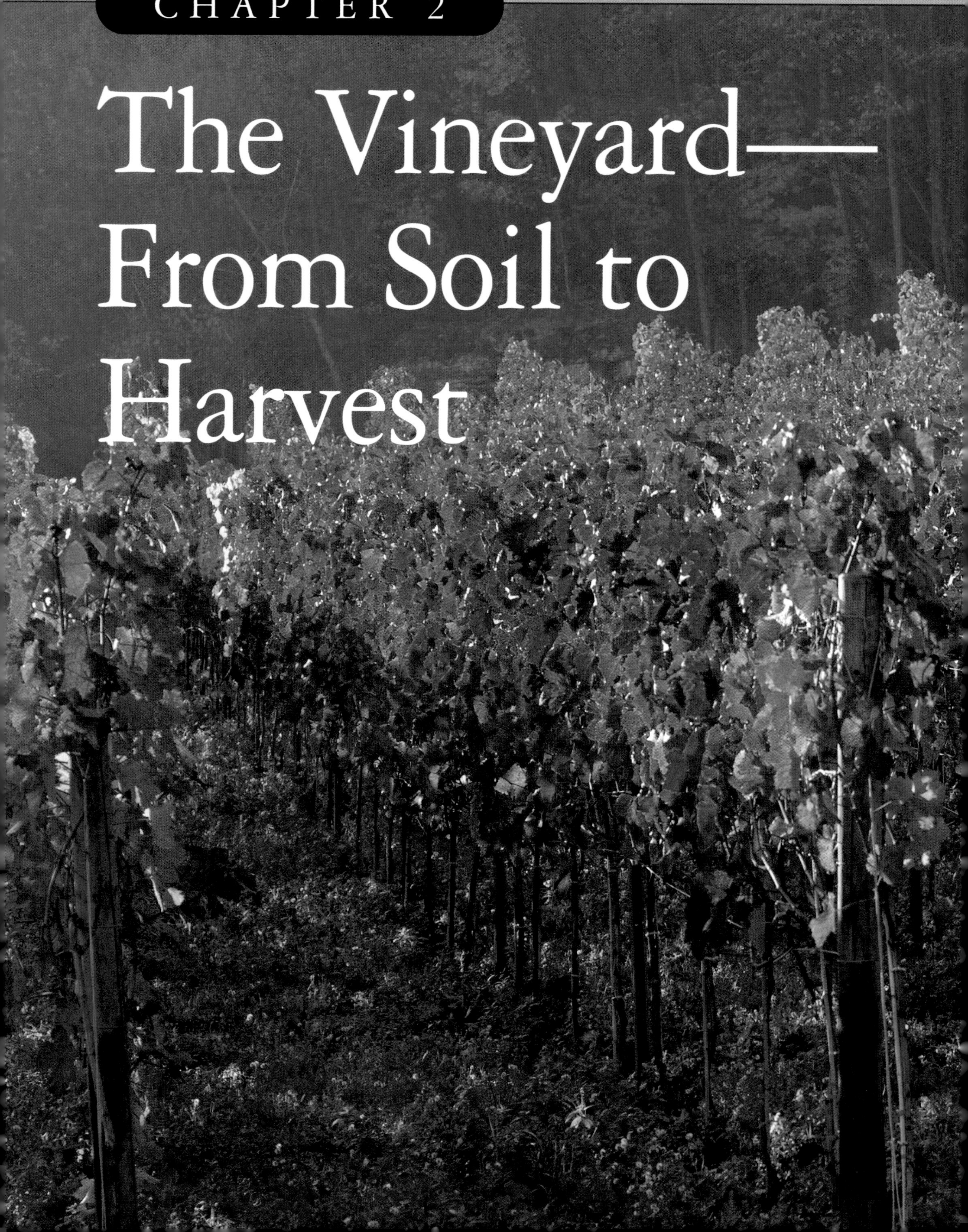

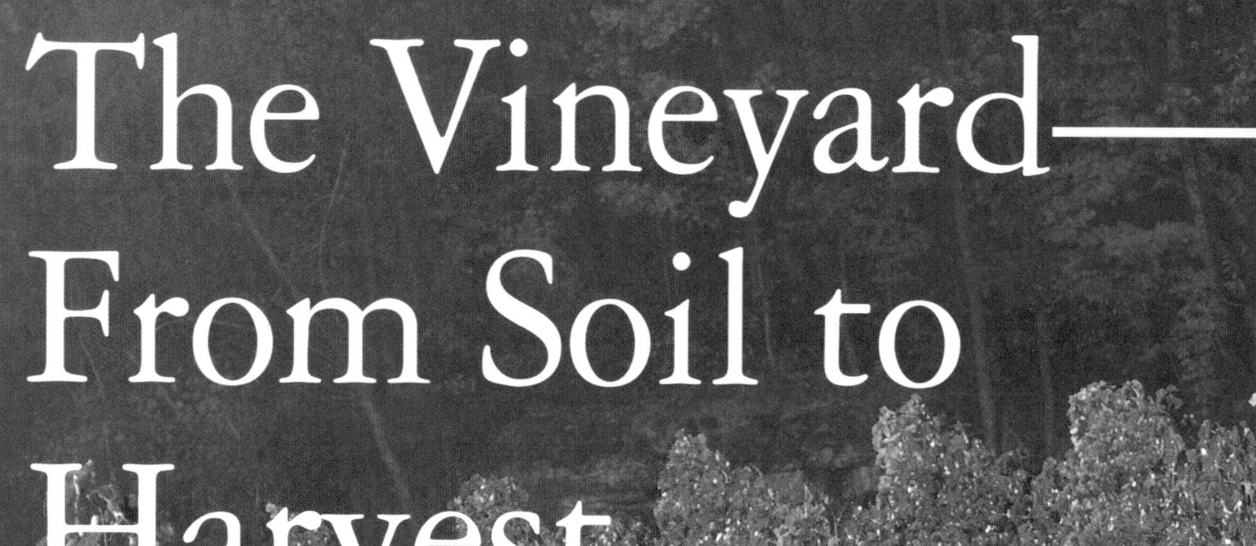

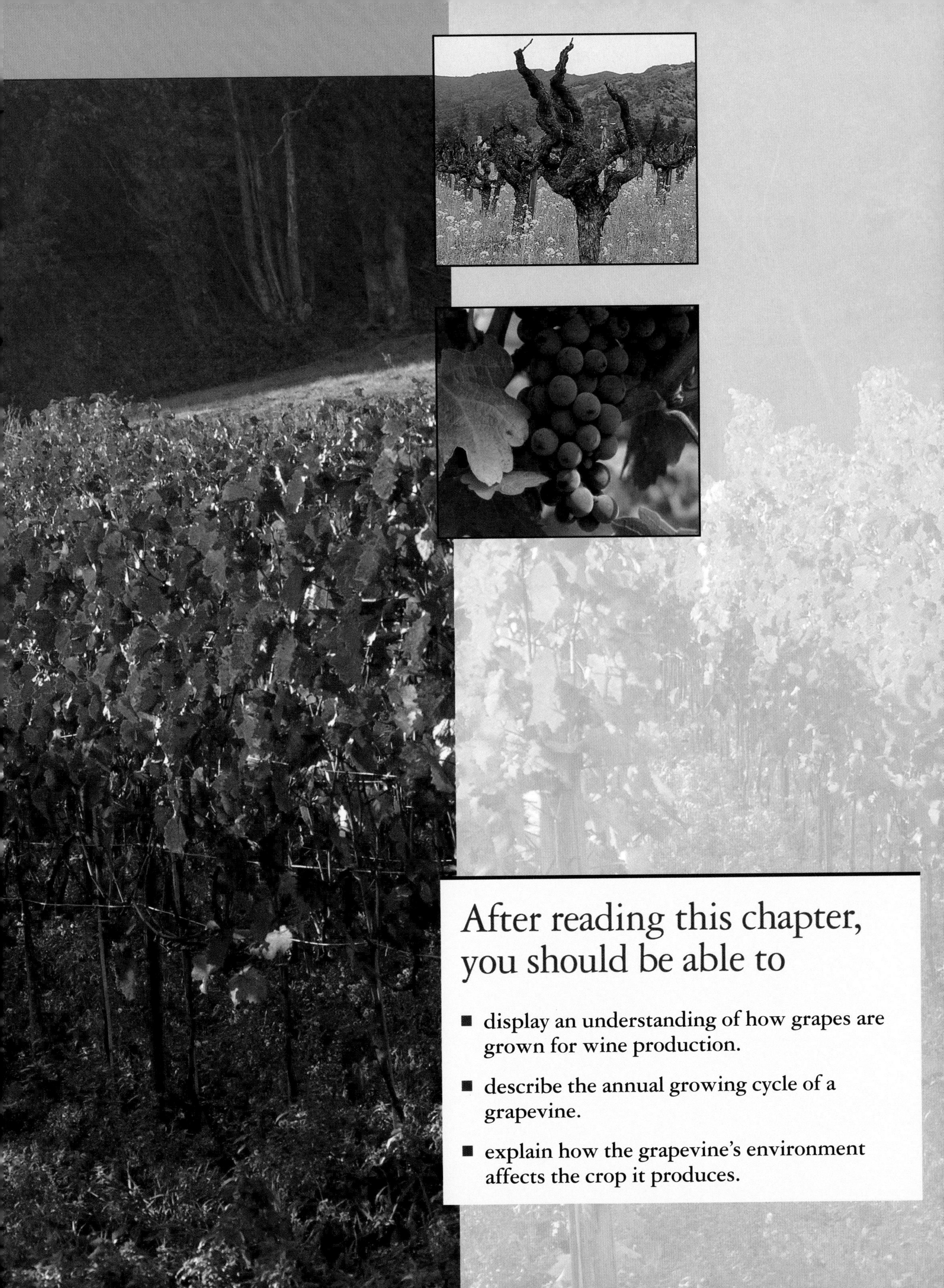

After reading this chapter, you should be able to

- display an understanding of how grapes are grown for wine production.
- describe the annual growing cycle of a grapevine.
- explain how the grapevine's environment affects the crop it produces.

KEY TERMS

alluvial	loam	terroir
bloom	macroclimate	trellis
canes	mesoclimate	véraison
clone	microclimate	*Vitis labrusca*
cordons	*Muscadinia rotundifolia*	*Vitis vinifera*
grafting	rootstock	winterkill
heavy soil	scion	
light soil	shatter	

Great wines are made from great grapes, and the ultimate quality of a wine is determined in the vineyard as much as it is at the winery (Figure 2–1). Most fruit crops are grown with an emphasis on appearance, and picked before ripeness so they can survive the trip to the market without deterioration. Additionally, many agricultural crops are grown as a commodity, where the quality of the product and prices are uniform among growers. In contrast, premium wine grapes are one of the few crops that are still grown primarily for their flavor, and there is greater variation in prices that growers receive based on quality of their fruit. There are two factors that influence the character of grapes from a given vineyard, environmental and cultural. Environmental factors are all of the natural attributes of the vineyard site, including climate, soil, and drainage, while cultural practices are all of the actions performed by the grower, such as pruning, trellising, and selection of grape variety.

FIGURE 2–1 Pinot Noir grapes. The dark purple color of the skins is one indication that they are ready for harvest.

Grapes Used for Winemaking

Before covering the types of grapes that are used for winemaking, it is important to consider why grapes are the preferred fruit for wine production. While it is possible to make "wine" out of many fruits and berries, it cannot be done without adding amendments to the juice before it can ferment, for example: sugar, water, and nutrients. Grape juice has all of the attributes, high concentration of sugar and enough nutrients to support yeast growth, necessary for fermenting the juice into a beverage with enough alcohol to inhibit microbial spoilage. There is also an inherent association between grapes and yeast. The outside of the grape berry is covered with a waxy layer that contains naturally occurring yeast. Therefore, it is possible to make a rudimentary wine by simply crushing grapes into a vessel and letting the natural yeast ferment the juice. The great majority of wine produced in the world is from grapes, so much so that the term "wine" has become synonymous with wine made from grapes.

There are many indigenous species of grapes worldwide but the overwhelming majority of wine produced is from the species ***Vitis vinifera.*** Native to Asia Minor it has been spread by humankind throughout the Old and New Worlds. Within the species *Vitis vinifera* there are over 5,000 named cultivars or cultivated varieties (Boulton, Singleton, Bisson, & Kunkee, 1996); of these only a fraction are grown commercially. These cultivars, such as Cabernet Sauvignon and Chardonnay, exhibit great latitude in growing characteristics, appearance, and flavors. The difference in grape varieties is analogous to those in other fruits, where a single species like *Malus domestica,* or the common apple tree, has a number of varieties that taste very different, such as Red Delicious and Pippin.

There is also variation within a single grape variety. Grapevines can be propagated by taking cuttings off the parent plant and grafting them onto another grapevine, or if the cuttings are planted, they will form roots to become a separate vine that is a **clone** (genetically identical) of the original vine. Different clones have different growing and flavor characteristics. In the Burgundy region of France viticulturists have developed more than 50 certified clones of Pinot Noir (Entav, Inra, Ensam, & Onivins, 1995), each with its own unique properties (Table 2–1).

Although the *vinifera* grape is very versatile, it is not well suited to cold or humid climates. In the eastern United States, native grape varieties such as Concord ***(Vitis labrusca)*** and Scuppernong (*Muscadinia rotundifolia,* also known as *Vitis rotundifolia*) are grown for winemaking. Since these species are indigenous to the east coast of North America, they evolved to thrive in the climate as well as being naturally resistant to pests of the region. In an effort to combine the hardiness of native American varieties with the flavors of European (*vinifera*) varieties, the two were

Table 2-1 Examples of Multiple Clones of a Single Grape Variety Pinot Noir

CLONE NUMBER	LOCATION OF ORIGIN	CLONE NUMBER	LOCATION OF ORIGIN
Pinot Noir 01A	Sel B111, Wadenswil, Switzerland	Pinot Noir 31	Roederer, France 236
Pinot Noir 02A	Sel Bl 10/16, Wadenswil, Switzerland	Pinot Noir 32	Roederer clone, France 386
Pinot Noir 09	Jackson, CA	Pinot Noir 37	Mt. Eden, CA
Pinot Noir 13	Martini 58, CA	Pinot Noir 38	France 459
Pinot Noir 15	Martini clone 45, CA	Pinot Noir 39	France 386
Pinot Noir 16	Jackson, CA	Pinot Noir 40	France 236
Pinot Noir 18	VEN, UC Davis, GB type	Pinot Noir 44	France 113
Pinot Noir 19	VEN, UC Davis, GB type	Pinot Noir 46	France 114
Pinot Noir 22	VEN, UC Davis, GB type	Pinot Noir 47	France 114
Pinot Noir 23	Clevner Mariafeld, Switzerland	Pinot Noir 48	France 162

FIGURE 2–2 The root louse phylloxera feeding on a grape vine root.

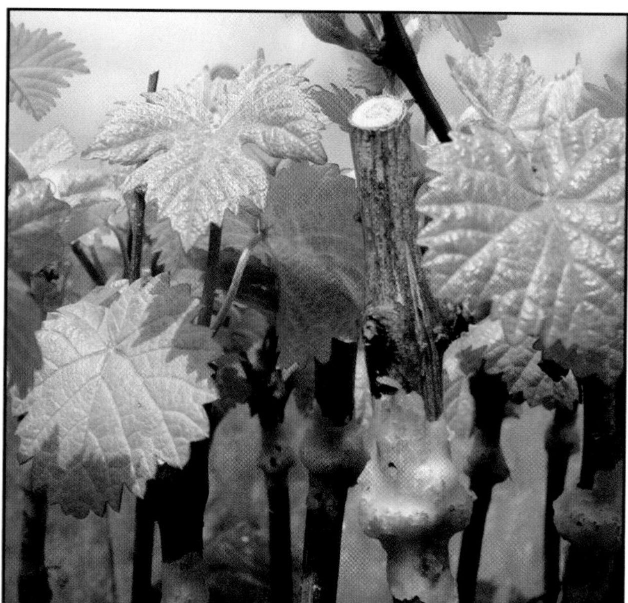

FIGURE 2–3 A young bench-grafted grapevine showing the union between the rootstock and the scion covered with wax to protect it while the graft becomes established.

crossbred to produce hybrids. Examples of these French-American hybrids are Delaware and Vidal Blanc. These native and hybrid varieties are planted throughout the eastern United States but have exotic flavors and have not found wide acceptance with wine consumers.

In the second half of the 1800s European interest in developing new grape varieties for winemaking led to the importation of native American grapes. When native American grapevines were transported to Europe, they inadvertently brought with them several grape pests and diseases that had not been known in Old World before. Downy mildew, powdery mildew, black rot, and phylloxera quickly spread throughout the wine growing regions of Europe in part because the *vinifera* grape has little natural resistance to these pests. The root louse phylloxera was by far the most devastating of these pests, destroying nearly all the vineyards of France by 1900 (Figure 2–2). Vines brought from France further spread phylloxera to California and other parts of the world. To combat the problem of phylloxera, viticulturists developed the technique of **grafting.** Grafting is the process of taking a cutting, or **scion,** from a *vinifera* variety and affixing it to a **rootstock** that is bred from native American grapes that are phylloxera resistant (Figure 2–3). The result is a vine that has roots that are resistant to phylloxera and produces *vinifera* fruit for winemaking. This is still the universal method for growing grapes where the root louse is present or likely to come. There are many different rootstocks that are available to vineyard managers that are adapted to a variety of growing conditions.

Soil and Site

The most fundamental aspect of any vineyard is the ground in which it is planted, and the qualities of the soil affect the character of the wine that the vineyard produces. Grapevines are not demanding when it comes to the types of soil that they will grow in and the vineyards of Europe were originally planted in areas where soils were not fertile enough to grow other foodstuffs. In fact, there is a belief that vineyards that are stressed by environmental conditions produce more flavorful grapes. There is a saying in Bordeaux, "If these were not the best soils in the world they would be the worst." Soils that are shallow or low in nutrients will put pressure on the grapevines resulting in smaller berries and lower cropload with less vegetative growth. Here vegetative growth refers to the vines' production of leafs and **canes;** it is not to be confused with "vegetative aroma," a green bean/bell pepper-like aroma that is sometimes found in wines. For example, Cabernet Sauvignon with small berries has a higher skin-to-juice ratio and will produce a wine with deeper color and more tannin. If there is less vegetative growth, the grape clusters will not be as shaded by the leaves; sun exposure on Cabernet Sauvignon clusters will give them more fruity aromas and less of a vegetative aroma. Grapes grown in fertile soils can also produce high quality fruit but they must be managed in such a way as to permit the proper amount of sunlight and air on the grape clusters and leafs and not be overcropped.

The composition of the soil a vine is planted in is important to the plant's health as well as the quality of the fruit that it produces. There are a number of parameters that go into a soil's makeup:

- The parent material or rock that the soil is composed of
- The size of the particles the soil is made of: clay has very fine particles, silt has larger particles than clay, and sand has larger particles than silt
- The chemical composition and pH (acidity) present in the soil
- The organic matter and nutrients that are present in the soil
- The depth of the soil

Soils are characterized by the ratio of sand, silt, and clay that are present in them. Soils that have a high proportion of clay are said to be **heavy** and have a great capacity to hold water and generally contain more nutrients. Sandy soils are called **light;** they hold less water and usually are lower in nutrients. **Loam** is a mixture of clay, silt, sand, and organic matter that is fertile and drains well. Grapevines generally prefer light or rocky soils that drain well, keeping the roots from being waterlogged. Rocky soils also warm up more quickly in the spring, allowing grapes to be grown in cooler climates with shorter growing seasons. **Alluvial** soils lie in the floodplains that flank rivers and streams; they are usually a mix of silt, sand, gravel, and loam. Grapevines need adequate, but not excessive, nutrients from their soil to sustain healthy growth. The minerals and nutrients that are present in the soil have a limited effect on the flavor of the grapes unless they are present in quantities that are too little, deficient, or too great, toxic, to support healthy vines.

A vineyard's topography is as important as its soil. A vineyard on a south-facing hillside will absorb more sunlight and be warmer than its counterpart on the north side of the hill. Furthermore, hillside vineyards will have better drainage but will be more susceptible to erosion and wind damage. Cold air will settle into valleys and low-lying areas on still mornings without wind making these locations more susceptible to spring frost and **winterkill,** the death of vine tissue from excessive cold. The Rheingau region of Germany is at the northern limit of where grapes can be grown, yet by planting in the rocky soils of the south-facing river valleys, grapes can still get ripe (Figure 2–4).

Climate

Climate has an even greater influence on wine grape quality than soil. Grapevines do best in temperate zones between 30° and 50° in latitude north or south. In this zone, winters are sufficiently cold to allow the vines to drop their leaves and go dormant, but do not often get below 0°F (−18°C) and cause winterkill. Grapes also require adequate rainfall to support growth and crop development, but this can be augmented by irrigation. Vines grown in dry areas with drip irrigation have an advantage because rain and humid conditions promote mildew and rot as well as other diseases. These broad weather conditions of a particular wine-growing region are defined

FIGURE 2–4 Vineyards grown on south-facing hillsides, such as these in Germany's Neckar River Valley, absorb more energy from the sun, allowing them to obtain full ripeness in the cool northern climate.

Terroir

Terroir is the French term to describe all of the environmental factors that nature imparts to a given vineyard. It is a common misconception that when viticulturists speak of terroir they are referring only to the soil. While the proper soil is very important to growing wine grapes, the soil of a vineyard has less of an effect on the flavor of a wine than weather. Terroir is a holistic philosophy and relates to all of the properties of the soil—composition, drainage, mineral content, the topography, orientation, and direction of the slope; and all of the climatic conditions such as rainfall, temperature, and humidity.

Vintners and wine writers sometimes elevate terroir to almost magical proportions, and make comments like "a truly great Pinot Noir can only be grown in the Côte de Nuits" (Figure 2–5). The Côte de Nuits district of Burgundy does grow excellent Pinot Noir, but there are many other grape-growing regions of the world that have a terroir that is similar to the Côte de Nuits and also produce excellent wines from Pinot Noir (Figure 2–6). Part of the concept of terroir is not only having the proper environmental conditions but also matching the choice of variety and vineyard management to suit the terroir. A vineyard that grows great Cabernet Sauvignon would be unlikely to produce great Chardonnay; and even if the terroir and variety are perfectly suited to each other, if the vineyard is poorly managed, the crop will be of inferior quality. In Europe, where cultivation of grapevines has been carried on for over a thousand years, there is a great deal of tradition of which appellations are appropriate for certain varieties. These selections have been worked out by trial and error over hundreds of vintages, and in some areas the choice of variety has been codified into law. In the New World, there is generally less regulation, this allows more innovation and flexibility, but it also means that sometimes varieties are planted in inappropriate terroirs with less than ideal results.

Figures 2–5 and 2–6 show vineyards in separate wine regions, one in Burgundy France and one in the Russian River appellation of California. Although they are located on different continents both regions have terriors that are known for producing excellent wines from Pinot Noir grapes.

FIGURES 2–5 The Burgundy region of France.

FIGURES 2–6 Russian River Appellation of California.

as the **macroclimate.** Local conditions that influence the weather in a particular vineyard or portion of a vineyard are referred to as the **mesoclimate** or **microclimate.** While the term microclimate technically refers to the conditions around a particular vine, it is used much more frequently than mesoclimate to describe the conditions at a given vineyard. Grape varieties not only have a wide range of color and flavors, they are also suited to a diversity of climates. As an example, Pinot Noir tastes better when grown under cooler conditions then Cabernet Sauvignon.

Grapevines are often grown in coastal areas where the ocean has a moderating influence on the climate keeping it from getting too warm in the summer or too cold in the winter. Warm nights will increase the metabolism of malic acid in ripening grapes so inland areas that do not have the benefit of evening sea breezes will have a lower acid level than those grown near the coast. This is evident in California where the inland Central Valley is known for producing large crops of lower priced, less flavorful grapes, while the coastal valleys such as Sonoma and Napa produce a higher quality and higher cost product.

Techniques of Grape Growing

Vineyard managers rarely get the amount of recognition that winemakers do; however, their contribution to making good wine is every bit as important. If they are not managed properly, even the best vineyards will produce mediocre grapes. Growers must always be aware of the status of their grapevines and monitor them throughout the growing season. Water stress, pests, diseases, nutrient deficiencies, and extremes of temperature can all be moderated if the grower is observant and reacts to the problem as soon as it develops. The modern viticulturist has many tools that he can use to influence the development of his crop and correct small problems before they get out of hand.

The most important decisions that a grape grower will make all take place before the vineyard is planted. First the site must be chosen and prepared for planting the grapes. The soil can be tested for its composition and amendments to the soil can be tilled into it to get the mineral content to the necessary level. If the area has been used to grow grapes before, grape pests and diseases may be present in the soil (Figure 2–7). There are a number of soil pests such as nematodes (a microscopic worm that can spread grapevine disease) and fungi that will grow on the roots of grapevines and diminish their productivity. These factors can be controlled by using a rootstock that is resistant to them or by fumigating the soil before planting. Fumigation with methyl bromide is an effective preplanting treatment and is used for many different crops, but it is in the process of being phased out and replaced with other fumigants due to environmental concerns.

After the vines are planted and begin to grow, there are a number of options for the trellising of the vines. Young grapevines cannot support themselves and if left on their own will grow spread out along the ground, or in the wild, they grow using nearby trees for support. Both of these options are impractical so growers use an artificial support called a **trellis.** There are many different types of trellis systems from the simple to the complex, and they are used according to

FIGURE 2–7 This old-vine Zinfandel vineyard in the Sonoma Valley has red leaves in the fall, caused by the presence of viral diseases infecting the vines.

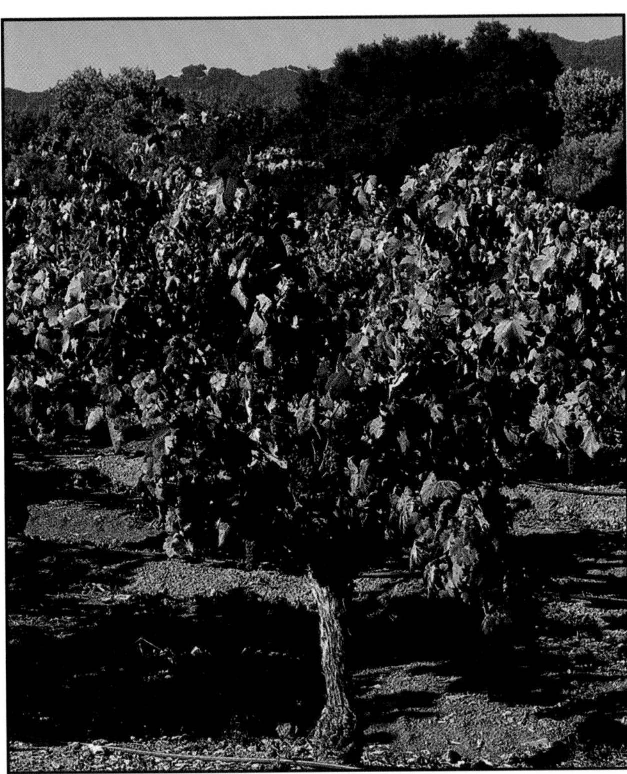

FIGURE 2–8 A head-trained Zinfandel vine. Head training and spur pruning was the most popular method of growing grapes in California until the 1970s when more elaborate trellis systems were introduced.

FIGURE 2–9 A Sauvignon Blanc vineyard trained in a two-curtain trellis called a U-system. How light penetrates the divided canopy is evidenced by the double shadow cast by the row of vines on the right side.

varying viticultural situations (Figure 2–8). High vigor vines with lots of vegetative growth will benefit from a complex trellis that spreads out the canes in an orderly fashion, providing a maximum of sun exposure to the leaves and adequate ventilation to the grape clusters (Figure 2–9). A low vigor vine will not have an enough growth to fill a large trellis so a simpler system is more appropriate. Grapevines can be thought of as a type of solar collector that uses water, carbon dioxide, and sunlight to produce sugar for the ripening grapes. The more efficiently the leaves collect the sunlight, the more easily that they will ripen the fruit. A vineyard's vigor is also directly related to its yield. While non-irrigated Chardonnay, planted in a cool area with poor soils, may be able to produce only 2 tons per acre, the same vines planted in a warm area with deep fertile soils and plenty of irrigation may produce more than 10 tons per acre. It is worth noting that in many countries, particularly in Europe, yields are expressed in terms of the quantity of wine the vineyard produces per hectare or hectoliters/hectare (hl/ha). Depending on the variety, a ton of grapes yields approximately 175 gallons (662 liters or 6.62 hectoliters) of wine, so 1 ton per acre is roughly equivalent to 16 hectoliters per hectare.

The Growing Season

Grapevines are deciduous, meaning that they lose their leaves in the fall and go dormant over the wintertime (Figure 2–10). In the fall when the vine is going dormant, the canes harden and become woody in texture; and with the leaves gone there is no green tissue on the vine, so photosynthesis does not take place. Because there is no green tissue on the vine, it is more tolerant of cold temperatures than at other times of the year. This dormancy creates an annual cycle of the growing season that begins in the spring and ends in the fall after harvest.

Budbreak

The growing season begins in the early spring—usually between February and April in the Northern Hemisphere depending on latitude. In the Southern Hemisphere this occurs six months later, in August to October. When the average temperature reaches 50°F (10°C), the vines end their winter dormancy and the buds formed during the previous year's growing season begin to swell. High soil moisture will keep the root zone cool so in a wet year budbreak will be delayed more than in a dry year. Soon tender green shoots, or canes, sprout from the buds and begin to grow quickly (Figure 2–11). At this point, the new shoots are very delicate and sensitive to subfreezing temperatures.

Spring frosts are common in low-lying vineyards where the cold air can settle in the early morning hours. Growers pay attention to frost warnings and sleep with a temperature alarm on

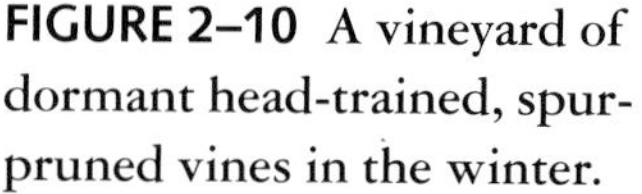

FIGURE 2–10 A vineyard of dormant head-trained, spur-pruned vines in the winter.

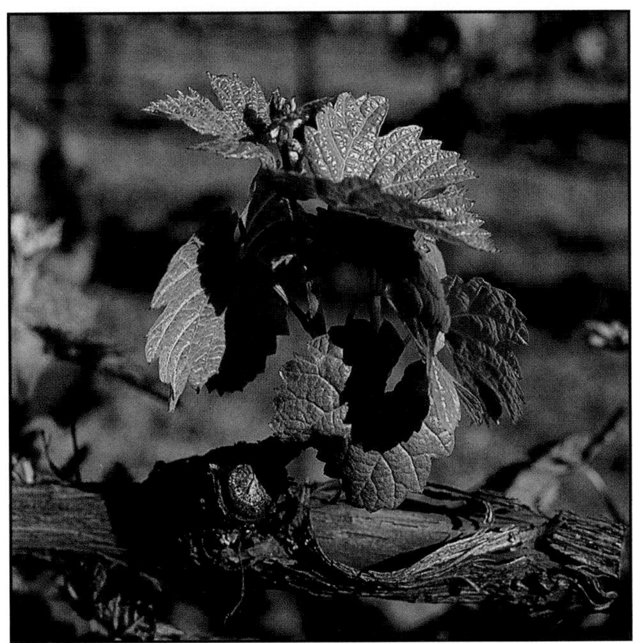

FIGURE 2–11 A young shoot about 10 days after budbreak. As the season progresses this shoot will grow to become a cane with developing grape clusters.

FIGURE 2–12 A wind machine in a Chardonnay vineyard in the Carneros region of Sonoma County. On frosty spring nights these large fans stir up cold air on the surface of the ground, mixing it with warmer upper air to keep the tender young shoots from freezing.

FIGURE 2–13 Overhead sprinklers being used on a frosty spring morning to protect tender shoots from freezing. As water freezes it actually liberates heat, and as long as there is a new layer of ice constantly forming on the vines, they will not fall below 32°F (0°C).

their nightstand to wake them when a frost is approaching. Once awakened they go to their vineyard and start large wind machines to stir up the cold layer of air along the ground and mix it with the warmer air off the surface to keep the shoots from freezing (Figure 2–12). Another method is to put overhead sprinklers in the vineyard and begin watering as soon as the temperature falls close to freezing (Figure 2–13). It is counterintuitive, but as water freezes it actually liberates heat, and as long as there is constantly a new layer of ice forming on the vines they will not fall below 32°F (0°C).

Once the shoots begin to grow they are also susceptible to rot and other diseases. To control this, growers keep a close watch on their young vines and spray with sulfur or other man-made fungicides to prevent them. Later in the year, bunch rot can also grow on ripening fruit and lower its quality, so it is important that it is not established early in the season. Weed control is also done in the form of mowing, tilling, or herbicide application. After the young canes reach about 18 inches (45 centimeters) in length, field workers go through the vineyard and tie the shoots to the trellis to keep them growing in the proper direction.

Bloom

Flower clusters look like miniature clusters of grapes and are located at the base of the young shoots (Figure 2–14). About eight weeks after budbreak, they begin to **bloom.** Grapes are self-pollinating and do not require the action of bees to become fertilized. Once fertilized, a grape flower will begin to develop into a berry. If a flower is not fertilized, it will drop off the cluster in a process called **shatter.** By this time, the danger of frost is usually past, but growers are still very concerned about the weather. For optimum pollination, warm, even temperatures are desired, without too much wind or wet weather. Hot weather or rain will increase the incidence of shatter and if the weather is too cold, bloom will be prolonged, making the ripening of the crop uneven. Some years it is necessary for the grower to go through the vineyard and thin the fruit clusters so

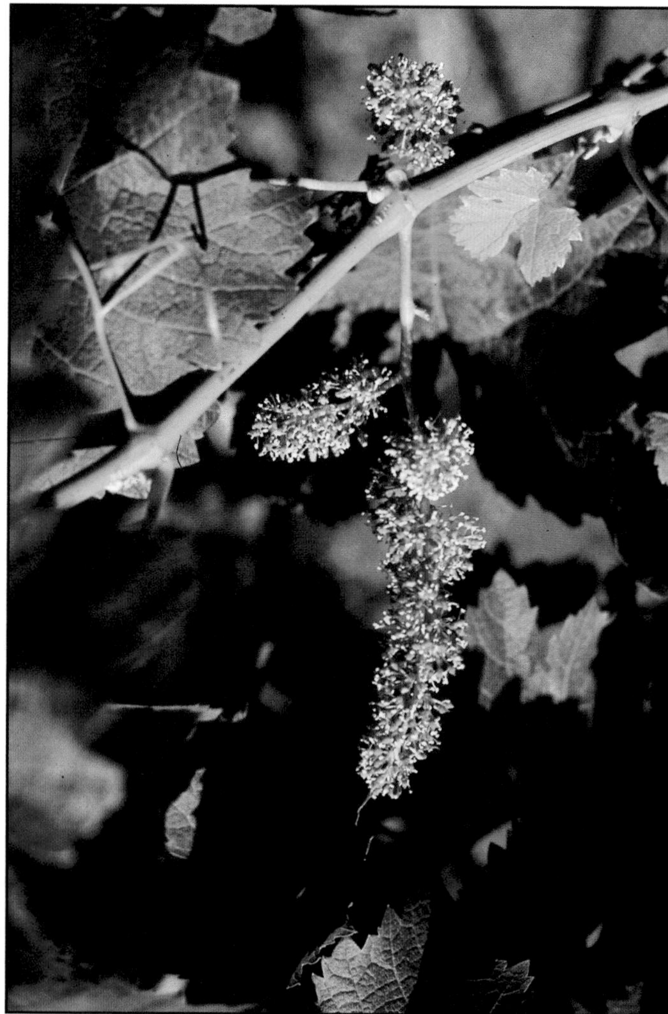

FIGURE 2–14 A grape flower cluster. After they are fertilized, these individual flowers will develop into berries, forming a cluster or bunch of grapes.

there will be less cropload. This is done when flowering clusters set too heavy and the vine has more cropload than the grower thinks will ripen.

After bloom, the canes will continue to grow, but the fruit clusters enter a lag phase and remain green and hard. If tasted they are very sour with no perceptible sugar. As the summer progresses vineyard workers continue to spray sulfur and tie up the growing canes as needed. In vigorous vineyards, some of the leaves that are at the base of the canes are pulled off to allow more sunlight and air to the fruit clusters. In growing regions that do not have adequate summer rainfall, irrigation can be used to keep the vines from becoming water stressed. Grapevines have deep root systems and can survive on very little water once they are mature. Dry-farmed vineyards are known for having low yield and intensely flavored fruit. This being said, a vine that is limited by its water supply would not be able to produce as much fruit as one that is not water stressed. The key is for the grower to balance properly the water demand of his vines so they get just the right amount of water for healthy, but not excessive, vine growth.

Véraison

Véraison is the beginning of ripeness and starts in mid to late summer about 8 to 10 weeks after bloom. At this time, the vines have begun to slow their vegetative growth and the canes are at their maximum length. This is also the point where dramatic changes begin to take place in the fruit clusters. Up until now the berries have remained hard and green but at véraison, they swell and start to change color (Figure 2–15). The sugar that the leaves are producing through photosynthesis is now going into fruit development instead of producing more leaves and canes. Irrigation is diminished to help the fruit ripen by slowing vine growth. Spraying is also discontinued both because after véraison rot is less likely and because no residual sulfur or vineyard chemicals should remain on the grapes when they are harvested.

Six to 10 weeks after véraison the grapes will be ready for harvest. The amount of time depends on the variety, the weather conditions, and the degree of ripeness that the winemaker desires. Both growers and vintners keep a close watch on the vineyard, taking samples often and analyzing them for acid and sugar content. In addition to the chemical parameters

FIGURE 2–15 A grape cluster at véraison, the beginning of ripening.

that are measured, the berries are tasted for flavor and observed for signs of ripeness such as the seeds and stems turning from green to brown. The decision to harvest is based on the maturity of the fruit as well as operational concerns. Sometimes crops must be brought in before they are ripe to beat an approaching storm, or may become too sweet before a crew of pickers is available during hot weather. When logistics and the weather allow the vineyard to be picked at the optimum sugar and acid levels at the same time that it is at its peak of flavor, it is considered a "vintage year." The term "vintage year" means an exceptionally good harvest; it is not to be confused with the term "vintage" that refers to the particular year a grape crop is produced.

Harvest

In the traditional method of harvest, grapes are picked by hand into boxes or baskets that are then carried to the end of vineyard rows to be loaded into trailers or trucks for transport to the winery (Figure 2–16). This method is still popular today because it is very gentle to both the fruit and the vines. It also has the advantage of being selective because only the healthy ripe fruit is picked. In some vineyards on steep slopes or with limited access, it is the only way to bring the crop in. However, as is the case with many other agricultural products, labor shortages and high costs are causing mechanization to play an increasingly larger role. Mechanical harvesters are designed to straddle a row of vines and shake them vigorously to dislodge their fruit (Figure 2–17). The grape clusters are collected below and carried to a bin on the back of the harvester. Vineyards that are to be mechanically harvested must be trellised with sturdy wires and stakes to keep the machine from damaging the vines during harvest. Mechanical harvesters can also be operated at night so they will bring in cooler fruit. This is particularly an advantage with white grapes in warm climates, because the cooler the fruit is, the less it will degrade on the trip from the vineyard to the winery. Both methods have their advantages and disadvantages, and if done properly they each can provide the winery with high quality fruit.

FIGURE 2–16 Pickers harvesting Zinfandel grapes in California's Sonoma Valley. The boxes of fruit are placed in the bin being towed by the tractor. After the bin is full, it is lifted onto a truck to be carried to the winery.

FIGURE 2–17 A mechanical grape harvester. These machines straddle a row of vines and shake them, dislodging the clusters of fruit, which are collected at the base of the machine. An example of a mechanical harvester in action is in Figure 11-4 on page 338.

After picking, the grapes are weighed and brought to the winery for processing into wine. If there is mild weather after the harvest, some photosynthesis will occur and the sugar that is produced is stored in the trunk and root system of the vine for use when it comes out of dormancy the next spring. At the first frost, the leaves will turn brown and fall off the vine, marking the beginning of the winter dormancy period.

Dormancy

While the vine is dormant, no new growth occurs, and as previously stated it is much less sensitive to cold weather. However, vineyard operations do not stop in the cold weather; fertilizers and soil adjustments will be made to prepare the plant for the upcoming year. A cover crop of grass or clover may be planted to control erosion. However, the most labor intensive and important task to be completed is pruning. After a vine is established, the process of pruning removes almost all of the new growth from the previous year. Each bud left on the vine will produce a new shoot in the spring that will have one to three clusters on it, and the amount of the next season's crop is determined by the number of buds. The vineyard manager evaluates the previous year's growing season and cropload and will make adjustments to have more or less in the upcoming season. No matter what trellis system is used, vines are either cane pruned or spur pruned. In cane pruning, healthy canes are selected from the past season's growth and trained along wires; each bud on the cane will grow a new cane of its own in the spring (Figure 2–18). With spur pruning, the grapevine is grown with permanent arms or **cordons** that have spurs located about every 6 to 8 inches (15 to 20 centimeters) along their length. Each spur will have one to several buds on them for the next year's growth (Figure 2–19). Cane pruning is more difficult to perform but it is preferred for varieties that have small clusters like Chardonnay. Cane pruning generally leaves more buds and clusters than spur pruning on cordons, and this allows grape varieties with small clusters that do not weigh very much to achieve an adequate crop.

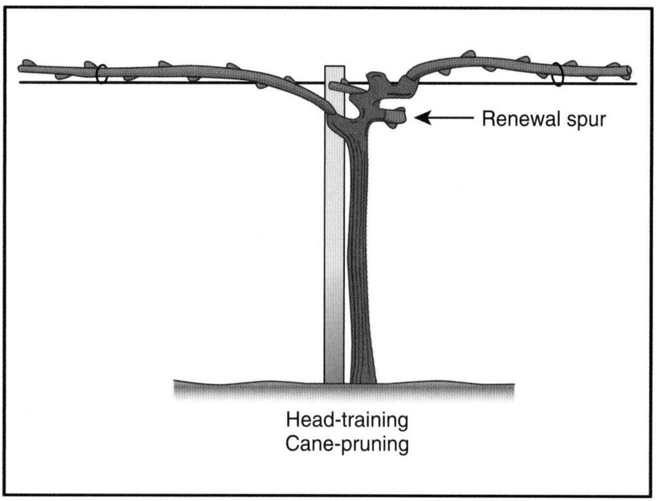

FIGURE 2–18 A dormant head-trained, cane-pruned vine after pruning.

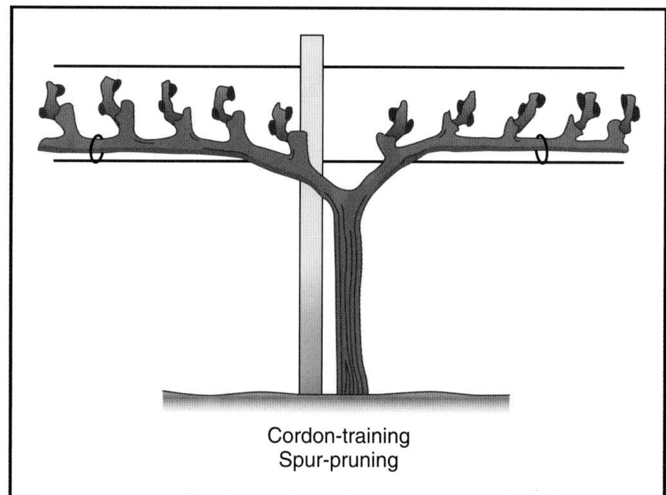

FIGURE 2–19 A dormant cordon-trained, spur-pruned vine after pruning.

The most important concept in pruning is balance. The vine must be left with the appropriate amount of buds to produce the correct cropload. If a vine is overcropped, it will have a difficult time producing enough sugar to get the grapes ripe. Overcropping will also stress the vine, leaving it with fewer reserves stored up for the following year. Undercropping also has undesirable consequences because if there is not enough crop, the vine will put too much energy into vegetative growth, making a large canopy of leaves that are hard to manage and shade the fruit clusters. The idea of balanced pruning is to leave the maximum amount of fruit that can achieve the proper level of maturity without weakening the vine. This depends on the vine's vigor, which results from the combination of terroir, variety, rootstock, and management. Grapevines that are grown on fertile soils with good weather can support larger croploads than those grown on less fertile soils in cool climates.

Major Grape Varieties

Although there are thousands of varieties of *Vitis vinifera* that are grown for winemaking, only a few make up the vast majority of production. We will go over 20 of the most widely planted varieties.

Barbera

Known for producing intensely colored, tart wines with moderate tannins, the vigorous Barbera grapevine is native to the Piedmont region of Italy (Figure 2–20). The grape's tendency to hold on to its acid in warm climates made it popular as a "blender" in jug wines in warm areas like Argentina and California's Central Valley. Interest in growing premium Italian varietals such as Barbera and Sangiovese have led to increased plantings in the coastal growing regions in California.

Cabernet Franc

This variety from the Bordeaux and Loire regions of France has small berries and loose to compact clusters (Figure 2–21). Related to Cabernet Sauvignon, it typically produces wines with less complexity and lighter tannins and color than its relative does. Although some wineries,

Organic Viticulture

Organic viticulture is the practice of growing grapes without the use of any man-made substances. While simple in concept, it requires a great deal of skill and concentration by the vineyard manager. In an organic vineyard, weed control is done by tilling or planting of cover crops. Elemental sulfur can still be used to combat rot, but synthetic chemicals cannot be used. An organic grower must keep very alert for any developing problems in his vineyard and react to them quickly because the natural alternatives are often less powerful than man-made pesticides. A small but growing number of growers in both Europe and the United States are raising their grapes organically. In California, it is currently less than 2 percent (Ness, 2002). Some growers do this out of a belief that it produces healthier, better tasting fruit, others out of a commitment to protect the environment. A number of organizations exist to help promote organic agriculture and certify vineyards that have not used man-made pesticides. Excellent grapes can be grown organically but the extra handwork required makes them more expensive to grow.

Wine grapes do well organically because they are grown for flavor and not for appearance. Blemishes and marks from insects significantly devalue a fresh fruit crop such as peaches. However, with wine grapes minor cosmetic imperfections do not matter as much since the fruit is crushed when it arrives at the winery. Some vineyards are more suitable than others are for going organic, and a vineyard's terroir plays a huge part in determining if it will be a success. A vineyard grown in a dry area with low humidity will have less pressure from rot and mildew as well as needing less weed control. Above all, an organic grower who is not attentive to their vineyard will have a great deal of difficulty producing high quality fruit.

Sustainable Viticulture

Even more growers use a technique called sustainable agriculture. This concept promotes agricultural practices that allow the minimal use of pesticides in the vineyard. This encourages the development of a natural vineyard ecosystem with predatory insects to help combat grape pests. Since the most expensive aspect of organic farming is hoeing at the base of the vine to control weed growth, in a sustainable vineyard a small amount of herbicide is often sprayed at the base of the vine. If a problem develops that the grower cannot control through natural means, he or she has the option to use man-made chemicals. Sustainable grape growing is a popular option because it provides much of the benefits of organic farming, with less risk and at a lower cost of labor.

Organic Winemaking

Organic practices can be taken into the winery along with the grapes. If a wine is produced and bottled from organically grown grapes and without the use of any man-made additives in the cellar it can be called Organic Wine. The major implication of this is that sulfur dioxide cannot be used as a preservative before bottling. Sulfur dioxide, not to be confused with vineyard use of elemental sulfur, inhibits oxidation and the growth of microbes that can spoil wine. The lack of sulfur dioxide is most notable in white table wines; if they are bottled without it they will quickly turn darker in color and lose some of their fruity aromas. For this reason there is much more wine bottled with the phrase "made from organically grown grapes" than labelled as "organic."

particularly in the Loire, produce excellent varietal Cabernet Francs, it is most often used for blending with Cabernet Sauvignon and Merlot in a Bordeaux-style blend. Its soft qualities make it useful in toning down the sometimes harsh tannins in Cabernet Sauvignon and adding fruity character.

FIGURE 2–20 Barbera

FIGURE 2–21 Cabernet Franc

Cabernet Sauvignon

The classic variety of Bordeaux, it is one of the most popular varieties grown worldwide. It is a late season ripener with loose clusters and thick-skinned berries that make it resistant to rot (Figure 2–22). It is known for its excellent color and good tannins combined with complex flavors. At its best Cabernet Sauvignon exhibits a strong cassis (black currant) aroma combined with interesting aromatic notes such as cedar, pipe tobacco, and mint. When grown in cool areas, or on vines with too much canopy so the fruit clusters are shaded, it can have a distinct vegetative or green-bean character.

In 1997 researchers studying the DNA of grapevines determined that Cabernet Sauvignon is a cross between Cabernet Franc and Sauvignon Blanc (Bowers & Meredith, 1997). While 100

FIGURE 2–22 Cabernet Sauvignon

percent Cabernet Sauvignons are often made with great success, blending with other Bordeaux varieties such as Merlot, Cabernet Franc, Malbec, and Petite Verdot can make a wine that is even more balanced and complex. In the New World, Cabernet has gained wide popularity. In California it is one of the most commonly produced varieties by wineries, and there are also extensive plantings in the Southern Hemisphere. The combination of plentiful tannins and fruit flavors allow most Cabernet Sauvignons to improve for 10 to 20 years in the bottle.

Chardonnay

One of the best known white varieties, Chardonnay comes from the Burgundy and Chablis regions of France. It budbreaks early in the spring, which makes it susceptible to frost damage, but it is a mid-season ripener, allowing it to be grown in cool regions (Figure 2–23). This versatile variety can be made in a number of styles. Its aroma can be described as green apple, pear, or citrus, and depending on how it is made, it can have a body that is anywhere from crisp and tart to soft and viscous. When it is fermented in small oak barrels and undergoes malolactic fermentation, it takes on toasty vanilla and butter flavors. If a winemaker is not careful, it is easy for the flavor of the oak to overwhelm the delicate fruit aromas.

Chardonnay's popularity resulted in extensive planting in California throughout the 1980s and 1990s with great success in the cooler coastal valleys. Previously it was always grown as a premium varietal, but in the 1980s, a surplus of Chardonnay wine led to the introduction of Chardonnay as a "fighting varietal." This category of wine was priced similarly to jug wines but had a varietal identity and was packaged in a 750 ml bottle. As part of this market segment, Chardonnay also replaced many of the more neutral varieties like Chenin Blanc and French Colombard that were grown in warmer regions to produce bulk wines. The large quantity of

FIGURE 2–23 Chardonnay

mediocre Chardonnay released on the American market in the late 1990s has resulted in a backlash against the varietal and consequently falling grape prices (California Department of Food and Agriculture, 2003).

Chenin Blanc

Chenin Blanc is native to the Loire Valley of France where it is called Pineau de la Loire (Figure 2–24). It is a prodigious producer and adapts well to a number of different soils and climates. For these reasons, it is popular throughout the world, although in most areas outside of France it is considered a simple grape and used for making inexpensive wines. Until the late 1980s, it was the most widely planted white grape in California with most of the vineyards located in the Central Valley. Since that time it has lost more than half of its acreage, much of it to Chardonnay for "fighting varietal" wines. Usually made into a clean, crisp, wine with a minimum of oak aging, it can be made in either sweet or dry styles and it is known for being easy to process in the wine cellar.

Gewürztraminer

Gewürztraminer is a white grape, but unlike most other white varieties, it turns a deep russet color at ripeness instead of staying green or yellow (Figure 2–25). It has thick bushy vines and does best in cooler growing areas; it is one of the most popular grapes grown in the Alsace district of France. Gewürztraminer is also grown in New Zealand, the northwest United States, and the cool coastal valleys of Northern California. Cool growing conditions help to bring out the distinct floral-spicy aroma for which the variety is famous. Gewürztraminer makes a delicious dry wine; however, it is best known for its sweeter styles including late harvest dessert wines made from clusters infected with botrytis or "noble rot."

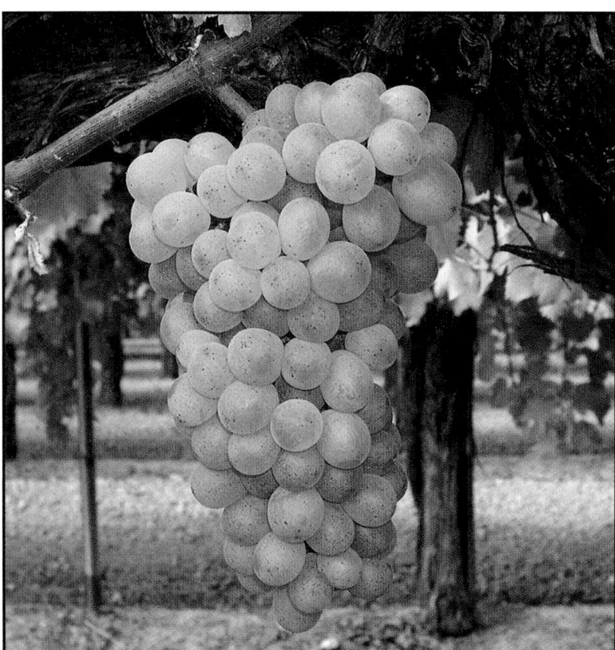

FIGURE 2–24 Chenin Blanc

FIGURE 2–25 Gewürztraminer

Grenache

Grenache is the most popular grape in the southern Rhône Valley, where it is the mainstay of the popular Rhône blend Châteauneuf-du-Pape (Figure 2–26). It has ripe, fruity, plum-like flavors with moderate tannins—qualities that make it useful for blending with Syrah, which can be more tannic. It thrives under warm growing conditions and can support a large cropload on fertile soils. In California, it is often grown in the Central Valley where it makes soft, early maturing red wines that are used mainly in generic blends. In cooler areas with a lighter crop, it produces much better wine, but its acreage is limited in California's coastal valleys.

Merlot

Merlot is from the Bordeaux region, where it is sometimes made into a wine by itself but more often is used as a blender with Cabernet Sauvignon (Figure 2–27). It has similar flavors to Cabernet but has a softer mouthfeel and gets ripe earlier in the season. The latter quality makes it even more useful in cooler vintages where Cabernet may not be able to attain full ripeness. In California and Washington, Merlots have become very popular because the good flavors and lighter body make it more approachable for wine drinkers that are making the move from white to red wines. Its consumer acceptance has made it one of the most widely planted grapes in California. In addition, like Chardonnay, its success has led to overplanting in some areas, resulting in lower grape prices.

Muscat Blanc

Also called Muscat Canelli, it is a member of the Muscat family of grapes (Figure 2–28). There are more than 200 different varieties of Muscat with varying skin color and flavor (Robinson, 1986). They have in common a distinct "Muscat aroma" that is described as intensely fruity and floral. It can be made in a variety of styles from a light-bodied and dry table wine to a sweet dessert wine that is fortified with alcohol in the style of a white port. It can also be made into sparkling wine as in the style of Asti Spumante. Muscat does well in a diversity of conditions, with cooler areas producing the best dry styles and warmer areas making the best dessert wines.

FIGURE 2–26 Grenache

FIGURE 2–27 Merlot

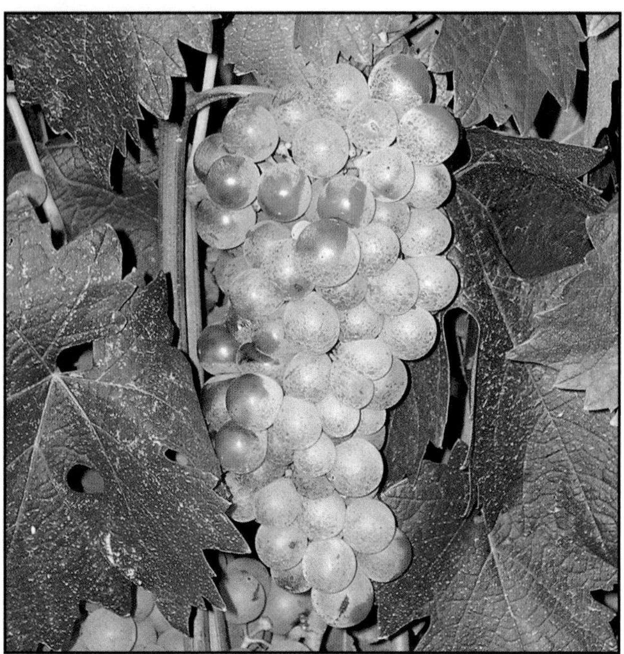

FIGURE 2–28 Muscat Blanc

FIGURE 2–29 Petite Sirah

Petite Sirah

Called Durif in France, it is descended from a cross between the Rhône varieties of Syrah and Peloursin (Figure 2–29). Although native to the Rhône, Petite Sirah has found a great deal of popularity in the coastal valleys and Sierra Foothills of Northern California. It makes a deeply colored, full-bodied wine with lots of fruity aromas such as raspberries and plums. While it makes an excellent varietal wine, it is often blended with other reds, particularly Zinfandel, to provide them with additional color and body.

Pinot Blanc

Known as Pinot Bianco in Italy and Weissburgunder in Germany and Austria, it is a mutated clone of the grape variety Pinot Gris. Pinot Blanc does best in cooler areas and it has small pale green clusters that have flavors that are similar to Chardonnay, but more delicate in nature (Figure 2–30). In Europe, it is generally used to make crisp, light-bodied wines with a minimum of oak aging. In California, a riper style is produced with more body and often more oak. In addition to its uses as a table wine, it can be used along with Chardonnay and Pinot Noir in sparkling wines.

Pinot Gris/Pinot Grigio

The parent of the variety Pinot Blanc, Pinot Gris itself is mutated from the red variety Pinot Noir. Although it produces a white wine, the clusters have a light pinkish/brown color (Figure 2–31). It is widely grown in Alsace and northern Italy, where it is called Pinot Grigio. It is an early season ripener and is popular in cool regions with short growing seasons such as Oregon. It makes a slightly more full-bodied wine with less of a fruity aroma than Pinot Blanc does. It is currently one of the fastest growing varieties in America in terms of consumption, due to imports as well as new plantings.

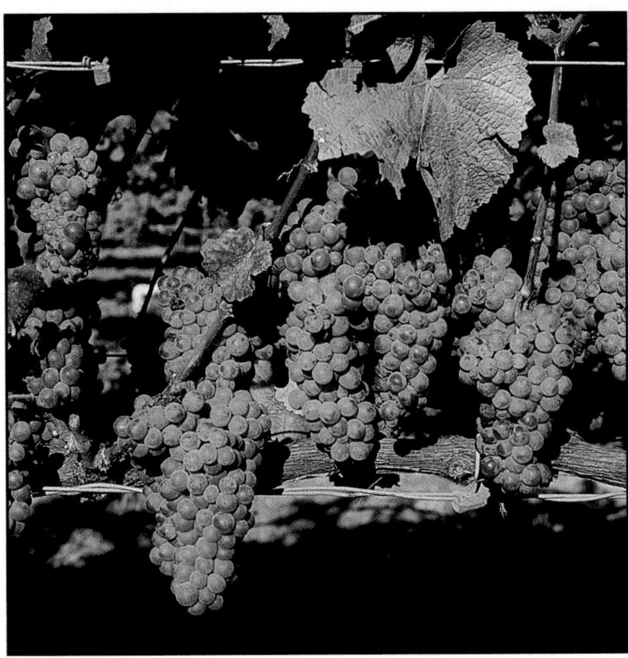

FIGURE 2–30 Pinot Blanc

FIGURE 2–31 Pinot Gris

Pinot Noir

The primary grape of Burgundy, its reputation for producing excellent wines is probably only surpassed by Cabernet Sauvignon (Figure 2–32). Although it makes long aging, complex fruity wines, it has a well-deserved reputation for being both difficult to grow and troublesome in the winery. There are a number of clones of Pinot Noir available to grape growers, from large-clustered, heavy-producing vines that are suited to sparkling wine production to small-clustered "Burgundy" clones with dark skins that are better for table wine production. It is an early ripening variety that does best when grown under cool conditions such as Oregon and the Carneros

FIGURE 2–32 Pinot Noir

and Russian River appellations in California. However, even grown under the best conditions there can be problems with obtaining good color and flavor from the fruit. It is a very delicate wine and must be treated very gently while processing at the winery so that the balance and flavor is not lost.

Riesling

Called White Riesling or Johannisberg Riesling in the United States, it is the most famous variety grown in Germany and is planted throughout its steep river valleys (Figure 2–33). Riesling is similar in character to Gewürztraminer with strong floral and fruity notes but less spicy. When it is grown in cool areas, the fruity qualities and tart acid that the grape is known for are preserved. It can be made in a number of styles from a dry, tart wine that is low in alcohol to the famous German dessert wine Trockenbeerenauslen, or TBA, that is very concentrated with flavor and sugar. In the 1970s it was the most expensive grape grown in California, and usually made in a sweet style. However, it fell out of favor as the public started to drink more dry wines.

Sangiovese

Sangiovese is the classic grape of the Tuscany region of Italy, and is the major variety used in Chianti wines. The variety's thin-skinned berries leave it vulnerable to rain and high temperatures at ripeness, and depending on the clone, it can sometimes have light color (Figure 2–34). It produces tart wines with medium body and cherry flavors. Sangioveses do well on their own or blended with other red varieties such as Merlot or Cabernet Sauvignon. It is increasingly planted in California and Australia where the climate is similar to that in Tuscany.

Sauvignon Blanc

Sauvignon Blanc, also known as Fumé Blanc, grows bushy, vigorous vines that produce tight clusters of thin-skinned pale green berries (Figure 2–35). These thin skins and tight bunches also make it very susceptible to rot. It has a distinct varietal aroma that runs a spectrum, including vegetative, grassy, gooseberry, and melon. When grown under cool conditions, the varietal character can become very intense. Because of the vigorous nature of the vines when grown in

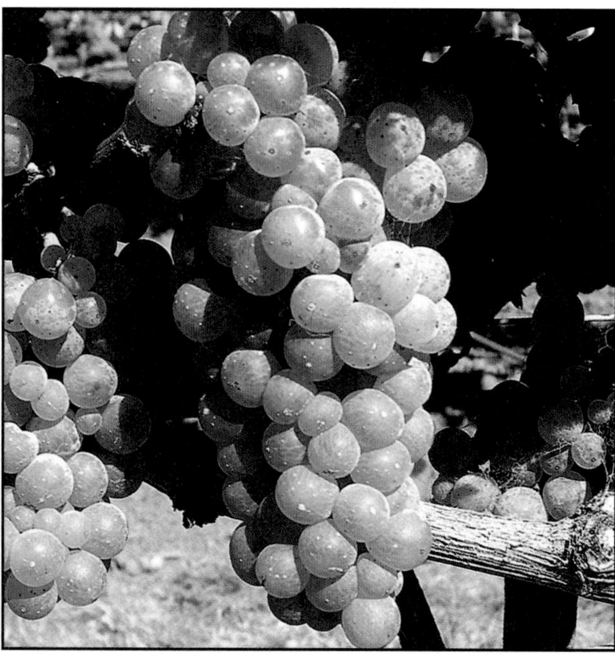

FIGURE 2–33 Riesling

FIGURE 2–34 Sangiovese

FIGURE 2–35 Sauvignon Blanc

FIGURE 2–36 Syrah

fertile soils, it is necessary to carefully trellis the vine and practice leaf pulling to expose the clusters to the sun to get optimal flavors. The classic white variety of the Graves district in France, it has also found acclaim in New Zealand and California.

Syrah

From the northern Rhône Valley, Syrah produces a meaty wine with lots of tannins and good acid (Figure 2–36). It has a complex aroma that includes fruity flavors of blackberry and plum balanced out with spicy-peppery and earthy-leathery notes. The powerful tannins combined with plentiful fruit allow Syrah wines to age for a very long time. In the southern Rhône, it is often blended with other Rhône varieties such as Grenache and Mourvèdre to balance the tannins and add more complexity. It has great popularity in Australia where it is called Shiraz; this name is also used at some wineries in California. In the New World it is sometimes blended with Cabernet Sauvignon and generally is made in a more fruit-forward style (a wine with a predominately fruity character).

Tempranillo

The dominant grape used in Rioja wines of Spain and widely planted in Argentina, Tempranillo has vigorous vines that ripen early in the season and produce thick-skinned berries (Figure 2–37). It makes an intense wine with excellent color and tannins, and its aroma typically has notes of strawberries and plums with earthy overtones. It is also grown in Portugal where it is known as Tinta Roriz.

Viognier

This distinctive grape from the Rhône region is difficult to grow and has low yielding vines (Figure 2–38). It makes a relatively low acid wine with very intense tropical and floral fragrances. Sometimes it is blended with other varieties to add structure to the body and tone down the strong aromas. It is becoming increasingly popular in California, but is still not widely planted.

FIGURE 2–37 Tempranillo

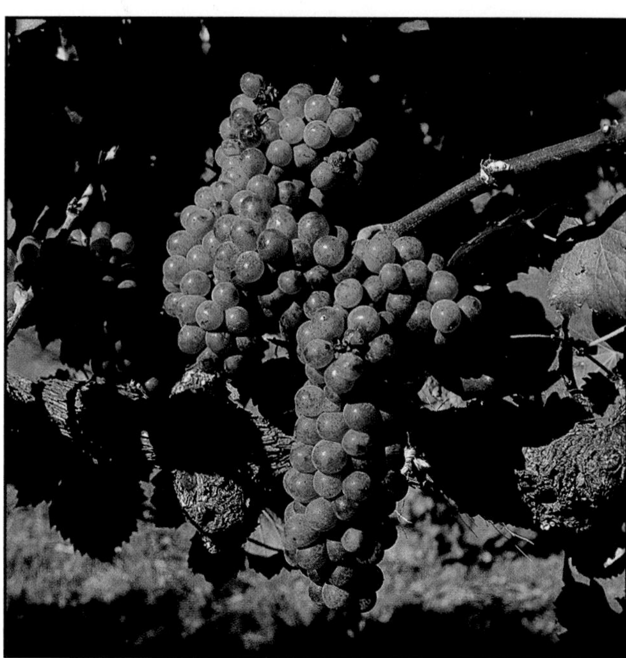

FIGURE 2–38 Viognier

Zinfandel

Like all *vinifera* grapes, Zinfandel is native to Europe; however, it is best known in its adopted home of California where it is widely planted throughout the state. For years it was not known what country Zinfandel came from or what its original name was. This lack of a provenance was cause for a great deal of speculation. Its origins remained unknown until recent DNA analysis has determined it to be native to Croatia where only 20 vines were found to be still in production. In Croatia, it is called Crljenak Kastelanski, pronounced "tsurl-Yen-ahk-kahstel-AHN-ski" (Smith, 2002). Not surprisingly, it became known as Zinfandel when it was brought to the New World. It has large, tight, thin-skinned clusters that have a tendency to become overripe in hot weather, which can result in a high alcohol wine that has a "raisiny" character (Figure 2–39). It makes a full-bodied wine with blackberry and pepper flavors and light tannins.

FIGURE 2–39 Zinfandel

Summary

Growing premium wine grapes is a collaboration between the winemaker and grower where the vintner lets the grower know what qualities are wanted in the fruit and the grower manages the vineyard in a way that will deliver them. To ensure production of the best wine grapes, the grower should be rewarded with higher prices for producing the best fruit and paid less for fruit of lower quality. Market conditions in the wine business also affect prices where popular varieties in high demand are worth more than varieties that have fallen out of favor with consumers. The best grapes are usually produced from vineyards that have a long and mutually beneficial relationship with the winery. In these situations, a trust between the grower and the winemaker develops and the grower knows exactly what kind of fruit the winemaker wants.

Review Questions

1. What are the attributes of grapes that make them the ideal fruit for wine production?
2. What species of grape is used for most wine production and where did it originate?
3. What risks does the vineyard manager face in the early spring and how are they dealt with?
4. What is the beginning of ripening called?
5. What factors are considered when deciding the proper time to harvest?
6. What is the difference between "organic grapes" and "organic wines"?
7. Explain the relationship between grape species, variety, and clone.

CHAPTER 3

The Winery—From Grapes to Bottle

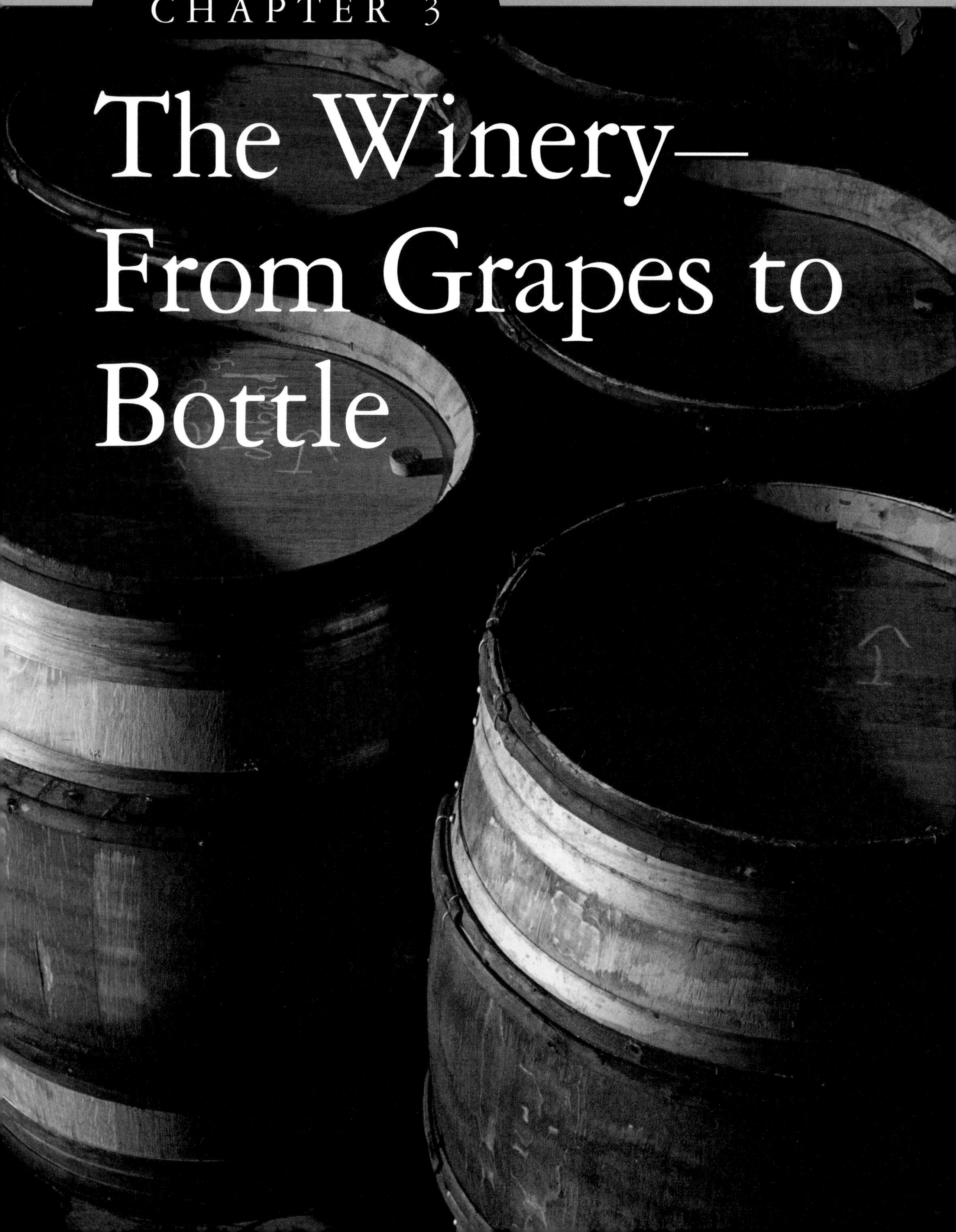

After reading this chapter, you should be able to

- display an understanding of how both red and white table wines are produced.
- describe how sparkling wine is produced.
- explain how dessert wines are produced.

Key Terms

autolysis
Botrytis cinerea
cap
carbonic maceration
Charmat process
coopers
cuvée
degrees Brix (°Brix)
dessert wines
disgorging
dosage
dry wine
Eiswein
ethanol
extended maceration
fining
fino
flor yeast
free run
lagar
late-harvest
lees
malolactic bacteria
manzanilla
marc
méthode champenoise
must
oloroso
polymerize
pomace
Port
Port-style
press fraction
primary fermentation
primary lees
pumping over
punching down
quinta
racking
riddling
rotary fermentors
Saccharomyces cerevisiae
Sherry
solera
sparkling wine
stemmer-crusher
still wine
sur lie
sur pointe
table wine
tirage
ullage

Great wines begin in the vineyard, but they are finished at the winery. Much like a chef preparing a fine meal, the winemaker takes the produce of farmers and converts it into a food that is both nourishing and delicious. Like a chef, the winemaker works with flavors and aromas to create a wine that will give the consumer the maximum amount of sensory pleasure. To the uneducated observer the choices and decisions the winemaker makes may appear random in nature, but in reality they are based on a scientific understanding of the ingredients and techniques used to produce wine. In this way, winemaking is a craft that is a combination of art and science. Complicating the winemaker's quest to create great wine is the fact that people have different tastes and preferences, and there is no one "ideal" style wine. This, of course, is why there are so many different types of wines in various styles. It is also what makes wine such a diverse and interesting subject of study.

As mentioned in the previous chapter, wine is simply grape juice that has been fermented by yeast. Although this definition is quite simple, in the more than 6,000-year history of winemaking, wine production has evolved into a number of complex procedures that produce a variety of wines. **Table wine** is a wine designed to accompany food. It is produced in numerous forms, both red and white, and is the most common type of wine consumed in the United States, making up over 90 percent of the market (Adams Beverage Group, 2004). A table wine is a **still wine** (a wine without effervescence) and is also a relatively **dry wine** (without sweetness) having a moderate alcohol content of about 11 to 15 percent. To achieve this percentage of alcohol content, grapes are picked between 22 and 25 **degrees Brix,** or **°Brix** (the percentage of sugar by

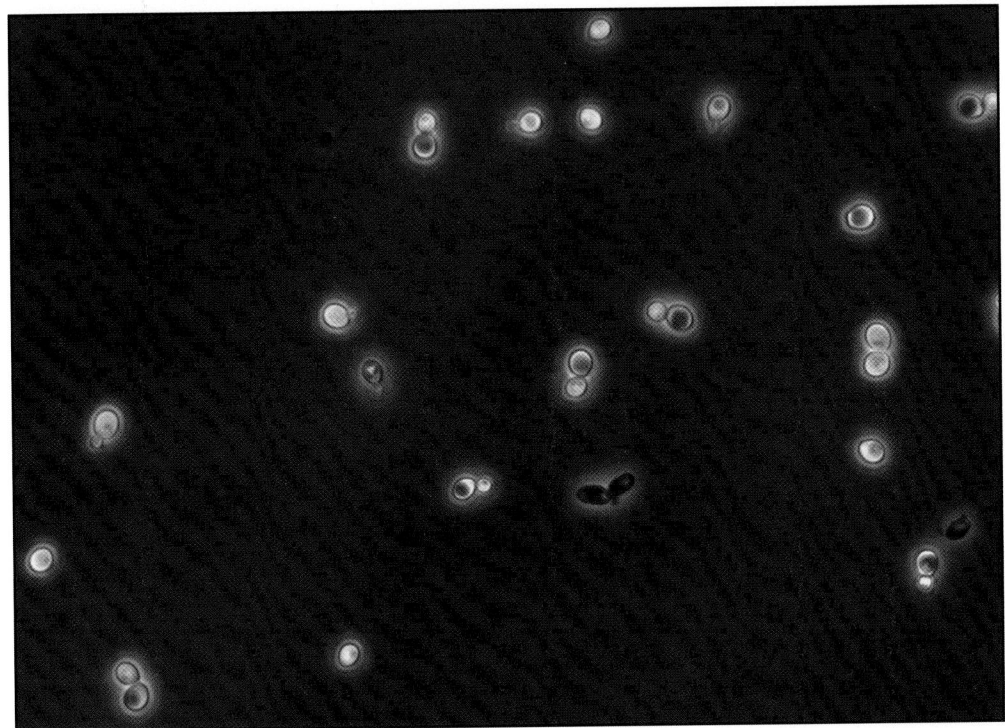

FIGURE 3–1 Microscopic photograph of cells of the wine yeast *Saccharomyces cerevisiae.*

weight). In the United States, a wine must be less than 14 percent alcohol to be labeled a "table wine." This is an arbitrary number that was chosen for reasons of tax collection; wines with a higher alcohol content are taxed at a higher rate. The exact alcohol content has little to do with the true definition of being a wine made to complement food, and there are many table-style wines that are bottled at over 14 percent alcohol. Some table wines are also made with a small amount of residual sugar in an "off-dry" style.

The Process of Fermentation

Fermentation is the process of yeast (unicellular or one-celled fungi) (Figure 3–1) converting the sugar in grape juice to alcohol and carbon dioxide, releasing some heat during the process. Yeast ferments sugar to produce energy to sustain life and reproduce. Other microorganisms can do this but yeast ferment with the most efficiency and can survive in the higher alcohol at the end of fermentation (Figure 3–2). The species of yeast that is best suited for winemaking is called ***Saccharomyces cerevisiae.*** The name *Saccharomyces* is derived from the Latin sugar fungus, while *cerevisiae* refers to grain. This is not surprising because the most common use of *Saccharomyces cerevisiae* is in bread making. When bread dough rises, it is from the bubbles of carbon dioxide that are produced inside the loaf during fermentation. Alcohol is also produced during bread making but it is baked off while the loaf is in the oven. This is what gives baking bread its distinctive smell. While the yeast used for winemaking and bread making is the same species, different strains are used which are adapted for their individual roles.

While this formula looks simple, it is actually a biochemical pathway with 12 separate reactions that are controlled by different enzymes in the yeast (Figure 3–3). The rate of fermentation is affected by a number of factors, including:

- temperature. The warmer the juice, the faster it will ferment; however, at temperatures above 100°F (38°C) yeast will die off.
- acidity. The higher the concentration of acid (lower the pH), the slower the rate of fermentation.

FIGURE 3–2 Stainless steel wine tanks in a fermentation cellar.

- nutrients. If the juice is low in nutrients, the yeast may not be able to ferment to dryness.
- alcohol. At higher concentrations, 13 to 16 percent depending on strain, yeast begin to die.
- sugar. Although sugar is required for yeast growth, if the sugar concentration is greater than 30 percent, it inhibits yeast growth.

Winemakers use these factors to control the fermentation and make different styles of wine. As an example, Port wine is made by adding brandy to fermenting wine to kill the yeast before it can ferment to dryness. This way a stable, sweet wine can be bottled without further risk of fermentation.

Wine was made for thousands of years before anyone knew how fermentation worked or that there were such things as microscopic organisms called yeast. The conversion of grape juice into wine was considered a miracle of nature. Although early winemakers did not understand the mechanism, they knew how to use fermentation to produce good wine. There are still a few wineries that use this method of fermentation with natural or "wild" yeast to make wine. The winemakers at these wineries feel this method can give their wine more complexity, but there also is a higher risk of off-flavors or an incomplete fermentation. Today, most winemakers use commercially available strains of yeast that have been isolated from different wineries and manufactured for sale. These yeasts are usually sold in an "active dry" form that has a similar appearance to bakers' yeast, and they give the winemaker a clean, efficient fermentation with no off-aromas.

Alcoholic Fermentation

$$C_6H_{12}O_6 \rightarrow 2\ C_2H_5OH + 2\ CO_2 + \text{Energy}$$

(Sugar) Alcohol Carbon Dioxide

FIGURE 3–3 The chemical equation for alcoholic fermentation in which one molecule of sugar is converted to two molecules each of ethanol (wine alcohol) and carbon dioxide. Byproducts of this fermentation include heat and energy for the yeast cell.

Red Wine Crush and Fermentation

The harvest is the busiest time of year at the winery because the grapes must be harvested and processed as soon as they reach their peak of flavor. The weather conditions set the pace of harvest and it is not uncommon for winery workers to be on the job 12 hours a day for 7 days a week. Once the grower and the winemaker have determined that the grapes have reached their optimum ripeness and flavor, they are picked and brought to the winery. When the crop arrives at the winery it is weighed, inspected, and analyzed before being processed (Figure 3–4). If the grapes are being purchased and are not grown on the winery's estate, the results of inspection are very important. This is because grape contracts between growers and vintners often include bonuses and penalties that depend on the analysis at harvest and the overall quality of the fruit. Particularly at larger wineries, this inspection and analysis is performed by an independent third party to avoid conflicts of interest.

After the grapes are weighed and inspected, they are brought to the receiving hopper and unloaded. At the bottom of the hopper, there is either a screw or a belt conveyor that is used to transport the fruit to the **stemmer-crusher** (Figure 3–5). The stemmer-crusher has two func-

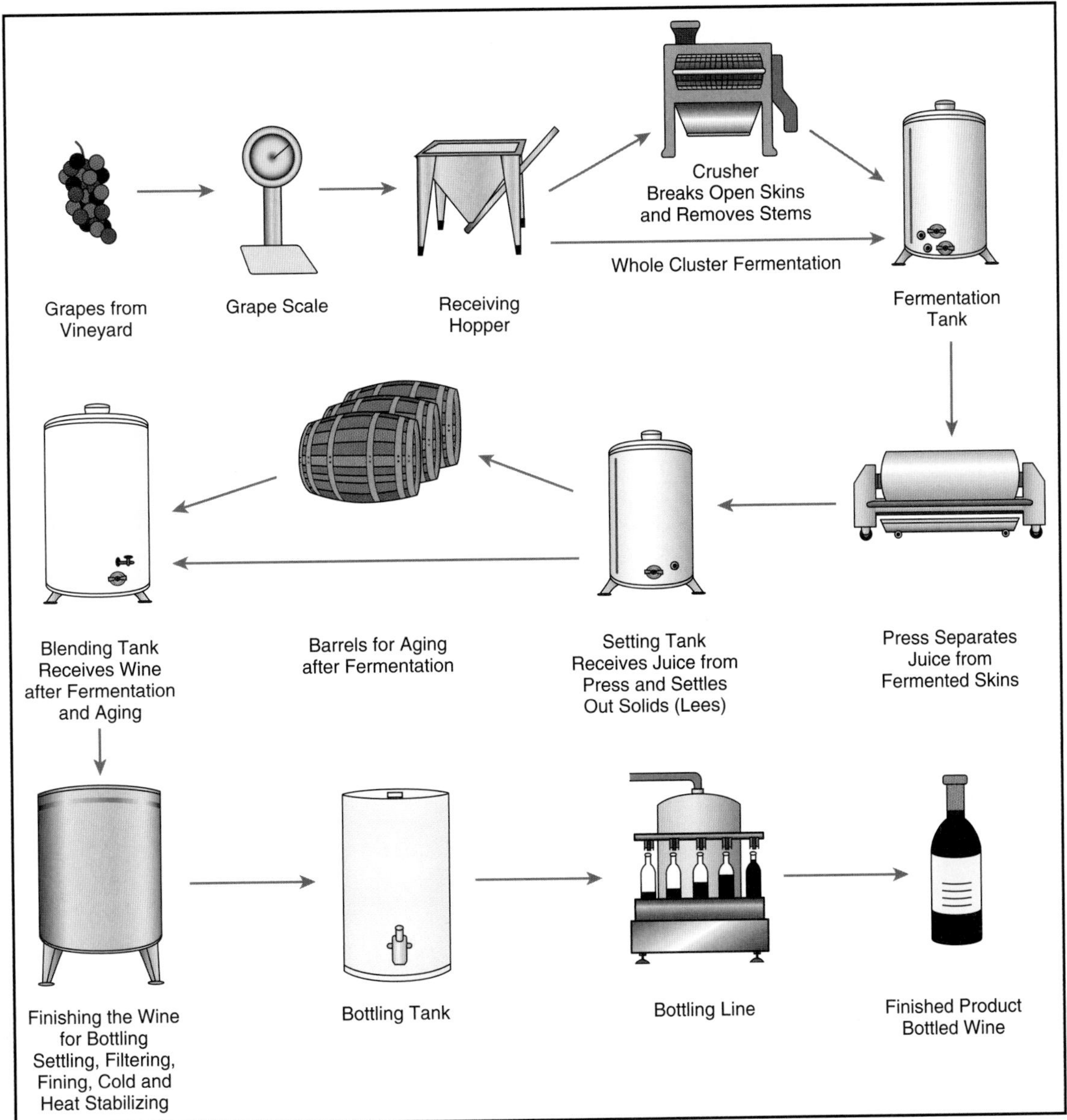

FIGURE 3–4 Flowchart of operations in making red wine.

FIGURE 3–5 A grape stemmer-crusher with side panels removed to show the perforated stainless steel drum and bars.

tions: first it takes the berries off the stems, and second it breaks the berries open to release the juice. Stemmer-crushers are made up of a perforated stainless steel cylinder or drum that is 1 to 4 feet (0.3 to 1.2 meters) across. The perforations are holes that are large enough to let the individual grapes through, but not whole clusters or stems. Inside the cage is a set of bars that are arranged in a helix pattern. When the crusher is started, the bars begin to rotate at several hundred revolutions per minute (RPM), while the cage rotates at a much slower rate. The clusters of grapes enter in through the back of the cage and, when they come in contact with the bars, the berries are knocked loose and fall through the holes in the cage. The stems, once they have lost their grapes, are pushed out the front of the machine by the helix pattern of the bars (Figure 3–6).

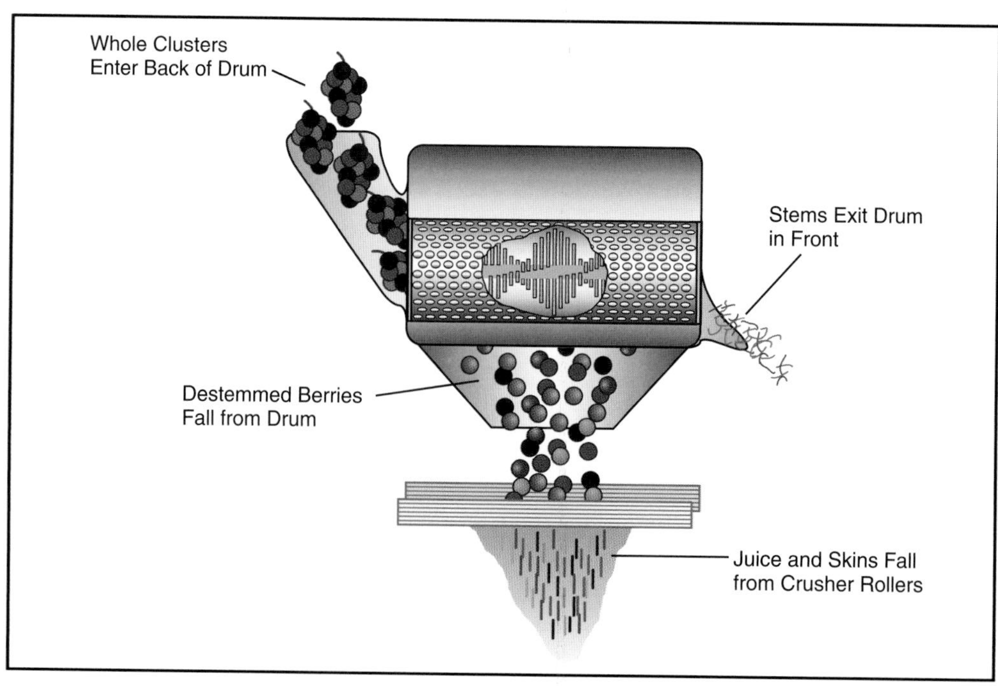

FIGURE 3–6 Diagram of a grape stemmer-crusher.

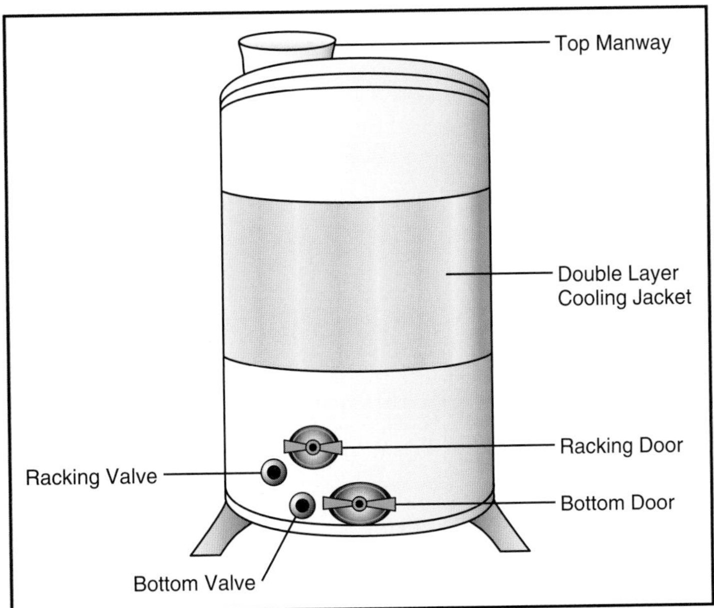

FIGURE 3–7 Diagram of a stainless steel fermentation tank showing doors, or man ways, for access to the inside of the tank, valves for the transfer of wine into and out of the tank, and a cooling jacket, a section of double walled stainless steel that cooling fluid is circulated through to maintain temperature during fermentation.

After the berries are destemmed, they fall to the second part of the machine—the crusher. The crusher is a set of rollers designed to break open the berries and release the juice. In modern crushers, the gap between the rollers can be adjusted to provide a greater or lesser degree of crushing. On some models the rollers can be removed entirely to allow whole berries to pass through. The mixture of approximately 80 percent juice, 16 percent skins, and 4 percent seeds produced by the crusher is called **must.** At this point, the must is liquid enough to be pumped to a tank for fermentation.

In modern wineries fermentation tanks are most often made of stainless steel (Figure 3–7), although vats made of wood or concrete are also still in use. The tank is filled to three-quarters capacity to allow room for expansion during fermentation and the must is analyzed and adjusted, if necessary. Usually, with the exception of the preservative sulfur dioxide, the compounds that are added to adjust the must, such as sugar, acid, nutrients, and yeast, are natural and already present in the must to some degree. Additives to wine are regulated and vary from region to region. For example, it is legal to add sugar to must in France but not acid, while in California the opposite is true. This is not a hindrance, however, because grapes grown in California seldom need additional sugar and French musts seldom need additional acid.

When the yeast is first added the must is homogenous; however, once fermentation begins, the carbon dioxide that evolves causes the skins to float to the top of the tank and form a **cap** (Figure 3–8). In large tanks, the cap is several feet thick and very firm. The juice from most red wine varieties is clear; therefore, to produce a red wine it is necessary to extract the red color out of the skins. If the skins are in a cap that is floating above the juice, very little extraction will take place. To combat this, the cap is mixed into the juice several times a day. There are many ways to do this, and the manner in which it is done, and the frequency, have a major effect on the overall style of the wine being made. If a cap is mixed in vigorously, and frequently, the result will be a wine with more color, body, and astringency than one with a more gentle treatment.

Methods of Cap Management

Punching down (Figure 3–9) is the oldest, simplest, and most gentle method of mixing the cap of skins and the juice. A punch-down device is used to press down the cap into the juice. Done by hand, it works well on smaller tanks with an open top. In larger tanks, pneumatically powered plungers are used.

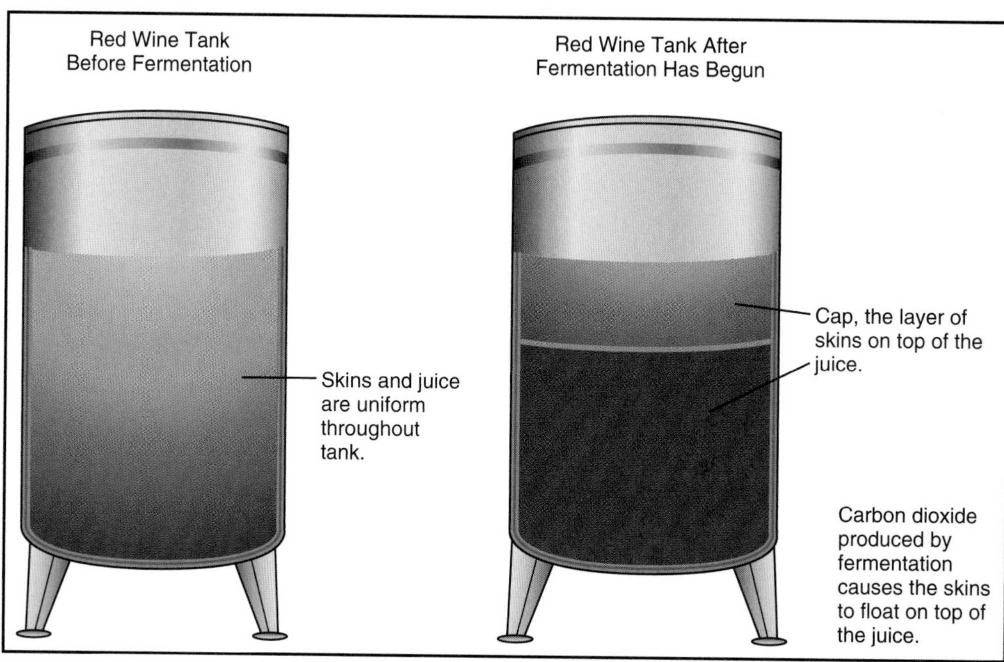
Red Wine Tank
Before Fermentation
Skins and juice are uniform throughout tank.
Red Wine Tank After
Fermentation Has Begun
Cap, the layer of skins on top of the juice.
Carbon dioxide produced by fermentation causes the skins to float on top of the juice.

FIGURE 3–8 At the beginning of fermentation the must is homogeneous and the skins and juice are distributed evenly throughout the tank. When fermentation begins, the carbon dioxide gas that is produced causes the skins to separate from the juice and float to the top of the tank forming a *cap* of skins.

Pumping over (Figure 3–10) is the method by which the juice is taken from beneath the cap and irrigated over its top. As the juice percolates through the skins it extracts the color and flavor, similar to the way a drip coffee maker uses hot water to extract flavor from ground coffee.

Rotary fermentors (Figure 3–11) are the most modern and least labor-intensive way of dealing with the cap. They are large horizontal tanks that have fins along the inside, similar to a cement mixer, and when they are rotated, the cap is rolled over into the juice. The main advantage of rotary fermentors is that they make it very easy to extract the skins after fermentation by opening the door at the end of the tank and rotating it. The disadvantage is their high cost.

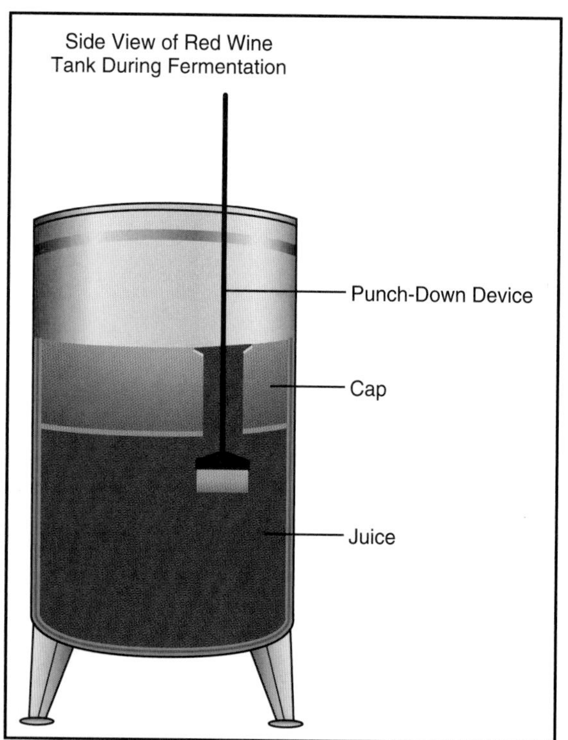
Side View of Red Wine
Tank During Fermentation
Punch-Down Device
Cap
Juice

FIGURE 3–9 Breaking up the cap by hand or punching down.

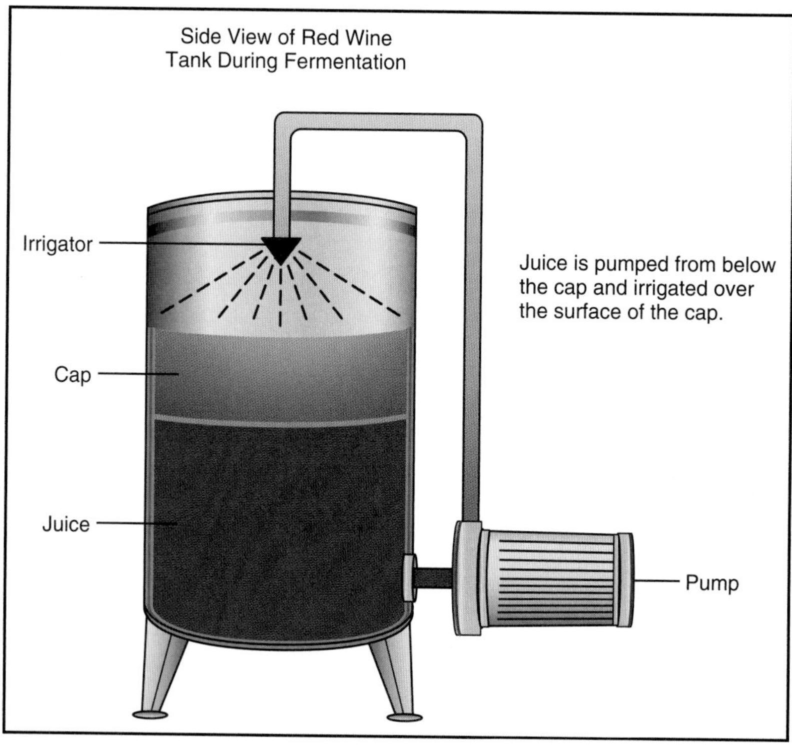
Side View of Red Wine
Tank During Fermentation
Irrigator
Juice is pumped from below the cap and irrigated over the surface of the cap.
Cap
Juice
Pump

FIGURE 3–10 Pumping over or irrigating the cap.

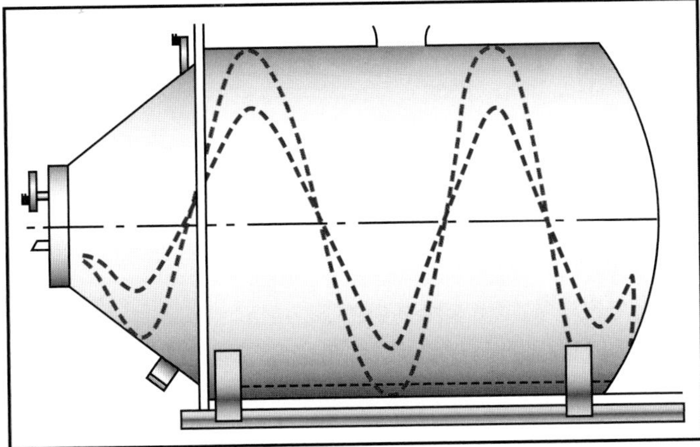

FIGURE 3–11 A rotary fermentation tank for red wines. The dotted lines show the location of the helix-shaped stainless steel that mixes the cap and must during fermentation, and also removes the skins after fermentation is complete and the wine has been drained off.

From the time the yeast is added, the fermentation usually takes about 1 to 3 weeks. This depends on several factors: the amount and type of yeast added, the nutrients in the must, and the temperature. Most red wine fermentations will peak at about 85°F (30°C); at this temperature there is good color extraction without the yeast becoming too hot. When the yeast has fermented all of the sugar in the must to alcohol or, in the case of sweet wines, as much sugar as the winemaker wants to be fermented, then the must is considered wine. At this point, the juice is drained off the skins and the skins are removed from the tank for pressing.

Pressing the Skins

In red wines, when fermentation is complete and the winemaker is satisfied with the flavor extraction, it is time to separate the wine from the skins. The majority of the wine is simply

Carbonic Maceration and Extended Maceration

In red wine production, the most important stylistic decision a winemaker has to make is the manner in which the skins are handled during fermentation. How this is done will determine most of the flavor components in the finished wine. There are many ways to influence extraction from the skins beyond how the cap is punched down or pumped over. Two of the most common procedures are carbonic maceration and extended maceration.

Carbonic maceration is the process whereby either a portion or all of the grapes are not crushed but loaded into the tank as whole clusters. The weight of the fruit crushes some of the berries at the bottom of the tank and releases juice. A small amount of fermenting must is added to begin fermentation and to fill the tank with carbon dioxide. As the fermentation in the juice progresses, it also begins to take place within the cells of the intact grape berries. This intercellular fermentation produces soft tannins and a unique strawberry or bubble-gum aroma. This technique works well with both Pinot Noir and Gamay and is the trademark characteristic of Beaujolais Nouveau.

Another method of production, **extended maceration,** is more suited to big-bodied red wines such as Cabernet Sauvignon. With this technique the fruit is crushed and fermented with typical cap management; at the end of fermentation, however, the must is not pressed. Instead, the tank is topped off and the skins are left in contact with the young wine for 1 to 8 weeks. At first, the young wine becomes more bitter and astringent from the increased skin contact, but after several weeks, the tannins begin to **polymerize.** This is the process whereby small, harsh tannins join together and become so large that they are no longer soluble and begin to drop out, leaving the finished wine softer and more drinkable.

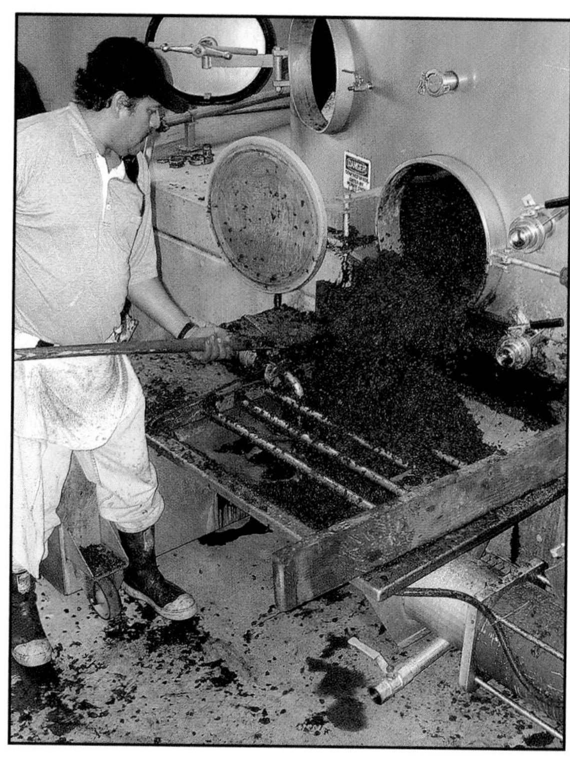

FIGURE 3–12 Removing skins from a tank into a portable must pump that transfers the fermented must to the press for separation of the juice from the skins.

drained out of the tank by gravity. The remaining wine, 10 to 20 percent, is held within the skins still inside of the tank. The skins are then removed and loaded into a press, which squeezes out their remaining liquid. The removal of the skins from the fermentation tank is one of the most labor-intensive aspects of winemaking (Figure 3–12). Great care must be taken when entering a tank that has just finished fermentation due to the danger of asphyxiation from the residual carbon dioxide. Before the tank can be entered, it must be properly ventilated and the atmosphere tested to make sure it is safe.

There are a number of types of presses, but they all work in the same manner. Force is applied to a layer of skins against a screened or slatted surface which allows the juice or wine to drip through, but holds the skins and seeds back. After pressing, the compressed layer of skins is called a cake. To extract the maximum amount of liquid from the skins it is necessary to break up the cake and re-press it a number of times at progressively higher pressures. The first wine to come off is usually combined with that which was dejuiced from the tank and is called the **free run.** As the cycles of pressing continue, the quality of the juice diminishes and becomes more astringent and bitter. Often, the wine that is removed at the end of the press cycles is kept separate from the free run and is called the **press fraction.** The young wine is then collected in a sump at the base of the press before being pumped into a receiving tank. After the skins dry they are called **pomace** or ***marc*** (the French term for pomace), and are removed from the press and used for compost in the vineyard.

The basket press is the oldest and most simple design (Figure 3–13). It is a vertical cylinder made of slats of wood arranged with small gaps in between. The fruit is loaded into the top and a plate is pushed down by mechanical means, which causes the juice to drip out through the gaps in the slats. Traditional basket presses are gentle but require the cake to be broken up by hand in between press cycles. More modern basket presses are made of fiberglass and mounted horizontally; the cake can then be both broken up and unloaded by simply rotating the press.

FIGURE 3–13 Diagram of a basket press.

Another type of press uses air pressure, or pneumatics, to squeeze the juice out of the skins. There are a number of designs for pneumatic presses; one of the most common is the tank press. Tank presses are cylindrical steel tanks that are 3 to 8 feet (1 to 2.4 meters) in diameter and are mounted horizontally. On one side of the interior there is an inflatable bag or membrane, and on the other side is a series of perforated screens or channels. Once the press is loaded with grapes and the door is closed, it rotates so that the screens are down and the bag is above. The bag then inflates, squeezing the skins against the screens and removing the juice. There is less chance for contamination or oxidation with tank presses because they extract the juice inside the press. Their efficiency and gentleness toward the grapes make them the workhorses of most modern wineries. Yield after fermentation is typically about 170 gallons of wine per ton of grapes (700 liters per metric ton).

After pressing, the wine is pumped to a tank in the winery cellar for storage. At this point the new wine is very turbid and full of suspended solids that are primarily yeast cells and particles of grape skins and pulp. After several days, the suspended solids begin to settle out to the bottom of the tank, forming a layer of thick, mud-like material or dregs called **lees.** After a week or two, the clean wine is decanted off the layer of lees in a process called **racking.** This process of settling and racking can be done once, or repeated several times, to clarify the wine before it is transferred to the aging cellar and placed into barrels.

White Wine Crush and Fermentation

It is no surprise that white wines are made from white grape varieties. However, since the juice of most red grapes is colorless it is also possible to make a white wine from red grapes, as is done with White Zinfandel and Blanc de Noir sparkling wine. So, white winemaking is defined not only by the color of the grapes that are used but also by how they are processed (Figure 3–14). The major difference between white and red wines is that reds receive most of their flavor from the skins and whites get their flavor from the juice. Therefore, in processing, the most important

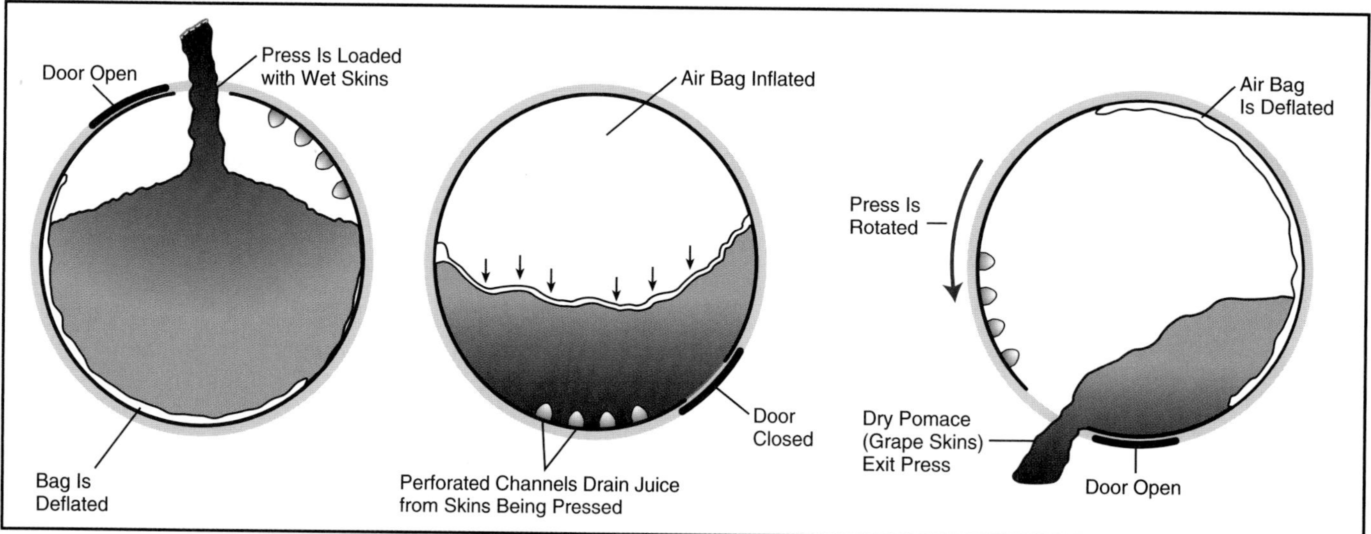

FIGURE 3–14 Diagram of the operation of a tank press.

difference is that red wines are pressed *after* fermentation and white wines are pressed *before.* Because the flavor of white wines is not as dependent on what is extracted from the skins, the grapes are usually picked early in the morning and brought to the winery while they are still cool in order to preserve their fresh fruit flavors.

White winemaking begins in much the same way that red winemaking does—the grapes are picked, weighed, inspected, and delivered to the receiving hopper in much the same way they are for red wine production (Figure 3–15). However, since the delicate character of white grapes is more sensitive to warm temperatures than red grapes, a special effort is made to pick white wine grapes in the morning when temperatures are cooler and to quickly transport them to the winery. Once the grapes are unloaded, red and white winemaking techniques diverge and for white wine production the juice is separated from the grapes before fermentation. The winemaker has several options on how to accomplish this. The fruit can be (1) crushed and pressed; (2) it can be crushed, dejuiced, and pressed; or (3) it can be whole-cluster pressed.

In the first option, the grapes are destemmed and crushed and the must is pumped into the press for the juice to be separated. In the second option, the must is dejuiced before being loaded into the press. This is done by having a slotted screen to drain the juice inline on the way to the press. A more gentle method of draining the juice is the dejuicing tank. These tanks are mounted above the press, and the must is pumped into them directly from the crusher. A screen is located on the inside of the tank and the force of gravity helps the grape juice to drain through it. The third option of processing white grapes, whole-cluster pressing, is also the most gentle. In this method, the stemmer-crusher is bypassed entirely and the whole clusters are loaded directly into the press. This minimizes the amount of skin contact the juice receives, and since the grapes are not macerated by the crusher, it produces a juice with lower solids and a more delicate flavor. Whole-cluster pressing, however, is more difficult and expensive, because it requires a larger press and more time to load as whole clusters of fruit do not dejuice as readily as crushed fruit.

After pressing, the juice is pumped to a settling tank in the fermentation cellar. In white grape pressing the difference in quality between free run and press juice is even greater than it is with red wines, so the press juice is usually kept separate from the free run. The juice is

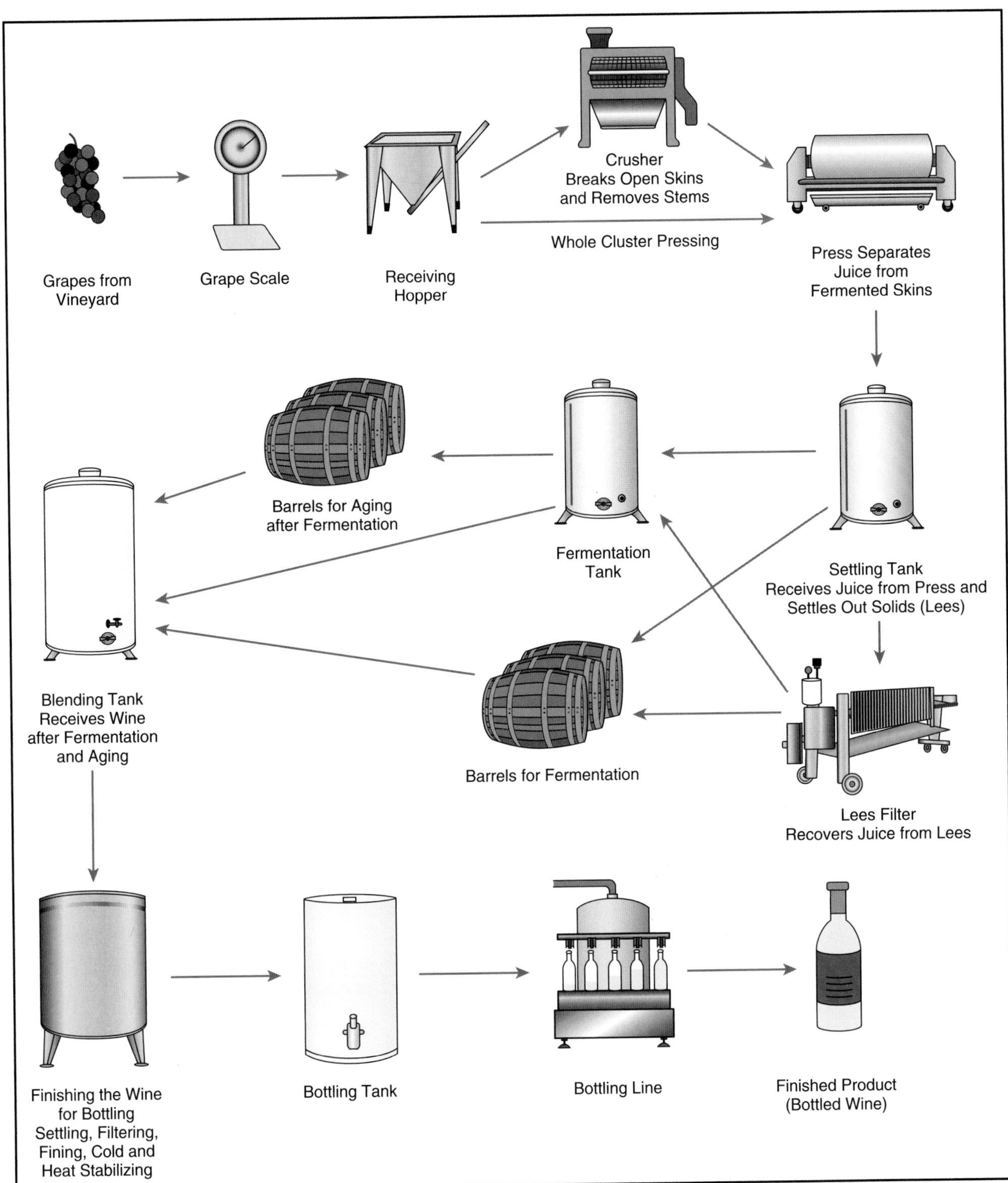

FIGURE 3–15 Flowchart of operations in making white wine.

kept cool, at around 50°F (10°C), and held in the settling tank for 12 to 72 hours to allow the lees to form. It is necessary to separate the juice from the grape solids, or **primary lees,** to avoid the production of undesirable flavors during fermentation. After settling is complete, the clean juice is racked off into the fermentation tank (Figure 3–16) where it is adjusted with yeast and fermentation additives, if needed. Similar to red winemaking, the tank is not

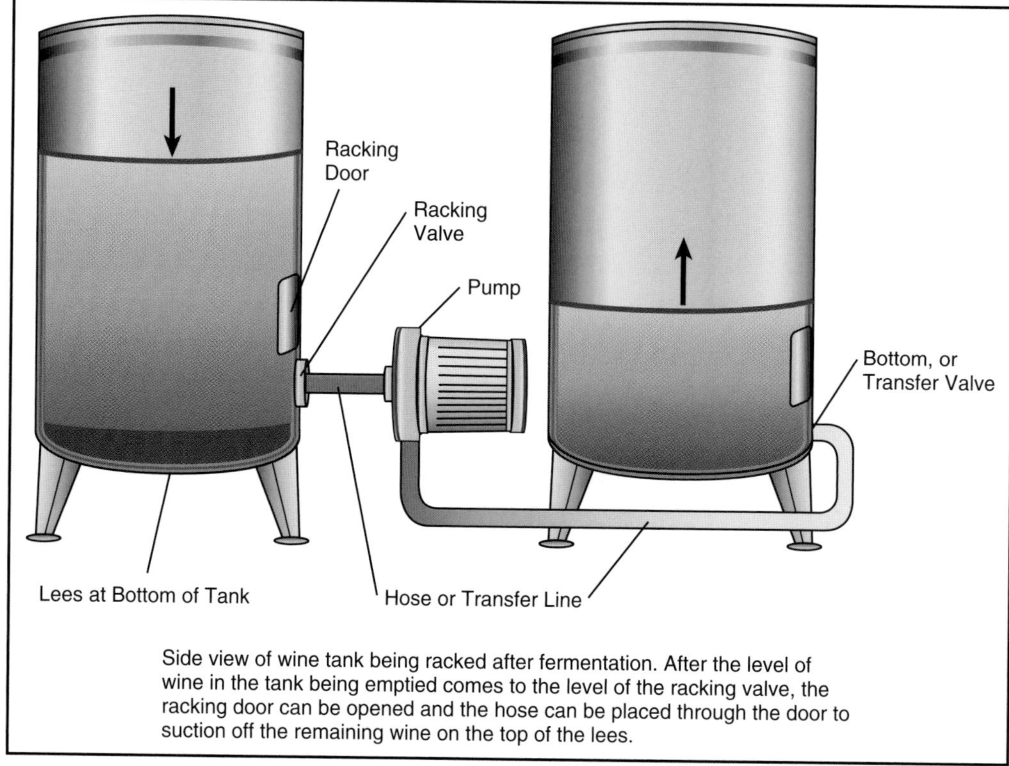
Racking Door
Racking Valve
Pump
Bottom, or Transfer Valve
Lees at Bottom of Tank
Hose or Transfer Line
Side view of wine tank being racked after fermentation. After the level of wine in the tank being emptied comes to the level of the racking valve, the racking door can be opened and the hose can be placed through the door to suction off the remaining wine on the top of the lees.

FIGURE 3–16 Diagram of racking a tank of clean wine off the lees that have settled to the bottom of the tank.

filled to capacity in order to allow room for the foam that forms during fermentation. White wine fermentations take place at a cooler temperature, 45° to 60°F (7° to 15°C), because there is no need to extract color from the skins as in red wine fermentations. The cooler temperature helps the juice to retain its fruity aromas. Because it takes place at a cooler temperature, white fermentations take two to three times longer than red fermentations, about 3 to 6 weeks. After fermentation, the new wine is racked off the yeast lees into a holding tank in preparation for aging and processing.

Some white wines are transferred to barrels just as they are starting to ferment. Barrel fermentation of white wine gives it a distinctly toasty aroma and is very popular with Chardonnay. After the fermentation is finished, some of the barrels are used to top off the rest of the lot and the wine is left in contact with the yeast lees at the bottom of the barrel. This technique of aging is called ***sur lie*** (French for "on the lees"), and it gives the wine more of a yeasty-bready aroma and more viscosity. The young wine can be left *sur lie* for many months; sometimes the yeast is periodically stirred to intensify the character.

Barrels and Aging

The first **coopers,** or barrel makers, were the Romans over 2,000 years ago (Jackson, 2000). The Romans used barrels to store and transport a variety of goods, including wine. Winemakers soon discovered that storing wine in barrels had positive effects on the wine's flavor and body. The qualities that barrel aging gives to a wine are so positive that barrels are still used for winemaking today, long after their other uses have been discontinued. Although there has been some mechanization, barrels are still made by hand in the method they have been for hundreds of years. There are two types of reactions that take place during aging: the wine undergoes a slow oxidation and it absorbs flavor components from the wood. Both of these make significant contributions to a wine's flavor. Aging a wine in small, 60-gallon (225-liter) barrels is both

Malolactic Fermentation

In addition to being very sweet, grape juice is also quite tart. This natural acidity primarily comes from the presence of two types of acid: tartaric and malic. Malic acid is found in many fruits, while tartaric acid is unique to grapes (Yair, 1997). There is a group of microorganisms, called **malolactic bacteria,** that can use malic acid as an energy source for growth. They do this by converting malic acid in wine or grape juice into lactic acid, the type of acid found in milk (Figure 3–17). Malolactic fermentation usually takes place after the **primary fermentation,** or alcoholic fermentation, and occurs at a much slower pace. Often the fermentation takes place in barrels, sometimes not completing until the spring following the harvest.

Malolactic fermentation has several effects on the wine, the primary effect being deacidification. Since malic acid is stronger than lactic acid, a wine will taste less tart and have a higher pH (lower acidity) after fermentation. Malolactic fermentation also makes wine more microbiologically stable. If malolactic fermentation finishes during aging, it will not be able to spoil the wine by taking place after the wine is bottled and the wine will not have to be filtered as tightly as a nonmalolactic wine.

Finally, malolactic fermentation produces a compound called diacetyl that has a distinct buttery character. The presence of diacetyl is more noticeable in white wines than reds; Chardonnays often go through malolactic fermentation to get this aroma. Winemakers can encourage malolactic fermentation by adding cultures of the bacteria after primary fermentation or by placing the wine into barrels that have previously been used for wines undergoing malolactic fermentation. Malolactic fermentation is usually encouraged in red wines for reasons of stability, and because it is difficult to prevent it from spontaneously occurring during the long barrel-aging process. With white wines, it is a stylistic concern; in a light-bodied, fruity wine like Riesling, it is usually avoided, while in a rich, oak-aged Chardonnay it would be more appropriate.

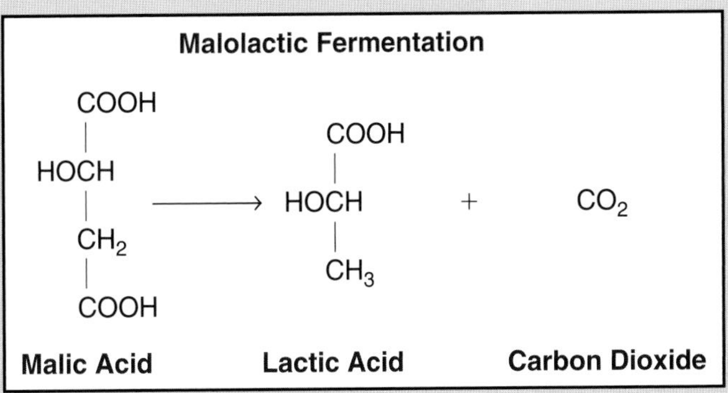

FIGURE 3–17 The chemical equation for malolactic fermentation in which one molecule of malic acid is converted to one molecule each of lactic acid and carbon dioxide.

expensive and labor intensive, but the positive effect that barrel aging has on wine makes it worthwhile (Figure 3–18).

Barrels can be made out of many types of trees, however, oak is the chosen wood for wine-barrel production. In addition to being strong and durable, it is also nonporous so the barrels will not leak (Figure 3–19). Most importantly, it has excellent flavor and aroma compounds that are extracted into the wine during storage. Although oak is the wood of choice for winemaking, there are many different types of oak from which to choose. The two major categories of oak are European and American. There are two species of European oak that are used for making wine barrels: *Ouercus sessilis* and *Quercus robar,* and they are grown throughout France and central

FIGURE 3–18 Barrels of red wine aging in a winery cellar.

FIGURE 3–19 Toasting the inside of wine barrels during production at a cooperage, or barrel-making facility. Toasting helps give the wood the proper flavor for aging wine, and also softens the wood staves so they can be bent into the characteristic barrel shape without breaking.

Europe. European oak is known for giving wine a rich, toasty vanilla aroma. In the United States *Quercus alba,* or white oak, is used for barrel making and has a stronger, more woody flavor than European oak. Beyond the type of oak used, a barrel's flavor varies depending on the forest the wood is from, how the wood is seasoned, and the various methods of production that different coopers use. This variety in styles gives winemakers a wide selection of flavors that they choose to put into their wine by aging.

Much of the flavor obtained from aging wine in barrels comes from what is extracted out of the oak, however, the softening of the wine's texture that comes with aging is due to the process of slow oxidation. Oxidation can be a vintner's enemy, spoiling the wine's aroma and color as well as promoting the growth of bacteria that produce vinegar. Oak has the quality of being semipermeable to oxygen, allowing it to be incorporated into the wine at just the right rate. A small amount of oxygen in an aging wine helps tannin molecules to polymerize and settle out, softening a wine's body and making it less bitter. Furthermore, a small amount of alcohol and water in the wine can evaporate through the oak of the barrel. This evaporation causes the remaining wine in the barrel to become more concentrated with acid and flavor. From time to time, the **ullage** (headspace in the barrel) that is produced by this evaporation must be displaced by topping the barrel up with some wine from the same lot. The period of time that a wine spends in oak depends on the tastes of the winemaker and the body of the wine being made. A big-bodied red such as Cabernet Sauvignon or Syrah may need 2 or more years in oak before it has sufficiently mellowed for bottling. A fruity, light-bodied wine like a Beaujolais Nouveau or Gewürztraminer may be bottled with little or no oak aging. Wine can also be aged in stainless steel tanks or after it has been bottled. Under these conditions there is much less exposure of the wine to oxygen than there is in barrels, so the aging process is slower and has less of an effect on the flavor of the wine than barrel aging. In addition, during tank or bottle aging there are no flavor compounds being extracted into the wine from oak. Some wineries place oak wood in stainless steel tanks to get the flavor of the wood without getting the aged quality of barrel storage.

Finishing a Wine

After aging is complete, the wine is pumped out of the barrel and sent to the tank cellar for preparation for bottling. Wines can be bottled from a single fermentation lot but more often different lots are blended together (Figure 3–20). Blending can combine lots from different vineyards, even

FIGURE 3–20 An enologist, or wine chemist, making a laboratory trial blend at Kenwood Vineyards in Sonoma County, California. When the favorite trial blend is selected, the same proportions of different lots of wine in it will be used to make the final blend in the wine cellar.

different regions and varieties, each with its own attributes. The art of blending lies in putting different combinations of these lots together in trial blends to find the combination that has the most balance and complexity. After the favorite trial blend is selected, its proportions are used to assemble a bottling blend in the cellar. Having a wide selection of wine lots with different flavors gives a winemaker many options to fine-tune the blend and achieve the desired style. Sometimes winemakers will blend before or in the middle of the aging process to give the blend time to harmonize in the barrel. After the blend is selected, two more steps must be completed before wine is ready to be bottled: clarification and stability. Clarification produces a wine that is brilliant and free of suspended solids, while stability operations are performed to ensure that a brilliant wine stays so. These operations are closely linked, and often one will complement the other.

The simplest and most gentle form of clarification is settling and racking. As wines age in barrels, particles that are suspended fall out and accumulate at the bottom of the barrel. If the wine is carefully pumped out, the solids remain behind, sending a clean wine to the tank. Two more active methods of clarification are fining and filtering. **Fining** is the process of adding a compound called a fining agent to the wine that will react with compounds in the wine causing the two materials to combine and become insoluble. After the wine settles, the fining agent and the wine component that it removed are left behind in the lees when the wine is racked. Most fining agents are proteins although some, such as bentonite (a type of clay) and carbon, are inorganic. Fining not only helps to clarify and stabilize a wine, it can also affect the flavor. A classic example is egg white fining whereby egg whites, which contain the protein albumin, are added to a red wine. The albumin reacts with tannin molecules, causing them to drop out and make the wine softer in character.

Another fining agent that is used for clarity and stability is bentonite. Bentonite is commonly used in white wines to make them protein or "heat" stable. All wines contain some residual grape protein; this protein can denature (lose its shape) over time and become insoluble. If this happens after the wine is bottled, it will form a milky haze on the bottom of the bottle. To combat this, bentonite is added to white wines to remove the protein, and in the process it also helps to clarify the wine. Red wines have a much higher level of tannins. Since tannins react with proteins in a manner similar to bentonite, it is seldom necessary to fine with bentonite to make red wines protein-stable.

In addition to protein or "heat" stability, a wine is also "cold" stabilized to remove excess potassium bitartrate before bottling. Potassium bitartrate, or cream of tartar, is a salt comprised of two natural constituents of wine: potassium and tartaric acid (Figure 3–21). Potassium bitartrate is semisoluble and forms crystals over time, especially under cold conditions. These crystals will form in bottles or in tanks and have an appearance of ground glass. To avoid an excess of tartrates crystallizing in the bottle, wines are chilled in the cellar to just above the freezing point. The crystals then settle to the bottom and the walls of the tank (Figure 3–22).

Filtering is another way to obtain clarity in a wine prior to bottling. There are many types of filters designed for different winemaking applications (Figure 3–23). They all work by using pressure to force the wine through a porous substance that allows the liquid to go through but holds solid particles back. Filters are available in many grades of "tightness" that retain larger or smaller particles. Filtration is very important when making a wine that has the presence of residual sugar or malic acid. In such cases, if all of the microbes are not removed before bottling, they can begin to ferment in the bottle and spoil the wine. In any case, there are no human pathogens that can tolerate the alcohol in wine, so it is important to keep out microbes only because of their effect on wine stability and quality. Most winemakers use some form of fining or filtration to ensure the quality of their wine, however, others prefer a wine that is unfined and unfiltered. The philosophy here is that although a wine that is not fined or filtered may be less brilliant and less

FIGURE 3–21 Crystals of potassium bitartrate formed on the cork of a bottle of Syrah during aging.

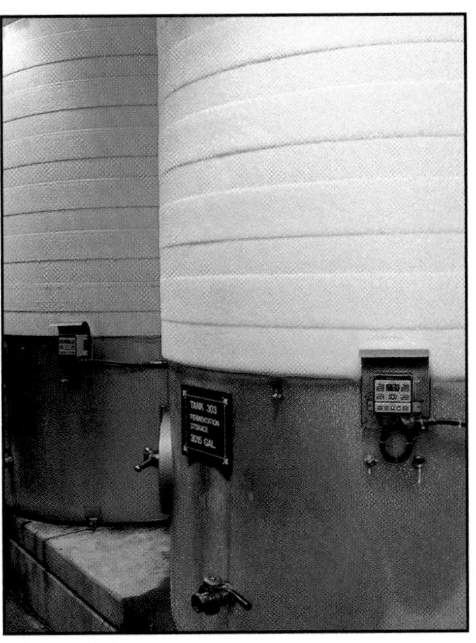

FIGURE 3–22 A stainless steel wine tank with its cooling jacket covered with ice during cold stabilization.

FIGURE 3–23 A plate and frame filter filtering a red wine before bottling.

stable, it retains more of its natural flavor. While there is some truth to this argument, if fining and filtering are properly handled, they will have very little effect on a wine's flavor.

Bottling

Bottling, the final step in winemaking, must be done with great care because it is not easy to rectify mistakes after the wine is in the bottle. Before bottling, the wine is analyzed and checked for stability one final time and any necessary adjustments are made. The wine is then sent to the bot-

FIGURE 3–24 Bottles of Cabernet Sauvignon on a filling machine at Valley of the Moon Winery in California.

tling room where a filler machine distributes it to the bottles (Figure 3–24). Immediately after being filled, the bottles are corked or capped to protect the wine from contamination. The bottles are then sent to a capsule machine to have a capsule applied to cover the neck and the cork. The final steps of applying the label and packing the bottles into cases then take place. Bottling is some of the roughest treatment a wine will receive and can leave a wine with less fruity aromas and body for a period of time. This condition is called *bottle shock,* and will go away if the wine is allowed to have some bottle age before consumption. While this is a real condition, bottle shock is often used as a scapegoat for anything a winemaker does not like about a new wine.

Sparkling Wine

Sparkling wine is defined as wine with bubbles or effervescence (Figure 3–25). It was first developed in the Champagne region of France in the 1700s, and was the result of two seventeenth-century winemaking inventions: the cork and the wine bottle. These innovations provided, for the first time, an airtight package for wine. Inadvertently, young wines were bottled before they had finished primary fermentation. Because of the tight seal, when the wines finished their fermentation in the bottle, the carbon dioxide was trapped inside, giving them effervescence. Over the next hundred years this accident was developed into the elaborate procedure used to make sparkling wine called ***méthode champenoise,*** or the Champagne method (Figure 3–26). There are other processes used to make sparkling wine; however, the original *méthode champenoise* is still considered to yield the highest quality product. The term "Champagne" refers to sparkling wine made in the Champagne region of France. In the United States, "champagne" is often used as a generic term to mean any sparkling wine, and it is legal to use the term on the label as long as the region of origin is listed (e.g., "California champagne").

Since the Champagne region is very cool, the grapes used for making sparkling wines are early ripeners. Pinot Noir, Chardonnay, and Pinot Meunier are three of the most common grapes

FIGURE 3–25 The glassware that is used for sparkling wine is called a flute. It is tall and narrow to prolong the evolution of bubbles.

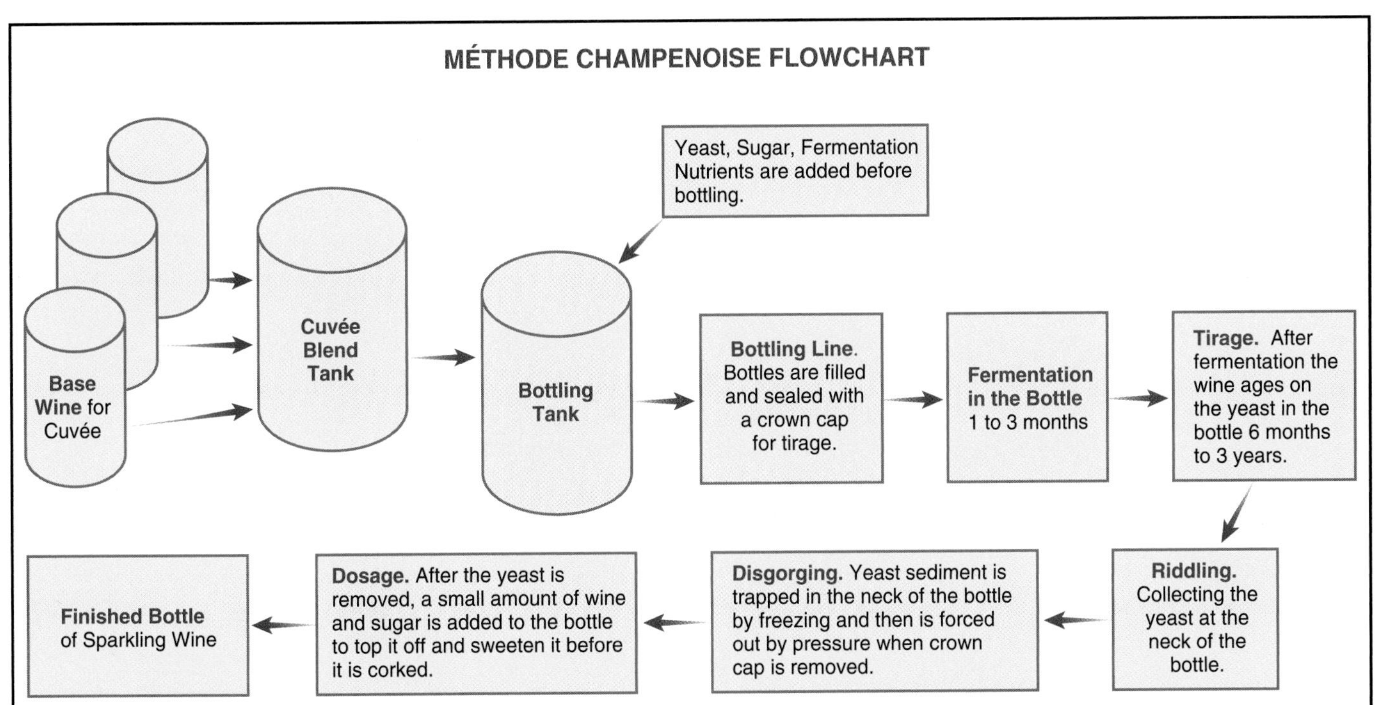

FIGURE 3–26 Flowchart of operations in making *méthode champenoise* sparkling wine, the traditional method used for making French Champagne.

used for sparkling wine; in California, the Pinot Blanc grape is also popular. The grapes used for sparkling wine are picked even earlier than those used for still wines for several reasons. The base wine used for sparkling wine should be low in alcohol and should not have a lot of varietal character. This is because the secondary fermentation will increase the alcohol content, and the finished wine should exhibit the flavors produced from the *méthode champenoise* process. Furthermore, the grapes are handled very gently during picking and pressing to avoid extracting too much flavor or color from the skins. Press cycles for sparkling wine are longer and more press fractions are taken to ensure the best juice is kept separate. After fermentation, the wine is racked and stored until blending.

In the winter following harvest, the winemaker tastes the various lots of wine produced and puts together the base blend called the ***cuvée.*** *Cuvée* is a French word that translates literally to "tub full" or "vat full." The *cuvée* is low in color and alcohol, but high in acid, and it takes considerable talent as a taster to see how the flavors in the *cuvée* will translate into the finished wine. After the blend is made, it is bottled with a small amount of sugar and actively fermenting yeast, and sealed with a crown cap. The wine is then stored in a cool, dark place during the fermentation in the bottle. As the yeast ferment, they produce about 1.5 percent more alcohol and about 90 pounds per square inch (6 atmospheres) of carbon dioxide. The bottles used are much heavier than those for still wine in order to hold back the pressure; they are also dark green because sparkling wine will develop off-flavors if exposed to excessive light. After fermentation, the bottles undergo ***tirage,*** whereby they are aged on the yeast cells for a period of several months to many years depending on the style of sparkling wine being made. During this time, the yeast cells begin to break down in a process called **autolysis,** which is what gives *méthode champenoise* sparkling wines their unique flavor.

After tirage, it is necessary to remove the yeast from the bottle before the wine can be finished. The bottle is taken from storage, mixed to loosen the yeast from the sides of the bottle, and placed in a riddling rack. **Riddling** is a process used to accumulate the yeast at the end of the neck of the bottle. The bottle is placed horizontally in the riddling rack and every day it is twisted and pushed back into the rack at a slightly steeper angle. After several weeks, the yeast has settled at the end of the neck and the bottle is upside down, or ***sur pointe.*** Hand riddling is still practiced at some wineries but at most wineries the process is now done by machines.

FIGURE 3–27 Bottles of sparkling wine entering a disgorging machine. The plug of ice visible in the neck of the bottle contains the residual yeast from fermentation in the bottle. When the crown cap that seals the bottle during tirage is removed, the pressure of the sparkling wine in the bottle will eject the ice plug leaving the clean wine behind.

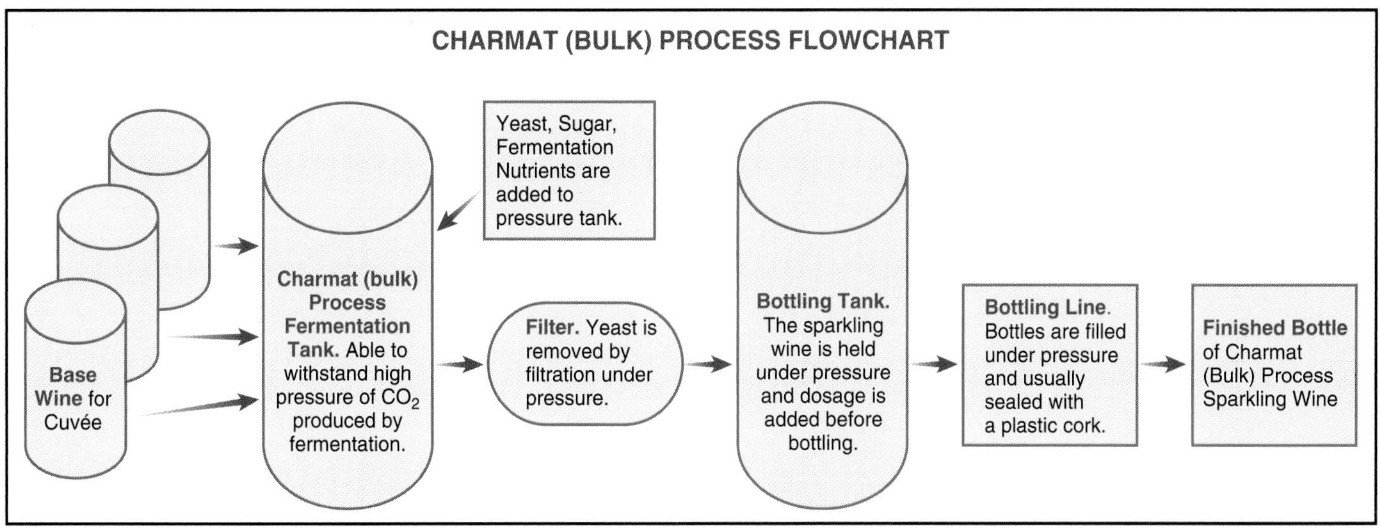

FIGURE 3–28 Flowchart of operations in making Charmat or "bulk" process sparkling wine, the method used for the majority of sparkling wine produced in the United States.

Following the riddling process, the yeast is ready to be removed by **disgorging** (Figure 3–27). The bottles are chilled to just above the freezing point and placed upside down in a brine bath to freeze the wine in the neck of the bottle. This traps the yeast, and when the crown cap is removed, the pressure of the wine expels the plug of frozen wine, taking the yeast with it. The bottle is then topped off with a small amount of base wine called **dosage.** Since sparkling wine is quite sour, the dosage often has a small amount of sugar to balance out the acid. The bottle is then finished with a wide-diameter agglomerate cork that is only inserted halfway to give it its mushroom shape. Sparkling wine produced by *méthode champenoise* is labeled as such or as "fermented in this bottle."

Other Methods of Sparkling Wine Production

Although the finest sparkling wines are made by *méthode champenoise*, this accounts for only a small portion of production. The majority of sparkling wine is made by the **Charmat process** or "bulk" process developed by the French winemaker Charmat in 1907 (Figure 3–28). In the Charmat process, instead of having the secondary alcoholic fermentation take place in the bottle, it takes place in large steel tanks that are specially designed to withstand the pressure produced by fermentation. After fermentation, the wine is racked off and the yeast is filtered out under pressure. Once the dosage is added, the wine is bottled and usually sealed with a plastic mushroom-shaped cork. The ability to filter the yeast out saves the effort of riddling and disgorging, making these bulk-processed sparkling wines much less expensive to produce. Without the extended time on the yeast during tirage however, these wines do not have the same character as those produced by *méthode champenoise*. The grapes that are used for the Charmat process are typically less expensive varieties such as Chenin Blanc and French Colombard.

There are two other methods used to make sparkling wine: the transfer method and artificial carbonation. In the transfer method, the *cuvée* is fermented in bottles and aged in tirage for a time. At the end of tirage, the wine is transferred from one bottle to another, being filtered in the process. Sparkling wine made this way is labeled "fermented in *the* bottle" instead of "fermented in *this* bottle." The transfer method, because it incurs extra expense without adding significantly to the quality, is not as popular as the *méthode champenoise* or the Charmat process. In artificial carbonation, a still base wine is injected with carbon dioxide, carbonating it before serving, much the same way a soda pop dispenser works. These wines are usually served at large banquets and have little of the qualities of natural fermented sparkling wine.

Dessert Wines

There are multitudes of unique dessert wines that are produced throughout the world's wine-making regions. **Dessert wines** are made with appreciable sugar and often have higher alcohol to stabilize the wine and prevent it from fermenting in the bottle. Dessert wines make an excellent dessert in themselves, can be offered as a digestive after a meal, or can be a complement to a sweet dessert course. Although the classic definition of a dessert wine is a wine that is sweet, for purposes of taxation the United States Government classifies all wines that are fortified with additional alcohol as "dessert wines" whether they are sweet or not, and they are taxed at a higher rate than table wines. We will examine the production methods used in some of the most common types of dessert wines: late harvest, Port, and Sherry.

Late-Harvest Wines

Late-harvest wines are made from grapes picked at a much higher sugar level than grapes used for table wines. Through photosynthesis grapevines can ripen the crop up to about 28°Brix, while late-harvest wine grapes are frequently picked at 35°Brix or more. Late-harvest wines achieve this higher level of sugar concentration due to the fruit partially dehydrating on the vine. Under the right conditions, water will evaporate through the skin of the berry, concentrating the sugar that is left behind. This high sugar means that the yeast will have a difficult time fermenting due to the combined inhibitory effects of alcohol and sugar concentration. Late-harvest fermentations progress at a very slow rate and are unable to ferment to dryness. When the fermentation eventually slows to a stop, a microbial stable, sweet wine is produced.

This dehydration is increased by an infection of a mold that is usually considered a vineyard nuisance, ***Botrytis cinerea*** or "noble rot" (Figure 3–29). This mold is a common problem in vineyards and is normally discouraged by applying sulfur dust; however, under the right conditions with the right varieties, it has the ability to make some of the world's best wines. *Botrytis cinerea* infects ripe grapes that are exposed to high humidity; the growth of the mold perforates the skin of the grape, opening a path for the water to leave. When wet weather is followed by dry, warm weather, the berries then dehydrate to reach the high sugar levels needed for late harvest. Two excellent examples of wines made under these circumstances are the Trockenbeerenauslese, or TBA, of Germany and the Sauternes wines of France.

FIGURE 3–29 Growth of *Botrytis cinerea* on clusters of Sauvignon Blanc.

The growth of botrytis is sometimes encouraged by artificial means, such as watering the grapes with overhead sprinklers, to get the needed humidity to start growth. In addition to the concentration of sugar, botrytis produce a number of compounds that affect the flavor of the wine. One of these, botrycine, has a distinctly apricot aroma. Thin-skinned grape varieties like Zinfandel will shrivel up in hot weather during the harvest season and significantly concentrate the sugar without the presence of mold. However, these late-harvest wines have a different, more "raisiny" character than botrytis-affected wines. Late-harvest grapes, because of their high solids and sugar, are notoriously difficult to press, and fermenting and clarifying the wine is no easier. The unique weather conditions that are required, combined with the difficulty of their production, make late-harvest wines both rare and expensive.

Late-harvest wines can also be made without the growth of *Botrytis cinerea*. In Germany and other cold-climate growing regions, the grapes can be left on the vine until freezing weather sets in at the end of the fall. Wines that are made from frozen grapes are called ***Eiswein,*** or ice wine. As the water in the berries freezes the remaining juice is concentrated, increasing the sugar level to about 35°Brix. The grape clusters are then picked, transported, and pressed while they are still frozen. The pressing is done very slowly and, as the juice is removed from the grapes, some of the water in the berries remains behind as ice. Like botrytized wines, the fermentation proceeds slowly and stops before it can complete, resulting in a sweet dessert wine.

Port-Style Wines

Port wines are full-bodied red wines with about 10 percent sugar and 20 percent alcohol, and are native to the Douro River wine region in northern Portugal. **Port-style** wines are wines made in the style of Port, but produced outside the Port region. Historically, in the Douro River wine region, brandy was added to red wine to stabilize it for export. After a time, the practice of adding the brandy, or fortification, in the middle of fermentation was developed. This had the effect of killing the yeast while the must was still quite sweet. Because a deep-red wine with lots of tannins is desired for Port, and the time of the fermentation is limited, winemaking practices are designed to maximize extraction from the skins. There are a number of deep-colored red grape varieties grown in the Douro River region for Port production; some of the most popular are Touriga Nacional, Touriga Francesa, Tinta Cão, and Tinta Roriz.

The terrain of the Douro River is steep and rocky, conditions that stress the grapevines and help to intensify the character of the fruit. When the grapes are ripe, they are picked and brought to small wineries located near the vineyards called ***quintas.*** Traditionally, Port wines were fermented in a shallow stone trough called a **lagar.** The fruit was placed in the troughs and crushed by being trodden upon by cellar workers. The treading would continue, mixing the cap and juice throughout the fermentation, until the sugar had fallen to 12° to 14°Brix. At this point, the must would be pressed and brandy would be added at the ratio of one part brandy to three parts new wine in order to kill the yeast and preserve the residual sugar in the must. Although treading is still practiced at some wineries today, most Port is mixed during the fermentation by mechanical means.

After fermentation, the Port is transported down the river gorge to the city of Oporto on the coast. Here "Port houses" assemble the wines from many different vineyards and quintas into blends, based on their quality. Many of the Port houses have British names, which indicate their origins as export companies. Some quintas hold on to their product, bottling it under their name as a "single quinta" port. Port-style wines are made around the world and, like "champagne," "port" has become a generic term for the style of sweet red wine produced in the Port region of Portugal.

In California, sweet dessert wines are often made from the Zinfandel variety. Called "Zinfandel Port" the winemaking is a hybrid of the late-harvest and Port winemaking methods.

The Zinfandel grape has thin skins and, when allowed to hang on the vine during warm fall weather, it can achieve sugar levels of about 30°Brix through dehydration. The grapes are then harvested and fermented on the skins and when the must reaches the desired sugar level it is pressed and the juice is fortified with alcohol to arrest the fermentation.

Sherry

Sherry originated in Spain and, like Port, it is produced in a variety of styles. The Spanish have a saying that "there is a Sherry for every occasion." This reflects the wide range of sherries from light and dry table wines to the more common rich and sweet dessert wines. This is also an indication that Sherry was so important to the region that different styles have been designed to complement many types of food. The defining characteristic of Sherry is that it is purposely oxidized, making it high in acetaldehyde, which is the result of the reaction of ethanol (wine alcohol) and oxygen. This gives Sherry wines their distinctive roasted nut aroma. Wine drinkers who are not accustomed to Sherry can sometimes find this aroma unsettling because it is the same compound found in a table wine that has been spoiled by oxidation. Sherry was once one of the most popular wines in the United States, however, in recent years its consumption has declined significantly.

The flavor of Sherry is produced during the aging process so, like sparkling wines, a fairly neutral wine is desired as a base for Sherry. For this reason, neutral grape varieties like Palomino are used for its production. Sherry production starts by fermenting the base wine to dryness and fortifying to achieve an alcohol content of about 15.5 percent. The high alcohol level inhibits the growth of vinegar bacteria. To reach this level, the grapes must be very sweet and sometimes they are dried on mats after picking to reach the appropriate sugar concentration. After the base wines are made, they are graded by color, taste, and body to determine which type of Sherry they will be used to make. The lighter wines are inoculated with **flor yeast** and called **fino** or **manzanilla** (when aged in the coastal region of Sanlúcar de Barrameda, Spain). The more full-bodied wines are fortified with brandy to 18 to 20 percent alcohol and called **oloroso.** The wines are then placed in partially full barrels to expose the wine to oxygen.

Table 3-1 Types of Port

TYPE	WINEMAKING METHOD	WINE STYLE
Ruby	Blended from several vintages and aged from 2 to 3 years in oak casks before bottling.	Bright ruby-red in color with a fruity aroma. Meant to be consumed soon after bottling.
Vintage	The highest quality Port, blended from lots all produced in the same harvest; only the best years are selected to be "vintage." Aged from 1 to 2 years in casks, it is then bottled without filtration.	Deep red color and more complex flavors and aroma then Ruby Port. It should be aged for 10 to 30 years in the bottle before consumption. Due to long bottle aging, there is usually a great deal of sediment in the bottle.
Late Bottled Vintage	Ruby Ports blended from a single non-vintage year, aged in casks for 4 to 6 years before bottling.	Good quality wine that does not have the intense flavors found in vintage Port. Meant to be consumed soon after bottling.
Vintage Character	Ruby Ports blended from several vintages and aged 4 to 5 years in oak casks before bottling.	Good quality wine that does not have the intense flavors found in vintage Port. Meant to be consumed soon after bottling.
Tawny	Port wine that has been aged 10 to 40 years in casks before bottling.	Extended aging gives them amber, "tawny" color and a slightly oxidized character. Meant to be consumed soon after bottling.
White	Port wine made from white grapes, usually has limited aging in casks.	Light gold in color, it is made in a variety of styles from dry to sweet.
"Zinfandel Port"	Late-harvest Zinfandel grapes are fermented then fortified to arrest the fermentation before aging and bottling.	Lighter in color than traditional Port, dehydration on the vine gives it a distinctive "jammy" aroma.

Table 3-2 Types of Sherry

TYPE	WINEMAKING METHOD	WINE STYLE
Fino (Manzanilla)	Aged in partially full barrels under a surface layer of flor yeast growth.	Pale yellow in color with a distinctive nutty aroma. Dry and served slightly chilled. Alcohol 15.5 to 18%
Amontillado	A mature fino Sherry that is allowed to oxidize as it ages in partially filled barrels after the flor yeast is removed.	Darker than fino Sherries with a richer flavor and nutty aroma, dry with alcohol from 16 to 22%
Oloroso	Fortified before aging to prevent flor yeast growth, aged in partially filled barrels.	Dark amber in color and dry to slightly sweet (1 to 3% sugar), alcohol from 17 to 22%
Cream Sherry	Fortified before aging to prevent flor yeast growth, aged in partially filled barrels. Sweetened before bottling.	Dark amber in color and sweet (7 to 10% sugar), alcohol from 17 to 22%
Baked (California Style)	Flor yeast is grown throughout the wine (submerged culture) by bubbling oxygen through the tank. After fermentation the wine is heated to 130°F (55°C) for 1 to 3 months.	Dark amber in color and usually sweet (7 to 10% sugar), alcohol from 17 to 22%

In the fino Sherries, the flor yeast begins to grow, using the alcohol that is present as an energy source. As it grows it forms a thick film on the surface of the aging wine because it can grow only in the presence of oxygen. Sherries made in this style are light and dry and are an excellent table wine to accompany savory foods. In the United States, however, the more popular style Sherry is the full-bodied oloroso. After the oloroso is fortified, it is aged in partially full barrels but without flor yeast. Oloroso Sherries are often sweetened before bottling to make a dessert wine.

The traditional method of aging Sherry is also unique; it is done in a fractional barrel system called a ***solera.*** The barrels of a *solera* are set up in five to twelve tiers. When it is time to bottle, one-quarter of the wine on the bottom level is removed for bottling. It is replaced by one-quarter of the wine from the next highest tier. This process goes on until one-quarter of the wine from the top tier is moved to the next level to make room for the wine from the new vintage. *Solera* aging provides a great deal of consistency from year to year, and as a *solera* system matures, the average age of the wine that is bottled gets older. There is another saying in Spain: "No one sets up a *solera* for themselves, they do it for their grandchildren."

In California, methods of production differ significantly from those in Spain. Flor yeast is often grown in a submerged culture made possible by bubbling oxygen through the tank until it has reached the desired flavor. California sherry is often finished by aging in barrels at an elevated temperature, more similar to the production of Madeira than Spanish Sherry.

The Attributes of Wine

Wine is a complex mixture of nearly 1,000 different, naturally occurring chemical compounds. These constituents come from three sources: (1) the compounds that are present in grape juice, (2) the compounds that are produced by microorganisms fermenting the grape juice, and (3) the compounds that are added by the processing and aging of the wine. In addition to the natural chemicals in wine there can also be a small amount of man-made materials that are added to wine, usually in the form of sulfites used as a preservative (Figure 3–30).

The major component of wine is water, making up 80 to 90 percent of the solution. Water content affects the chemical and sensory qualities of wine, but its most important role is as the solvent in which all other wine constituents are dissolved. After water, alcohol, or more specifically ethyl-alcohol, or **ethanol,** is the next most prevalent compound. It has a significant role in

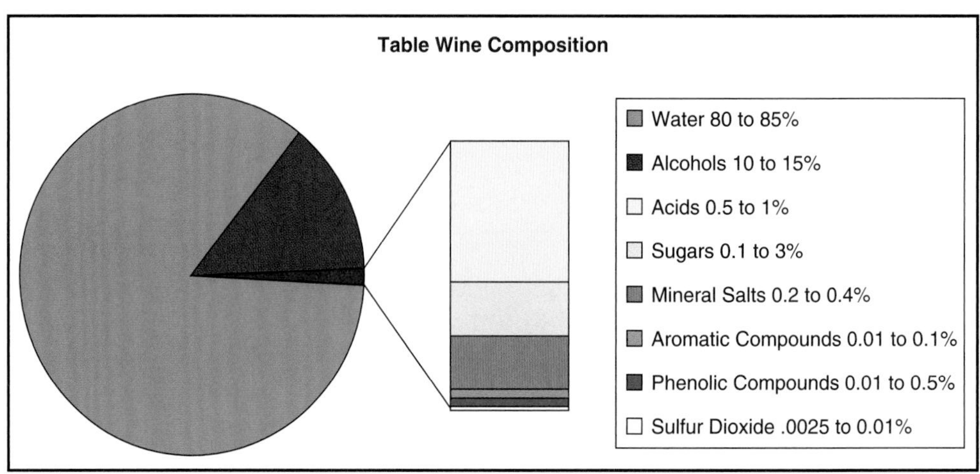

FIGURE 3–30 Table wine composition.

the sensory and stability aspects of wine, as well as many physiological effects. Glycerol is another type of alcohol that is produced by yeast. Unlike ethanol, it is nonintoxicating but it does make sensory contributions to the viscosity, or body, of the wine. Organic acids are present in about the same quantity as glycerol but have much more of a sensory effect. A wine's natural tartness is one of the qualities that make it an excellent accompaniment to food. The acids in wine also contribute to its microbial stability by inhibiting the growth of bacteria.

Some of the most important flavor compounds in wine are present in very small amounts. Trace constituents such as phenols, esters, and sugars each represent groups of complex chemical compounds with similar structures. Each of these groups has many members; for example, there are ten different alcohols found in wine besides ethanol and glycerol. Each individual wine has a unique combination of these chemicals that gives it a distinctive character. The various amounts of these compounds present are determined by factors such as grape variety, the vineyard's *terroir* (total environment, including soil, climate and location), and the production decisions that the winemaker and the grower make.

Summary

A wine's sensory qualities are determined by its chemical makeup, and the chemical makeup of a wine is influenced by a vineyard's *terroir* and the actions of the grape grower and the winemaker. As described in the previous chapter, the grower sets the stage for a wine's flavor by controlling factors such as selection of a clone and how the vineyard is pruned. Once the fruit is delivered to the winery, the winemaker takes over. Winemaking decisions including when to press, and what type of barrels to age in, build upon the flavors that the grape grower established in the vineyard. The great complexity of a wine is ultimately shaped by numerous choices available to the people who produce it. This is why there are so many different types of wines made around the world and also why they are made in such a variety of styles. The interpretation of what a Cabernet Sauvignon should taste like varies from region to region, winery to winery, and vintage to vintage. In the end, the consumer of the wine makes the ultimate decision on which interpretation is the proper one.

Review Questions

1. Discuss the three major products of alcoholic fermentation and explain their significance in the winemaking process.
2. What is the defining sensory characteristic of sherry wines? Explain how that characteristic is produced.
3. What is the difference in the pressing of red and white wines?
4. What is the definition of a table wine?
5. How does aging a wine in oak barrels affect its flavor?
6. Describe the various methods of sparkling wine production.
7. What effects does malolactic fermentation have on wine?
8. What processes does a wine undergo to be finished for bottling?

CHAPTER 4

Tasting Wines

Of the five senses—sight, smell, taste, touch, and hearing—only the first four are significant to evaluating a wine. One can hear the pop of a cork, or the fizz of a glass of champagne, but to assess a wine's flavor, hearing is not required. Of the four senses that are used the flavor of a wine is defined by the impression it makes on one's sense of smell (aroma), sense of taste, and sense of touch (texture). Sight is important to evaluate the aesthetic visual aspects of a wine, such as color or turbidity, which are part of the sensory evaluation but do not directly play a role in a wine's taste.

The Sense of Sight—Appearance

Although sight does not play a direct role in determining flavor, the visual appearance of a wine is a very important part of sensory evaluation. The color of the wine is observed for the **hue** (shade) along with its **depth** (intensity) and how appropriate these are for the type of wine being tasted (Figures 4–1 and 4–2). The clarity of the wine is also observed and noted whether it is **brilliant,** clear of any defects, or **dull,** turbid and cloudy. Looking at a glass can give the taster clues on what to expect when the wine is consumed. If a wine is an inappropriate color, or is cloudy or turbid, it will have a negative effect on its visual appreciation. The positive or negative aspects of appearance play a role in how the taster will perceive the wine when it is consumed. We are all familiar with how a food can look appetizing and stimulate hunger or look unappetizing and have the opposite effect. While a wine's appearance gives us clues on what it might taste like before it is consumed, it is important not to let the expectation of how a wine will taste prejudice your judgment. We have all had the experience of sampling a food that looks tempting but tastes terrible.

The Sense of Smell—Aroma

The sense of smell is the oldest and one of the most highly developed senses. The sense of smell is also much more acute than the sense of taste, being able to detect many more compounds at much lower concentrations. The human nose can identify thousands of different types of aromas, some at levels as low as several parts per trillion. For a compound to have an aroma it first must be **volatile,** or able to evaporate and be carried by air. Inhaling through the nose carries the air with any volatile compounds present into the upper sinus where there are two membranes called the **olfactory epithelium;** each is about the size of a postage stamp and is located to either side of the nasal septum (Figure 4–3). Volatile chemicals in the air react with receptor neurons located in these membranes. There are more than 200 different types of receptor neurons, and the degree a certain compound reacts with different types of receptors is what is responsible for a

FIGURE 4–1 One-year-old Sauvignon Blanc (left) and ten-year-old Sauvignon Blanc (right). As a white wine ages, the phenolic compounds present in the wine oxidize and turn brown, giving it an amber color.

FIGURE 4–2A 6-month-old Cabernet Sauvignon. Young red wines have a bluish/purple tint to their red color.

FIGURE 4–2B 5-year-old Cabernet Sauvignon. After a few years, the wine becomes more red as the bluish/ purple tint is lost.

FIGURE 4–2C 20-year-old Cabernet Sauvignon. After many years, the oxidation of phenolic compounds gives red wines an amber/tawny color.

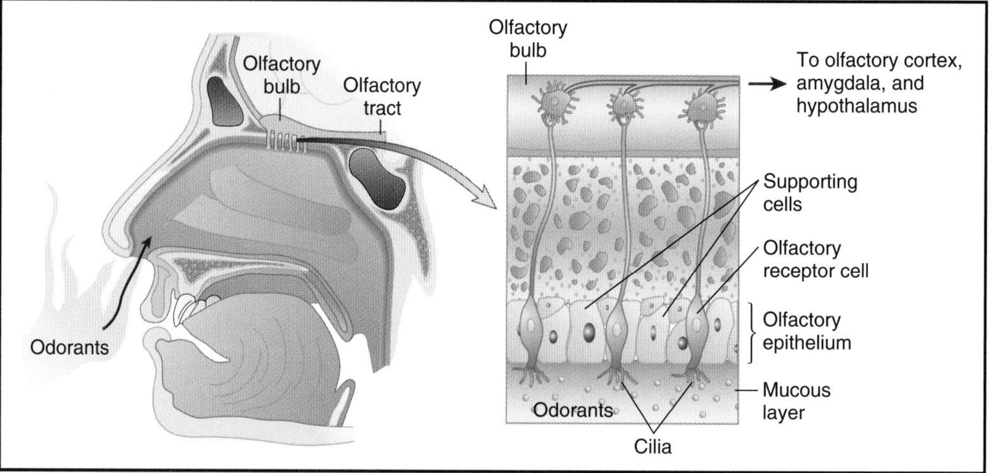

FIGURE 4–3 Diagram of sinuses, retronasal pathway, and olfactory organs.

particular smell. Neurons in the olfactory epithelium transmit signals from the receptors to the **olfactory bulb** above the nasal cavities before these signals are sent on to the brain where the information is processed. Smell is one of the most evocative senses, and how we describe an aroma is usually based on comparisons to aromas we remember having smelled before. There is a great deal of variation in sensitivity to smell in the general population; however, with training it is possible to increase one's ability to distinguish different aromas.

Wine has a very complex chemical composition with volatile compounds that originate from the grapes as well as from processing activities such as fermentation and aging. Wine has as many as 800 volatile compounds, some present in very low quantities; this high number is what is responsible for the great complexity of a wine's aroma. When describing a wine's aroma it is common to isolate and identify the different aromas present and describe what they smell like; this is called **descriptive analysis.** A taster might mention detecting the aroma of black pepper, raspberries, and vanilla in a wine. This does not mean that these flavors are from ingredients added to the wine by artificial means, but rather that some of the same compounds that are responsible for these aromas in other products are also present in wine. Descriptive analysis works particularly well because of the evocative nature of the sense of smell. Whenever we encounter a new aroma we naturally try to categorize it by what other smells it is reminiscent of.

The Sense of Taste—Flavor

The sense of taste is very simple when compared to the sense of smell. Most people identify only four flavors that can be discerned by taste alone. They are bitter, salty, sweet, and sour. Some flavor chemists also include the tastes of both metallic and umami, neither of which has much of a role in wine. Most people are familiar with the concept of a metallic taste but umami is less well known. Umami was first described in Japan and is the savory character that is found in broths and meat as well as monosodium glutamate, or MSG (Jackson, 1994). The sense of taste comes from receptor cells located within the taste buds. Taste buds lie within small fleshly protuberances called papillae that are located throughout the soft tissue of the mouth and upper esophagus but are concentrated primarily on the tongue. For many years it was thought that different areas of the tongue were sensitive to different tastes resulting in "tongue maps" being published in many texts. Recently this theory was discredited; the different tastes can be identified on all parts of the tongue (Smith & Margolskee, 2001). In spite of the limitations of the sense of taste it is possible to experience more than the four basic flavors when tasting a food or beverage. This comes from the interaction of the senses of taste and smell working together. Much of the flavor one perceives when tasting something comes from the aroma that enters the sinuses through the retronasal pathway at the back of the mouth. These aromas stimulate the olfactory nerves at the same time the taste is being perceived. This is evidenced by how bland food tastes if the sinuses are congested and the sense of smell is blocked.

The Sense of Touch—Texture

The tactile sensations or **mouthfeel** that is produced when one drinks a glass of wine are integral to describing its flavor. The nerve endings of the mouth and tongue detect parameters such as

- temperature. The temperature a wine is served at has a great effect on how the wine is perceived. At warmer temperatures, the aroma will become more intense because of the greater volatility of the aromatic compounds in the wine. Cooler temperatures will give wine a more refreshing quality but will diminish the aroma, making the tastes of acid (sour) and sugar (sweet) more prevalent. Red wines are traditionally served at room temperature—61° to 68°F (16° to 20°C); while white wines are usually served at a lower temperature—50° to 59°F (10° to 15°C). Sparkling wines are served at the coolest temperature—around 41°F (5°C), which helps to slow the evolution of bubbles, prolonging the effervescence in the glass.
- viscosity. The body or "thickness" of a wine is influenced by its temperature and composition. Alcohols, acids, tannins, and sugar all play a role in a wine's body.
- effervescence. This is the prickly sensation from the carbonation or dissolved CO_2 present in the wine that is residual from fermentation. Its qualities are readily apparent in sparkling wines as well as young white wines where CO_2 contributes to the tartness and "fresh" taste.
- alcohol. Wines with higher alcohol have a burning or "hot" character that is reminiscent of distilled spirits (Figure 4–4).
- astringency. Of these sensations astringency is the most poorly understood. It is the drying or "puckery" sensation that is often confused with bitterness. Astringency is a tactile drying sensation, whereas bitterness is a flavor. It is produced by the reaction between phenolic com-

FIGURE 4–4 Tears, also called "legs," on a wine glass are formed by the interaction between alcohol and water in the wine as it evaporates from the side of the wine glass. Tears are often incorrectly thought to be an indication of viscosity but actually are a function of the wine's alcoholic strength.

pounds, such as tannins, with proteins in the saliva. Astringency is perceived more slowly than the other mouth-feel sensations and is important to the aftertaste of a wine as well as how it will complement food.

The physical sensations that are perceived when drinking a glass of wine interact with its aroma and taste to make up a wine's overall impression of flavor. The perception of a wine's flavor can also be influenced by its appearance. Among wine tasters there is a great deal of variation in their natural ability of sensory perception. This can be compensated for by using proper tasting techniques combined with training and experience.

Organizing a Tasting

There are many different ways to taste wine, from the formal and analytical to relaxed and social. Less structured tastings often offer a number of wines from a variety of different regions and producers. Tasters usually walk around the room sampling the wines they wish to try as well any food offered as an accompaniment (Figure 4–5). While informal tastings of wine with pleasant conversation and good food can be a very enjoyable experience, it is difficult to evaluate wine carefully under these circumstances. For purposes of serious sensory evaluation, there is a certain procedure for setting up a tasting that minimizes distractions and allows tasters, also called wine judges in this context, to concentrate on the wine. While social tastings are common and very popular with the public, more analytical tastings are used by students, enologists, and judges of wine for more careful evaluation. In this section of the chapter, we will focus on the proper protocol for setting up an analytical tasting.

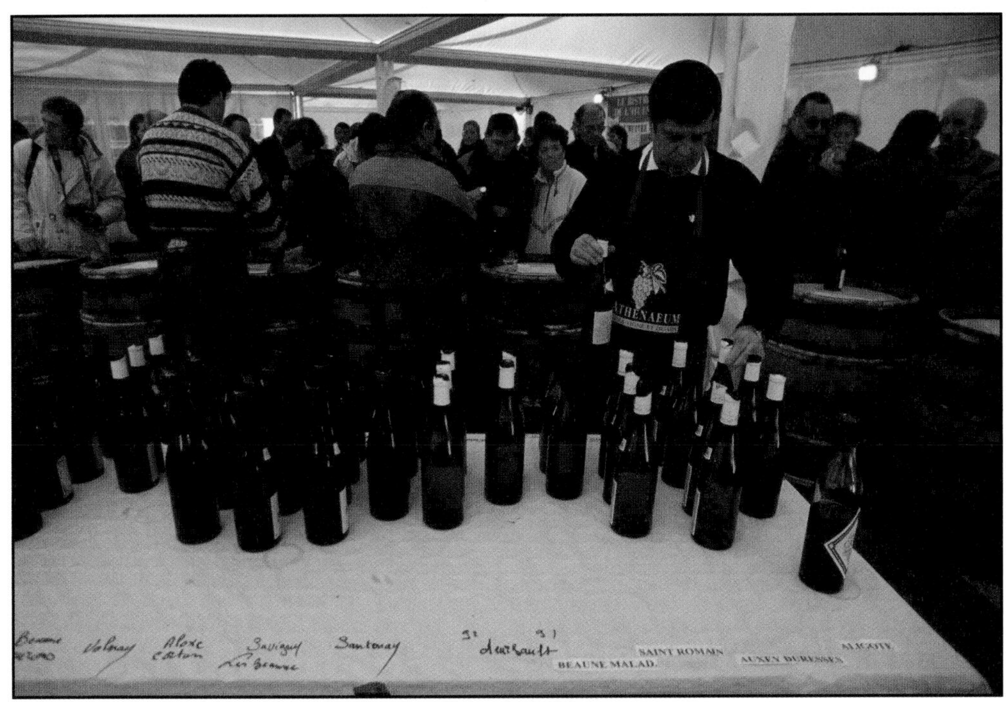

FIGURE 4–5 A large wine tasting in Beaune, France. Large public tastings offer an enjoyable opportunity to taste a number of wines. However, the fact that the wines are usually not tasted blind, combined with the distracting atmosphere, makes them difficult places to perform careful sensory evaluation.

The Proper Setting for Tasting

Like any task that requires concentration, sensory analysis should take place in an environment that has a minimum of distractions. The location that the tasting is to take place in should be quiet and without any distracting activities taking place nearby. The room should be at a comfortable temperature and without drafts. Since part of sensory evaluation is to appraise the color and clarity of a wine, good lighting is also important. Natural light from the sun is best, but since sunlight is not always available, care must be taken when selecting indoor light sources. It should be bright enough to illuminate deeply colored red wines as well as have a neutral white hue. Some florescent lights can have a blue tint to them and incandescent lights have a yellowish character, while wide-spectrum bulbs provide a more natural color.

Most importantly, the tasting area should have no distracting smells that would interfere with the sometimes delicate aromas of the wines to be evaluated. Cleaning products, fresh paint, and smoke from fireplaces all affect a taster's ability to discern aromas in wine. Even pleasant aromas such as those from food preparation, flower arrangements, or scented candles can interfere with the tasting process.

The importance of limiting distractions applies not only to the setting of the wine tasting, but to the tasters themselves as well. Tasting etiquette requires that people who attend the tasting be conscious of personal hygiene and any odors that might be coming from them or their clothes. Obviously one should be clean and free of body odor; however, care must also be taken that they do not use any strongly scented soap or shampoo that could distract fellow tasters. Perfumes, colognes, and aftershave must also be avoided. Additionally, some types of lipstick can be difficult to remove from fine glassware and considerate tasters should not use them.

Sensory evaluation requires concentration; therefore, participants should arrive at the tasting well rested and fresh so they can remain focused on the task at hand. It is also advised to have a meal an hour or two before, but not immediately prior to, the tasting. If food is consumed right before evaluating wines, its aftertaste can interfere with the wine's flavor. Tasting on an empty stomach should also be avoided because this allows the alcohol in the wine to be absorbed more quickly by the body. Tasters should also be punctual because the commotion caused by setting up and serving the wines for a late arrival is very distracting to the other judges.

Tasting protocol also dictates what the proper behavior is during formal tastings. In most wine judgings and critical tastings one group, also called a **flight,** of similar wines is evaluated at a time. When the flight is presented, the wines are first assessed by the tasters, and afterward they are discussed. During the first phase, individual tasters carefully appraise each wine's attributes and take note on their impressions. While this is going on, conversations and talking should be kept to a minimum so that fellow tasters are not distracted. To avoid influencing fellow judges, opinions should also be kept to oneself and not shared at this time. After all judges have finished their own evaluations and recorded their impressions of the wine, then a lively discussion of each wine's qualities can proceed.

Presenting the Wines

The table setting for a formal tasting should be simple yet provide everything that the taster needs (Figure 4–6). It begins with a basic white linen tablecloth that will absorb spills as well as provide a good backdrop for evaluating a wine's color. Place settings should also include a napkin, water glass, and a small cup or bucket to be used as a spittoon. Tasters may bring their own notebooks for recording observations, but it is always a good idea to have pen and paper available in case they are needed.

One of the most important aspects of setting up for a tasting is selecting the proper glassware. Wine glasses come in a diverse array of sizes and styles. Some glasses incorporate elaborate colors and patterns into their designs that are attractive to the eye, but interfere with the evaluation of the wine and should be avoided. There are also a number of wine glasses available that are designed for use with a particular type of wine such as Cabernet Sauvignon or Chardonnay. With these types of glasses the shape and size of the bowl is designed to emphasize the positive aspects of the wine they are made for. Matching the glass to the wine being poured is an important part of serving fine wine and is discussed in chapter 16; however, for critical evaluation of wine, a more basic tasting glass is all that is needed.

A glass for general tasting should have a capacity of about 10 to 14 ounces (300 to 425 ml), and when the wine is served, it should be filled about one-quarter to one-third full. This allows the taster to swirl the glass without spilling and provides enough wine for several tastes. The glass should be wider near the base of the bowl and narrower at the top, forming the shape of a tulip

FIGURE 4–6 A proper place setting for an analytical wine tasting, featuring glasses that are clearly marked with letters representing the different wines, notepaper and pen for writing down impressions, and water with unflavored crackers to refresh the palate. Although it is not pictured, a spittoon should be made available to tasters as well.

FIGURE 4–7 A wine tasting glass suitable for the evaluation of table wines. The curved-in rim at the top of the glass helps to hold in the wine's aroma and prevent spills as the glass is being swirled.

(Figure 4–7). This curved-in shape at the top makes swirling easier and helps to concentrate aromas and keep them in the glass. Most wine glasses have their bowls connected to their base with a glass stem. This provides a convenient place to hold onto the glass without your hands warming up the wine. There are also several styles of glasses designed specifically for wine evaluation that do not have stems. On these models, the bottom of the bowl on the glass is flattened to form the base. For sparkling wines a narrow "Champagne flute" should be used, as illustrated in Figure 3-25.

Another factor to consider when selecting a wine tasting glass is whether to use one made out of crystal or glass. Crystal stemware is usually handmade and considered a finer product. The stem and sides of the bowl are thinner on fine crystal, and this both gives the glass an elegant appearance and allows the taster to clearly view the wine in the bowl. Additionally, the slender surface at the rim of the glass makes it more pleasant to drink from. The thin construction, however, makes crystal stemware much more fragile than glass stemware. This, coupled with their more expensive price tag, means that some wine tasters prefer to use glass stemware because of its durable and inexpensive nature. Whatever the wineglass is made of it should be very clean before it is used for tasting and great care should be taken that it has been well rinsed so that no residual soap remains to alter the taste of the wine. Most crystal stemware is lead crystal, which contains some lead. The amount of exposure is very small if the wine is only served in the crystal. However, wine should never be stored in lead-crystal containers for more than a short period of time.

Water should always be available to wine tasters for rinsing out their mouths as well as drinking. Occasional sips of water will help to prevent tasters from becoming fatigued or getting dehydrated while consuming alcohol. Food is also a useful tool while tasting wine; a small bite of French bread, plain crackers, or rare roast beef between sips of wine will help to keep the taster's palate fresh. While good food complements the taste of wine, it can also distract from the qualities of the wine itself. Care must be taken, however, to select foods that do not have strong aromas that will interfere with the scent of the wine. Strongly flavored or spicy foods will overwhelm a wine's taste making it more difficult to evaluate. Aromatic foods such as soft cheeses and smoked meats, while popular at social tastings, should be avoided when performing serious sensory evaluation.

Other Considerations

It is human nature to have preferences and prejudices; however, they should not be allowed to influence one's judgment when evaluating a wine. If a wine judge is biased towards a particular wine region or producer, it can affect either consciously or subconsciously their opinion of that wine. It is important to always judge a wine based on its attributes and not by its reputation or price. The simplest and best way to eliminate any potential bias it to taste the wines blind. A **blind tasting** is set up with the tasters not knowing specific information about the wines that they evaluating. They may be informed of the variety or vintage of the flight of wines they are tasting, but the appellation and producers should remain anonymous. After the evaluation of the wines is completed and ranking or scoring is done, they then can be unveiled while they are being discussed. To hide a wine's identity the bottles can be covered with paper bags and labeled with numbers or letters before serving (Figure 4–8). If a bottle has a distinctive shape that might give clues to its identity, it can be decanted into plain bottles or a carafe.

The number of wines in a flight can vary anywhere from 4 to 12, with 6 being a good average number. This allows the judges to easily compare the different wines to one another and rank them in order of preference. Giving wines an absolute score on a scale of 1 to 20 or 1 to 100 is another way of reviewing a wine. This method is simple to understand and popular with consumers, but often is arbitrary and difficult to standardize among judges. After the judges complete their evaluations, their scores or rankings can be compiled to find the overall group

FIGURE 4–8 To hide the identity of wines being tasted, they can be placed in bags marked with letters or numbers.

ranking of the wines. Using statistics to evaluate the data from tastings is essential for making accurate conclusions about the wines being tasted. Because statistics are difficult to perform, this important step is often neglected, except by enologists doing research on wine.

No matter how one ranks or scores a wine, it is essential to be able to describe the wine's particular attributes and explain why the wine received the score that it did. In this way if one carefully describes the impressions of the flavors and aromas they perceive in a wine, it will help others in the group to understand the judge's viewpoint, regardless of whether or not they agree with the opinions. Accurate and carefully worded descriptions are also very useful when reading reviews of a wine. If the description is well done, a reader can get an impression of what the wine being reviewed tastes like and get a better understanding of how the reviewer arrived at their opinion. Novices at wine tastings are often intimidated by the serious setting and by other tasters with more experience. Beginners should remember that because taste is subjective there are no wrong answers when it comes to which wine they prefer. It is important to be honest when expressing your opinions and not change them to suit others.

Flights are usually composed of similar wines of the same variety or style; however, there are some variations. A **vertical flight** is a series of consecutive vintages of the same grape variety or type of wine from a single winery. Vertical flights can be very instructive about how a particular wine will age, or how a winery changes its style over time. Tastings also can be set up with a number of different wines from a single producer. While these tastings are not always done blind, they can still be very informative. When tasting multiple flights of wine, or different types of wine within a single flight, there are several basic rules in how to set up the order of the tasting:

- White wines should be consumed before red wines.
- Dry wines should be consumed before sweet wines.
- Light-bodied wines should be consumed before full-bodied wines.
- Young wines should be consumed before older wines.
- Table wines should be consumed before dessert or fortified wines.

This method of tasting wines with more delicate flavors before those with stronger flavors prevents the flavors of the first wines from overwhelming those that come later. This helps to prevent tasters from becoming fatigued and allows them to have impressions that are more accurate.

Proper Tasting Techniques

There is a systematic procedure a taster uses for the sensory evaluation of a group of wines. This method is designed so that the taster is less likely to become fatigued while tasting, and to make sure each wine in the flight gets an equal treatment. All the wines are first appraised by their appearance, then by their aroma, and then by their taste and mouthfeel. By using the senses in order of sight, smell, and then taste, the taster's senses will not tire as quickly. This is because the sense of sight seldom becomes fatigued and the sense of smell remains more acute than the sense of taste as multiple wines are sampled. Additionally, by evaluating the wines by sight, then smell, before moving on to taste, it gives each of the wines in the group more equal treatment. If each of the wines of the group were completely evaluated by sight, smell, and taste before moving on to the next wine, the last wine in the flight would not have as equal a treatment. This is because the taster could become less focused as the tasting goes on and the aroma of the wine can change as the wine sits in the glass. While the wines are being tasted, it is important to keep track of your impressions by taking notes on the sensory attributes of the wines. These will be useful later in the tasting when the judge is ranking the wines and discussing them with others in the tasting group.

Evaluation by Sight

When a taster approaches a group of wines the first attribute that is obvious is their appearance. To begin the assessment of the wines, the taster selects one and observes it for its clarity and color. The easiest way to view the clarity is to hold the glass up to a light source and see how clear the image of the light passing through the wine is. This step can be difficult to perform with a tasting area that is not well lit or with deeply colored red wines. In these cases, the wine glass can be held at an angle to reduce the layer of wine the light has to pass through. The wine is studied to see if there is any turbidity or haze present, or if the wine is free of any particulate matter and is brilliant. Most modern commercial wines are brilliant; exceptions are wines that are bottled unfined or unfiltered, as well as older wines that have dropped sediment as they aged. In older wines the sediment can be removed before tasting by decanting as outlined in Chapter 16.

To observe color the glass should be held at a 45° angle and viewed against a white surface such as the tablecloth or a napkin (Figure 4–9). The hue and depth of the wine's color should be observed, compared to the other wines in the flight, and recorded. In red wines, looking at the edge of the wine in the glass can show subtle differences in color that are otherwise not as obvious. Young red wines have a bluish/purple tint to them that changes to a brick red/orange tone as the wine ages. White wines can have a light-yellow/greenish tint when they are young that becomes more golden as they age. Dessert wines have their own color standards, with late-harvest whites tending to have a golden hue and tawny ports an amber color. In sparkling wines, note the color as well as the size and quantity of the bubbles. All of the wines in the group should be judged for appearance and observations noted before moving on to the next step.

Evaluation of Aroma

The second step in wine tasting, appraising a wine's aroma, is the most important part of the sensory evaluation of a wine. An experienced wine judge can tell much about a wine's identity by simply smelling it. To begin, the glass is selected and then, while held by the stem, briefly swirled to concentrate the wine's aroma (Figure 4–10). After this is done, place your nose inside the glass and

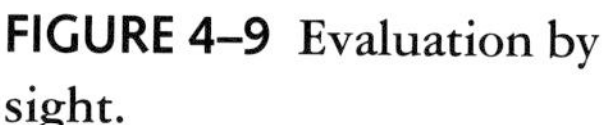

FIGURE 4–9 Evaluation by sight.

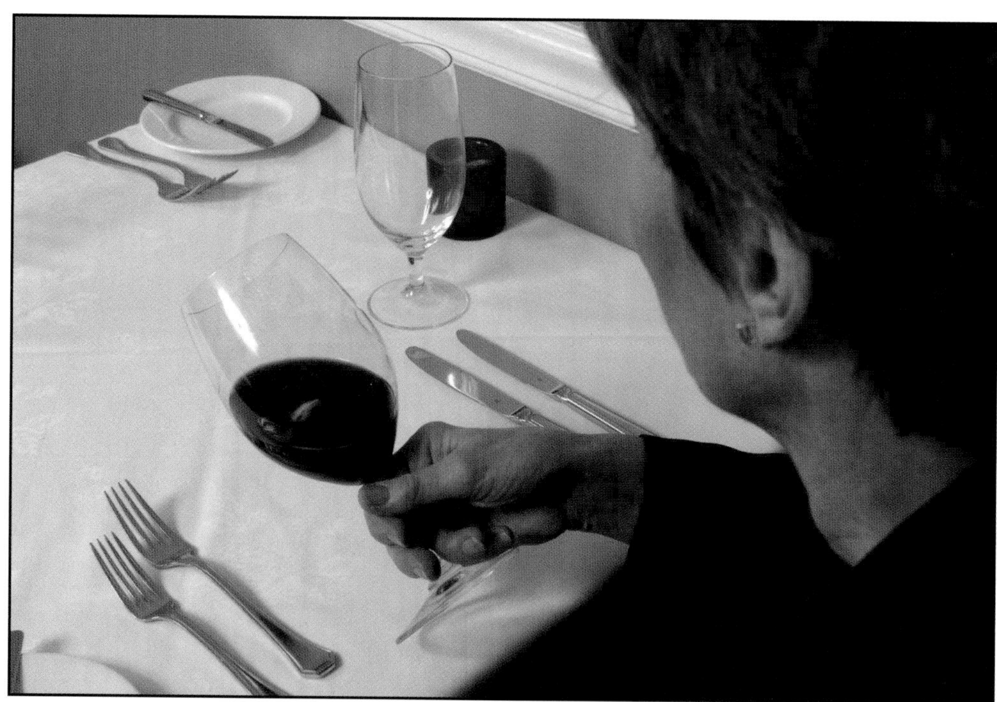

inhale deeply (Figure 4–11). First, note the aromas that are present, what types of smells are detected, and whether they are pleasant or unpleasant. Assess the intensity of these aromas, making note of which are more obvious and which are more subtle. The Wine Aroma Wheel can be a useful tool when looking for descriptors, especially for novices. This is discussed further on page 92. For sparkling wines there is no need to swirl the glass. First impressions tend to be the most accurate, but remember a wine's aroma can change over time. Some unpleasant aromas, such as the rotten egg aroma from hydrogen sulfide, can dissipate quickly and "blow off," so the wine's aroma will improve later in the tasting. While resting for 15 to 30 seconds write down your observations on

FIGURE 4–10 Swirling the glass.

FIGURE 4–11 Evaluation of aroma.

the aroma of the wine. After this, repeat the swirling and sniffing, record any changes in the aroma or qualities that you may not have observed when smelling the wine the first time.

It is important to resist sharing your observations, good or bad, with the other judges at this time so they are not prejudiced by your opinion. One exception to this rule is if the wine smells as if it has cork taint, a musty, mildew smell from a bad natural cork. If a wine has this character and a second bottle is available, a new glass can be poured. If a second bottle is not available, the wine can be withdrawn from the flight. Cork taint is described in more detail in chapter 18. The aroma will be reduced if the wine is too cool; in incidences such as these the wine glass can be warmed with the hands to increase the aroma. After all of the wines in the group have been judged for aroma, then the tasting or evaluation by mouth can begin.

Evaluation by Mouth

Before the wine is tasted, it is once again swirled and smelled but instead of stopping at this point, a small sip of wine is taken immediately after inhaling the wine's aroma (Figure 4–12). Hold the wine in your mouth for a few seconds examining its acidity, sweetness, bitterness, and astringency as well as any flavors or new aromas that are perceived. While the wine is in your mouth, appreciate the tactile sensations it makes such as viscosity, alcohol content, and astringency. Observe how the sensory qualities of the wine develop over time, swallow or spit out the wine, and then note the aftertaste. As in the previous steps, record your thoughts on the wine for later discussion. Some tasters like to swish the wine around in their mouth, coating all of the taste buds to intensify the flavors and the texture that is experienced. Another technique that tasters use is to draw a small amount of air through the wine in their mouth or "gurgle" the wine. This increases the concentration of volatile aromatic compounds and intensifies the aroma that is perceived. Both of these practices are popular, but by no means universal, and individual tasters can try them and determine what works best for them.

After the wine is tasted and the observations recorded, this step is repeated by taking a second sip. This gives the wine judge another chance to confirm their impressions and look for flavors or aromas they may not have noticed the previous time. One of the most important aspects of evaluation of a wine by mouth it to appraise its overall balance. Does the acid, bitterness, or

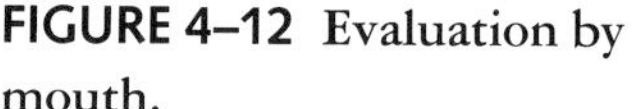

FIGURE 4–12 Evaluation by mouth.

astringency seem insufficient or too strong? It is important to remember that many wines that taste out of balance when consumed alone taste much better when consumed with food. A red wine that seems to be too astringent may taste much better when accompanied with roast meat; likewise a white that tastes too sour may be the perfect complement to seafood. To appreciate fully the aftertaste it is a good idea to swallow a small amount of the wine. This is usually not a problem if there are only a few wines being tasted but can be more troublesome with large or multiple flights. Professional judges at competitive tastings often are required to sample several hundred wines a day. By being very careful with what they swallow and spitting samples, they can do this without their judgment or senses becoming impaired.

Group Discussion of the Wines

After all of the wines in the flight have been evaluated for appearance, aroma, and taste you can go over your notes and see if there any of them that you would like to revisit. Since aromas change over time it is a good idea to re-smell any of the wines that had an off character to see if it persists. If the wine has improved, it should be noted but do not discount the fact it was originally flawed when it was first tasted. This is also the time to select which were your most and least favorite in the group and assign them their scores or rank. If you are finished before the others, wait quietly until everyone else is done. After everyone has completed tasting and the ranks or scores of the judges are compiled, the discussion of the wines can begin.

When discussing the wines do not be shy about sharing your observations and opinions with the others in the group; your opinion is as valid as theirs. This being said, be sure to give others the chance to share their thoughts as well. Novice wine tasters must not give in to the temptation to change their scores or reviews to match the more experienced members of the group; just be prepared to explain the reasons you arrived at your decisions. If the wines were tasted blind, they can be unveiled as they are being discussed or after the discussion is completed. In professional competitive tastings, it is not uncommon for the judges never to know the identity of the wines that they reviewed.

Many tasters will have a tasting notebook to record their observations so that they can refer to it at a later date. Notebooks are useful to keep all of the taster's notes in the same place, as well as to refer to when the taster is shopping for the wines that they liked. Tasting sheets as in Figure 4–13 are also a useful method of taking notes and can be copied and handed out before the tasting. The steps to evaluate a wine are useful in formal settings, as outlined above, in addition to situations that are more informal. Whether one is tasting wines in a class or at the table while enjoying a meal at a fine restaurant, the procedure for evaluating a wine is the same. If the basic steps for tasting a wine are followed, it does not detract from the enjoyment of wine, and it allows tasters to turn every wine they sample into a learning experience.

Discussing and Evaluating Wine

When studying any subject it is crucial to be able to describe and categorize it. Moreover, while tasting and evaluating a wine is initially a solitary undertaking, after one's observations are recorded it is always beneficial to discuss them with others. By expressing your point of view and listening to others' opinions, one can gain a better understanding of wine. The best method of improving your sensory evaluation ability is practice. The best method of improving one's skills in evaluating wine is to taste frequently with a group. This way a larger sampling of different wines can be obtained at a lower per person cost, and a great deal of knowledge comes from discussing the wines that you taste.

FIGURE 4–13 Sheet for tasting notes. Here, six sauvignon blancs from a single vintage were evaluated. The judge records his comments, including the personal and group rankings.

Wine Tasting Sheet

2004 CALIFORNIA SAUVIGNON BLANCS

Wine #	Comments	Personal Rank	Group Rank
A	SIGHT: BRILLIANT PALE STRAW COLOR SMELL: FRESH CUT GRASS AROMA TASTE: TART AND A LITTLE THIN - SHORT FINISH OVERALL: AVERAGE QUALITY - DOESNOT STAND OUT	3	4
B	SIGHT: BRILLANT SMELL: LOTS OF GRASSY VARIETAL CHARACTER WITH TROPICAL FRUIT AROMAS, OF MANGOS PINEAPPLES TASTE: CRISP ACIDITY - LONG FINISH OVERALL: HAS THE MOST FLAVOR OF ANY OF THE WINES	1	2
C	SIGHT: SLIGHT HAZE SMELL: NOT MUCH FRUIT OR GRASSY CHARACTOR SLIGHT ROTTEN EGG SMELL TASTE: LOW ACIDITY AND SHORT FINISH	6	6
D	SIGHT: GOOD CLARITY BUT DARKER COLOR SMELL: HERBACEOUS AND SLIGHT TOASTY-VANILLA AROMA. TASTE RICH BODY AND SLIGHTLY SWEET — OVERALL - NOT MY FAVORITE STYLE	4	5
E	SIGHT BRILLANT SMELL - VERY STRONG HERBACEOUS-BELL PEPPER AROMA TASTE - VISCOSITY IS GOOD BUT NOT MUCH ACIDITY — AROMAS TOO STRONG FOR MY TASTE	5	3
F	BRILLIANT CLARITY SMELL: FRESH MELON AND GOOSEBERRY AROMA TASTE: GOOD ACID BALANCE - REFRESHING LONG FINISH OVERALL: ALMOST AS GOOD AS WINE B	2	1

Although sensory evaluation is the definitive method to evaluate what a wine will taste like, it is not without its complications. In laboratory analysis of wine it is important that the data be reproducible by others as well as quantifiable. By contrast, in sensory analysis of wine these objectives are much harder to obtain. The sensory evaluation of wine is by its very nature a very personal experience, and because the experience of tasting wine relies on the interpretation of individual tasters, the results can be as varied as the wine tasters themselves can. Furthermore, when evaluating a wine for its quality, the results are dependent on the preferences of those doing the judging.

Difficulties in Evaluating Wine

Evaluating wine does have challenges, including the following:

- individual sensitivities. Among wine tasters there is variation in ability to differentiate aromas and tastes. The level at which a taster can detect a given flavor is called the **threshold.** For example, a level of 0.5% residual sugar in a wine may taste sweet to some, while other judges may not be able to identify the character as sweet. Because sugar level also affects viscosity, to them the wine might taste more "full bodied." Through tasting experience and training, it is possible to hone one's skills of identification and to lower the threshold level for identifying wine aromas and tastes.

- definitions. The lexicon of wine terms is often obscure and full of jargon. It is not uncommon to find terms being used that are unfamiliar to most people such as hazelnut and cassis to describe a wine's aroma. In addition to this problem, different judges may use different terms to describe the same flavor or aroma. What one person calls fruity another might identify as floral. Another example is the aroma of vanilla and French oak. French oak barrels impart a vanilla quality to a wine that is aged in them. An inexperienced judge will describe the aroma as vanilla and a judge that is familiar with the effect of barrel age will describe it as "oaky." Like thresholds this problem can be overcome by training and tasting experience, making standards of unfamiliar smells. Use of The Wine Aroma Wheel outlined in Figure 4–14 is particularly helpful.
- preferences/prejudices. This is perhaps the most difficult problem to overcome when evaluating a wine. When performing sensory evaluation it is important to be able to describe objectively the characteristics of what the wine smells and tastes like without letting your opinion of their quality affect your results. After one has fully evaluated the wine's sensory characteristics, one can then complete a review on your opinion of the flavors. Different judges will always have different preferences, and there is "no accounting for taste." However, if the flavor of a wine and the judge's reasons for liking or disliking it are accurately portrayed, people reading the review will be able to determine how they might like the wine and whether or not they would agree with the reviewer. A taster's prejudices are not limited to certain flavors; they might also include certain wine varieties, regions, and producers. This can easily be overcome by always reviewing wines "blind," where the identity of the wine is not known until after the review is completed.
- fatigue. Wine tasting, like any reasoned activity, requires concentration, and it is easy to become fatigued. Fatigue can also be compounded with the ingestion of alcohol. When tasting red wines, repeated sips of the same wine will taste increasingly astringent; this is called **tannin buildup.** To combat these effects it is important to taste in an area free of distractions and taste a reasonable amount of wines, 4 to 10, per flight. Aroma is less likely to become fatigued than taste and gives you more information, so it can be relied on more readily when tasting large numbers of wines. Having water available with neutral flavored crackers or plain French bread will help to keep the palate fresh in between wines. Particularly when tasting a large number of wines, spitting out the wine rather than swallowing will prevent the taster from becoming intoxicated.

Understanding Wine Descriptors

As previously mentioned the sense of smell is a particularly evocative experience. This attribute allows even the inexperienced wine taster to describe a wine's aroma by how it reminds him of other things he has tasted before. Wine often has a very complex aroma made up of many different types of aromas without one smell dominating the others. Familiar aromas that are reminiscent of pears, green apples, melon, and bell peppers are all commonly found in wine. When smelling a wine, examine the qualities that remind you of items you have smelled before. As the taster gains experience, he or she will learn how to identify some of the aromas and recognize and identify the source of those aromas. For example, the aroma produced by the growth of the ***Brettanomyces*** yeast has a smell reminiscent of a barnyard, leather, or horse sweat. When the taster learns to identify these particular smells he or she may just refer to them as being "*Brett*" or *Brettanomyces.*

Table 4–1 is a glossary of wine descriptors that are used on wine that may be unfamiliar to beginners in wine tasting. There are literally hundreds of aromas and tastes that can be found in wine, and this is by no means a complete list.

The Wine Aroma Wheel

Beginning wine tasters often have a difficult time coming up with terms to describe the aromas that they find in wine. Furthermore, experienced tasters often use obscure terms or jargon to describe wine aromas, or individual tasters may use different terms to describe the same aroma. In an effort to combat these difficulties Dr. Ann Noble at the University of California at Davis developed The Wine Aroma Wheel. The Wine Aroma Wheel categorizes the most common aromas that are found in wine, organizing similar aromas together on the different tiers, or rings, on the wheel. For example, the first tier term of fruity is divided into six different categories of fruity aroma on the second tier. These second tier terms are further divided into nineteen individual types of fruity aromas on the third tier.

In addition to categorizing the aromas that are commonly found in wine, The Wine Aroma Wheel also provides procedures for preparing aroma standards for the

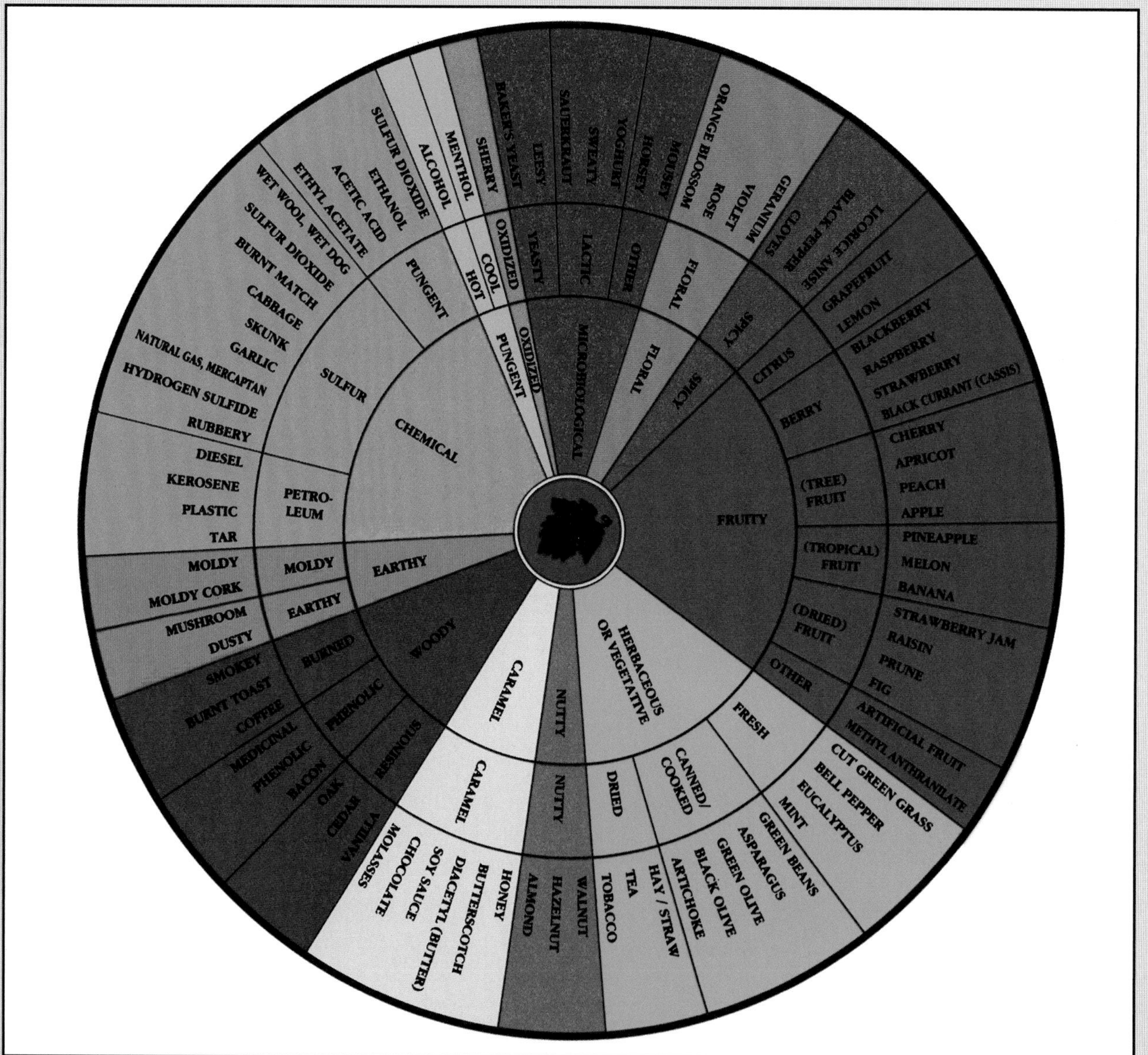

FIGURE 4–14 The Wine Aroma Wheel organizes smells commonly found in wines into groups that have similar qualities, and is a useful tool for experts and novices alike. The Wine Aroma Wheel, Copyright 1990 A. C. Noble. Colored laminated plastic wine aroma wheels may be obtained from www.winearomawheel.com.

The Wine Aroma Wheel (continued)

smell of third tier terms in wine. The third tier term of eucalyptus aroma is found under the first and second tier terms of vegetative and fresh. The aroma of eucalyptus is unfamiliar to many wine tasters who live in areas where eucalyptus trees do not grow. A standard for this term can be made to train tasters by placing one crushed eucalyptus leaf in a glass of base wine (Noble, et al. 1987). Tasters can then use the eucalyptus standard to become familiar with what the character smells like in wine. Procedures for preparing The Wine Aroma Wheel standards are available at her Web site http://wineserver.ucdavis.edu/Acnoble/home.html or in her *American Journal of Enology and Viticulture* article, mentioned above.

Table 4-1 Fifty Common Wine Descriptors

Acetaldehyde	A component of wine that gives off a nutty smell, found in sherry and oxidized wines.
Acetic	Vinegar smell, referring to the chemical name for vinegar, acetic acid. See Volatile Acidity.
Aftertaste	The flavor of a wine that lingers after it is swallowed. See Finish.
Astringent	The drying or "puckery" sensation that is produced by tannins, most obvious in red wines.
Attack	The first impression a wine's flavor makes.
Backbone	The tannin structure of a red wine.
Balanced	A wine that has a harmonious balance of flavors, such as acid and sweet, or tannins and fruitiness.
Barnyard	A slightly earthy smell that is reminiscent of manure, sometimes found in wines that have had the growth of *Brettanomyces.* See Horse Sweat.
Black Currants	A berry fruit popular in Europe that has a flavor often found in Cabernet Sauvignon, also called cassis.
Botrytized	The apricot-like aroma that is produced from the growth of *Botrytis* mold on grapes.
Bouquet	An old term for aromas that are produced by wine processing, such as fermentation bouquet or bottle-aged bouquet.
Burnt Match	The smell of sulfur dioxide, a preservative that is used in wine.
Cloying	Overly sweet taste.
Complex	A wine that has a number of different aromas and tastes.
Corked	The musty-wet newspaper smell of compounds that are produced by mold growth, most often caused by bad corks. See Musty.
Crisp	A wine that has a high level of acid and a light body.
Diacetyl	A compound that has a buttery smell produced by malolactic fermentation.
Distinctive	A wine that has a strong aroma, the opposite of Vinous or Dull.
Dry	A wine that has no perceptible sugar.
Dull	A wine that has little aroma or flavor. See Vinous.
Earthy	The smell of freshly turned earth, similar to barnyard.
Finish	The flavor of a wine that develops after it has been sipped and as it is being swallowed. See Aftertaste.
Flat	A low acid wine that is out of balance.
Foxy	The distinctive varietal aroma of native American grape varieties such as Concord.
Herbaceous	The herbal/vegetable-like aroma that is often found in Cabernet Sauvignon as well as Sauvignon Blanc and Merlot.
Horse Sweat	The distinctive aroma of the growth of *Brettanomyces,* usually reserved for wines that have a very strong "barnyard character."

Table 4-1 Fifty Common Wine Descriptors (continued)

Hot	The tactile sensation that is produced by wines with high alcohol.
Hydrogen sulfide	A compound that has a strong rotten egg smell.
Jammy	A cooked fruit smell that usually results from very ripe grapes.
Legs	The small rivers of wine that roll down the sides of a glass after it has been swirled. See Tears.
Linalool	A compound that has a floral aroma often found in Gewürztraminer and Muscat wines.
Mercaptan	A strong unpleasant aroma that can have a variety of characteristics, such as onions, garlic, or rubbery.
Methoxypyrazines	A group of compounds that produce herbaceous and vegetal aromas.
Monoturpenes	A group of compounds responsible for the spicy character in Gewürztraminers, Muscats, and Rieslings.
Mousse	A French term for the bubbles in sparkling wine.
Musty	A moldy smell associated with cork taint. See Corked.
Nutty	The smell produced by acetaldehyde. See Oxidized.
Oaky	The woody smell produced by aging in oak barrels.
Oxidized	The nutty smell produced by acetaldehyde, often accompanied by a yellowish (in whites) or brownish (in reds) color change in the wine.
Raisiny	The raisin-like smell produced by making wines from grapes that have been partially dehydrated, often found in Zinfandels.
Sec	A French term for "dry," in regards to sparkling wine it refers to a wine with a low level of residual sugar.
Spicy	A distinct aroma found in Gewürztraminers, Muscats, and Rieslings. See Monoturpenes.
Spritzy	A wine with a detectable amount of dissolved CO_2, usually found in young white wines.
Tears	The small rivers of wine that roll down the sides of a glass after it has been swirled. See Legs.
Thin	A wine that has a light body, not very viscous.
Vegetal	A wine with an aroma of fresh vegetables. See Herbaceous.
Vinous	A term that literally means "smells like wine," sometimes used to describe a wine that has little identifiable character. See Dull.
Volatile Acidity	A term used to describe vinegar or acetic acid. See Acetic.
Woody	The smell of a wine produced by aging in oak barrels. See Oaky.
Yeasty	A baking-bread smell produced by the breakdown or autolysis of yeast, found in méthode *champenoise* and *sur lie* aged wines.

Wine and Food

When done properly, the marriage of wine and food is a mutually beneficial relationship, with the qualities of each partner befitting the other (Figure 4–15). Water, being neutral in flavor, does not affect the taste of food; it merely serves to moisten the mouth during ingestion. Wine, in contrast, can have a profound effect on the flavors of the food it is consumed with. Previously in this chapter we have discussed the role of analytical tasting where wines are tasted by themselves so one can concentrate on their flavors and aromas. In the study of food and wine pairing the opposite is true, where individual foods and wines are matched together so that their flavors complement each other. Most people are familiar with the overused and oversimplified rule of "white wine with fish and red wine with meat" but not surprisingly, the subject is much more complex than the rule implies. While it is the job of the vintner to produce a quality wine, the role of the sommelier and the chef is to match the wine with the proper food.

The topic of food and wine pairing is an extensive one that goes beyond the scope of this book. However, there are a few basic principles that will aid consumers and wine servers alike in

FIGURE 4–15 Serving wine with food improves the taste of both items. While there are many books written on the subject of wine and food pairings, there are no absolute rules of what wine must be served with a certain dish. Wine drinkers should feel free to experiment with different combinations and come up with matches that they themselves enjoy.

selecting what type of wine is best suited to accompany a meal. When eating food, wine can serve to freshen the palate; a sip of wine between bites of food will help to cleanse the aftertaste of the food out of the mouth and make your senses ready to fully appreciate another mouthful. An example of this principle is the interaction of the astringency of red wine with the rich "fatty" taste of red meat or cheese. Another illustration of this is a crisp white wine with a food with a "creamy" character such as avocado. Young crisp wines also do well with spicier dishes where their acidity stands up to the food's strong flavors.

Another basic principle of food and wine pairings is when the textures and body of the two complement each other. Ideally the flavors and the textures of the two should be somewhat evenly matched and not overwhelm each other. For example, light-bodied foods such as fish and chicken are better paired with light-bodied wines. Alternatively, full-bodied "big" wines do better with rich foods and red meats. Tart wines generally do not do well with sweet foods because the sugar content of the food brings out the acidity in the wine, making it appear too sour. Salad courses prepared with vinegar also present a problem because the acidity and vinegar aroma clash with most wines. Sparkling wines are usually paired as a light-bodied white wine would be. The older and more complex a sparkling wine is, the better it goes with heavy foods. A dry wine paired with a dessert course will make the dessert seem too sweet, and a wine with higher sugar content would be more appropriate. Table 4–2 provides a set of basic guidelines for pairing wine and food.

The indigenous cuisine of many wine-producing regions is particularly well suited to the types of wine they produce. In these countries, years of experience and trial and error have produced exceptional combinations that bring out the best qualities in both the food and the wine. While these established matches such as a fine Bordeaux and filet mignon are excellent, they are by no means the only combinations that work. Traditional foods from one part of the world readily make exciting new combinations with wines from another part of the globe. As an example, Champagne and sushi make a delicious pairing. When deciding what wine to have with a meal, the diner will usually be well served by following the basic rules of food and wine pairing. However, do not feel limited by these rules; experimentation with different types of food and wine will lead to a greater understanding of how the flavors of the two come together.

A Simple Experiment in Pairing Wine and Food

Chefs and wine researchers at Beringer Vineyards in California have developed a simple experiment that demonstrates how the basic flavors found in wine and food interact. This experiment is easy to perform and can be done by individuals in a home setting but even more instructive when done with a tasting group so the wine tasters can discuss how the flavors of the food and wine in the trial interact. To conduct the experiment assemble the following materials.

EACH TASTER WILL NEED:

1 Glass of a light-bodied off dry white wine (such as an early-harvest Gewürztraminer)
1 Glass of a dry, full-bodied, oaky white wine (such as Chardonnay)
1 Glass of a medium-bodied red wine (such as Merlot)
2 Wedges of apple
2 Wedges of lemon
2 Slices of cheese (mild flavor with firm texture)
Several olives (unpitted)
Saltshaker

STEP ONE

Assemble the materials and then taste each of the wines by itself, starting with the light-bodied white wine, then the full-bodied white wine, then the medium-bodied red. With each wine, write down your impression of its flavors, particularly how the qualities of sweet, sour, and bitterness/astringency taste.

STEP TWO

Taste the slice of apple followed by the off dry white wine, then taste the apple again followed by the full-bodied white wine, and then taste the apple once again before sampling the red wine. Record how the sensory qualities of the wine change from food to food. Repeat these steps for the lemon slice, the cheese, and the olives. Each time you taste the wine be sure to write down how the taste is affected by the food that it is sampled with.

RESULTS

Apple: Notice how the sweet fruity taste of the apple goes best with the off dry white wine, while it makes the dry white taste tart and the red wine taste bitter and astringent.
Lemon: The acidic taste of the lemon makes the off dry white wine taste sweet and the dry white wine taste more balanced. The tart character of the lemon also makes the red wine taste less bitter.
Cheese: The rich flavor of the cheese has a subtle effect on the flavor of the off dry white wine. However, notice how the cheese lowers the perceived astringency and bitterness in the dry white and red wines.
Olives: The bitter and slightly tart nature of the olives makes the off dry wine taste sweeter and the dry white taste less acidic. In the red wine, notice how the taste of the olive make the flavors appear less bitter and astringent

STEP THREE

To show how flavors can be used to bridge the gap between food and wine, start by tasting the apple slice again followed by a sip of the red wine. Notice how the wine tastes bitter and astringent after sampling the apple slice. Now take another slice of apple and squeeze a bit

(continues)

A Simple Experiment in Pairing Wine and Food (continued)

of lemon juice over it followed by a sprinkle of salt. Taste the apple with the lemon and salt again following it with a sip of the red wine. This time notice how the tart flavor of the lemon juice combined with the savory character of the salt makes the red wine taste more balanced and less harsh than when the apple is tasted by itself. This effect can easily be demonstrated at the dining room table where a squeeze of lemon and a pinch of salt can help many light-bodied foods to stand up to the tannins that are found in red wine.

Table 4-2 Basic Guidelines for Pairing Food and Wine

Rich foods are complemented by full-bodied wines.
Light-bodied foods are complemented by light-bodied wines.
Sour foods decrease the perception of acid in wine and are best paired with tart wines.
Sweet foods accentuate the perception of acid and are best paired with wines that are slightly sweeter than the food.
Foods with fruity flavors go best with wines that also have a fruity character.
Complex foods with intricate flavors go best with simple wines; conversely, wines with complex flavors go best with simple foods.
Spicy foods bring out the bitterness and astringency in wine and are best paired with light off dry wines.
Salt in food decreases the perception of bitterness and astringency in wine.
A sauce or glaze can be used to bridge the gap between flavors to allow a particular food to go with a certain wine.

Wine and Health

The role of wine on human health is influenced by two contradictory concepts. The first: wine is a beverage that should be considered a food; the second: alcohol is a drug and should be regulated and controlled. Society's attempts to come to terms with these competing philosophies have resulted in a wide variety of customs and laws that affect wine consumption. In some parts of the world consuming wine is not only prohibited but is considered sinful; in other areas wine is thought of as an essential part of a healthy diet. The era of Prohibition began in the United States as a popular movement by those who saw the damage that was caused by alcohol abuse and thought that outlawing alcoholic beverages would eliminate the problem. However, during Prohibition, alcohol abuse did not diminish, and many more problems were created because it was illegal and its production and distribution were unregulated. Today scientific studies show that moderate consumption of wine can have positive health effects, and moderate drinkers outlive those who abstain. Whenever one is producing or serving alcoholic beverages, it is important to be cognizant of its role in human health.

Negative Effects from Excessive Alcohol Consumption

Alcohol in its many forms is involved to some degree in the deaths of more than 100,000 Americans a year (National Institutes of Health, 2001). The majority of these deaths are due to driving under the influence and cirrhosis of the liver from alcoholism. The harmful effects of excessive consumption can occur chronically over many years, or acutely in a single "binge drinking" episode. The long-term consequences of overconsumption include liver damage, as well as an increased risk of cancer and heart disease. Those who chronically overconsume are usually referred to as alcoholics or problem drinkers; however, the term problem drinker can also apply to someone who rarely drinks but when they do, they drink in excess. Becoming intoxicated affects one's judgment and behavior. Although it is possible to drink so much that the level of alcohol in the body becomes toxic, many of the deaths from binge drinking are associated with risky behavior. Drunk driving, or driving under the influence, is by far the most common form of accidental death while intoxicated. Despite more awareness of the problem it continues to kill more than 17,000 Americans a year (National Center for Statistics and Analysis, 2001), many of them innocent bystanders.

Positive Effects from Moderate Wine Consumption

While excessive consumption of alcohol results in higher mortality rates, epidemiologists have known for many years that people who consume wine in moderation have a longer lifespan than both alcoholics and those who abstain from alcohol (Figure 4–16). In recent years, numerous scientific studies have shown a direct link between modest wine drinking, particularly red wine, and increased cardiovascular health. This research shows that people who consume an average of one to two glasses of wine per day have a 30 to 50 percent reduction in mortality from heart disease (Goldfinger, 2003). This is surprising to many people, and health officials are reticent to recommend alcohol consumption for fear that citizens would use it as a license for excess. This effect is not due to the alcohol in wine because laboratory studies show that mice given a solution of alcohol and water did not have the same lower rate of heart disease as was seen in mice that were given wine. The mechanism for these results is still unknown but it is believed to be a product of naturally occurring substances such as **quercetin** and **resveratrol** that are present in wine.

There are many opinions of what is considered a "moderate" amount of wine to consume, but most researchers considerer one to two 6-oz (175 ml) glasses per day with meals to be moderate, depending on one's circumstances. The positive effect on cardiovascular heath is often referred to as the **"French paradox."** This term was coined by Boston University epidemiologist Dr. Curtis Ellison who noticed that despite the fact that French citizens smoked more and had a much higher intake of saturated fat then Americans, they enjoyed a much lower level of heart disease.

Special Considerations for Women

The health benefits of moderate wine drinking are much more clear cut for men than they are for women. This is primarily due to the relationship between alcohol consumption and breast can-

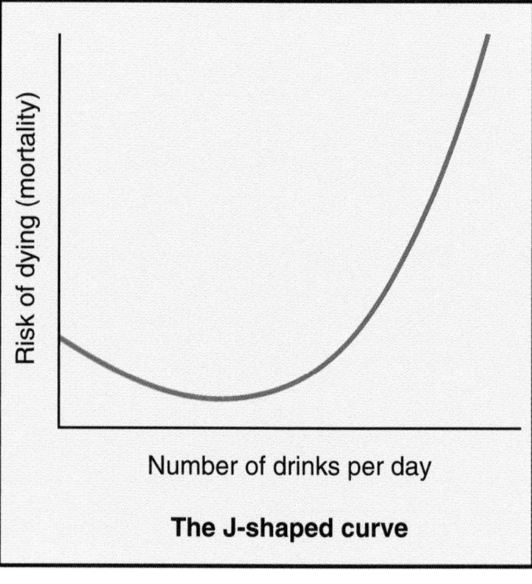

FIGURE 4–16 The "J"-shaped curve illustrating how mortality rates decrease for moderate wine drinkers and increase for excessive wine drinkers.

cer. Although women receive the same benefit that men do in cardiovascular heath, some studies show a correlation between moderate consumption of wine and an increased risk of breast cancer. Women also have to take into account the effects of alcohol consumption on pregnancy. About 1 in 20 women who are heavy drinkers during pregnancy give birth to babies with a form of mental retardation called fetal alcohol syndrome, or FAS (Abel, 1995). While there seems to be little evidence that moderate wine drinking plays a role in FAS, most obstetricians recommend that their patients abstain during pregnancy.

It is important for both men and women to evaluate honestly their personal heath and family history when deciding what is healthful consumption for them. The role of wine and health can be summarized with the statement that the majority of scientific studies show that moderate consumption of wine increases the lifespan for most individuals with the exception of premenopausal women with a family history of breast cancer.

Summary

The ability to taste a wine and enjoy its flavor requires very little skill and can be done by almost anyone (Figure 4–17). However, tasting a wine and being able to evaluate it critically is a talent that can take many years to master. It is by gaining this knowledge of how to taste and understanding the sensory qualities of a wine that we are able to fully appreciate and share with others how the many components present in wine combine to make up a wine's flavor. The protocol for setting up a tasting to evaluate wines is designed to help minimize distractions so the tasters can concentrate on the qualities of the wines. Whenever possible wines should be evaluated blind so that preconceived notions do not affect the taster's opinions. When one has a better understanding of the sensory characteristics of different wines it is easier to match them with food in a way that enhances the qualities of both. This knowledge, combined with an awareness of how wine has an effect on the body, allows the consumer to enjoy fully its consumption in a healthful manner.

FIGURE 4–17 While analytical tasting methods are useful for understanding the qualities of a wine that make it appealing to the senses, it is important not to forget that a wine is best enjoyed as part of a delicious meal with friends.

REVIEW QUESTIONS

1. What considerations are important to take into account when setting up a formal tasting?
2. How do the different senses respond to wine?
3. Explain how The Wine Aroma Wheel can be used to train wine tasters.
4. Discuss the various positive and negative health effects of wine consumption.
5. Prepare a list of different types of foods and wines that would accompany them.

SECTION II

Wine Regions of Europe

Wine Regions of Europe are covered in five chapters: France; Italy; Spain and Portugal; Germany; and Other European Regions and the Mediterranean. Europe is the birthplace of modern winemaking and most of the styles of wine and grape growing throughout the world have their origins on this continent. The chapters cover individual regions within the country, their history of winemaking, and the wines they produce.

CHAPTER 5

France

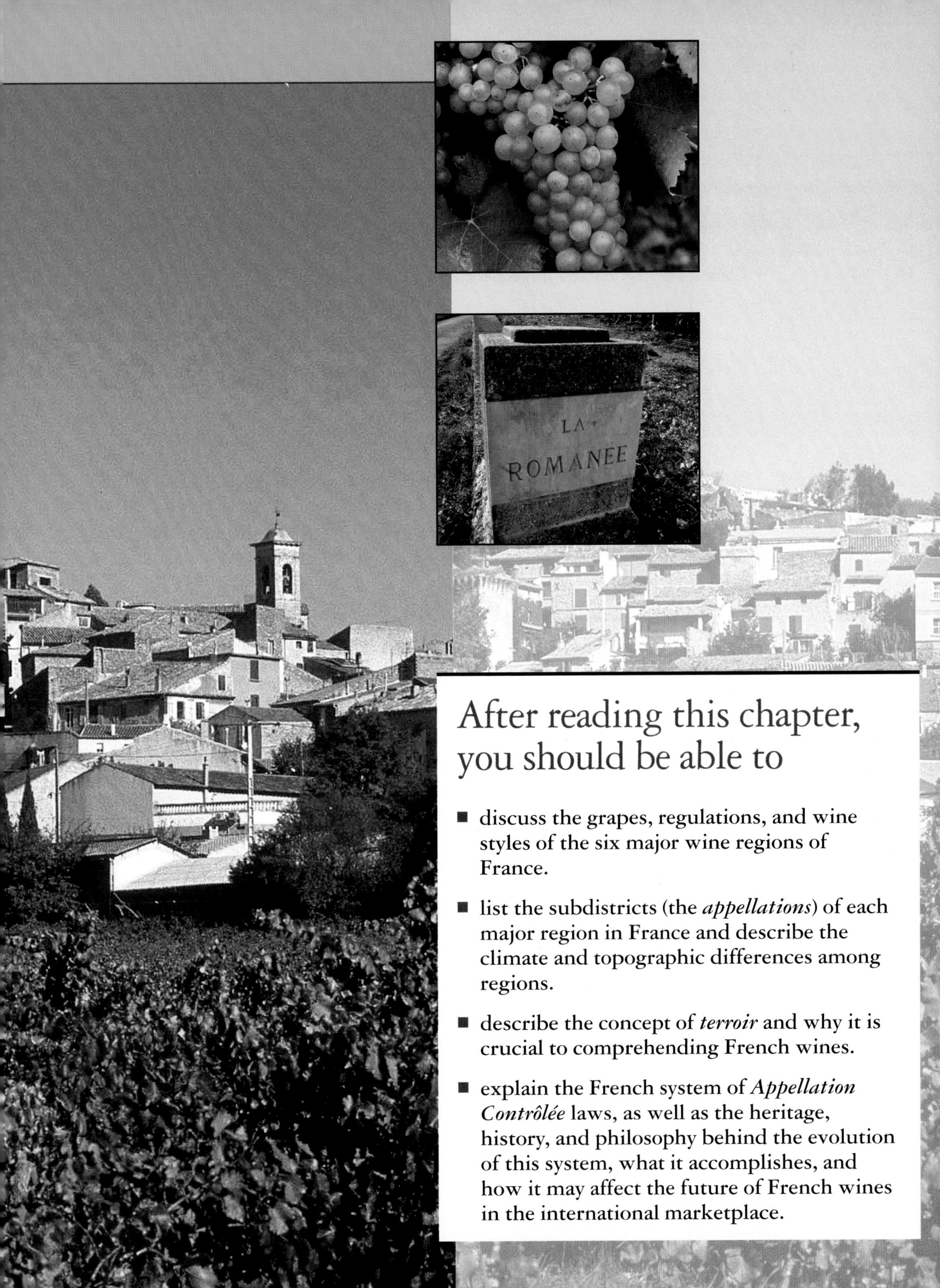

After reading this chapter, you should be able to

- discuss the grapes, regulations, and wine styles of the six major wine regions of France.
- list the subdistricts (the *appellations*) of each major region in France and describe the climate and topographic differences among regions.
- describe the concept of *terroir* and why it is crucial to comprehending French wines.
- explain the French system of *Appellation Contrôlée* laws, as well as the heritage, history, and philosophy behind the evolution of this system, what it accomplishes, and how it may affect the future of French wines in the international marketplace.

KEY TERMS

- appellation
- *Appellation d'Origine Contrôlée (AOC)*
- *blanc de blanc*
- *blanc de noir*
- brut
- carbonic maceration
- *clos*
- *cru*
- *cuvée*
- *demi-sec*
- *doux*
- extra brut
- extra dry
- *moelleux*
- *mousseux*
- *négoçiants*
- noble rot or *pourriture noble (Botrytis cinerea)*
- phylloxera
- *propriétaire*
- *sec*
- terroir
- *tête de cuvée*
- triage
- *vin délimité de qualité supérieure (VDQS)*
- *vin de pays*
- *vin de table*
- *vin doux naturel*
- vintage

In addition to producing great wines in nearly every category, France is also the original home to most of the "noble varietals," the grapes from which the best wines are made. Of the twelve most important noble varietals, eight are indigenous to France: Chardonnay, Riesling, Sauvignon Blanc, and Chenin Blanc for whites, and Cabernet Sauvignon, Merlot, Pinot Noir, and Syrah for reds. Although Riesling is also indigenous to Germany, and the Sangiovese and Nebbiolo of Italy and the Tempranillo of Spain are also counted among the noble varietals, the majority of important grape varietals originated in France.

The French also demonstrated important initiative in the creation of a country-wide system of laws to control viticultural practices and the production of wines, along with a federal-level government agency to oversee the wine trade and enforce the regulations. One of the primary purposes of these laws is to protect the geographic names of the places of origin of specific wines. This protection is very important as French wines (like most European wines) are named for the region where the grapes were grown. This geographic designation of origin is called the appellation of the wine.

The French are passionate about wine, and are understandably proud of the wines they produce. As proof that the French believe in the quality of their wines, one need look at just one statistic: only 3 percent of all wine consumed in France is imported (Osborne, 2004). Although great wine is being produced elsewhere in the world, to truly understand wine, one must understand French wines.

French Wine—Historical Perspective

The history of wine production in France is inextricably intertwined with the politics and sociological development of the country. The first wine grapes were planted in the southern part of what is now France by Greek traders as far back as 600 BC. As the Romans spread into Gaul (as France was then called) and colonized the country, the planting of grapes and the production of wine increased. By the time of the birth of Christ the exporting of wine from Gaul to Rome was well established. When the Roman Empire began to crumble in the second century AD, the expansion of viticulture ceased, although wine continued to be produced, often by the monasteries and abbeys of the Christian Church that had been established in Gaul by the Romans. Barbarians from the North invaded Gaul and caused the collapse of the Roman Empire by AD 400. The Dark Ages set in across Europe. During this time, it was the Christian Church that kept viticulture and enology alive in Gaul and elsewhere in Europe.

Charlemagne brought stability to Gaul during his reign which began in AD 768. He introduced the first laws on wine production. Although he was based in the north of Gaul, in the Champagne region, his influence was felt as far south as the Mediterranean. Charlemagne and his successors encouraged the export of wine (Figure 5–1).

In 1152 Eleanor of Aquitaine married Henri of Anjou. An important trade alliance was established when Henri ascended the English throne as King Henry II. The combination of land owned by this couple and the taxes they collected from their domains on either side of the Channel allowed active exchange of wine and other goods. English entrepreneurs came to France, especially to Bordeaux, and played a crucial role in building a long-lasting English appreciation of the red wines of this region.

In subsequent centuries, the Dutch played an increasing role in the shipment of Bordeaux wines to other Northern ports, like Amsterdam. The production of wine in the Bordeaux region increased considerably over the next two hundred years. In other sections of France, wine production also grew, as did the influence of the Christian Church. As the Church acquired land, often in the form of gifts from wealthy aristocrats, the importance of the monasteries as winemaking centers increased. The monks and priests had the time and resources to develop better vineyards and to perfect winemaking procedures. They also had the ability to record their successes (and failures), thus helping other vintners to learn. Wine was one of France's most important exports during the Middle Ages.

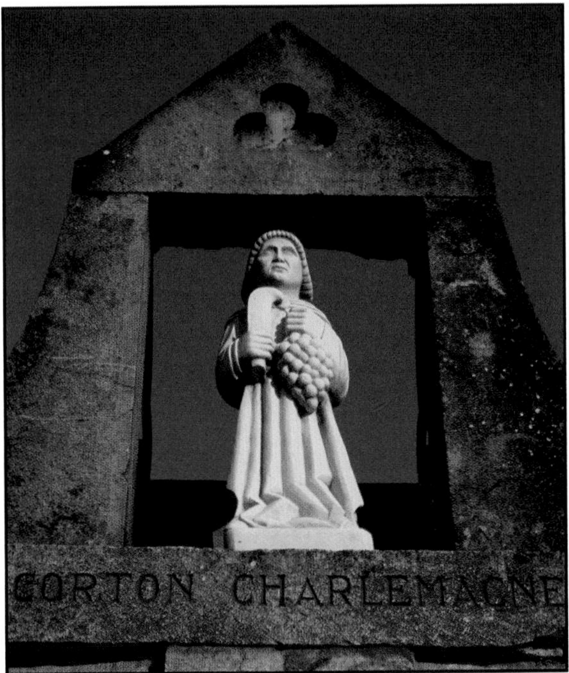

FIGURE 5–1 The gate to a great vineyard in Burgundy, showing a statue of a man holding a bunch of grapes, typifies the French tradition of, and dedication to, the production of fine wine.

After the French Revolution (1789–1791) and the rise of Napoleon, the church and the aristocracy lost a great deal of their power. Land was taken by the government and given to the farmers. Large land-holdings that did remain with wealthy families became fragmented over time as the Napoleonic Code did away with the medieval concept of primogeniture; that is, the practice of a rich man leaving all his holdings to his oldest son. Now all children of a landowner, including daughters, inherited equal amounts. Fortunately, the change in land ownership patterns did not adversely affect the quality or the popularity of French wines. The production and exporting of French wines, especially those of Bordeaux, continued to increase until the **phylloxera** epidemic in the late 1800s. This insect attaches itself to the roots of a vine and literally sucks the life out of it, eventually causing the vine to die. As discussed in Chapter 2, the epidemic spread throughout the vineyards of Europe before it was eventually halted when phylloxera-resistant rootstocks from North America were grafted to the classic varietals.

Appellation Contrôlée Laws

As French wine production recovered from the phylloxera epidemic, a new, man-made problem arose: fraud. As certain French place-names acquired panache, the demand in the international marketplace for the products of that region naturally increased. If, as demand increases, the supply remains constant, the natural economic tendency is for price to rise. When the price for a certain wine rose, some wine producers and merchants could not resist the temptation to increase the supply through fraud. Unscrupulous producers simply attached that region's name to their bottles in order to get a higher price. Or producers within a famous region would expand their saleable inventory by buying grapes grown outside the region, blend them with their legitimate crop, and label the whole batch by the regional name.

The need for government intervention to protect the authenticity of geographic names of origin became evident as early as the late 1890s. Fraud became so widespread in France that some place names on bottles became essentially meaningless. The problem was particularly evident in the Champagne region. It has been estimated that by 1911, the Champagne houses were selling at least 11 million more bottles of wine than their region's vineyards could possibly have produced (Kramer, 1989). This blatant fraud caused the Champagne region to explode into violence that year. The grape growers rioted to protest the practice of the large Champagne producers buying grapes outside the region to meet the ever-expanding demand for their product. Angry grape growers rampaged through the street, broke into warehouses, and destroyed hundreds of cases of wine. The country was so shocked by the violence and waste that the government immediately passed legislation defining the boundaries of the Champagne region and decreed that the valuable name "Champagne" on a label could be used only if all grapes used in the production of that batch were indeed grown inside those boundaries. This was the first step towards a system that guarantees the authenticity of specific geographic locations.

At this point it would be beneficial to take a close look at the concept of **terroir.** The full term in French is *gout de terroir,* which translates literally as "taste of the soil." However, in the context of wines, the definition of terroir is the unique and distinctive character a specific wine will exhibit due to the fact that it was grown in a specific vineyard. The French place enormous importance on the vineyard, the site where grapes were grown. It is the specific location, with its unique soil, be it limestone, chalk, or slate, as well as the mineral content and drainage capacity of that soil, along with the location's unique microclimate (average temperature, winds, exposure to sunlight, precipitation, etc.) that is the primary influence on the quality of the grapes. As discussed in Chapter 2, the term terroir encompasses the entire physical environment in which the grapes were grown. What the French care about more than anything else in their wine is that it reflect the terroir of its region, that it be typical of that region, and that it be *authentic.*

The importance of protecting the authenticity of wines being labeled with any of the famous wine regions was not lost on the French government, but before the work that began in Champagne could proceed to other regions and a nationwide system of legislation could be created, the First World War intervened.

After the war, in 1923, a revolution took place in the Châteauneuf-du-Pape region of the southern Rhône Valley. This incident was similar to, but more orderly than, the uprising in Champagne in 1911. Châteauneuf-du-Pape was famous for its red wine and for its memorable name with a quaint story behind it. The name came about when Bertrand the Goth was elected Pope Clement V in 1305. At the time, the relationship between the King of France and the papacy in Rome was badly strained, and Italy was in political turmoil. Clement chose to stay in France, and established his papal court in Avignon, an ancient city on the Rhône River. His successor, Pope John XXII, improved the papal finances sufficiently to build himself a summer palace outside the city, on the foundations of an old castle (Fig. 5-2). This palace became known as Châteauneuf-du-Pape, "the new castle of the Pope." Of course, the old castle's vineyards came with the property and Pope John made sure these vines were well tended so that he could produce his own wines. Thus began the Châteauneuf-du-Pape wine-producing region.

By the 1920s the demand for the red wines of this historic region was very high. Fraud in the form of inflated production numbers had been going on for years, and now became widespread throughout the region. A group of producers, under the expert guidance of Baron LeRoy of Château Fortia, set out to define their own boundaries and to set prescriptions on which grapes could be used in wine to be labeled as Châteauneuf-du-Pape. They decreed what viticultural practices were to be allowed and spelled out specific techniques that were banned. The vintners also set out strict standards for minimum ripeness of grapes at harvest, minimum alcohol level in wines, and other factors that are critical to quality and authenticity. The system devised by this dedicated group of Châteauneuf-du-Pape vintners eventually became the model for the national system of quality control laws.

In other regions, especially Burgundy, trouble in the form of fraud and blackballing of recalcitrant growers continued for many more years. Most of the wrongdoing was at the hands of

FIGURE 5–2 The historic town of Châteauneuf-du-Pape, showing the ancient castle on the hill.

négoçiants, businessmen who acted as the middlemen between grape growers, and the producers and shippers of wine. Due to the Great Depression, demand for wine was down severely, and many growers could not afford to protest against *négoçiants* who bought cheaper grapes from outside areas and then sold the wine as something far more expensive than its quality merited. Authenticity and quality took a serious step backward in Burgundy and other premier wine regions of France.

Finally, in 1935, the French government passed legislation creating the *Institut National des Appellations d'Origine des Vins et Eaux-de-Vie* (INAO) under the Ministry of Agriculture (Figure 5–3). The charge given to the *INAO* was to work with local growers, to establish legally defined appellation boundaries, along with a codification of grape-growing and winemaking practices appropriate to each area. The system has continued to evolve and is continually under review.

All wine regions of France are classified into one of four levels of quality. Wine coming from each region also carries that classification. The four levels are, in descending order of quality, ***appellation d'origine contrôlée*** or ***AOC*** (higher quality wines from one of the better limited areas of production.); ***vin délimité de qualité supérieure*** or ***VDQS*** (quality wines from a limited area); ***vin de pays*** (country wine); ***vin de table*** (table wine).

Appellation d'Origine Contrôlée (AOC)

In order to carry the name of an *AOC* region, a wine must meet very specific criteria:

- The wine must be made 100 percent from grapes approved for that appellation.
- The grapes must have all been grown within a limited zone or area of production. In general, the smaller that geographic designation, the better and more distinctive the wine. Some *AOC* wines attain even higher recognition of quality if the vineyard or estate where the grapes were grown is further rated by the authorities as being a particularly impressive location. Rated vineyards are usually designated as *grand cru* or *premier cru* or some comparable term indicating high quality.
- The grapes must have been picked at the minimal level of sugar specified for that appellation. A minimal alcohol content must be achieved after fermentation.
- The amount of grapes harvested must not exceed a certain amount per hectare. In general, the smaller or more specific the area, the smaller the yield allowed. If all the vigor of the vine goes into fewer bunches, those bunches will have more concentrated flavors.

FIGURE 5–3 These bottles from a producer in the Sauternes area of Bordeaux are ready to be sent to the local authorities of the *INAO*, where they will be analyzed and tasted to be sure they meet all standards and reflect the terroir of Sauternes.

- The methods used in the vineyard and in the winery must conform to the regulations of the region.
- The wine must be bottled in the same region as the appellation.
- The wine must pass a tasting test by the local branch of the *INAO*. What the tasters are judging is not the quality of the wine so much as its terroir, that is, they are determining if the wine reflects the character of the appellation. In other words, does the wine have typicality?

Presently over one-third of the wine produced in France is designated as *appellation d'origine contrôlée* (Figure 5–4).

Vin Délimité de Qualité Supérieure (VDQS)

This designation was begun in 1949. These wines are also produced according to *INAO* guidelines, and producers are supervised by the local bureau. However, standards are not as strict nor as numerous as at the *AOC* level. Growers and producers in these regions often aspire to have their area elevated to *AOC* status. At this time, only about 1 percent of French wines are designated *VDQS*.

Vin de Pays

Higher yields and a higher percentage of nonindigenous grapes are allowed at this level. Since 1979, wines at this level have been permitted to be labeled by varietal (although region of production must also be listed). Varietal labeling has been a boon to French entrepreneurs wanting to sell their products in the New World, especially the United States. *Vins de pays,* most of which come from the south of France, differ considerably in quality, style, and price.

Vin de pays regions can fall within three different types.

1. *Regional.* There are three of these. They are very large, covering wide swaths of land with many different soil types and microclimates.
2. *Departmental.* This covers an entire département, the French equivalent of an American state or Canadian province.
3. *Zonal.* This is the smallest type of region, often just one district or even one town. There are over 100 zonal *vin de pays* regions.

Today approximately 25 percent of French wine is designated as *vin de pays.* With the success of marketing these wines into the United States, that percentage may well increase.

Vin de Table or Vin Ordinaire

This type of wine can be made from grapes grown anywhere in France. There are no limits on yield and no specifications on varietals. Wine that is fermented purely for the purpose of being

FIGURE 5–4 This label confirms that the wine in the bottle meets all standards for its region, and, therefore, is entitled to carry the controlled appellation Bordeaux.

distilled into spirits fits into this category. The European Commission is putting pressure on France to decrease the amount of acreage dedicated to this level of wine, as the glut of bulk wine and wine grapes causes prices to fall.

Weaknesses of the System

The French system of wine laws is one of the most comprehensive and strict in the world. These laws have done a great deal to guarantee the authenticity of wine names, and thus, to protect the prestige of the finest wine appellations. The purpose of the laws is not to guarantee quality. The government feels that that is up to individual producers, and that the open market will determine a wine's success or failure. Rather, what the wine laws are intended to do is to assure that each wine carrying a region's name will be typical of that region. This way the consumer will know the essential style and character of the wine when purchasing it. In meeting this objective the wine control laws of France are successful. Moreover, the system does rate regions (the highest rating being *AOC*), and also rates some of the highest quality locations within *AOC* regions. These ratings also assist the consumer in making purchasing decisions.

However, despite its successes and strengths, the system does have its weaknesses, the worst being that in some of its applications, the system of laws protects the grower and producer more than it does the consumer. Changes advocated by experts, including Clive Coates, a leading authority on French wines, include adding consumer representation to the local *INAO* commissions. In other words, each tasting panel and regulatory body should have an objective observer, with a vote, who has no direct involvement with any facet of the wine trade, but will speak simply as a consumer, *for* consumers.

The tasting and analysis of *AOC* and *VDQS* wines should be done with an eye to quality, not just to typicality. True, the open market will eventually eliminate low quality wines that are not worth the price being asked, but it seems the authorities should step in before consumers have wasted money on a low quality wine.

Another suggested improvement has to do with the matter of yields. As the law now stands, if a grower exceeds the yield allowed for his appellation, the amount by which he exceeds the allowed amount per hectare has to be downgraded. Logically, the entire crop should be downgraded, for if the yield for a vineyard has been too high, all the grapes from that vineyard will be of inferior quality.

Labeling laws could also be improved. The requirements for the use of words like *domaine* or *clos,* and phrases like *mis en bouteille à propriété* (estate bottled) need to be made stricter so as not to mislead the consumer. Moreover, from a marketing viewpoint, the French authorities and wine producers should expand the use of explanatory back labels. Consumers in the United States like to have helpful information on the style of the wine in the bottle, how to serve that wine, what grapes it was made from, and so forth. Explanatory labels have greatly helped in the sales of Australian and Californian wines. The French should follow suit.

Wine Regions of France

The major wine regions covered in this chapter are Bordeaux, Burgundy, Côtes du Rhône, the Loire Valley, Champagne, and Alsace. These six *AOC* regions account for less than 20% of France's total wine production, but their wines are the country's most famous and most impressive wines. We will also look at some of the promising regions in the South of France.

Map of France

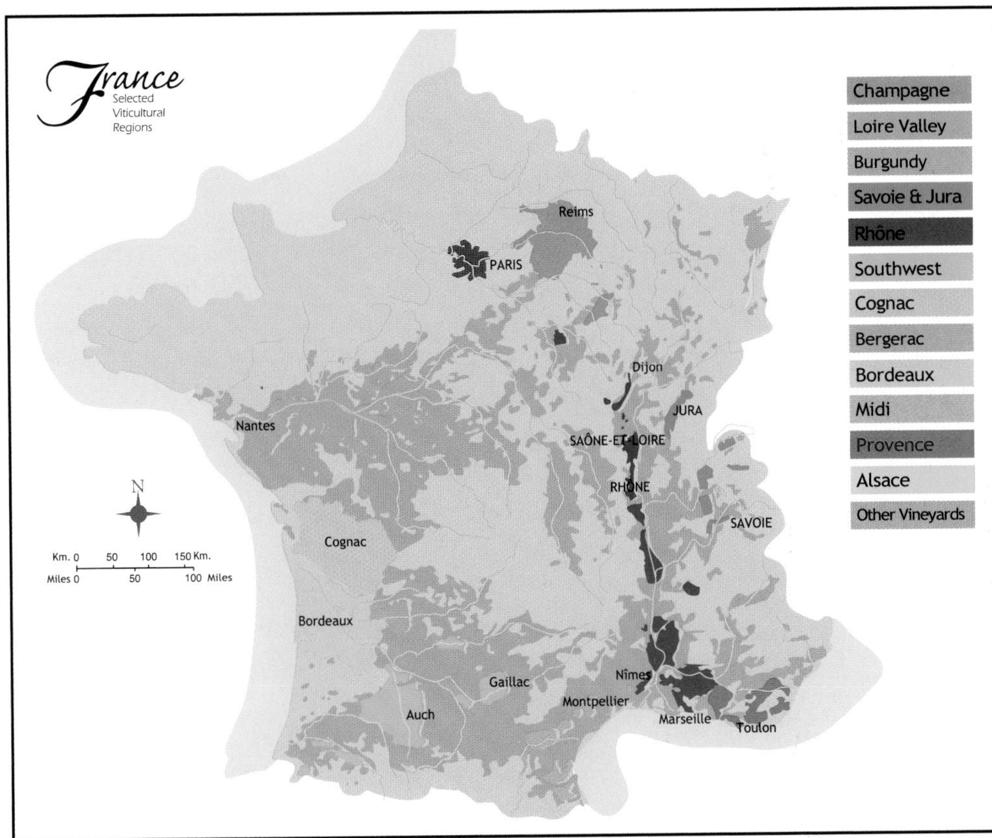

Bordeaux

Bordeaux is one of the world's largest and most diverse wine-producing regions. There are almost 304,000 acres (123,077 hectares) under vine, and annual production is over 660 million bottles of wine. Fully 22 percent of France's total *AOC* production is from Bordeaux, a region that produces fine wine in three major categories—red table wines, dry white table wine, and luscious dessert wine. Many of Bordeaux's wines rate among the very best: elegant, complex reds; honeyed sweet whites; and well-balanced, crisp dry whites.

Bordeaux is a city and a wine region as well as the name of a wine. The city of Bordeaux, eighth largest in France and for centuries an important port, is the capital of the *département* of Gironde, the largest of France's 95 *départements.* Excluding a section along the coast, the *département* of Gironde essentially encompasses the wine region, or appellation, of Bordeaux. This is a region of large, self-sufficient estates in which the vineyards, the winemaking facilities, and often the owner's house are all under one ownership and located in close proximity. Many of these estates have been under the same ownership for centuries. This uninterrupted proprietorship has allowed development of high quality vineyards, confident winemaking skills, and a pride in name, heritage, and product that results in extraordinary wine repeatably. There is a distinguishable style for each major estate that stays the same year after year.

The late seventeenth and early eighteenth centuries were the period in which many of the great estates developed. Prior to that time, a high quantity of wine was exported, but much of it was not of high quality. After the marriage in 1152 of Eleanor of Aquitaine (which included Gascony, where the city of Bordeaux was located) and Henry Plantagenet, Duke of Normandy and Count of Anjou, the planting of vineyards in Bordeaux expanded. The extent of trade with England also expanded enormously when Henry ascended the throne of England two years after his marriage to Eleanor. However, most of the vineyards were not in the Médoc, which was much too swampy to be useful for growing grapes. The grapes were grown further inland, in the Dordogne region; the valleys along the Garonne, Tarn, and Lot Rivers; and area known as the "high country." Bulk wine was brought into Bordeaux from as far away as Cahors, well to the southeast, to be blended into bottles being sold as Gascon wine.

After the English were expelled from Gascony in 1453, the French kings were wise enough not to disrupt the Bordeaux wine trade. The privileges and favors granted under the English monarchy to Bordeaux wine producers and merchants remained in place. Trade with England continued, and business with the Dutch expanded.

The quality of wine from this area took a major step forward in the seventeenth century due to the ingenuity of Dutch entrepreneurs who had become increasingly involved in the exporting of wine from Bordeaux. Long familiar with marshy low lands, the Dutch businessmen brought in engineers from their homeland who were able to drain the marshes of the Médoc peninsula. This process exposed gentle hills of very gravelly soil, perfect for *vinifera* vines. Many of Bordeaux's great estates are now located in the Médoc. Winemakers in Médoc and adjacent growing areas took steps to protect their products and passed strict regulations against bringing "high country" wines into Bordeaux.

As the wine trade grew, a new social class emerged and became the new aristocracy. The merchants who attained success in trading and exporting wine began to purchase land and build châteaux. This moneyed class replaced the old nobility. The merchant families invested resources into improving their vineyards. After the French Revolution, some Bordeaux vineyards owned by the church were confiscated by the government and turned over to peasant families. However, many of the wealthy merchant families escaped that fate, and the top properties remained largely intact. Their estates became the great, highly rated estates of modern Bordeaux.

As in other regions of France and Europe, production of wine in Bordeaux was set back by the infestation of phylloxera in the late nineteenth century. By 1869, land under viticulture in Bordeaux had decreased by over a third, with many more hectares dying each year. At a conference called in the city of Bordeaux in 1881 to study the problem of phylloxera, the Bordelais vintners agreed to accept the proposed solution of grafting their vines onto American rootstock.

The process of replanting vineyards proceeded slowly, partly due to a fear on the part of Bordeaux landowners that American rootstock would adversely affect the flavor of their wines, and partly due to an infection of the vineyards by downy mildew, a disease that primarily affects the leaves of the plant. This scourge was quickly eliminated by the spraying of copper sulfate solution. By the early twentieth century, the vineyards of Bordeaux were well on their way to recovery.

In the first half of the twentieth century, the tribulations of the Bordeaux wine region, as was true throughout France, were not at the hands of Mother Nature, as in the previous half century. Rather, the ensuing years saw an unprecedented string of man-made disasters: The First World War, the Great Depression, Prohibition in the United States and, of course, the Second World War. The production of wine fell drastically during World War II and the German occupation, partly due to lack of manpower, and partly to German forces seizing supplies of wine. Many Bordeaux producers used ingenious methods to hide their wine from the Nazis. Fortunately, most of the German occupying forces had the foresight to realize it was in the long term best interest of Germany to allow the Bordeaux trade to remain as undisturbed as possible. When the war ended, they wanted there to be Bordeaux wine to import into Germany (Kladstrup & Kladstrup, 2001).

In the second half of the twentieth century and on into the twenty-first century, the Bordeaux wine trade grew and strengthened. A rising standard of living throughout the Western world, an increasing appreciation for fine wine as an inherent part of cuisine, and the emergence of the United States as a particularly important and sophisticated market for wine have all worked to widen the consumer base for Bordeaux's wines.

Soil and Climate—The Terroir of Bordeaux

The *département* of Gironde is located on the west coast of France, on the Atlantic Ocean. Exactly halfway between the North Pole and the Equator, extending about 65 miles (105 km) from north to south and 80 miles (129 km) from east to west, the Gironde is spared any temperature extremes. A thick pine forest along the coast protects the vineyards from cold ocean breezes. The region contains many different soil variations that can nourish a wide variety of grape types. The soil composition is a major factor in deciding which vine shoots will be planted. The style of wine produced within each appellation of Bordeaux is a direct reflection of the proportion of each varietal planted there.

In Bordeaux, the grape varietals allowed by *AOC* laws are

Red	*White*
Cabernet Sauvignon	Sauvignon Blanc
Merlot	Semillon
Cabernet Franc	Muscadelle
Malbec	
Petit Verdot	
Carmenère	

(For regional white wines up to 30 percent of lesser grapes such as Colombard, Merlot Blanc, and Ugni Blanc is allowed.)

From this line-up of varietals, one can easily surmise that Bordeaux wines are *not* single-varietal wines. Rather, winemakers are free to blend the allowed varietals together to obtain the most complex and interesting combination possible. French wines made in Mediterranean-influenced zones tend to be blends, whereas wines from cooler, continentally influenced regions tend to be single varietal.

In Bordeaux, red varietals take up 89 percent of total acreage, and Merlot is the most widely planted red varietal with 162,000 acres (65,587 hectares) as opposed to 70,000 acres (28,340 hectares) for Cabernet Sauvignon, and 32,110 acres (13,000 hectares) for the third most important grape, Cabernet Franc (Figure 5-5). The lesser red varietals take up only about 5,400 acres (2,186 hectares). Each of the lesser red grapes can contribute necessary characteristics to the final blend. For instance, Carmenère adds deep color, and Malbec adds additional body. Petit Verdot is a late-ripening varietal, and when harvested at the same time as the other grapes, tends to be higher in acidity.

For the high quality, dry white grapes, Sauvignon Blanc is the most important. However, the most widely planted white grape is Semillon (Figure 5–6) (18,387 versus 11,367 acres—7,444 versus 4,602 hectares—for Sauvignon Blanc). Muscadelle is third with 2,341 acres (948 hectares). This grape gives high yields and adds fruity flavors and floral aromas to the wine (Acreage figures courtesy of *SOPEXA*, August 2004).

Within a region as large as Bordeaux, there are many different terroirs, each favorable to different varietals. In general, Merlot and Cabernet Franc are dominant on the right bank of the Gironde River, in St. Émilion and Pomerol, and Cabernet Sauvignon is dominant on the left bank of the river, in Médoc and Graves.

The Classifications of Bordeaux Estates

The tendency to rank wine-producing estates has become quite prevalent in recent times. During the late nineteenth and on through the twentieth century, as the market for wine became less regional and eventually international in nature, the need arose for a simple and understand-

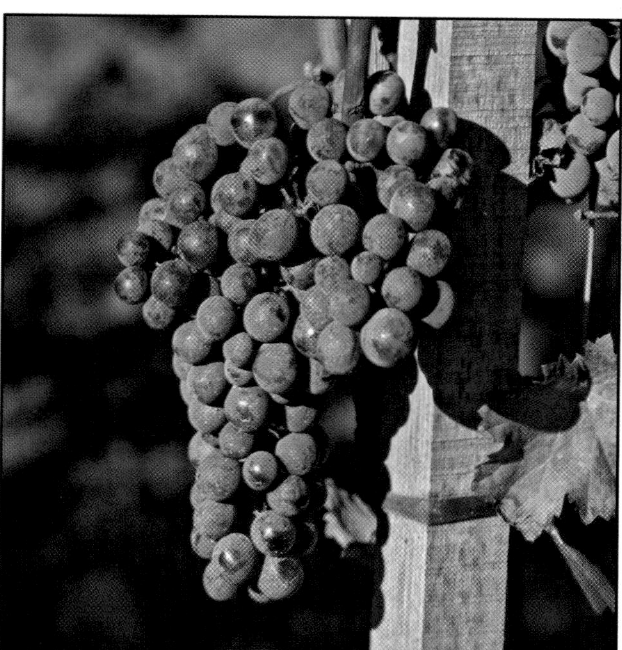

FIGURE 5–5 Merlot grapes ripening in the sun. Merlot is the most widely planted varietal in Bordeaux.

FIGURE 5–6 Semillon bunch almost ready to be picked. In Bordeaux, 55 percent of acreage devoted to white grapes is planted to Semillon.

able rating for the many diverse wines available. The most famous rating, and the most enduring, was the *Classification of 1855* for the wine-producing estates of the Médoc. In that year, the Exposition Universelle (the World's Fair) was to be held in Paris. To be sure that only the very best of France's great wines would be shown to visiting dignitaries, Napoleon III asked the wine merchants of Bordeaux to rank the wine-producing estates of that region. Even before then, there were already several informal rankings of Bordeaux's châteaux by the businessmen and interested connoisseurs (including one listing by the United States' first serious wine collector, Thomas Jefferson). Moreover, the market reflected the comparative worth of different estates' wines by the price consumers were willing to pay. However, what Napoleon III had requested was a formal, quantified ranking of the recognized top estates.

The merchants (also called brokers or *négoçiants*) took their task very seriously, and proceeded to formalize the ranking that they, and the open market, had been using for Bordeaux's wines. Referring back to prices fetched over the previous century, the brokers were able to divide the top Médoc estates into five tiers of quality. They issued their final classification through the Bordeaux Chamber of Commerce in time for the Exposition, and it remains the official ranking to this day, with only one change. In the top tier, called first growth or *premier* ***cru,*** there were only three Médoc estates, Lafite, Latour, and Margaux, as well as one estate that, although located in the Graves region, was of such a high caliber and its wines were so highly regarded, that it could not be omitted from this ranking of the Médoc. This estate was Château Haut-Brion. In 1973, Château Mouton-Rothschild was elevated from second growth to first growth. An additional 56 estates from the Médoc were rated at *deuxième cru* or second growth, and on down to *cinquième cru,* or fifth growth. Since there were thousands of properties producing wine at the time, it is indeed impressive to be included in the Classification of 1855. These 61 châteaux continue to be regarded as among the world's very best wine-producing estates. Even today the classification done so many years ago affects the pricing for Bordeaux wines in the highly competitive international marketplace. Every year, the demand for the 61 classified growths, especially the five *premier crus,* far exceeds the supply, thus driving the prices up.

In 1855, the wine brokers of Bordeaux also classified the estates of Bordeaux that produced sweet white wines. They ranked these estates into two classes, again based on market demand, price, and quality of the wines. These estates are all within the appellations of Sauternes and Barsac.

The wine-producing estates of the Graves region were not officially classified until 1953 for the red wines and 1959 for the white wines. Both lists consist of one class. It is worth noting that Château Haut-Brion is the only estate to be included in three classifications—the 1855 Classification of the Médoc, the 1953 classification of Graves reds, and the 1959 classification of Graves whites.

The estates of St. Émilion on the right bank of the Gironde River were first classified officially in 1955. In an effort to assure that their ranking be always current, the vintners of St. Émilion arranged for periodic reassessments of the classification, supposedly every 10 years. This plan makes sense for, although the vineyards themselves may be immutable, there are often changes in ownership or other human influences that need to be factored in. This system allows poorly managed vineyards to be demoted, while promising, well-cared-for estates can be promoted. The first modification took place in 1969 and was followed by further modifications in 1986. The list was again updated in 1996. The two top estates, Châteaux Cheval Blanc and Ausone are listed as *premiers grand cru classé A,* while the next 10 (increased to 11 in 1996) are listed as *premiers grand cru classé B.* Below that level there are an additional 55 estates all listed as *grand cru classé.* There is considerable variation in quality among these estates. The most recent list of classifications is in Appendix C.

The estates of the other famous appellation on the right bank, Pomerol, have never been officially classified. It is widely accepted, however, that the best wines from this region rank among the world's very best red wines.

The Classifications of Bordeaux estates can be found in Appendices C-F.

The Wine Regions of Bordeaux

The Médoc peninsula lies between the Atlantic Ocean and the muddy estuary of the Gironde River. For any wine lover, driving along the D2 highway, the *Route du Vin,* that wends its way up the peninsula from the city of Bordeaux is a magical experience. The landscape is not particularly spectacular. It is a bit flat, and in the southern portions, there are signs of urban sprawl. What *is* magical are the names one sees on the signs at the entranceways to the various wine-producing estates along the way. Château Margaux, Château Brane-Cantenac, Château Gruaud-LaRose, Château Lafite-Rothschild, Château Mouton-Rothschild, Cos d'Estournel—these are words any wine connoisseur has seen on bottles of extraordinary wines. To see these names on signs designating the vineyards from which the grapes come, and be able to glimpse, across those acres of vines, the facilities where these legendary wines are made, is surreal. Some of the châteaux are simple country homes. Some are large, beautiful mansions (Gruard-LaRose). Some have an unexpectedly exotic look to them (Cos d'Estournel's chai resembles a Chinese pagoda.) There are even former priories (Château Meyney). What ties these diverse estates together is the quality of the great red wines made here, in Bordeaux's Médoc region (Table 5–1, page 125).

The Haut-Médoc

Most of the very best of Bordeaux's wines come from famous estates in the lower two-thirds of the Médoc peninsula. This subregion, known as the Haut-Médoc, begins in the suburbs just north of the city of Bordeaux, and continues on up the peninsula through the small rural hamlet

Map of Bordeaux

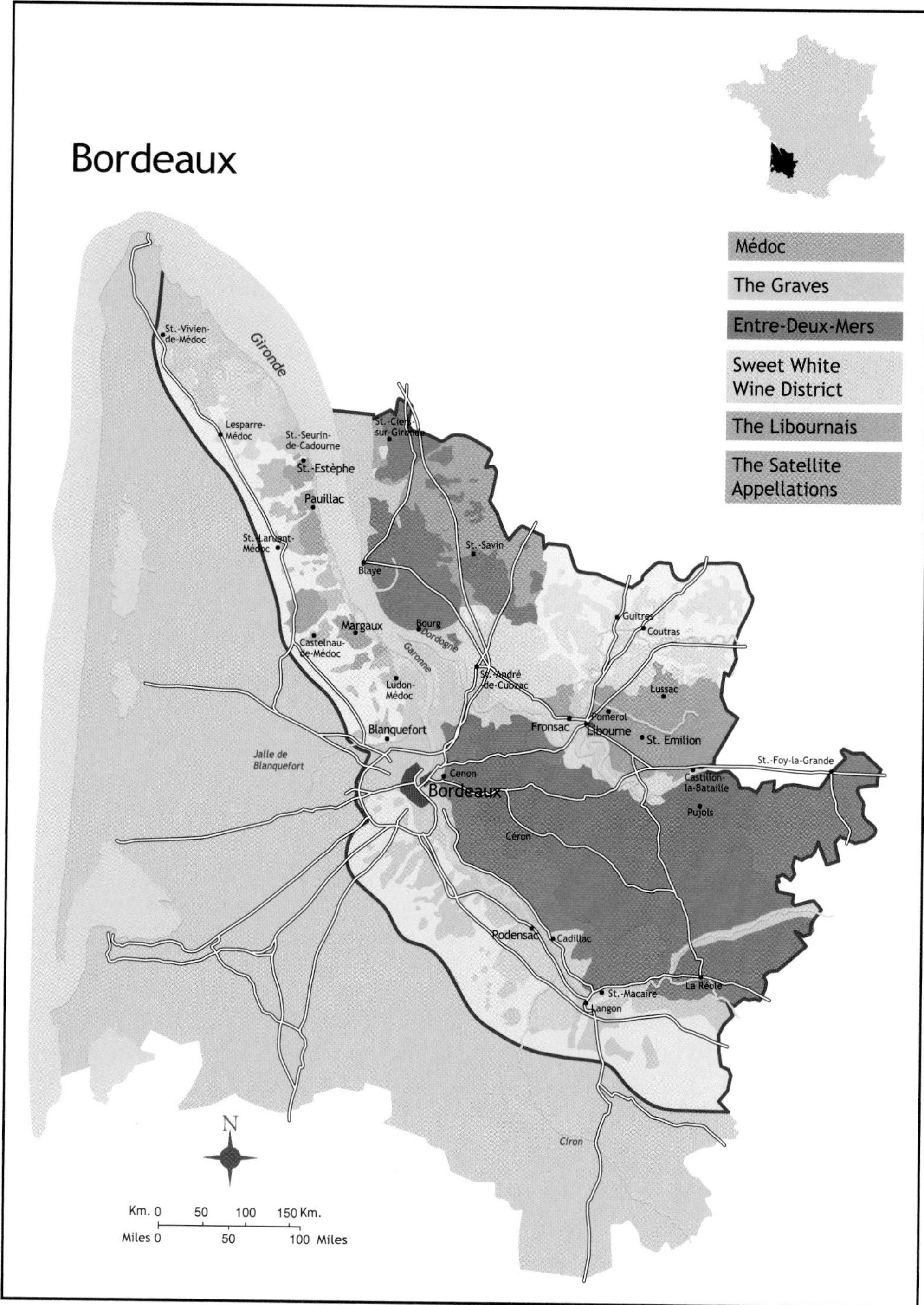

of Cadourne. This is just over 30 miles (4.8 km) as the crow flies. There are 29 communes (towns or villages) and a total of 25,000 acres (10,121 hectares) of vineyards within the Haut-Médoc. The greatest estates have been classified, that is, officially rated as superior. Most of these classified estates are located within the boundaries of four villages. These villages, listed from south to north, are Margaux, St. Julien, Pauillac, and St. Estèphe. Each of these towns is a separate appellation (as are the villages of Moulis and Listrac.)

One physical characteristic common to great estates throughout the Haut-Médoc is a good balance between water stored in the soil, and the depth of roots. In an article (Smart, 2004), Australian viticulturist Richard Smart refers to the famous study of terroir by Gerard Seguin of the University of Bordeaux. Dr. Seguin found that all the great estates of the Haut-Médoc had water tables within reach of vine roots. The water level drops progressively from spring into late summer. As Dr. Smart explains, by August the water table has fallen below the level at which the roots can reach it, and vine growth stops just as the berries are changing color, a process referred to as *véraison.* This allows the plants' energy to be concentrated in the berries, producing more concentrated flavors. Despite the similarity of water supply patterns, other factors of soil content and microclimate are diverse within the Haut-Médoc. In general, wines from the southern communes are softer, a bit richer and more accessible than the more tannic, elegant, and restrained wines from further up the peninsula.

Margaux The appellation Margaux actually encompasses five villages: Labarde, Arsac, Cantenac, Margaux, and Soussans. The soil varies considerably throughout the Margaux appellation, but it is essentially sandy gravel, quite thin and light in color. In the town of Margaux, the gravel lies atop a base of clay and marl. (Marl is a geological term for the conglomerate of magnesium and calcium from the shells left behind when the sea water drained out of this part of the peninsula.) In the surrounding villages, the base is sometimes gravel, sometimes iron-rich sandstone, and in some places even sand and grit. The percentage of plantings to Merlot is higher in Margaux than in the communes further north in the Haut-Médoc.

The wines of Margaux tend to be raspberry scented, smooth and medium-bodied on the palate, and redolent of rich, ripe berry flavors. The Margaux appellation is home to 20 classified estates, more than any other appellation in Bordeaux. After 20 years or so of declining quality in the 1960s and 1970s, Château Margaux, a *premier cru* estate, began producing stunning wines again. The international wine guru and writer, Robert Parker, has declared the wines of Château Margaux from 1978 on to be consistently outstanding (Parker, 1991).

St. Julien North of Margaux there is a wide stretch of land unsuitable for grapevines because the land is too marshy and flat. The next great vineyards appear as one comes into the commune of St. Julien. This is the smallest and most compact of the Haut-Médoc appellations, with only about 2,200 acres (891 hectares) under vine. The average quality of wine in this commune is very high. The vintners of St. Julien take great pride in the quality of their winemaking. Eleven estates in St. Julien are classified (rated).

The soil is gravelly with some clay; the subsoil has more limestone than Margaux's. Drainage is good. The vineyards are planted primarily to Cabernet Sauvignon such as these being picked in Figure 5–7. The wines of St. Julien have more tannic backbone and are fuller-bodied than those of Margaux, but still elegant.

FIGURE 5–7 Picking Cabernet Sauvignon grapes at Château Finegrave in St. Julien.

Another well-known estate, and justifiably so, is the beautiful Château Gruard-Larose. Situated off the main *Route du Vin* (D2), the stately château is surrounded by its meticulously maintained vineyards, 64 percent of which are planted to Cabernet Sauvignon. The wines produced here are very full-bodied; "massive" is the word chosen by some wine writers. One must be patient with Gruard-Larose, and not open the wine until years after the vintage, when the tannins will have softened and the fruit developed. Patience will be amply rewarded!

Pauillac This is perhaps the most famous of the communes in the Haut-Médoc. Virtually all of its 2,916 acres (1,180 hectares) of vines belong to or are controlled by its 18 classified estates. The soil throughout has the gravelly composition that permits excellent drainage, and retains the sun's heat and reflects it back on the vines in the cool evening, thus assisting ripening. Despite the relative uniformity of the top levels of soil, the subsoils differ from vineyard to vineyard, thus allowing for noticeable differences in style. In general, however, one can say that the wines of Pauillac tend toward full-bodied, smooth texture, exhibit a distinctive lead pencil/cedar combination in the bouquet, and are very long lived.

Three of the very top-rated estates are in Pauillac—Lafite-Rothschild, Mouton-Rothschild, and Latour (Figure 5–8). Each of these estates has a recognizable style. For instance, Lafite-Rothschild's vineyards, in the northern part of the appellation, have a limestone base, resulting in a particularly complex bouquet and subtle flavors of currants. Mouton-Rothschild sits on a gravelly ridge looking down on the small town of Pauillac. Its vineyards have more sandstone in their base soil than Lafite, and its wines are more opulent and complex, as well as very structured. (The fact that Mouton uses as much as 85 percent Cabernet Sauvignon also contributes to its distinctive style.) Latour, a grand old estate located in the southern reaches of Pauillac next to St. Julien, produces wines that are more supple and open. Latour's vineyards are entirely on loose, fine gravel, affording excellent drainage and heat retention. Latour's style is unmistakable. As Robert Parker succinctly puts it, "Latour is simply Latour, and . . . there are no 'look-alikes' in style or character" (Parker, 1991).

The famous trio of top-rated estates (first growths) are just the beginning of great Pauillac wines. Among the second growths, perhaps the most distinctive is Pichon-Longueville-Comtesse

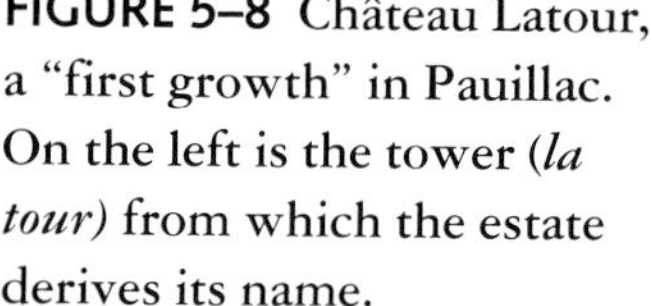

FIGURE 5–8 Château Latour, a "first growth" in Pauillac. On the left is the tower *(la tour)* from which the estate derives its name.

de Lalande, usually referred to as Pichon-Lalande. Located next to Latour on the St. Julien border, this estate produces wines that are soft, supple, and full of appealing fruit. An undervalued property that has begun to be noticed more and more by wine aficionados is Grand-Puy-Lacoste. Placed at the fifth level of quality in the official rating, this estate turned out wines in the past few decades of such high quality and distinctive character that some experts say this estate should be upgraded.

St. Estèphe Past Latour over a small man-made drainage ditch, the land suddenly rises in the commune of St. Estèphe, perhaps the least lauded of the Haut-Médoc's appellations. There is not as high a percentage of rated estates here, but there are many very fine properties, such as Château Meyney, Château Haut-Marbuzet and the rapidly improving Château Les Ormes-de-Pez, that, although unrated, produce attractive, agreeable, balanced (and affordable) wines that are deservedly popular in export markets. The style of these wines, as well as those coming from the five rated estates located here, is more tannic and backward than that of other communes. Despite an effort on the part of vintners in St. Estèphe to make more approachable wines by increasing percentages of the less-tannic Merlot and by allowing longer ripening and less maceration, the presence in most of St. Estèphe of a thick, dense claylike soil with inferior drainage and lower heat retention results in wines that are chunkier, unyielding, more acidic, and a bit awkward compared to wines that come from finer, gravelly soils.

This is not to imply that there are no world-class wines in St. Estèphe. Wine experts and consumers agree that Cos d'Estournel is indeed one of the world's great red wines. It is also the first estate one sees after crossing over into the commune from Pauillac. The strange pagoda-style chai (wine-making facility) sits on a slight ridge, overlooking Pauillac's famous Lafite-Rothschild. Because of the high percentage of Merlot (40 percent) (Figure 5–9), extensive use of new oak, and very careful attention to quality, the wines of Cos are fleshy, full-bodied, and complex. Usually austere when young, they are very long-lived and can be rich and lush with black fruit flavors once they mature.

Impressive wines have also been made for many years at other rated estates in St. Estèphe, most notably Calon-Ségur. Located on a bed of sandy gravel and iron-enriched limestone in the

FIGURE 5-9 Harvesting Merlot grapes at Cos d'Estournel in St. Estèphe.

northernmost reaches of the commune, Calon-Ségur is an ancient property. Hidden behind its high stone wall, the château is a lovely white building sporting two towers. The unusual drawing of a heart on the label stems from a charming story about the eighteenth century Marquis de Ségur who has been quoted as saying, "I make my wine at Lafite and Latour, but my heart is in Calon" (Parker, 1991).

The Médoc

North of Calon-Ségur, the land dips down and becomes too marshy for quality vineyards. This is the beginning of the Bas-Médoc, a low-lying area viticulturally inferior to its famous neighbor, the Haut-Médoc. Much of the land is dedicated to pasture rather than grapes. The soil here is sandy and has poor drainage. There are 14 wine-producing communes within the Bas-Médoc (often called simply the Médoc) and a total of 11,600 acres (4,696 hectares) of vines, mostly planted to Cabernet Sauvignon and Cabernet Franc. Some very decent and affordable red wines are made in the Médoc, and the adventuresome buyer can be rewarded with some exciting finds from this region.

Graves

Unlike the appellations Médoc and Haut-Médoc, which can be applied only to red wine, the appellation Graves applies to both reds and whites. A large area that runs about 34 miles (55 km) along the southern edge of the Garonne River (one of the two tributaries to the Gironde), Graves' 8,255 acres (3,342 hectares) of vineyards are planted 4,540 acres (1,838 hectares) to red wine grapes and 3,715 acres (1,504 hectares) to white wine grapes. Just over 45 percent of Graves' production is white wine. The dry whites of Graves can be among the most elegant, complex, and food-friendly wines based on the Sauvignon Blanc grape. The Semillon that is blended in softens the acidic edge and makes the wines rounder and smoother, as well as adding complexity through complementary flavors. The wines are fragrant with appealing citrus, gooseberry, and fresh grassy aromas.

The best red wines of the Graves region are velvety smooth, full of ripe berry flavors. They are not as full as some Haut-Médoc reds and mature more quickly, primarily because of the good dose of Merlot in most Graves reds. The leading estates are planted anywhere from 25 to 40 percent Merlot and 50 to 65 percent Cabernet Sauvignon, with the balance being the three lesser varietals (Cabernet Franc, Malbec, and Petit Verdot).

The soil of the Graves region is different than in other parts of Bordeaux. The region actually gets its name from the gravelly, pebble-strewn soil (Figure 5–10), the vestige of ancient Ice Age glaciers. This top level of gravel allows for excellent drainage and heat retention that helps the grapes to ripen fully. The gravel sits on base soils of sand and clay. Pine forests to the west afford considerable protection from the ocean's cool winds, just as in the Médoc.

The finest vineyards in Graves are in the communes of Pessac and Léognan in the northern section. The soil here is more alluvial where sediment has been deposited by the river over the millennia. In recognition of the fact that the best reds and the best whites of Graves come from these two towns, they were granted a separate appellation in 1986, Pessac-Léognan. The appellation covers ten communes and essentially divides Graves in two, with all the classified estates being in Pessac-Léognan. The reds from this appellation have complex bouquets of berries, earth, chocolate, and minerals. They feel full and firm, yet supple, on the palate and exhibit delicious flavors of ripe berries.

Among the most acclaimed producers in Graves is Château Haut-Brion, located in the commune of Pessac, which is really a bustling suburb of the city of Bordeaux (Figure 5–11). This château, now one of the premier wine-producing estates in the world, was purchased in 1935 by an American family, the Dillons. It was the first important Bordeaux estate to be owned by

FIGURE 5–10 This pebble-strewn soil in a Graves vineyard is typical of the region. Graves takes its name from the gravelly nature of its soil.

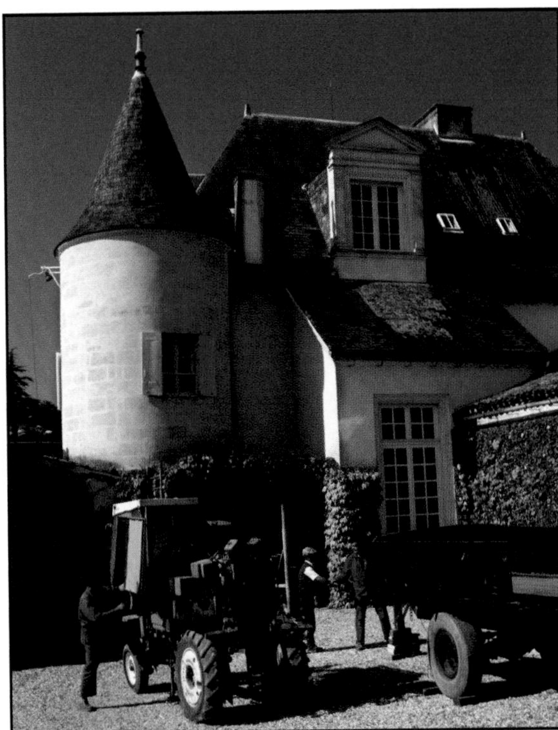

FIGURE 5–11 Château Haut-Brion in the commune of Pessac was included in the 1855 Classification of the Haut-Médoc. When Graves was classified in the 1950s, Haut-Brion was classified for its red wine. Although it is widely accepted that Haut-Brion makes the best white wine from Graves, the estate was not included in that classification at the request of the owners.

Americans. Because the winery and vineyards were in deplorable condition at the time, the Dillon family had to invest a large amount of money to return the estate to its full potential. The Duchesse de Mouchy, the present proprietor and a Dillon descendent, maintains the highest level of quality to this day.

Right across the road from Haut-Brion sits its rival, Château La Mission Haut-Brion, an extraordinary property of just under 50 acres (20 hectares). In 1919 La Mission was bought by Frederic Woltner. In the following decades, he and his family oversaw the ascendancy of this estate. Frederic's son, Henri, was widely recognized as one of Bordeaux's leading experts in viticulture and enology. He did invaluable research into varietal clones. His property was famous for both its rich complex reds and its fresh elegant whites. At the time of Henri's death in 1974, management of the property was taken over by Françoise and François Dewavrin-Woltner. After less than a decade the decision was made by the Dewavrin-Woltners and the other family members who still had an interest in the estate that, because of many internal disagreements, it was best to sell the property. Their neighbor, the Duchesse de Mouchy of Haut-Brion, purchased La Mission in 1983, freeing up Françoise and François Dewavrin-Woltner to pursue their dream of making wine in the New World. Château Woltner on Howell Mountain in Napa Valley, California, for many years produced some of that state's most elegant Chardonnays.

South of the busy towns of Pessac and Léognan, the region of Graves becomes more rural and even bucolic. Although not classified-growth territory, many lovely wines are produced in this part of Graves. It is possible to find some attractive, well-made reds, most of them primarily

Cabernet Sauvignon, which certainly rival the bourgeois-level Médoc reds. Some of the estates in this part of the Graves making very good (and affordable) red wines are Châteaux Barat (owned by the famous Lurton family), Cabannieux, d'Archambeau, Rahoul, Roquetaillade La Grange, and Sansay.

Many properties also produce clean, fresh white wines. In this southern part of Graves, as is typical in the entire region, almost 45 percent of the wine produced is white. The predominant white grape here is Semillon, taking as much as 70 percent of the acreage at some estates. Many experts feel that Semillon, when handled correctly and allowed a judicious amount of time in oak, can be made into as good a wine as the more illustrious Sauvignon Blanc. Some of the better estates for white wine are Clos Floridene and Châteaux de Carrolle, Roquetallade La Grange, and Sansay.

Sauternes

The appellation of Sauternes is restricted to sweet white wines. The appellation actually encompasses five villages—Sauternes, Bommes, Fargues, Preignac, and Barsac. Barsac is an appellation in its own right and can be sold either as Barsac or as Sauternes. The communes lie on the south bank of the Garonne River, in the southern part of Graves. The wines from Sauternes and Barsac, which by law must be botrytized, are widely regarded as the most luscious, rich dessert wines in the world. (If there is no **noble rot,** or ***pourriture noble,*** and dry wines are produced, they can be sold only as Bordeaux blanc.)

The appellations of Sauternes and Barsac contain less than 5,500 acres (2,227 hectares) of vines. The grapes planted are Semillon, Sauvignon Blanc, and Muscadelle. The most widely planted is Semillon, as it is the most susceptible to the noble rot. The climate here is perfect for the botrytis fungus, as the air is very damp. The vineyards lie along the tiny Ciron River, a tributary of the Garonne. The waters of the Ciron are very cold, and when they flow into the warmer Garonne, a gentle mist rises that often lasts through the morning. The mists create the humidity the fungus needs to thrive. As the fungus grows on the grapes, it causes them to shrivel up and eventually crack. The watery juice of the grapes escapes through the cracked skin, but the grapes continue to hang on the vine and ripen, producing more sugar. When harvested, these shriveled up, fungus-infected grapes have extremely concentrated sugars. The resulting wines are simply ambrosial, redolent with aromas and flavors of dried apricots and clover honey.

The most famous of the estates in the Sauternes appellation is the legendary Château d'Yquem. The wines from this property are so extraordinary that a special category had to be created for d'Yquem when the wines of this region were rated in 1855. As the only *premier grand cru* (first great growth), Château d'Yquem is literally in a class by itself. In the next level, *premier cru,* there are 11 estates, and at the *deuxième cru* (second growth) level there are an additional 15 properties. A combination of successful vintages (due to good weather) and increased interest in sweet wines during the decade of the 1980s led to an infusion of badly needed capital into this region. In the subsequent decades, with their newly updated equipment and clean new barrels, the rated estates of Sauternes and Barsac (and some of the as yet unrated estates) have continued to produce lovely dessert wines, rich but balanced.

The Libournais

The Libournais, named for the simple small town of Libourne, lies across the Gironde River from the Médoc and extends along the opposite bank of the Gironde and its tributary, the Dordogne River. This wine-producing region is often referred to as the "Right Bank." This is a very old wine-producing area, steeped in tradition and history. Many of the grape growers here have handed down their land, their expertise, and their way of life from generation to generation. Driving into one of its villages, full of simple stone houses with red tile roofs, ancient little churches, and quaint marketplaces, one has the feeling of having stepped back in time a hundred years or more (Figure 5–12). Because this part of Bordeaux was still isolated and considered a backwater at the time of the 1855 classification, no estates from the Right Bank were ranked. Even though the Libournais may still appear sleepy and dusty, it now produces some of the best red wines of France, particularly from the two most famous villages, St. Émilion and Pomerol.

The soils on the Right Bank are quite different from those in the Médoc, tending more toward clay and limestone. In this soil base, Merlot does very well and is, not surprisingly, the predominate grape planted. The second most widely planted varietal is Cabernet Franc. Cabernet Sauvignon, which excels in the gravelly soils of the Médoc, does not do well in the clay/limestone combination and is planted in relatively small quantities on the Right Bank.

St. Émilion Archaeological evidence shows that wine was made around the walled village of St. Émilion during Gallo-Roman times. A wine-making history stretching back two millennia makes for a great deal of tradition, and the current day wine producers of this area demonstrate considerable pride in their heritage. The small village has such a medieval air to it that sometimes one gets the feeling the local people are making a conscious effort to maintain a sense of history.

There are few grand châteaux in and around St. Émilion. Rather, this is an area of small properties and unpretentious houses and chais, once owned by peasants and bourgeois families. St. Émilion is one of the most compact and densely planted appellations in France. Close to 13,000 acres (5,263 hectares) of vineyards are packed into a relatively small appellation. The vineyards are planted primarily to Merlot, which ripens earlier than Cabernet Sauvignon, and due to its softer tannins, matures more quickly in the bottle. The wines of St. Émilion, in other words, are softer and more fruit-forward than wines of the Médoc. The blackberry aromas are tempered by an appealing violet or rose-petal nuance. The best St. Émilions will age amazingly well, taking on additional complexities as they mature.

From locations in and around the town of St. Émilion come many fine wines, especially from the classified properties atop the large plateau and its slopes south of the town. Here the vineyards look down on the valley of the Dordogne below. The soil is a thin layer of limestone debris on top of a solid limestone rock base. The vineyards receive bountiful sunshine tempered by

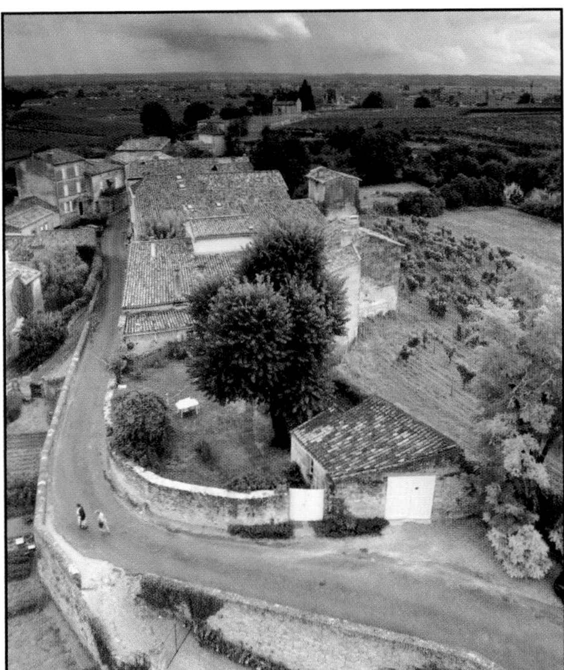

FIGURE 5–12 This view of a village on the outskirts of the town of St. Émilion shows the old stone houses, tiled roofs, and ancient walls typical of this historic region.

cooling breezes. Of the 11 estates presently included in St. Émilion's classification, eight have at least part of their vineyards on this plateau and its slopes.

One of the most famous of these classified growths is Château Ausone, named for the Roman poet and statesman, Ausonius, who was rumored to own a vineyard nearby. The caves of Ausone are dug right into the limestone hillside (Figure 5–13), and its vineyards are filled with gnarled old vines of Merlot and Cabernet Franc. There are less than 20 acres (8 hectares) of vines, producing a minuscule 2,200 cases of wine a year. These wines are very expensive and difficult to find commercially.

The other *premier grand cru classé* A estate is Château Cheval-Blanc. It lies to the west of town, near the border with Pomerol. This small section is called the Graves-Saint-Émilion, named for the gravel in its soil. There is less limestone and more sand here than on the plateau above the river. This difference in terroir creates a noticeable difference in the wines of these two great estates. Cheval Blanc is more approachable when young than Ausone, and due to its high percentage of Cabernet Franc (66 percent) the bouquet has none of the floral hints of Ausone, but tends instead towards minerals and spice. The flavors of black fruit are intense and concentrated.

Near Cheval Blanc lies another famous St. Émilion property, Château Figeac. Here the vineyards are planted almost equally to Cabernet Sauvignon, Cabernet Franc and Merlot, making a complex wine, quite different from the two A-level *grand crus.* (Figeac is also a classified estate.) The range of styles among the great St. Émilion estates is impressive. Some of the other best classified estates are Châteaux Pavie, Canon, and La Gaffelière (all *premier grand crus).* There are also many very fine, up and coming estates, as yet unclassified, which may well be included in future revisions. Most notable of these is Château Tertre-Roteboeuf. (The list of St. Émilion's most recently updated in 1996 classified growths is found in Appendix F.)

Pomerol Pomerol is a much smaller grape-growing region than its neighbor St. Émilion, having only 1,900 acres (769 hectares) of vines versus St. Émilion's 12,800 acres (5,182 hectares). Fully three-quarters of the vineyards in Pomerol are planted to Merlot with Cabernet Franc playing a supporting role. The soils vary throughout Pomerol, with a mixture of sand, clay, and gravel over a base of either sedimentary rock or iron. The presence of iron is one reason the wines of Pomerol are rich and concentrated, with a distinctive aroma of minerals and pencil lead. Pomerol may be the smallest of Bordeaux's important wine regions, but its wines

FIGURE 5–13 The caves (aging cellars) of St. Émilion's famous Château Ausone are carved directly into the stone hillside. This same limestone is found throughout the vineyards of the estate and contributes to Ausone's unique terroir.

are among the most impressive in the world. They are also among the most expensive because of the combination of superb quality and very limited production. Even though the wines of Pomerol have never been officially rated, their reputation is such that demand will always outpace supply.

The undisputed star of Pomerol is Château Pétrus. Some experts will state unequivocally that Pétrus is the best Merlot-based wine made anywhere. (Vineyards are planted 95 percent to Merlot). A very small estate, just 28.4 acres (11.5 hectares), Pétrus year after year turns out wonderfully rich, smooth, complex wines that spend more than two years in barrels. Demand for Pétrus always exceeds supply, making this the world's most expensive wine.

Other notable Pomerol producers are Châteaux La Conseillante, Le Pin, and Trotanoy. Also superb is Vieux Château Certan.

Lesser Appellations Beyond the five regions of Bordeaux that produce her undisputed champion wines—the Haut-Médoc, Graves, Sauternes/Barsac, St. Émilion, and Pomerol—several other *Appellation d'Origine Contrôlée* districts produce admirable wines. Fully 24 percent of France's *AOC*-level wines come from Bordeaux.

Entre-Deux-Mers: This fairly large appellation, whose name means "between two seas," lies between the two tributaries of the Gironde, the Dordogne and the Garonne. The appellation is restricted to dry white wine. Any red wine made from grapes grown here can be labeled only as Bordeaux Rouge. Although small quantities of lesser grapes, such as Ugni Blanc and Colombard, are allowed in Entre-Deux-Mers wines, most producers use high proportions of Sauvignon Blanc, which imparts racy citrusy flavors. This region produces large quantities of very affordable, clean, crisp white wine, light in body and straightforward in flavor. Entre-Deux-Mers whites are very food-compatible, particularly good with seafood.

Premières Côtes de Bordeaux: Stretching along the northern bank of the Garonne for 37 miles, the Premiéres Côtes de Bordeaux produces mostly red wine. Due in part to the high amount of gravel in the soil, some of these reds are quite distinctive.

Fronsac: Across the Dordogne River, on the Right Bank, one of the more important districts is Fronsac. The vineyards are west of the commune of Pomerol, mostly on limestone bluffs and are planted primarily to Cabernet Franc. Cabernet Sauvignon is also found in many of the wines. The result is complex, fruity wine at very reasonable prices. Some of the better wines come from the sub-appellation of Canon-Fronsac.

Table 5–1 Appellations of France: Bordeaux

REGION	SUBREGION	PRINCIPAL VARIETAL*
Haut-Médoc	Margaux	Cabernet Sauvignon, Merlot
	St. Julien	Cabernet Sauvignon, Merlot
	Pauillac	Cabernet Sauvignon, Merlot
	St. Estèphe	Cabernet Sauvignon, Merlot
Libournais	St. Émilion	Merlot, Cabernet Sauvignon
(The "Right Bank")	Pomerol	Merlot, Cabernet Sauvignon
Graves		Sauvignon Blanc, Semillon
	Pessac-Léognan	Cabernet Sauvignon, Merlot
Sauternes/Barsac		Semillon (botrytized)
Entre-Deux-Mers		Semillon, Sauvignon Blanc
Médoc		Cabernet Sauvignon, Merlot

*The varietal mentioned first is the prevalent one for that region. Remember, all Bordeaux wines are blends.

Lalande de Pomerol: This small satellite appellation lies just north of Pomerol. The wines, understandably, are like lesser Pomerols—full of Merlot, soft, fruity, and approachable even when young. They often represent excellent value.

Bourg and Blaye: On the right bank of the Gironde lie the large region of Côtes de Blaye, and its smaller neighbor, Côtes de Bourg. These appellations are allowed on both white and red wines, but the majority of the production is red (90 percent). Merlot dominates, and the wines are medium bodied, aromatic, and pleasantly fruity.

Bordeaux: The most general appellation, Bordeaux, can be used for white or red wines made from grapes grown anywhere within the boundaries of this large region. This is also the appellation used if grapes from two or more subdistricts of Bordeaux are blended together and for wines that do not conform to the restrictions of the appellation in which the grapes were grown (e.g., red wine from Entre-Deux-Mers). Even at this regional level, the law spells out specific requirements such as minimum alcohol content and yield per acre.

Bordeaux Supérieur: Bordeaux Supérieur has 0.5 percent higher minimum alcohol requirement that wines labeled as Bordeaux, and must have lower yields. Moreover, lesser grape varietals are excluded whereas in Bordeaux appellation whites, up to 30 percent can be from these subsidiary grapes.

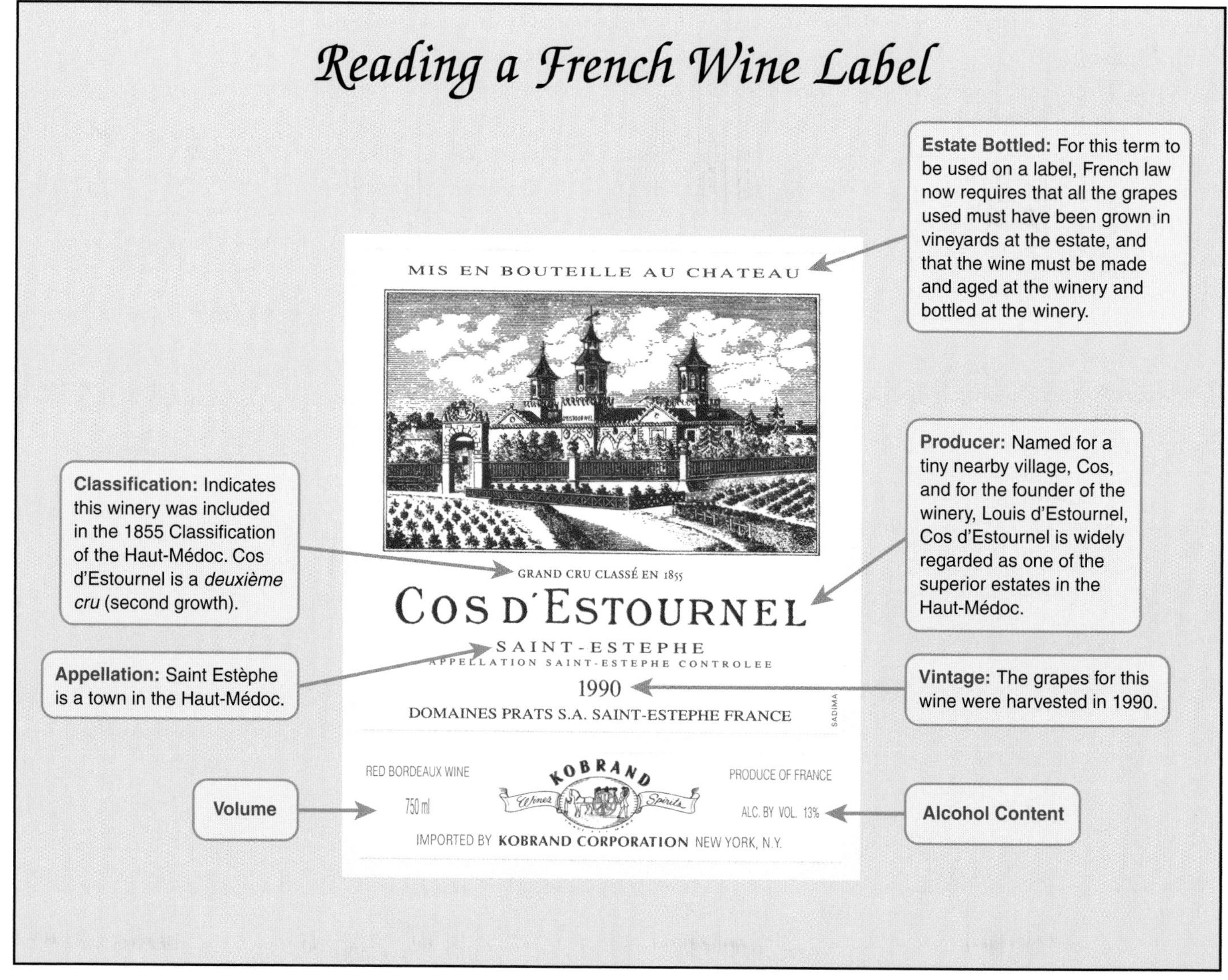
Reading a French Wine Label
Estate Bottled: For this term to be used on a label, French law now requires that all the grapes used must have been grown in vineyards at the estate, and that the wine must be made and aged at the winery and bottled at the winery.
MIS EN BOUTEILLE AU CHATEAU
Producer: Named for a tiny nearby village, Cos, and for the founder of the winery, Louis d'Estournel, Cos d'Estournel is widely regarded as one of the superior estates in the Haut-Médoc.
Classification: Indicates this winery was included in the 1855 Classification of the Haut-Médoc. Cos d'Estournel is a *deuxième cru* (second growth).
GRAND CRU CLASSÉ EN 1855
COS D'ESTOURNEL
SAINT-ESTEPHE
APPELLATION SAINT-ESTEPHE CONTROLEE
Appellation: Saint Estèphe is a town in the Haut-Médoc.
1990
Vintage: The grapes for this wine were harvested in 1990.
DOMAINES PRATS S.A. SAINT-ESTEPHE FRANCE
SADIMA
RED BORDEAUX WINE
KOBRAND
Wines
Spirits
PRODUCE OF FRANCE
Volume
750 ml
ALC. BY VOL. 13%
Alcohol Content
IMPORTED BY KOBRAND CORPORATION NEW YORK, N.Y.

The region of Bordeaux is an immensely complex and varied wine region, with a long history of wine production and a stellar reputation in the international marketplace. The producers of Bordeaux have become fiercely competitive with each other, and with wine producers the world over. They are astute enough to realize that they must maintain high levels of quality if they are to hold their prominent position in the eyes of wine consumers. As observed by the American wine critic, Robert Parker, "never in the history of Bordeaux have so many estates been making so much fine wine" (Parker, 1991).

Burgundy

Burgundy is much smaller than Bordeaux, produces only half as much wine, but is far more complicated. Burgundy is difficult to comprehend because of the plethora of appellations, maze of ownership patterns, and prevalence of *négoçiant* labels. The main complicating factor is the pattern of land-ownership. In Bordeaux, the wine-producing estates are self-sufficient entities in that they grow their own grapes, have the winemaking facility and aging caves on the property (and in many cases, the proprietor's dwelling also), and market the wines under the name of the estate. This is not the case in Burgundy. Each village here has its own appellation, and the vineyards within that village may each have their own individual appellations. Those vineyards may have several owners. For instance, Clos Vougeot, a single 123-acre vineyard of high quality, is subdivided into 100 parcels and has 80 owners (Figure 5–14). Moreover, the winemaking facilities are located in the towns, away from the vineyards. The name under which a wine is marketed may be that of a merchant or *négoçiant,* who is in no way connected to the owner of the vineyards where the grapes were grown. The *négoçiant* buys grapes or juice from several different growers and blends them together, often to the detriment of distinctive character due to terroir.

The effort to learn about Burgundy's wines is well worth it, however. This region produces elegant and complex whites based on the noble varietal Chardonnay and beautiful, refined reds made entirely of Pinot Noir. Sadly, one must shop for Burgundies carefully for despite high prices, due to limited supply and considerable demand, the quality is not consistent. Part of the problem is the northern location of the region, where continental weather patterns can make grape growing problematic. Another important factor, at least for the reds, is that Pinot Noir is a

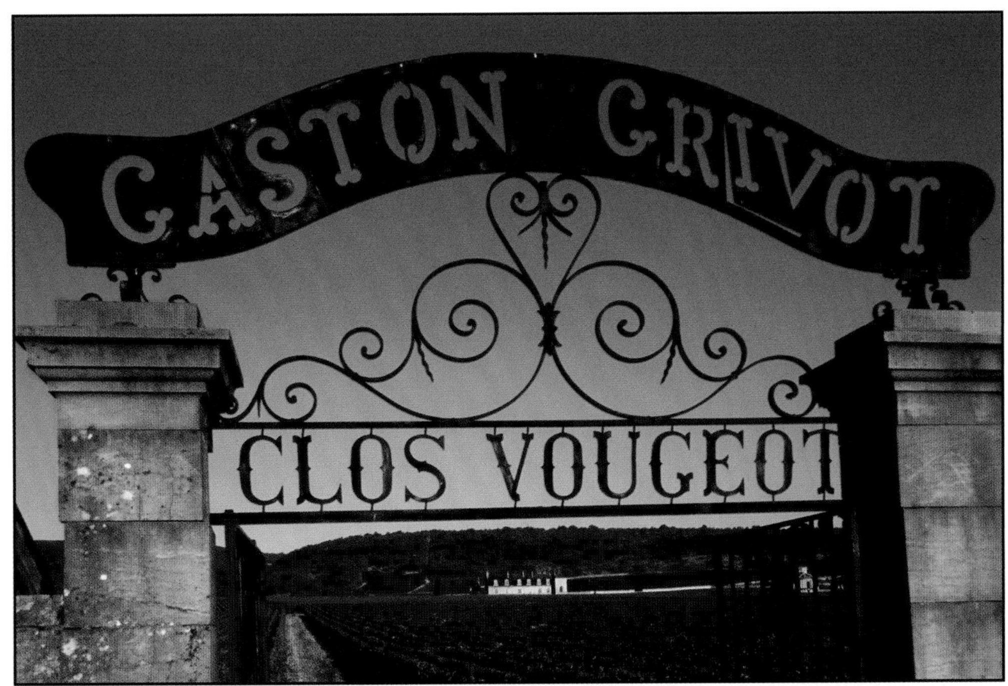

FIGURE 5–14 Clos Vougeot is one of the most famous vineyards in Burgundy's Côtes du Nuits. The term *clos* refers to a vineyard enclosed by walls.

Map of Burgundy

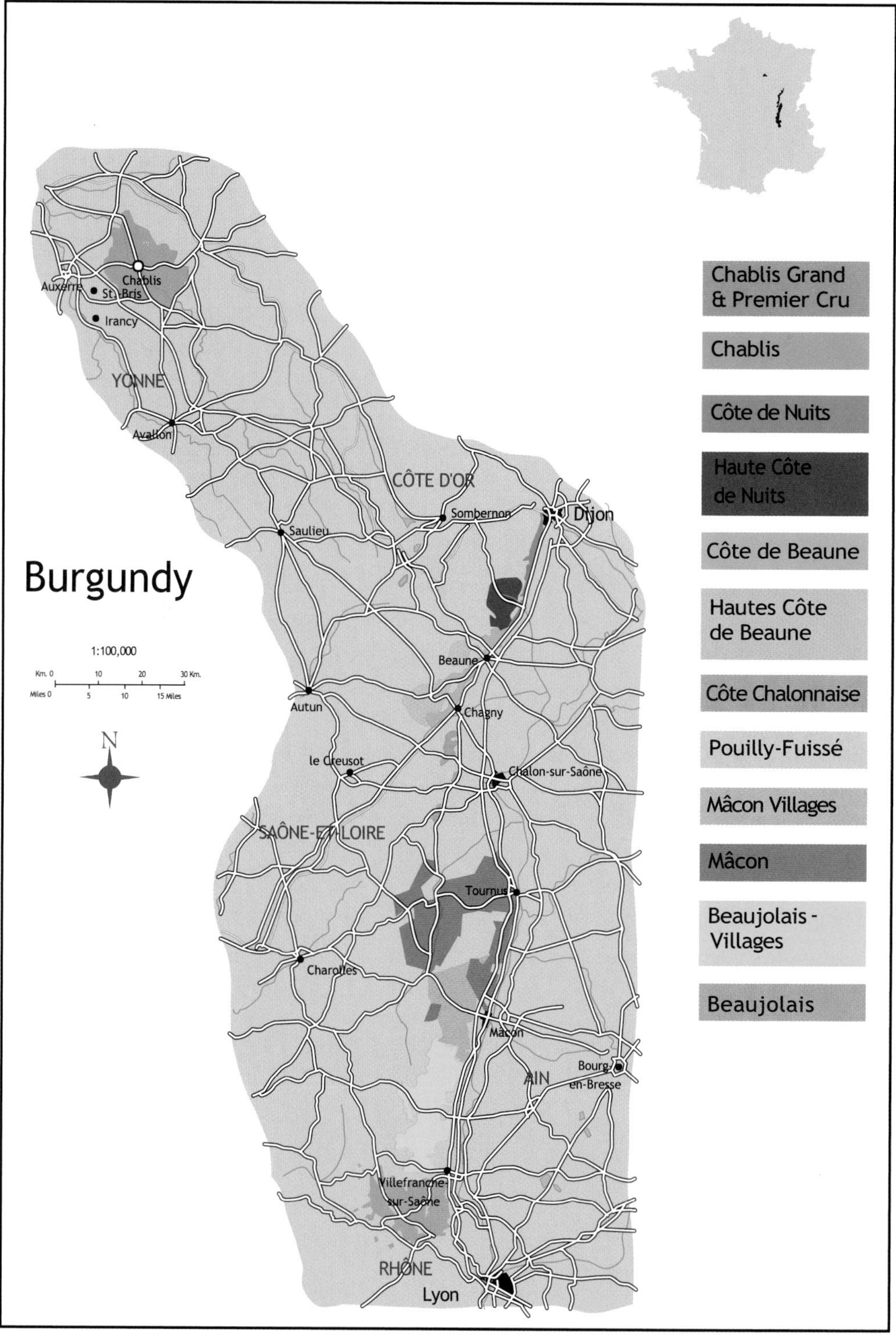

notoriously finicky and difficult varietal to grow. There is real variation from vintage to vintage and from producer to producer. The wise consumer does his homework and buys Burgundies from good vintages and only from reputable producers. Fortunately, in the past few decades, more moderately priced wines of good quality are being produced in Burgundy. The trend is away from selling to *négoçiants* and toward ***propriétaire*** labels, that is, wines for which the wine-making, bottling, and marketing are all done by the growers themselves.

History

The history of wine production in Burgundy precedes the Roman Empire. There is clear evidence that viticulture was well established here by the second century AD. The region survived the collapse of the Roman Empire and the invasion of barbarian tribes with little disruption to wine production. In fact, the name of the region evolved from one of those tribes. The Burgondes were a little-known people who migrated from Germany in the second half of the fifth century and stayed in the area well into the next century. At that time (approximately AD 530), they were absorbed into the Frankish kingdom after being defeated in battle (Coates, 2000).

Over the next thousand years, as Burgundy evolved first into an independent kingdom which lasted until the early eighth century, and then an autonomous duchy enlarging its boundaries and its power into the Middle Ages through carefully negotiated dynastic marriages, the most important single factor in the development of the region was the ever-increasing influence of the Catholic Church. In no other region of France did the church play such an important role vis-à-vis wine production. The church's vineyard holdings in Burgundy were enormous. Much of the land owned by monasteries and parishes was acquired as gifts from knights of the aristocracy as they left to fight in the Crusades. The knights' hope was that the monks and priests would pray for their souls should they die in battle far from home.

During the Middle Ages, as its landholdings increased, the church played a crucial role in perfecting techniques of viticulture and winemaking. The Cistercian order, for instance, which by 1336 owned over 123 acres (50 hectares) of prime vineyards in the northern part of Burgundy, did extensive systematic research into the relationship among grape varietal, soil and climate conditions, and the wine that resulted. These monks were among the very first to investigate and define the concept of terroir. From their meticulous work evolved the idea of *crus* (growths), the dividing of vineyards into sections each with its own distinct character. Many of the viticultural steps now practiced in Burgundy, such as pruning, grafting, and soil preparation, were developed by the Cisterians, as were important wine-making techniques (Phillips, 2000).

The invaluable contributions of the Catholic Church continued, as did the expansion of its landholdings, up to the time of the French Revolution and the abolition of the monarchy, at which time the pattern of land ownership in Burgundy changed. After the Revolution, the new government confiscated the lands of the Church and aristocracy, and sold them to the bourgeois families and to the peasants. Shortly thereafter, in 1790, the Napoleonic Code contributed even further to the fragmentation of landholdings by abolishing primogeniture, the age-old custom of leaving all one's holdings to one's oldest son. All children, including daughters, were to receive equal portions of an inheritance. It did not take many generations for a family's landholdings to become very small indeed. In modern-day France, one individual's holding can be as small as a few rows of vines.

After the Napoleonic Wars came to an end in 1815, and new patterns of land ownership were established, while economic and political conditions stabilized throughout France, wine production in Burgundy expanded. With the rise of the bourgeoisie, France's middle class, a new market for Burgundy's wines opened up. Unfortunately, attention to quality and authenticity was not always maintained. It was not unusual for a vintner to expand his production by blending in juice from grapes grown in inferior vineyards, or even grown outside of Burgundy. This type of fraud became even more prevalent as increasing numbers of *négoçiants* emerged. With the completion in 1856 of the railroad line that connected northern France to the Midi in the south, it became even easier for unscrupulous businesses to bring in inferior wines, expanding the amount of wine they had to sell but seriously eroding quality.

Before Burgundy could correct the problem of fraud, the region was hit hard by phylloxera. Many of Burgundy's vineyards were wiped out. The one benefit of the phylloxera epidemic was that, when replanting was undertaken on American rootstock, only the most suitable locations were planted, thus eliminating inferior vineyards. Most vineyards are now on slopes leading up from the river valley. The upper reaches of the hills are too exposed and cold for vines, and the low-lying sites along the valley floor are too alluvial and marshy.

The passage of the *Appellation d'Origine Contrôlée* laws in 1935 eliminated the worst of the fraud and gave protection to place names within Burgundy. The *AOC* laws also established standards of viticulture and winemaking. Since the 1980s, there has been a trend away from small growers selling their grapes to *négoçiants,* and instead the number of *propriétaire* labels has increased.

The Classification System of Burgundy

When learning to decipher Burgundy's classification system, it is helpful to think in terms of concentric circles (Figure 5–15), while bearing in mind our maxim about European appellations: "The smaller and more specific the geographic designation, in general, the better and more distinctive the wine." In the case of Burgundy, the outermost concentric circle is the *general appellation,* Burgundy. The label will say simply "Bourgogne Rouge" or "Bourgogne Blanc" (Figure 5–16). Grapes for this level of wine may be grown anywhere within the region of Burgundy. Burgundy is a small region, with only 98,000 acres (39,676 hectares) under vines, and the grapes used must be the approved varietals of Pinot Noir (for reds) or Chardonnay (for whites). Therefore, a wine labeled as generic Burgundy can still be quite distinctive (Table 5–2).

The next circle in our hypothetical "target" is that of the *regional appellation.* For these wines, the grapes must all be grown within a specific subregion of Burgundy. An example would be Côte de Beaune or Côte de Nuits-Villages. Sometimes a regional appellation signifies that grapes from vineyards located in two or more villages have been blended together.

The next smaller circle is the *commune appellation.* A commune is a village or town. All the grapes used in a wine labeled with the name of a specific commune must come from vineyards located within the boundaries of that village or town (Figure 5–17).

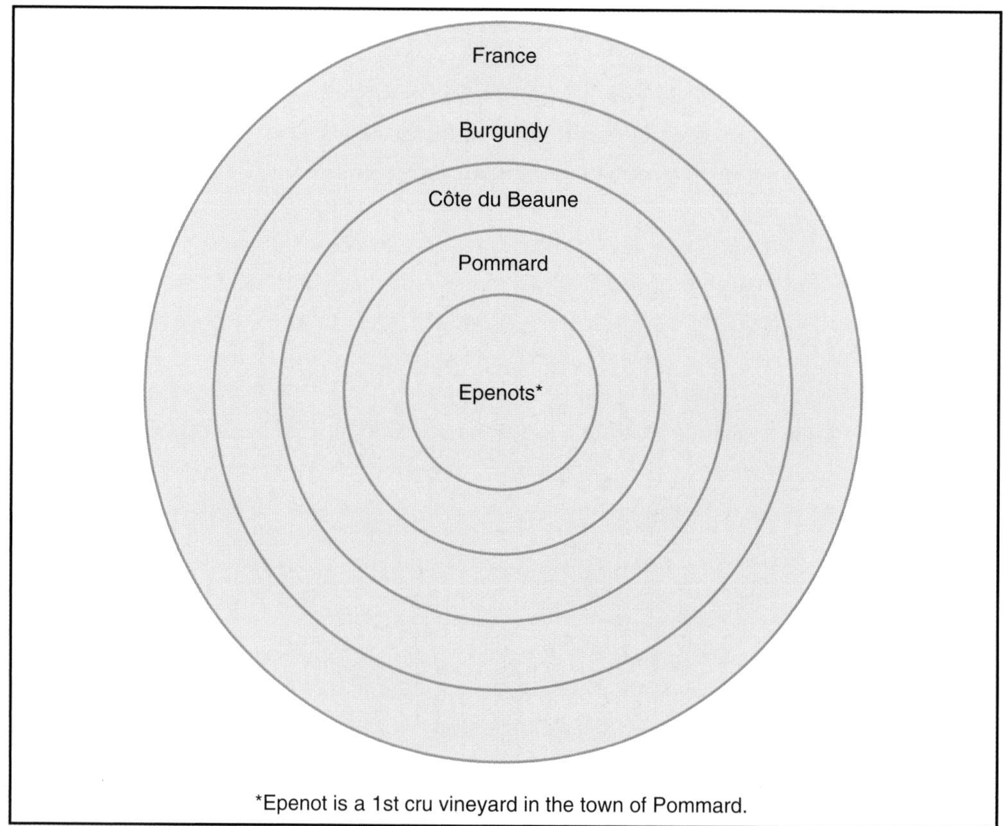

FIGURE 5–15 The analogy of concentric circles illustrates how French appellations fit one inside the other as the geographic designation gets smaller. Generally, the smaller the appellation, the better and more distinctive the wine.

FIGURE 5–16 An example of a general appellation. These Chardonnay grapes could have been grown anywhere within the boundaries of the Burgundy region. ("Burgundy" is the English spelling for the Bourgogne region.)

Table 5–2 Appellations of France: Burgundy

REGION	SUBREGION	PRINCIPAL VARIETAL*
Chablis	None	Chardonnay
Côte de Nuits	Fixin	Pinot Noir
	Gevrey	Pinot Noir
	Morey-St. Denis	Pinot Noir
	Chambolle-Musigny	Pinot Noir
	Vougeot	Pinot Noir
	Vosne-Romanée	Pinot Noir
	Nuits-St. Georges	Pinot Noir, Chardonnay
Côte de Beaune	Aloxe-Corton	Pinot Noir, Chardonnay
	Pommard	Pinot Noir
	Volnay	Pinot Noir
	Beaune	Chardonnay, Pinot Noir
	Meursault	Chardonnay
	Puligny-Montrachet	Chardonnay
	Chassagne-Montrachet	Chardonnay
Côte Chalonnaise	Rully	Chardonnay, Pinot Noir
	Mercurey	Pinot Noir, Chardonnay
	Givry	Pinot Noir
	Montagey	Chardonnay
Mâconnais	Pouilloy-Fuissé	Chardonnay
	St. Véran	Chardonnay
	Mâcon-Villages	Chardonnay
Beaujolais	Morgon, St. Amour, Fleurie	
	Moulin-à-Vent, Brouilly, Côte de Brouilly	Gamay
	Julienas, Chenas, Regnie	
	Beaujolais-Villages	Gamay

*Burgundy wines are *not* blended. The grape mentioned for each village indicates whether that village produces red wine or white. When both varietals are mentioned, the one shown first indicates which type of wine the village is more famous for.

FIGURE 5–17 An example of a commune label. Chassagne-Montrachet is a village in the Côte de Beaune, famous for the quality of its white wines. *Vielles Vignes* means "old vines." This is not an official designation, but is used by vintners to indicate the grapes come from older vines that usually impart more intense flavor to wine.

The next two levels are for specific single vineyards, often very small indeed. The vineyards that carry their own individual appellation are those that have been officially rated by the authorities. Only in the best regions of Burgundy are there any rated vineyards. Single vineyard appellations are found only in Chablis and the Côte d'Or. The first level of rated vineyards is the *premier cru appellation* (first growth designation) (Figure 5–18). The label for *premier cru* wines will show both the name of the vineyard and the name of the commune in which it is located. For instance, "Pommard *(the commune)* Epenots (*the rated vineyard*)" or "Beaune (*the town)* Clos de la Mousse *(the rated vineyard*)."

The final level of quality for Burgundy, the "bull's eye" of our concentric circles analogy, is the *grand cru appellation,* or "great growth designation" (Figure 5–19). The *grand cru* vineyards have been rated by the authorities as the very best sites for growing Pinot Noir or Chardonnay. These vineyards are capable—due to immutable physical factors such as hours of sunshine, protection from cold winds, drainage, and unique make-up of the soil—of producing, year after year, grapes that are superior to those harvested from other vineyards. Wines from *grand cru* vineyards carry just the name of that vineyard. An example would be "Grands Echézeaux" or "Le Corton." The commune name is not mentioned. *Grand cru* wines are the *crème de la crème* of Burgundy's wines. There are only seven *grand cru* vineyards in Chablis, and 30 in the Côte d'Or.

FIGURE 5–18 A *premier cru* (first growth) label will show both the specific rated vineyard (in this case Boucherottes) and the commune (Beaune). Note the full official appellation reads *"Appellation Beaune 1er Cru Contrôlée."*

FIGURE 5–19 A *grand cru* label shows only the name of the specific rated vineyard, in this case, Grands Échezeaux, one of Burgundy's most acclaimed sites for the Pinot Noir grape.

The Classification System of Burgundy

Burgundy has five different levels of classification, the top two of which are relevant only to Chablis and the Côte d'Or. These quality levels are:

1. Non-specific general appellation with no geographic definition, i.e., Bourgogne.
2. Regional appellation. Usually a blend of one or more commune wines made by a merchant, or *négoçiant*. Example: Côte de Nuits-Villages.
3. Commune appellation. All the grapes used in the bottle were grown within the boundaries of one town, or commune, but not from vineyards that are rated as superior by the authorities. Example: Vosne-Romanée.
4. First growths (*premier cru*). The label shows the name of the commune *and* the name of the rated vineyard. Example: Vosne-Romanée "Les Malconsorts."
5. Great growths (*grands crus*). These are the very finest vineyards. The label will show the name of the vineyard *only*. Example: La Tâche.
   In Chablis the label will say Chablis Grand Cru, followed by the name of the rated vineyard, e.g., "Blanchots."

These are the 14 most important communes of the Côte d'Or, moving from north to south. The best-known *grand cru* vineyard of each commune are also shown.

Cote de Nuits:

1. Fixin. Red wine only.
2. Gevrey. Red. Eight *grand crus,* two of which are extraordinary:
   Chambertin — Clos de Beze
3. Morey-St.-Denis. Red.
   Bonnes Mares (a small portion)
4. Chambolle-Musigny. Mostly red.
   Bonnes Mares — Musigny
5. Vougeot. Mostly red.
   Clos de Vougeot
6. Vosne Romanée. Red only.
   Échezeaux — Grand Échezeaux
   La Tâche — Richebourg
   Romanée-Conti — La Romanée
7. Nuits-St.-George. Mostly red.

Cote de Beaune:

8. Aloxe-Corton. Some white; mostly red.
   Le Corton (red) — Corton (white)
   Corton-Charlemagne (white)
9. Pommard. Red only. No *grand crus.* Six *premier crus.*
10. Volnay. Red. No *grand crus.* Thirteen *premier crus.*
11. Beaune. Red and white. No *grand crus.* 34 *premier crus.*
12. Meursault. White only. No *grand crus.* 17 *premier crus.*
13. Puligny-Montrachet. Mostly white; some red.
    Bâtard-Montrachet — Bienvenues-Bâtard-Montrachet
    Chevalier-Montrachet — Le Montrachet
14. Chassagne-Montrachet. Mostly white; some red.
    Bâtard-Montrachet — Criots-Bâtard-Montrachet

These vineyards are small in area and their production limited. Needless to say, *grand cru* wines are very expensive.

THE WINE REGIONS OF BURGUNDY

Burgundy is divided into six main regions: Côtes de Nuits, Côtes de Beaune (which together are often referred to as the Côte d'Or), Chablis, Côte Chalonnaise, Mâconnais, and Beaujolais. Chablis lies geographically separate from the rest of Burgundy, some 81 miles (131 km) to the northwest. The remaining regions are spread in a contiguous line along the Saône River valley, from the city of Dijon in the north to the city of Lyon in the south. The vineyards are not contiguous, however. Burgundy is not a tightly planted region, as vineyards are planted only on the slopes where vines can flourish.

Map of Chablis

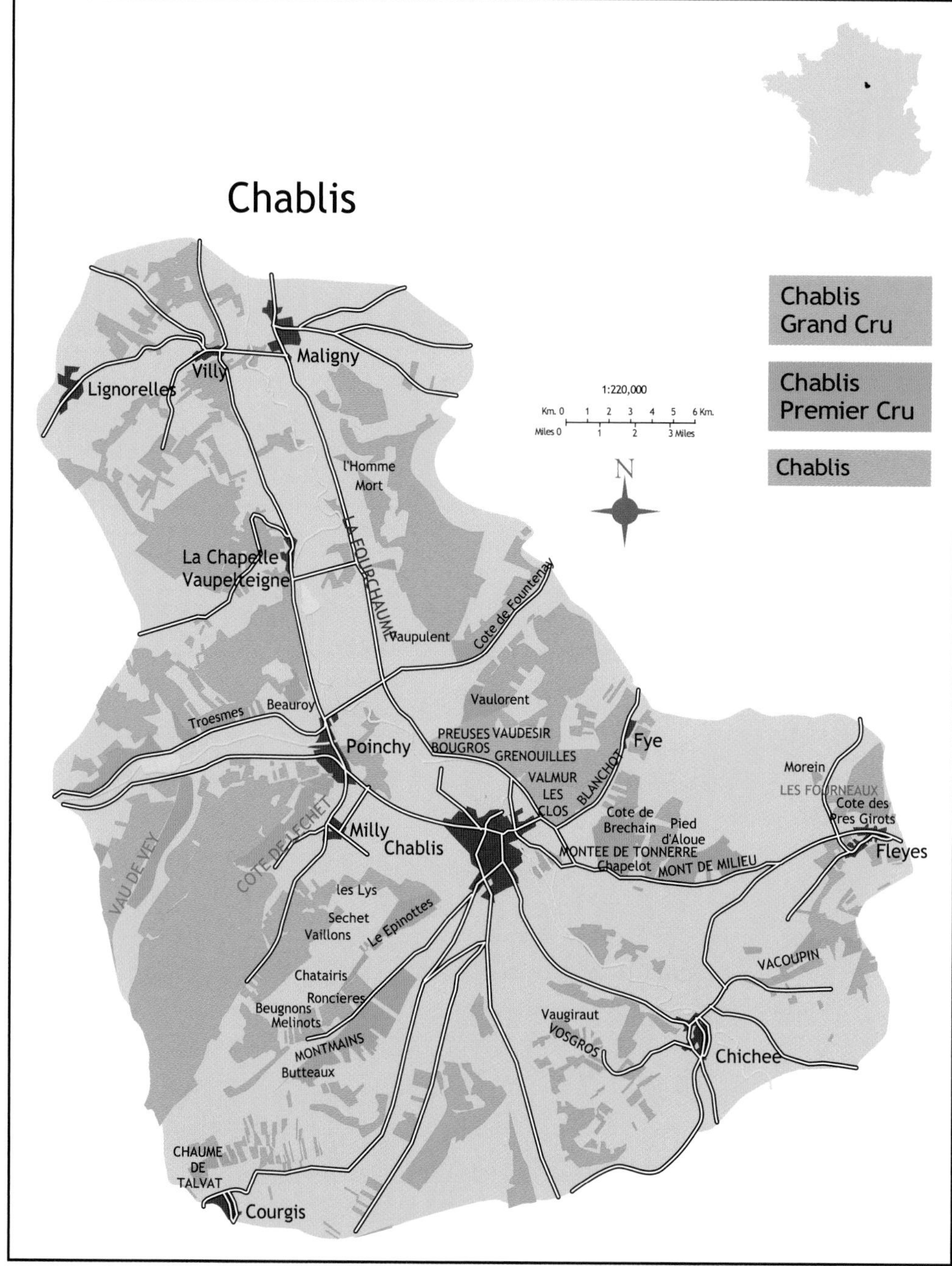

Chablis

Chablis is an appellation restricted to dry white wine. These are among the dryest and most elegant wines made from the Chardonnay grape. The climate here is cool enough that the grapes maintain an excellent crisp acidity. The flavors fully evolve because the grapes enjoy a lengthy ripening period as they hang on the vines into fall. However, the vintners must be constantly alert to the danger of frost.

Chablis is a fairly small region, with fewer than 7,000 acres (2,834 hectares) under vines. The soil throughout Chablis is uniform, a unique mix of chalky limestone and clay. This mixture imparts distinctive aromas (hay, apples, wet slate) and flavors unlike those found in Chardonnay grapes grown even a few miles away. There is a minerally, almost flinty, edge to Chablis that perfectly sets off its subtle flavors and crisp acidity. Visually, Chablis whites have a vibrant yellow color with a touch of green at the edge. These wines are superb companions to a wide variety of foods, especially seafood, poultry or pasta in creamy sauces, and certain veal dishes.

Classification within Chablis In 1936, the French authorities began the process of rating the vineyards of Chablis. The very best vineyard sites, on a slope above the town of Chablis, face southwest, thus benefiting from more sun exposure than other locations. On this slope the soil also has a higher level of fossilized oyster shells, which add a subtle but highly desirable extra dimension of complexity to the wines. In 1938, seven vineyards on this slope were awarded the *grand cru* rating (Figure 5–20). These seven vineyards are very small, averaging less than 40 acres each. Wines made from grapes grown in these vineyards are labeled "Chablis Grand Cru," with the specific vineyard also listed (Table 5–3).

Next in quality are those vineyard sites designated *premier cru* (Figure 5–21). The original group of 11 *premier cru* vineyards were classified in 1967. Some of these vineyards actually encompass several subsidiary vineyards, but the authorities streamlined the original list of 26 vineyard sites down to the more comprehensible eleven. In 1986, an additional seven sites were designated *premier cru,* while one of the original 11 was absorbed into its neighbor. Consequently, there are now a total of 17 *premier cru* vineyards. The wines from these sites are labeled as "Chablis Premier Cru," with the specific vineyard also listed (Table 5–4).

Table 5–3 *Grand Cru* Vineyards of Chablis

Bougros	Vaudésir
Preuses	Grenouilles
Valmur	Les Clos
Blanchot	

FIGURE 5–20 The soil of Chablis' best vineyards is white because of the high concentration of chalk and limestone. Stretching over the gently rolling hills in this photograph are portions of four of the seven *grand cru* vineyards.

Table 5–4. *Premier Cru* Vineyards of Chablis

CLASSIFIED IN 1967	
Fourchaume	Montée de Tonnerre
Monts de Milieu	Vaucoupin
Les Fourneaux	Beauroy
Côte de Léchet	Vaillons
Montmains	Vosgros
Mélinots (absorbed into Vaillons)	
CLASSIFIED IN 1986	
Vaudevey	Vau Ligneau
Côte de Vaubarousse	Chaume de Talvat
Les Landes	Les Beauregards
Berdiot	

Wine made from Chardonnay grapes grown anywhere else within the official boundaries of the Chablis appellation is simply labeled *Chablis*. These wines must have a minimum of 10 percent alcohol. Approximately 85 percent of the wine produced in the region is classified as generic Chablis.

Several *negoçiant* firms produce consistently fine Chablis. Especially reliable sources include Joseph Drouhin, Henri Laroche, and Christian Moreau. Among the best of the *propriétaire* labels from Chablis are William Fèvre, René Dauvissat, and Gerard Tremblay.

FIGURE 5–21 A bottle of *premier cru* Chablis.

Map of Côte d'Or

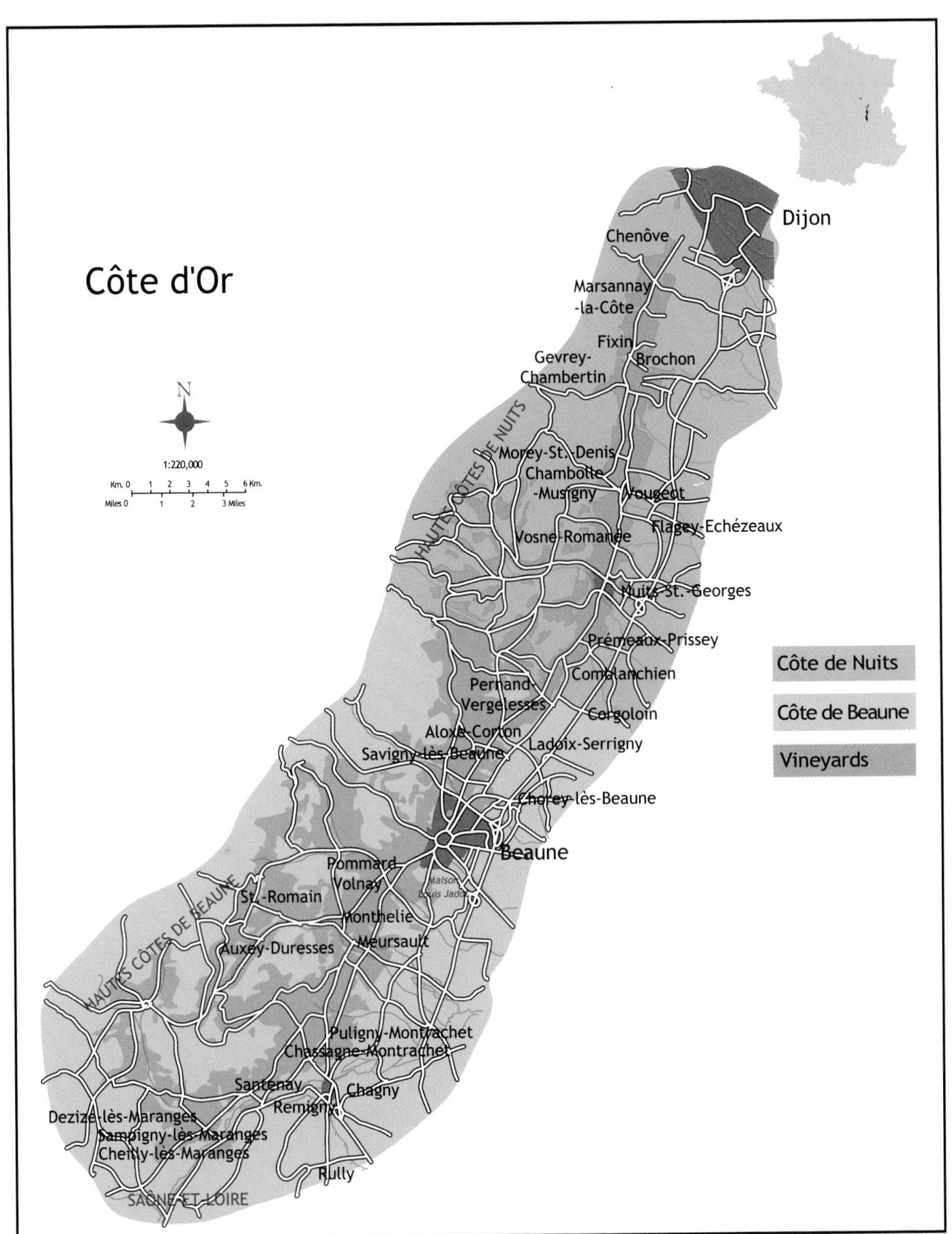

Côte d'Or

Burgundy's Côte d'Or, or Golden Slope, is widely regarded as one of the world's best areas for growing cool-climate grapes. It is only about 30 miles (48 km) long and less than 2.5 miles (4 km) at its widest. Elevation is between 720 and 1,000 feet (219 and 304.8 meters). The hills protect the vineyards from excessive rain and provide south and east facing slopes which catch more sunlight.

The Cote d'Or is divided into two subregions. The northern portion is the Côte de Nuits (named for the town of Nuits-St.-George). The southern portion is the Côte de Beaune (named for the city of Beaune). In general, the Côte de Nuits is famous for its red wines and the Côte de Beaune for its whites. The reds of the Côte de Nuits are big but not tannic, elegant with solid structure, and incredibly complex. The complex mélange of aromas and flavors is all the more surprising when one realizes that the wines are made from only one grape, the Pinot Noir. The bouquet is earthy, displaying enticing aromas of mushrooms, root vegetables like beets, or even the typical barnyard aroma of old manure. The fruit is reminiscent of cherries or strawberries. These wines can be consumed as early as 2 to 6 years after the vintage year, but can age very well, often not reaching maturity until 15 or 20 years old.

The whites of the Côte de Beaune are, in the opinion of many experts, the world's most elegant and complex. Made entirely from Chardonnay grapes, Côte de Beaune whites are noted for their complex bouquets of hazelnuts or blanched almonds, apples, appealing vegetable tones of fresh cabbage, and a hint of toast. The flavors are of ripe fruit and toasty oak, perfectly balanced by fresh acidity. The lesser whites of Burgundy are fermented in stainless steel and bottled young, never spending any time in barrels. But the great whites of the Côte de Beaune are both fermented and aged in oak, which adds to their nutty/buttery richness.

The Côte de Nuits

The Côte de Nuits starts in the north with the village of Marsannay, just south of the city of Dijon and continues for 14 miles (22.6 km). There are small quantities of rosé and whites wines made here, but the Côte de Nuits is famous for its world-class reds. The important communes of the Côte de Nuits, from north to south, are Marsannay, Gevrey-Chambertin, Morey-St.-Denis, Chambolle-Musigny, Vougeot, Vosne-Romanée, and Nuits-St.-George.

Marsannay This small village just south of Dijon is famous for its rosé, a dry wine made from Pinot Noir grapes. The bouquet is of strawberries, and the flavors are clean and fresh. These wines are best served young. Like all rosés, Marsannay rosés are very good with salty foods, such as ham or anchovies.

Gevrey-Chambertin The name of this commune reflects a common practice in the Côte d'Or: Hyphenating the name of the commune's most famous *grand cru* vineyard (in this case, Chambertin) (Figure 5–22) to the original name of the commune. Gevrey-Chambertin is an appellation for red wine only. There are eight *grand cru* vineyards in the commune. Besides Chambertin, the most famous vineyard is Clos de Beze. (A ***clos*** is a small walled-in vineyard. It is a common way of naming vineyards in Burgundy.) The soil of the *grand crus* vineyards varies depending on how high up the hillside the vineyard is located. The primary component is limestone, mixed with some clay and flint. The amount of clay decreases in sites higher up the hills. Gevrey-Chambertin also has 24 *premier cru* vineyards.

The wines from Gevrey-Chambertin are among the best of Burgundy's reds. They are full-bodied, smooth and very complex. They can age extremely well, often not reaching their prime until fifteen years after the vintage.

Morey-St.-Denis The commune of Morey-St.-Denis produces primarily red wine. Lying between the more famous villages of Gevrey and Chambolle, Morey-St.-Denis is often overlooked. This is a shame as its wines can be as concentrated and refined as the best produced in either of its neighbors. There are four *grands cru* vineyards in Morey-St.-Denis (plus a tiny portion of one of the most prestigious of Burgundy's *grand crus,* Bonnes Mares). Clos de la Roche is considered the best of the four, although it is Clos St. Denis that the village of Morey chose to hyphenate to its name. There are also 25 *premier cru* vineyards in Morey-St.-Denis, located down the hillside from the *grand crus.* For the most part these vineyards are very small, averaging under 3 acres (1.2 hectares) each.

Chambolle-Musigny A small amount of white wine is made in Chambolle-Musigny, but the commune is renowned for its great reds. The outstanding characteristic of these wines is

FIGURE 5–22 The historic village of Gevrey-Chambertin in the Côtes de Nuits. The old wooden wine press stands in front of a castle that was built in the tenth century.

their aromatic bouquet, reminiscent of strawberries and roses, as well as their finesse and delicacy. Both of Chambolle's *grand crus,* Bonnes Mares and Musigny, are of very high quality. The soil in both vineyards contains very little clay, a factor that contributes to the delicacy of the wines. There are also 22 *premier cru* vineyards in Chambolle, covering a total of just over 150 acres (61 hectares). The best of these vineyards is Les Amoureuses, which is at the northern edge of the commune. Also very good is Les Charmes, further south.

Vougeot Vougeot is a tiny little village with only a couple of dozen inhabitants. The village is dominated by its one *grand cru* vineyard, Clos de Vougeot, which, at 124 acres, is one of Burgundy's largest rated vineyards. Clos de Vougeot was originally planted by monks in the early twelfth century, and by 1340, the high stone wall enclosing the vineyard was completed. Fully four-fifths of Vougeot's wine production is red wine from this one *grand cru* vineyard. Because it is so large and is owned by many different entities, the wines of Clos de Vougeot can differ considerably in style.

A tiny amount of white wine has the commune appellation. The remainder is red wine from the 44 acres of *premier cru* vineyards. Almost no red wine from Vougeot carries the commune appellation. The reds of Vougeot have a distinctive truffle or mushroom hint to their bouquet, and have concentrated flavors.

Flagy-Échezeaux No commune wine is made in this village. Production is almost entirely from its two *grand cru* vineyards, Échezeaux and Grands-Echézeaux, which together cover 113 acres (45.7 hectares). The smaller one, Grands-Échezeaux, is regarded as the superior vineyard. In both vineyards there is more clay than in other important Côte de Nuits villages, a factor that gives more weight and density to the wines.

Vosne-Romanée This village produces red wine only. According to some wine experts, the Pinot Noir grape achieves its absolute pinnacle of quality in the *grand cru* wines from Vosne-Romanée. There are six *grands cru* vineyards, all of them famous and justly celebrated. These wines are among the world's most expensive. The *grand crus* are Romanée-Conti, La Romanée, La Tâche, Richebourg, Romanée-St.-Vivant, and La Grande Rue. In 1650, a vineyard formerly known as Le Cloux was renamed Romanée, because Roman artifacts had been found nearby. When the vineyard was purchased in 1760 by the Prince of Conti, it was given the name Romanée-Conti. Lying on the slope right above it is La Romanée (Figure 5–23). In both vineyards, the soil is primarily calcareous with up to 45 percent clay. The resulting wines have a superb balance of concentration and refinement. Fewer than 1,000 cases of wine from these two legendary vineyards are made each year.

In a pattern unusual for Burgundy, several of Vosne's *grand crus* have only one owner. Romanée-Conti is entirely owned by Domaine de la Romanée-Conti, and La Romanée by the Liger Belair family. La Tâche (15 acres/6.1 hectares) is also entirely owned by Domaine de la Romanée-Conti. The tiny vineyard of La Grande Rue (3.5 acres/1.4 hectares) is owned by Domaine Lamarche. More typically, the other two *grands crus* each have several owners, six in the case of Romanée-St.-Vivant and 10 in the case of the even smaller (22 acres/8.9 hectares) Richebourg.

Nuits-St.-Georges In the industrialized town of Nuits-St.-Georges there are no *grand cru* vineyards, but there are many fine wines made at the *premier cru* and village levels of quality. There are an impressive 28 *premier crus* within the boundaries of the town, plus an additional 13 in the adjoining town that are entitled to use the Nuits-St.-Georges appellation. Many of the *premier crus* lie north of the town, near Vosne. These vineyards cover a wide swath, north to south, on top of a steep slope. The soil here is primarily iron-rich limestone, like that of the Vosne *grand cru* vineyards. The best of these *premier crus* produce elegant, subtle wines similar in character to their *grand cru* neighbors in Vosne. South of the town of Nuits-St.-Georges, the

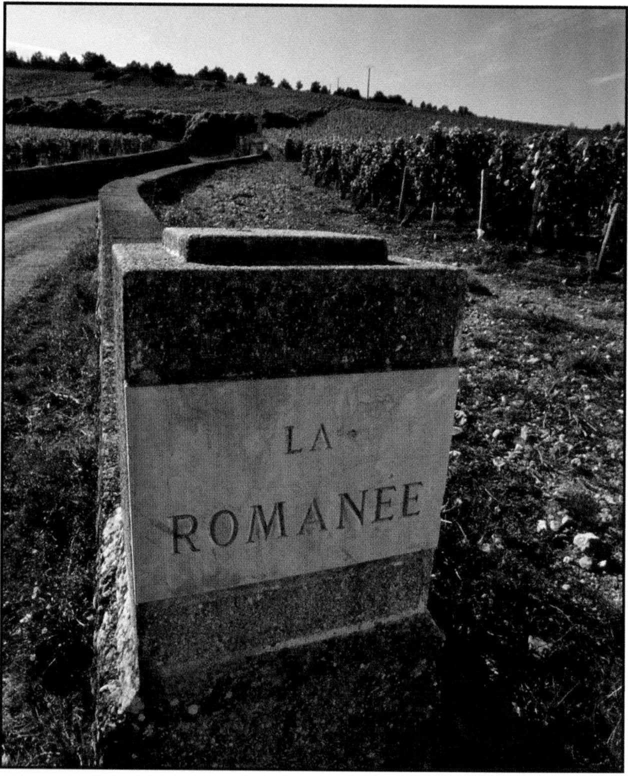

FIGURE 5–23 This stone pillar marks the corner of the world famous La Romanée vineyard outside the town of Vosne-Romanée. The name derives from "Roman" because it is believed the Romans may have first planted grapes on the site.

vineyards are planted on gentler slopes, where the soil tends to have a higher clay content. The wines from these vineyards are a bit fuller bodied and more concentrated.

Côte de Beaune Since the Côte de Beaune is so famous for its elegant, complex whites—with one exception, all the Grand Cru vineyards are white—it is often forgotten that three-quarters of the production here is red wine. These reds are not as full, velvety, refined, and multifaceted, nor as long-lived, as those of the Côte de Nuits. However, Côte de Beaune reds can be very appealing, with vibrant fruit, silky texture, and quiet refinement. The Côte de Beaune is a large region, more than twice the size of the Côte de Nuits, stretching some 71 miles (114.5 km) from north to south. The hills here have gentler slopes and face primarily southeast.

Côte de Beaune-Villages In the French appellation system, the word *villages* affixed to a regional appellation indicates that the vineyards are of higher quality than those in surrounding areas. Thus, Côte de Beaune-Villages is a more distinctive appellation than Côte de Beaune. In fact, only 74 acres (30 hectares) of vineyards are designated Côte de Beaune, and the wine is not particularly good. Sixteen villages are included in the Côte de Beaune appellation. Many of these villages are of high enough quality that they also have their own appellation, but when grapes from two or more villages are blended together, the only appellation allowed is Côte de Beaune-Villages. Many of these wines, both red and white, represent very good value.

Aloxe-Corton This commune produces primarily reds, but some superb whites are made here. It is home to Burgundy's largest *grand cru* vineyard, Corton. This famous vineyard, which is spread over three communes, is close to 400 acres (162 hectares) in size. Its ownership is spread among many entities, including the *négoçiant* firm, Louis Latour. To further complicate matters, this vineyard produces both *grand cru* red, usually labeled as Le Corton, and *grand cru* white, which is labeled as Corton.

The Pinot Noir in the Corton vineyard is planted on heavier soil with a preponderance of clay. The resulting red wines seem to have the heft of Côte de Nuits reds combined with the grace and vibrancy of Côte de Beaune reds. The Chardonnay grapes are planted in higher sections of the Corton vineyard and its adjacent *grand cru,* Corton-Charlemagne, where the soil is lighter and finer, full of chalk and pebbles. These whites have wonderfully appealing aromas of almonds, fruits, and delicate flowers, and are rich and weighty on the palate.

There is a minuscule amount of *premier cru* Aloxe produced from the 100 acres (40.5 hectares) so classified. An insignificant amount of commune-level Aloxe is also produced.

Savigny-lès-Beaune The commune of Savigny is one of the larger villages in the Côte de Beaune. It has nineteen *premier cru* vineyards and produces mostly reds that are fruity and charming.

Beaune The city of Beaune is the center of the Burgundy wine trade. It is an ancient walled city with the distinct feel of medieval times, but it is bustling with commerce and wine-oriented activities. The appellation to which the city lends its name is a large one, with 13,300 acres (5,263 hectares) of vines. Over 90 percent of production is red wines with pleasing bouquets of cherry, medium-bodied with lively fruit. There are 42 *premier cru* vineyards, covering almost 800 acres (324 hectares), but there are no *grand crus.*

The soil content among these many *premier crus* is complicated and varied. The soil structure is based on limestone, but to the north, near Savigny, the soil is very thin and the vine roots have to reach deep for nutrients. The result is wines that are intensely flavored and concentrated, with good structure. Among the best of these vineyards north of the city are Marconnet and

Bressandes. As one moves into the middle section of *premier crus,* to the west of the city, the soil becomes more gravelly and less thin. The wines from these vineyards are riper, rounder, and more succulent. The best-known vineyard here is appropriately named Grèves, meaning gravel. To the south of the city, the soil has less gravel and more clay. The wines are fuller-bodied, softer and fruitier, and quicker to mature. The best of these vineyards are Clos de Mouche (which produces both whites and reds) and Chouacheux.

Pommard Pommard and its neighbor to the south, Volnay, are home to the best reds in the Côte de Beaune after those from the *grand cru,* Le Corton. The name Pommard comes from the French word for apple, but there are almost no apple orchards left. All available agricultural land has long been planted to wine grapes. There are 832 acres (337 hectares) of vines, of which 275 acres (111 hectares), or about one-third, are rated as *premier cru.* The reds of Pommard are among Burgundy's most popular because they are easy to drink, silky and full of fruit, and show lively acidity. They can be enjoyed while still quite young. Among the best of Pommard's *premier crus* are Épenots and Rugiens.

Volnay The reds of Volnay are justly famous for their charm and seductive fruit. Some writers refer to Volnays as among the world's most feminine red wines. A perfect illustration of the overall quality of Volnay's vineyards is that, of its 527 acres (213 hectares), over half (284 acres/115 hectares) are rated at the *premier cru* level. Like Pommard, Volnay is well balanced and approachable when young, full of vibrant, berrylike fruit. Of the 34 *premier crus* in Volnay, the best known are Clos de la Bousse d'Or and Les Caillerets.

St. Aubin This often overlooked village lies to the west of Chassagne and Puligny, up behind a little hill. It is gaining quite a reputation for its wines, especially its appealing whites with their distinctive nutty, appley flavors. These wines are considerably less expensive than the whites of its prestigious neighboring villages to the east. The reds of St. Aubin are pleasant and full of pleasant cherry/berry fruit.

Santenay Santenay is the southernmost village of importance in the Côte de Beaune. It is a fairly large commune with 975 acres (395 hectares) under vine. In most vineyards the soil has considerable marl mixed in with the limestone. There are eleven *premier cru* vineyards, spread over 346 acres (140 hectares). Santenay's production is 90 percent red. Santenay reds are charming: medium bodied, fruity, and very pleasant, albeit a little more rustic than Volnay. Joseph Drouhin, the *négoçiant,* makes a particularly appealing Santenay.

Côte Chalonnaise

The southern edge of the Côte de Beaune marks the end of Burgundy's prestigious appellations, with its world-class wines. That does not mean, however, that there are no more wines worth seeking out. Many excellent wines are produced in the southern regions of Burgundy. Immediately south of the Côte de Beaune, the region of Côtes Chalonnaise begins. The region is named for the town of Châlone on the Saône River. The vineyards are planted on hillsides a little east of where the vineyards of the Côte de Beaune end. The soil on these hills is similar to the soils of the southern communes of the Côte de Beaune—a mixture of gravel and marl on limestone. The wines of the Chalonnaise lack the elegance, depth, and longevity of those from the Côte d'Or, but they can be charming, balanced, and appealing. These wines are also excellent values.

In the Côte Chalonnaise, there are four commune appellations of particular importance. Moving from north to south these villages are Rully, Mercurey, Givry, and Montagny.

Rully The village of Rully produces approximately equal quantities of red and white wines. Its vineyards are also a good source for the sparkling wine of the Burgundy region, *Crémant de Bourgogne.* There are 19 vineyards rated as *premier cru.* The whites of Rully are fresh and clean with apple flavors, and can be drunk young. Several *négoçiants* are using oak aging to round out their Rully whites. One example is Joseph Drouhin, whose Rully Blanc is particu-larly well made. Rully reds are medium-bodied and have strawberry aromas and pleasant fruit.

Mercurey The vineyards of Mercurey begin just south of Rully. There is more clay and iron in the soil here, and the resulting Pinot Noirs are fuller, rounder, and have more cherry/berry nuances than Rully reds. Mercurey was granted *AOC* status in 1936 and has maintained its reputation for solid, attractive reds that are eminently affordable. They age well and should be allowed to mature before being drunk. Four to eight years after the vintage is recommended. Some white wine is also produced in Mercurey.

Givry This small village, which produces mostly red wine, lies a few miles west of the town of Châlone. The soil here begins to shift away from the clay and marl which sits atop the lime-stone base of the Cote d'Or. At Givry a sandy, lighter limestone is evident. The Pinot Noirs are simpler than those of Mercurey, but have more depth than the reds of Rully. There are 540 acres (219 hectares) of vineyards in the commune of Givry, one-sixth of which were recently designated as *premier cru.*

Montagny All wines from Montagny are white. The wines are light to medium bodied, have crisp acidity, and are made to be drunk young. The designation *premier cru* on a Montagny does not indicate a superior vineyard site, but rather is assurance that the wine has attained an alcohol level of at least 11.5 percent.

Mâconnais

The Mâconnais region, surrounding the small town of Mâcon, marks the transition, climatically and geologically, from northern to southern France. Although winters can be very cold and spring chilly enough that frost can be a concern, the summers are sunny and balmy. The lime-stone base of farther north is still present, but the topsoil is more sandy and less chalky, with patches of granite. The majority of Mâconnais wines are white. They are primarily Chardonnay, but another grape, Aligoté, is also allowed. Red wines, which represent only about 15 percent of production, are made from Pinot Noir or Gamay. Gamay-based red wines are fruitier, softer, and lighter bodied than Pinot Noirs. The vineyards all lie to the west of the Saône River, which flows south past the town of Mâcon and through the Beaujolais region.

Mâcon-Villages Grapes grown in any of 43 villages can be blended together to make Mâcon-Villages. If a wine is made exclusively from grapes grown in one village, the label can show that village's name (e.g., Mâcon-Viré or Mâcon-Lugny). Mâcon-Villages is a pleasant, light wine, perfect for everyday consumption. Several *négoçiants* make nice Mâcon-Villages, including Louis Jadot, Joseph Drouhin, and Georges DuBoeuf.

Pouilly-Fuissé In the United States, Pouilly-Fuissé is perhaps the most recognized of any Burgundy wine (Figure 5–24). It is now ubiquitous on wine lists and in liquor stores, probably because it is made in large quantities, is affordable, and can be quite distinctive with its nutty aromas, fresh apple/lemon flavors, and good acidity. There are 2,100 acres (850 hectares) in this appellation, and they are planted exclusively to Chardonnay. There are no officially rated *premier cru* vineyards, although a producer can list a specific vineyard on the label if desired. One of the most renowned producers of Pouilly-Fuissé is Château de Fuissé, a *propriétaire* label.

Map of The Mâconnais

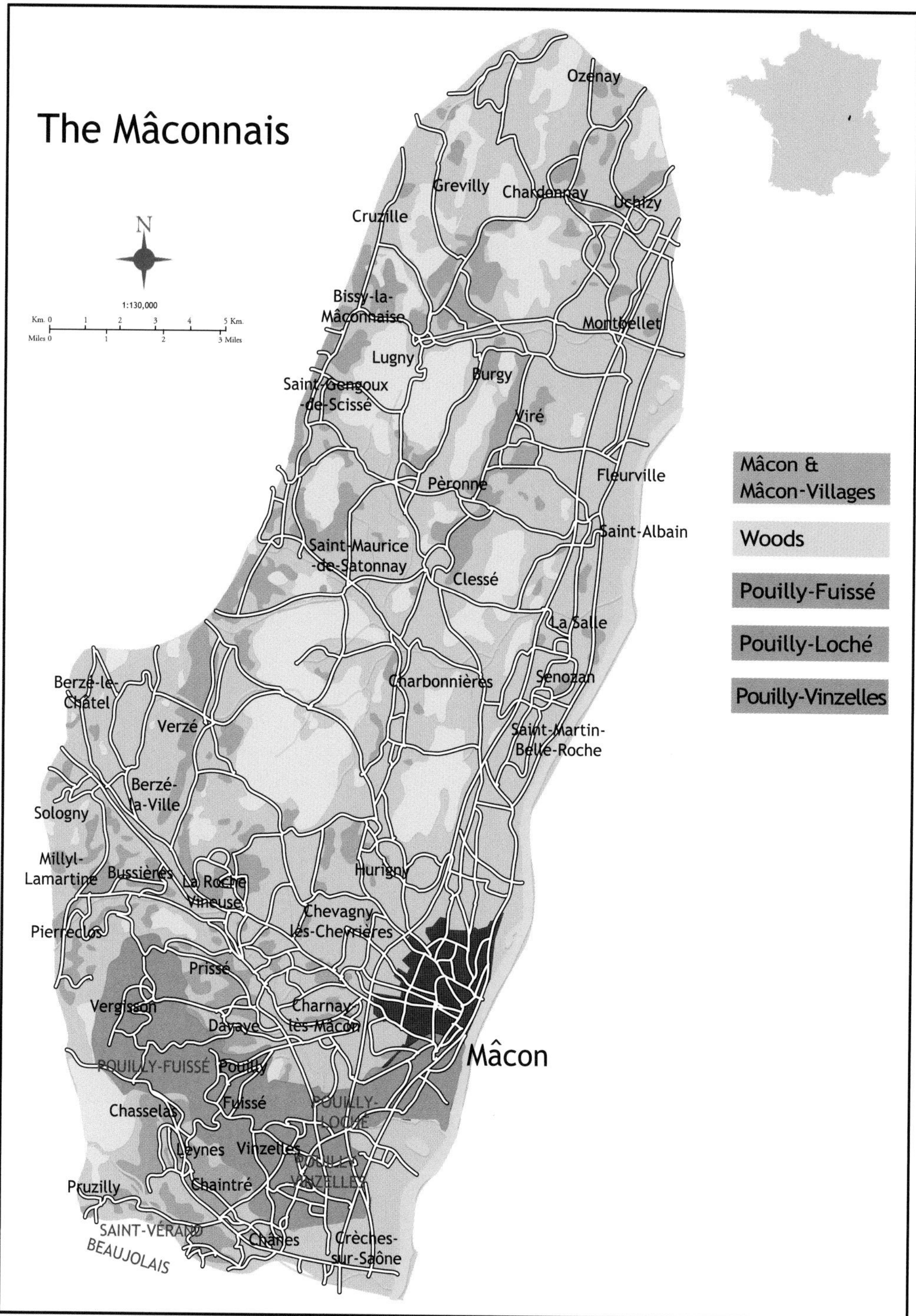

At the town of Pouilly, the limestone plateau on which all Burgundy vineyards to the north are located starts to level out and disappear. From here on to the south, the base is granite rock, with topsoil of chalk or clay.

St. Véran The appellation St. Véran was approved in 1971. It was carved out of areas that previously made Mâcon-Villages or Beaujolais Blanc. The soil is chalky. The wine, all white, is light and crisp and best consumed young. St. Véran is very affordable.

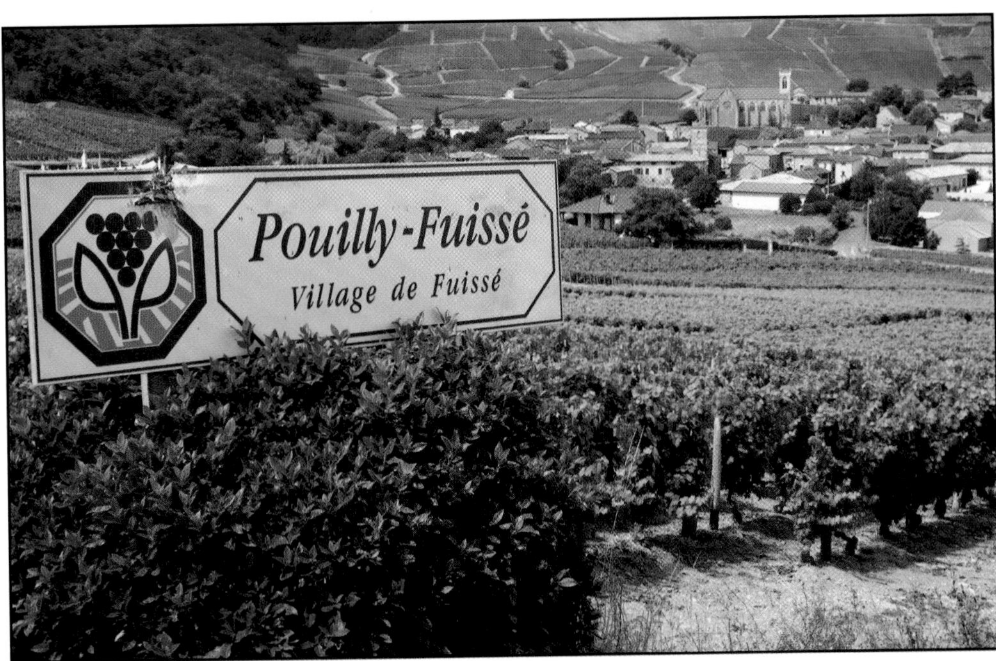

FIGURE 5–24 A Chardonnay vineyard looks down on the town of Fuissé in the Mâconnais. The wines produced here are among the best-known French wines in the United States.

Beaujolais

Beaujolais is classified as part of Burgundy, although the climate and soil are different, it is in a different *département* than the rest of Burgundy, and the primary grape is Gamay, not Pinot Noir. The style of the wine is entirely different. However, this large region is treated as a subdistrict of Burgundy.

Beaujolais is one of the most popular red wines in many countries around the world. One reason, of course, is that Gamay makes easy-to-drink reds with cherry/raspberry fruitiness, soft tannins, and light body. Another is that it is widely available—12.5 million cases of Beaujolais are produced annually from 49,540 acres (20,057 hectares) of vines. Also helpful is the annual widely publicized release of Beaujolais Nouveau (Figure 5–25), a very light, simple wine that, by tradition, is released by mid-November. Since the wine is only a few weeks old at the time of release, it is termed *nouveau,* or new. As much as 50 percent of a Beaujolais producer's wine is released as Beaujolais Nouveau, which is helpful to the producer from a cash-flow viewpoint. Most producers export as much as one-half of their nouveau.

The portion of any vintage year's wine that is not sold as nouveau is released starting the next spring. In ascending order of quality, Beaujolais is classified as *AOC Beaujolais, Beaujolais Supérieur* (higher minimum alcohol content; rarely exported), *Beaujolais-Villages,* and *Cru Beaujolais* (from specific villages whose vineyards have been judged to be better than those in surrounding areas).

The Winemaking Process in Beaujolais As stated earlier, Gamay grapes make soft, easy-to-drink wines with lots of fresh fruit flavors. However, Beaujolais is even softer and fruiter than Gamay-based wines made elsewhere because of a unique winemaking process employed in this region, **carbonic maceration.** With this method, whole bunches of uncrushed grapes are placed in a container from which oxygen has been removed by pumping in carbon dioxide. A mini-fermentation takes place inside each berry, producing very small quantities of alcohol while also reducing the tart malic acid and releasing aromatic flavorful compounds. Eventually the weight of the bunches at the top of the fermentation tank will crush the grapes on the bottom, thus releasing their juices. Normal fermentation will com-

Map of The Beaujolais

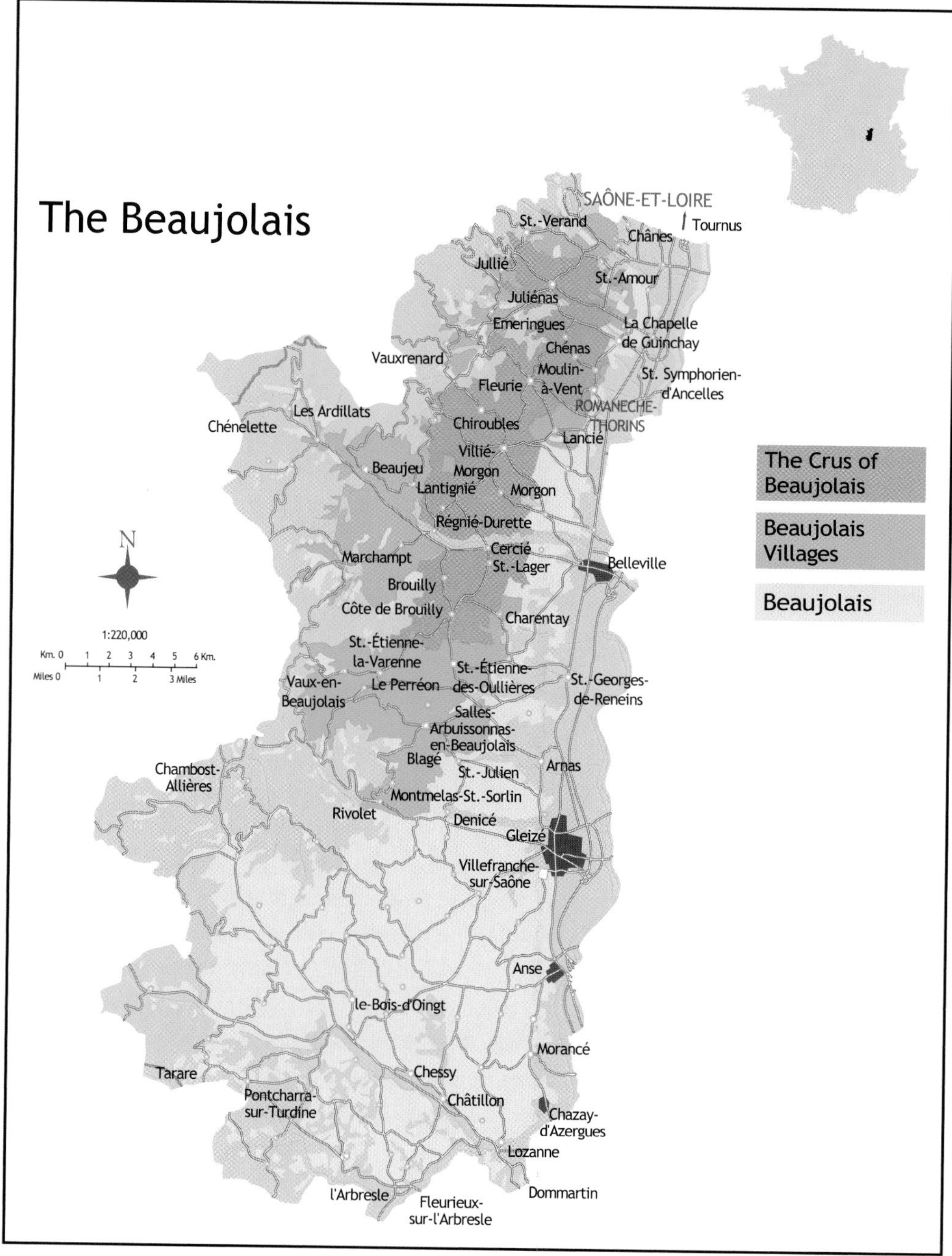

mence at that level, while carbonic maceration continues at the upper levels. After a short time, the winemaker will allow oxygen into the tank, and the entire batch will complete the sugar-to-alcohol fermentation. The portion that went through carbonic maceration allows the whole batch to be softer and more vividly fruity than would have otherwise been the case. The Beaujolais (other than nouveau) is then allowed to age in oak barrels for anywhere from a few weeks to several months before being released for sale.

Cru Beaujolais In the Beaujolais region there are ten villages where conditions are judged to be ideal for Gamay grapes. The wine from these towns is deeper in color, has more weight and substance, and is longer-lived than wine at the other levels of Beaujolais. The soil in these ten towns is granite-schist based with topsoil containing varying amounts of sand, clay, and chalk. The wines from these top vineyards have bouquets of flowers and fresh berries. The color

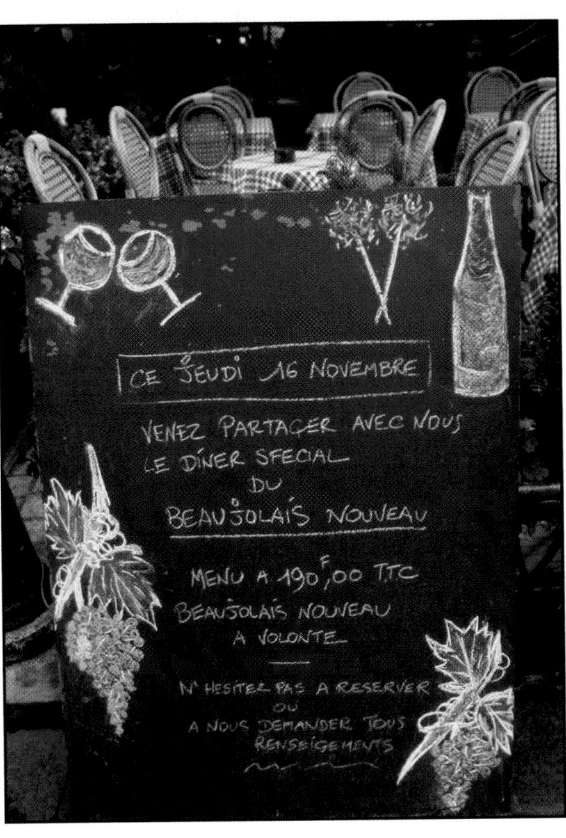

FIGURE 5–25 The release of Beaujolais Nouveau wine is eagerly anticipated each fall.

is cherry red with hints of violet. They are fuller bodied and deeper in flavor than other Beaujolais, but still much lighter than Pinot Noir–based wines. A Cru Beaujolais shows the name of the village on the label.

The northernmost village designated as Cru Beaujolais is St. Amour, only a few miles south of St. Véran in the Mâconnais. Its wines are light and pretty. Further south and to the west is Juliénas, whose wines are often overlooked, but which can be very good, with strong hints of raspberry in the nose and on the palate. Just south of Juliénas is Chénas. Because the soils of these two villages is almost identical, their wines are quite similar in character, with Chénas wines having slightly less weight. Next are the vineyards of Moulin-à-Vent, named for the large ancient windmill nearby (Figure 5–26). The Moulin-à-Vent wines are the most

FIGURE 5–26 A Cru Beaujolais label from the commune of Moulin-à-Vent. Notice that the label mentions only the town.

impressive, fuller, more complex, and able to take a surprising amount of aging. They also tend to be the most expensive of the Crus. The next villages south are Fleurie and Morgon, two villages very close to each other, and with similar wines—very fruity and nicely structured. Lying a few miles to the west are the villages of Chiroubles and Régnié. The latter is the newest Cru, having been elevated to that designation only in 1988. South of Régnié is the village of Brouilly, the largest of the Crus. Its wines are the lightest in body and fairly simple. Slightly to the east, on the slope of Mount Brouilly, are the vineyards of Côte de Brouilly. This is the southernmost of the Crus. The wine from these vineyards has pleasant flowery aromas and a bit more heft than Brouilly.

A comparison of the wines of Bordeaux and Burgundy can be seen in Table 5–5.

Côtes du Rhône

The region along the Rhône River in southern France is an ancient wine-producing area (Figure 5–27). It is believed that vines were planted here as early as 600 BC by Greeks. The valley of the Rhône extends from the city of Lyon in the north where it is joined by the River Saône, and extends south for approximately 120 miles (194 km) to the city of Avignon. For much of the length of the river the valley is bursting with commercial activity and is heavily industrialized. The areas down by the river banks are not promising for growing quality grapes. However, if one climbs up the slopes (the *côtes*) on either side of the river, the topography changes drastically. Along the northern section one discovers open rolling agricultural land, while in southern sections, an ascent of the slopes reveals rugged, dry quiet open spaces of scrub oak and heather. This is where the vineyards are located.

Table 5–5 Comparison of Bordeaux and Burgundy

	BORDEAUX	BURGUNDY
Size	Very large; 304,000 acres	Small; 98,000 acres
Climate	Warm and dry, good for early-ripening grapes	Cool and moist, perfect for late-maturing grapes
Viticulture	Red: Cabernet Sauvignon, Merlot, plus three others	Pinot Noir
	White: Sauvignon Blanc, Semillon	Chardonnay
Winemaking	All wines are blends	No blending
Landownership	Large, discrete, self-sufficient properties	Many small holdings. Few proprietary bottlings
Classification	General appellation	General appellation
	Regional, e.g., Graves	Regional, e.g., Côte de Beaune
	Commune, e.g., Margaux	Commune, e.g., Volnay
	Single estates, non-rated	Rated vineyards: *premier cru*
	Classified estates	*grand cru*
Styles	Whites: Very dry, crisp, herbaceous bouquet	Whites: elegant, complex, dry, nose of nuts/butterscotch/fruit
	Reds: Full-bodied, tannic, very complex; nose of cedar/coffee/blackberry	Reds: Medium-bodied, elegant, earthy nose, strawberry/cherry flavors
Bottle Shape	Shoulders	Sloping sides

FIGURE 5–27 The Rhône river flows for 120 miles through countryside and small towns. The *côtes* (slopes) along its length are often covered with grape vines.

The History of the Rhône Valley

Although introduced by the Greeks, viticulture did not take hold in the valley along the Rhône until many centuries later, in the early Christian period, when wine was exported to Rome. After the decline of the Roman Empire, winemaking essentially disappeared until the popes moved to Avignon, as discussed earlier in this chapter.

Historically, perhaps the most significant contribution to French wine production to emerge from the Côtes du Rhône is the work done by Baron LeRoy of Château Fortia on quality control laws. It was the baron who, after the First World War, led an orderly revolt against the desecration of the appellation Châteauneuf-du-Pape. The region was famous for its big, robust red wines, and the name, with its quaint history of popes and summer castles, was well known and memorable. The temptation to falsely label inferior wine as Châteauneuf-du-Pape in order to receive a higher price was hard to resist. This type of fraud was widespread when the baron assembled his fellow vintners. Under his expert guidance, the group set out to define their own boundaries and to set prescriptions on which grapes could be planted in Châteauneuf-du-Pape. They also decreed certain viticultural pratices that they knew worked well in their region. They outlined minimum ripeness at harvest and many other factors that were critical to quality and authenticity. The system of laws and regulations devised by Baron LeRoy's group eventually became the model for the national system of quality control laws adopted in 1936.

The Terroir of the Côtes du Rhône

Geographically and climatically it makes sense to separate the Rhône into two regions, the Northern Rhône and the Southern Rhône. The entire region is a warm, dry region whose climate is influenced by the Mediterranean Sea. But the North is definitely cooler, and the vineyards there cling to the stony soil of steep hillsides. The narrow northern section extends from Lyon to the village of Valence, a distance of about 45 miles (72.6 km). For the next several miles, there are no vineyards. The soil is not suited for wine grapes. The southern section begins south of the town of Montélimar and continues on south of Avignon, into the delta of the river. The climate is definitely Mediterranean, very warm and sunny and dry. The soil is more alluvial with a complex mixture of gravel, sand, clay, and limestone left as glaciers receded millions of years ago, and then moved and ground up and redeposited as the river changed course repeatedly over the centuries.

There is considerable variation among soil types and climatic patterns in the northern section of the Rhône Valley. In the northernmost sites, vineyards are protected from drying hot winds by tall ridges on the west side of the river. Cooler temperatures make for higher acidity and lower alcohol in the wines. The soil on these sheltering ridges is mostly granite. Further south, the prevailing winds are a little warmer, and the soil less stony, and more calcareous. From vineyard to vineyard, there are varying amounts of sand, clay, and chalk. The differences in terroir come through in the widely varying character of the fine wines made here. In the southern Rhône, there is even more variation among soil types from commune to commune.

The principle grape varietals of the Northern Rhône are Syrah for reds and Viognier for whites. The Syrah grape is one of the noble varietals, producing full-bodied wines famous for their deep color, tannic structure, and glorious aromas of blackberries, spice, and tar. Tight and austere when young, Syrah-based reds will open up to show warm, accessible flavors when mature. Viognier is considered by some to be one of the noble white varietals. The grapes are a deep yellow color, and the resulting wines are vivid in color and high in alcohol, possessing an intriguing bouquet of peach, almonds, and spring flowers. The wines are usually very dry and show excellent acidity.

The southern section of the Rhône valley is much larger than the northern one. The total acreage for the entire Rhône appellation is almost 150,000 acres (60,728 hectares). Of that, only 5,900 acres (2,389 hectares) are in the nine communes and *crus* of the Northern Rhône. The rest is in the very large, highly varied region of the Southern Rhône. The variation in soil among the many communes and individual vineyard sites of the Southern Rhône is tremendous, and will be pointed out in the discussion of appellations below.

The vineyards of the Southern Rhône support a much more complex array of grape varietals. Whereas the wines of the Northern Rhône, both reds and whites, are mostly single-varietal, those of the Southern Rhône are blends of several varietals. In Châteauneuf-du-Pape, for instance, 13 different grapes are authorized. The principle red grape of the southern appellations is the Grenache, a noble varietal that thrives in warm, sunny climates. It produces soft, mellow, round reds with succulent ripe plum flavors and a distinct aroma of fresh-ground black pepper. Other varietals used for blending in the Southern Rhône are Mourvèdre and Cinsault, both red grapes, and Marsanne and Roussanne, white grapes. Syrah is also often blended into Southern Rhône reds.

THE APPELLATIONS OF THE CÔTES DU RHÔNE

The appellations of the Rhône fall into three quality levels.

Côtes du Rhône: Almost 98,000 acres (39,676 hectares), scattered in peripheral sections of the Southern Rhône, are classified simply as Côtes du Rhône. With 7 million cases of generic Côtes du Rhône produced annually (the vast majority of it red), quality can vary widely. Some wines are heavy, dense, highly alcoholic, and even a bit rustic. This style is disappearing, however, as producers, most of whom are either *négoçiants* or local cooperatives of growers, improve their wine-making equipment and techniques and use lower percentages of the coarser local grapes in favor of higher quantities of noble varietals to turn out softer, smoother, classier wines of medium body, good balance, and nice black fruit. The newer style of wine labeled as plain Côtes du Rhône can be an excellent value.

Côtes du Rhône-Villages: The standards at this level are higher, and certain requirements must be met. Most importantly, the yield of grapes per acre must be lower, and the minimum alcohol content is higher. There are 30 villages included within this appellation, sixteen of which are authorized to add their village name to the label.

Commune: The best wines from throughout the Rhône valley carry the name of the commune or village where the vineyards are located. Twenty-five percent of the Rhône's wines are labeled by commune. In some cases, a specific vineyard will also be included. Although there is no system for rating vineyards as *premier* or *grand cru,* the very best vineyard sites, especially in the Northern Rhône, are well known, and can be shown on the label, thus adding considerably to the value of the bottles so labeled.

The Northern Rhône

Moving from north to south, the important communes of the Northern Rhône are Côte Rôtie, Condrieu, Château-Grillet, St. Joseph, Crozes-Hermitage, Hermitage, Cornas, and St. Péray (Table 5–6).

Côte Rôtie This commune's name translates as "the roasted slope," an apt name as the vineyards here receive excellent sunshine from their southeastern exposure on the steep ridges. The combination of hours a day of sun, stony minerally soil, and moderate temperatures results in the great red wines of unusual power and finesse for which the Côte Rôtie is justly famous. The world has discovered these wines, and they are not as affordable as they were 15 or 20 years ago.

The 320 acres (130 hectares) of vineyards in the Côte Rôtie are planted primarily to Syrah, with some Viognier also showing up. As much as 20 percent Viognier is allowed to be added to the red wines for aromatics and delicacy. However, most producers add less than that. The wines are typically medium to full bodied, deep in color, redolent of blackberries and currants, and incredibly long-lived. The best of these big red wines will hold their own for 20 years or more.

The best-known producer in the Côte Rôtie is the Guigal company, now headed by Marcel Guigal (Figure 5–28), the son of the founder, Étienne. Two of the very best vineyards in the region are owned by the Guigals, La Landonne and La Mouline. Some feel that the high quality of the Guigal wines and the great reviews writers have given these wines, coupled with the ener-

Table 5–6 Appellations of France: Côtes du Rhône

REGION	SUBREGION	PRINCIPAL VARIETAL
Northern Rhône	Côte Rôtie	Syrah
	Condrieu	Viognier
	Hermitage	Syrah
	Crozes-Hermitage	Syrah
	St. Joseph	Syrah, Viognier
	Cornas	Syrah
Southern Rhône	Coteaux du Tricastin	Grenache, Syrah, Mourvèdre
	Châteauneuf-du-Pape	Grenache, Syrah, Cinsault, Mourvèdre, Marsanne, 5 others
	Vacqueryas	Grenache, Cinsault
	Gigondas	Grenache (red and rosé)
	Tavel	Grenache, Cinsault (rosé only)
	Beaumes-de-Venise	Muscat (sweet only)
	Côtes du Ventoux	Grenache, Syrah, Mourvèdre
	Côtes du Lubéron	Grenache, Syrah (reds), Ugni Blanc, Marsanne (whites), Grenache (rosé)

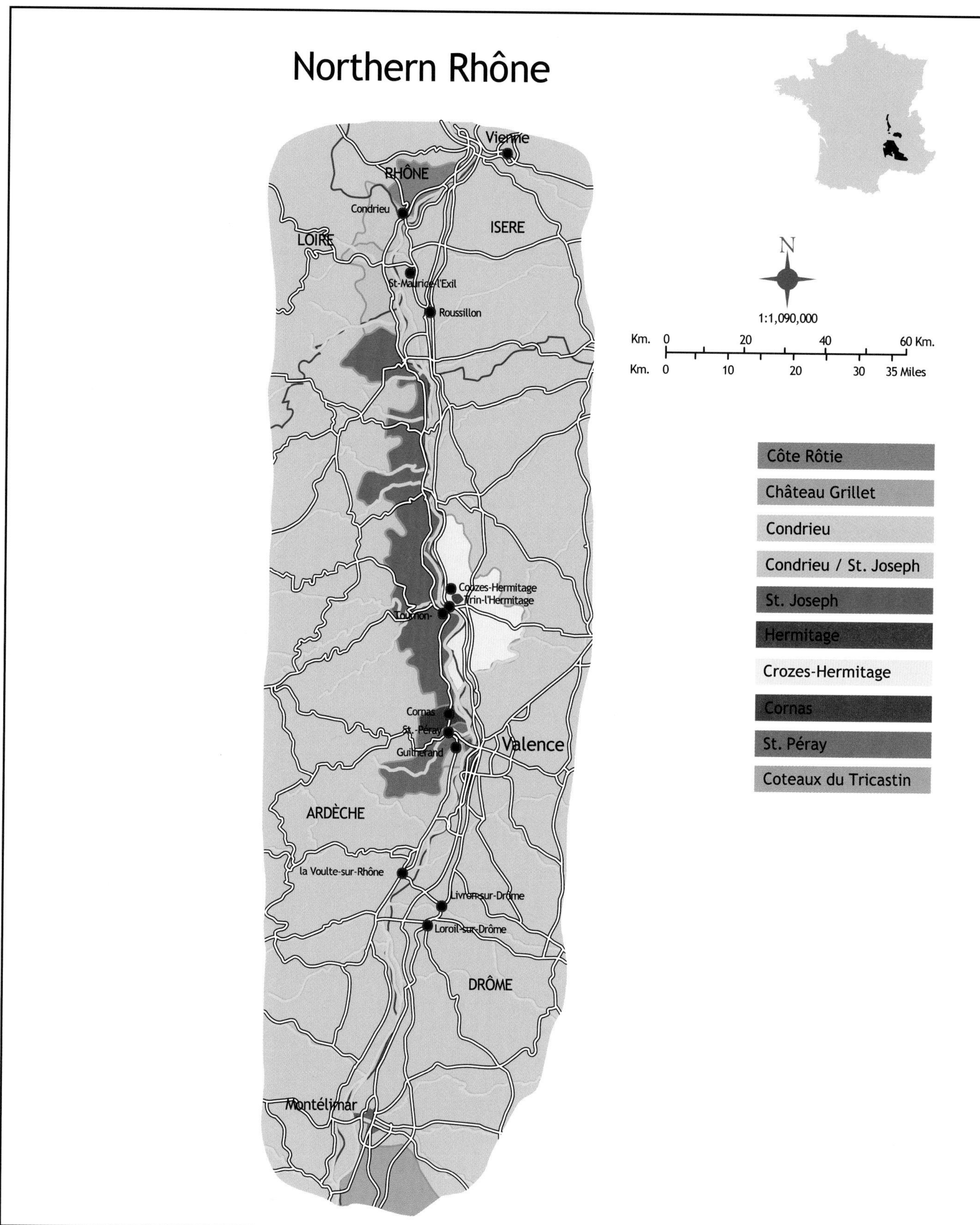

Map of Northern Rhône

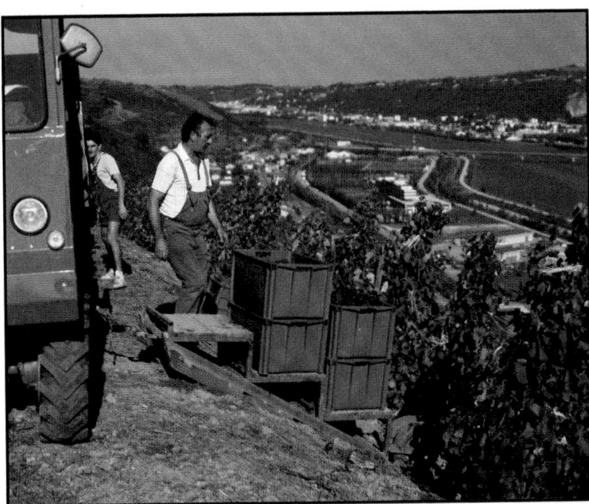

FIGURE 5–28 Harvesting grapes in one of the vineyards in the Côte Rôtie owned by the famed Guigal family. The workers need a winch to haul boxes full of grapes up the steep hill.

getic and competent marketing efforts of Étienne and his son, were the impetus behind the international rediscovery of the Côte Rôtie region in the past two decades.

Condrieu The elegant, flinty, and highly aromatic Condrieu whites have become some of the world's most expensive dry white wines. This is as much because of scarcity as quality. The Condrieu appellation is small, only 200 acres (81 hectares) of Viognier vineyards clinging to the west bank of the Rhône River. The ripe peach, melon, and honey flavors are held together by good acidity. However, there is not adequate acidity to allow the wines to age well. Therefore, Condrieu whites are best consumed between 18 months and 4 years after the vintage.

Château-Grillet Château-Grillet is France's smallest appellation (fewer than 10 acres/ 4 hectares) and one of the very few single-owner appellations (the Neyret-Gachet family has owned Grillet since its inception). Château-Grillet is an enclave within the Condrieu appellation. Château-Grillet's vineyard is a bowl carved out of a granite slope and carefully terraced to afford the Viognier vines a foothold. Château-Grillet is somewhat less perfumed and a bit more acidic than Condrieu, and thus can age longer.

St. Joseph

This is a fairly large appellation, stretching 40 miles (64.5 km) along the west bank of the Rhône. The production of St. Joseph is 80 percent red wines. These reds, made from Syrah grapes, are medium bodied and full of berry and red currant flavors. Their acid and tannic structure is not adequate to allow long-term aging; the wines are at their best within 3 to 6 years of vintage. The *propriétaire* Andeol Salavert makes a particularly warm and flavorful St. Joseph.

Crozes-Hermitage

This is the largest appellation in the Northern Rhône, with just over 3,000 acres (1,215 hectares) of vines. Annual production is over 400,000 cases of wine, 90 percent of which is red. These wines are made from Syrah grapes to which two white varietals, Marsanne and Roussanne, are added to make the wine more delicate. There are 11 communes within this appellation. (The name is from one of these villages, Crozes.) Crozes-Hermitage could be called the poor man's Hermitage, for it is more affordable than its prestigious neighbor. Although lighter than many Syrahs, Crozes-Hermitage can be quite good, with ripe berry fruit mingled with peppery spiciness. Given 5 years to mature, a Crozes-Hermitage can be a fine accompaniment to a full-flavored dinner.

HERMITAGE

The story behind the name Hermitage is another example of a predilection for folklore, or at best, a tendency to embellish history. Apparently, a French knight who went to fight in the Middle East during the Crusades came home to the south of France in 1220 feeling so guilty about the ravages imposed on the conquered people and their countries that he vowed to spend the rest of his life as a hermit, praying for redemption. He bought some land on the east slopes of the Rhône, built a small chapel (Figure 5–29), and prayed devoutly. Eventually, the hermit wanted wine to drink, so he planted grapevines and began to make very good wine. The peasant families nearby followed suit, and the region around this hermit's chapel eventually became known as Hermitage. In fact, one of the region's better vineyards, now owned by the *négoçiant* Paul Jaboulet is named La Chapelle (the chapel).

Regardless of the truth of the legend, it is an indisputable fact that the wine from Hermitage is superb. The hermit chose his site well, for the hill on which he planted is solid granite covered with a thin but complex layer of chalk and decomposed flint. To some connoisseurs, Hermitage is as good, as elegant, as complex and noble as the best reds from Bordeaux or Burgundy. In Hermitage, Syrah reaches its pinnacle of quality. Hermitage is deeply colored, beautifully structured, and full bodied. The nose is redolent of blackberries and raspberries, and on the palate the multiple layers of flavor open. Although the wine is a big, muscular wine, the impression is one of finesse and mystery. Able to age incredibly well, the best Hermitage is not really mature and fully accessible for 20 years or more.

CORNAS

Located south of St. Joseph on the west banks of the Rhône, Cornas has 173 acres (70 hectares) of Syrah planted on steep granite slopes. The hills give shelter from the cold north winds while affording the vines plenty of sun from the southeastern exposure. The terraced vineyards have a top

FIGURE 5–29 La Chapelle Vineyard in the Hermitage appellation is named for the small stone chapel built by the hermit who lived here after returning from the Crusades. This site is now famous for the quality of its Syrah grapes.

layer of limestone. The best Cornas is deeply flavored, full bodied, and harmonious. It can hold its own against the wines from the more illustrious commune across the river, Hermitage, and is far more affordable.

The Southern Côtes du Rhône

Whereas the communes and *crus* of the Northern Rhône are compact and dense, the Southern Rhône's appellations spread out in a huge lopsided circle from the town of Montélimar in the north along the Rhône and its tributaries all the way south to the city of Avignon, with a long bulging arm reaching east and south along the Durance River. In this enormous region of almost 100,000 acres (40,485 hectares) of vineyards, there is tremendous variation in terroir and styles of wine. Approximately 85 percent of the wine made here is red. About 5 percent is dry white. There is also some very good rosé made, and very small quantities of fortified dessert wine. From north to south, the most important appellations of the Southern Rhône are Coteaux de Tricastan, Gigondas, Muscat Beaumes-de-Venise, Vacqueyras, Châteauneuf-du-Pape, Tavel, Côtes du Ventoux, and Côtes du Lubéron.

Coteaux de Tricastan

This region on the eastern fringe of the Rhône valley produces pleasant reds and dry rosés. The reds are blended from the traditional grapes of Grenache, Syrah, Mourvèdre, and two lesser varietals. Several white varietals are also allowed to be added. The rosé is primarily Grenache with white grapes blended in. The climate is decidedly Mediterranean, but because of altitude and exposed terrain, the grapes do not ripen as much as in surrounding areas classified as Côtes du Rhône. Tricastan wines, therefore, are not as forward in their fruitiness, and are more acidic than generic Côtes-du-Rhône.

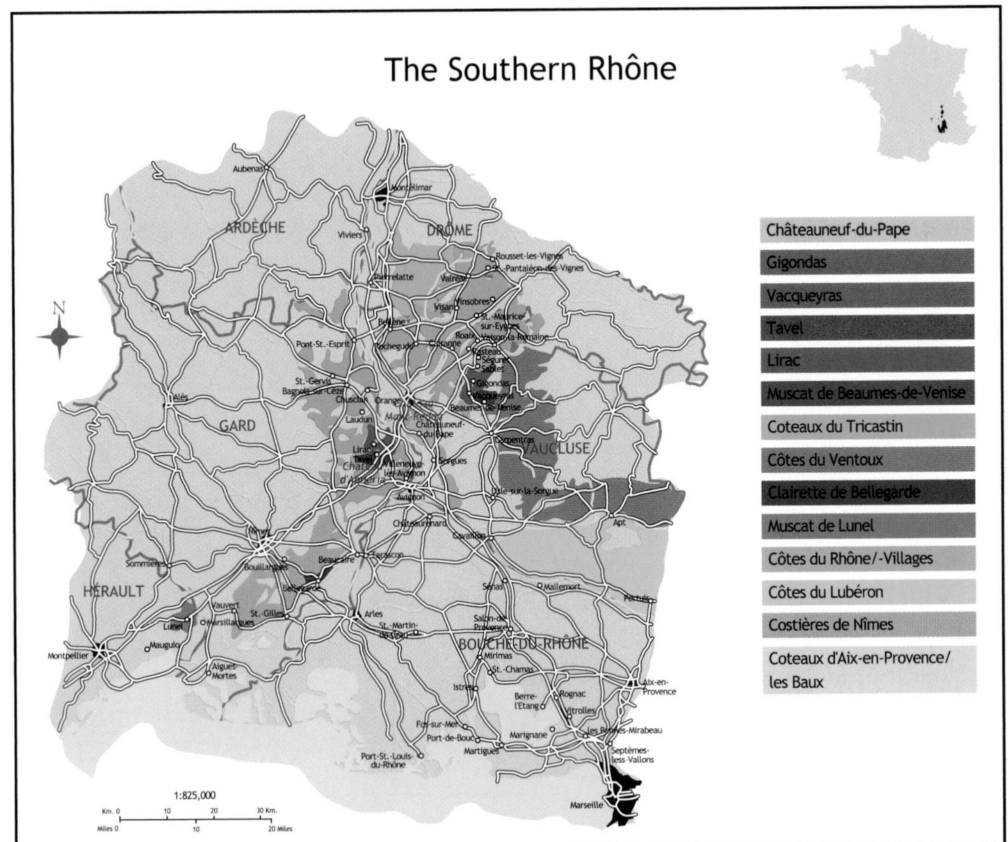

Map of The Southern Rhône

GIGONDAS

Located at the foot of the limestone-rich Dentelle mountains, Gigondas produces primarily red wines from its red clay soils. Despite having adopted some of the same stringent requirements as Châteauneuf-du-Pape, Gigondas spent several decades as part of the general Côtes du Rhône appellation, much to the commercial disadvantage of the district. In 1966, it was elevated to the Côtes du Rhône-Villages classification. Not until 1971 did Gigondas receive its own *AOC* designation. The regulations stipulate that Gigondas reds can be no more than 80 percent Grenache. The balance is Mourvèdre and/or Syrah. Rarely as distinguished a Châteauneuf-du-Pape, Gigondas can represent good value.

MUSCAT BEAUMES-DE-VENISE

The ***vin doux naturel*** Beaumes-de-Venise is one of France's prettiest dessert wines. It is made by arresting the fermentation of the Muscat grapes with neutral alcohol, which kills the yeast cells. The residual, or unfermented, sugars give the wine a delightful natural sweetness. The wine is medium bodied, honeyed, and fairly high in alcohol (15 percent minimum).

VACQUERYAS

Elevated to *AOC* status only in 1990, Vacqueryas makes reds, whites and rosé. Only a tiny portion of the 1,700 acres (688 hectares) is planted to white varietals. The reds are medium-bodied blends with pleasant berry, spice, and floral tones. At least half the blend must be Grenache. Vacqueryas resembles concentrated Côtes du Rhône Villages.

CHÂTEAUNEUF-DU-PAPE

The most celebrated of the Southern Rhône appellations, Châteauneuf-du-Pape has become increasingly popular in the past two decades. Although this is a large appellation—over a million cases each year—the standards are high. The vintners of Châteauneuf imposed on themselves stringent requirements including a very low yield per hectare, a high minimum alcohol content (12.5 percent), and the practice during harvest of the grape-sorting technique known as **triage,** by which at least 5 percent of the grapes must be rejected before fermentation begins. This way the substandard grapes, whether under-ripe, too ripe, or diseased, never find their way into the final batch. Triage is an expensive and labor-intensive process, but it is helpful in maintaining the highest quality.

The vineyards of Châteauneuf-du-Pape are on gently sloping hillsides that provide good drainage and exposure to the sun. The soil of reddish clay is covered with large pebbles that retain the day's heat (Figure 5–30) and reflect it back onto the grapes in the cool evenings, thus helping them to fully ripen. Only 3 percent of Châteauneuf-du-Pape is white, made from white Grenache, Roussanne, and several lesser grapes. The reds are immensely complex wines blended typically from 50 to 70 percent Grenache, 10 to 30 percent Syrah, up to 20 percent Mourvèdre and other reds, and up to 10 percent white varietals. These are deeply colored, full-bodied wines with incredible bouquets reminiscent of everything from blackberries, figs, cinnamon, and cloves to tar, coffee, and cedar. Many of the better wines can age for decades.

The producers of Châteauneuf-du-Pape are usually self-sufficient estates similar to Bordeaux's, in which the vineyards, wine-making facilities, and aging cave are all at one location and under one ownership. Each of the famous estates has its own recognizable style, some more fruit-driven, some austere and age-worthy. Among the best producers are Château Fortia, Château de Beaucastel, Chante-Perdrix, Château la Nerthe, and Domaine de Mont Redon.

TAVEL

The appellation Tavel is restricted to rosé made from Grenache (no more than 60 percent) and Cinsault (at least 15 percent) and several local grapes. The clay-based soil and warm climate

FIGURE 5–30 The Châteauneuf-du-Pape region has unique soil: red clay under large stone pebbles.

produce medium-bodied, flavorful rosés. The wine is surprisingly dry and fruity, and a versatile food wine.

Côtes du Ventoux

The large sprawling region of Côtes du Ventoux takes its name from the 6,500-foot (1,981-meter) high Mount Ventoux which towers over the area. The appellation contains over 18,000 acres (7,287 hectares) of vines, the best of which are on the flanks of the mountain. Elevated to *AOC* status in 1973, Côtes du Ventoux produces mostly red, plus some rosé, wines made from Grenache, Syrah, Mourvèdre, and Cinsault. A typical Côtes-du-Ventoux is a pleasant lighter wine intended for early consumption. One of the best examples available is the brand La Vielle Ferme which seems to have more extraction and more care in the making.

Côtes du Lubéron

Stretching from the southern boundary of the Côtes du Ventoux to the banks of the Durance River in the south, the Côtes du Lubéron appellation was created in 1988 and contains 7,410 acres (3,000 hectares) of vines. The wines are of all three colors. The reds, light and juicy, must contain some Syrah, blended with Grenache, Mourvèdre, and other traditional varietals. The whites, dry and clean, cannot contain more than 50 percent of Ugni Blanc (a lesser grape used to make brandy) and can have five other grapes blended in. The rosés are mostly Grenache and are soft, fruity, and pretty. Stylistically, the Côtes du Lubéron creates a bridge between the wines of the Rhône and those of Provence.

Champagne

No appellation in the history of wine has been more misused than the term *Champagne.* Champagne is not merely a type of wine. It is a geographic region in France, and only wine made in a specific method from specified grape varietals grown inside the boundaries of that region is technically Champagne. The Office of Champagne USA, a branch of the Comité Interprofessionel du Vin de Champagne, has devised, with the help of a New York agency, a very clever print advertising campaign to make the point that not all sparkling wine is Champagne. "Valencia oranges from

Maine?" asks one of the ads. "Alaska salmon from Florida?" The essential point the ads make is that Champagne is from Champagne. Sparkling wine made elsewhere is not Champagne.

The advertising campaign may be fun and playful, but it does make a solid point: Geographic factors, unique to specific locations, do affect the character of the products raised or captured there. In the case of the Champagne region, the differentiating characteristic are its unique soil, a mixture of clay and chalk, and its climate as one of the northernmost fine wine regions in the world. There is no terroir quite like Champagne's. So although very fine sparkling wines are made elsewhere in the world, they will not be quite like Champagne.

THE HISTORY OF CHAMPAGNE

Champagne was not always famous for its sparkling wine. Rather it started out in the time of the Roman Empire as a producer of still white wines most of which were consumed by Roman legions. After the decline of the empire, communications with other regions deteriorated, commerce was interrupted, vineyards were destroyed, and winemaking disappeared. As Christianity moved into northern Europe, winemaking re-emerged, receiving a huge boost when Clovis, King of the Franks, was converted to Christianity by the Bishop of Rheims, the major city of Champagne. The baptism of Clovis in AD 460 is said to have taken place where the magnificent Cathedral of Rheims now stands. With the support of the church, the vineyards of Champagne flourished, wine-making techniques were perfected, and markets for the wine were expanded. (Paris is only 90 miles, 145 km away.)

It was monks who rescued the vineyards of Champagne, and it was a monk who developed the style of wine for which the region has become famous (Figure 5–31). Dom Pérignon did not invent Champagne. What he did do was lend his viticultural genius to sorting out which grapes to plant and how best to tend them, and he perfected numerous wine-making procedures.

FIGURE 5–31 The monk Dom Pérignon is often cited as the inventor of Champagne. Although he did not actually invent this style of wine, he made crucial contributions to the viticulture of the Champagne region, and perfected the methods of producing the famous sparkling wine.

Pierre Pérignon was born in 1638 and, at a young age, entered a monastery near Épernay in Champagne, where he stayed until his death in 1715. During his long tenure as cellar master, Dom Pérignon greatly improved the quality of the wine made at the monastery. His greatest discovery was a process for capturing effervescence in the glass bottles by allowing a second sugar-to-alcohol fermentation to proceed. The by-product of fermentation is carbon dioxide, which remained trapped in the bottle. It can not be proven that Dom Pérignon was the very first to bottle an effervescent wine, but there is no doubt that he greatly improved the quality of wines from the Champagne region.

The sparkling wine of Champagne did not find immediate favor, but once it was discovered by the fun-loving aristocracy surrounding the royal court in the late eighteenth century, Champagne soon became the wine of celebration. New markets opened in continental Europe, the British Isles, and even as far away as Russia. Over the next 150 years, demand for Champagne increased at such a rate than demand could not keep up. Some producers started expanding production with inferior grapes brought in from other growing areas. Fraud became so widespread that the legitimate growers of the Champagne region revolted in 1911, demanding protection of their place name.

In 1927 the French government did implement laws spelling out the exact boundaries of the Champagne region. With the passage in 1935 of the national *Appellation d'Origin Contrôlée* laws, the Champagne name received full protection, as did the quality of the product. In 1941, even with the Second World War raging, the vintners of Champagne formed a trade association, the *Comité Interprofessionel du Vin de Champagne (CIVC),* to give further protection to the authenticity and prestige of the appellation. The *CIVC* remains active today, both in the Champagne region overseeing production and pricing, and in markets around the world, assisting with marketing and promotion.

Viticulture in Champagne

Three grapes are allowed in Champagne: Chardonnay, Pinot Meunier, and Pinot Noir. The latter two are red grapes, but the juice of these grapes is white. There are more than 72,000 acres (29,150 hectares) of vineyards, owned by 19,000 individual growers. Physical conditions determine which varietal to plant in each location. Legislation, based on centuries of close observation of weather and soil patterns by *vignerons,* now dictates what is to be planted in each site. For instance, in the Vallée de la Marne, the hardy Pinot Meunier is widely planted. Throughout the Champagne region, more acres are dedicated to this relatively easy-to-grow varietal than to Chardonnay or Pinot Noir. Near the Montagne de Marne, Pinot Noir predominates. The Côtes des Blancs, as the name implies, are exclusively for the growing of Chardonnay.

As stated earlier, the terroir of the Champagne region is truly unique. The most distinguishing characteristic of this environment is the high concentration of chalk in the soil (Figure 5–32). Essentially two types of chalk (which is a form of limestone) can be found in soil. One of these types, belemnite chalk, is found only in Champagne. The topsoil is a thin layer of clay with some chalk. The underlayer of chalk reaches as deep as 800 feet (244 meters).

The roots of the vines can push through the soft crumbling chalk to great depths in order to reach water. Chalk does an effective job of draining rainwater so that it does not pool near the surface, risking rotting of the vines. Moreover, as the water drains downward, the chalk forms natural storage chambers for it, where the deeper roots can reach. A further advantage of heavily chalky soil is heat retention. The warmth of the sun is absorbed by chalk, and radiated back onto the vines in the cooler evenings. Also, the poor nutritional content of chalk discourages the growth of leaves on the grapevines, thus allowing more sunlight to hit the grape bunches, which assists ripening. A meager canopy also allows better circulation by the breezes, which reduces the likelihood of mildew forming on the bunches.

FIGURE 5–32 This vineyard in the Côte des Blancs is typical for the depth and density of the chalk soil that is unique to the Champagne region.

Soil is only one part of terroir (albeit a critical one). The other determinant is, of course, climate. The climate of Champagne holds many perils for grape growers. Most obvious is cold. Champagne is farther north than any other important wine region, and the damp cold weather patterns of the North Atlantic (only 110 miles, 77 km away) are not blocked by any mountain range or other natural barrier. The average temperatures in Champagne are barely enough to allow grapes to ripen. Acidity levels stay high in such a cool climate, which is desirable in any sparkling wine. However, a minimum sugar level (set by law) must be reached, and if the temperatures stay too cool, the grapes have a difficult time reaching the necessary ripeness. Another constant concern for the *vignerons* is frost, either in the spring when the vines are budding or in the fall before the grapes are harvested. Growers in Champagne use the method called *aspersion* to protect their vines. This method, also used in other very cool wine regions, consists of spraying the vines with a fine mist of water which freezes on the buds or the grape bunches, forming a protective shield from the cold.

The weather in Champagne presents still another danger for wine grapes, and that is moisture. Grapes are susceptible to mildew if they remain damp for extended periods. With a high average annual rainfall of 25.6 inches (65 cm), Champagne is often very damp indeed. Fungicides are sometimes employed to decrease this danger. Other potential problems in years of heavy rainfall are swelling of the grapes with too much water and a reduced rate of pollination.

Champagne Producers and the Style of the Wine

There are approximately 110 companies, called houses (or, in French, *marques*), that make Champagne. Because these companies own only 10 percent of the vineyards in Champagne, they buy the vast majority of their grapes from growers. The oldest, most established houses are called *grands marques.* They are required by law to maintain a presence in export markets. Exports of Champagne play a vital role in the French economy, accounting for 25 percent of wine and spirits exports.

The various vineyards of Champagne have been officially rated, but these ratings do not show on a label. Rather, the rating of a vineyard was used to determine what price the houses would be required to pay for grapes from that vineyard. The most highly rated sites would act as benchmarks, and once their prices were set, other vineyards' grapes would be priced accordingly. The law was changed in 1990 so that even though prices are still officially determined according to the quality rating of a vineyard, the houses are not required to pay that amount.

Each of the major Champagne houses has a distinct style that it maintains year after year. Champagne is made from still wine that has been fermented dry. (See Chapter 3 for the

Winemaking Process.) After malolactic fermentation is complete, the wine is racked, and a careful blending of different batches and varietals *(assemblage)* is undertaken. The wine is put into thick glass bottles and a second sugar-to-alcohol fermentation is induced by adding the *liquer de triage,* a mixture of reserve wine, sugar, and cultured yeasts. As the second fermentation occurs, the secondary by-product of carbon dioxide is trapped in the bottle, giving the wine its effervescence. After the wine is aged, the yeast cells are removed from the bottle by a careful disgorging (Figure 5–33). The *dosage* of sugar and water is then quickly added to give the finished product the amount of sweetness its classification requires.

FIGURE 5–33 A cellar worker riddles bottles in preparation for the disgorging, which will remove all dead yeast cells from each bottle.

Classification of Champagne Styles

Extra Brut: Bone dry. Residual sugar is less than 0.6 percent per liter. At this level there is usually no dosage.

Brut: This is the most common classification, and forms the backbone of any house's line. Residual sugar is 0.5–1.5 percent per liter (Figure 5–34).

Extra Dry: These Champagnes are off-dry, with residual sugar 1.0-2 percent.

Sec: Although *sec* means "dry," these Champagnes have noticeable sugar—between 2 and 3.5 percent. They are rarely seen in the United States.

Demi-Sec: The literal translation is "off-dry" but these Champagnes are quite sweet. The dosage causes residual sugar to be between 3.5 and 5 percent. These Champagnes are meant to be served with dessert.

Doux: The sweetest form of Champagne has a minimum of 5.5 percent sugar, and in some cases contains as much as 8 percent.

FIGURE 5–34 Label for nonvintage Brut from the house of Taittinger.

Other important terms show up on Champagne labels. Most of these types will be made in a Brut style, even if that word does not show on the label.

- **Nonvintage:** A nonvintage wine is not made exclusively from grapes grown in one vintage year. Grapes from several different years are blended together to get consistency of quality, even in years when weather patterns are less than ideal. A portion of each year's wine is held back for this purpose (Figure 5–35).
- **Vintage:** When conditions are favorable, the winemaker can choose not to blend in wine reserved from lesser vintages. To be declared a vintage Champagne, the wine must contain at least 80 percent grapes from the declared year. Not every year is a vintage year. The winemaker at each house decides whether to declare a vintage.
- ***Blanc de blanc:*** Literally, "white from white" is a Champagne made exclusively from chardonnay grapes. Since only 25% of the vineyards in the region are planted to chardonnay, the grapes are expensive, and therefore, so is this type of wine. These are the most delicate and lightest of Champagnes.
- ***Blanc de Noir:*** Literally, "white from black." The wine is made exclusively from the two allowed red varietals. These are the fullest of Champagnes, with considerable fruits and complexity.
- ***Rosé:*** If some red wine is added to a **cuvée** of white wine, or if the juice of the red grapes is given some skin contact, the resulting Champagne will be a rosé. Most rosés are in the brut style and are full-flavored and elegant.
- ***Tête de cuvée:*** Most *marques* have a prestige label, the top of the line. These bottlings are almost always made from vintage brut. Each *marque* has a name for their *tête de cuvée.* For instance, Veuve Cliquot names its prestige label La Grande Dame to honor the Veuve ("widow") Clicquot.

Each *marque's* distinct house style guarantees consistency of quality and recognizable character in every release. This is very important in building a brand and assuring repeat customers that they will get the same Champagne every time they buy a bottle from that house. The factors that influence the distinct style of a Champagne house are numerous. Among them are the proportion of Chardonnay to the red grapes; the vineyard sites from which grapes are purchased; the blending *(assemblage);* the amount of time the wine spends aging on the lees. The major Champagne houses take great pride in the distinctiveness of their region, and in the quality and unique styles of their companies.

FIGURE 5–35 Label for nonvintage Brut Rosé.

Some of the major Champagne houses are listed below, with the name of their *Tête de Cuvées* listed to the right.

Champagne House	*Tête de Cuvées*
Billecart-Salmon	Cuvée Columbus
Bollinger	Année Rare R.D.
Charles Heidsieck	La Royale
G. H. Mumm	Grand Cordon Rouge
Gosset	Cuvée Grand Millesième
Krug	Grande Cuvée, Clos de Mesnil, Blanc de Blancs
Moët et Chandon	Dom Perignon
Piper Heidsieck	Cuvée Florens-Louis
Pol Roger	Cuvée Winston Churchill
Taittinger	Comte de Champagne
Veuve Clicquot	La Grande Dame

Alsace

Two natural barriers define the small region of Alsace. On the west the Vosges mountains separate Alsace from France. On the east, the Rhine runs between Alsace and Germany. Forced by conflicts between these two powerful nations to change political affiliation many times over 1,000 years, the people of Alsace have absorbed the best of each culture. The language, the arts, the cuisine, and certainly the wine of Alsace reflect the best of both French and German influences. For instance, the wine produced here (90 percent of which is white) are named for the varietals, mostly of German origin, from which they are made. But the wines are made in a quintessentially French style, dry and elegant, intended to complement food, never overpower it.

Despite, or perhaps because of, the many conflicts and opposing national pressures, the heritage of Alsace is first and foremost Alsatian. These are fiercely independent people, proud of their heritage and history. There is a historical, medieval look to this region, with its ruined castles and its charming small villages of half-timbered cottages and stone churches, surrounded by bucolic meadows. But winemakers here are thoroughly modern in their approach to grape growing and winemaking. By combining ancient traditions with up-to-date technology, Alsace has emerged as one of the world's premier regions for white wine.

History

Clovis, King of the Franks, established Alsace as a Frankish territory in AD 496, when Christianity was taking hold in northern Europe. As in other parts of France, the Catholic Church played a crucial role in the wine-making tradition of the region. Monasteries came to own and plant many of the sites best suited to grapevines. The monks and priests produced wine for the sacraments, for use as medicine, and for their visitors.

The first change of sovereignty took place in 843 when Charlemagne's kingdom was divided, and Alsace was ceded to Louis the German. In the seventeenth century, Alsace was annexed to France at the end of the Thirty Years' War. During this time several of the leading families of Alsace started their wine companies—Beyer in 1580, Dopff around 1600, Hugel in 1639. In 1870, Alsace was reclaimed by the new German empire. Soon the plight of phylloxera wiped out the vineyards. The Germans replanted only the easier-to-work sites on the flat plains, with inferior but hardy hybrids to use in blending. Most of the more desirable locations on the rocky hillsides were left fallow. Noble varietals essentially disappeared.

In 1918, after the defeat of Germany in the First World War, Alsace was again part of France. Vineyards, even the inaccessible sites on the steep hillsides, were replanted to noble *vinifera* grapes, including Riesling, Pinot Gris, and Gewürztraminer. The quality of the wines increased dramatically. Sadly, Alsace suffered a devastating setback during the Second World War, when the Nazis occupied the region. The oppression of Alsace under the Nazis was far more severe than during the 1870–1918 period. Everything French was forbidden, even the language, and many young men were conscripted into the German army to fight on the Russian front. Exporting was forbidden and all wine produced was shipped to Germany (except the sizable cache that the Alsatians concealed from their occupiers). After the Allied victory in 1945, Alsace returned to France. Again the residents put strong emphasis on quality. The vintners worked with the authorities to forbid inferior grapes, define boundaries, establish regulations, and set quality standards. In 1962, Alsace was granted *AOC* status.

Geography and Climate: The Terroir of Alsace

To understand the unique soil composition of Alsace, it is necessary to go back 60 million years to the period when the mighty Alps were forced up out of surrounding waters by the violent collision of tectonic plates. At the same time, the Black Mountains and the Vosges chain were formed, but with a large fault between the two chains. This fault flooded and became a huge inland sea, which was further eroded as glaciers came and then receded. Eons of accumulating sea skeletons, of rocks being pulverized, of deep erosion, of volcanic eruptions, and of geological debris left by the movement of water and ice have given Alsace an incredibly complex pattern of soil composition. There are at least 20 different soil formations in this small area.

The upper steep reaches of the Vosges mountains have thin topsoil on a base of well-worn granite, schist, and volcanic sediments. The gentler slopes further down the hillsides have deeper topsoils derived from the delta of the Rhine, and subsoils of clay, marl, limestone, and sandstone. In general, grapes planted in clay and marl will yield wines with more heft and broader flavors, whereas those planted in the finer soils of limestone and sand will be lighter, more subtle in flavor, and more elegant. The plains below the foothills of the Vosges have alluvial soil, rich and fertile, and better suited to the growing of produce than wine grapes.

Although Alsace lies quite far north (of French wine regions, only Champagne is more northerly), it enjoys a far milder climate than other regions at the same latitude. The warmer temperatures and lower rainfall are due primarily to protection from the prevailing westerly winds by the Vosges mountains. Winters can be quite cold, but spring is mild, allowing for good bud-set, summers are usually warm and sunny, and very importantly, fall stays sunny, dry, and frost-free on into October.

The noble varietals of Alsace (listed here roughly in order of "nobility") are Riesling, Gewürztraminer, Pinot Gris, Pinot Noir, Muscat, Pinot Blanc, Chasselas, and Sylvaner. Riesling takes just over 20 percent of vineyard acreage, and that is increasing as Sylvaner, a blending grape, is being removed. Sylvaner still accounts for 20 percent of acreage. Also losing ground is Chasselas, another blending grape. Pinot Blanc is also widely planted, which, along with the lesser varietal Auxerrois, which is sometimes blended into it, accounts for another 20 percent of acreage. Gewürztraminer can be a picky grape to work with, being slow to ripen, but it accounts for 20 percent of the vineyards space. Pinot Gris currently accounts for only 5 percent of acreage, but it is slowly increasing in favor as consumers discover its spicy flavors and crisp acidity. More rapidly increasing in plantings is Pinot Noir, Alsace's only red varietal, which now covers about 5 percent of acreage. The remaining vineyard space is divided among Muscat, Chasselas, and the ubiquitous Chardonnay, not yet an approved varietal.

The great vineyards of Alsace have long been recognized by producers and consumers alike. These vineyards are those with superior terroir. The names of these finer vineyards have traditionally been shown on labels. For two decades after being awarded *AOC* status, the *vignerons* of Alsace saw no need for a system of classification of their vineyards. However, in the early 1980s a cooperative effort between the government authorities and landowners to give official recogni-

THE HOUSE OF LÉON BEYER: Fifteen Generations of Winemakers

When it comes to the wines of Alsace, Marc Beyer has many strong opinions. He is entitled to speak out on this topic because his family has been growing grapes and making wine near the village of Equisheim since 1580. Presently Marc is the President of the house of Léon Beyer, and both his father, Léon, and his son, Yann, are also involved with the business (Figure 5–36).

One topic about which Marc Beyer has strong opinions is what he terms the "classic style" of Alsace's wines—dry and elegant. He regrets the recent trend among wine producers in Alsace (and elsewhere in France) to leave residual sugar in the traditionally very dry white wines. In the opinion of Mr. Beyer and his father (both of whom are widely recognized experts in gastronomy), wine's *raison d'être* is to complement food. As Marc puts it, "We believe that the best way and the most frequent way to enjoy wine is in partnership with food." The Beyers believe that wine complements food best when there is no residual sugar in the wine to fight with the flavors of the dish it accompanies. For this reason, all Beyer wines are fermented entirely through to dryness (except *Vendage Tardive* and *Sélection de Grains Noble*). Marc Beyer makes the analogy between a sugary Alsace wine and a woman who is overly made-up with rouge, lipstick, etc., which creates only the illusion of beauty while hiding the woman's natural loveliness. Marc Beyer surmises that some producers are leaning toward increased sugars not to improve their wines' ability to complement food, but rather to win additional points in blind tastings by judges and journalists. "A classic Alsace white wine has structure, body, alcohol, acidity, and complexity of flavors, but *not* sugar," says Mr. Beyer.

FIGURE 5–36 Winemaker Yann Beyer uncorks a bottle to be tasted by his father, Marc (seated left), president of the winery, and his grandfather, Léon, chairman of the company. The Beyer family has been making wine in Alsace since 1580.

Another issue on which Marc Beyer has expressed heartfelt opinions is that of the classification of his region's vineyards. The process of officially declaring Alsace's best vineyard sites to be *grand cru* vineyards, which began in 1983, is superfluous, according to Marc. His objections are myriad. First, many of the *grand cru* vineyards have several noble varietals planted in them, as many as six to eight different grapes. A site that is excellent for one varietal may produce only mediocre specimens of another varietal. Second, with 50 vineyards already approved, and another 40 or so under consideration, there will soon be too many vineyards in a relatively small region for the designation to carry much validity. Third, and perhaps most important, the array of soil types within Alsace is extremely varied. Within one vineyard, there could be dozens of different terroirs. As Marc Beyer says, "We would need over 500 *grand crus* to demarcate every site in Alsace that has a unique terroir."

THE HOUSE OF LÉON BEYER: Fifteen Generations of Winemakers

Marc Beyer's advice to consumers of Alsatian wines is to become familiar with the styles of different houses, as one would when buying Champagne (Figure 5–37). When you find a style you like, stick with that producer and do not worry about vineyard designations or sugar levels. He advises, "Stick with the producers who allow the varietal character and the natural terroir to speak for themselves."

FIGURE 5–37 The Léon Beyer winery labels its top-level wines as Comtes d'Equisheim, named for the village located near their vineyards and wine-making facility.

tion to superior properties was begun. Regulations were written requiring that only the four truly noble varietals—Riesling, Gewürztraminer, Pinot Gris, and Muscat—could be planted in classified vineyards; that the yield not exceed four tons per acre; and that wine to be labeled with the classification must be made from one varietal only.

The *grand cru* appellation was created in 1983, and has been creating controversy ever since. Of the 94 sites originally considered for designation as *grand cru,* 25 were chosen in 1983. However, other sites continued to be added, and by 1990 there were 50 *grand cru* vineyards. Some Alsace producers feel standards have gotten too lax in that even mediocre vineyards are now classified. Furthermore, some of the classified vineyards are very large, spreading over differing terrains, and thus possessing several different terroirs. Another problem is that some of these sites are appropriate for only one or two of the specified varietals, but growers want to plant what they think will sell so they plant popular varietals where they do not excel. Moreover, growers do not want to stop using vineyard designations they have used for generations, so they push for *grand cru* designations for the vineyards they own, whether or not the property is of superior quality.

While the disagreement over the validity of the classification system goes on, some producers continue to use the traditional name for their reserve-level wines while ignoring the *grand cru* designation. For example, the Beyer family has always labeled its top wines "*Comtes des Equisheim.*" They continue to use that designation, not mentioning the *grand cru* vineyards they own. The famous Riesling "Clos Ste.-Hune" made by Trimbach is not labeled as Pfersigberg, the *grand cru* vineyard from which it is made.

THE WINES

Many Americans shy away from Alsace wines in their tall green bottles and Teutonic-looking labels, assuming that the wines are German and will, therefore, be too sweet. People who make that assumption are wrong on two counts: Alsace wines are French in style, and German wines are not all sweet (see Chapter 8). In passing by the Alsace wines, the consumer is depriving herself of an extraordinary experience. The whites of Alsace are fermented dry and are extremely versatile food wines.

- *Riesling:* Usually bone dry, Alsace Rieslings have superb steely acidity to hold up their flavors of green apple and minerals. When young, the wines often have a floral aroma, which with age (and these wines can age very well), evolves into gunflint and wet slate.
- *Gewürztraminer:* With its unique nose of lychee nuts and its racy acidity and pronounced spice flavors overlaying ripe forward fruit, Alsace Gewürztraminer is the ideal accompaniment to highly spiced food, especially Asian cuisine.
- *Pinot Gris:* Traditionally known as Tokay d'Alsace or Tokay Pinot Gris, but since an agreement with Hungary in 1993, now called just Pinot Gris, this is perhaps the most underrated of Alsace's noble varietals. Pinot Gris combines some of the spice of Gewürztraminer with the steely acidity of Riesling. The aromas and flavors are reminiscent of peaches or ripe melon and are perfectly balanced by firm acids.
- *Pinot Blanc:* Perhaps the lightest and least complex of Alsace whites, Pinot Blanc can nonetheless be a charming and appealing wine. It is clean and dry, with crisp acidity and can be sipped alone or matched to a variety of light, simple dishes.
- *Pinot Noir:* In vineyards this far north, red grapes have a hard time fully ripening. Rouge d'Alsace has traditionally been light in color and body, full of young strawberry aromas and flavors, and soft on the palate. These are pleasant quaffing wines.

There are other styles of wine produced Alsace in addition to the dry table wines described above.

- *Crémant d'Alsace:* Based primarily on Pinot Blanc, with Pinot Noir and Riesling sometimes blended in, Alsace's sparkling wine is made in the *méthode champenoise.* It now accounts for about 10 percent of total production in the region, and that number is increasing as the wine achieves commercial success. The wine is fairly light in body, has excellent *mousse* (fizziness) and pleasant fruit, and is quite dry.
- *Vendage Tardive:* The French term for these wines means "late picked." Left on the vine to develop additional sugars, these grapes result in delicious wines, usually off-dry in style. The flavors of late-picked wines are very rich and deep. To be labeled as *vendage tardive,* a wine must be from one vintage of the approved varietals—Riesling, Gewürztraminer, Muscat, or Pinot Gris—and cannot be enriched with additional sugar. The grapes have to be picked on a date set by the authorities, when the natural sugars have reached a specified level. They do not have to be botrytized.
- *Sélection de Grains Noble (SGN):* Wines at this level almost always contain some grapes infected with botrytis, the noble rot. This makes them sweeter, richer and heavier than *vendage Tardive* wines. *SGN* wines are made from the same four permitted varietals as *vendage tardive.* A rich, unctuous *SGN* with just enough acidity to hold up its complex mélange of apricot, ripe peaches, and honey can be an unforgettably ambrosial wine. A good *Sélection de Grains Noble* is extremely difficult to make, but when done well, it can be superb.

The Loire Valley

There is a large, regional appellation of the *vin de pays* level that encompasses all of the Loire Valley and some of its surrounding areas: *Vin de pays du jardin de France,* "Wine from the Garden of France." This is a beautiful name, and an appropriate one. The region is truly beautiful, like one very large, plentiful, and well-tended garden. The Loire Valley used to be the center of power in France, for it was in this wealthy, bountiful region that the French royal family had its roots. The region reached its pinnacle of power and influence in the late 1400s to early 1500s. In 1589, Henri IV moved the royal court, and the sphere of influence shifted to Paris and its environs.

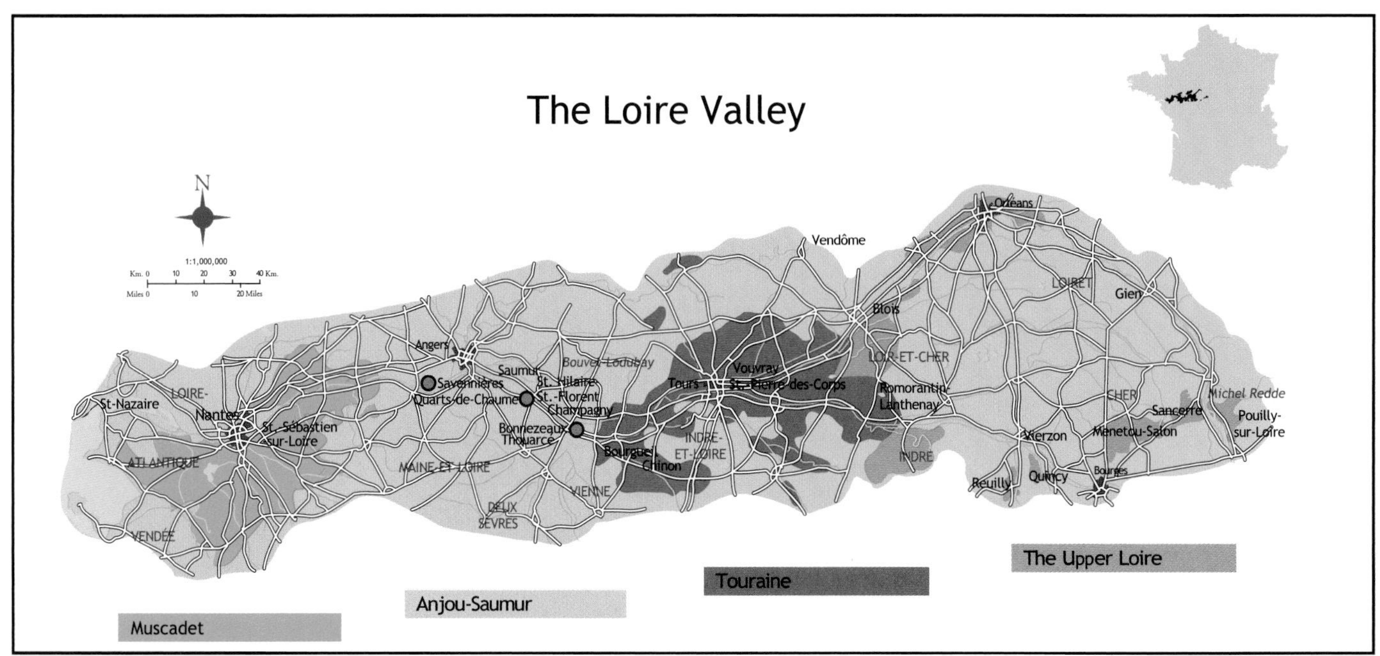

Map of the Loire Valley

Wine had been produced along the Loire River for centuries but the region's wines sank out of favor when the Court moved. For hundreds of years, Loire valley wines were not seen outside of the area. Only in the past 50 years have the Loire's more important appellations been rediscovered in Europe, and only in the past 20 years have Loire wines been widely available in the American market.

The Loire is a very long river, the longest in France. It starts in the south and flows north for 635 miles (1,024 km) before spilling out into the Atlantic Ocean. The Loire and its tributaries drain a quarter of the land mass of France. The *jardin de France* is a huge area, where a variety of fruits and vegetables is grown, livestock and dairy cows graze, and a total of almost 440,000 acres (178,138 hectares) of grapevines is planted. However, the fine wines of the Loire *AOC* appellation are found only in the final third of the area, after the river takes a turn and starts its westward journey to the sea.

HISTORY

Along the Loire River stand numerous ruined fortified castles, staring down from rocky precipices at the river far below. These once-magnificent châteaux were the homes of powerful aristocratic families that exerted considerable influence on the royals who ruled the country. Culturally, the Loire Valley was a trendsetter in the arts, in architecture, and in cuisine. Most important for French cuisine was the arrival of Catherine de Médicis when she married Henri II. She brought Italian chefs with her who taught the basics of fine cooking to the French. The heyday for the Loire was the period of the later Valois kings from the ascension of Charles VIII in 1483 to the assassination of Henri III in 1589. His successor, Henri IV, moved the court to Paris; the Loire's period of decline began.

Viticulture in the Loire Valley has been traced back as far as the eighth century AD. Many of the aristocrats who built their châteaux along the river during the next several hundred years also planted grapevines. By the late eleventh century, the wines of the Loire were highly regarded in France. That fame soon spread outside of the country, and exportation to northern markets was simplified by the ease of transport by boat along the river itself and its tributaries. Demand for Loire wines was particularly strong in the cities of Flanders. Some wine was shipped even farther, to England. Commerce in the fine wines of the Loire continued to grow and its reputation spread

FIGURE 5–38 The Loire River flows for 600 miles through the French countryside. There are many different terroirs along its banks.

until the move by King Henri IV to Paris. The royal trendsetters turned their attention to other wines, and Loire wine production was cut back. Most wine was consumed locally.

Geography and Soil: The Terroirs of the Loire

The Loire River stretches for such a long distance that no generalizations can be made about the appellations along its bank (Figure 5–38). Viticulturally it makes sense to divide the Loire into four distinct regions. First is the Upper Loire, where the river makes its jog to the west. The climate is continental, with cold winters and hot dry summers. The primary grape is Sauvignon Blanc. Moving down river into the Central Loire, the climate becomes gradually more temperate, and the soil less rocky. In the two subregions of the Central Loire, Anjou/Saumur and Touraine, the principle white grape is Chenin Blanc. Some red grapes are also planted—most importantly, Cabernet Franc. Here spring frost can still be a problem, as can drought. In the westernmost reaches of the river valley, the Loire is joined by its tributaries, the Sèvre and the Maine, near the city of Nantes and soon spills into a bay of the Atlantic. The climate has a maritime influence, warmed by the Gulf Stream and more humid than farther upriver. Occasionally the weather is too damp and overcast, hindering efforts to ripen the grapes.

The Appellations

The appellations of the Loire are not divided into levels, but rather are characterized solely by location (Table 5–7).

The Upper Loire

The majority of wine made here is dry, zesty white made entirely from Sauvignon Blanc grapes. Many culinary experts believe that Sauvignon Blanc is the most food-friendly white because of its excellent clean acidity and restrained citrusy fruit. A little red wine is made from Pinot Noir. Appellations are the names of individual communes.

Pouilly-Fumé: The village of Pouilly-sur-Loire dates from Roman times. Its wine is named Pouilly-Fumé, not because there is a "smoky" component to the bouquet, as some have believed. Rather, *fumé* refers to the grey-green nuances of color on the ripening Sauvignon grapes (Coates, 2000). The area received *AOC* status in 1937, and today there are close to 1,600 acres (648

Table 5–7 Appellations of France: Loire Valley

REGION	SUBREGION	PRINCIPAL VARIETAL
The Upper Loire	Sancerre	Sauvignon Blanc
	Quincy	Sauvignon Blanc
	Pouilly-Fumé	Sauvignon Blanc
	Menetou-Salon	Sauvignon Blanc
	Reuilly	Sauvignon Blanc
Touraine	Vouvray	Chenin Blanc (dry, semi-sweet, and sparkling)
	Chinon	Cabernet Franc
	Bourgueil	Cabernet Franc
Anjou/Saumur	Savennières	Chenin Blanc
	Coteaux du Layon	Chenin Blanc (sweet and semi sweet only)
	Quarts de Chaume	Chenin Blanc (sweet and semi-sweet only)
Nantes	Muscadet	Melon de Bourgogne

hectares) of vines, virtually all Sauvignon Blanc. Most vineyards are owned by farming families. There are few large land-holdings. The soil is mostly a clay-limestone mix. The wines are very dry and have forward grassy bouquets with often a whiff of what the French call *pipi du chat.* The acidity is crisp and clean, the flavors of citrus fruit.

Sancerre: Across the river from Pouilly lies the village of Sancerre. In 1936 the white wines were granted *AOC* status, but it was not until 1959 that the Pinot Noir–based reds and rosés achieved equal status. Most of the 5,800 acres (2,348 hectares) of vineyards are held by small growers, with some families owning plots as small as six acres (2.4 hectares). The vineyards, mostly on limestone mixed with some clay, are on gentle slopes facing east, south, and west. The white wines are very similar to those of Pouilly, perhaps a bit more pronounced in their aromas and flavors and showing a little more finesse. The reds are light in body and color, with strawberry aromas and flavors.

Menetou-Salon: Located just 18 miles (29 km) southwest of Sancerre, the village of Menetou-Salon was granted its own appellation in 1959. The wines are similar to Sancerre, but even more focused and elegant, with a slight floral hint to the bouquet.

Quincy: Twenty-five miles (40.3 km) southwest of Menetou-Salon, across a tributary river, the Cher, is the town of Quincy. Its chief claim to fame is that it was the second appellation in France (after Châteauneuf-du-Pape) to be granted *AOC* status in 1936. The soil here is a little more calcareous than in other villages, the average temperature is a little cooler, and frost is more of a problem. The wines are very racy and zesty, and in some years, can taste a bit unripe.

Reuilly: This is a small appellation, only 150 acres (60.7 hectares) of vines, very near to Quincy. The soil and climate are essentially the same. The wines are quite austere and very dry.

Touraine

Named for the city of Tours, this region is home to a variety of wines—white, red, rosé, and sparkling. Wine made from approved varietals grown within the boundaries of the Touraine region, but outside any of the commune appellations, or a blend of grapes from two or more communes, is given the generic appellation of *Touraine.* This appellation can be applied to both white and red wines, which are produced in almost equal quantities. The name of the varietal can show on a generic Touraine label. There is also a small amount of *Touraine Mousseaux* made primarily from Chenin Blanc.

Vouvray: In the commune of Vouvray, the Chenin Blanc grape reaches its zenith of quality. There are 5,000 acres (2,024 hectares) of vines in Vouvray, virtually all of them Chenin Blanc. This lovely varietal is often overlooked, but with its delightful aromas of ripe pear, its round mouthfeel, and its smooth but lively acidity and ripe fruit, Chenin Blanc is capable of making delightful wines in a wide variety of styles. In some cases, Vouvray is dry and crisp. Sometimes it is made off-dry (*demi-sec*) with a hint of residual sugar complementing the ripe pear nuances. Vouvray can also be sweet, in which case it is labeled as *Vouvray* ***Moelleux***. Chenin Blanc is also made into a sparkling wine, *Vouvray* ***Mousseux***.

Chinon and Bourgueil: The commune of Chinon and its neighbor across the Loire to the north, Bourgueil, are the two major regions for red wine within the Loire. The wine is primarily Cabernet Franc, although Cabernet Sauvignon is also authorized. Loire reds can be described as charming—light, pleasant, fruit-forward, soft, easy to drink, and easy to like. These are not wines made to age or to seriously contemplate. The wines from Bourgueil are a bit fuller in body, since the soil in that commune contains more clay, whereas the soil of Chinon is mostly lighter sand and gravel.

Anjou/Saumur

The large province of Anjou contains nineteen appellations at the *AOC* level, including generic Anjou (Figure 5–39) and generic Saumur. The vineyards of Anjou cover 35,600 acres (14,412 hectares). A variety of wines is made here, including dry whites, reds, rosés, sparkling wine, and sweet whites. *Rosé d'Anjou* is made in copious quantities from a lesser grape. Much better are the dry and semisweet rosés, *Cabernet d'Anjou* and *Cabernet de Saumur,* which must be made from the Cabernet Franc varietal.

Also prevalent are *Saumur Mousseux* (also called *Saumur d'Origine*) and *Anjou Mousseux,* sparkling wines made in the *méthode champenoise.* The variety of grapes allowed is impressive. For both the rosé and white styles, red grapes can be used: Cabernet Franc, Cabernet Sauvignon, Gamay, and three others. The white grapes allowed are, naturally, Chenin Blanc, but also small percentages of Sauvignon Blanc and Chardonnay. The wines are usually made in the brut style, although Loire bruts are sweeter than those of Champagne.

Savennières: This elegant Chenin Blanc–based wine is a rarity among Loire whites in that it is made to age. When young, the wine is austere and closed, very dry. Given time, Savennières can evolve into a complex, full, round wine with delicious flavors of Bosc pear, lime, and hints of

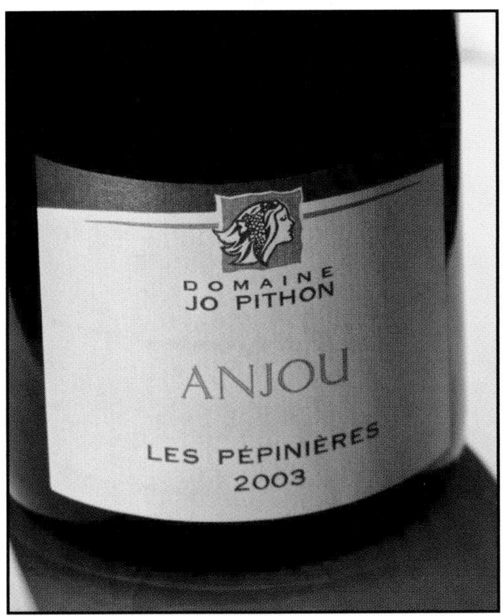

FIGURE 5–39 A label for generic Anjou wine from the Jo Pithon estate. The grapes were grown in the Les Pépinières vineyard.

honey. Savennières is on the north bank of the Loire, where weather is cool enough that keeping the grapes on the vines longer into the fall in order to fully develop that panoply of flavors is sometimes risky due to frost. One of the more important producers of Savennières is Nicolas Joly of Château de la Roche-aux-Moines (Figure 5–40). Mr. Joly is one of France's leading advocates for the concept of biodynamic viticulture, a totally organic and natural way to care for the vines in harmony with Nature's cycles. Another source for consistently high quality Savènnieres is Clos du Papillon, owned by the Baumard family.

Coteaux du Layon: This is an appellation restricted to sweet and semisweet wines based on botrytized Chenin Blanc grapes. The wines are rich and honeyed, but never cloying, thanks to their acidity. These wines can age for many years and gain further complexity and fullness. The dessert wines of Coteaux du Layon are among Europe's best, and are comparatively affordable.

Quarts de Chaume: This is a tiny appellation within the Coteaux du Layon. It is essentially a single-vineyard appellation that is rated as one of the best sites for the sweet, honeyed, floral botrytized wines. The standards, including yield, sugar content, and alcohol level, are more stringent than for Coteaux du Layon.

Bonnezeaux is another enclave within Coteaux du Layon that is allowed to use its own appellation. Larger than Quarts de Chaume but with similar standards, Bonnezeaux is not yet as well known as its neighbor to the northwest.

NANTES

The Nantes, or Atlantic region, of the Loire is home to the bone-dry white wine, Muscadet. There is more Muscadet made each year than the wines of Touraine and Anjou combined. The grape, which has come to be called Muscadet, is actually Melon de Bourgogne. This varietal was brought to the Nantais region of the Loire from its native home of Burgundy in the eighteenth century by Dutch traders. The grape is easy to grow, has high yields, and produces a clean, fresh, uncomplicated wine that perfectly complements the seafood, shellfish, and freshwater fish that are such an integral part of the diet of the Atlantic section of the Loire.

Over three-quarters of Muscadet comes from vineyards in the Sèvre et Maine district, named for the two rivers that flow through it to join the Loire. The soil here has a good amount of clay mixed in with the sand and gravel, so its wines are a tad less tart than wine labeled as Muscadet. *Sur lie* is a technique widely used in Muscadet. In this process, the wine is left on the lees until it is ready to be bottled, rather than being racked to an interim container. The time on the lees gives the wine a prickly feel and adds complexity.

Muscadet is now widely distributed around the world, where it has found favor in many markets, including the United States, as a refreshing aperitif and a perfect companion for light seafood dishes.

FIGURE 5–40 This Savennières from Nicolas Joly is from a vineyard, La Coulée de Serrant, that is entirely organic and cultivated in a method consistent with the tenets of biodynamic viticulture.

The South of France

Having covered the six major wine regions of France, where her world-class wines are made, we have touched only one-third of the *AOC* production, and barely 15 percent of total wine production, for the country. There are a great many other regions producing very nice wine. Many of the best of the lesser-known *appellation d'origine contrôlée* regions are found in the South of France. In the past the south was known for rugged, sometimes coarse wines, mostly red, made from indigenous varietals like Mourvèdre, Cinsault, and Carignan. In the past several decades, however, there have been remarkable improvements in the quality of wines. Part of the reason is a trend to planting more of the noble varietals that can survive the heat and small amount of rainfall, such as Cabernet Sauvignon and Syrah. These are being planted on superior, cooler sites farther up the hillsides. The best of these face north or east so that exposure to sunlight is reduced, thus lessening the chance of over-ripening. Another important change has been the modernization of winemaking techniques. For instance, many winemakers are now fermenting at cooler temperatures in stainless steel tanks; this protects the aromas and natural flavors of the grapes. In the South of France, one can now find some impressively elegant and balanced wines.

PROVENCE

In southeastern France, Provence extends from the delta of the Rhône east to the border with Italy (Figure 5–41). This is beautiful, rugged country, extremely hot in the summer and rather desolate in the winter, but fertile and well suited to the vine. For decades, Provence was known for its large quantities of light rosé made from the high-yielding Carignan and some Grenache. Production is still 60 percent rosé, but the percentage of Carignan has decreased, and the wines have more depth. Reds account for 30 percent of production and are made with Mourvèdre, along with increasing percentages of Syrah and Cabernet Sauvignon. White wine is also made in Provence, but native varietals like Ugni Blanc and Clairette are being replaced with Semillon.

There are seven *AOC* appellations in Provence, two of which, Bellet and Palette, are so small as to be inconsequential. There is also a large section, the *Coteaux Varois,* which is rated *VDQS*.

Côtes de Provence: With 44,500 acres (18,016 hectares), most of them in the easternmost section of Provence, this is the largest of the three generic *AOC* appellations. Production is mostly red wine, which typically is a blend of Carignan (no more than 40 percent by law), Cabernet Sauvignon, and Syrah. Several of the better producers are giving their red wine more time in new oak barrels, which allows the wine to mellow out and become smoother than was previously the case.

Coteaux d'Aix-en-Provence and *Coteaux d'Aix-en-Provence-les-Baux* are the other two generic AOC appellations. They encompass 10,550 acres (4,271 hectares) of vines in the western section

FIGURE 5–41 Terraced vineyards in Provence soak up warm sunshine above the Mediterranean Sea.

of Provence and are named for the city of Aix-en-Provence. The wines, mostly red, are similar to those of Côtes de Provence.

Bandol: The old fishing port of Bandol is right on the coast of the Mediterranean and is now a popular tourist destination. It is also home to the best and most interesting wines of Provence—big, structured, and full reds based on Mourvèdre (by law at least 50 percent) blended with Grenache and Cinsault. The regulations also allow up to 20 percent white varietals to be blended in. Bandol reds have a unique spiciness sprinkled in with the ripe plummy fruit flavors. They can age for at least a decade. Some whites are also made from local varietals, with Sauvignon Blanc adding additional acidity.

Cassis: Not to be confused with the black currant liqueur of the same name, Cassis is a small region unique in that it is more famous for white wines than red or rosé. The production of Cassis is 75 percent white, whereas in most of Provence, indeed, in most of the south of France, reds or rosés predominate. The 400 acres (162 hectares) of vines, protected from the mistral winds by a massive stone cliff, are planted to indigenous white varietals like Ugni Blanc and Marsanne with some Sauvignon Blanc. Very little of this perfumed, sprightly wine ever makes its way to export markets as the influx of summer tourists consumes most of it.

LANGUEDOC-ROUSSILLON

The very sizable region of Languedoc-Roussillon, also known as the Midi, is a popular tourist destination on the Mediterranean. The region produces ever-improving wines as investment in the area and awareness of its wines in foreign markets have increased. This varied region reaches from the western side of the Rhône delta along the coast to the border with Spain at the Pyrénées. The Midi was long famous for its simple, rustic table wines, and though quality of many of its wines has certainly improved, the Midi is still the source of much of France's *vin ordinaire,* as well as copious quantities of *vin de pays.* Eighty percent of the country's *vin de pays* is from this area. Languedoc and Roussillon contain 40 percent of the vineyards in the nation of France, yet account for only 10 percent of *AOC* production.

Throughout the Midi, the climate is Mediterranean with mild winters and warm, sunny, dry summers, perfect for ripening a wide variety of varietals, from Chardonnay to Viognier, from Cabernet Sauvignon to Syrah. The soils are consistently limestone-based, with enough variation in topsoils to allow discernible if subtle differences in terroir among the regions.

The largest *vin de pays* appellation is the regional Vin de Pays d'Oc, and its subregions. Other important *vin de pays* appellations include l'Hérault and Aude. The trend in these and other *vin de pays* regions within the nearby *départements* of Gard and Pyrénées Orientales is toward varietally named reds and whites, produced by large international *négoçiant* firms that have invested considerable amounts of capital to replant vineyards with noble varietals, install giant stainless steel fermentation tanks, and equip their caves with oak barrels for aging the wines. These mass-produced, computer-monitored wines can be quite pleasant and very affordable. Fortant de France, a brand owned by the Kobrand Corporation of New York, is one of the better-known brands. Also successful in the United States market are brands like Domaine de la Baume (previously owned by the Hardy's company of Austalia), Val d'Orbieu (originally launched by importer Martin Sinkoff of Dallas), and Réserve St. Martin (also part of the Sinkoff portfolio).

AOC Appellations of the Midi

Côtes du Roussillon, Côtes du Roussillon-Villages: Between them, these two appellations cover 19,000 acres. The smaller one, Côtes du Roussillon-Villages, is located in the northern portion along the Argly River and its tributaries. The vineyards here are superior, with a complex mix of topsoils (schist, sand) on limestone. Regulations require a lower yield in *villages*-level wines, and a higher minimal alcohol level. Twenty-five communes are included in this appellation. The wines of both Roussillon and Rousillon-Villages are primarily red, blended from Syrah, Cinsault, and Grenache. They show considerable depth and character.

Corbières: Corbières is the largest appellation in the Midi, with over 35,000 acres (14,170 hectares) of *AOC*-rated vineyards (Figure 5–42). This is primarily a red wine appellation, although a small amount of rosé and a minuscule amount of white are also made. The reds are solid, perhaps a bit dense, but with appealing aromas of lavender and plum. They are widely available in the United States, and are affordable everyday wines.

Minervois: Just northwest of St. Chinian, across the valley of the Aude River, lies the Minervois. There are 45,000 acres of wine grapes, but only a small percentage of these vineyards (11,500 acres/4,656 hectares) are rated at the *AOC* level. The remainder are *vin ordinaire* or *vin de pays.* The *AOC* vineyards, on terraced slopes in the eastern part of the region, are planted mostly to red varietals. The wines produced have a definitive acidity to them and are medium bodied and fairly fruit-forward.

St. Chinian: Once a commune within the Coteaux du Languedoc appellation, St. Chinian received *AOC* status in 1982. This appellation can be applied to red and rosé wine only. Noble varietals—Syrah, Grenache, and Mourvèdre—are slowly displacing the rougher Carignan as the maximum amount allowed of this traditional grape is decreased. These red wines are similar to those of the Minervois, but with more weight and depth.

Banyuls: Banyuls is one of six appellations for *vin doux naturel* within the Midi. *Vins doux naturel* in the Midi can be made from either Muscat, like Beaumes-de-Venise in the Rhône, or for the delicious richly flavored red *vin doux naturel*, from the Grenache grape. Banyuls is a red *vin doux naturel*, which by law must be 50 percent Grenache. The wine spends many months in oak barrels. It is aromatic, rich, intensely fruity, and ages very well.

The other Midi *vin doux naturel* appellations are *Muscat de Frontignan, Rivesaltes, Muscat de Lunel, Muscat de Mireval,* and *Maury.*

Coteaux de Languedoc: The vineyards of Coteaux de Languedoc lie on the hills that run in a line behind the city of Montpellier. From vineyards on the lower levels, where the soil is alluvial and fertile, come vast quantities of *vin ordinaire,* most of which will be distilled into spirits. Further inland, as elevations increase, the soil becomes rockier. From these better vineyards come the wines, predominately red, that are labeled with the *AOC* Coteaux de Languedoc. (The name comes from the time this land was inhabited by a people whose language was Occitan. In that ancient language, *oc* was the word for "yes," hence *langue d'oc.*)

Large quantities of *vin de pays* wine is produced in Languedoc, much of it under the regional *vin de pays* appellation of Vin de Pays d'Oc. Some of these wines are very good indeed, but cannot carry the *AOC* appellation because they are made from noble varietals not authorized for this region.

FIGURE 5–42 Very old vines are gnarled and stubby like those in this vineyard in the picturesque region of Corbières in the Midi.

The quality of both the *vin de pays* wines and the *AOC* wines is increasing steadily in this huge, wild region of small family growers and large cooperatives. As outside investment continues to grow, and better viticulture, more noble varietals and more modern equipment result, the wines of the Coteaux de Languedoc will become substantially more attractive while maintaining, one hopes, their eminently affordable prices.

THE SOUTHWEST

The catchall term "the Southwest" encompasses a huge part of France, including all viticultural areas south of Bordeaux and east of the Midi. In such a large area, there is obviously a huge variety of terrains, microclimates, soil types and wine-making preferences. Of the 70,000 acres (28,340 hectares) of vines in the Southwest, only about half produce *AOC* wines (Table 5–8).

Madiran: In the extreme southwest, the most important appellation is Madiran. The principle grape is Tannat, which is made into big heavy complex wines, quite tannic (the name of the grape derives from the same root as the word "tannin"), and, therefore, age-worthy. In many Madiran reds, Tannat makes up 40–60 percent of the blend, the balance being Cabernet Sauvignon, Cabernet Franc, and a lesser red grape. One of the most exciting producers in the region is Château Montus, where Alain Brumont, one of France's leading enological consultants, is in charge of winemaking.

Bergerac: Just east of France's largest fine wine region, Bordeaux, in the *département* of Dordogne lies Bergerac, with vineyards planted along the banks of the Dordogne River. Source of a variety of wines—dry whites, reds, rosé, sparkling and sweet whites—Bergerac has long been eclipsed by its more prestigious neighbor. This is a shame, for there are some splendid wines being produced here. The varietal mix is the same as in Bordeaux. The reds are of medium-to-full body, fresh and nicely balanced. The whites have lively citrus and herbal aromas and flavors and sprightly acidity. Rosés are fruity and pretty. One of the premier producers of Bergarac appellation wines is Château de la Jaubertie, near the tiny town of Colombier. Bought by Englishman Nick Ryman in 1973, and raised by him to high standards of excellence, the estate is now run by Mr. Ryman's son and various partners.

Within Bergerac are several subdistricts of *AOC* status.

Monbazillac: Located on the south side of the river, Monbazillac produces a sweet wine, made in the same manner as Sauternes and from the same grapes. It can be delicious, if well made, with just enough acidity to hold up the lush honeyed flavors. Unfortunately, many producers in Monbazillac do not leave the grapes on the vine to become fully botrytized for fear of losing them to a killing frost. If *vignerons* would take that risk, the wine would be even richer and more ambrosial.

Pécharmant: North of the busy market town of Bergerac lies the region of Pécharmant, home to some of Bergerac's best red wines. A blend of Cabernet Sauvignon, Cabernet Franc, Merlot, and Malbec, Pécharmant resembles a lighter St. Émilion.

Table 5–8 Appellations of France: The South

REGION	SUBREGION	PRINCIPAL VARIETAL
Provence		Mourvèdre, Cinsault, Grenache
Languedoc-Roussillon (The Midi)	Corbières	Mourvèdre, Cinsault, Grenache, Carignan
	Minervois	Mourvèdre, Cinsault, Grenache, Carignan
	Banyuls	Grenache (*vin doux naturel* only)
The Southwest	Madiran	Tannat, Cabernet Franc
	Bergerac	Cabernet Sauvignon, Merlot, Cabernet Franc
		Semillon, Sauvignon Blanc
	Cahors	Malbec

Montravel: This region produces Bergerac's best dry whites from Sauvignon Blanc, Semillon, and Muscadelle. Similar to Entre-Deux-Mers, these can be pleasant everyday whites.

Cahors: South of Bergerac, where the small Lot River flows into the Garonne, is the old town of Cahors, an important trading center in the late Middle Ages. The town lends its name to the surrounding wine-producing area. The principle wine is a big, deeply colored (almost inky), tannic but balanced red made from the Malbec grape, mellowed with the addition of some Merlot and Tannat. Not a wine to be drunk young, it needs time for the plumlike fruit to open up and the fierce tannins to soften.

Corsica

The island of Corsica, off the coast of southern France, has always produced wines. By some accounts, Corsica is Europe's oldest wine-producing region, dating from 570 BC when Phoenicians first settled there. This mountainous island in the Mediterranean is actually closer to Italy than to France, but has been under French jurisdiction since 1768.

Corsica produces a wide variety of wines—red, white, rosé, still, sparkling, and sweet. Most is *vin de pays* and *vin ordinaire.* Very little Corsican wine, even the minuscule amount that is *AOC,* is exported off the island.

Summary

The incredible variety of wines from France is quite mind-boggling. Although the task of becoming familiar with France's many different wines may seem daunting, it is worth the effort. The best of French wines will provide a benchmark against which all other wines can be measured. Moreover, an understanding of France's *Appellation d'Origine Contrôlée* laws is helpful in understanding the quality control laws of other European wine-producing countries, as most of them modeled their systems on the French system.

Review Questions

1. What are the varietals allowed in red Bordeaux wines? What varietals are used to make white Bordeaux?
2. Name the six subdistricts of Burgundy.
3. What was the Classification of 1855?
4. What is *la méthode champenoise?* Are all wines made by this process entitled to be called "Champagne"?
5. What is the dryest style of Champagne?
6. Are dry white wines the only style of wine made in the Loire Valley?
7. What is the most important grape in Alsace? In that region, what does the term *Vendage Tardive* mean?
8. What was the primary purpose of the *Appellation d'Origine Contrôlée* laws enacted in 1935?
9. What is the principal red varietal of the Northern Côtes du Rhône?
10. What is Beaujolais Nouveau?

CHAPTER 6

Italy

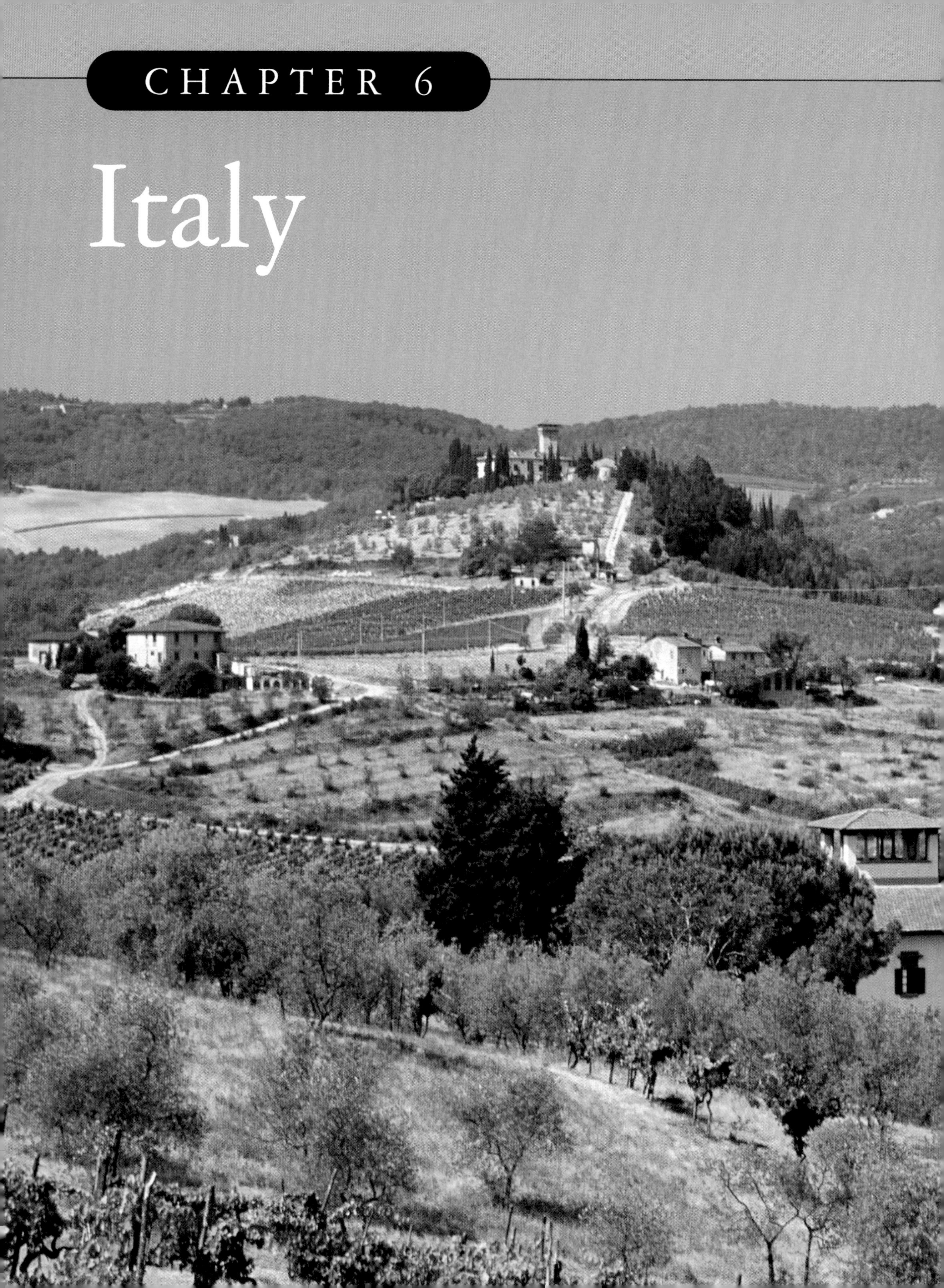

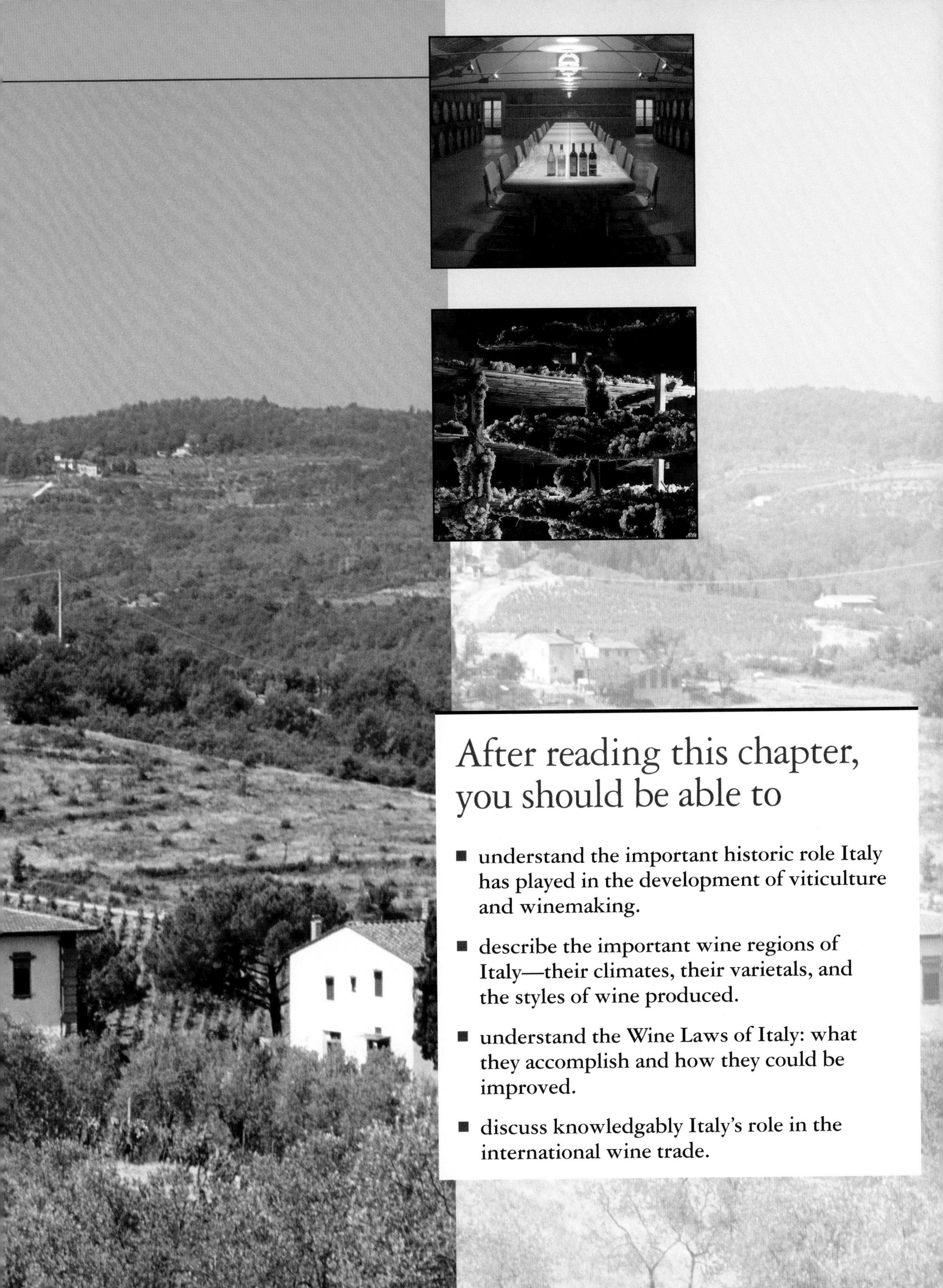

After reading this chapter, you should be able to

- understand the important historic role Italy has played in the development of viticulture and winemaking.
- describe the important wine regions of Italy—their climates, their varietals, and the styles of wine produced.
- understand the Wine Laws of Italy: what they accomplish and how they could be improved.
- discuss knowledgably Italy's role in the international wine trade.

KEY TERMS

- Aglianico
- *amabile*
- *classico*
- Cortese
- *Denominazione d'Origin Controllata e Guarantita (DOCG)*
- *Denominazione d'Origin Controllata (DOC)*
- *dolce*
- *frizzante*
- *imbottigliano dal produttore all'origine*
- *Indicazione Geografica Tipica (IGT)*
- *metodo classico/metodo tradizionale*
- Moscato
- Nebbiolo
- Recioto
- *riserva*
- Sangiovese
- *secco*
- *spumante*
- *superiore*
- Tocai Friulano
- Verdicchio
- *vino da tavalo*

It would be difficult to overemphasize the significance of Italy as a wine-producing country. It has played a vital role for thousands of years in the development of effective viticultural practices, cultivation of new varietals (there are currently over 400 grape types grown in the country), perfection of wine-making techniques, and the shipping and selling of wines. Modern Italy is the world's largest producer of wine. In all phases of the wine business, Italy's government and her vintners are working hard to increase the country's presence in the global marketplace while maintaining time-honored traditions. This chapter will explore Italy's wine heritage, the emergence of the modern wine business, (including the development of the wine laws), and the present conditions in her 20 wine regions. Based on the information presented, we will venture some predictions as to the future of Italy's wine trade.

Italian Wine—Historical Perspective

For almost as long as there have been people in Italy, there has been wine made in Italy. Evidence of grape growing and of wine consumption (earthenware jars with stains and residues of wine) dating from the Neolithic period, or late Stone Age, have been discovered in northern Italy near the city of Venice. This would date winemaking in Italy to as far back as 4000 BC. Central and southern parts of the Italian peninsula were colonized by the Greeks starting in 1000 BC as part of the expanding Greek Empire. With the colonizing forces came increased knowledge of viticulture (along with new varietals) and improved methods of making and storing wine. By the fifth century BC a true wine industry had evolved in Greece, and trading became prevalent throughout the Greek colonies around the Mediterranean Sea (Figure 6–1). It was during the Greek domination of the region that viticulture was perfected, and wine became, along with olives and grain, one of the mainstays of Mediterranean agriculture. So successful was the production of wine in Italy that the Greeks called their colony *Oenotia,* or "the land of trained (trellised) vines." The Greeks did not colonize northern Italy, for the Etruscans already had a long history of vine cultivation. The Etruscans (believed to have come originally from Asia Minor) were trading their wine as far north as modern-day Burgundy. The Etruscans did, however, pur-

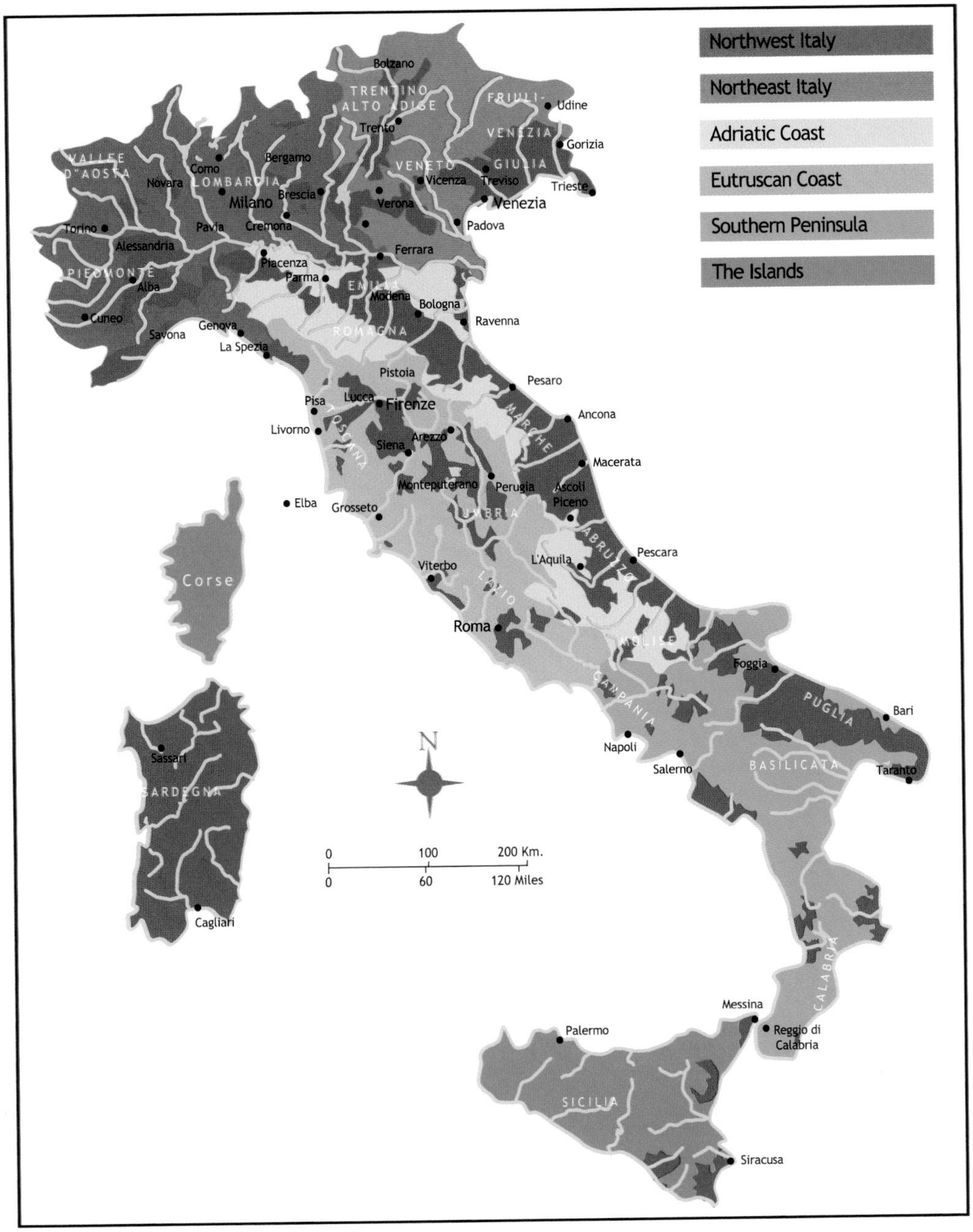

Map of Italy's Wine Regions

chase from the Greeks vessels for the drinking and storage of wine, and like the Greeks, they came to consider consumption of fine wine one of the sure signs of a civilized people.

A new empire began to control southern Italy when the Romans emerged as the dominant force in the region between the third and second centuries BC. The Romans' interest in wine is proven by the many mentions of wine in their writings. The oldest surviving Latin work on wine, Cato's *De agri cultura,* dates from 200 BC and suggests that wine was no longer a rare luxury in the Roman world, but a crucial ingredient in the growing commercial activities of the Empire. As the population of the city of Rome grew, reaching one million inhabitants by the start of the Christian era, it became the single most important market for the wine made from vines throughout the peninsula and in outlying regions of the Empire. Romans were consuming so much wine that in AD 92, the emperor of Rome decreed that too much arable land was devoted to the wine grape. Frustrated with having to import grains from north Africa, the emperor

FIGURE 6–1 This antique wine vessel is typical of the ones used by the Greeks to transport wine.

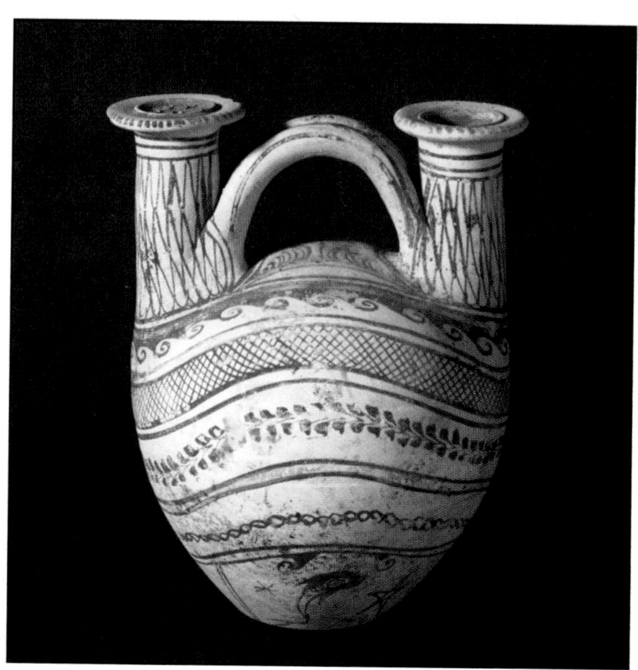

ordered many vineyards uprooted to make room for edible crops. But viticulture persisted, and the Romans continued to develop their vineyards, and to trade with wine in all parts of the known world. Evidence of Roman wine vessels have been found as far north as England, as far west as the Iberian Peninsula, and as far south as northern Africa. The Romans also introduced viticulture to any colony where climatic and soil conditions were appropriate. It is under the Romans that vineyards were planted throughout Gaul (modern-day France), reaching into Bordeaux by the first century AD and into Alsace and Burgundy by the third century. It is fair to say that most of the great wine regions of Europe were originally planted by the Romans.

When the Holy Roman Empire fell to the combined invasion of barbarians from the North and the Turks from the East in approximately AD 300, wine production in Italy essentially came to a standstill. Although wine continued to be produced in former colonies, within Italy virtually the only wine produced was for the sacraments of the Christian Church and to fill the daily allocated ration for members of religious orders living in monasteries. Wine's integral role in Christian (and Jewish) religious ceremonies was a strong factor in its survival of Europe's Dark Ages. In Italy, as elsewhere in Europe, monasteries became the owners of large swaths of vineyard land, and monks became proficient winemakers. The sale of their wines became an important source of income for religious orders.

As Europe emerged from the Dark Ages into the later Middle Ages, wine once again became an important commodity for trade. The demand for wine was growing as population increased across the European continent, and vibrant urban centers of commerce grew, especially in northern Italy (Venice, Milan, Genoa, and Florence). As trade increased, a wealthy middle class of merchants and businessmen developed, with an interest in, and the money for, luxury items like wine. No longer was wine merely for the aristocrats and the church. The vineyards of France, Iberia, and Germany could not supply enough wine to meet the new demand. Into the breach stepped the vintners of Italy. From the south of Italy, from farmlands in northern Italy, and from the provinces in central Italy, especially Tuscany, came domestic wine to fill the demands of markets in the northern urban centers. Moreover, because wine from Italy's warm climate was hardy, sweet, and high in alcohol, it could travel well. The enterprising merchants and traders of Italy's wine business soon found new markets for their product in Paris, the British Isles, and as far away as Eastern Europe and the Baltic area. As Europe emerged from the Middle Ages at the time of the Renaissance, more people acquired a taste for the beverage, and that taste became more discriminating. Consumers began to differentiate among wines from different regions, and developed preferences. Winemaking techniques were improved during the sixteenth century, and in the seventeenth and early eighteenth centuries, Tuscany became the center of Italy's wine trade. Vintners in Tuscany formulated their first wine laws in the early 1700s. During this time, Italy was still fractured into many small competing provinces, some independent states, some under the control of Austria's Hapsburg Empire. The unified country of France, with the central government lending support to the wine industry, became the leading force in the international wine trade.

By the 1800s, the various provinces of the Italian peninsula began their steady emergence into the modern wine trade. A stronger insistence on quality among wine consumers across Europe and in emerging markets in the New World forced Italian winemakers to be more conscientious in their viticulture and vinification methods. Most Italian wine at this time was of average quality and was marketed locally. However, certain wine regions became recognized as producing wine superior to the everyday wine being made elsewhere in the country. Barolo, Barbaresco in Piedmont, Valpollicella in Veneto, and Chianti in Tuscany became known for the quality of their wines, thus opening up foreign markets to Italian wines. When Italy finally united as one country in 1860–1861, progress toward a modern wine trade gathered real momentum.

In the late nineteenth century Italy, like the rest of Europe, suffered a serious setback in the production of quality wine when phylloxera, and later the powdery mildew, decimated vineyards up and down the country. The process of replanting vines on American root-stock was expensive, and many small vintners who could not afford to replant ceased operation. Also lost to phylloxera were many of Italy's indigenous varietals. These traditional varietals were often expensive to replant and to trellis, required labor-intensive care, and were low yielding. Cash-strapped landowners instead replanted with hardy, disease-resistant, high-yielding varietals that produced acceptable wines, lacking in complexity, but plentiful and less expensive to produce. Italy acquired a reputation over the ensuing decades as a producer of large quantities of undistinguished, affordable wines (Figure 6–2). This reputation was to continue on into the 1970s. Only recently

FIGURE 6–2 Early twentieth-century Italian women picking grapes in the time-honored method. After the phylloxera invasion many of Italy's vineyards were planted on American rootstock.

has the country been able to shake this image. One important factor was the adoption of meaningful quality control laws in 1963. Also significant was the leadership, in various regions, of conscientious, dedicated vintners determined to put the emphasis on quality.

The Denominazione d'Origine Controllata (DOC) Laws

After the Second World War, Italian vintners and government officials agreed that if the country was to continue its progress into the modern commercial world of wine, there would need to be standards set for the production of wine. Regions that had already acquired a reputation for quality wine, like Barolo and Chianti, needed regulations to protect the authenticity of wines bearing their names, and regions that were struggling to improve the quality of their wines needed guidelines and standards to assure progress. Italian leaders also recognized that it would be wise to work as closely as possible with the French who already had a system of quality control laws in place, and who also had earned, over the previous hundred years, the respect of wine consumers throughout Europe and the rest of the world. Accordingly, the Italian Parliament passed, in 1963, the *Denominazione di Origine* laws, based on the French *Appellation d'Origine Contrôlée* laws. The laws, which became effective in 1966, created the ***Denominzione D'origine Controllata (DOC)*** designation, which guarantees the place of origin of any wine bearing the name of a region that held the *DOC* designation. The laws also established basic standards of quality for *DOC* regions. The first *DOC* designation was granted in 1966 to Tuscany's white Vernaccia di San Gimignano. The wine laws included a top category of classified wines, the ***Denomiazione d'Origine Controllata e Garantita (DOCG),*** that not only regulated the production of wine in regions so designated, but further guaranteed they would be of high quality. These would be Italy's most elite wines, more impressive than even *DOC* wines. At both levels, the laws specified which varietals could be grown, what the yields could be, minimum alcohol levels, and even acidity levels.

From the time the *DOC* laws were first enacted, there was resentment on the part of many vintners who did not want the government dictating how they should tend their vineyards or make their wines. As a result, there was minimal immediate effect on the overall quality of Italian wines. Some vintners reverted to cheaper methods or easier-to-cultivate vineyards sites or higher-yielding varietals to produce wine that met the basic standards for their respective regions, but was really no better than what had been made in the past. Other more adventuresome and innovative producers began to make wine outside the parameters of the new laws, bypassing the *DOC* designation to make interesting, exciting wines. The legitimacy of the wine laws was further eroded after 1980, when the authorities began elevating to the *DOCG* level regions whose wine was clearly not among Italy's finest.

In 1992, Italy's wine laws were revised under the leadership of the newly elected prime minister, Giovanni Goria. He was determined that the laws set legitimate guidelines and be respected as a reliable indicator of quality. In the overhaul of the laws, new designations were created and existing regulations were tightened. The tougher, fairer laws earned the respect and support of powerful landowners and businesspeople within the wine industry. With government and vintners and corporations now working together, Italy has improved the quality of its wines considerably in the past decade, and as a result, is again recognized as an important producer of a wide variety of fine wine.

Quality Designations

The new *DOC* laws encompass all of Italy's many wines, even those that do not meet any specific standards. The laws divide all of Italy's numerous wine regions into four levels of quality.

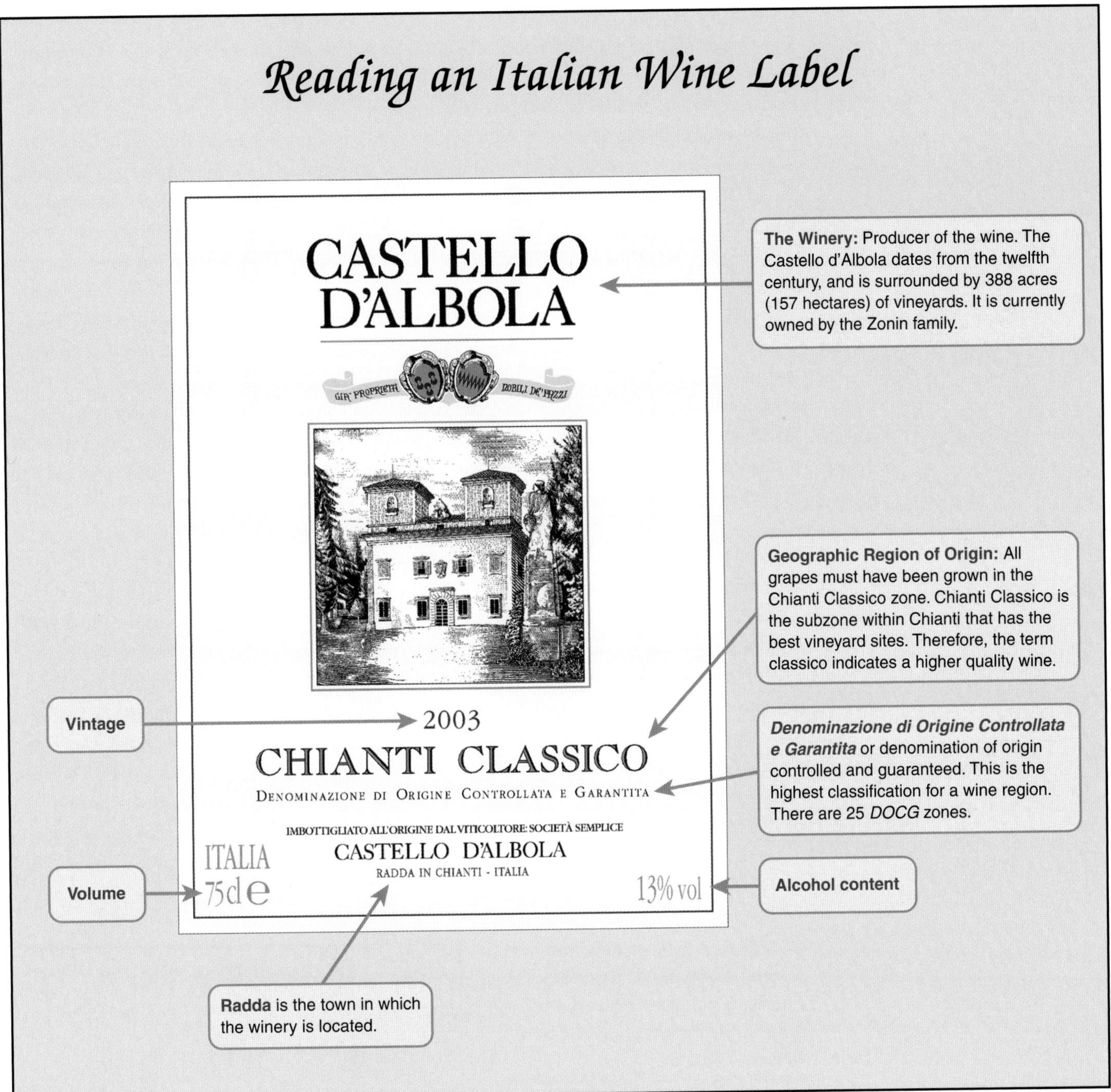

Reading an Italian Wine Label
CASTELLO D'ALBOLA
2003
CHIANTI CLASSICO
Denominazione di Origine Controllata e Garantita
IMBOTTIGLIATO ALL'ORIGINE DAL VITICOLTORE: SOCIETÀ SEMPLICE
CASTELLO D'ALBOLA
RADDA IN CHIANTI - ITALIA
ITALIA
75cl e
13% vol
The Winery: Producer of the wine. The Castello d'Albola dates from the twelfth century, and is surrounded by 388 acres (157 hectares) of vineyards. It is currently owned by the Zonin family.
Geographic Region of Origin: All grapes must have been grown in the Chianti Classico zone. Chianti Classico is the subzone within Chianti that has the best vineyard sites. Therefore, the term classico indicates a higher quality wine.
Vintage
Denominazione di Origine Controllata e Garantita or denomination of origin controlled and guaranteed. This is the highest classification for a wine region. There are 25 DOCG zones.
Volume
Alcohol content
Radda is the town in which the winery is located.

Vino da Tavola

"Table wine." At this level no geographic place of origin can be named. There are essentially no regulations imposed on producers except those required for health and safety standards. If the wine is bottled, only the color of the wine (e.g., *vino da tavola rosso*) and the name of the producer can show on the label. The only other conditions to be met are that the wine be made from grapes recognized by the European Union community, and that alcohol content and volume of wine per bottle show on the label. Even the vintage date is not required. This category includes wine that will be sold in bulk or distilled into spirits. Approximately 77 percent of wine produced in Italy is in the *vino da tavola* category.

INDICAZIONE GEOGRAFICA TIPICA (IGT)

This category was newly created under Goria to include wines that were made in a *DOC* region, but not according to the laws of that region. This way some of the very good wines that had been made in previous decades by imaginative vintners like Piero Antinori in Tuscany, who blended the native **Sangiovese** grape with Bordeaux varietals, could have a designation higher than just *vino da tavola* so long as they reflected the terroir of the larger region whose name was indicated on the label. In other words, if a wine is typical of one of the twenty provinces of Italy, it can show the name of that region. An IGT label cannot, however, show a smaller subregion or specific village or an individual vineyard. Since 1996, when some of the first Sangiovese–Cabernet Sauvignon blends from Tuscany, the so-called Supertuscans, were granted IGT status, the number of IGT wines in Italy has steadily increased. These wines, most of which are made in Tuscany, cover a wide range of styles from solid, good-value Sangiovese-based brands to very fine proprietary wines, usually Sangiovese blended with Cabernet or other French varietals. Some of the wines in this latter category, such as Antinori's Solaia and Tignanello, Ornellaia from the Marchese Lodovico, and Sassicaia from the della Rocchetta family, are among the most expensive wines produced in Italy, but also among its most impressive. Understandably, they have had tremendous success in export markets.

DENOMINAZIONE D'ORIGINE CONTROLLATA (DOC)

As in the original 1963 law, a *DOC* wine must be made from specified grape varietals grown within a delimited geographic area, according to prescribed methods of viticulture. The major improvement of the 1992 overhaul was to allow the dividing of a large *DOC* region into more specific subzones, townships, villages, microzones, or even into individual vineyard sites. The recognition of variability of terroir within a *DOC* gave credibility to the system that it had previously lacked. There is now a hierarchical basis to the geographic delimitations, mandating that the smaller the zone, the more rigid are the standards, especially on production limits. This hierarchy actually allows more flexibility to vintners as they can choose to declassify a wine from a smaller, more restricted *DOC* to a larger, in a sense "lower" *DOC*. A consumer can now assume

FIGURE 6–3 IGT wines can have a brand name like Il Bastardo. The label will also indicate the geographic region of origin, here Toscana (Tuscany).

that a wine from a restricted, homogenous zone with a proven track record for producing fine wine will be more distinctive than one from a larger region that may incorporate the small zone, but will also contain grapes from a variety of other, lesser zones. The "concentric circles" hierarchical concept that holds true for French appellations is now applicable also to Italy's classified wines.

Denominazione d'Origine Controllata e Garantita (DOCG)

This category was first created in 1963 to designate the most prestigious subregions within the *DOC* regions. The four subregions that were designated in the next few years, Piedmont's Barolo and Barbaresco, and Tuscany's Chianti and Vernaccia di San Gimignano, are indeed among Italy's best. However, over subsequent years some questionable designations were made. By 1992 there were eleven *DOCGs,* seven red and four white. Among the revisions of the wine laws that year was a tightening of the requirements for *DOCG* status. As of 2005, there were 32 *DOCGs*, 20 red and 12 white. Wines from these *DOCG* regions are given more stringent taste analysis than wines from the 300 *DOC* regions. To qualify for *DOCG* status, a *DOC* zone must have at least five years' record as a recognized demarcated zone, its wines must have established a reputation as of distinctive style and high quality, and the wines must have attained a measurable commercial success domestically and in export markets. The criteria may be difficult to quantify, but there is general consensus that *DOCG* zones that have been elevated in the past 10 years are indeed worthy of the honor. For a summary of the current *DOCGs,* refer to Tables 6–1 through 6–4.

At both the controlled levels of Italy's wine laws, very specific requirements are listed and closely regulated. For each *DOC* and *DOCG* in Italy the following factors of wine production are spelled out:

- Grape varietals allowed and the percentage of each that must be used, usually listed as a minimum and maximum allowed
- Yield per hectare that can be harvested and allowable pruning methods
- Total amount of wine to be produced
- Vinification methods (for instance, chapitalization is not allowed in any Italian *DOC*)
- Aging requirements and methods (the use of the term *riserva* is carefully controlled).

Additionally, these wines undergo taste analysis to confirm that they meet standards for their denomination. The quality analysis of *DOCG* wines is more stringent than for *DOC.* Moreover, each bottle of wine within an approved *cuvée* of *DOCG* wine is sealed with a numbered government seal over the cap to prevent any further manipulation. Currently, 14 percent of wine produced in Italy is from a controlled denomination, at either the *DOC* or *DOCG* level.

Naming of Italian wines

Within Italy's *DOC* laws there are several ways a wine can be named. At the *vino da tavola* level a label will carry only a brand name (or the producer's name) and the color of the wine. No geographic location, no vintage date, and no grape varietal can be named. For a wine of the *indicazione geografica tipica* category, a brand or proprietary name is often used. However, at this level the label can provide information on the varietal used and the geographic region. Refer to the IGT label in Figure 6–3. It shows both the varietal (Sangiovese) and the geographic region (*Toscano,* or Tuscany).

At the classified levels, that is, *DOC* and *DOCG,* there are two ways a wine can be named. First, the name of the wine could be just the geographic region of origin—for instance, Barolo or Chianti. A second method of naming a classified wine is by region of origin and varietal (Figure 6–4). Barbera d'Alba (red wine made from Barbera grapes grown in the subregion of Alba in

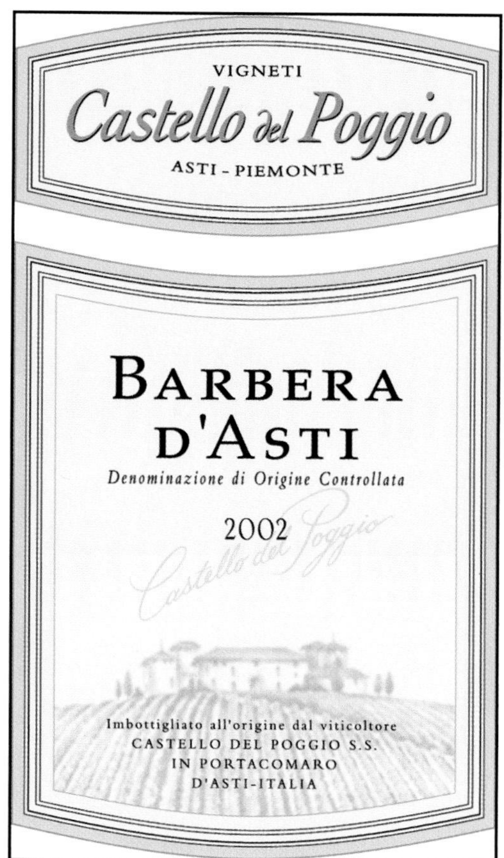

FIGURE 6–4 The wine here is named both for varietal (Barbera) and region (the commune of Asti in Piedmont).

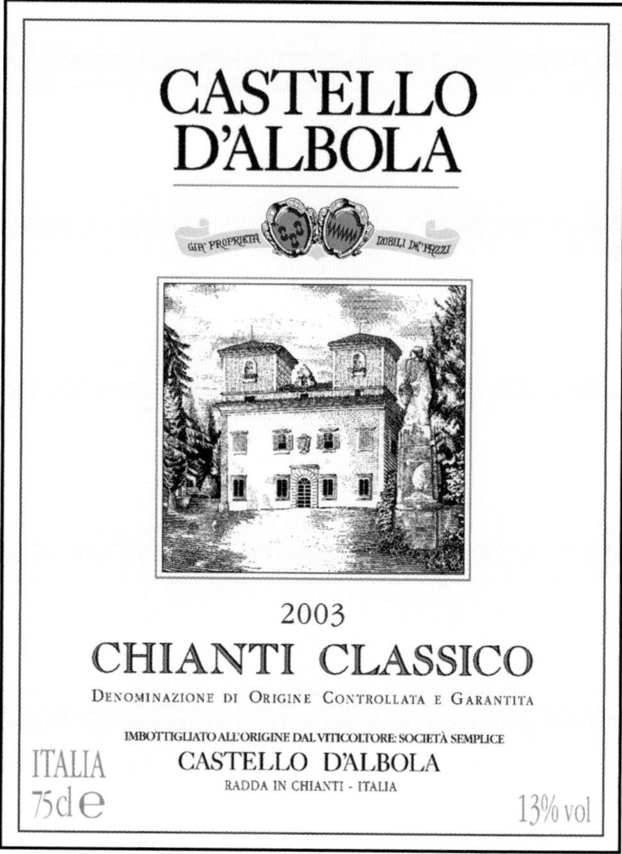

FIGURE 6–5 Wine can be named for the region where the grapes were grown, in this case Chianti Classico.

Piedmont) and Cortese di Gavi (a white wine made from **Cortese** grapes grown in the subregion of Gavi in Piedmont) are both good examples. If a wine is to be named in the latter method, the varietal must be one approved for use in that region. If the wine is a blend of two or more varietals, the wine can be named for its color. Examples are Bianco di Custoza (a blended white from the Custoza subregion of Veneto) or Rosso di Montalcino (a blended red from the Montalcino subregion of Tuscany).

Other words that may be part of a wine's name could indicate information about the style of the wine, its vinification methods, or additional information about its place of origin. Some terms that can be incorporated into a wine's name are the following:

Classico is a geographic term indicating that the vineyards where the grapes were grown are in the portion of the wine region which traditionally has produced better, more distinctive grapes (Figure 6–5). Example: Chianti Classico

Riserva is a winemaking term indicating additional aging. Each classified wine region has its own minimal aging requirements regulating the use of the term *riserva.*

Example: Chianti Classico Riserva

Amabile means semisweet. Example: Orvieto Amabile, an off-dry white wine from Orvieto. (***Dolce*** indicates a truly sweet dessert wine)

Secco means dry. Example: Orvieto Secco is a dry version of the white wine, Orvieto.

Superiore indicates the wine is at least 1 percent higher in alcohol than the minimum required for its *DOC.* Moreover, it may have received extra time in cask.

One other term that can show up on an Italian label, but is not part of the wine's name, is *imbottigliato dal produttore all'origine,* which means the wine was bottled by the producer at the source of the grapes. This is the equivalent of France's *mis en bouteilles au château* or *au domaine.*

Wine Regions of Italy

There are twenty political regions in Italy. Each of these regions produces wine, and each has delimitated wine zones within it. There are over 300 *DOCs* in Italy, and literally thousands of different wines produced. We will not attempt to cover every region, but instead will concentrate on the five regions whose wines are most likely to be found in the North American market: Piedmont in the northwest; Veneto, Friuli-Venezia, and Trentino-Alto Adige in the northeast (these three regions are often grouped together and referred to as Tre Venezie due to their proximity to the city of Venice); and Tuscany in central Italy. This is not to imply that other regions

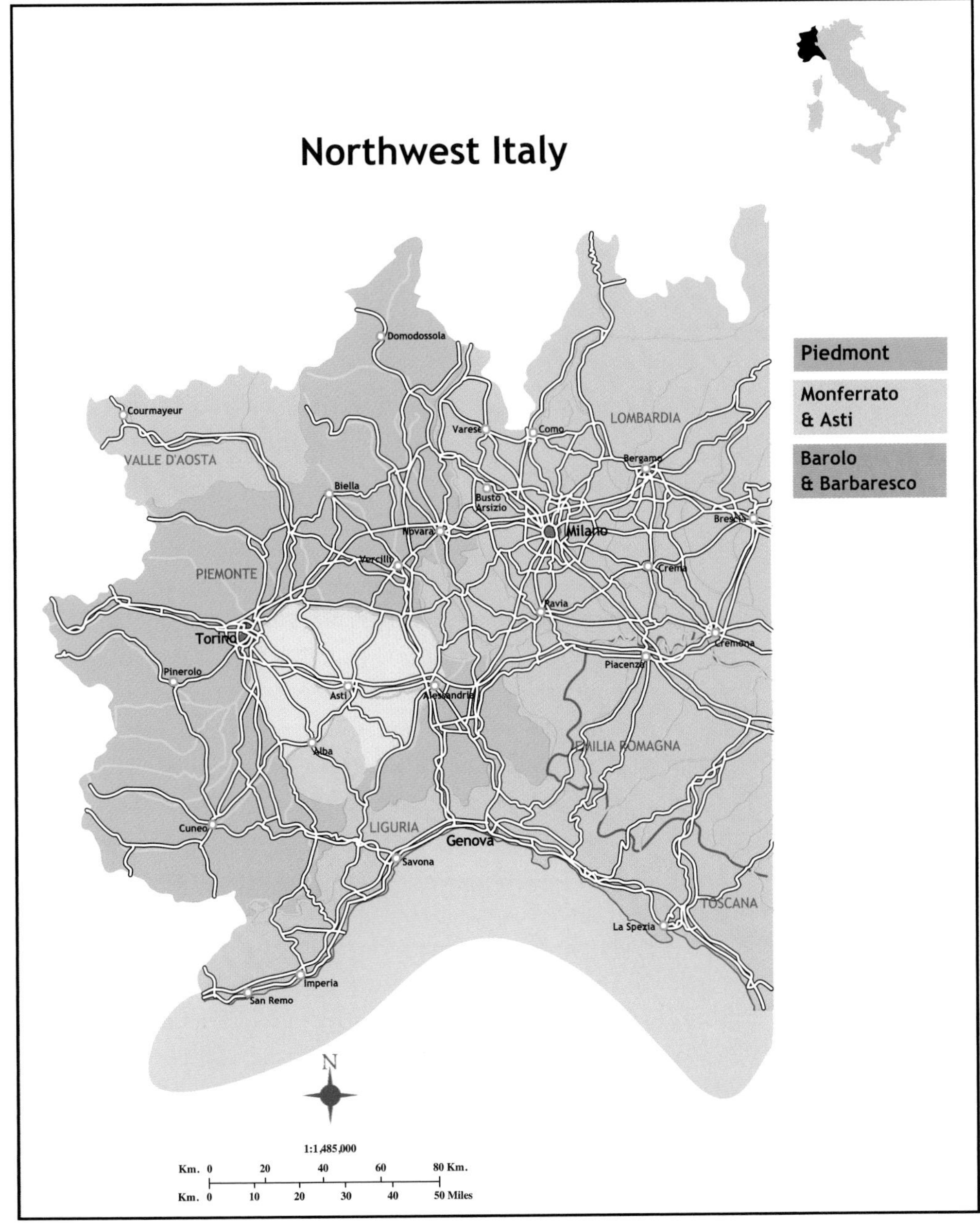

Map of Northwest Italy

of Italy do not produce good wine. Indeed, there is a wide variety of fine wines from other areas, especially in the south, many of them packed with unusual, even exotic, aromas and flavors, but eminently affordable. Knowledgeable wine professionals often refer to the south of Italy as one of the most exciting, up-and-coming wine regions of the world. We will give an overview of southern Italy, touching on its most promising wines that are imported into the United States and Canada.

Piedmont

The large region of Piedmont in northwest Italy is nestled up against the Alps. (Piedmont, or *Piemonte* in the Italian spelling, means "foot of the mountain.") It is the largest region on the mainland, and there is a considerable variety of mesoclimates. However, many of the foothills and ridges are too steep for vineyards. So, despite its size, Piedmont is not Italy's largest producer of wine. However, it does have the highest percentage of classified wines—fully 17 percent of Italy's *DOC* and *DOCG* wines are from Piedmont. This is not to imply that Piedmont is the only region of northwest Italy to produce fine wines. Two of Piedmont's neighbors, Lombardy and Emilia-Romagna, have equally long histories of wine production, and are still turning out impressive wine today.

Albana di Romagna, for instance, was one of the first white wines to be granted *DOCG* status. It is made from an indigenous grape in Romagna, the eastern side of this hyphenated region. Within the other half of the region, Emilia, there are four subzones that produce prodigious quantities of the red, frizzy, fruity, innocuous wine called Lambrusca. Lombardy is home to many *DOCs,* and is particularly well known for the quality of its sparkling wine, Franciacorta Spumante, which is made in the same method as Champagne (here called ***la metodo tradizionale***) and from the same grapes, Chardonnay and Pinot Noir (although here Pinot Gris and Pinot Bianco are also allowed to be blended in). So impressively elegant is this sparkler, with its austere flavors and tight frothy mousse, that it has been elevated to *DOCG* status. Lombardy also has two other *DOCG* wines, Valtellina Superiore and Sforzato di Valtellina, both full-bodied reds made from local indigenous grapes in the subzone of Valtellina. For a summary of the DOCs of Lombardy and Emilia-Romagna, see Table 6–1.

Piedmont is unquestionably the source of the best wines from the northwest. It is the home of two of Italy's most esteemed wines, Barolo and Barbaresco, the second and third *DOC* regions, respectively, to be elevated to *DOCG* status. Both wines are made entirely from the region's famous red grape, the **Nebbiolo.** This grape is site-specific in that it easily reflects the terroir of whatever vineyards it is planted in. Both Barolo and Barbaresco are big, rich, tannic, concentrated wines with full bouquets, but because of the differences in terroir, they are different, even though the two subregions are only a dozen or so miles apart. Connoisseurs claim the aromas of Barolo tend more toward tar, licorice, and truffles, while Barbaresco sends forth nuances of rose petals and plums. Both wines are extremely long-lived.

The two big Nebbiolo-based reds of Piedmont are not the everyday wines of the people of Piedmont. They are too expensive for that. Fortunately, there are other more affordable wines available. The most widely planted grape in the region is the Barbera, less tannic, lighter, and more acidic than Nebbiolo, it is made into pleasant, zippy, cherry-flavored wines that accompany tomato-based pasta sauces well. Also prevalent is the Dolcetto grape, naturally fruit-driven and juicy, softer at the edge than Barbera. Piedmont is also home to Italy's most famous sparkling wine, Asti (formerly called Asti Spumate), easy and pleasant to drink, in its soft, off-dry style. There are also attractive dry whites made in Piedmont from the Arneis and Cortese grapes, as well as a delightful sweet wine made from the **Moscato** grape (Italian name for Muscat). Piedmont undoubtedly produces a wider variety of fine wines than any region of Italy.

Table 6–1 Wine Regions of Northwest Italy

REGION	SUBREGION	CLASSIFICATION	VARIETAL	STYLE
Piedmont	Barolo	*DOCG*	Nebbiolo	full-bodied red
	Barbaresco	*DOCG*	Nebbiolo	full-bodied red
	Barbera d'Alba	*DOC*	Barbera	medium-bodied red
	Gattinara	*DOCG*	Nebbiolo (Spanna)	medium red
	Ghemme	*DOCG*	Nebbiolo (Spanna)	medium red
	Roero	*DOCG*	Nebbiolo	full/medium red
	Dolcetto	*DOC*	Dolcetto	light red
	Brachetto d'Acqui	*DOCG*	Brachetto	light, *frizzante* red
	Cortese di Gavi	*DOCG*	Cortese	dry white
	Arneis de Roero	*DOCG*	Arneis	dry white
	Asti	*DOCG*	Moscato	sparkling
	Moscato d'Asti	*DOC*	Moscato	semi-sparkling; off-dry
Emilia-Romagna	Lambrusca	*DOC*[1]	Lambrusca	frizzante red
	Albana di Romagna	*DOCG*	Albana	dry white
Lombardy	Valtellina Superiore	*DOCG*	Nebbiolo	medium reds from 4 subzones
	Sforzato di Valtellina	*DOCG*	Sforzato	full-bodied red
	Franciacorta Spumante	*DOCG*	Chadonnay/Pinot Noir	sparkling

[1]Lambrusca is made in four separate *DOCs*.

Barolo

Named for the village at its center, the region of Barolo produces the most impressive and powerful expression of the Nebbiolo grape (Figure 6–6). Planted on hillsides around the village of Barolo and surrounding townships, the Nebbiolo grape takes its name from the Italian word *nebbia,* the fog, because it ripens late in the autumn when the hills are shrouded in mist. These cool, misty hillsides mean that Barolo is the more austere, restrained, structured, and muscular of

FIGURE 6–6 Vineyards surround the town of Barolo.

Piedmont's two famous Nebbiolo-based reds. Barolo has been made for hundreds of years, and vintners of the area used to advise buyers to hold the wines for up to 10 years before even attempting to consume them. The extra aging time would give those ferocious tannins a chance to mellow out. Once mature, these big aggressive wines were the perfect accompaniment to the substantial meats, pungent cheeses, and hearty risottos, often made with local truffles, which were all intrinsic components of the regional cuisine.

A significant development in Barolo is the move since the early 1990s toward estate-bottled and vineyard-designated wines. Many landowners, after decades of selling their grapes to *négoçiant* houses to be blended with grapes from other sites, now want to make wines that reflect the distinct character of their own small plots. Although there is no official system in Piedmont (indeed, in any region of Italy) of rating specific vineyards, several sites are recognized by the cognoscenti as capable of producing superior wine. The owners of these sites, usually families that have been making wine in Barolo for generations, are including these sites' names on their small bottlings.

In the past 10 years there has been another movement in the Barolo region among younger winemakers who are moving away from the traditional style of Barolo, which was unapproachable when first released. Recognizing the fact that the modern consumer is not usually willing to lay a wine away for decades before consuming it, many of the new vintners of the region are making their Barolo in a softer style. While honoring the traditions of the region, they are now practicing better vineyard management, avoiding too much extraction of phenols during the crush, limiting the fermentation and maceration periods, and investing in more new oak barrels each year. These new winemakers are turning out excellent wines that require shorter aging time but still reflect the concentrated flavor profile of the Nebbiolo. These newer, more modern Barolos make their forerunners, like pioneers Elio Altare, Luciano Sandrone, Roberto Voerzio, and Paolo Scavino, proud. Mr. Sandrone has observed that the new generation of winemakers in Barolo are showing attention not only to the "finer aspects of winemaking but also to the global aspects of the territory of Piedmont" (Cooke, 2002).

BARBARESCO

If Barolo is the king of Italian wines, then Barbaresco is surely the queen. A little softer, less tannic, more refined and subtle, with more flowers and less spice in the bouquet, Barbaresco is another powerful and true reflection of the Nebbiolo, but of different character. Its gentility meant that Barbaresco could be consumed younger than its compatriot from across the valley. Named for the village of Barbaresco, the region is smaller than Barolo, with only about 40 percent as much vineyard acreage. The Barbaresco is a younger wine than Barolo, not made in the current dry style until the mid-1890s, when a professor at the enology school in Alba succeeded in coaxing the grapes to fully ferment. Barolo had already been made for over 50 years, and had the financial support of the nobility in nearby Turin. Subsequently, Barbaresco had little commercial success until the groundbreaking work in the 1960s and 1970s of two skilled and dedicated local vintners, Giovanni Gaja and Bruno Giacosa. They both made superb Barbarescos that proved their region did indeed have a distinct character. They then worked tirelessly to promote the wines of the region locally and in export markets. Giovanni Gaja's son, Angelo Gaja, continues his father's work, traveling around the world to sing the praises of Barbaresco, and of Barolo, where he now also owns vineyards. Gaja was the first producer in Barbaresco to vineyard-designate his wines. He was also one of the pioneers who brought more modern winemaking technology to Barbaresco. Dedicated to the uniqueness and power of the Nebbiolo, Angelo Gaja is Piedmont's most effective spokesperson.

Nebbiolo ripens earlier in Barbaresco, where flat vineyards stretch through the valleys, as opposed to the steep, windy slopes in Barolo. This topography allows for more sunshine to reach the ripening bunches hanging on the vines. Barbaresco is generally a lighter-bodied wine, but it certainly does not lack for tannins and acidity. These wines have plenty of structure and are capable of a long life, and even though they are not pleasant sipping wines when young, they can be served alongside a hearty dinner soon after being released. Wine-making techniques here, as in Barolo, are moving away from extended maceration and prolonged cask aging to brief fermentation in temperature-controlled steel tanks and short maturation in oak to assure a softer, more approachable, style. Barbaresco is usually at its best between 5 and 10 years after the vintage date.

Barbera

Nebbiolo may be the noble grape of Piedmont, but Barbera is definitely the region's workhorse. Widely planted throughout Piedmont, Barbera produces the majority of the wine consumed by local folk. It accounts for 55 percent of the red wine produced in the region (Robinson, 1994). Barbera is a hardy varietal, not fussy about location, and very productive. It is planted throughout Italy and has been widely exported. (It grows extensively in California, for instance.) The grape ripens comparatively late, a good two weeks after the lesser red grapes of Piedmont, which allow plenty of time for flavors to evolve. (Despite that long hang-time, Barbera still is picked well before the noble Nebbiolo). Barbera is made into an incredible variety of styles, from light and fizzy to dense and complex. Certain characteristics are constant, no matter what the style: deep ruby color, fairly full body but low tannins, and, most characteristic, very high acid levels. When a Barbera vineyard is allowed to produce a high yield, the varietal's tendency toward pronounced acidity becomes exaggerated, resulting in thin and unpleasantly sharp wines.

Unfortunately the *DOC* regulations for Piedmont allow for high yields of Barbera. In the opinion of one expert on Italian wines, American Daniel Thomases, the quality of Piedmont's Barbera-based wines could be considerably improved if the authorities would change the regulations, and lower the allowed yield per hectare, while also raising the minimum alcohol level and lowering the high minimum acid level. In his opinion, these changes would better allow the desirable characteristics of the varietal, such as full body and deep cherry-like fruits, to show through (Robinson, 1994).

There are three *DOC* zones within Piedmont that lend their names to Barbera-based wines—Alba, Asti, and Monferrato. Many viticulturists believe that the Barbera grape originated in the latter zone, as archives from the mid-1200s found in the cathedral of Monferrato refer to a contract for leased vineyards planted to Barbera (Robinson, 1994). Today, however, it is the Barberas of Asti and Alba that are held in the highest esteem. Although there are semi-sparkling, or ***frizzante,*** Barberas made in these two *DOCs,* most Barbera d'Asti and Barbera d'Alba imported into this country is of the dry, still, fairly full-bodied style with fruit-forward flavors. Many of the better Barberas are now spending time in small oak barrels, which adds another layer of spicy complexity to the wine while also allowing the acids to mellow.

Gattinara and Ghemme

In the dialect of the eastern provinces of Piedmont, the Nebbiolo grape is called Spanna. There are seven *DOC* wines made here either partly or fully from Spanna, but the only two found in any quantity in the American market are Gattinara and Ghemme. Both of these wines were recently elevated to *DOCG* status, perhaps in hopes that the increase in stature would encourage local vintners to strive to upgrade the quality of these two historically important wines. In the past decade, progress certainly has been made to correct weaknesses of the past. Through the 1980s, many of the wines from these two *DOCs* and from their neighboring *DOCs* were sloppily made, with inadequate or incomplete malolactic fermentation, or excessive time in oak barrels leading to oxidation. Today, many versions of Gattinara and Ghemme are big, sturdy, fragrant wines that show strong Nebbiolo character, even though their blends include two lesser red grapes. This blending is necessary because high elevation vineyards in the hills of these provinces are cool enough that the grapes have a difficult time ripening. Without the addition of the fruitier, less tannic grapes, the wines would be too harsh and thin. Although Gattinara and Ghemme do not attain the nobility and great intensity of Barolo and Barberesco, it is possible to find well-made, very attractive versions of each.

DOLCETTO

Dolcetto translates literally as "sweet little thing." The wine made from this grape is not actually sweet, but rather a soft, fruity pleasure, often compared to Beaujolais. The comparison is apt, although Dolcetto does have more of a bitter bite from tannin. This serves to nicely balance the savory fruit. The better Dolcettos come from *DOC* zones around the town of Alba. Best consumed young and fresh, they are admirably versatile food wines, matching nicely to roast poultry, lighter veal dishes, many pasta sauces, and risottos.

WHITE WINES OF PIEDMONT

Although renowned for the variety and quality of its red wines, from the light charming Dolcetto to the massive, multifaceted elegance of Barolo, the region does also produce several whites worth seeking out. The best known of these is Cortese di Gavi, which was awarded *DOCG* status in 1999. Made entirely from the Cortese grape, the wine is flinty-edged and full of mineral nuances. Very dry and crisp, Cortese di Gavi (which can also be labeled as Gavi di Gavi or simply as Gavi) has been dubbed "this nation's Chablis" by no less an authority than Hugh Johnson. Mr. Johnson, who is the world's best-selling wine writer, describes Gavi as "Italy's most prestigious white wine."

Of increasing importance are the wines made from the Arneis grape. For many centuries it was used solely as a blending grape for overly-tannic Nebbiolo-based reds, but Arneis is now made into pleasant, round, straightforward whites reminiscent of Pinot Blanc from Alsace. (See Chapter 5). Arneis exhibits a unique hint of blanched almonds in the finish. The best of these whites is the Arneis di Roero, a *DOCG* made from grapes grown in the Roero Hills northwest of the town of Alba.

While demand for Gavi and Arneis is increasing, the most famous white wine from Piedmont remains the sparkling wine, Asti (Figure 6–7). This wine was known as Asti Spumante since it was first made in 1850, but after its elevation to *DOCG* status in 1995, only

FIGURE 6–7 This Asti is in the *dolce,* or sweet, style. Zonin is the producer of the wine.

the word Asti shows on the label. Asti is not made in *la méthode champenoise,* nor is the Charmat process or the bulk method used. Asti, made from the Moscato grape, often does not go through a second fermentation at all. Instead, after the crush, the must is kept in stainless steel tanks at a temperature low enough to prevent fermentation. Batches of the must is racked to new tanks and put through a cool fermentation during which the carbon dioxide is captured inside the sealed tanks. When the required level of alcohol is reached, the temperature is quickly lowered to allow the proper amount of residual sugar. The wine is bottled only as needed. This way the Asti, with its delicate flavors, is always fresh when bottled. Its light, off-dry style has proven very popular in export markets. In sheer volume, it is now one of Italy's most important wine exports.

Moscato d'Asti is a *frizzante* wine, made from the same grape grown in the same region as Asti, but produced in a different style. There is less of a mousse, the alcohol content is lower, and there is more residual sugar. With its light froth, floral bouquet, and touch of sugar, Moscato d'Asti is not a wine to be taken seriously, but rather is a simple charmer to be enjoyed with informal fruit-based desserts.

SUMMARY

There is little doubt that the wines of Piedmont will continue to be evident in the North American market. The easy-to-like and highly affordable Asti and Moscato d'Asti are made primarily by large *négociant* companies and shipped regularly, and in considerable quantities, into U.S. and Canadian markets. The demand for the increasingly elegant Cortese di Gavi is improving, especially in fine restaurants where alert wine stewards have recognized its food compatibility. The soft, likable and reasonably-priced reds of Piedmont such as Barbera and Dolcetto continue to carve out market niches. Unfortunately, the high prices of the great reds of Piedmont will preclude their becoming any more prevalent in North America. Barolo and Barbaresco are made in small quantities and are expensive to produce, so it is not likely those prices will come down. Nonetheless, most producers of Piedmont's noble reds sell everything they can make every vintage.

Tuscany

The Appennines Mountains descend down the boot that is Italy, effectively dividing the country in half, geographically and culturally. In the central portion of the country, east of the mountains and over to the Adriatic Sea, lie the Marches, Abruzzi, and Molise. On the other side of the mountain range, on the western coast, are Tuscany, Latium, and Umbria. All six of these regions produce great quantities of fine wine (Table 6–2) The people of the Marches are proud of their wines and export a great deal of their bone-dry white Verdicchio. The *frizzante* red wine of the Marches, Vernaccia di Serrapetrona, was recently elevated to *DOCG* status. (Do not be confused by the name of the grape. There are several grape varietal in Italy called Vernaccia. Most are white; this one is red.) Considerable quantities of the blended red Montepulciano d'Abruzzo are also exported to the United States where its lush, smooth, fruity appeal is appreciated. (Molise is a very small, rugged, and sparsely populated region. Most of the tiny amount of wine produced there is consumed locally.)

Latium, despite the fact that the metropolitan area of Rome lies at its center, manages to produce close to 90 percent of central Italy's white wines, the most popular in this country being the fresh, crisp Frascati. Landlocked Umbria produces several *DOC* wines with distinct regional character, the best known being the dry white Orvieto. Made in the western section of Umbria, on the border with Tuscany, Orvieto is made in various styles, from *secco* to *amabile* and even the truly sweet *dolce.* Umbria also has two *DOCG* wines. One, Torgiano Rosso Reserva, is a medium-bodied, spicy wine with snappy acids and good structure. The blend is similar to Chianti's

Map of Tuscany

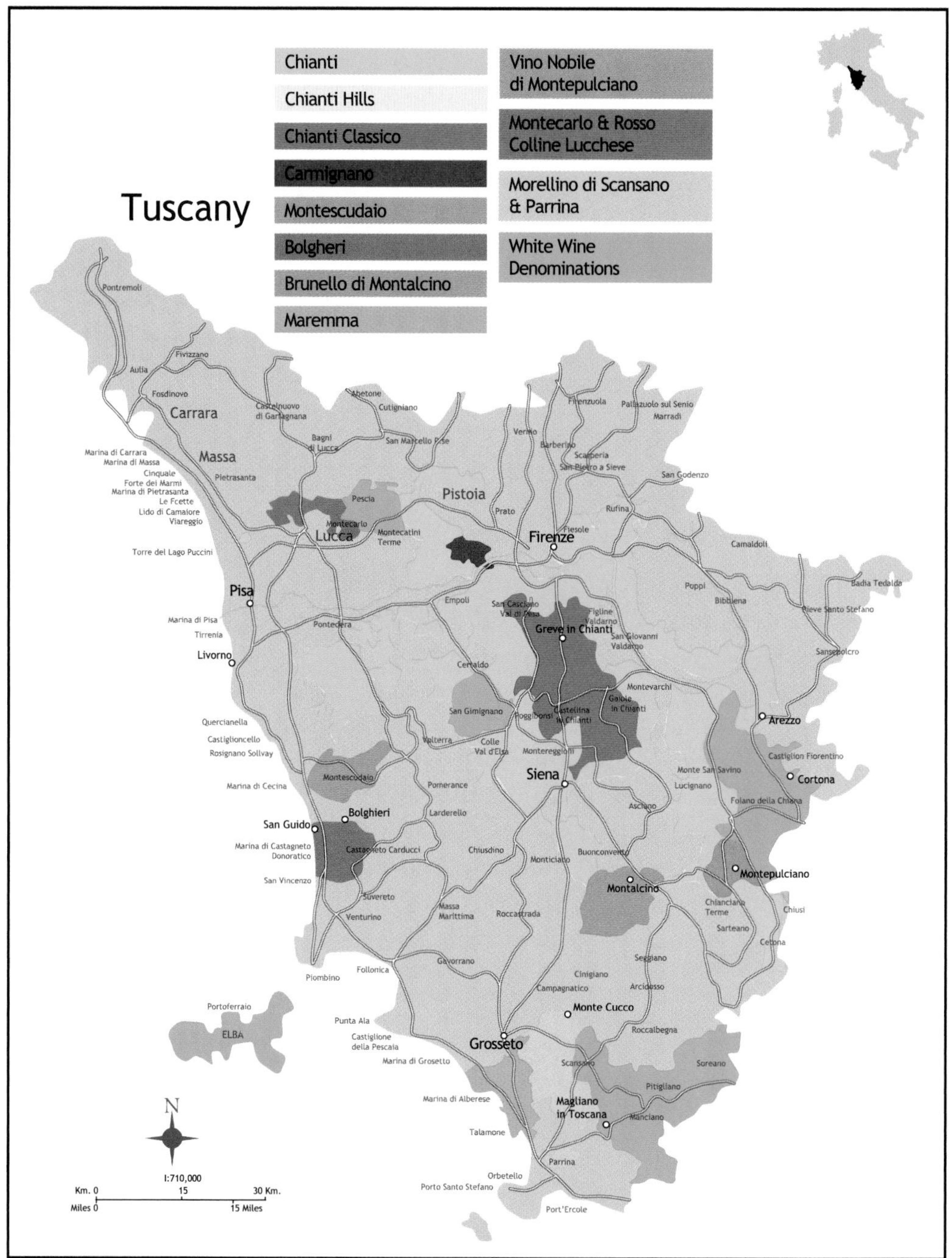

Sangiovese, Canaiolo (a lesser red grape), and small amounts of the simple white grape, Trebbiano. The other *DOCG* from Umbria is Sagrantino di Montefalco, a tannic, full-bodied red made from the local varietal, Sagrantino, grown in the small subzone of Montefalco in central Umbria. The wine can sometimes be a little harsh or awkward, but many producers, and the government authorities, appear optimistic about the potential of this wine.

Though many of the wines from the six regions of central Italy may be popular and successful in the global market, the undisputed leader in wine production is Tuscany. Tuscany, along with Piedmont, is certainly one of Italy's premier wine regions. Its *DOCG* red wines—Chianti, Brunello di Montalcino, Vino Nobile de Montpulciano, and more recently, Carmignano—are well known and widely appreciated around the world. In addition to these proud reds, Tuscany also produces many appealing whites.

Table 6–2 Wine Regions of Central Italy

REGION	SUBREGIONS	CLASSIFICATION	VARIETAL	STYLE
The Marches	Verdicchio	*DOC*	Verdicchio	very dry white
	Vernaccia di Serrapetrona	*DOCG*	Vernaccia	*frizzante* red
	Conero	*DOCG*	Montepuliciano	medium red
Abruzzi	Montepuliciano d'Abruzzi	*DOC*	Montepulciano	light red
	Colline Teramane	*DOCG*	Montepulciano	full-bodied red
Latium	Frascati	*DOC*	Malvasia, Trebbiano	very dry white
Umbria	Orvieto	*DOC*	Trebbiano, 3 others	*secco, amabile,* or *dolce*
	Sagrantino di Montefalco	*DOCG*	Sagrantino	full-bodied red
	Togiano Rosso Riserva	*DOCG*	Sangiovese, Trebbiano, Canaiolo	medium red
Tuscany	Chianti	*DOCG*	Sangiovese, plus 7 others	full-bodied red
	Chianti Rufina	*DOC*		
	Chianti Colli Senese	*DOC*		
	There are five other subregions within the Chianti DOC, but these names rarely show on labels.			
	Chianti Classico	*DOCG*	Sangiovese, plus 7 others	full-bodied red
	Brunello di Montalcino	*DOCG*	Sangiovese	full-bodied red
	Rosso di Montalcino	*DOC*	Sangiovese	medium-bodied red
	Vino Nobile de Montepulciano	*DOCG*	Sangiovese, plus 3 others	full-bodied red
	Carmignano	*DOCG*	Sangiovese & Cabernet	full-bodied red
	Vernaccia di San Gimignano	*DOCG*	Vernaccia	dry white
	Bolgheri	*DOC*	Several are allowed	rosé; some dry white
	Vin Santo	*DOC* [1]	Trebbiano, Malvasia	sweet white

[1]There are 12 *DOC* subregions that can produce Vin Santo.

As we saw earlier, Tuscany was one of the first wine regions in Europe. The Etruscan people started making wine in what is now Tuscany shortly after arriving there from Asia Minor. By 1000 BC the Etruscans were trading their wine with the Greeks. The Etruscans (from whom the Etruscan Coast and Tuscany itself got their names) flourished in central Italy until their territories were absorbed into the Roman Empire in the third century AD. Viticulture has been a proud tradition ever since. Some wine-producing estates in Tuscany have been owned by the same families since the Middle Ages. For instance, the Mazzei family has made wine in Tuscany for over 1,000 years, and has owned Castello di Fonterutoli in Chianti since 1435. The first recorded mention of Chianti wine was in a letter written by one of their ancestors in 1398 (Schoenfeld, 2005). The pattern of landownership in which the majority of land in Tuscany was owned by a few wealthy, noble families and the Roman Catholic Church, while the land was worked by sharecroppers, started to die out in the 1950s and 1960s. The ensuing two decades saw a decrease in investment and thus a deterioration of vineyards and cellars, and a plummeting in the quality of wine. Starting in the 1980s, landowners became more directly involved with the management of their estates (as is the case with the Mazzei family) or the estates were sold to new owners, either individuals or companies, that had the capital to upgrade vineyards and winemaking facilities, as well as the business acumen to effectively promote their wines in domestic markets and abroad. The new breed of Tuscan vintners are showing an admirable respect for age-old traditions while still engaging in creative and exciting innovations. The improvement in the quality and variety of Tuscan wines in the past 20 years is nothing short of remarkable.

The Tuscan countryside is famously hilly—a mere 8 percent of the land is flat (Figure 6–8). The hillside vineyards, with their elevation and good exposure to sunlight during the day, but chilly and calm evenings, produce very high quality grapes with good aromatics and a solid backbone of acidity. At the core of Tuscan winemaking for centuries has been the Sangiovese grape. The undulating hills of Tuscany and its fertile rivers valleys contain many different zones where the Sangiovese and several other varietals, including the whites Trebbiano and Malvasia, can thrive in the temperate climate and the calcium-rich clay and loam of the hillside plots.

CHIANTI

In the United States, the best-known Italian wine is undoubtedly Chianti. It may also be the most misunderstood. Because much of the Chianti imported during the 1960s and 1970s, in those ubiquitous straw-covered flasks (Figure 6–9), was insipid, sharp, and boring, many Americans came to think of this as a mediocre, affordable wine to be served at little cafés, or around the kitchen table, with simple pizza or pasta. Actually the Chianti of today is not the Chianti of your parents' era. Elevation to *DOCG* status in 1984 started the turn-around, with the most important change being the mandated lowering of yields. Vintners, perhaps inspired by the new prestige that the *DOCG* designation brought to their appellation, invested in improved facilities and new technology. Now Chianti made by reputable producers with grapes from the best vineyard sites can be truly world class, exhibiting bright berry/cherry aromas, complex hints of anise or licorice, solid backbone, and suave elegance,

The Chianti region is divided into seven official subregions, each with its own slightly different terroir. These subregions are Chianti Colli Aretini, Chianti Colli Fiorentini, Chianti Colline Pisane, Chianti Colli Senesi, Chianti Montalbano, Chianti Rufina, and Chianti Montespertoli (the most recent). Not all Chianti labels specify the subregion. Rufina is the one most often seen.

Traditionally Chianti was a blend of Sangiovese, four lesser red grapes, and two whites, Trebbiano and Malvasia. The *DOC* laws of 1963 mandated that all these grapes be used. The objective was to have the whites lighten the wine and decrease the impact of its tannins. Fortunately, the laws were changed in 1999, and now stipulate that Chianti can be between 75 and 100 percent Sangiovese. The wine may continue to include the four other reds and up to 10 percent of the whites, but this is no longer required. Moreover, the use of up to 10 percent other reds is also now allowed, which frees vintners to blend in non-indigenous varietals such as Bordeaux's Cabernet Sauvignon, which flourishes in the terroir of Tuscany. These changes have added immeasurably to the quality and complexity of the better Chiantis.

FIGURE 6–8 The vineyards of Tuscany are often steep, rolling hills.

FIGURE 6–9 Italian women making straw flasks the traditional way for wine bottles.

Chianti Classico is a separate region, and is itself now a *DOCG.* It is not a subregion of the general Chianti *DOC.* Located at the very heart of the Chianti region, indeed in the very heart of Tuscany, Chianti Classico is in the central hills, north of Siena (Figure 6–10). The geographic delimitation of the region was first mentioned in an edict from the Grand Duke of Tuscany in 1716, one of the first wine regions to be so defined. The boundaries of Chianti Classico have been made more inclusive of surrounding areas with the same terroir, and Chianti Classico is now larger than any of the Chianti subregions. The Classico region is itself further divided into smaller zones, grouped around the nine communes, or villages, located there. These villages are said to each have a distinct terroir. Since labels do not always indicate the village, the distinctiveness of the various terroirs is difficult to confirm. It is true, though, that wines from the southern towns, nearer to Siena, are fuller and more complex, while those from the towns further north tend to be less fruit-driven and have firmer structure.

FIGURE 6–10 Vineyards in the central hills of Chianti.

The traditional blend for Chianti Classico was the same as for the rest of Chianti: Sangiovese, the lesser red Canaiola to soften the acids, two other reds in smaller parts, and a small percentage of Malvasia and one other white grape to lighten the wine. The composition and quality of the wine changed significantly after the enactment of the *DOCG* legislation of 1984. Those laws, first mandated a much smaller yield, second, allowed the amount of white to be vastly reduced, and third, permitted up to 10 percent of "other red varietals." This latter regulation was interpreted by vintners in the Classico region as a green light to blend in some of the Cabernet Sauvignon that had been originally planted in the late 1940s. As was seen earlier, this practice later became prevalent throughout Chianti.

Many of the producers in Chianti Classico have banded together in a consortium that is pledged to maintain the highest standards of winemaking. Wines produced by members of the consortium display the black rooster that is the group's logo. The consortium is now working closely with the Italian government to further upgrade the quality of their wine through research and experimentation into improved viticultural practices, such as better matching of clones to vineyards' terroirs, more effective trellising, and vine density. The consortium remains more committed than ever to keeping their region in the vanguard of Tuscan wine production.

SUPERTUSCANS

In 1948, a gentleman vintner in the Maremma region on the coast of Tuscany did something unheard of at the time. The Marchese Mario Incisa della Rochetta planted Bordeaux varietals in the vineyards around his estate near the village of Bolgheri. He had received Cabernet Sauvignon and Cabernet Franc cuttings from his friend who owned Château Lafite-Rothschild. The Marchese made small batches of deep red wine from these vines and aged them in small oak barrels. He called his creation Sassicaia, and shared it with family and friends. Many of them were very impressed that such a nontraditional wine, made from non-Italian grapes in a non-Italian style, could be so good.

One of the Marchese's cousins, himself a vintner, saw the commercial potential for Sassicaia. The cousin, Marchese Niccolò Antinori, and his son, Piero Antinori, brought Sassicaia to market in 1968. They released it in Italy and in selected export markets, where it was received enthusiastically. Soon the Antinori family started producing their own wine from an untraditional blend, this time from grapes grown at their estate in Chianti Classico (Figure 6–11). Their Tignanello, a blend of 80 percent Sangiovese and 20 percent Cabernet Sauvignon, was also aged

FIGURE 6–11 A large tasting room in an Antinori cellar. The Antinori family has been making wine in Chianti for generations.

in small barrels. It, too, was well received, both by critics who gave it high grades similar to those awarded to Sassicaia, and by consumers who seemed willing to pay whatever necessary to acquire the new wine. Sassicaia and Tignanello were soon being referred to as "Supertuscans." They, and the similar blends that followed, were among Italy's most expensive wines, although they were officially designated as mere *vino da tavola* because they were made from varietals not allowed by the *DOC* laws. When the *indicazione geografica tipica* classification was created in 1992, the Supertuscans at last were above the lowest *vino da tavola* level and could show the name of their home region on their labels. Producers of the best Supertuscans are hopeful of attaining eventual *DOC* classifications for their individual wines. The only Supertuscan to achieve that honor to date is the original: Sassicaia was elevated to *DOC* status in 1998. This is fitting recognition for an extraordinary wine that began an amazing trend towards increased quality and a more modern style in Tuscan wines and that brought international attention to that quiet hilly home-base where most of the Supertuscans are made—Chianti Classico.

Brunello di Montalcino

In 1888 Ferruccio Biondi-Santi, who owned an estate near the small town of Montalcino in southern Tuscany, made a wine from what he claimed was a superior clone of Sangiovese. He called the clone Sangiovese Grosso or Brunello, and named his new wine Brunello di Montalcino. To this day, this is the only important Tuscan red made entirely from Sangiovese (Figure 6–12).

More important to the wine's superior quality than clonal choice is the climate around the historic town of Montalcino (Figure 6–13). In this area south of Florence, the temperatures are warmer and precipitation lower than in other sections of central Tuscany, while the open topography and gentle winds assure cool evenings. These factors allow the Sangiovese to ripen fully while maintaining solid acid balance. Typically, the better producers do not pick until well into October. The wine is then put through a slow fermentation with extended maceration to extract maximum color and flavor. *DOCG* regulations mandate a minimum of 36 months of oak aging,

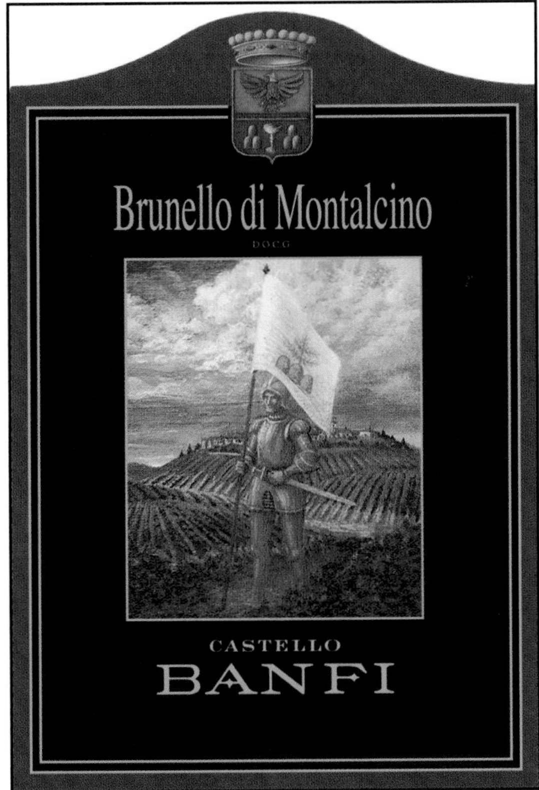

FIGURE 6–12 Brunello di Montalcino is one of Italy's wines made of 100 percent Sangiovese grapes. Brunello is the local name for the Sangiovese grape.

FIGURE 6–13 In the hills of Tuscany near the town of Montalcino stands an eleventh-century castle that the wine producer Banfi has transformed into a state-of-the-art winery.

but some producers exceed that time. Brunello di Montalcino is the fullest, richest, and most long-lived of Tuscany's Sangiovese-based wines.

The long aging requirements for Brunello di Montalcino can put a wine producer into a cash-flow bind, so as the late twentieth century saw an increase in the production of the great Brunellos, there was a corresponding increase in the production of the lesser *DOC* wine, Rosso di Montalcino. This light, fruity wine can be released after only one year, thus generating revenues while producers give their Brunello the extra time in oak that makes it so smooth and rich.

Vino Nobile di Montepulciano

Montepulciano is a quaint village, due east of Montalcino, across a wide swath of the Chianti *DOC.* Here vintners for hundreds of years have made a red wine based on their clone of the Sangiovese, called locally Prugnolo. The same grapes allowed in Chianti are blended here with the Prugnolo to produce a deeply-colored, quite tannic wine whose initial hardness and restraint gives way to deep, warm flavors. Oak aging accelerates the process of softening and opening. A minimum of two years in cask is required, and three for *riserva.* Fortunately most producers have rejected the traditional Slovenian oak and substituted small French *barriques.*

Vino Nobile di Montepulciano is often referred to as being somewhere between plain Chianti and Brunello di Montalcino in quality. The producers of this historic proud wine resent that comparison and are determined to raise the overall quality of wine from their region. The best of these wine unquestionably do deserve the *DOCG* classification, and as improvements are made, more of the wines made here will be worthy of the adjective "noble" that is part of its name.

Carmignano

Tuscany's newest *DOCG* was elevated to that status in 1988. Like so many of Tuscany's wines, Carmignamo has a long history. It was mentioned in the edict of 1716 by the Grand Duke of Tuscany that defined the boundaries of the region's first delimited wine zones. Wine has been made here since the Middle Ages, perhaps because there is a ready market for wines in the large city of Florence, only 10 miles to the east. The hills here are not as high as in nearby sections of Chianti. Lower elevation means that the average temperature is a little warmer, thus allowing the Sangiovese grapes, which are the basis of Carmignano as they are for Chianti, to ripen more fully. Therefore, Carmignano tends to be less acidic than Chianti, with fruit flavors more fully developed. Adding to the complexity and depth of the wine is the Cabernet Sauvignon that has been allowed since the region became an independent *DOC* in 1975, previous to which it was part of one of Chianti's subregions. Carmignano, in other words, was the first Tuscan *DOC* to allow Cabernet Sauvignon as part of the blend. Presently up to 15 percent Cabernet Sauvignon and/or Cabernet Franc is allowed. Production is small, with only 35,000 cases being made each year. As demand for Carmignano increases, it is hoped that other producers will join the eleven estates presently making the wine.

Vernaccia di San Gimignano

Italy's first *DOC,* Vernaccia di San Gimignano (Figure 6–14), became the country's first white *DOCG* in 1993, but Vernaccia has a long history. There are references to the wine in the archives of the town of San Gimignano dating from 1276 (Robinson, 1994). The Vernaccia grape does very well in the sandstone-based soils in the province of Siena, where temperatures are moderate enough to allow the grapes to retain crisp, vibrant acidity. The wine does not exhibit particular depth of flavor, but has enticing hints of citrus and is a perfect companion to fish or pasta in cream sauce. Little of the wine is exported, as droves of tourists consume most of the wine in the cafés and country inns of this historic corner of Tuscany.

Vin Santo

Translated, Vin Santo is "holy wine." The traditional explanation for the name of this sweet amber wine is that it was originally made in that style so as to be palatable to children during the

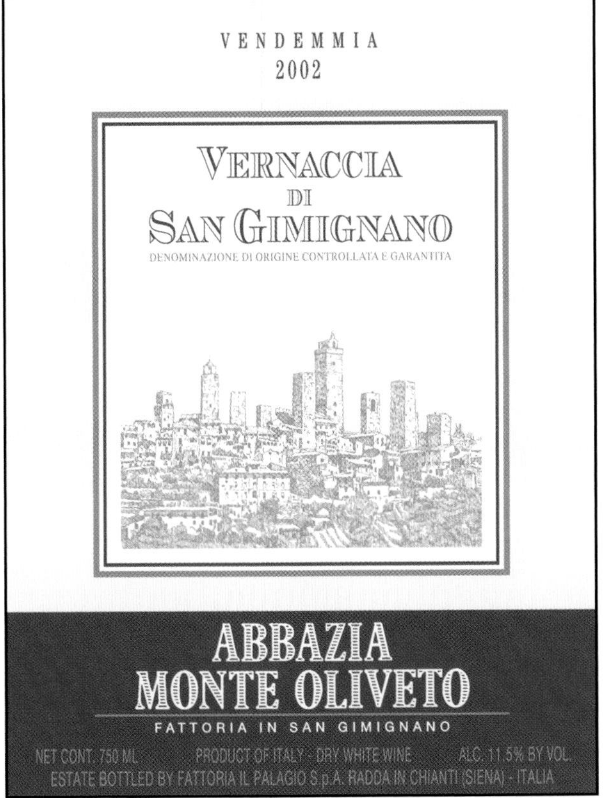

FIGURE 6–14 Vernaccia di San Gimignano was Italy's first white wine honored with *DOCG* status.

Mass. Vin Santo is made throughout Tuscany. There are 12 *DOC* zones where production of Vin Santo is controlled. If made outside those zones, it is classified simply as *vino da tavola.*

Traditionally Vin Santo is made from the local white grapes, Trebbiano and Malvasia, which have been partially raisined, or dried on straw mats, in a room that is carefully ventilated (Figure 6–15). After the juice of the grapes has adequately evaporated from drying out (a process that can take several weeks), the grapes with their high concentration of sugar are crushed and allowed to ferment. After a lengthy fermentation, the wine is aged for as long as 5 years in small oak barrels. The style of the wine varies considerably depending on the extent of the raisining and the length of the fermentation. The dry style, which resembles Fino sherry (see Chapter 7), is served as an aperitif. The sweet style makes a scrumptious dessert wine, rich but balanced, full of hints of toasted nuts and honey.

SUMMARY

The improvement in the wines of Tuscany in the past 15 to 20 years is nothing short of phenomenal. With the every increasing understanding of ancient vineyard sites, improved viticultural techniques, modern wine-making technologies and equipment, more sophisticated marketing, and close cooperation between vintners and government authorities, Tuscany, at the opening of the twenty-first century, is poised to cement its emerging reputation as one of the true stars of the international market for fine wines.

Tre Venezia

In the early Middle Ages, the ancient city of Venice was at the center of the cultural and commercial development of northern Italy. So crucial was the Republic of Venice as a trade partner that the Byzantine Empire granted it a full tax exemption. Located on the Adriatic Sea, Venice was well positioned to receive goods, including sweet wines, from countries in the Byzantine Empire at the eastern edge of the Mediterranean, and move them to markets of northern Europe, just over the Alps. To satisfy the demand for dry wine within the city, its merchants purchased

FIGURE 6–15 Trebbiano and Malvasia grapes drying for Vin Santo at the Isole e Olena winery.

Map of Northeast Italy

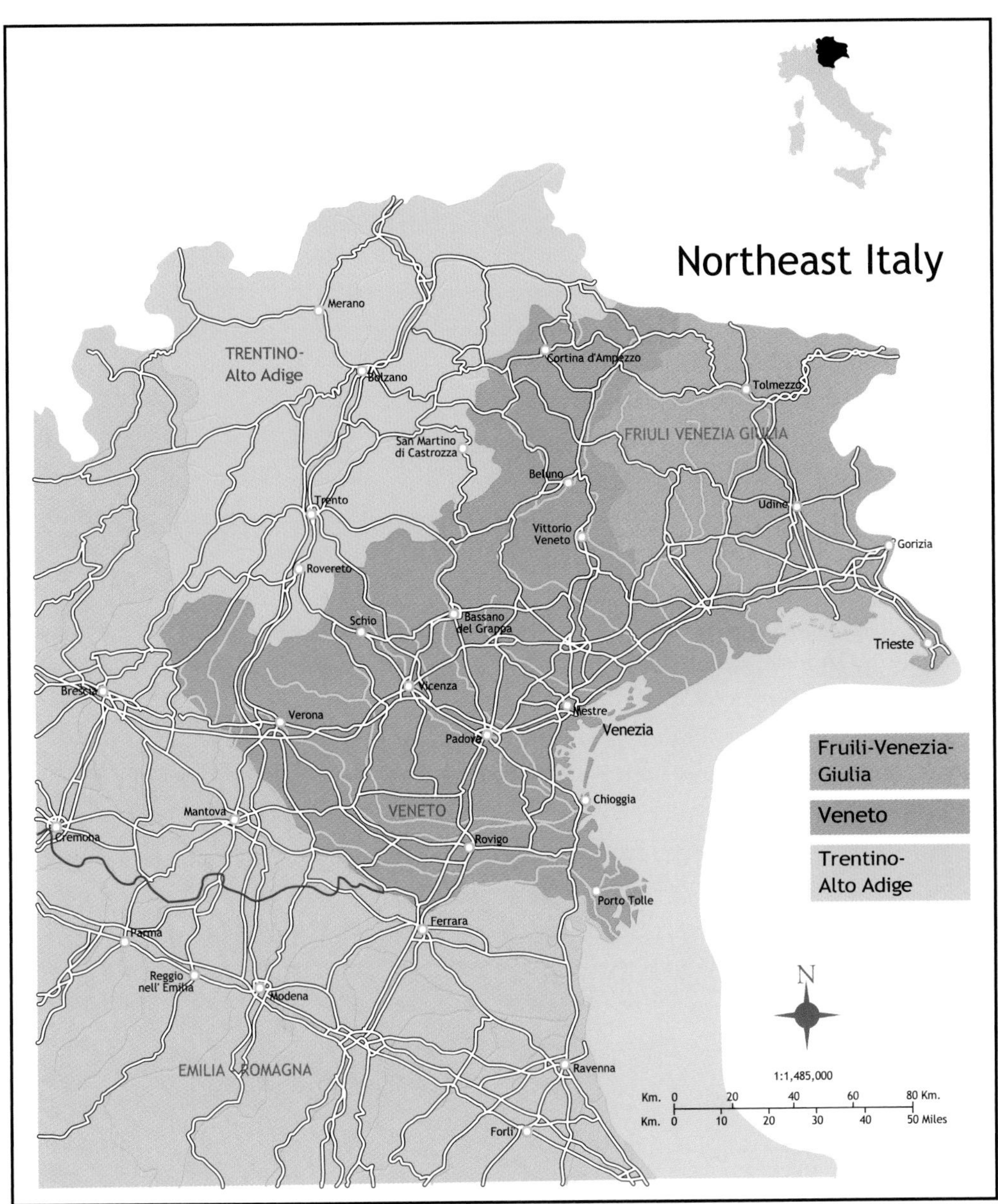

great quantities of locally made wines from the three regions surrounding it. By the late Middle Ages, Venice was the richest and most vibrant port city on the whole peninsula.

Due to their strong historical connection to the city of Venice, the three wine regions of the northeastern corner of Italy are referred to collectively as the Tre Venezia, the "Three Venices." The three regions, Friuli-Venezia Giulia, Trentino-Alto Adige, and Veneto, lie north of the great River Po and south of the Alps. These three regions are not large producers of wine, accounting for only abut 15 percent of Italy's total production. However, a large share of their wines are in the *DOC* category—fully 30 percent of the regions' wine qualify for that status.

Friuli-Venezia Giulia

Friuli-Venezia Giulia, usually referred to simply as Venezia, is one of the few Italian wine regions in which white wine plays a more important role than red. Tucked up against the Alps in the far northeastern corner of Italy, Friuli lies just across the border from Austria. The culture, language, and cuisine, including wine, of Friuli reflect that influence, and that of Austria's neighbor to the north, Germany. The vibrant white wines of Friuli are made with the same attention to detail as are the wines of those two countries. One technique imported from Germany in the late 1960s was cold fermentation for white wines, which greatly extends the duration of the process and allows the aromatics and inherent flavors of the grapes more opportunity to develop (See Chapter 3). There is a precision and a focus to the wines of Friuli that seems far removed from the carefree, sensorial, almost joyful approach to food and wine that we associate with Italy.

Most of Friuli's wines are named for the varietal and the region (Figure 6–16). Because the northern part of the region is right up against the foothills of the massive Alps, it is not suited to viticulture, being too high in elevation, too rocky in soil composition, and too rugged to be feasibly cultivated. There are some excellent vineyard sites in the sloping foothills of the Alps, but most of Friuli's vineyards are located on the flat plains extending inland from the Adriatic Sea. The unique combination of mountain air and maritime breezes and humidity make an ideal situation for viticulture: warm sunny days, cool evenings, and adequate precipitation. This mesoclimate explains why Friuli's whites are so zippy and flavorful. Grapes get the sun and moisture they need to ripen up and evolve their flavors, while the drop in temperature each evening during the growing season assures that the grapes' acids retain their vibrancy.

The varietals that do particularly well in this sunny but cool region are Pinot Grigio, Pinot Bianco, and Sauvignon Blanc. The white grape that Friulians think of as their own is the Tocai, here called the **Tocai Friulano.** "Tocai" is a Slavic word, which is not surprising since Friuli is bordered on the east by Slovenia, Croatia, and further east, Hungary. Tocai Friulano is not the same grape used in Hungary to make the famous Tokaji Aszú dessert wine. There is some speculation, however, among leading botanists that Friuli's Tocai may be the same grape as Hungary's Furmint (Robinson, 1994), but that genetic connection has not been definitely proven. No matter what its origin, there is no doubt that Tocai Friulano is the region's most popular white grape; fully 20 per-

FIGURE 6–16 This Pinot Grigio is named for its varietal (Pinto Grigio) and Region (Friuli).

cent of total vineyard acreage is devoted to this grape (Robinson, 1994). Tocai is used on its own to make varietal wines, and it is often blended with other varietals to make some of the many enticing blended whites. Also widely planted and popular for blending, is the indigenous varietal, Ribolla Gialla. Red grapes are also planted in Friuli and make up almost 50 percent of total production (MacNeil, 2001). Among the more popular red varietals are the local Schioppettino and Refosco, along with the Bordelais varietals that are better suited for export markets. Merlot, Cabernet Sauvignon, and Cabernet Franc are all made into varietally named wines.

Friuli-Venezia Giulia is divided into seven *DOC* subzones, two of which are in the hilly eastern area near the Alps. These two, Collio (the hills) and Colli Orientali (the eastern hills) have calcareous stony soil. In the flatlands below the foothills, the soil is alluvial, with quantities of sand and pebbles having been deposited over the millennia by the numerous rivers that slice through the plains. The five *DOCs* of the plains are, moving from west to east, Lison-Pramaggiore, Latisana, Grave del Friuli, Aquileia, and Isonzo. Of these seven *DOCs,* only Collio, Colli Orientali, Grave, and Isonzo are likely to be seen in North America (Table 6–3).

The Wines of Friuli-Venezia Giulia

The wine producers of Friuli-Venezia are committed to protecting their region's reputation for racy, balanced, clean white wines, in which varietal character is allowed to shine through and

Table 6–3 Wine Regions of Tre Venezia

REGION	SUBREGION	CLASSIFICATION	VARIETAL	STYLE
Friuli-Venezia Giulia	Friuli Grave	*DOC*	Sauvignon Blanc, Tocai	dry white
	Isonzo	*DOC*	Tocai, Pinot Grigio, Pinot Bianco, Sauvignon Blanc Cabernet Franc, Merlot	dry white light red
	Collio	*DOC*	Tocai, Pinot Grigio, Pinot Bianco	dry white
	Colli Orientali	*DOC*	Tocai, Sauvignon Blanc, Verduzzi, Merlot, Cabernet Franc, Refosco	dry white light red
	Ramandola	*DOCG*	Verduzzo	white *(dolce)*
Veneto	Soave	*DOC*	Garganega, Trebbiano	dry white
	Soave Superiore	*DOCG*	Gargenaga, Trebbiano	dry white
	Bianco di Custoza	*DOC*	Trebbiano, Garganega, 5 others	dry white
	Bardolino	*DOC*	Corvina, Rondinella, Molinara	light red
	Bardolino Superiore	*DOCG*	Corvina, Rondinella, Molinara	light red
	Valpolicella	*DOC*	Molinara, Rondinella, Corvina	light red
	Amarone	*DOC*	Molinara, Rondinella, Corvina	full red
	Recioto di Soave	*DOCG*	Garganega, Trebbiano	sweet white
	Proscecco	*DOC*[1]	Prosecco	sparkling
Trentino-Alto Adige[2]	Alto Adige	*DOC*	19 varietals are allowed Schiava is predominant red; Silvaner is predominant white	dry white medium red sparkling
	Santa Maddelena	*DOC*	Schiava	full red
	Trentino[3]	*DOC*	Schiava, Lambrusca, Lagrein Pinot Bianco, Pinot Grigio, Müller-Thurgau	medium red dry white
	Teroldego Rotaliano	*DOC*	Teroldego	light red

[1]There are two *DOC* regions within Veneto that produce Prosecco.
[2]There are twelve *DOC* in this region, but the four listed are the only ones often seen on labels.
[3]A total of twenty varietals are allowed. Those listed are predominant.

food-compatibility is evident. Friulians, both small private winemakers and larger cooperatives, are resisting the temptation to make the creamy, rich, oaky style of white that seems to have found such acceptance in the U.S. market and in other countries. White wines here are not fermented or aged in oak barrels, so there is no layer of vanilla and toast to smother the natural varietal aroma and tastes. Friuli winemakers also shun malolactic fermentation, believing that racy fresh acids increase the versatility of white wine as a food match. The emphasis here is on indigenous varietals, even though Chardonnay, the most popular of the international varietals, has been planted here since the late nineteenth century. The Friulians make a delightful fruity wine, bursting with peachy aromas and additional layers of citrus flavors, from Ribolla Gialla, a native grape that surprisingly is grown nowhere else, except Greece and parts of Slovenia.

One native varietal the Friulians have helped to create international demand for is Pinot Grigio, which is the Pinot Gris of Alsace in France. Many Pinot Grigios from northern Italy are truly one-dimensional wines, simple and even boring. The best Pinot Grigios from Friuli can have forward, enticing bouquets reminiscent of almonds, pears, and ripe peach. On the palate they have weight and are smooth, almost unctuous, with layers of fruit flavors beautifully supported by racy acidity.

The varietal which the people of Friuli-Venezia consider the most important is Tocai, here called Tocai Friulano. They make their favorite grape into lovely wines, each with its own personality reflecting the terroir of its subzone. Favorite of some congnesenti are the Tocais from the Colli Orientali *DOC,* which exhibit a pronounced peppery spiciness. Others prefer the aromatic Tocai from the Isonzo *DOC,* more lush in texture and more exotic in its spiciness.

The *DOC* region Grave del Friuli (which takes its name from the same root as the Graves region of Bordeaux—both have gravel-based soils) is renowned for the distinctiveness of its wines made from Sauvignon Blanc, shortened here just to Sauvignon. The Sauvignons from Grave are intensely aromatic, redolent of fresh green herbs, and fairly dance with lively acidity.

Although renowned for its whites, Friuli-Venezia also produces red wine. Actually almost 40 percent of its production is red. The predominant red varietal is Merlot, brought here from Bordeaux over a century ago, along with Cabernet Sauvignon and Cabernet Franc. The Merlot and Cabernets, made in Friuli are soft, light and fruity, and a bit lean. More successful are the reds made from indigenous varietals, especially Schioppettino, which is made into a muscular wine with aggressive spice and intense cherry flavor.

Friuli also has a renowned dessert wine, Ramandola. Ramandola is a village in the Colli Orientali *DOC.* The native white grape, Verduzzo, flourishes in its sunny, hilly vineyards to a sugar-laden ripeness. After being harvested, the grapes are allowed to raisin for a short period before being fermented into a lovely sweet wine. Although not as unctuous or weighty as some very sweet dessert wines, Verduzzo di Ramandola can be exquisite: golden in color, emanating inviting honey aromas, full of delicious honey/dried fruit flavors, and beautifully balanced. Verduzzo di Ramandola has recently attained *DOCG* status.

Although the overall production of Friuli-Venezia Giulia is quite modest in size, the percentage of that production that is of high quality and distinctiveness is high. Forty-five percent of wines from this region are *DOC* or *DOCG,* a large number by Italian standards.

THE VENETO

Veneto is the largest of the regions in the northeast and by far the most prolific in wine production. In some years, the Veneto has the highest production of all of Italy in sheer volume. Many of the wines made here are eminently forgettable—simple, straightforward, and totally unremarkable. In the 1960s and 1970s, it was impossible to walk into a liquor store or wine shop anywhere in the United States and not see shelf after shelf in the imported section dominated by the white Soave and the two reds, Bardolino and Valpolicella, from Veneto. Millions of cases of these wines are still produced in the Veneto, many of them by large corporations. Because these corporate brands are so ubiquitous in this country, it is hard for Americans to be aware of the strides being made by smaller producers in the Veneto to improve the quality of their Soave, Bardolino, and Valpolicella. Also lost in that ocean of wine are the truly impressive wines of the Veneto, like Amarone, an intense big red, and the genuinely appealing wines, like Prosecco, a delightful *frizzante* white. The mediocrity of Veneto's *DOC* regions was due primarily to two factors. First, within a decade of the passage of the *DOC* laws, the boundaries of Valpolicella and Soave were hugely extended, from the traditional hillside vineyards out onto the vast plains that previously had been used to grow grains. The second misjudgment on the part of the authorities was to set the yields permitted in these *DOC* regions, as well as in Bardolino, too high. The resulting wines, although ostensibly *DOC,* are too often without character or distinction. Fortunately, there are producers within the Veneto who will not settle for mediocrity in their wines and who are setting their own standards higher than those established by the authorities.

Soave

Made from Garganega and Trebbiano, Soave is a very light, straightforward white wine. Not meant to age, it is best drunk while young and fresh. Soave from better vineyard sites can be smooth and suave (*soave* is Italian for suave), but never great or even particularly complex. The denomination is named for the small town of Soave, which is located among the hills east of Verona in the western section of the Veneto. The Soave region was originally defined in 1927 and was limited to 3,500 acres (1,417 hectares) in the hillsides surrounding the town. The soil is volcanic rock. At the time the official Soave *DOC* was created in 1968, a large section of alluvial plains along the Adige River valley were tacked on to the classic Soave region. This original region is now the Soave Classico *DOC,* and its wines are a definite step up in quality from simple Soave, most of which is made by large cooperatives. Even further up in quality is Soave Classico Superiore, which must be aged at least 8 months in oak before being released. The best Soave Classico Superiore is made by small conscientious producers who keep their yields below the allowed maximum, shun use of the highly productive but inferior Trebbiano Toscana, using instead the more concentrated Trebbiano di Soave, and age their wine in small *barriques* of French oak. Because of the efforts of small producers like Anselmi, Bertani, and Pieropan, Soave Classico Superiore was recently elevated to *DOCG* status.

Recently there has been renewed interest in **Recioto** di Soave, a sweet wine made from raisined Garganega grapes. This time-honored method of concentrating the grapes' sugars, a technique called *passito* in Italian, was widely practiced in Soave. Production, however, was very small, and most of this lovely dessert wine was consumed locally or at least domestically. However, interest in what some consider to be Italy's best dessert wine, and the fact that Recioto di Soave is now a *DOCG,* has encouraged some wineries to explore the commercial viability of the wine for export markets.

The only other dry white wine from the Veneto to have attained any recognition is the Bianco di Custoza, a very light and simple wine. It is a blend of Trebbiano di Toscana (20 to 45 percent), Garganega (20 to 40 percent), and Tocai Friulano (5 to 30 percent) with four other white grapes also allowed. Because such a large percentage of the inferior Trebbiano di Toacana is allowed, the wine tends to be thin and lacking in character, albeit refreshing.

Bardolino

Named for the town of Bardolino on Lake Garda, this very light, fun little wine is made from the same grapes as Valpolicella, but with less Corvina with its structure and weight and more neutral, simple Rondinella (Figure 6–17). The result is an imminently quaffable red, best served slightly chilled. As in adjoining regions, the better, original Bardolino region is now the

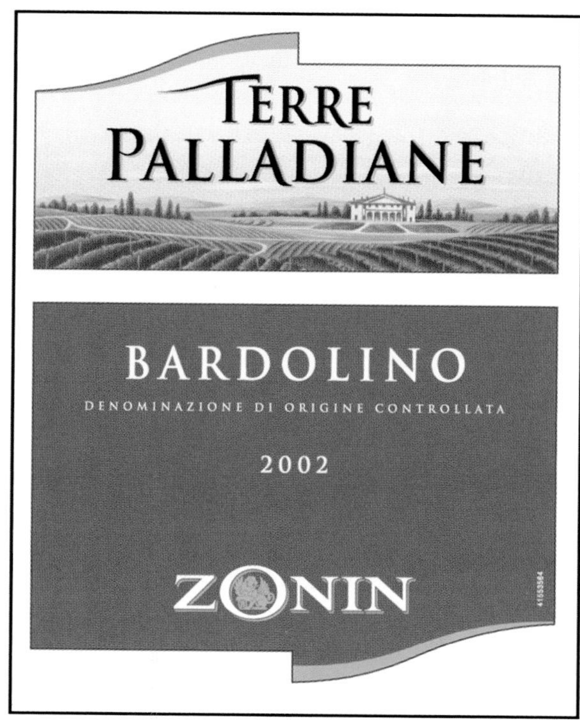

FIGURE 6–17 Bardolino is a light, fruity red wine from the Veneto.

Bardolino Classico *DOC.* Bardolino Superiore, now a *DOCG,* a slightly fuller-bodied and more concentrated wine, requires 1 percent higher alcohol content and at least one year of aging before release. There is also a rosé style called Bardolino Chiaretto. It is made either as a still rosé or is slightly *frizzante.* In the late 1980s, Bardolino Novello was created to compete with Beaujolais Nouveau.

Valpolicella

Like Soave, Valpolicella's defined area was greatly expanded in 1968 when it achieved *DOC* status, more than doubling in size and stretching eastward all the way to the boundary of Soave. The original defined section is now the Valpolicella Classico *DOC,* and its wines are considerably more interesting than simple Valpolicella. Better yet is Valpolicella Classico Superiore, which must have a minimum alcohol content of 12 percent (the requirement is lower for plain Valpolicella) and must receive a minimum aging period of one year. (Plain Valpolicella has no aging requirement.)

The blend specified by *DOC* law for any Valpollicella wine is a minimum of 70 percent Corvina, which has distinctive cherry/smoky bouquet and lively cherrylike flavors. To that basis are blended Molinara for acidity, and Rondinella, which provides a neutral backdrop for the flavors of the Corvina. Sadly many producers go overboard on the high acid Molinara and bland Rondinella. *DOC* law does not allow for a wine from this *DOC* to be 100 percent Corvina, and the better vineyards of this varietal are usually saved for use in Valpolicella Superiore. So the majority of simple Valpolicella, especially that made by large cooperatives or bottom-line oriented corporations, remains an undistinguished wine, often thin and too acidic. Worth seeking out, though, is Valpolicella Classico Supiore, the best of which are excellent wines with full open bouquets of cherry and distinct flavors of cherries with a hint of licorice.

Amarone

Within the Valpolicella region there is a truly impressive wine made that has its own official *DOC.* This wine is Amarone. Highly aromatic, intensely flavored, very full bodied and heady, this is a big sturdy wine to serve alongside braised beef or venison steak or pungent cheeses, or even at the end of the meal, in place of Port. The original name of the wine, Recioto di Valpolicella Amarone (Figure 6–18), has been shortened, but the original name does give insight into how the wine is made. Essentially, Amarone is made from grapes, primarily Corvina, that are allowed to hang until they become very ripe. When picked in late fall, the bunches of grapes are spread on straw mats for two to three months and allowed to dry out, or raisin, slightly to concentrate the sugars as the water of the grapes evaporates. Fermentation is allowed to continue until all sugars are converted, which leads to a natural alcohol level of between 14 and 16 percent. This is a labor-intensive method, with inherent risks—for instance, a spell of wet weather could cause the grapes to rot as they lie in the drying sheds. Consequently, Amarone is not cheap. It is worth the price, though; big but graceful, intense by elegant, redolent with earthy aromas and mocha/chocolate flavors. This is easily the most impressive wine from the Veneto, and it is probably only a matter of time before Amarone is a *DOCG.* Some of the best Amarone is made by the Masi winery, and by Zonin.

Prosecco

One of the most popular wines in the Veneto is its own *frizzante,* the delightful Prosecco. Made primarily from the grape of the same name, with some Pinot Grigio and a little Pinot Bianco blended in, Prosecco is frothy, charming, and usually slightly off-dry. The best Prosecco is made from grapes grown in the hills just north of Venice. By law, Prosecco must be made in the ***metodo classico*** (*la méthode champenoise*) and be labeled by its official *DOC* designations, either Prosecco di Conegliano or Prosecco di Valdobbiadene.

Trentino-Alto Adige

Unlike the other two "Venezias," Trentino-Alto Adige is landlocked. Nestled right up against the Alps, it is also the northernmost of Italy's wine regions (Figure 6–19). Politically and culturally, this region is as infiltrated by northern influences as by Italian ones. This is especially true in Trentino, where German is the primary language. In both parts of this region, the approach to winemaking reflects that Teutonic influence. The wines have the precision and focus that is typical of Germany's and Austria's finest whites. While northern climates are usually more conducive to nice white wine, there is more red wine made in Trentino-Alto Adige than white.

Vineyards are planted everywhere that conditions allow—in the south-facing foothills of the Alps, in alpine meadows, in the steep valleys of the river Adige, which flows through the center

FIGURE 6–18
Amarone is made from partially raisined grapes in the region of Valpolicella.

FIGURE 6–19 St. Madalena in the center of Alto Adige is surrounded by sheltering mountains. The vineyards have soil of volcanic rock and limestone, perfect for the Pinot Grigio planted there along with several local red varietals.

of the region. The Alps provide enough protection from cool continental weather patterns to allow grapes to ripen well, despite the northern location. Soil content is also close to ideal, with a mixture of well-drained volcanic rock laced with limestone in higher elevations and, in lower meadows and along the valleys, clay and sand left by retreating glaciers in ancient times and by the flowing rivers of more modern times. The white grapes that are prevalent are an interesting mix of indigenous varietals, such as the Pinot Grigio and Traminer, and international varietals, like Müller-Thurgau from Switzerland and Burgundy's Chardonnay, which was introduced in the nineteenth century. The array of red varietals is similar. There are local grapes, the most prevalent being Schiava and Lagrein, planted alongside imports. Cabernet Sauvignon, Cabernet Franc, and Merlot are all successfully cultivated. The most important of the white varietals is Chardonnay, which is made into crisp attractive still wines and is the basis for the lively ***spumante.*** The *spumantes* that are made in the *metodo tradizionale,* with Chardonnay as the base with some Pinot Bianco blended in, can rival the best sparkling wines made in the world. The ubiquitous Pinot Grigio can be unusually interesting , with nice depth and roundness and pleasant melon tones. The Müller-Thurgau, a rather ordinary grape elsewhere, can be made into nice zippy wines here. Of the reds, the indigenous Schiava produces the most exciting wines, medium to full bodied, well structured, and packed with black fruit overlaid by hints of licorice.

The wines are named for the varietal and for the specific *DOC* region. There are 12 *DOC* zones, but many of them do not show on labels, since varietal differentiation is more relevant that place of origin. There are four *DOCs* that are likely to show on wines that are exported. (Fully a third of the region's *DOC* wines are exported). The four relevant *DOCs* are Alto Adige; Santa Maddalena, a subzone at the southern part of Alto Adige reputed to be the source of the best Schiava; Trentino; and Teroldego Rotaliano, a red wine made of the local Teroldego grape in Rotaliano, a subzone of Trentino where gravelly vineyards in the flat plains seem ideal for bringing this somewhat rustic grape to an adequate ripeness level. The wine is considered the best red from the whole region, with its ruby color, full body, and ripe fruit. It ages very well. There is now a lighter style of Teroldego that is released early as a novello. To admirers of the old, deep style of Teroldego di Rotaliano, the Novella di Teroldego may seem blasphemous, but it has been received well and may give Beaujolais Nouveau some serious competition. Pinot Noir vineyards were planted in cooler zones in the late 1980s, and many of them show real potential, especially those given oak aging. The Cabernet Francs of the Trentino are also showing increasing promise. As small producers in both Alto Adige in the north and Trentino to the south continue to more carefully match appropriate varietals to the various mesoclimates of their region, the overall quality of the wines is sure to climb.

Southern Italy

The southern part of Italy, much of which is rugged, sparsely populated, and economically disadvantaged, has not been prominent in international wine markets. Campania, Apulia (Puglia in Italian), Basilicata, Calabria, and the islands of Sicily and Sardinia have made wines, like the rest of the country, for hundreds of years. The amount of wine coming out of these regions is prodigious, especially from Apulia and Sicily. Acreage devoted to vineyards is 406,315 acres (164,500 hectares) on Sicily, which is exceeded only by Apulia where 420,000 acres (170,000 hectares) are devoted to wine grapes. However, only 2.5 percent of the wine from Sicily, and 2 percent from Apulia is classified. Because much of the wine from those regions, as in other parts of southern Italy, is used primarily for blended bulk wine or is distilled into spirits, the entire area accounts for only 10 percent of the country's total *DOC* production (Italian Trade Commission Web site).

In the belief that the flood of mediocre wine from southern Italy has contributed to a worldwide situation in which the supply of wine now exceeds demand, thus leading to a general lowering of prices, the Italian government and the E.U. are offering incentives to landowners to reduce production by tearing out vineyards, especially in questionable areas. Some landowners are complying, reducing their acreage, and more carefully managing their crops. Other vintners, whose families have been making quality wine for generations, continue to produce quality wines, never having fallen into the pattern around them of making large quantities of high-alcohol, overly astringent, and highly extracted wine of little finesse or character. It appears that the standards of these few pioneers is spreading to other producers around them, with better wines appearing in many sections of the South.

It is certainly possible to make good wine in the South, despite the impression that most of the southern peninsula and the two islands, with abundant sunshine and hot Mediterranean temperatures, would be incapable of producing quality grapes because there would be inadequate hang-time. However, there are numerous sites well-suited to viticulture because of cooling sea breezes and/or elevation. If producers can continue the trend towards reducing acreage in less-suited areas and concentrate on the better vineyard sites, while working to increase the overall quality of their wines, the south will greatly improve its chances of becoming a player in the export market for wines (Table 6–4).

Campania

When one thinks of Campania, one is likely to think of the lovely old city of Naples or the stunning beauty of the Amalfi coast more than of wine. Despite the grinding poverty of much of the interior, there is some fine wine produced here, and has been for a very long time. It is well documented that one of the most popular wines in the time of the Roman Empire came from Campania. Named Falernian, this big red wine was made from grapes grown in the foothills of Mount Falernus south of Naples. It is claimed that even Julius Caesar enjoyed this wine (Standage, 2005).

In more modern times, the most famous wine from Campania, indeed one of the most famous from all of southern Italy, has long been Taurasi, a full-bodied red made from Aglianico, which thrives in the volcanic soil of the hillside sites where sunshine is abundant but temperatures are moderated by breezes. Recently awarded *DOCG* status, Taurasi has unquestionable

Table 6–4 Wine Regions of The South

REGION	WINE	CLASSIFICATION	VARIETAL	STYLE
Campania	Taurasi	*DOCG*	Aglianico	full red
	Greco di Tufo	*DOCG*	Greco	dry white
	Fiano di Avellino	*DOCG*	Fiano	dry white
Puglia (Apulia)	Locorotondo	*DOC*	Merdicchio, plus 4 others	dry white
	Salice Salentino	*DOC*	Negroamaro	full red
Basilicata	Aglianico del Vulture	*DOC*	Aglianico	full red
Sicily	Bianco d'Alcamo	*DOC*	Verdello, plus 3 others	very dry white
	Marsala	*DOC*	Grillo Plus 2 others	fortified
Sardinia (Sardegna)	Cannonau di Sardegna	*DOC*	Cannonau (90 percent)	medium red
	Vermentino di Sardegna	*DOC*	Vermentino	dry white
	Vermentino di Galluria	*DOCG*	Vermentino	dry white

character. Increasingly evident on wine lists in the United States is another Campanian *DOCG* wine, this time a white, Greco di Tufo. Greco, as can be seen from the name, was brought to Italy by the Greeks. When grown around the village of Tufo and nearby communes, the grape produces an impressively full-bodied dry wine of considerable depth and rich flavors. Another *DOCG* white from Campania is also attracting attention. The Fiano di Avellino is full and rich. It perfectly complements the great mozzarella cheeses and pizzas for which Campania is famous.

PUGLIA

Italy's most prolific wine region, Puglia (Apulia) covers the heel of Italy's "boot." The majority of its wine is distilled into industrial alcohol or is concentrated down into a potent must used to strengthen thinner wines in the north. There are 24 *DOC* zones in Puglia, but few of them are seen in North America. One standout is widely available, however. Salice Salentino, from the Salento peninsula, is a full, fruit-driven, aggressively flavored red that has found favor in the United States. It is made from the local grape, Negroamaro. The Primitivo grape, believed by many botanists to be the forebear of California's Zinfandel, is also widely planted in Puglia, where it is made into several interesting *IGT* wines, and into the *DOC* Primitivo di Manduria. The best white made in Puglia is the *DOC* Locorotondo, a vivacious wine made primarily from the **Verdicchio** grape, with four other grapes allowed to be blended in.

BASILICATA

From this hardscrabble region comes one wine worth seeking out, Aglianico del Vulture. **Aglianico** is an ancient grape believed to have been cultivated in this area for over 3,000 years. The grape does especially well in the foothills of Mount Vulture, where volcanic schist imparts a mineral nuance without burying the grape's natural berry aromas. The wine is highly structured and very dry. The *riserva* must age for five years before release, and two of those years must be in oak. When released, the mature smooth wine is a perfect accompaniment for game meats or pungent cheeses.

The mountainous region of Calabria, south of Basilicata, is more famous for its cheeses, like pecorino and ricotta, than for its wines. There are some respectable red wines made from the native varietals, Primitivo, Aglianico, and Gaglioppo, as well as some pretty whites from Greco and Malvasia.

Sicily and Sardenia

The two islands off the coast of Italy, Sicily which is right off the "toe" of the boot, and Sardenia to the west of the peninsula in the Tyrrhenian Sea, both have long histories of wine production. The best-known wine from the islands is Sicily's Marsala, a justifiably popular fortified wine, which has been made for over 200 years (Figure 6–20). It can be fortified with both additional alcohol and with concentrated must made from very ripe grapes. For several decades in the mid-twentieth century the quality of Marsala suffered a downturn. It became common practice to sweeten the wine with various syrups, thus hiding its character. That type of degradation has almost disappeared, and several producers are returning to the traditional style of naturally sweetened and well-matured wine. The best Marsala, designated Vergine, is made without the heavy must and, after fortification with pure alcohol, is aged in wooden casks for up to 10 years. Obviously popular with cooks for the delicious sauces made from it, Marsala could regain its status as a delicious beverage if *DOC* laws governing its production can be streamlined and tightened.

Another Sicilian wine of a very different style is finding its way into foreign markets. Bianco d'Alcamo, a dry white, is made from the same grapes as Marsala. It is medium bodied and has unique almond aromas. There are several other *DOC* whites made on the island (95 percent of Sicily's *DOC* production is white, including Marsala) (Figure 6–21).

Sardinia is 150 miles off the coast, thus isolated from the rest of Italy. Culturally and viticulturally, the island has been heavily influenced by Spain, its neighbor to the west. Both its most famous red, Cannonau di Sardegna, and its most famous white, the *DOCG* Vermentino di Gallura, have Spanish roots. The Cannonau grape is the Garnacha of Spain, from whence it was brought to Sardinia centuries ago. (The Garnacha of Spain is originally from France's southern Rhône, where it is called Grenache.) The *DOC* Cannonau di Sardegna covers the whole island. The wine can be made in a dry style or can be semisweet and fortified. The best is designated *riserva.* The Vermentino grape of Sardinia's dry white wines was brought from Spain to Corsica in the fourteenth century, and from there to the narrow coastal region of Liguria, just south of Piedmont. It soon became widely planted throughout southern France and parts of coastal Italy.

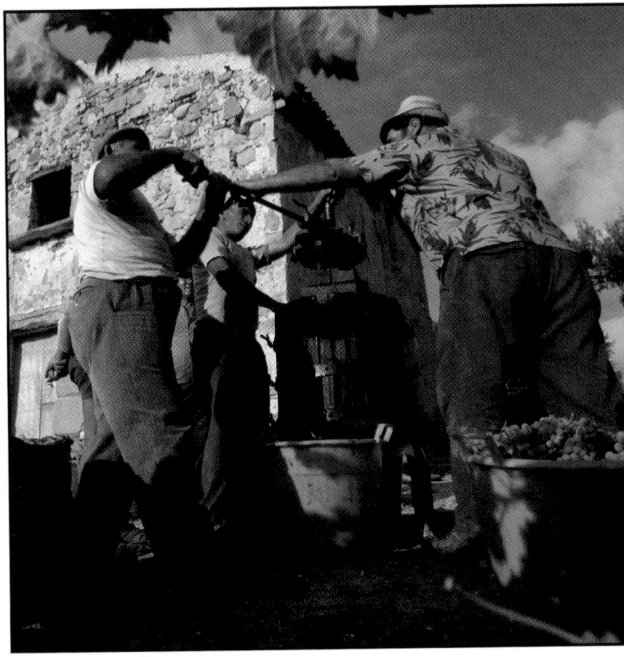

FIGURE 6–20 Workers at the Speranza family winery press grapes for Sicily's famous fortified wine, Marsala.

FIGURE 6–21 The climate of this vineyard in Sicily with dry pebbly soil and baking hot sun is moderated by ocean breezes.

But the grape did not show up on Sardinia until the late nineteenth century when it was brought from Liguria to Gallura, the island's northernmost tip (Italian Trade Commission Web site.) Although the grape is planted throughout the island, the best Vermintino is still grown in Gallura, where fierce winds from the Alps keep temperatures moderate and the granite-based soil imparts a distinctive mineral character. Vermentino di Gallura has good body and vibrant acidity well balanced by fruit. Locally it is served with calamari fritto, baked cod, and other seafood dishes.

In looking to the future of wine production in the south of Italy, one can only hope that the trend away from quantity and toward quality will continue to gather steam. Many experts are optimistic that the enthusiastic reception in the United States to Campania's Taurasi and Greco di Tufo, Puglia's Salice Salentino, Sardinia's Cannonau and Vermentino, and the many other exciting wines made from native varietals will encourage producers in the southern peninsula and the islands to become ever more responsive to consumers' sophisticated tastes and the increasing demand for interesting wines. In the words of one expert, Deidre Magnello, a District Manager for Zonin USA, a major importer of Italian wines, "After 4,000 years of producing wines that never left home, southern Italy is just now receiving deserved recognition for its varied and value-priced wines" (Deidre Magnello, personal communication, January 9, 2006).

Summary

Italian wines have long been popular in the United States, with a real surge in the 1990s, when imports of Italian wines first outpaced those from France. The increase in sales of Italian wines coincided with the wider acceptance of and respect for Italian cuisine during that same period. Perhaps Americans felt the Italian culture was more *simpatico* with our lifestyle, as French cuisine, wines, art, and even language are perceived as elitist, formal, elegant, and sophisticated, and the Italians represent a more relaxed approach. As one observer, Professor Sean Shesgreen of Northern Illinois University has put it, "Italy stands for naturalness, informality, accessibility, practicality, spontaneity, optimism, intuitiveness and family feeling" (Shesgreen, 2003). These qualities all reflect the American approach to entertaining and to dining out.

Whatever the explanation, Italian wines are steadily increasing their share of the U.S. market for wines. For the year ending in June 2005, Italian imports were up 8.5 percent over the comparable figure for 2004 and now hold a healthy 8.3 percent market share for all table wines sold. The only country whose wines were imported in higher volume was Australia (Nielsen Report, reprinted in *Wine Business Monthly,* 2005). Pinot Grigio leads the growth in sales of Italian wines (it is now the fourth most popular varietal in this country), followed closely by Chianti and various *spumantes* and *frizzantes.* However, the perennial favorites are being joined more by other, lesser-known Italian wines, such as the top quality wines Barolo and Brunello di Montalcino, and the excellent value, but even lesser-known wines, especially from the south. With the *DOC* laws of Italy becoming steadily more strict, and with more producers aware of the importance of quality control versus mass production, it is safe to predict that Italian wine will improve its share of the North American market in the years ahead.

Review Questions

1. What is the principle red grape of Piedmont, and what style wine is made from it?
2. What is the principle red grape of Tuscany? What style wine does it make?
3. Is there any region in Italy in which white wine plays a more important role than the reds? Which region(s)?
4. What is the difference between a *spumante* wine and a *frizzante* wine? Are they made in the same method?
5. Is Cabernet Sauvignon allowed to be blended into Chianti and Chianti Classico?
6. List the two methods by which a *DOC* or *DOCG* wine may be named. Give an example of each way of naming these classified wines.
7. What is the meaning of the term *indicazione geografica tipica?*
8. What people originally brought the Greco grape to southern Italy? What is the most important wine now made from that grape? What region is it from?
9. How many *DOCG* zones are there presently? How many are red? White?
10. Define the term Supertuscans.

CHAPTER 7

Spain and Portugal

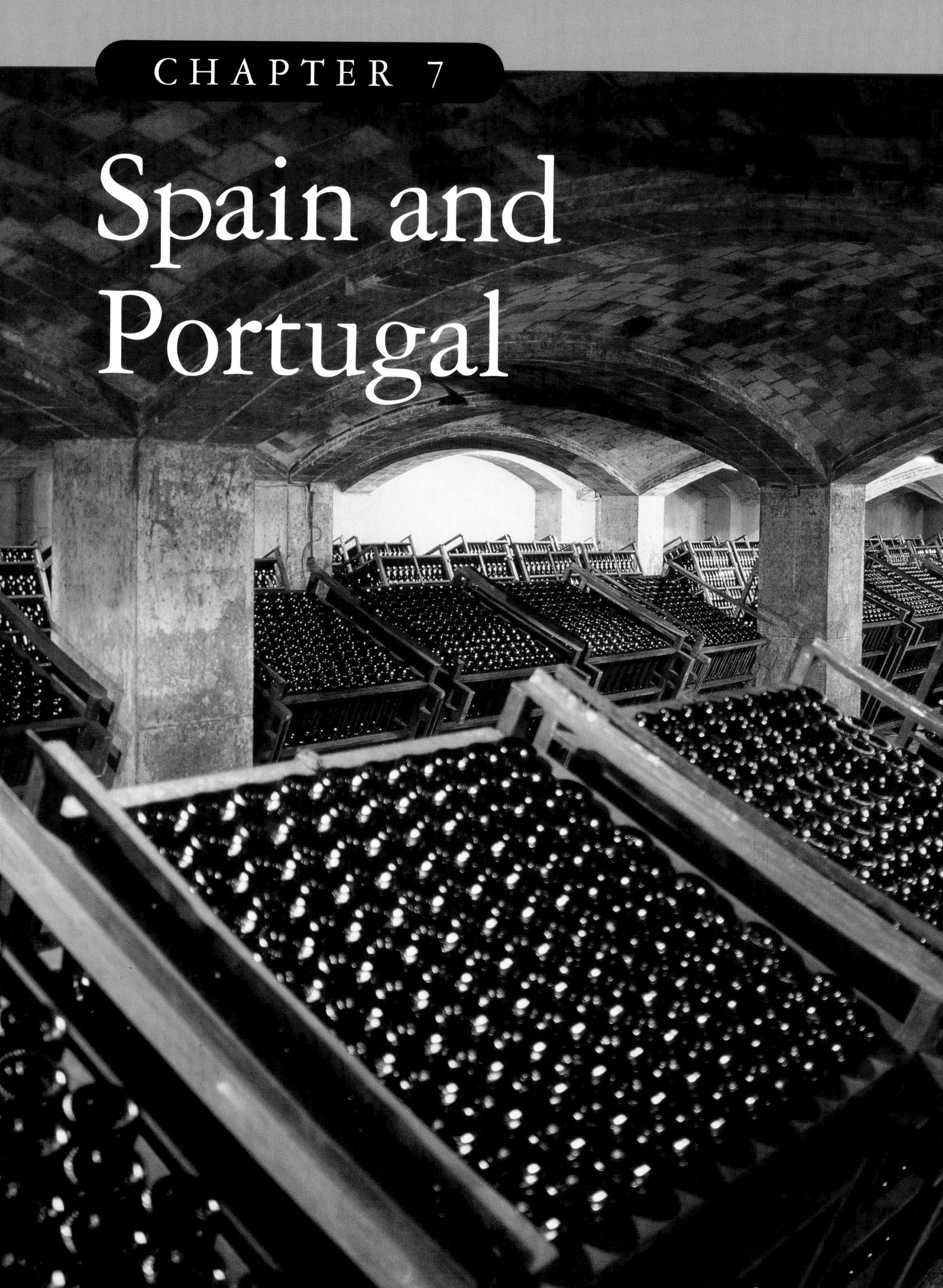

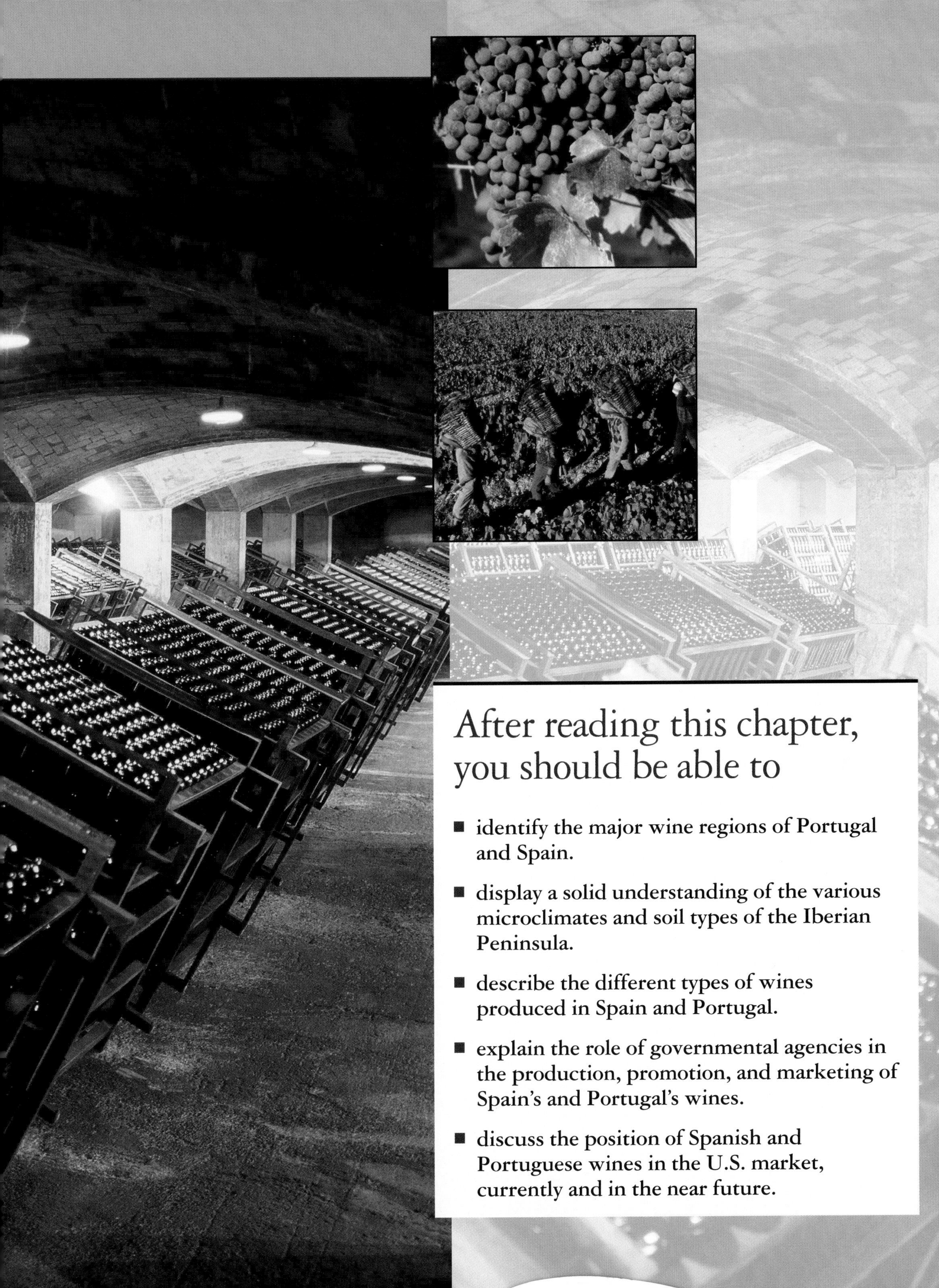

After reading this chapter, you should be able to

- identify the major wine regions of Portugal and Spain.
- display a solid understanding of the various microclimates and soil types of the Iberian Peninsula.
- describe the different types of wines produced in Spain and Portugal.
- explain the role of governmental agencies in the production, promotion, and marketing of Spain's and Portugal's wines.
- discuss the position of Spanish and Portuguese wines in the U.S. market, currently and in the near future.

KEY TERMS

Airén
Albariño
Alvarinho
bodega
Cava
Colheita Selecionada
crianza
Denominação de Origem Controlada (DOC)
Denominación de Origen (DO)
estufa
fino
flor
fortified wine
Garnacha
garrafeira
gran reserva
Instituto Nacional de Denominaciones de Origen (INDO)
Instituto da Vinha e Vinho (IVV)
oloroso
Palomino
Pedro Ximénez
Port
quinta
reserva
ruby Port
solera
tawny Port
Tempranillo
Verdejo
Viura

The *vinifera* grape is indigenous to the Iberian Peninsula. Archeological evidence shows that grapevines were growing here millions of years before *Homo sapiens* ever set foot on the rugged mountains and arid plains of this large peninsula at Europe's southwestern edge. From the time of the Roman Empire wine has been produced and exported from the two countries in what was once known as Iberia. (The Latin word for the Ebro River in the northeast section of the peninsula was Iberus.) These countries remain important wine producers to this day, Spain ranking third in the world for total production and Portugal sixth.

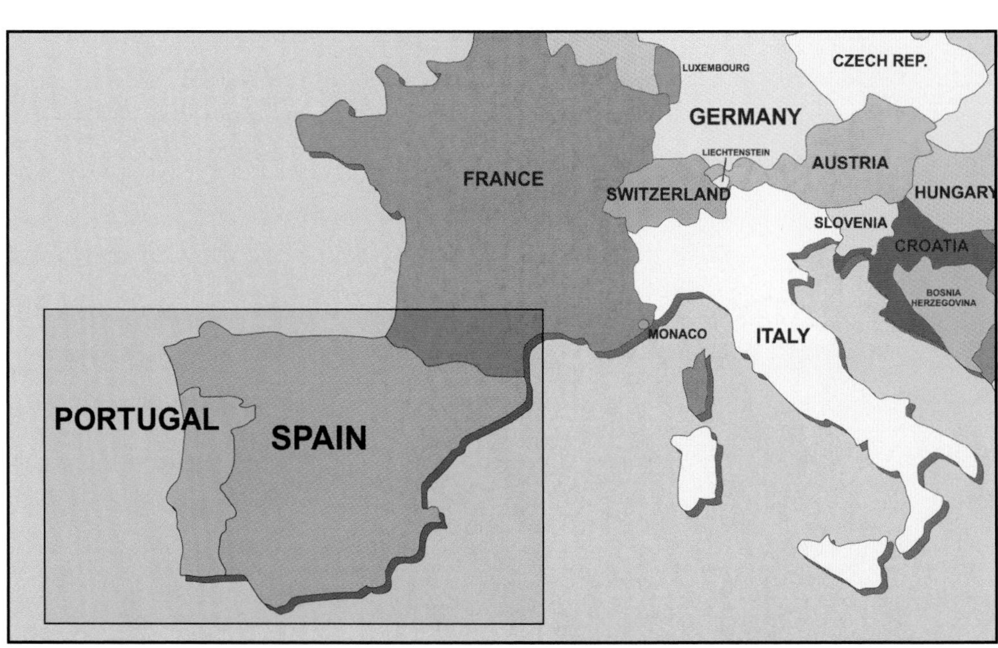

Map showing the Iberian Peninsula.

Both countries have suffered through natural disasters and political upheaval, all of which adversely affected wine production. By the 1960s and 1970s, the image of wine from the Iberian Peninsula was not one of quality. Spain was known for over-oaked reds and oxidized whites, and the production of its most popular wine, the fortified Sherry, was in turmoil. Outside of its own borders, Portugal was known only for sweet rosé, exported in large quantities by Mateus and Lancer's, and at the other end of the price spectrum, the **fortified wine,** Port. In the past 30 years both countries have made extraordinary progress in modernizing the production of wine, in upgrading the quality of their table wines, and in increasing penetration into foreign markets. In both countries these improvements have been the result of close collaboration between the private sector and government agencies. The close cooperation among property owners, vintners' groups, national and regional governments, and the European Union will continue to reap impressive results. The story of wine from Spain and Portugal, begun so many thousands of years ago, is still unfolding.

History of Wine Production

The Phoenicians did not introduce grapevines to the Iberian Peninsula; the vines were already growing there when they arrived around 1100 BC. It was the Phoenicians, however, who engaged in the first commercial winemaking on this vast peninsula. The Carthaginians (named for their home city of Carthage in North Africa) invaded Iberia around 250 BC. They ruled for 200 years, coexisting with the Romans, who had possessions east of the Ebro River. The Carthaginians greatly expanded the production of wine. At one point they were shipping wine to all parts of the extensive Roman Empire. The Romans wrested control of Iberia from the Carthaginians and colonized the whole peninsula under Emperor Augustus in the first century BC. Wine production continued, and Iberian wines were sent as far away as Normandy, England, and even the Roman frontier in Germany. The Roman Empire began its fall in the second century AD. Germanic tribes invaded the peninsula. In the fifth century AD, the Visigoth tribe came to control most of what is now western Spain and Portugal, and established a kingdom there. Records are scarce from this period, but enough have survived to offer proof that viticulture continued under the Goths.

In 711 AD the Visigoth kingdom was overthrown by the Moors, an Islamic tribe from North Africa. The Moors ruled peacefully for over 600 years and did not demand the cessation of viticulture. Even though the Prophet Mohammed forbade the consumption of wine (or any alcohol), the Moors did not impose their ways on the local culture. Wine was taxed, and perhaps the conquerors were wise enough to recognize the need for these revenues. At any rate, Christians and Jews were allowed to continue making and consuming wines under the Moors.

In the early twelfth century, Christians began to rise against the Moors to drive them from the peninsula. During this time, Navarra, Aragon, Castilla y León, and Barcelona in Spain were under Christian rule, and in 1136 Portugal declared itself a Christian kingdom. By 1320, Christians were largely successful in their efforts to reconquer Iberia. Only Granada in southern Spain remained under Moorish rule. Between 1300 and 1500, trade with the rest of Europe increased considerably.

Spanish Wine—Historical Perspective

In January 1492, Spain became a united Christian country under one crown when the Spanish Army drove the Moors from Granada. In October of that same year, Christopher Columbus discovered the West Indies, opening up a new world for Spanish and Portuguese trade, including wines (Figure 7–1). In 1494, under the arbitration of Pope Alexander VI, the whole of the New World was divided between Portugal and Spain. The entire continent of South America became Spanish, except for Brazil, which became a colony of Portugal.

In 1492 there was yet another historic (albeit very unfortunate) event that affected the wine trade. The Spanish government decreed that all Jews who refused to be baptized as Christians must leave the country. The Spanish Inquisition, although condemned by other Europeans, did open the way for English, Dutch, and French merchants to come into Spain and build up the wine trade. The Spanish wine that attracted the most attention from these foreign traders was Sherry. By the late sixteenth century it was the best selling wine in England. Trade with other parts of Europe also greatly expanded at this time, partly because Spain, through a fortuitous royal marriage, became closely aligned with the Hapsburg Empire. This alignment gave Spanish wine producers access to Holland's exporters and their ships.

Unfortunately, the period of peaceful commercial enterprise did not last. Relations between Spain and England began to deteriorate after Henry VIII divorced Catherine of Aragon in 1533. Tensions escalated into war within a few decades. Trade declined and remained sporadic even after the English defeated the Spanish Navy in 1588. The English imposed heavy excise taxes on Spanish wine, so the Spaniards concentrated on building their wine business elsewhere in Europe and in the New World, while the Portuguese stepped in as primary supplier of wine to England. Unfortunately for the Spanish, their colonies in South America turned out to be a less lucrative market than hoped. Peru and Chile were so successful at viticulture that they were soon meeting all domestic demand for wine themselves and imported very little from Spain.

Demand for wine did pick up throughout Europe during the seventeenth century as the population increased during a time of relative peace. Gradually wine became an integral part of daily life for many Europeans, especially in cities. The increase in sales of Spanish wines into other parts of Europe continued on into the eighteenth century. A small portion of Spanish wine was also exported to the British colonies in the New World. By the late eighteenth and early nineteenth centuries the Spanish wine trade was well established, both to the domestic market and to export

FIGURE 7–1 This old engraving depicts Queen Isabella and King Ferdinand of Spain as they send Columbus off on his trip that led to the discovery of a whole new world outside Europe for Spanish and Portuguese trade.

markets. By 1825, for instance, two-thirds of wine imported into England was Spanish, most of it Sherry (Phillips, 2000, p. 221). During the early nineteenth century, vineyard acreage in Spain increased fourfold. By 1850, wine constituted fully one-third of all Spanish exports.

The lucrative expansion of wine production in Spain was dealt a severe blow with the arrival of the devastating phylloxera. The vine-destroying louse made its first appearance on the Iberian Peninsula in 1878, and by 1901 had spread to the vineyards of Rioja. From there the infestation moved across Spain, and the wine business was decimated. The only silver lining is that after phylloxera was controlled many inferior vineyards were not replanted to grapevines. Moreover, on the advice of French vintners who had come to Spain after French wine regions were destroyed by phylloxera, many vineyards were replanted, not to the lesser grapes that had been there, but to higher quality varietals.

Spain was unable to continue the improvement of its wine production and its expansion into foreign markets because of the political turmoil created by the Spanish Civil War of 1936–1939 and the isolation resulting from the dictatorial regime of General Francisco Franco, who ruled the country from the end of the Civil War until his death in November of 1975. In the years since Franco's death, Spain has recovered remarkably and is now a thoroughly modern and prosperous nation. As the nation has progressed into the twenty-first century so has its wine trade. Partly because of government controls, which were instigated as early as 1926 with the demarcation of Rioja and continued with a nationwide system of quality control laws in 1972, and partly because of the introduction of modern technology like stainless steel fermentation tanks, Spain is now producing a range of table wines, sparkling wines, and fortified wines for which the demand around the world is steadily growing.

Portuguese Wine—Historical Perspective

Portugal's King Alphonse established the country's first Parliament in 1249, thereby reducing territorial disputes and opening the way for this small, seafaring nation to build commercial relationships with other European countries. England was a particularly important trading partner, shipping grain, salt cod, and other commodities to Portugal in exchange for wine. In 1386, the two countries ratified the Treaty of Windsor, which strengthened the commercial and political ties between them. When England and France went to war in the seventeenth century and French wine became unattainable, the natural beneficiary was Portugal. Wine shipments to England increased dramatically. Many English wine merchants went to Portugal seeking alternative sources of wine. In the Douro region of inland Portugal, the merchants found very deeply colored, intensely tannic red wines. To ensure that these wines would make the sea crossing to England unspoiled, the English began fortifying these red wines with additional brandy. Legend has it that it was a wine merchant from Liverpool who, in 1678, first fortified his wine during fermentation rather than after fermentation was completed, thus leaving the touch of natural sweetness for which Port is now famous. In 1703 England and Portugal further strengthened their trading partnership with the Treaty of Methuen which gave tariff advantages to wines imported from Portugal. By that time, a thriving community of English and German merchants had established itself in the city of Oporto, on the coast of the Douro region, from where they controlled the production and shipping of the increasingly popular Port.

The Treaty of Methuen gave protection to another, very new industry in Portugal—the cork industry. In the late 1600s English shippers of wine discovered that cork stoppers provided an airtight closure that greatly reduced spoilage of wine. The invention of glass bottles and cork stoppers made long-term storage of wine possible. Cork stoppers are made from the bark of cork trees, most of which grow in Spain and Portugal. The Treaty of Methuen gave Portugal's wines further protection against tariffs in England in exchange for Portugal promising English merchants a supply of corks for their shipments of wine to foreign countries.

During the eighteenth century and into the nineteenth century, conflicts between France and England continued to benefit the Portuguese wine trade. The island of Madeira, a Portuguese colony off the coast of Africa, became an important trading post for passing ships. That island began shipping its own fortified wine to England and her colonies. The British colonies in North America were also an important and growing market for the wines of Portugal, especially her fortified wines.

In the late nineteenth century, Portugal's vineyards were struck by the same natural disaster as those in the rest of Europe—phylloxera. Some wine regions have never recovered. At the beginning of the twentieth century, Portugal turned its back on the rest of the world, operating in relative isolation during a 20 year period of political and economic disruption. The king was assassinated in 1908, and the monarchy was eliminated shortly thereafter. The last king of Portugal, Manuel II, started the process of developing a system of quality control laws for wine production. Manuel II was able to obtain international agreement in 1916 granting Portugal the sole right to use the terms "Port" and "Madeira" shortly before he was deposed.

Stability began to return when Antonio de Oliveira Salazar, the son of a small vineyard owner, stepped forward in 1926 to take the position of Finance Minister. He became Prime Minister in 1932. Salazar's regime lasted for 40 years during which time his centralized one-party system of government encouraged the modernization of Portugal's economy. Salazar completely reorganized his country's disorderly and anarchic wine business, creating the *Junta Nacional do Vinho* in 1937 to oversee wine production. The JNV encouraged grape growers and small producers to work together. In the next 20 years over 100 cooperatives were built, mostly in northern Portugal, to produce wine and sell it domestically and abroad. The system imposed by the central government was very inflexible, not conducive to experimentation or improvement of techniques. With little incentive to improve their wine, vintners allowed their standards to deteriorate, resulting in mediocre wine.

In 1974 Portugal once again descended into chaos, with a military-led revolution that lasted for 2 years. In 1976 Portugal held a successful democratic election and began its ascent into modern Europe. In 1986 Portugal was admitted to the European Union, which greatly benefited the wine industry. The E.U. gave generously of its expertise and capital to help elevate the standards of wine production in Portugal. The improvement in the quality of table wines—red, white, and rosé—can be attributed to improved technology such as stainless steel tanks and controlled fermentation, and to the innovative leadership and vision of younger winemakers. One group of these younger winemakers, the G7, is made up of seven vintners from seven different regions working together since 1993 to promote awareness of Portuguese table wines in foreign markets. These leading winemakers are indicative of the trend throughout Portugal to improve the overall quality of wines and to incorporate modern, farsighted marketing strategies to move the Portuguese wine business forward.

Government Involvement

The contemporary governments of both Spain and Portugal have been actively involved with bringing their respective country's wine trade into the increasingly competitive international marketplace. The support of the Portuguese and Spanish governments takes three forms.

1. Quality control laws that spell out boundaries of regions, regulate the production and naming of wines, and create regional agencies to oversee production and enforce regulations.

Reading a Spanish Wine Label

This is the name of the winery (**bodega** in Spanish) that made the wine. Montecillo is one of the oldest (founded 1874) and most respected bodegas in Rioja. In Portugal the word for winery or estate is **quinta**.

Vintage Year: The Spanish *DO* laws require that at least 85 percent of the grapes be harvested in that year.

Gran Reserva: Indicates that this wine received additional aging. Each *DO* region has its own requirements for the gran reserva and reserva designations. In Rioja the requirements for gran reserva are 5 additional years of aging, 3 in oak and 2 in bottle.
Portugal has comparable terms: Garrafeira signifies a wine with 2½ years in oak casks and 1 year in bottle. The term reserva on a Portuguese wine shows that the wine is a garrafeira from an exceptional vintage.

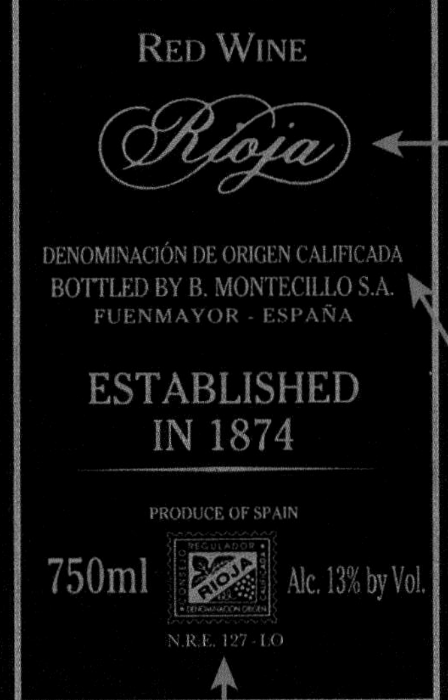

Appellation: Designated wine region, in this case, Rioja.

***Denominación de Origen (DO)*:** Spanish term for a defined region that produces quality wine. Note that in this case the label says *Denominación de Origen Califacada (DOCa)*. This is the highest classification for a wine region. Rioja is the only region to be granted this designation thus far.
In Portugal the comparable term for a designated wine region is *Denominação de Origem Controlada (DOC)*.

Seal: The official seal of the local *Consejo Regulador,* the agency that oversees wine production in that region. The presence of such a seal from any *DO* guarantees that the wine conforms to the regulations of that region.

2. With assistance from the European Union, research and development of improved viticultural and enological technologies, and monetary investment in training and physical equipment.
3. National marketing programs that promote their countries as world-class wine regions and assist individual producers to devise sophisticated marketing strategies for foreign markets.

Wine Laws

As with other countries, both Spain and Portugal have laws controlling wine production.

SPAIN'S *Denominación de Origen (DO) Laws*

Spain's efforts to control the production of wines and wine regions began in 1926 with the official demarcation of the Rioja region. The process was completed for the rest of the country in 1972 with the passage of legislation that created the ***Instituto Nacional de Denominaciones de Origen (INDO)*** and established a system of ***Denominacion de Origen***, the equivalent of France's *appellations d'origines*. There are presently 55 *DOs* in Spain (20 of which have been approved in the last 10 years) (Figure 7–2). Each of the *DOs* has its own *consejo regulador,* the local authority that oversees viticulture, production, labeling, and distribution of wine for that region. The activities of the 55 local *consejos reguladores* are coordinated by the INDO. The regulations overseen by *INDO* include boundaries of wine regions, allowed varietals, yield per hectare, pruning and trellising methods, vinification and aging requirements, minimum alcohol content, and labeling information. Furthermore, since joining the European Union in 1986, Spain has had to conform to all regulations of that body that mandate continent-wide standards for winemaking, land use, and marketing and distribution of alcoholic beverages.

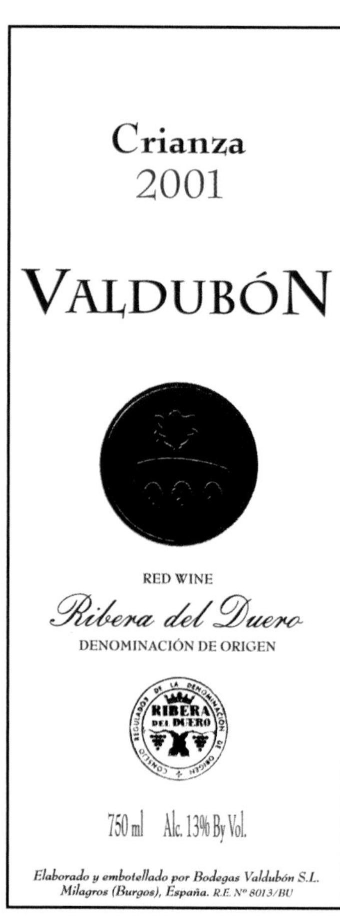

FIGURE 7–2 A typical Spanish label under the *Denominación de Origen* laws. Ribera del Duero is one of the 55 DOs; *crianza* refers to aging requirements for that DO.

QUALITY DESIGNATIONS OF THE INDO

Within the Spanish system, all wine regions are designated at one of four levels of quality, and all wines coming from each region carry that same designation. In ascending order of quality the designations are vino de mesa, vino comarcal, vino de la tierra, and quality wines, which are subdivided into different levels of quality.

VINO DE MESA

Basic table wine, the equivalent of France's *vin de table.* These wines are often blends of various grape varietals, and may come from several different regions. No vintage date is shown, nor may any region of origin be mentioned on the label.

Within the *vino de mesa* designation is a second tier that equates to Italy's *Indicazione Geografica Tipica (IGT)* designation. At this level, winemakers who produce a wine within a *DO,* but from grapes unauthorized for that region, or who use vinification methods outside the parameters allowed for that region, may call their wine *vino de mesa* followed by the name of the province and a vintage year. They are not allowed to show the registered *DO.* For instance, a white wine made from Chardonnay grapes (an unauthorized varietal) in the Rías Baixas *DO* in the province of Galicia in northwestern Spain (see map on page 230) may say on its label "Vino de Mesa de Galicia, 2003," but the words Rías Baixas may not be mentioned.

VINO COMARCAL

A regional wine designation, similar to France's *vin de pays.* There are 21 regions in Spain, all of them quite large, with diverse growing conditions. The label of a wine at this level would read "Vino Comarcal de [name of classified region]." If the producer feels it would be beneficial from a marketing viewpoint, he can choose to designate his wine as "Vino de Mesa de [the name of the province]," followed by the vintage year.

Vino de la Tierra

The same as France's *vin délimité de qualité supérieure (VDQS),* essentially a stepping stone to the highest designation. Producers of wines at this level hope that their district will someday receive *DO* status. The wines are made entirely from authorized grape varietals. The grapes are grown within one district and exhibit the character of that district.

Quality Wines

Wines from official *Denominaciones de Origen,* made from authorized varietals and vinified and aged according to the regulations of that *DO*. An additional level of quality was created by law in 1986, when Spain passed a law specifying that the most prestigious wine districts would be designated as *Denominación de Origen Calificada (DOCa),* "calificado" being Spanish for "eminent" or "distinguished." To date, only Rioja has been elevated to *DOCa* status.

The quality level of a *DO* or *DOCa* wine is indicated on its label by a term that is based primarily on the amount of aging the wine received. The requirements for aging are spelled out separately for each district. The terms used are Vino de Cosecha, Crianza, Reserva, and Gran Reserva.

- *Vino de Cosecha:* Vintage wine. At least 85 percent of the grapes used must have been harvested in the year shown. These young wines are offered for sale as early as the spring after harvest. The style is soft and fruity.
- *Crianza:* From the Spanish word for a child's room or nursery, a **crianza** wine is released in its third year, after spending at least 6 months in small oak barrels and an additional 2 years in the bottle. Some districts specify even more aging. For instance, in Rioja a Crianza wine must receive a full year of oak aging plus the 2 years in bottle. (Figure 7–2 shows a Crianza wine from the Ribera del Duevo region.)
- *Reserva:* Red wines aged at least 3 years, with one of those years spent in oak barrels. Most producers of fine reds will exceed the minimum requirements for their **reserva** wines. There are also stipulations for white wines at the Reserva level; however, few of these wines are exported because most consumers outside of Spain do not like the oxidized character of these aged whites.
- *Gran Reserva:* Produced only in the finest years, and only with the approval of the local *consejo regulador,* **gran reserva** wines must be aged for a minimum of 3 years in barrels and an additional 2 years in bottle. Most producers exceed these minimums to give their wines more richness and smoothness. Some Gran Reservas are not released until 8 or more years after the vintage year. These reds are wonderfully complex and subtle. Figure 7–3 shows an example of a gran reserva label.

Wine Laws of Portugal

Portugal created the world's first demarcated wine region in 1756. At that time demand for Port from Portugal's Douro region was so strong in England that some unscrupulous producers and shippers began stretching the supply by adulterating the wine, either by adding elderberry juice or by bringing in inferior grapes from outside Douro. The Portuguese government realized that these fraudulent practices could badly damage the reputation of their valuable export. Therefore, officials reacted quickly and created a body to oversee the Port trade. Exact boundaries of the Douro region were confirmed, and the government authorized the new agency to supervise each step in the production of Port. Between 1908 and 1929 further work on regulating the wine trade was done with official geographic demarcations being drawn up for other major regions, such as Vinho Verde, Dão, Madeira, and Setúbal.

Because of political upheaval, Portugal did not finalize the creation of quality control laws for its wine trade until its admittance into the European Union in 1986. The system of laws for

FIGURE 7–3 Gran Reserva wines from Bodega Montecillo are made from 95 percent Tempranillo and often receive more than the required 3 years of aging in oak.

Portugal's ***Denominação de Origem Controlada (DOC)*** is based on France's *Appellation d'Origine Controlée* system. The regulations spell out boundaries for wine regions, viticultural practices (including allowed grape varietals and yields per hectare), vinification techniques (including minimum and maximum alcohol content), labeling requirements, and distribution practices. All laws are overseen by the ***Instituto da Vinha e Vinho*** (Institute of Viticulture and Wines), which works closely with local authorities in each province. The only local agency that does not report to the *IVV* is the *Instituto do Vinho do Porto,* which supervises the production and selling of Port.

Within Portugal's *DOC* system, wine regions (and their wines) are classified as *Vinho de mesa, Vinho regional,* and *Denominacào de Origem Controlada.*

- *Vinho de Mesa:* Table wines are the lowest level of quality and are produced with minimal regulations and oversight. No vintage year may be stated. Most of these wines are consumed locally.
- *Vinho Regional:* There are eight regional wine areas. Moving from north to south, these *VR*s are Minho, Trás-os-Montes, Beiras, Ribatejo, Estremadura, Alentejo, Terras do Sado, and Algarve. At least 85 percent of the grapes that go into a wine at this level must be grown in the region shown on the label. The grape varietals used must be authorized for that region.
- *Denominação de Origem Controlada:* Quality wines made under specific requirements and high standards from authorized grapes grown entirely within the *DOC* specified. To date there are 40 *DOCs.* At this highest level of quality, additional information on the wine is found in the

terms on the label. For instance, **garrafeira** signifies a red wine aged a mimimum of 2½ years, including a year in the bottle. For whites and rosés, the aging requirement is a mimimum of 1 year. for all wines, a minimum alcohol content of 11.5 percent is required. The term reserva is used for high quality wines from superior vintage years. The alcohol content must be 0.5 percent above the minimum stipulated for that *DOC*. **Colheita selecionada** signifies a very high quality wine, from excellent vintages and often from prime vineyards. The alcohol content must be 1 percent above the requirement for that *DOC*.

Wine Regions

The Iberian Peninsula is the westernmost outpost of continental Europe. Its climate is strongly influenced by the Mediterranean Sea to the east and by the Atlantic to the north and west. Four of Iberia's five major rivers (including the Duero/Douro) flow westward and drain into the Atlantic; the fifth, the Ebro, flows southeast to the Mediterranean. Iberia is a large and diverse area, with numerous climatic and cultural differences.

Spain encompasses most of the Iberian Peninsula. This country has 3.5 million acres (1.4 million hectares) of grapevines, more than any other country in the world. Because of a very arid climate in most of the country and a ban on irrigation, yields are very low, averaging 1.4 tons per acre (3.1 metric tons per hectare). In the vast central plain and in the southernmost sections, rainfall is minimal and temperatures are high during the long growing season. Only in the north, at the foot of the Pyrenees Mountains that separate Spain from France and in Galicia on the northwestern coast, are temperatures more moderate and rainfall adequate. In these cooler regions, the primary grape varietals for white wines are **Viura** (also called Macabeo), **Albariño** (especially prominent in Galicia), and **Verdejo**. For red wines the most important varietals are **Tempranillo,** which is the most-widely planted of the quality wine-producing varietals. In some regions it is called Tinto Fino. **Garnacha** (the Grenache of southern France) is the second most planted red varietal. As we move inland and south, conditions are less suited to the early ripening varietals. In the sun-baked Central Plain the drought-resistant white grape, **Airén,** is widely planted, covering three times the acreage of any other varietal. In Spain's warmest and dryest section, the southern Andalucía province, most vineyards are planted to the grapes from which Sherry is made—**Palomino** and **Pedro Ximénez.**

Portugal occupies the western flank of the Iberian Peninsula. The culture and the climate of this sea-going country are strongly influenced by the Atlantic Ocean. Portugal is a small country—only 360 miles (581 km) long and 120 miles (194 km) wide, approximately the size of the state of Indiana. However, within that space is an amazing diversity of soils and microclimates, so it is not surprising that Portugal has a considerable range of varietals and wine styles. In the north's flat littoral areas along the Atlantic coast, the temperate maritime climate produces warm summers and wet, cool winters. Rainfall is plentiful, over 100 inches (256 cm) a year in parts of Vinho Verde. As one moves inland or to the south, rainfall is considerably lower and temperatures more extreme. In the southern plains it is not unusual for summer daytime temperatures to average 95°F.

Portugal developed its viticulture in virtual isolation. Very few grapevines were brought in from other European countries. Among the wide variety of indigenous grapes, the most important are Touriga Nacional, Touriga Francesca, and Tinta Roriz (the Tempranillo of Spain). In the cool damp north, the white grape Alvarinho (same as Spain's Albariño) is prominent, especially in Vinho Verde. In the warmer dryer south, a red grape that thrives is Castelão Frances, also known as Periquita. There has been very little experimentation in Portugal with internationally famous varietals, such as Chardonnay and Merlot. Rather, the Portuguese vintners and the government agencies that support them, such as *Comércio e Tourismo de Portugal (ICEP)*, concentrate their marketing efforts on promoting the unique character of Portugal's indigenous varietals.

The Wine Regions of Spain

Spain is divided into seventeen *autonomías*, or "autonomous communities," analogous to the 50 states in this country. Most of these regions have demarcated wine zones, or *denominaciones de origen,* within their boundaries. (See map of Spain.) The majority of *DO*s are in the cooler, more humid northern regions. We will cover in some detail the most important *DO*s of the north, then touch briefly on the large central plain before moving to the south, where Spain's famous fortified wine, Sherry, is made.

GALICIA: RÍAS BAIXAS AND RIBEIRO

The northwest region of Galicia is Spain's coolest and most humid section. The Atlantic Ocean's damp breezes assure moderate temperatures and plentiful rainfall. As mentioned earlier, the white grape Albariño flourishes in these conditions. Rías Baixas, a *DO* located on Galicia's coast and named for the oft-flooded, deep coastal valleys *(rías)* that extend inland for miles, produces one of Spain's most sought-after white wines from this grape. Albariño covers 90 percent of vine-

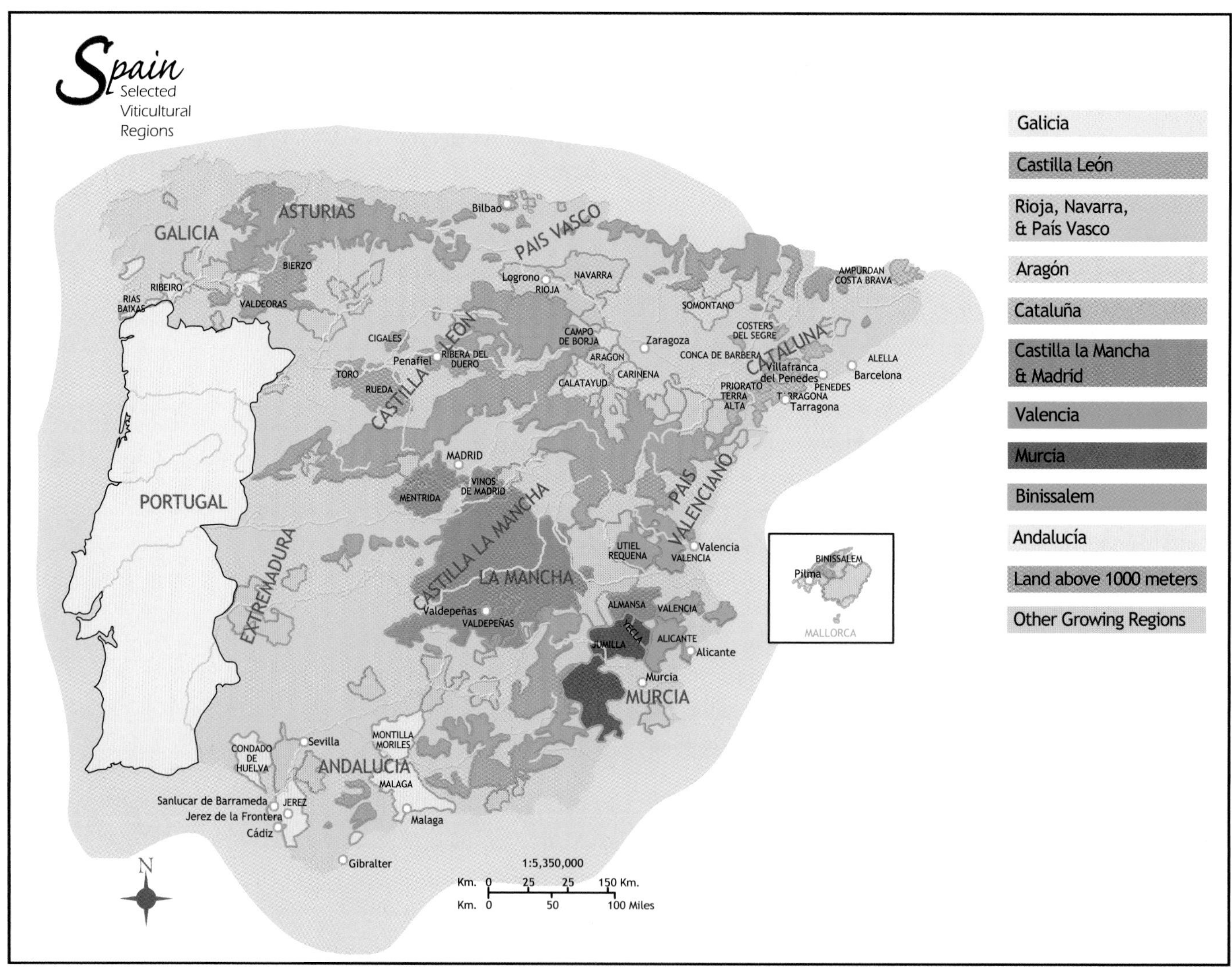

Map of Spain's Wine Regions

yard acreage throughout Rías Baixas, even though the *DO* law allows 11 different varietals. Although Galicians had been making and exporting wine for centuries, after the phylloxera devastation in the late 1900s many of Rías Baixas's vineyards were torn out. Those that were replanted went to inferior hybrid vines that produced mediocre wine. Fortunately, with advice and financial support from the European Union, many property owners upgraded their vineyards and equipment, and replanted their land to the Albariño grape.

Rías Baixas is divided into three sections. The two southern sections stretch along the river Miño, which separates them from the Portuguese wine region Vinho Verde. The third section is a little farther north and centers on the town of Cambandos on the coast. All three sections concentrate on producing dry white wine from Albariño. The varietal name is usually shown on the label. With bracing acidity, distinctive floral, mineral, and fresh peach aromas, and intense fruit, these are excellent food wines. The wine can be quite expensive because Albariño is such a low-yielding varietal that demand is starting to exceed supply, thus driving the price up. Nonetheless, Albariños are rapidly gaining favor in the American market.

The *DO* Ribeiro extends along the Miño river ("*ribeiro*" means "riverbank" in Spanish). It also concentrates on white wine. The small quantities of light-bodied red wine (primarily Garnacha) are mostly consumed locally. After phylloxera wiped out Ribiero's vineyards, most farmers chose to replant with Palomino, a grape from the south used to make Sherry. It was totally unsuited to the cool maritime climate of northwest Iberia. Ribeiro was granted *DO* status in 1957, and with assistance from the *INDO,* landowners began to replant, using two white varietals better suited to their climate, Torrontés and Treixadura. Both can be made into dry, crisp wine, which, though lacking the fragrance and complexity of Albariño, can make an appealing match for shellfish and other seafood.

Castilla y León: Ribera del Duero, Rueda

The region of Castilla y León, Spain's largest, is a historic place to visit. The two most important of the five *DO*s in the region are near the ancient city of Valladolid. Located on opposite sides of the Duero River, they are Ribera del Duero and Rueda. Wine has been made in these regions for over 1,000 years, but fortunately wine producers here are thoroughly ensconced in the twenty-first century, as Ribera del Duero acquires an international reputation for its big but elegant reds, and demand for Rueda's stunning fresh whites continues to grow.

Ribera del Duero straddles the wide valley of the Duero River (called the Douro once it crosses into Portugal), east of Valladolid. The vineyards lie on the slopes of the hills rising from the river, in places reaching a height of 2,600 feet (792 meters) above sea level. At this altitude there is considerable temperature fluctuation, with temperatures in daytime summer reaching well over 100°F but plummeting at night, and cold weather posing a threat of frost far into spring. Although such conditions make viticulture arduous and risky, the reward is in the excellent acidity level maintained in the grapes as they ripen. The principal grape here is Tinta Roriz, (Tempranillo), which has adapted well to the climate and is made into a single-varietal wine of deep color, intense flavor, and firm structure. In the opinion of some experts, Ribera del Duero now rivals Rioja as Iberia's premier red wine region.

Ribera del Duero was granted *DO* status only in 1982, but the now-famous winery Vega Sicilia in the western edge of the region had been producing great wines for 100 years by then. Another pioneer emerged in the 1980s. Alejandro Fernández first released his red wine, Pesquera (named for a nearby village), in that decade. It was immediately praised by critics and continues to be eagerly sought by wine lovers in Spain, the rest of Europe, and the New World. This success encouraged his neighbors to upgrade their vineyards and wine-making facilities, and to concentrate on quality rather than quantity in their wines. Several properties have succeeded in producing wine very close in quality to Vega Sicilia and Pesquera.

Extending south from the Duero River, on a bleak flat plain, is the *DO* Rueda, not far from Valladolid. Although wine has been produced here since the Middle Ages, Rueda fell into mediocrity after phylloxera wiped out its vineyards. When some property owners did start to replant in the early years of the twentieth century, they made the same mistake as vintners in Galicia's Ribeiro district: They planted the Palomino grape from the southern *autonomía* of Andalucía which and used it to make undistinguished fortified wine.

Ironically, it was a vintner from the Rioja region who rescued Rueda from mediocrity. The Marqués de Riscal recognized the potential of the sunny but moderate region to produce attractive dry white wines. He released his first fresh, crisp white made from the Verdejo grape in the mid-1970s to wide acclaim. Rueda was granted *DO* status in 1980, and ever since the local *Consejo Regulado* has been urging landowners to plant more of this low-yielding, high-acid varietal. Fortunately the region now produces good quantities of fresh, crisp, citrusy Verdejo, fermented in stainless steel tanks and released young. Another varietal introduced by the Marqués de Riscal more recently also shows great promise. Sauvignon Blanc does very well in sunny Rueda, where the sand and clay in the soil and the abundant sunshine seem to bring out its best characteristics.

RIOJA

Rioja is Spain's best known and most prestigious wine region. In 1991, a ministerial decree granted Rioja the status of *Denominación de Origen Calificada,* the first, and so far only, wine region to be granted this ranking. Rioja is located in the northernmost part of the country, near the Pyrenees separating Spain from France. The majority of Rioja's 120,000 acres (48,580 hectares) of vineyards are in the *autonomía,* or autonomous community, of La Rioja, but small portions of it extend into the neighboring *autonomías* of Navarra and País Vasco (Basque Country) (Figure 7–4). Flowing the length of Rioja is the Ebro River and its tributaries, including Río Oja, from which the region's name is derived.

Although small, Rioja is divided into three subzones, each with a distinct microclimate and terroir. The northwest zone is called Rioja Alta. The soil is a combination of clay and limestone, and the climate is cool because of elevation at the foot of the Pyrenees and remnants of the Atlantic's damp breezes that work their way inland. The bustling small city of Haro is located in

Map of Rioja.

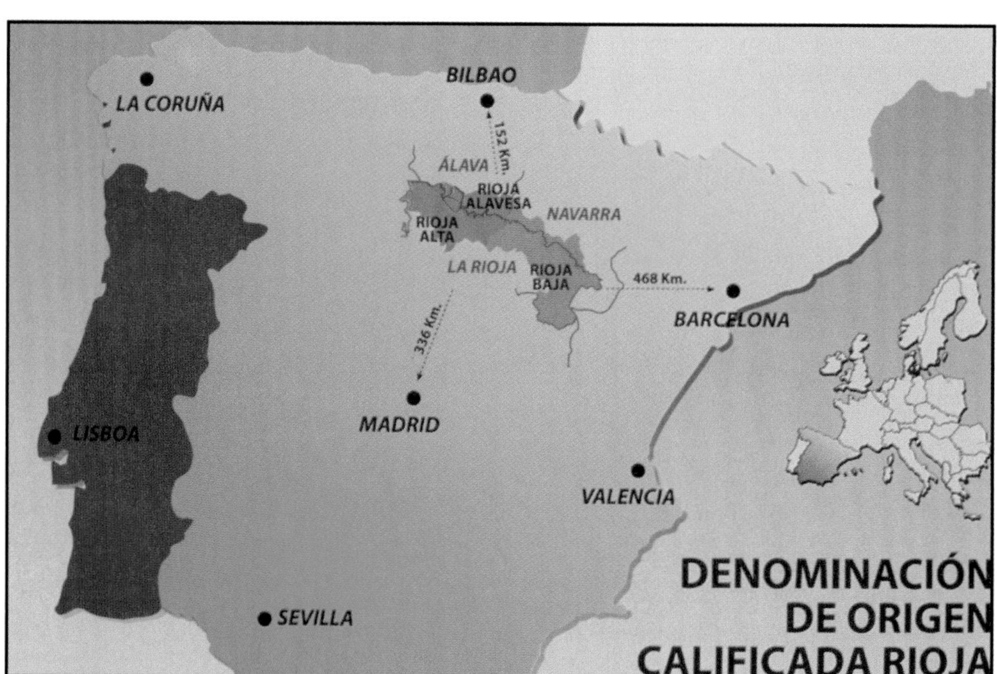

FIGURE 7–4 A typical winding road through Rioja's hillside vineyards, with a stunning view of the Pyrenees Mountains in the background.

Rioja Alta. The center zone, Rioja Alavesa, surrounds the provincial capital, Logroño. Rioja Alavesa has a climate that mixes Atlantic influences with warmer dryer air that flows up the Ebro River from the Mediterranean Sea. The soil is similar to the limestone and clay of Rioja Alta, with some ferrous clay. The eastern zone is Rioja Baja (lower Rioja). This low-lying zone is influenced by the warmer, dryer weather patterns of the Mediterranean, and its soil, especially along the Ebro, is mostly alluvial.

Seventy-five percent of the wine produced in Rioja is red (Figure 7–5). The principal grape is Tempranillo, which is believed to be indigenous to Rioja and ripens well in the clay and limestone soil and moderate temperatures of Rioja Alta and Alavesa. Three other red grapes are authorized. Garnacha does better in the dry, hot climate of Rioja Baja. It is often added to

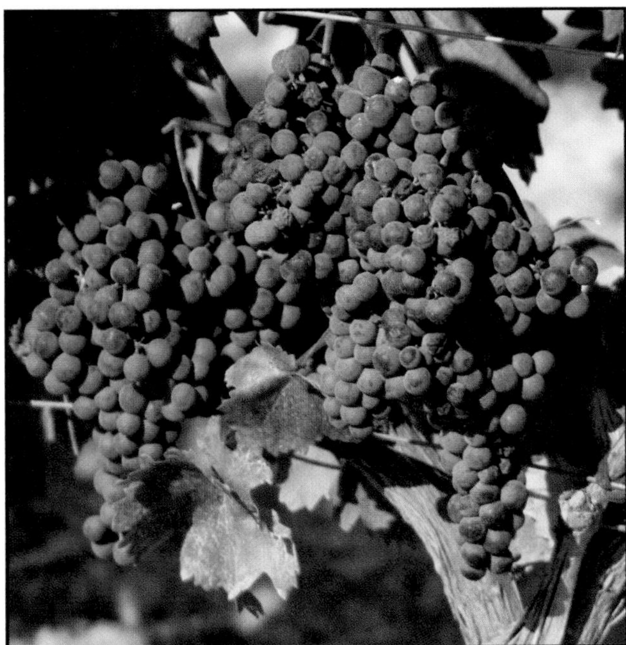

FIGURE 7–5 Tempranillo grapes ripening during a warm Rioja summer.

Tempranillo for additional body. Garnacha on its own does not produce impressive reds but is often used to make charming, dry but fruity rosé. The other two red varietals, Mazuelo (the Carignan of southern France) and Graciano are blended into red wine in small quantities, the Mazuelo contributing color and the Graciano attractive ripe aromas.

There are three authorized white varietals—Viura, Malvasia, and Garnacha Blanca. In the past, Malvasia was the favored grape. It produced big alcoholic wines that oxidized easily, and thus responded well to prolonged aging in oak barrels. Tastes have changed, however, and mature, oxidized, heavy whites have fallen from favor, especially in foreign markets. Since the 1980s demand is for fresh, cool-fermented dry whites with clean flavors. The current favorite is Viura, (called Macabeo in other parts of Spain) which has lively acidity, low alcohol, and pleasant fruit. It now accounts for ninety percent of white Rioja. Garnacha Blanca is light-bodied and low in acids. It contributes little, and is rarely used.

Rioja wines are classified according to aging guidelines. Forty percent of Rioja's wines fall into the three oak-aged classifications of crianza, reserva, and gran reserva. (The rest are sold, usually locally, as *joven,* young wine to be consumed early, or are white or rosé.) With so much wine aging at any one time, Rioja vintners need large quantities of barrels, as many as tens of thousands. At one time in the mid-1990s, a major producer of Rioja, Campo Viejo, was reported to have a stock of 45,000 barrels in its caves (Robinson, 1999). Bodegas are so dependent on a steady supply of aging barrels that some, like Bodega Montecillo, have their own resident coopers who, on the premises, make all the barrels needed (Figure 7–6). Rioja producers favor American oak because the hint of spice it imparts to wine blends nicely with the flavors of Tempranillo. It is also less expensive than French oak.

NAVARRA

Nestled next to Rioja in the foothills of the Pyrenees is the *DO* of Navarra, a beautiful mountainous area that has been producing wine since Roman times. Gracefully arched aqueducts and imposing stone bridges built by the Romans are still in use today. Although overshadowed by its neighbor Rioja, Navarra is producing increasingly attractive red wines, based primarily on

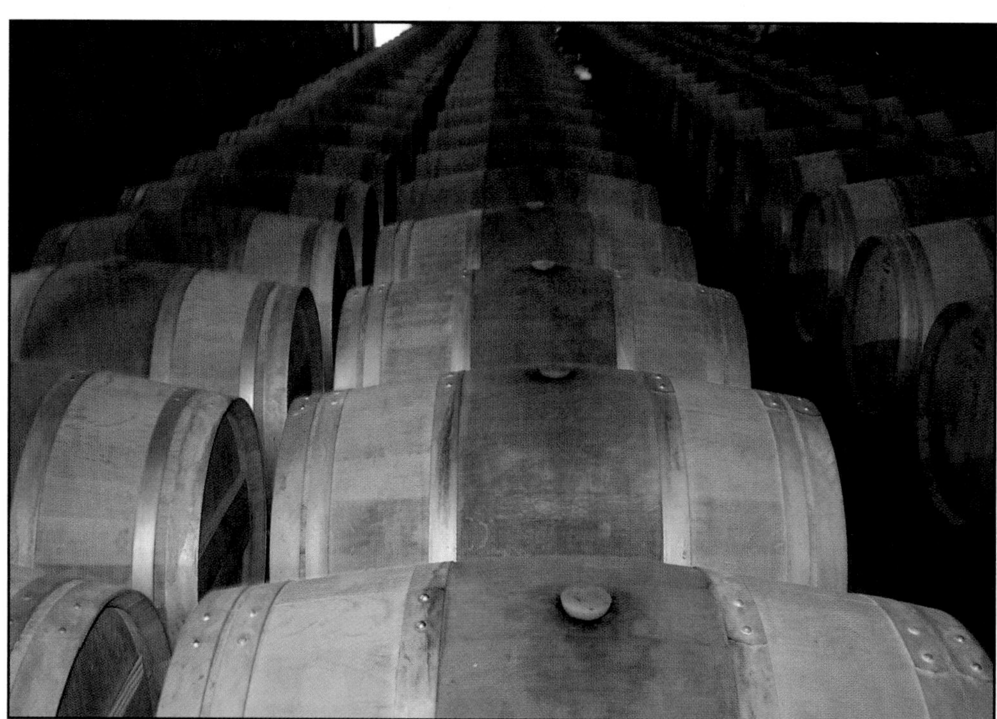

FIGURE 7–6 A typical aging "cave" in Rioja bodega with row after row of oak barrels full of wine that ages for years before being bottled and released.

Garnacha. Fully 80 percent of Navarra's vineyards are planted to this grape. Under proper conditions, Garnacha can produce soft, fruity and aromatic reds. Navarrans also make large quantities of pleasant dry rosé based on Garnacha. Tempranillo is the second most widely planted red varietal. White wines account for only 10 percent of Navarra's production. They are made primarily from Viura. Visiting French vintners introduced Cabernet Sauvignon, Merlot, and Chardonnay during the replanting that followed the devastation of phylloxera. These varietals did well, and are now authorized by the local *Consejo Regulado.* This agency in 1981 established an impressive station for viticultural and enological research in the town of Olite, where they continue research into suitable varietals and improved techniques. As the local agency and the handful of innovative private producers continue to experiment and cooperate on research, there is great promise for the future of Navarra's wines.

Cataluña: Penedès, Priorato

Cataluña on the Mediterranean Sea in northeast Spain is surely one of the most beautiful sections of this country, and its capital, Barcelona, is a busy commercial center for Cataluña's wine trade. A short drive inland from Barcelona lie the vineyards and bodegas of several exciting *DOs.*

Penedès has been described by British wine writer Tony Lord as the "heartland" of the Spanish viticultural revolution. He says, "the science of modern winemaking came to Spain via the Penedès" (Lord, 1988). One family has been particularly active in moving the revolution along. The Torres family became active in wine (and brandy) production in Penedès in 1870 when Jaime Torres built a winery at Vilafranca del Penedès, and swiftly made it and his shipping business, into large and successful enterprises. Jaime's heir was his nephew, Juan Torres, who continued expanding the production and exporting of wine. Juan's son, Miguel, rebuilt the businesses after the disruption of the Civil War, and by the 1950s the winery was back to its former level of production. Miguel began the practice of selling the family's wine in bottles rather than in bulk. He sent his son, Miguel, to study viticulture and enology in France.

It is difficult to overestimate the contribution to winemaking in Penedès that Miguel Jr. brought back with him after his years of study. He experimented with French and German varietals new to the region, including Cabernet Sauvignon, Chardonnay, Sauvignon Blanc, Riesling, and Gewürztraminer, and grew them successfully alongside indigenous varietals. He introduced the vine trellis system. He modernized the family's winery, adding such innovations as temperature-controlled fermentation tanks.

Much of what Miguel Torres introduced into Penedès was unknown anywhere in Spain at the time, and most of his innovations have now been adopted throughout the country. Miguel Jr. took over the presidency of his family's business in 1991 upon the death of his father. He now runs a large, thoroughly modern wine-making facility in Penedès, as well as properties in Chile and Sonoma, with his brother and sister. The Torres family's contributions to Spain's emergence as a producer of world-class wine is incalculable.

Despite the increasing number of fine varietal red wines coming from Penedès, as well as some balanced and fresh varietal whites, the wine most associated with Penedès is its sparkling wine. Sparkling wine made in the *méthode champenoise* was first produced in Penedès in 1872 when José Raventós, whose family owned the Codorníu firm, returned from a trip to France, fascinated with the festive wine of the Champagne region. He emulated the labor-intensive method of a second sugar-to-alcohol fermentation in the bottle to trap the side product, carbon dioxide, thus giving the wine its effervescence. (See "The Winery," Chapter 3.)

The term ***Cava,*** from the Catalan word for cellar (Figure 7–7), was adopted by the Spanish government in 1970, as an official *denominación.* Unlike other Spanish *DOs,* the term Cava did not signify a single delimited zone. European Union laws, however, stipulate that an official denomination refer to a specific geographic region. Therefore, the government of Spain has limited

FIGURE 7–7 The gyropalette was developed in Cataluña in the1970s as a substitute for the far more labor-intensive process of riddling in the production of Cava. This allows Cava to be sold at a much lower price than France's Champagne.

the use of the term Cava to sparkling wine made in the *méthode champenoise* from grapes grown in certain municipalities in Cataluña, Valencia, Aragon, La Rioja, Navarra, and País Vascos. However, Penedès remains the principal source for Cava, producing 90 percent of the country's sparkling wine. The allowed grapes are the traditional high-acid Viura, which accounts for half the makeup of most Cavas, Parellada, which adds body to the lighter Viura, and Xarel-lo which performs particularly well in higher altitude vineyards and contributes the earthy aromas connoisseurs now associate with Cava. Some major Cava houses are blending in more of the French transplant, Chardonnay. Total production of Cava is now over 12.5 million cases per year. Major producers include Codorníu, Freixenet (Figure 7–8), Paul Cheneau, and Segura Viudas (owned by Freixenet). Cava is popular in markets around the world for its fresh fruity flavors, delicate mousse, and very affordable price tag.

Priorato is a small *DO* (only 4,300 acres/1,740 hectares under vine) surrounded by the large *DO* of Tarragona. It is in the rugged mountains at the western edge of Tarragona. The soil is volcanic, full of flecks of mica that catch the sun's heat and reflect it onto the ripening grapes. The region has cold winters but long, hot, and very dry summers. The grapes, mostly Garnacha and Cariñena (called mazuelo elsewhere in Spain) get very ripe in these conditions. The result is a very full-bodied, intense red wine with high alcohol levels (the legal minimum is 13.5 percent

FIGURE 7–8 Freixenet is one of Spain's largest producers of Cava.

but the wine usually goes higher). Winemaking has changed little here since the Carthusian monks began making it at their priory (which gave the region its name) in the twelfth century. However, several producers, among them the Penedès firm of René Barbier and the local vintner Alvaro Palacios, are experimenting with modern technology, such as cool fermentation and aging in Limosin oak barrels, and careful blending with French *vinifera* grapes, especially Cabernet Sauvignon and Syrah. Some of the new reds coming out of Priorato are so good that they are putting this *DO* in contention with Ribera del Duero as the region most likely to topple Rioja from its traditional position as Spain's premier wine region.

La Mancha

The Central Plain of Spain is unimaginably vast and incredibly flat. The *DO* La Mancha covers most of the *autonomía* of Castilla-La Mancha. Vineyards stretch as far as the eye can see, broken only occasionally by small groves of olive trees and fields of cereal grains. There are over 1 million acres (405,000 hectares) of vineyard in the *DO*. La Mancha produces as much as one-third of Spain's total wine production each year. Most of the wine is sold in bulk, distilled into brandy, or used to make vinegar. Over 90 percent of the acreage is devoted to one grape, the white Airén, a varietal of little character and barely adequate acidity. However, it is one of the few varietals able to survive in the barren plain, where winters are long and bleak, and summers are short, arid, and relentlessly sunny and hot. Even with improved modern technology such as cool fermentation in stainless steel tanks, it is impossible, with such a harsh climate and an inferior varietal, to produce much wine that rises above the level of a solid low-alcohol everyday quaffer—light, clean, with a touch of pleasant fruit.

The local *Consejo Regulado* is encouraging growers to pull out Airén and replace it with the other allowed varietals including Tempranillo, which in this part of Spain is referred to as Cencibel, that can, with carbonic maceration, be made into decent soft wines. However, most growers are choosing to stay with the tried-and-true, so it appears that for the immediate future, La Mancha will continue to be the source of an seemingly endless supply of bulk white wine.

Andalucía: Jerez (Sherry)

The great fortified wine, Sherry, made in the small Jerez section of Andalucía in Southern Spain (Figure 7–9), is a wine misunderstood and underappreciated in the American market. (The word "Sherry" is an English corruption of "Jerez"). Few consumers here realize that Sherry can be a crisp bone-dry aperitif **(fino)** or a luscious deeply-flavored dessert wine **(oloroso),** with a range of styles in between. With governmental promotional efforts such as "Wines from Spain," a program organized under the aegis of the Commercial Office of Spain to educate foreign consumers, more Americans are discovering how versatile and delicious Sherry can be.

Sherry-making is a time-honored tradition in Jerez. Today the viticulture and winemaking, although adhering to tradition, are thoroughly modern. The grapes used, Palomino and Pedro Ximénez, are well adapted to the dry warm climate. Very little rain falls during the long summers, and the porous chalky soils allow vines to push deep to find the retained ground water.

Once harvested, usually in early September, the grapes are brought to the huge wineries owned by the major Sherry houses, where they go through a normal first fermentation until all natural sugars are converted. The new wine is run off into casks that are taken down to the cellars of the bodegas, where they are left as the famous **flor,** which is allowed to form on the surface of the wine (Figure 7–10). (See "The Winery," Chapter 3). The flor is the key to what style of Sherry will result. In some casks the flor develops like a thin white carpet over the entire surface of the wine, protecting it from excess oxidation while feeding off any remaining sugars. These casks will become Fino Sherry. The wine remains in the casks for several years, the flor receiving new nutrients upon which to feed as young wine is added each year. This is the basis of Jerez's

FIGURE 7–9 A view of Andalucía's beautiful sunny coast.

FIGURE 7–10 Large bodegas in Jerez, such as the one pictured, usually have several *solera* systems for aging and blending the different styles of Sherry.

solera system, an ongoing, rotating form of aging Sherry depicted in Figure 7–11. At the end of the aging period, Fino is removed from the older casks in the *solera* and fortified sparingly with additional alcohol. Amontillado, a style of Fino, is left longer in the *solera*. It is, accordingly, darker in color, richer, but dry, and nutty in flavor. It serves as a very good aperitif, excellent with salty nuts or briny olives.

In other casks, minimal flor will grow, thus exposing the wine to air inside the cask. These casks will be classified as oloroso, and fortified to 18°Brix, killing any remaining active yeast cells and thus leaving unfermented sugars. Oloroso Sherries spend longer in the *solera*, acquiring a deep color and richly concentrated, sweetly raisined flavors. Some olorosos are further sweetened through the addition of juice from sun-dried Pedro Ximénez grapes. The greatest of the

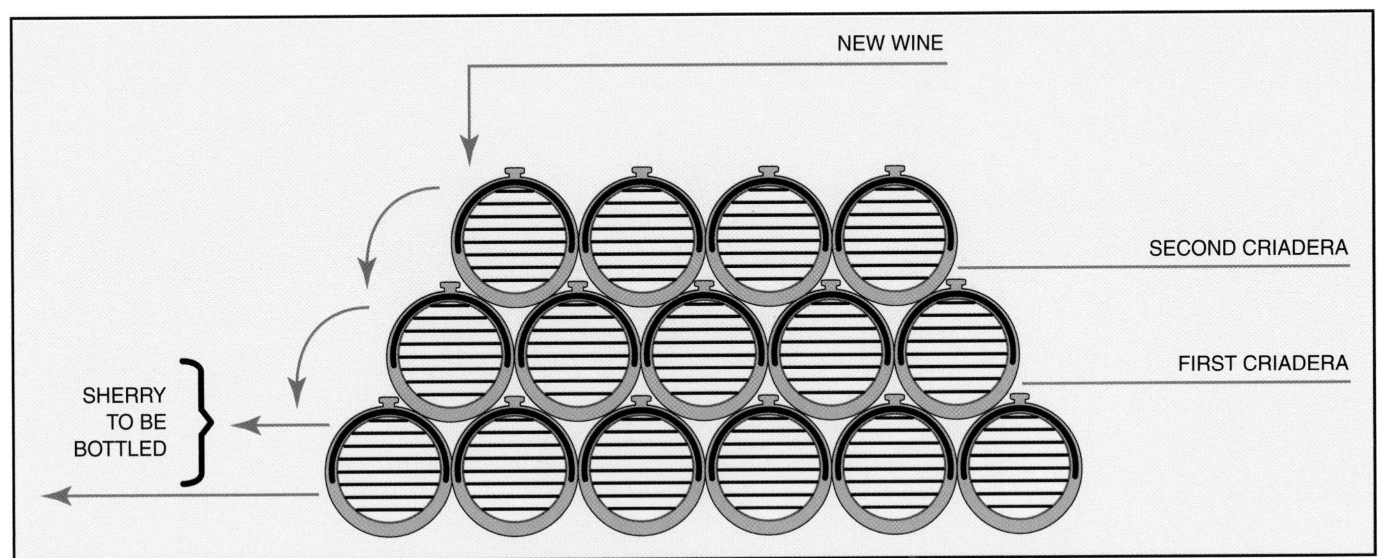

FIGURE 7–11 The *solera* system is the key to making Sherry. Each year, wine is siphoned from the bottom, or oldest, barrels to be bottled and released for sale. Wine is then racked from the next level of barrels, that is, the first criadera, to replenish the bottom level, although barrels are never completely filled. Next, some wine is racked from the second criadera to replenish the first criadera, and so on. The current year's wine is placed in the top, or youngest, level to begin its slow journey through the system.

sweet style Sherries are made exclusively of dried Pedro Ximénez grapes. After years in *solera*, they are labeled as Pedro Ximénez, a thick, syrupy, very sweet treat, with deep chocolate and raisin flavors.

A few Sherry houses own vineyards, but the majority of grapes are grown by small landowners and sold either to the major private producers or to one of the seven cooperatives. Among the most respected producers of Sherry are Sandeman, Domecq, Croft, Harvey, and Lustau.

Noteworthy wines are also being made in many other of Spain's *Denominaciones de Origen.* Among the more exciting emerging regions are Toro, Cigales, and Bierzo in Castilla y León; Tarragona and Terra Alta in Cataluña; Somontona in Aragon; and Jumilla in the Levante.

The Wine Regions of Portugal

There are a total of 988,000 acres (400,000 hectares) of vineyards planted in Portugal's eight *vinho regional* areas. Within those VRs are forty *Denominçãos de Origem Contralada.* The most important DOCs are:

Vinho Verde

The northwest *vinho regional* of Minho is famous for its bright, acidic, very dry white wines. The best-known *DOC* is Vinho Verde. (Literal translation: "green wine," despite the fact that half the production of this area is very dry, acidic, and slightly fizzy red wine, rarely exported.) Vinho Verde is Portugal's largest demarcated wine region, extending from the city of Oporto north to the border with Spain. The warm humid climate encourages intensive cultivation of vines despite the fact that north of Oporto is the most heavily populated part of rural Iberia. To get maximum use from the open land that remains, vintners here employ high trellises, allowing a second crop, usually vegetables, to be grown under the vines. The high trellises serve a second purpose: they discourage the growth of grey rot, a real danger in such damp conditions. The principal white grape is **Alvarinho**, which is made into a very popular, sprightly wine with truly distinctive flavors.

Map of Portugal

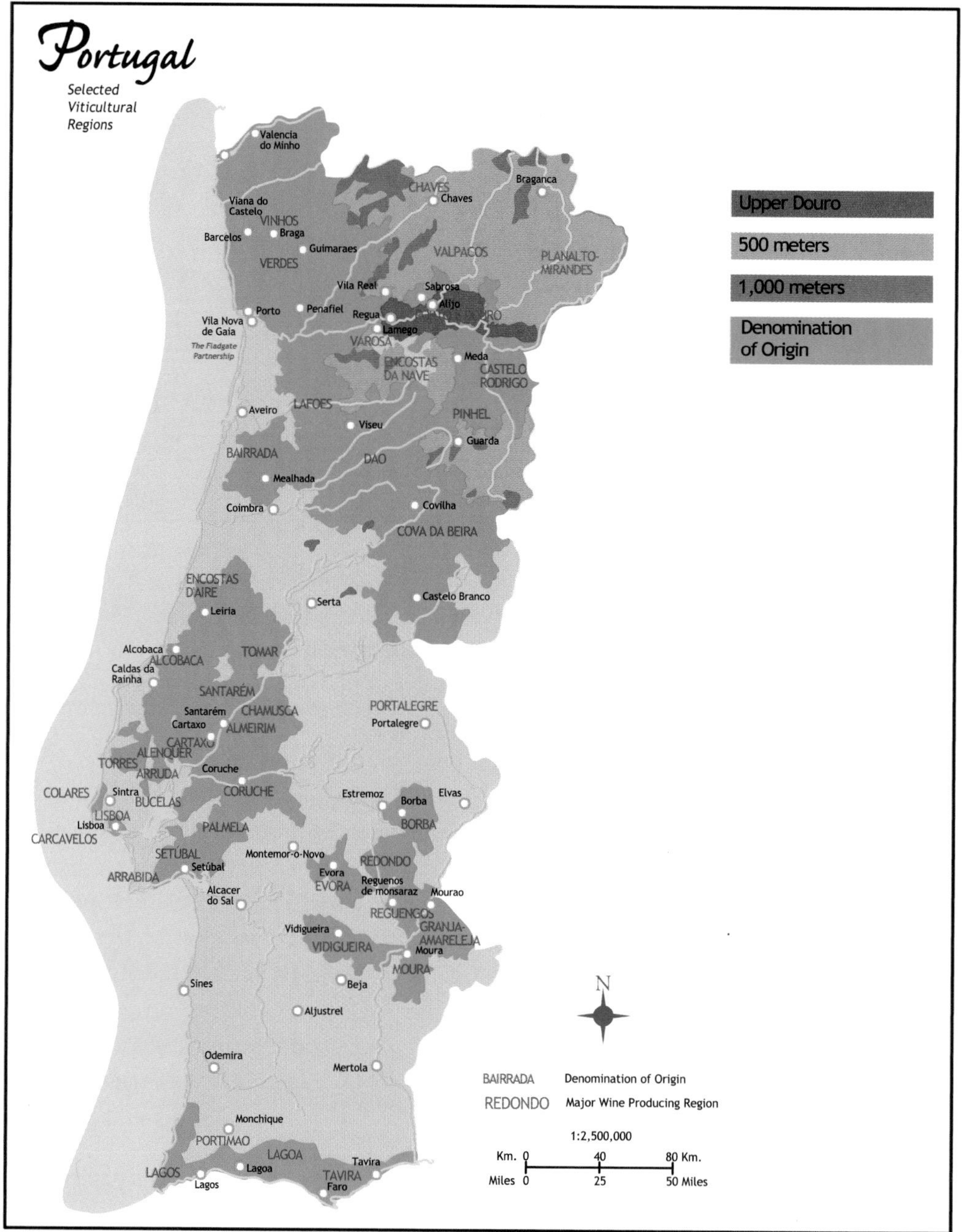

DOURO

This region is named for the Douro River which begins as the Duero in Spain and flows across Portugal to empty into the Atlantic. Famous for its great fortified wine, Port, the Douro region is rapidly acquiring an international reputation for solid, balanced red wines of considerable complexity. Table wines now make up half the production of Douro.

The vineyards for both Port and table wines are located inland, in the rocky, rugged hills of Trás-os-Montes ("behind the mountains") (Figure 7–12). The soil is primarily Precambrian schist, a very hard, mineral-laden rock that retains heat and is very difficult for roots to penetrate in their search for ground water. There are almost 90 approved varietals for Douro, but the favored grapes for both types of wine are Touriga Nacional and Tinta Roriz (the regional name for Tempranillo). These grapes impart to Douro table wine the same spicy, rich flavor, full body, and firm tannins found in a young ruby Port. In table wines Touriga Nacional con-

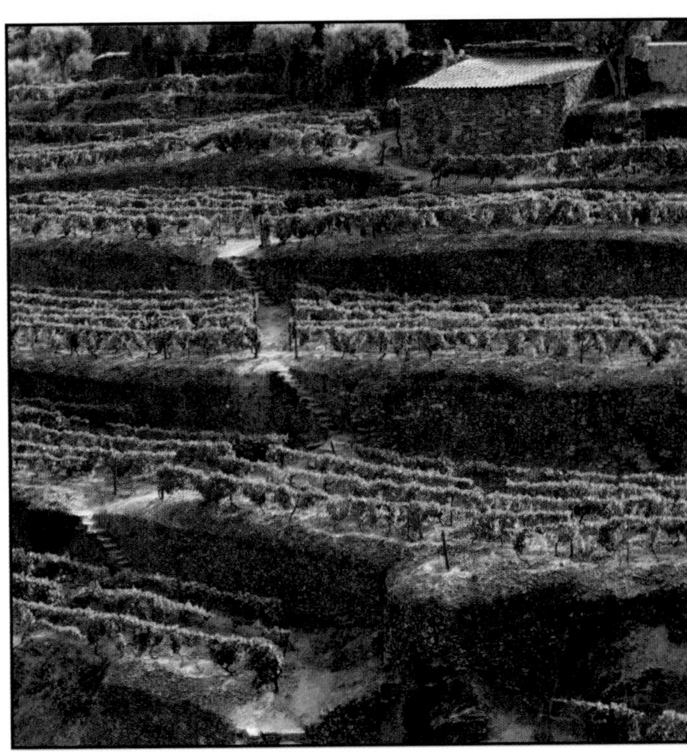

FIGURE 7–12 A terraced vineyard in the Douro region. Note the hard rock (schist) exposed at the edge of each laboriously constructed terrace.

tributes distinctive aromas of violets, berries, and smoked meat. Tinta Roriz is a hearty early ripening grape that likes lots of sunshine and thrives in minerally soil. Its yield is much higher than that of Touriga Nacional. Tinta Roriz has good acidity and tannic structure, allowing wines to age well. Its array of aromas revolve around its distinctive leather and floral components. The names of these two noble red varietals are showing up on labels more often.

For white Port, the favored varietals are Malvasia Fina and Gouveio (thought to be the Verdelho of Madeira.)

Porto

Unquestionably the most famous wine coming out of the Douro is still **Port**. As it has been for centuries, Port remains popular and revered the world over. Fortified with clear grape brandy whose high alcohol content (77 percent) kills the yeast cells before fermentation is completed, Port has a natural sweetness from the sugars that were prevented from fermenting. The final product is between 18 and 20 percent alcohol.

After an extended maceration to extract maximum color and tannins, the wine is pumped into barrels where it rests through the winter. The following spring the wine is shipped to the coastal city of Oporto where the Douro River spills into the Atlantic (Figure 7–13). Here the major Port companies, such as Croft, Fonseca, and Dow, have their caves, or aging facilities. In these facilities, under the watchful eye of the inspectors from the *Instituto do Vinho do Porto,* the wine is classified, aged, bottled, and eventually shipped to markets throughout Europe and the New World. By law no more than one-third of a company's stock can be released for sale in any year.

Styles of Port

The aging process determines a Port's style, and there are two basic categories. Wood-matured Ports are left for a short time in large wooden casks to age and are ready to be drunk when, after fining, filtering, and bottling, they are released. Bottle-aged Ports, on the other hand, are

FIGURE 7–13 For centuries, flat boats like these were used to transport casks of young Port wine from the wineries in the Upper Douro down the Douro River to blending and aging facilities in the city of Oporto. But since the river was dammed in the 1960s, this method of transport has disappeared in favor of other types of transport such as trailer truck.

FIGURE 7–14 Ruby Port is rich, smooth, and full of ripe berry flavors. From reliable producers like Osborne, ruby Port represents very good value.

intended to be aged further upon release. They are aged a short time in wood in the caves, then without filtration, are put into the bottles in which they may take up to 20 to 30 years to fully mature. Within these broad categories there are several different styles.

Ruby: This is the youngest and simplest style of Port (Figure 7–14). Named for the deep red color it retains upon release, **ruby Port** is meant to be consumed early. The berry flavors are robust and aggressive. In the making of ruby, wine from several vintages are blended together and aged briefly in large casks or perhaps steel tanks before being filtered and bottled.

Tawny: This term can be applied loosely to cover a variety of styles. One would assume that the amber color has evolved over years of barrel aging. However, most commercial tawnies are not much older than rubies, but without the fresh fruit flavors. The light color of **tawny Port** is not attained through patient aging in wood barrels thus allowing deliberate oxidation. Rather, the amber color of tawnies is achieved either by fermenting inferior, lighter-colored grapes or by blending in, after fermentation, some white Port. Approximately 80 percent of Port is simple ruby or commercial tawny.

Aged Tawny: This style of tawny comes by its color legitimately, as it must be aged in wood casks for six years or more. Aging not only imparts the golden color, but also gives the wine a smooth, soft texture as tannins polymerize. Aged tawny carries an indication of age, either 10, 20, 30, or over 40 years, which is an average of the ages of the various years' produce in that bottling. Aged tawnies are made from high quality grapes, and must pass a taste test by the IVP before being bottled. The best tawnies have delicate flavors of roasted nuts, honey, and dried fruit (Figure 7–15). The older the wine, the more subtle and profuse the flavors. Needless to say, as the age increases, so does the price.

Vintage Port: Occasionally an unusually warm and sunny summer will produce grapes of extraordinary character and ripeness. When this occurs, a Port maker will hold the resulting wine apart, rather than blending it in with the wine already aging in his cellars. After a year of careful aging, the winemaker will assess the wine again to determine if it is of quality to made into a vintage Port. If so, the company will send a sample to the *IVP,* along with a statement of their intent. If the wine passes the analysis of the *IVP,* the Port maker can declare a vintage. The

FIGURE 7–15 High quality tawny Ports exhibit appealing nutty aromas and delicate flavors.

single-vintage wine is then aged a further two to three years in wood barrels before being bottled and released. Thereafter the buyer takes over the aging, often holding the vintage Port for an additional 20 to 30 years. During this time the wine will throw considerable sediment, and it will need to be decanted before serving.

Port producers consider their vintage Port to be their flagship bottling. These are the rarest and most expensive of Ports. Only 1 percent of Port sold is from this category. Producers use only the best grapes from their finest vineyards. The greatest vintage Ports are velvety smooth and incredibly rich with deep but delicate flavors. In the past century the outstanding vintages for Port were 1963, 1966, 1970, 1977, 1982, 1985, 1990, 1995, and 1997.

Fine vintage Port, from quality producers such as Delaforce, Croft, Warre, Fonseca, Dow's, Osborne, González-Byass, and Taylor Fladgate & Yeatman, are in high demand in the United States, as well as in Great Britain and continental Europe. Vintage Port is usually served as the culmination of a fine meal, accompanied by sharply flavored cheeses such as Stilton or aged Cheddar, and walnuts or dried fruits.

LBV: Late-bottled vintage Port is a single vintage wine bottled between the fourth and sixth year after harvest. Most of these wines have been filtered and cold-stabilized before bottling, so they throw less sediment than vintage Port. Unfortunately, too often the filtration is excessive, stripping the wine of much of its character. Traditional-style LBV is bottled without filtration, and must be decanted before serving. LBV wines are made in good years that were not good enough to be declared a vintage. They need less time to mature than a vintage Port, and can be consumed within 5 years after bottling (Figure 7–16).

White: White Port is made in essentially the same method as red, except that the maceration period is shorter. Brandy is added to arrest fermentation at the same stage, leaving residual (or unfermented) sugars. White Ports, therefore, are medium-sweet, with fat, grape-like flavors. Alcohol content is between 16.5 and 17 percent as opposed to the 19 to 20 percent common in red Port. White Ports are usually aged no longer than 18 months, primarily in stainless steel tanks. White Port that does spend some time in wood has a golden color and nutty flavor. White Port is usually served chilled as an aperitif.

FIGURE 7–16 This 1999 Vintage Port was bottled 4 years after the harvest and is ready to be consumed, although it will improve for many years to come.

Bairrada and Dão

The *vinho regional* Beiras is a large region that stretches the width of Portugal, south of Vinho Verde and Douro, from the Atlantic Coast inland to the mountains that separate Portugal and Spain. There are two noteworthy *DOC*s in Beiras.

Bairrada is in the littoral region along the coast, extending from the city of Aveira south to the historic village of Coimbra. Wine had been produced in this region for centuries before the Marquis de Pombal, Portugal's powerful Prime Minister, in 1756 ordered all the vines in Bairrada ripped out as part of his effort to eliminate the fraudulent adulteration of Port. It took Bairrada two hundred years to recover, but after strenuous efforts on the part of its many small property owners, it was recognized as an official wine region in 1979, and now carries *Denominação de Origem Controlada* status.

The soils of Bairrada are primarily heavy but fertile clay. Over 70 percent of the vineyards are planted to one varietal, the hearty Baga, which produces the stout, dark, fairly tannic red wine for which the region is famous. All Bairrada wine must contain at least 50 percent Baga. Most growers sell their grapes to one of the six cooperatives. However, larger, private estates, such as that owned by Luis Pato, are emerging, where efforts are underway to produce Baga-based reds of more refinement. Two corporations based in Bairrada, Sogrape and Aliança, are also using modern vinification methods to make wines that are less rustic and more approachable.

In the hills at the eastern edge of Barraida, above the ancient village of Coimbra, lies the historic luxury resort, the Palace Hotel at Buçaco. Originally a monastery with its own vineyards and later a hunting lodge for the royal family, the hotel produces its own wines, red and white, from grapes purchased from local growers. Guests at the hotel can visit its dark cellars where these wines age in huge casks, then at dinner, order older vintages of these truly indigenous wines.

Dão lies inland to the east of Bairrada. Surrounded on three sides by high, granite-laden hills, Dão is protected from the winds and moisture of the Atlantic. It has the reputation of producing some of Portugal's finest red table wines. Unfortunately, the region was ill-served by the Salazar government's efforts to assist wine production by forcing the creation of many cooperatives and then restricting the sale of grapes to private producers. Standards fell as individual involvement was stifled. Such monopolistic practices were deemed inappropriate by the European Union and were discontinued upon Portugal's admission to that group in 1986. Now, with many talented individual vintners, Dão has re-emerged as a producer of fine table wines. Red grapes thrive in the granite-based soils, and 80 percent of the region's wine production is red. Many of the vineyards are on steep terraced vineyards in the hills. There are nine red grapes authorized, the most important being Touriga Nacional. According to a recently passed law, all Dão reds must now be at least 30 percent Touriga Nacional. These reds spend several months in oak barrels, and when released they are mellow, with vanilla tones mingling with natural red-berry and black pepper flavors.

White grapes are planted in the sandier, flatter area at the western edge of Dão. Most prominent of these is the Encruzado, a top grape with nice fruit and good acidity. It is unfortunately a low-yielding varietal. Other white grapes, such as Malvasia Fina (here called Arinto do Dão) and the ominously named *borrado das moscas* ("fly-droppings") are blended in. Dão whites, when not overly oaked or oxidized, can be fresh, crisp, and fragrant.

SETÚBAL

The Setúbal Peninsula, south of Lisbon, Portugal's capital city, protrudes into the Atlantic between the estuaries of the Sado and Tagus Rivers. The ocean and rivers provide moderating influences on the climate, and the warm temperatures and regular rainfall are excellent for growing grapes. The fishing town of Setúbal lends its name to the penisula, and Terras do Sado is the regional name for the wide range of table wines made on the peninsula. The most famous wine, though, is a sweet fortified wine that also carries the name Setúbal, now a *DOC.* In 1907, the region was demarcated as Moscatel de Setúbal for the principal grape, Moscatel (Muscat). However, E.U. regulations state that, to include the name of a varietal, a wine must be made at least 85 percent from that grape, and local customs allow as much as 30 percent of other grapes. Accordingly, since 1986 the *DOC* has been simply Setúbal.

Moscatel do Setúbal is made in the same manner as other fortified, naturally sweet wine, that is, fermentation is arrested with extra grape spirits before all natural sugars have a chance to ferment. The wine then has an extended maceration period with the Muscat skins, which gives a pronounced taste of fresh grapes to the wine. After as many as 5 months of maceration, the wine spends 4 to 5 years in large wooden casks. By the end of that time, the wine is deep gold in color, smooth and rich, and intensely flavored. The lively acidity prevents Setúbal from being cloying. A glass of this ambrosial wine, full of spice, caramel, honey, walnuts, and dried apricot flavors, is a dessert unto itself. The most prominent producer of Setúbal is the firm of J.P. Vinhos (formerly part of the José Maria da Fonseca company).

Alentejo

Alentejo is a huge agricultural region, stretching from the Tagus River east to the border with Spain and encompassing one-third of Portugal's land mass. It is known as the country's bread basket, covered as it is with grain farms. Portugal provides one half the world's cork, and Alentejo contains the majority of the country's cork forests.

After a period of disarray following the military-led uprisings of 1974 and 1975, Alentejo is again emerging as an important source of good table wines, with considerable help from the European Union in the form of financial investment and technical advice. The climate is not conducive to growing quality grapes, with very limited rainfall (as low as 23 inches/59 cm a year) and extreme temperatures that often soar to over 100°F in summer. Careful vinification with modern technology such as temperature-controlled fermentation tanks compensates for Nature's extremes. Red wines are made mostly from Aragonêz (local name for Tinta Roriz or Tempranillo), which lends elegance, Periquita with its blackberry and licorice aromas, and Trincadeira Preta which lends body and structure. After months in French oak barrels, the red wines emerge as complex, approachable, and long-lived, perfect accompaniments for the roasted meats and pungent cheeses of the local cusine.

Seven villages within Alentejo have been granted *DOC* status, including Borba and Redondo. The small rural village of Evora, better known for the breeding farms that produce the proud animals for Spain's bull-fighting rings, is also being considered for *DOC* status.

Madeira

The small island of Madeira (only 36 miles/58 km long and 15 miles/24 km wide) lies off the coast of Africa, a short plane ride from Lisbon. Its sparkling sunshine and white beaches make it a favorite vacation spot for European tourists. Claimed by Prince Henry of Portugal in 1420, the island was soon the site of vineyards. The soil is mineral-rich clay atop volcanic rock, sunshine and rainfall are abundant, and the terraced slopes of the southeastern facing hills, as seen in Figure 7–17, were soon producing high quality wine. With the help of British businessmen, Madeira wines were soon shipped to the Continent and the New World. By 1768, Madeira was demarcated as an official wine region.

In the mid-1800s shippers began to fortify the wine with additional alcohol so it could better withstand the long sea voyages. It was further discovered that the heat in the ships' holds gave the wine additional smoothness and richness. Today modern equipment is used, but the wine-making process is essentially the same. After vinification, during which brandy is added to arrest fermentation, the wine is placed for 3 to 4 months in an **estufa,** a heated vat which emulates the sun and shipboard heat of yore and imparts comparable qualities, including the distinctive nutty flavor. (A small percentage is heated naturally by being put in wood barrels and stored for up to 20 years in hot attics.) After the aging and heating are complete, sweetening in the

FIGURE 7–17 Harvesting grapes in the time-honored tradition at the type of sweeping, hilly, terraced vineyard found throughout rural Portugal.

form of caramel will be added to various degrees, depending on the style of Madeira being made. The whole process is carefully regulated by Madeira's quality control agency, *Insituto do Vinho da Madeira,* or *IVM.*

Even though 85 percent of the island's vineyards are planted to the lesser red grape, Tinta Negra Mole, quality Madeira is made from four premium white grapes. In accordance with E.U. regulations, each style of Madeira contains at least 85 percent of the grape for which it is named. In ascending order of sweetness, the styles of Madeira are Sercial, Verdelho, Bual, and Malmsey.

Sercial: The most delicate and lightest, with naturally high acidity, this dry, assertively flavored wine makes an excellent aperitif, especially good when matched with hors d'oeuvres of smoked fish or *paté de foie gras.*

Verdelho: With higher sugar content and lower acidity than Sercial, Verdelho is made in an off-dry style. It, too, is very good as an aperitif.

Bual (Boal): A heavier, richer wine, made in a sweet or semisweet style. The dark deep Bual with its intense raisin flavors can be served with caramel or coffee-flavored dessert. The Bual grape is the *vinifera,* Semillon.

Malmsey: This is the sweetest (and longest lived) of the Madeiras. It is made from Malvasia grapes grown in the warmest, sunniest vineyards where they achieve maximum ripeness and sugar levels. Additional richness and concentration of flavors are acquired during several years in wooden casks. The rich flavors and high sugar content are never cloying because the acidity levels remain high. These extraordinary wines can retain their beautiful complex flavors for as long as 100 years.

Summary

The progress made in the wine trade in both Spain and Portugal in the last half of the twentieth century, especially in the 20 years since both joined the European Union in 1986, is truly remarkable. Innovation in the vineyard; willingness to experiment with new varietals; modernization of facilities; utilization of computers to streamline all phases of vinification; research and teaching facilities where governments and private companies work together to seek new levels of quality; sophisticated marketing strategies; aggressive promotional programs representing close cooperation between government officials and businesspeople; and thoughtful and courageous investment at the continental (E.U.), national, provincial and local levels—all of these efforts combined have helped move Spain's and Portugal's many wines into competitive and constantly improving positions within the worldwide marketplace.

Review Questions

1. What is the principal red grape of Rioja?
2. Define the term "gran reserva."
3. Why is Ruby Port still red, while Tawny Port is a golden color?
4. List some of the responsibilities of the two government agencies, Spain's *INDO* and Portugal's *IVV*.
5. What is the wine-making process that is used to make Cava? Describe how Cava differs from Champagne.
6. What is the most widely planted grape in Navarra?
7. What is the highest level of classification within Portugal's quality control laws?
8. How has membership in the European Union affected the quality and marketing of wines produced in Portugal and Spain?
9. Sherry is produced in what region of Spain?
10. Name the eight *vinho regional* areas in Portugal.

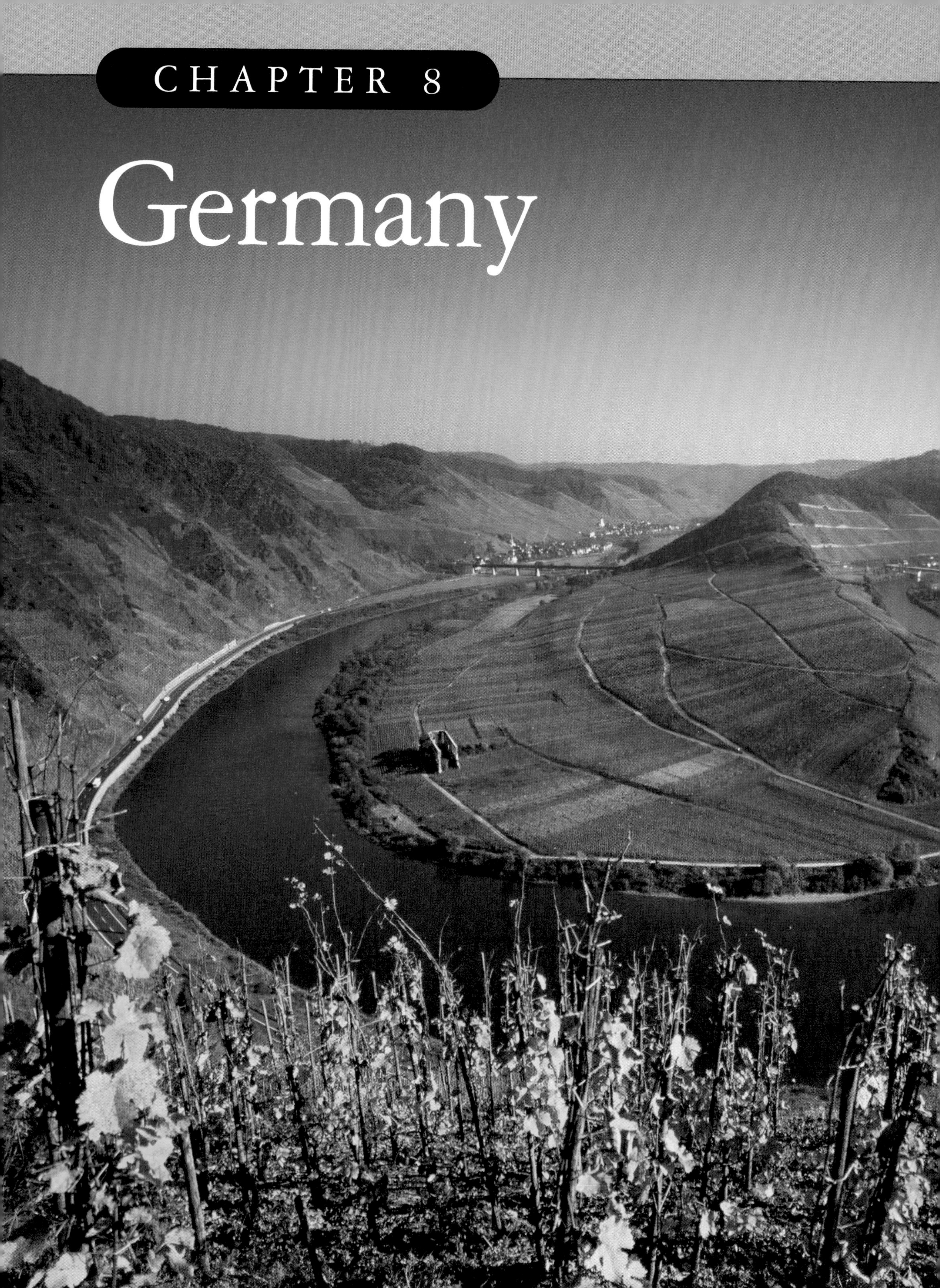

CHAPTER 8

Germany

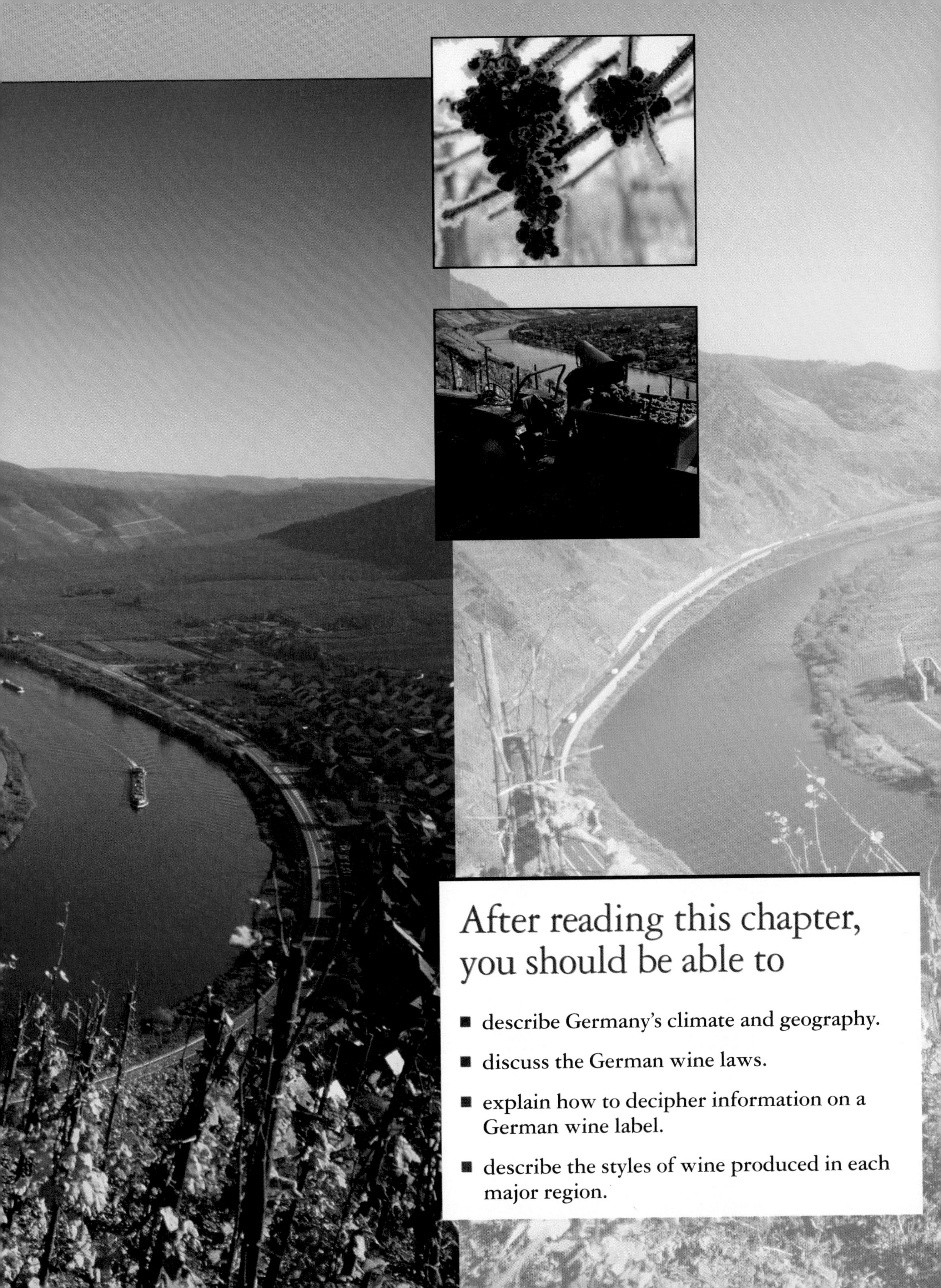

After reading this chapter, you should be able to

- describe Germany's climate and geography.
- discuss the German wine laws.
- explain how to decipher information on a German wine label.
- describe the styles of wine produced in each major region.

KEY TERMS

Anbaugebiete
Auslese
Beerenauslese
Bereich
chaptalization
Charta
cooperative
Einzellage
Eiswein
Erstes Gewächs
Grosslage
Gutsabfüllung
Kabinett
Landwein
Müller-Thurgau
Oechsle
Prädikat
Qualitätswein bestimmter Anbaugebiete (QbA)
Qualitätswein mit Prädikat (QmP)
Sekt
Spätburgunder
Spätlese
Tafelwein
Trockenbeerenauslese
Verband Deutscher Prädikatsweingüter (VDP)

In the United States, it seems that the least understood and most underappreciated category of wine is German whites. In the 1960s and early 1970s, when postwar production began to pick up in Europe, the bulk of German wine to reach these shores was made up of inferior, slightly sweet wines of the Liebfraumilch category. These heavily marketed brands dominated the U.S. market to such a degree that American consumers came to assume that all German wines are slightly sweet whites. In actuality, 15 percent of German wine production is red, and most German whites are dry or barely off-dry. The whites, especially those based on the noble varietals, are wonderfully food-compatible and lovely sipping wines.

The other factor that has worked against the wider acceptance of German wines in this market is the indecipherability of their wine labels. German wine labels provide more explicit information than those of any other country. Unless one knows

This meticulously tended vineyard is typical of many in Germany, with its rows closely spaced on a steep hillside. Notice how carefully the rows follow the slope of the land at the bottom of the photo, allowing maximum exposure to sunlight.

what to look for, however, it is intimidating to see all those words on a label, and the Gothic script does not help.

The effort to learn about the varying styles of German wines, the evolution of her wine laws (an ongoing process) and the philosophy behind the making of these versatile wines is well worth it. Germany is responsible for only about 5 percent of the world's total wine production, but its best wines are in the highest tiers of quality.

German Wine—Historical Perspective

Winemaking is an ancient tradition in what is now Germany. There is credible evidence to suggest that viticulture was brought to the region by the Romans during the time of the expansion of the Roman Empire. In AD 570 the Northern Italian poet Venantius Fortunatus mentioned steeply terraced vineyards along the Mosel River near the city of Trier (Robinson, 1991). Trier, the oldest city in Germany, was founded in AD 16 by the Roman emperor Augustus and archeological research near that city points to grape cultivation from that same era. Thus, grape growing and winemaking in Germany date from the first century.

Up until the time of Charlemagne, king of the Franks from 771 to 814, vineyards were concentrated on the west side of the Rhine (Rhein), in the region of Alsace and down the Rhein through the Palatinate (the Pfalz and Rheinhessen of today). Grape growing for wine extended along the Nahe River, a major tributary to the Rhein, and farther north along the banks of the Mosel and its two tributaries, the Saar and the Ruwer. The northernmost of these old Roman viticultural regions was the valley of the Ahr River.

Areas to the east of the Rhein, which were beyond the scope of Roman occupation, were not planted until later, as Christian monks moved into these districts to build their monasteries and spread the word of Christianity. In Franken, for instance, monks started planting vineyards in the eighth century, and Bavaria was widely planted to grape vines during the seventh century.

As in other parts of Europe during the Middle Ages, it was the Church and its monasteries and convents that owned many of the vineyards, and it was the monks, priests, and nuns who were responsible for keeping viticulture alive during that dark, unstable period (Figure 8–1). The members of religious orders maintained the tradition of making fine wines and perfected winemaking methodology. They also provided the principal market for wine, both for consumption with their meal and for use in the sacraments. Some of Germany's most famous vineyards were planted and tended by religious orders during the High Middle Ages, and still bear the names that signify their religious origins.

As viticultural practices improved and acreage under vine increased, production in medieval Germany eventually exceeded demand. Wine could be exported. Part of the commerce in wine was controlled by royalty and the aristocracy, and certainly princes stepped in to collect their share of excise taxes and tariffs, including tolls along the rivers which provided the main method of transport of wine from grape-growing districts to the ports of trade farther north. As trade with northern German cities, England, the Low Countries, and Scandinavia expanded, the Church and aristocracy could not handle every facet of the growing business. Expansion in trade gave rise to the *bourgeoisie,* or middle class. The wine merchants traded German wine for heavier red wines from France, and to Scandinavia for fish and grain, and to England for grain and other foodstuffs.

FIGURE 8–1 The Cistercian monastery in the town of Eberbach on the Rhein has been producing wine for more than 850 years. Pictured here is the room holding the original wine presses. In this room the laymen worked (supervised by the monks), and here they were served their meals.

The Thirty Years' War (1618–1648) dealt a severe blow to Germany's emerging wine trade. Many vineyards were torn out or allowed to deteriorate badly. However, by the late 1700s, winemaking and the concurrent trade were back on track in most regions. At this time governors, grape growers, and middle class merchants worked together to create a system of regulations that set quality standards for wine and simplified wine naming. Strong emphasis was put on planting more of the noble grape varietals that were proven performers in the cool climate, especially Riesling. Authorities and peers also discouraged landowners from planting vineyards in locations that could not produce quality fruit (Figure 8–2). In the early nineteenth century, as the tensions between the German states and France, and among the German states themselves, began to dissolve, internal customs and tariffs slowly disappeared. At the same time, transportation improved, especially as the railroads expanded. These two factors opened up the market for wine to such a degree that producers could not depend on selling their product locally. There was too much competition from the wines of other regions that were now available.

In the later nineteenth century the first attempts were made to produce dessert wines of specially selected or late-harvest grapes. The first designations of higher quality wines showed up at this time. By 1830 a system of quality control laws had begun. In that year, one region passed an impractical system defining 65 different levels of quality. Other states passed more workable ordinances defining quality levels with fewer tiers. Around this time several German states passed laws declaring that grapes within a vineyard were to be harvested at different times depending on their ripeness. Fortunately, a German scientist, Ferdinand Oechsle, had recently invented a method for measuring the sugar levels in grapes. The level of sugar is the clearest indication of the grapes' ripeness. The **Oechsle** system is still used today.

In 1892 the first national wine law was passed. It defined the borders of premier wine regions. It also spelled out which winemaking practices were to be controlled, such as **chaptalization,** the introduction of additional sugars during fermentation.

FIGURE 8–2 This historic village in the Mittelrhein, with its banked vineyards on steep slopes, old stone houses, historic cathedral, and commercial activity along the river, is typical of Germany's picturesque wine country.

Through the nineteenth century the emphasis on quality in wine production continued. The governments of the states became involved with the wine business, through legislation to assist in exporting of wines and in the establishment of state-sponsored institutions for training and research in viticulture and winemaking. As German wines became more competitive in the international market, the owners of small plots of land, many of them peasant families or members of the bourgeoisie, lacked the resources to maintain high levels of quality through improved technology and investment in modern equipment. Several small neighboring landowners teaming up into a **cooperative** allowed pooling of resources, which resulted in an improvement in the quality of wine being produced, as well as distribution outside their local community. The cooperatives remain an important influence in German winemaking to this day.

Germany's progress toward production of high quality wines and its participation in the international wine marketplace were dealt severe blows as the nineteenth century came to a close. Phylloxera destroyed vineyards throughout the region in the 1880s after being brought to Europe from North America. Downy mildew was also a problem in this cool damp climate, and no chemical yet existed to control it.

In the first half of the twentieth century, the German wine trade was set back by the two world wars, as was every aspect of Germany's economy and culture. The wars left vineyards decimated, labor supply limited as two generations of men were lost, and foreign markets adverse to German products. Fortunately, viticulture survived due to the dedicated efforts of private landowners and the increasing effectiveness of the winemaking cooperatives.

Between 1950 and 2000, the German wine trade made enormous strides forward. As the demand for the slightly sweet and very affordable Liebfraumilch increased in European and North American markets during the 1960s and early 1970s, the temptation to produce more wine led to an increase in allowable yields per hectare. Despite that weakening in quality control, Germany's best estates continued to produce some world-class wines. In the 1980s a more sophisticated base of wine consumers around the world turned away from semisweet whites and demanded dryer wines, particularly Chardonnay. Imports of German wine into the United States fell from almost 5,000,000 cases in 1980 to 1,800,000 cases in 2003. As we move into the twenty-first century, producers whose emphasis is, and always has been, on high quality

Reading a German Wine Label

Designation, or Prädikat: the designation indicates level of ripeness of the grapes at harvest, and therefore, the style of the wine. There are six Prädikat categories in chronological order of picking and in descending order of ripeness.
Kabinett: The lowest level of ripeness. Wines at this level are slightly off-dry.
Spätlese: Grapes for this category are picked later ("spät" means "late"). The wines are off-dry with more body.
Auslese: "Selected" grapes are picked later, and have higher sugar levels. The wines are definitely off-dry and quite rich.
Beerenauslese: Individual bunches are picked after being partially infected with *Botrytis cinerea,* the "noble rot." These wines are rich, sweet dessert wines.
Eiswein: Eiswein, like Beerenauslese, are partially botrytized bunches of grapes picked after the first hard frost in which the grapes are frozen.
Trockenbeerenauslese: Literally translates as "selected dried berries." These grapes are picked once they are fully botrytized, that is, shriveled up by the fungus. Sugars are very concentrated, and the resulting wine is very sweet and rich in texture.

Region: Mosel-Saar-Ruwer is one of the 13 official wine regions.

Gutsabfüllung: Estate-bottled. This means the Weis family owns the portion of the Goldtröpfchen vineyard from which these grapes were harvested.

Quality Level: "Qualitatswein mit Prädikat" indicates this wine belongs to the highest level of quality.

Government Approval Number: All Prädikat wines and QbA wines must display an official approval number on their label, indicating the wine has met all standards for its level of quality.

Alcohol Content: The percentage of alcohol by volume.

Village: All grapes were grown within the borders of the town of Piesport.

VDP Logo: The initials stand for "Verbands Deutscher Prädikats-und-Qualitats-weingüter," Germany's most prestigious growers' association for Riesling.

Varietal: 85 percent must be the grape indicated.

Winery Name: St. Urbans-hof, named for the German saint of wine, is owned by the Weis family.

Vineyard: Goldtröpfchen is a famous single vineyard in the town of Piesport. 85 percent of the grapes must come from this vineyard. In Germany, there are over 2,500 individual vineyards that can be indicated on labels.

Vintage Date: By German law, 85 percent of the grapes must have been harvested in this year.

wines compete with those producers who want to produce more and more mediocre wine to fill a growing demand in domestic and European markets. Germany may be poised to reclaim a place as one of the world's greatest wine-producing countries.

Wine Laws

Germany's quality control laws are among the most strict and thorough. However, the country's efforts to ensure quality and consistency in wine production and to increase the marketability of its wines have created a system that is so complicated as to be counterproductive. As Germany moves into the twenty-first century, its wine laws need to be streamlined and tightened, the labels need to be easier to read, and many groups, both private and governmental, will have to work more closely together so that their efforts are not redundant and confusing.

The original national Wine Law of 1892 defined boundaries of major regions and specified winemaking practices that were forbidden. It was amended in 1909. An important step forward for German viticulture was the Wine Law of 1930, in which many of the deficiencies of earlier laws were corrected. The definition of "quality wine" was refined, levels of quality were outlined, and certain winemaking practices that led to inferior wines were abolished. However, vineyard boundaries were often unclear, and quality levels were difficult to discern. Finally, in 1971, partly in response to pressures from the European Union and partly in response to market forces, the German government performed a major overhaul of its wine laws. The 1971 law greatly simplifies previous systems and is precise, definitive, and unambiguous. With only a few changes, this law has been the basis for producing and labeling German wines ever since.

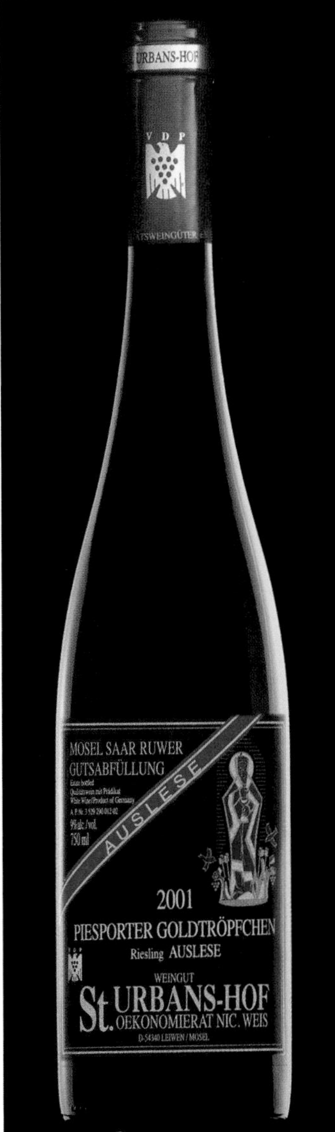

FIGURE 8–3 This bottle of wine is at the Qualitätswein mit Prädikat level (designation Auslese). It is from St. Urbanshof, one of the premier producers of Riesling in the Mosel region.

Wine Categories

Under the Wine Law of 1971, all vineyards are delineated and registered, and the definition of a wine's quality level depends not on vineyard location, nor on yield per hectare, but on the level of ripeness of the grapes at harvest. This method of measuring quality may be precise in that the sugar at harvest is easily quantified, but it has also caused controversy because it ignores the concept of terroir, the idea that different vineyard sites can impart distinct character and superior quality to wine. Critics further object that the law could be construed as encouraging production of sweeter wines, although the international market is leaning toward dryer white wines.

The quality control laws continue to evolve; see "Revisions to the Wine Laws" later. Current German wine law divides its wine production into four main groups:

- ***Tafelwein*** Table wine or ordinary wine
- ***Landwein*** Regional wine
- ***Qualitätswein bestimmter Anbaugebiete*** (often written as QbA) Quality wine
- ***Qualitätswein mit Prädikat*** (also written as QmP) Quality wine with designation (Figure 8–3)

Tafelwein

As in other countries of the European Union, "table wine" signifies everyday wine, produced with very few, if any, standards or stipulations. This is wine to be consumed locally and is rarely exported. In Germany, the average person is more likely to drink beer on a daily basis than wine. Therefore, the demand for ordinary wine is very low. In some years as little as 2 percent of German wine production is in this category. Plain *Tafelwein* can have grapes grown outside Germany blended in. For *Deutscher Tafelwein* four large regions are defined.

- Mosel und Rhein

- Bayern
- Neckar
- Oberrhein

One of these geographic designations must show on the label.

Landwein

This designation, added in 1982 to parallel France's *vin de pays* designation, is rarely used. Wine at this level is assumed to be higher quality than *Tafelwein* as some guidelines govern its production. The regions defined for *Landwein* are smaller than, and fit within, the four *Tafelwein* regions. There are 19 *Landwein* regions, and the region must show on the label.

Qualitätswein bestimmter Anbaugebiete

This is the lower level of Germany's quality wine production and in most years encompasses the largest percentage of German production. The term means "quality wine from a specified geographic location." Standards of quality must be met. The grapes must be of authorized varieties and must reach a specified level of ripeness that allows natural varietal character to show (Figure 8–4). Moreover, for any grape varietal to be mentioned on the label, the wine must contain a minimum of 85 percent of that varietal. The same percentage must also be true for any geographic designation of origin. The grapes must have been grown in one of the 13 approved regions or ***Anbaugebiete.*** These 13 regions are subdivided into smaller geographic designations of first, a district ***(Bereich)*** or village, and second, a group of vineyards ***(Grosslage)*** or specific vineyard (***Einzellage***) (Table 8–1). The name of the wine will sometimes incorporate the district (or village) and the vineyard site. However most QbA labels indicate the Anbaugebieten only (figure 8-5). Chaptalization, or the adding of natural sugars, is allowed at this level.

Qualitatswein mit Prädikat

This is the highest category of classified wines in Germany. Translated as "quality wine with designation," the term signifies the geographic location where the grapes were grown and their level of ripeness at harvest time. The name of the wine, therefore, will have at least three words

VON
OTHEGRAVEN
Maria v. O.
2002
RIESLING
MOSEL · SAAR · RUWER
Alc. 11% By Vol.
SOLE U.S. AGENTS:
CLASSICAL WINES - SEATTLE, WA
PRODUCT OF GERMANY
750 ml

FIGURE 8–4 This label from Maximillian von Othegraven is an example of a QbA label, that is, Qualitätswein bestimmter Anbaugebiete. The wine will reflect the terroir of the Mosel region, but does not indicate a style or level of ripeness.

Table 8-1 German Land Designations

DIVISION	EXAMPLE	DEFINITION
13 *Anbaugebiete*	Mosel-Saar-Ruwer	Specified wine regions
40 *Bereiche*	Saar	Smaller districts with a region
1,400 Villages	Ockfen	A specific village within the Bereich of Saar
163 *Grosslagen*	Scharzberg	A grouping, under one name, of several vineyards within one district or village
2,715 *Einzellagen*	Ockfener Bockstein	Individual small vineyard sites, each with its own name

in it: the village, the vineyard (either *Grosslage* or *Einzallage*), and the ***Prädikat,*** which shows the grapes' ripeness and hence the sweetness of the resulting wine (Figure 8–6). (No chaptalization of wines is allowed at the *Prädikat* level. The regulation on 85 percent minimum for grape varietal and geographic designation remains.)

There are six categories of *Prädikat,* or special designations. In ascending order of ripeness these designations are:

1. ***Kabinett:*** From the word for "cabinet," this category is the lightest and the driest of the *Prädikat* wines. Although there is residual sugar at this level, it is often undetectable due to the racy acidity of the Riesling grape. *Kabinett*-level wines can be matched to a wide variety of foods and also make excellent aperitifs.
2. ***Spätlese:*** From the German word for "late," these grapes are left on the vine longer and harvested when the sugar content is sufficiently high to make a slightly off-dry wine. Sugar

FIGURE 8–5 This is also a QbA label. Note, however, the term "Gutsabfüllung" on the St. Urbans-Hof label, which translates as "estate-bottled," signifying that the family that owns the St. Urbans-Hof winery also owns the vineyards in which the grapes for this wine were grown.

FIGURE 8–6 This label from the von Othegraven estate is more specific both as to place of origin and to style. Like all *Prädikat* labels, it indicates the *Bereich* or town in which the grapes were grown (Kanzem), the single vineyard or *Einzellage* (Altenberg), and the *Prädikat* level (*Auslese*). The wine will be distinct in character, reflecting the terroir of the vineyard, and will be definitively off-dry in style.

is usually detectable in Spätlese wines, but the acidity is still fresh enough that the impression is not one of "sweetness." These wines are excellent with foods that are either very spicy (e.g., chilli peppers, lemongrass), or tartly acidic (e.g., a lemon-based sauce), or have a lot of fruit (e.g., pork chops baked with apples).

3. ***Auslese:*** Meaning "selected," the grapes at this level are picked later still, when the sugar content has risen to a level that produces decidedly off-dry wine. Still nicely balanced with clean acidity to hold up the ripe fruit flavors and residual sugar, *Auslese* wines are nice as aperitifs. They can also be matched successfully to foods that also have a touch of sweetness, such as crab meat, lobster, or pâté.

So far the designations are all for table wines, meant to be served during the meal. (*Note:* Do not confuse the English/American use of the term "table wine," which merely signifies a non-sparkling, nonfortified, nonsweet wine meant to be served during meals, with the European Union's use of the term in its various wine laws. In those sets of wine laws, the term signifies a noncontrolled, lower quality category of wine.)

The last three designations in Germany's *Prädikat* tier of wine production definitely fall into the category of dessert wines and are truly sweet.

4. ***Beerenauslese:*** Translated literally as "selected berries," this designation signifies grapes with high levels of sugar and some botryitis (noble rot). Specific levels of sugar at harvest are spelled out in the laws for each region and each varietal. The resulting wines are gold in color, incredibly rich and ripe in flavor, with an impression of clover honey wrapping around the deep essence of the varietal. Risky to make and labor-intensive to produce, these wines are very expensive. In some drier years, noble rot does not occur and no *Beerenauslese* can be made at all.
5. ***Eiswein:*** Recently accorded its own *Prädikat, Eiswein* is also richly sweet, but it is not fully botryitized. The sugars are concentrated when grapes freeze while still on the vine, sometimes into late December or even early in the new year. The grapes are picked while still frozen (Figure 8–9). When the grapes are pressed, ice crystals are left behind and only the sweetest juice runs into the fermentation receptacle. *Eisweins* have a higher acidity level than boytritized wines, even though the minimum sugar content must be at least as high as that for *Beerenauslese.* They reflect varietal flavors more clearly because the cloak of clover-honey and dried-apricot nuances imparted by the noble rot is not present.
6. ***Trockenbeerenauslese:*** "Selected dried berries" are picked in late autumn or even early December. They are fully infected with botrytis, and have shriveled on the vine as the fungus

FIGURE 8–7 **Riesling grapes in the Schlossberg vineyard of the Rheingau region show the first signs of *Botrytis cinerea*, the noble rot.**

FIGURE 8–8 From one of the most famous vineyards in the Rheingau, Berg Schlossberg in the town of Rudesheim, this *Trockenbeerenauslese* is a lusciously sweet dessert wine. (For a picture of this vineyard see figure 8-23).

causes the skin to crack, allowing the water of the grape to evaporate (Figure 8–7). What is left is highly concentrated natural sugar. The wine that results is ambrosial—deep golden color, honeyed nose redolent of dried apricots and nuances of the varietal with rich deep raisiny flavors. In some years no *Trockenbeerenauslese* (often abbreviated to TBA) can be produced, and when it is made, the quantities are small (Figure 8–8). The price is very high, compensating the producer for the risk of losing the crop to grey rot or a killing frost, and for the difficulty of fermenting such a viscous juice.

Reading the Labels

Deciphering a German wine label is not as complicated as you may think although with wine laws currently in flux, there are some complicating factors. In general, however, the label of a *Qualitätswein* contains all the information the consumer could want in order to make a knowl-

FIGURE 8–9 These ice-encased grapes will be picked while still frozen. The ice will be separated from the grape at pressing, and the concentrated juice will become luscious *Eiswein.*

edgeable decision. Bear in mind the adage about wines getting better and more distinctive the smaller and more specific the geographic designation on the label. With German labels, this concept holds true as we move from *Tafelwein* up through *Landwein* to *Qualitätswein bestimmte Anbaugebiete* (often abbreviated to QbA), with each level being more specific and correspondingly higher quality. Within the QbA level, the adage continues to be true, as the geographic designation gets smaller and more specific.

Once we get to the *Prädikat* level, there is the added designation of quality, although here the quality is not based on the uniqueness of a specific vineyard or on a lower yield per hectare, as in some European countries. In Germany, that extra designation of quality is based on the level of ripeness of the grapes at harvest. In other words, the *Prädikat* label, like most European labels, tells you exactly where the grapes were grown, which gives you an idea of the character of the wine since the terroir of each *Bereiche* and village is unique. A German *Prädikat* label, however, goes further: It also gives the consumer a solid idea of the *style* of the wine because the designation indicates where the wine falls on the stylistic spectrum from very dry to very sweet.

Revision to the Wine Laws

As stated earlier, considerable dissatisfaction exists with Germany's existing wine laws and how those laws are interpreted on labels. Many different voices are protesting the status quo. Producers, wine consumers, the importing and marketing companies that handle German wines in foreign markets, and government officials are all raising concerns about the effectiveness of the present situation in ensuring continuing quality and stylistic integrity, and in improving marketability. In this section, we summarize the controversy and the various movements within Germany's wine trade that are working to improve the system.

New Terms and Ratings

The traditional style for German white wines has been off-dry, that is, with a degree of unfermented, or residual, sugar to play off the natural fruit flavors. Because of the piercing acidity of the Riesling grape and its offspring, German whites can carry off a less-than-bone-dry style without being cloying. While staying true to their long heritage, German producers have embraced the latest technology. Temperature-controlled, stainless steel fermentation tanks allow a slow, cool fermentation within an inert container, promoting an aromatic and complex, yet clean and fresh, style. Precision-engineered centrifuges are now common, especially in the production of bulk, or medium-quality, wines.

Many producers now use stainless steel containers in which the wine can be chilled to 33°F (0.5°C) to kill the yeast, thus stopping fermentation and leaving the desired amount of residual sugar (Figure 8–10). However, in response to perceived market demand for dryer wines, many producers in recent years wanted to reduce or eliminate residual sugar (Figure 8–11). They wanted a mechanism whereby their labels could indicate to the buyer that the wine was not in the traditional off-dry styles, but was still a *Qualitätswein.* In 1982, the government approved an amendment to the Wine Law of 1971 authorizing the use of the labeling terms *Trocken* ("dry") and *Halbtrocken* ("half-dry," i.e., a semidry wine with no more than 0.6 ounce of residual sugar per quart).

In 2000, further label terminology was approved and showed up for the first time on the 2001 wines of certain producers: Classic and Selection. Classic wines are made from only traditional grape varietals and a wine must be 100 percent one varietal. Twenty-two grapes are approved for Classic wines, specific ones for each of the 13 regions. The name of the vineyard is

FIGURE 8–10 These tall stainless steel tanks are typical of the highly mechanized, temperature-controlled fermentation tanks being used throughout Germany today.

FIGURE 8–11 A vintner extracts a sample of his wine from a cask with a glass siphon. He will taste it to assure himself that it is evolving according to his exacting standards.

omitted from Classic wines' labels. Selection wines are made from (to quote the German Wine Institute's Web site) "only the finest traditional grape varieties (which) meet the high quality standards prescribed" for this level. Twenty-two varietals are approved for Selection wines. The vineyard name is listed on the label.

Germany has never had an official rating of its vineyards. The sites that have for centuries been noted for producing the highest quality grapes were recognized in the Wine Law of 1971 by being designated as individual *Einzellagen.* Lesser sites were incorporated into *Grosslagen.* Ratings were deemed unnecessary anyway since Germany's *Prädikat* designations are based not on vineyard location but on level of sugars in the grapes at harvest.

However, one group of producers has banded together to protest the traditional *Prädikat* designations, and to work toward official recognition of prime vineyards through a classification system. This group of vintners, the First Growth Committee, are all from high quality estates in the Mosel, the Rheingau, the Pfalz, and Rheinhessen. They have persuaded the government to approve the use of the term ***Erstes Gewächs*** (first growth) to be used on the labels of a small group of selected vineyards. By 2000, the process of selecting the rated vineyards had been completed only in the Rheingau, where 35 percent of the vineyards now carry the *erstes Gewächs* designation. The process of rating vineyards in the other major regions is ongoing. Some of the owners of these now-rated vineyards are declassifying their wines from QmP level to QbA to drop the *Prädikat* designations (based on sugar content) from the labels. It will take several years before the task of classifying vineyards is completed in other prime regions.

FIGURE 8–12 An example of a label from a Charta estate. There is no indication of *Prädikat* level of ripeness. Rather, it carries the winery's own designation for its high quality Rieslings: "Terra Montosa."

TRADE ORGANIZATIONS

Other vintners feel strongly that their obligation is to continue to make the highest quality wines in the traditional style, while avoiding any measures that render the classification and labeling of German wines any more complicated than necessary. Many of these vintners have banded together to more effectively work toward their goals.

Charta Wines: A group of producers in the Rheingau who organized in 1984. Charta is dedicated to producing only the very highest quality wines that are typical of their region (Figure 8–12). Charta promotes the wines from their estates as the best from the Rheingau. Charta wines must be 100 percent Riesling and meet higher standards for minimum sugar content than is specified in national laws for *Kabinett, Spätlese,* and QbA wines. The back label of Charta wines shows twin Roman arches.

Verband Deutscher Prädikatsweingüter (VDP): Originally a collaboration of Mosel producers, the VDP was founded in 1908 by the Mayor of Trier, Freiherr von Bruchhausen. The group was dedicated to high-quality wines that "reflect the distinct character of their German origin" <http://www.VDP.de>. Later the Mosel group merged with similar associations from other regions. There are now 200 members from around Germany. The VDP is committed to the traditional style of German wines and to maintaining the *Prädikat* system. As stated by Nik Weis of St. Urbans-Hof in the Mosel, whose family has been active in VDP for generations, "The goal is to promote the German Estate Wineries, the quality of German wines, the traditions, and the best vineyards."

It may take another decade for all these pieces to fall into place, as Germany's producers and government officials work to clarify standards, classifications, and labeling practices. In the meantime, Germany continues to produce a wide range of fine wines, many of them of the highest quality. Although German wine laws are currently being reassessed, the example given here illustrates how to read a German wine label that conforms to all existing laws.

The Wine Regions

Climate

Germany is one of the northernmost fine wine regions in the world. Its climate is heavily continental; the climate is influenced by weather patterns within the land formation, rather than by oceanic patterns. A region with little mitigating maritime influence will exhibit wide variation in average temperatures each year. To produce quality wines in such a volatile and difficult

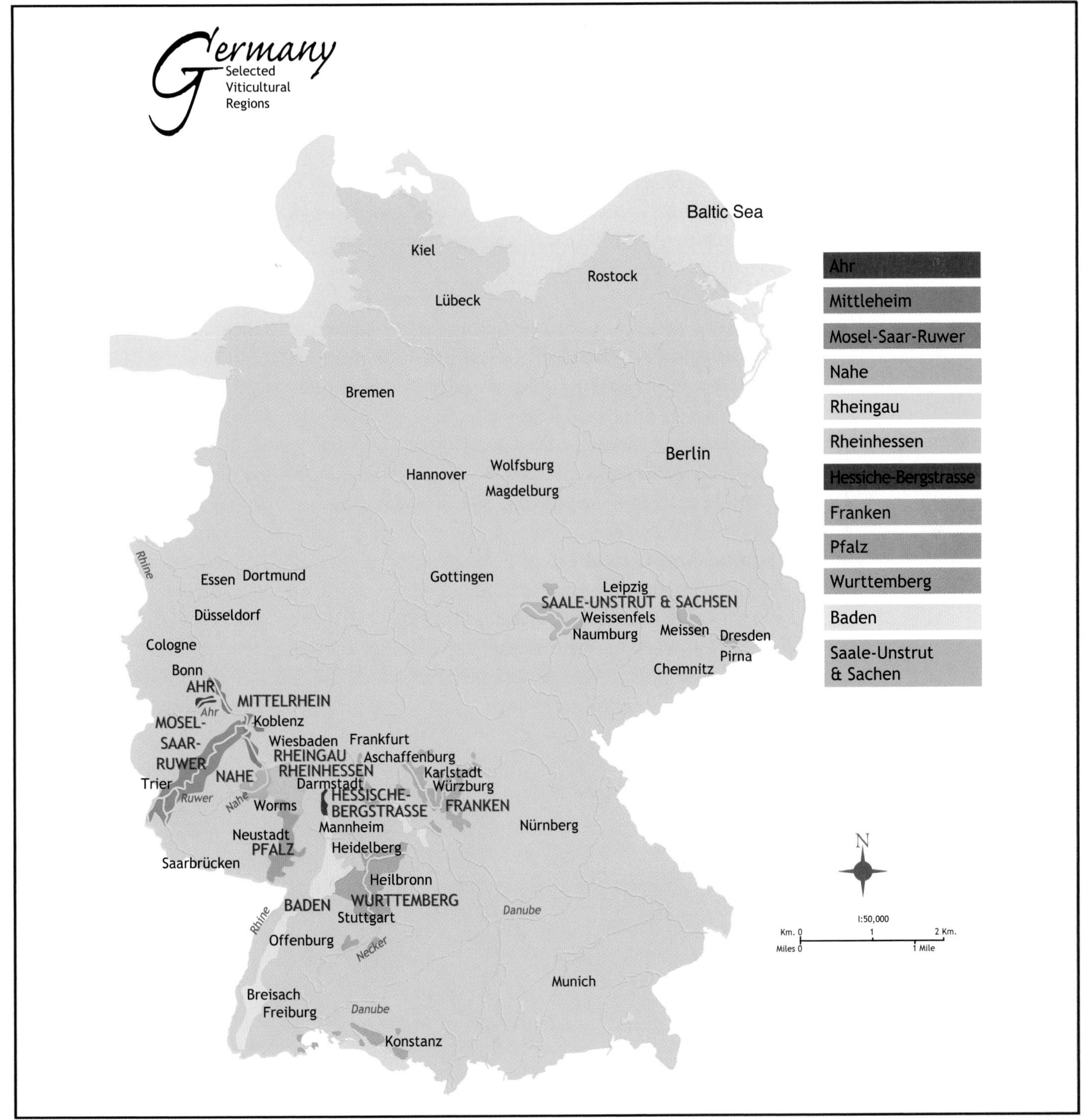

The wine regions of Germany

environment takes real dedication and skill. It is no accident that 80 percent of Germany's 260,000 acres (100,000 hectares) of vineyards are planted along steep river banks. The Rhein and its many tributaries provide a moderating influence on the climate, reflecting heat and light back onto the vines. The steepness of the slopes allows vines to capture as much sun as possible, while also providing some shelter from cool winds.

In most of Germany's vineyards, each individual vine is laboriously trellised onto its own stake rather than on the row-long trellises made of wire common elsewhere. The reasons are twofold, both reflective of the difficult terrain: First, when working these steep vineyards, the worker can reach any part of the vine he is working on without having to walk all the way around the row. Second, the individual stack allows the vine to hold its "heart" up to the sun, giving better exposure to the leaves. It is estimated that it takes three times as many man-hours to tend the vines on the steep terraced vineyards along Germany's rivers than is the case for flat rolling terrain (Figure 8–13).

Most of Germany's 100,000 grape growers (whose average holding is just 2.47 acres, or 1 hectare) do not make their own wines, but sell their grapes to the 26,000 winemaking facilities. The hard work and close collaboration required to make wine in a northern region like Germany is well worth the effort, though, for the grapes acquire enough ripeness to develop aromatic bouquets and appealing fruit flavors, while retaining excellent acidity. Moreover, the stony soils on the hillsides provide an additional quality of unique minerality to the wines, especially the Rieslings.

Because of the volatility of its climate and the difficulty of working the steep vineyards, Germany produces a very small percentage of the world's wine (about 5 percent). Not surprisingly, Germany is by far the largest importer of wine in the European Union. However, despite large imports of French and Italian wines, after exporting a portion of its own annual production, only 2 percent of domestic production is surplus each year. Obviously, then, Germany is not a contributor to the "ocean" of excess wine in the contemporary world that has led to the instability of wine (and grape) prices in the international marketplace.

Grape Varietals

As explained in the chapter on viticulture, different varietals require different conditions to thrive. Early ripeners tend to ripen quickly on the vine. If given too much sun and warmth, they will become ripe and have to be picked before having fully evolved their flavors. The resulting wines will be "green," thin and unpleasant. What early ripeners need is a long growing season and moderate temperatures. Germany—with its northern location, often overcast skies, and

FIGURE 8–13 The slope of the Doktor vineyard on the Mosel River near the town of Bernkastel is so precipitous that this worker needs a winch to help him back to the top, where he will again fill his basket with ripe grapes.

many rivers with their moderating influences—is perfect for these grapes. Most cool-climate varietals are white. Fully 85 percent of Germany's vineyards are planted to white grapes. Among the cool-climate grapes that thrive in Germany are the following:

Riesling: the most noble and important of Germany's white varietals. Riesling accounts for 25 percent of acreage under vine. In Germany's northernmost vineyards Riesling is at its best, shedding extraneous fat and richness and allowing sleek simplicity and elegant style to shine through.

Müller-Thurgau: This clonal offspring of Riesling was developed in Switzerland by botanist Professor Hermann Müller in 1893 by crossing the noble Riesling with a grape indigenous to his country. The resulting wine, which combined the delicate flavors and crisp acidity of Riesling with the hardiness of the local varietal, was named for Professor Müller and his home district. The **Müller-Thurgau** is now planted in approximately 20 percent of German vineyards. It is widely used in everyday blended wines, such as *Liebfraumilch*.

Silvaner: This workhorse of a grape is decreasing in importance. Useful as a blending grape, Silvaner contributes high acidity and a neutral background against which more distinguised varietals, such as Riesling, can show off their distinctive flavors and site-specific character. Presently less than 7 percent of total acreage.

Pinot Noir: This noble red grape, called **Spätburgunder** in Germany, is on the rise, presently accounting for 8 percent of vineyards. Many of Germany's Spätburgunders are anemic, too light in body and lacking in depth of flavor. Part of the problem, of course, is the cool climate in which the grapes struggle to reach adequate ripeness. The inadequacies could be somewhat alleviated, however, by lowering the allowable yield for this varietal (and other red grapes).

Other grape varietals: There are over 20 other grapes grown in Germany, from the truly obscure (e.g., Limberger, Gutedel) to the better known like Pinot Gris (Grauburgunder in German) and Pinot Blanc (Weissburgunder) (Figure 8–14 a&b). (Note: The German word *Burgunder* denotes a member of the Pinot family.) The older traditional grapes, many of which have long been associated with a particular region or district, are losing ground to the Pinot rela-

FIGURE 8–14 A and B These botantical cousins, the Pinot Gris (on the left) and Pinot Blanc (on the right), do well in Germany's cool climate, as do other members of the Pinot family.

tives. Even the ubiquitous Chardonnay, a Pinot cousin, was approved in 1991. It is not widely planted, however.

Two rare red grapes that have shown small increases in the past several years are the Dornfelder and the Portugieser. The origins of the latter grape are a mystery as no connection to any varietals grown in Portugal can be proven. The wines made from these two grapes have varied considerably in style and in quality.

Surprisingly, Gewürztraminer, which produces such delicious wines across the border in Alsace, is not widely grown in Germany. Perhaps its penetrating flavors are too pronounced for German producers who aim for refinement and delicacy.

The Mosel-Saar-Ruwer

The Mosel River curves and twists like a huge shimmering ribbon as it winds among the rocky hills that stretch across Germany's southwestern midsection (Figure 8–15). Over the millennia, the river has gouged out gorges through these hills, leaving incredibly steep banks on each side. The banks of the Mosel and its two tributaries are so steep, barren, and rocky that it is hard to believe anything can be cultivated there. But the Romans were cultivating wine grapes here sixteen centuries ago, and at one point only the highest, rockiest, and most inaccessible sections were left unplanted. Unfortunately there are more sections left bare today as fewer people can be found to undertake the backbreaking work of tending the terraced vineyards. There is an old saying in this region: "A true Moselaner has one leg shorter than the other." Working long hours in the sloping vineyards does not really lead to a shortening of the uphill leg, but it *is* incredibly hard work.

As the Mosel River has pursued its 150-mile route from France toward Koblenz where it feeds into the Rhein, it has deposited many layers of crushed rock in its winding path (Figure 8–16). These deposits have lent a distinct slate-like note to the wines made from grapes grown in the 31,530 acres (12,760 hectares) of vineyards along the Mosel and its tributaries. This note of slate, called Schieferton, gives such a unique edge to the wines of the Mosel that the government of Prussia drew up an extensive map of the region in 1868 specifying the degree of Schieferton to be found in each vineyard. That carefully researched map has recently been reprinted as a continuing guide to the terroir of the Mosel's vineyards. The touch of slate gives

FIGURE 8–15 The beautiful Mosel River twists and turns among the steep rocky hills it has gouged out over the many thousands of years that it has flowed through Germany's southwestern midsection.

FIGURE 8–16 The shale found throughout vineyards in the Mosel region soaks up the sun's warmth during the day, and reflects it back onto the vines after the sun sets, thus helping with the ripening process.

FIGURE 8–17 The Devonian slate found in the soil of many vineyards in the Middle Mosel has a slightly bluish tint, and imparts a unique aroma reminiscent of petroleum to the Riesling wines from this region.

the best Mosel Rieslings minerality that offsets the natural fresh green apple aromas and flavors (Figure 8–17). Racy acidity supports the fruit and slate, allowing these wines to exhibit an almost oxymoronic combination of delicacy and intensity.

The vineyards of the Mosel-Saar-Ruwer can be divided into five subregions: the Saar, Upper Mosel (between the village of Perl and the point where the Ruwer joins the Mosel), the Ruwer, the Middle Mosel (from the Ruwer to approximately the town of Zell), and the Lower Mosel (from Zell to Koblenz where the Mosel flows into the Rhein). These subregions correspond roughly to the six official *Bereiche:* Saar, Ruwertol, Obermosel, Moseltor, Bernkastel, and Zellwich. The six *Bereiche* are further divided into 19 *Grosslagen* and 525 *Einzellagen* (of which only about 60 vineyards are really noteworthy).

The westernmost vineyards are found along the Saar, which flows north into the Mosel River just 6 miles below Trier. Looking over the Saar in the village of Saarberg is a striking remnant of earlier times—a tall fortified stone tower built in the thirteenth century as a lookout against invaders. Many of the estates along the Saar also date from that early period. The *Grosslage* Scharzberg groups together many of the vineyards along the Saar. The most famous of these vineyards is the Scharzhofberg, an *Einzellage* near the village of Wiltingen (Figure 8–18). The historic village of Ockfen is also known for the quality of its Rieslings; its best known *Einzellage* is the hilly Bockstein vineyard. Sitting on a steep hill facing south, with no other hills blocking it, Bockstein receives excellent sunlight. Its soil is hard, gravelly slate that forms a blue-gray dust on the fingers when touched. This is indicative of how easily the soil mineralizes, allowing the vines' roots to pick up mineral components.

The Upper Mosel region, roughly the area between the two tributaries, is similar in terrain and climate to the Saar region. Outside of the city of Trier, the countryside is rural and picturesque. There are few private estates in this region. One large local cooperative does most of the cultivating and harvesting of grapes. The sparkling wine ***Sekt*** from this region shows promise.

FIGURE 8–18 The famous *Einzellage* Scharzhofberg in the village of Wiltingen looks down on the stately manor house of the Egon Müller winery. The estate and portions of the vineyard have been in the Müller family since the late eighteenth century.

The vineyards of the Ruwer cover about a 6 mile stretch, starting at the village of the same name and extending to the confluence with the Mosel River. The vineyards here are a little more sheltered than along the Saar. The soil is less gravelly and contains less slate, and there is a higher content of humus in the red soil. These two factors give Ruwer wines a little more ripeness and a more fully developed fruitiness than those from the Saar.

The subregion (and *Bereich*) of Bernkastel starts at the confluence of the Ruwer and Mosel and continues downstream almost to Zell. It is named for the town of Bernkastel which lies in the middle of the region, on the banks of the Mosel (Figure 8–19). The surrounding steep hills with their red slate soil contain vineyards whose wines are among Germany's best. Of the Mosel's approximately 60 quality *Einzellagen,* over half are in the *Bereich* Bernkastel. These exceptional vineyards are located in several villages along the river, starting with Treppchen in the village of Erden and progressing through Urzig where the Würzgarten vineyard is located and Wehlen

FIGURE 8–19 The beautiful and historic town of Bernkastel is located in the middle of the *Bereich* of the same name. Many of the Mosel's highest quality vineyards are in the *Bereich* Bernkastel.

with its pretty Sonnenuhr vineyard. The town of Bernkastel has two famous *Einzellagen,* Doktor and Graben (Figure 8–20). Moving further downstream, the village of Brauneberg is well known for the Juffer *Einzellage,* and Piesport has the superb Goldtröpchen (Figure 8–21). Each of these vineyards imparts distinctive terroir. It is important to recognize that the best wines from this region are bottled by quality-conscious private estates. The labels from these producers will con-

FIGURE 8–20 The *Einzellagen* Doktor and Graben lie on hillsides above Bernkastel. Note the precipitous gradient of the vineyards, both of which are famous for the quality of their Rieslings.

FIGURE 8–21 The *Einzellage* Goldtröpchen above the village of Piesport, with its perfect southern exposure, soaks up the afternoon sun, allowing its Riesling grapes to fully ripen while evolving unique mineral aromas and ripe flavors of stone fruit, counterbalanced by citrusy tartness. If Germany were ever to rate her vineyards, Goldtröpchen would surely be a *grand cru.*

tain the word ***Gutsabfüllung*** or "estate-bottled." There are also inferior wines from these villages made from grapes grown in less favorably located sections of the local vineyards and sold under a collective *Grosslage* designation.

At the town of Zell, one moves into the Lower Mosel subregion, which extends downriver to the town of Koblenz, where the Mosel meets the Rhein. The Lower Mosel has the highest percentage of Riesling in its vineyards of any section along the river. The river's slopes are particularly steep here, and there are virtually no sites that are smooth, flat, and easy to work (Figure 8–22). The wines from *Beriech* Zell are somewhat less delicate and refined than those from *Beriech* Bernkastel, but they can be very satisfying in their sturdy, true-to-varietal way. As one gets closer to Koblenz, the Reislings show some of the more muscular character of Rhein wines.

Throughout the Mosel-Saar-Ruwer there is an impressive commitment to quality and noticeable pride in heritage. Many of the new owners and managers of the established private estates work hard to reach standards of excellence that surpass those set by national quality control laws. Ernst Loosen of Weingut Dr. Loosen near Bernkastel, Manfred Prüm of J.J. Prüm in Wehlen, Johannes Selbach of Selbach-Oster in the little village of Zeltingen just downriver from Wehlen, Wilhem Haag of Fritz Haag near Brauneberg and Nik Weis of St. Urbans-Hof on the Saar—innovative, adventurous vintners like these are dedicated to producing the best Rieslings possible from the Mosel, combining meticulous viticulture, time-honored tradition, modern technology, and creative promotions.

The Rheingau

Historically, the Rheingau is the most commercially successful wine region of Germany (Figure 8–23). Over the centuries, the Church and the landed nobility gave direction and structure to wine production, allowing the region to recover quite quickly from the natural devastations of the late nineteenth century and self-induced tragedies of the first half of the twentieth century. The Rheingau is relatively small at 7,166 acres (2,900 hectares) of vineyards. Ninety percent of those acres lie on the right bank of the Rhein, from the boundary with the Mittelrhein on the north to the town of Wiesbaden in the south (Figure 8–15). The other 10 percent of Rheingau acreage is along the Main River which flows in from the east to meet the Rhein at the city of Mainz.

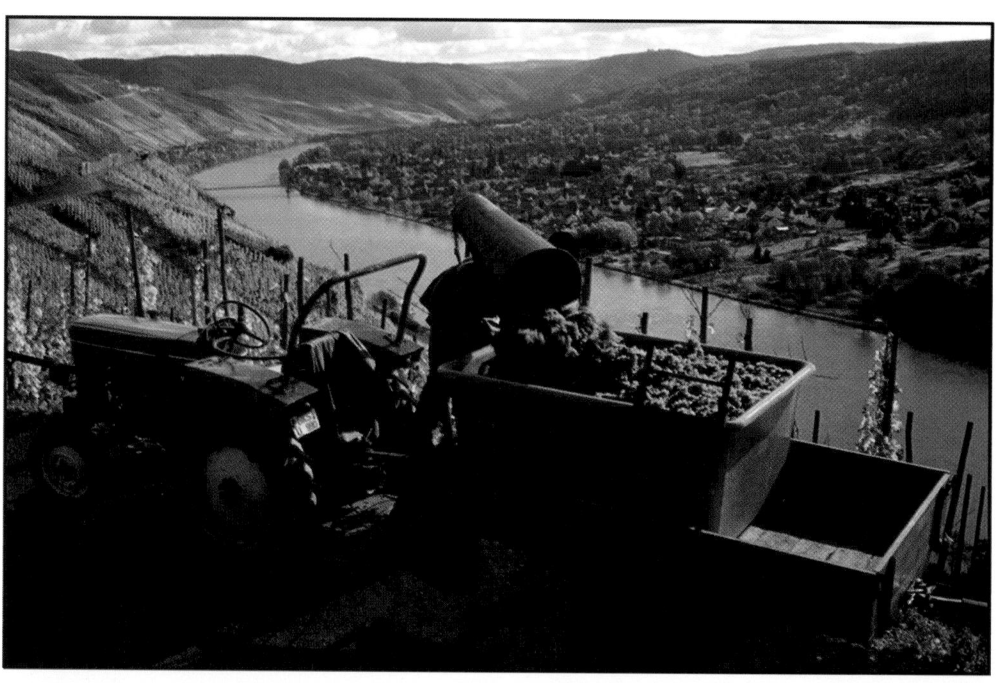

FIGURE 8–22 Harvesting grapes in the Zeltinger Sonnenuhr vineyard. Across the Mosel lies the village of Wehlen.

Nik Weis of St. Urbans-Hof

Nik Weis, co-owner and general manager of the St. Urbans-Hof estate in the small town of Leiwen on the Saar, is the third generation of his family to run the winery. Although still in his thirties, Nik has very clearly defined opinions on viticulture, oenology, and wine's place in world culture.

Nik's approach to vineyard management closely mirrors his family's longstanding concern for ecological balance. By limiting man's interference with the natural process of growing grapes, and using only gentle methods to help the grapes reach maturity, the Weis family concentrates its energy on producing wines that reflect the terroir of each vineyard. As Nik Weis says, "Great wines are the result of hard work over the whole year. Low crop yields, intensive soil work, leaf thinning, accurate trellising, late harvesting, and hand picking are the elementary steps that ensure that the wine reflects the original character of the land on which it is grown."

St. Urbans-Hof owns sections of two of the Mosel's premier *Einzellagen,* Piesporter Goldtröpchen (5 acres/2 hectares) and Ockfener Bockstein (12 acres/5 hectares). The estate also owns over 32 acres (13 hectares) of vineyards in three other villages. Surrounding the winery itself, there are an additional 31 acres (12.5 hectares) of vineyard. In total, the Weis family owns 82 acres (33 hectares) of vineyards, all of which are planted 100 percent to Riesling, which the family considers to be the traditional grape of Germany. St. Urban-Hof's Web site describes Riesling as "the most fruity and flavorful of all white varietals."

Once the grapes are harvested and brought into the winery, Nik Weis and his cellar-master, Rudolf Hoffman, follow traditional, time-honored methods of winemaking to allow the natural character of this noble grape to emerge. The grapes are crushed and the must is allowed a short time to macerate with the skins to bring out flavors. After a gentle pressing of the skins and pulp, the must is moved to stainless steel tanks for a cool fermentation. When fermentation is complete, the wine is racked into large neutral barrels to rest and harmonize before being lightly filtered and bottled.

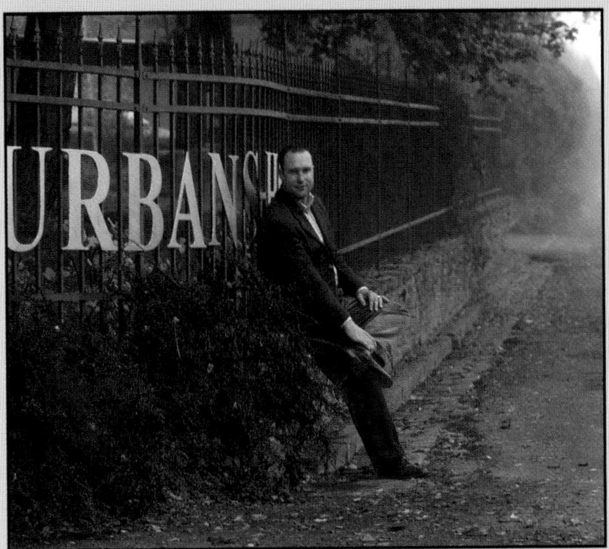

Winemaker Nik Weis sits outside his family's winery in the village of Leiwen on the Saar River.

Nik feels very strongly that when the noble Riesling is grown under the right conditions, allowed to reflect the natural terroir of its vineyard, and carefully handled in the winery, the result is the most food compatible of all wines. He has said, "Riesling's chameleon properties adapt easily to nearly any dish" (Bouchard, 2003, p. 4). The bracing acidity of a Riesling, coupled with its plentiful fruit, assertive flavors, and often a hint of residual sugar, make it the perfect foil for fatty meats, salty foods like anchovies or ham, or dishes in which there are natural fruits such as pork chops braised with chopped apples.

Flowing naturally from the Weis family's belief in traditional viticulture and winemaking and their dedication to Germany's traditional style of Riesling-based wines is their involvement for generations with the *Verband Deutscher Prädikats weingüter.* The group seeks to promote the German Estate Wineries, the quality of German wines, the winemaking traditions of the country, and its best vineyards.

Nik hopes that the traditional system of grading *Prädikat* wines by the level of ripeness in the grapes will continue. The touch of residual sugar in Rieslings is an intergral part of their appeal, he feels.

Nik has said, "Sugar belongs in German Rieslings the way bubbles belong in Champagne."

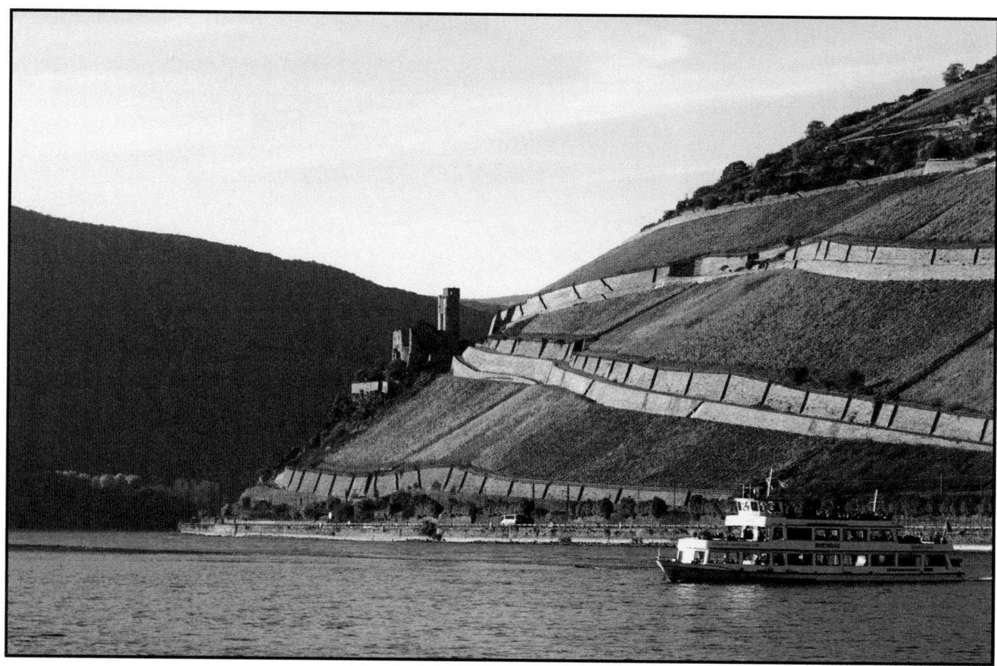

FIGURE 8–23 A medieval castle stands in one of the Rhein's most famous vineyards, Berg Schlossberg, near the town of Rudesheim in the Rheingau.

Except along the Main where the land is gentle and flat, the vineyards of the Rheingau are on steep sloping hills. The soil in the Rheingau is varied. Slate predominates. Downstream from Rudesheim, blue slate is prevalent whereas further upstream, there is more of the reddish minerally slate found in the Lower Mosel. The mesoclimate of the Rheingau is conducive to growing ripe, fully developed but nicely acidic, grapes. The Rhein (which for most of its path runs north or northwest) takes a jog at Mainz and runs west-southwest for about 18 miles. This variation in topography gives the river's banks a southerly slope, making this area marginally warmer than locations further inland in the Rheinhessen. Annual rainfall of approximately 20 inches ensures that there is adequate water for the vines. Although winters are very cold (the river often freezes over) temperatures stay warm long enough into the fall to allow the Riesling grapes to fully develop their flavors. The Rheingau is overwhelmingly a white wine region, and the majority of the vineyards are planted to Riesling. (In 2003 the percentage of acreage planted to Riesling was 78.2 percent.)

In general, wines from the Rheingau are fuller, firmer, and more assertive than wines from the Mosel. Some writers describe Rheingau *Rieslings* as Germany's most "muscular" wines. There is little dispute that these are among the most elegant, distinctive, and long-lived *Rieslings* produced. When fully mature, a Rheingau *Riesling* from one of the better private estates can proudly stand with the best white wines in the world.

There is only one *Bereich* in the Rheingau, Johannisberg, named for the town in the middle of the region. Sadly this name has been misused by wine-producers in several countries to try and attach legitimacy to their Riesling-based wines. The most famous *Einzellage* here is the historic estate Schloss Johannisberg (Figure 8–24). Other villages are Eltville with its *Einzellage,* Sonnenberg; Erbach with its incomparable Macrobrunn vineyard; and Rudesheim with several *Einzellagen,* the most famous of which is Rosengarten. The village of Winkel is home to one of the Rheingau's most famous, and oldest, single estates, Schloss Vollrands.

Besides the two great estates of Schloss Johannisberg and Schloss Vollrands, other very reputable producers in the Rheingau include Georg Breuer in Rudesheim, headed by the energetic Bernard Breuer; the house of Dr. Robert Weil; and Langwerth von Simmern.

FIGURE 8–24 The legendary estate of Schloss Johannisberg is one of only two *Einzellagen* in the village of Johannisberg.

Rheinhessen

The topography of the Rheinhessen is very different from that of the Rhiengau (or the Mosel). Instead of steep, rocky hills along river banks, the Rheinhessen is mostly flat rolling agricultural land. The soil is more alluvial and loamy. Few of the small landowners plant only grapevines. The area is very large, encompassing more than 61,460 acres (24,870 hectares). In general, the region is protected from cold winds and excessive rain by the hills on its western border, which rise as high as 2,000 feet (610 meters) in altitude. However, within such a large region, there are different microclimates.

Overall, the best vineyard sites are located on the eastern edge in an area known locally as the Rheinterrasse, where the vineyards (many of them around the town of Niersteiner) are on eastern-facing slopes. The soil is full of a reddish slate that imparts a distinctive mineral-like intensity to the wines. Vineyards here are planted primarily to Riesling. There is some talk among property owners about creating a new official region out of the viticulturally more favorable Rheinterrasse.

Other parts of the Rheinhessen are not dedicated to Riesling, but rather to hybrids like Müller-Thurgau or Scheurebe. The percentage of land dedicated to red varietals is also increasing. By 1992, 8 percent of acreage was occupied by red grapes such as Spätburgunder and the prolific Portugieser. There are over 400 individual vineyards in the Rheinhessen, very few of them of any merit. Most of the grapes from small vineyards are sold to cooperatives and made into pleasant, inexpensive still wines or increasingly, into decent, affordable Sekt (sparkling wine). One-third of all German exports come from the Rheinhessen. More than 50 percent of Liebfraumilch is made here.

The Pfalz

Formerly known as the Rheinpfalz, this is Germany's second largest wine region, with 58,060 acres (23,506 hectares) under vine (Figure 8–25). The Pfalz stretches along the left bank of the Rhein for about 50 miles (81 km). It lies on the eastern edge of the great Pfalz Forest and is well protected from the cold. Spring sets in early, and summers are long and warm. In July, the average temperature throughout the region exceeds the 64°F (18°C) considered the minimum for ripening grapes. The thick forest holds back the clouds so that rainfall is light. So favorable are conditions that in parts of the southern Pfalz fully 93 percent of arable land is planted to

FIGURE 8–25 Riesling vineyards surround the village of Deidesheim in the Pfalz. Conditions in this large region are so conducive to growing grapes that 90% of arable land is planted to vineyards.

grapevines. Of the region's 11,500 grape growers, about two-thirds deliver their grapes to cooperatives, producers' associations, or private cellars. For decades many of the bulk wines produced from these grapes were of mediocre quality, and in the 1970s and 1980s the reputation of the Pfalz was for inexpensive, pleasant, but unexciting wines. That is now changing.

Led by the 650 producers who estate-bottle their own wine, standards are rising fast. More acreage is being devoted to Riesling, and attention is being given to careful vinification methods. Because of the warmer temperatures, Pfalz Rieslings are generally fuller, riper, rounder, and higher in alcohol than Rieslings from other parts of Germany. Even with the fully evolved fruit flavors and higher sugar content at harvest, these wines retain lively acidity and an appealing elegance.

The best vineyard sites are in the north, incuding the famous Jesuitgarten in the village of Forst and Hohenmorgan in Deidesheim. Among the more important producers of high quality wines are historic old houses like Basserman Jordan, which owns vineyard sites around Deidesheim, and Bürklin-Wolf, which dates from 1875 and owns prime vineyard sites in Forst, Deidesheim, and Wachenheim. With all the changes taking place and the emerging emphasis on quality, the Pfalz is one of the more exciting wine regions in the world.

Other Regions

Ahr

The majority of vineyards in this small region are planted to red grapes, most importantly to Spätburgunder (Pinot Noir), which covers 40 percent of the acreage. Due to rocky heat-reflective soils, and protection by hills from the northerly winds, conditions here allow ripening of red grapes. Most of the wine is made by the five cooperatives in the region, and although cool fermentation and barrel-aging is rare, the wines are well received locally. There is also some Riesling made.

Mittelrhein

This area north of the Rheingau has fairly cool temperatures and produces primarily *Rieslings*. These wines have high acidity and are mostly sold within the region.

Nahe

The Nahe River flows in a north-northeastern direction and feeds into the Rhein where that river takes its easterly jog. Although Müller-Thurgau is the most widely planted varietal, 23 percent of acreage is Riesling, most of it on the more desirable hillside vineyards rather than on the flatlands. There are good quality, affordable Rieslings produced throughout the area, with some of the best coming from the section just north of the town of Bad Kreuznach.

BADEN

The longest (250 miles/400 km) and most southerly of Germany's wine regions, Baden extends from the border with Franken in the north at the River Neckar all the way south to the border with Switzerland (Figure 8–26). The region effectively divides into two regions, with the section south of the town of Baden lying just across the Rhein from France's Alsace region and closely mirroring that region in its terrain. In the northern sections where granite is common in vineyard soils, especially around the historic and very beautiful old city of Heidelberg, some attractive and graceful Rieslings with excellent acidity are produced. Most of the harvest here is handled by cooperatives.

In southern *Bereiche,* where temperatures are warmer, alcoholic strength is higher than in other German wines. Müller-Thurgau is the predominate grape. It is often blended with Riesling and Silvaner into pleasant, light whites for local consumption or for sale to supermarkets. Increasing numbers of acres are now being planted to Spatbürgunder, which can attain good ripeness at this southern latitude. As cooperatives and private owners continue to modernize their vineyards and their vinification methods, it is inevitable that more of Baden's whites, and her barrel-aged Pinot Noir–based reds, will find their way into export markets.

HESSICHE BERGSTRASSE

Hessiche Bergstrasse starts at the River Neckar, just north of the Baden region. In essence, Hessiche Bergstrasse is a continuation of the northern section of Baden and is very similar climatically and topographically. It is a very small region with only 1,000 acres (380 hectares) under vine, 53 percent of which is planted to Riesling. Most of the wine produced is consumed locally.

WÜRTTEMBERG

Directly east of Baden's northern section lies Württemberg, whose 24,000 acres (9700 hectares) of vines are planted along the sloping banks of the River Neckar and its tributaries. These vine-

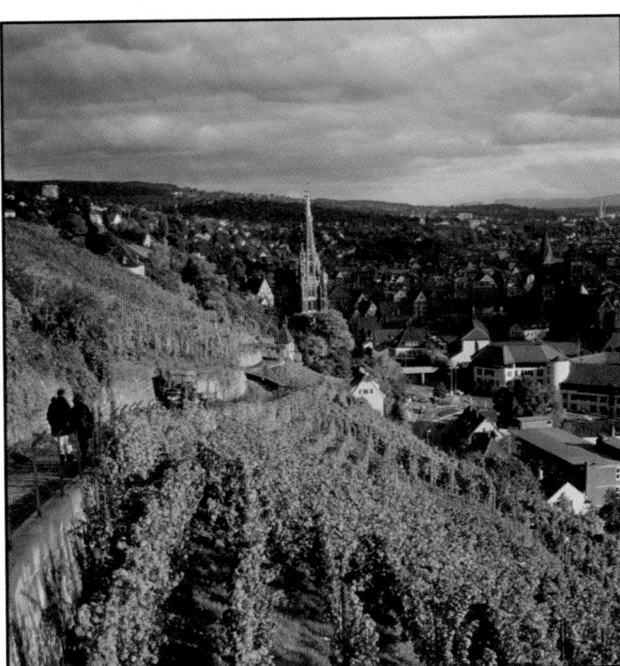

FIGURE 8–26 The vineyards of the Baden region, like this one above the town of Esslingen, are among Germany's southernmost. Their climate and soils are similar to those of France's Alsace region, which is located just to the west, across the Rhein River.

yards are very steep, and most are terraced. The climate varies from north to south, but most of the region is under continental influences, especially in its easternmost *Bereiche* where the danger of fall and spring frosts are high. Most landowners have very small plots. Over 80 percent of the region's harvest is handled by cooperatives. Württemberg is Germany's largest producer of red wine, most of it very light in body, low in tannins, and undistinguished. The vast majority of this wine is sold locally. (Citizens of Württemberg have the highest per capita wine consumption in Germany.)

FRANKEN

The Franken wine region is located east of Hessiche Bergstrasse and north of Württemberg. Being so far east, and removed from the moderating influences of the Rhein, Franken's climate is essentially continental. It is one of the coldest wine regions in Europe. Winters can be severe. Most of the region's 14,900 acres (6,050 hectares) of vineyards are along the Main River. They are planted primarily to lesser white grapes. The wines of Franken are sold in a traditional flagon-shaped bottle. The most distinctive Silvaner wines are produced in this region.

SAALE-UNSTRAT AND SACHSEN

The two wine regions in what was East Germany are just beginning to find their way into the modern world of winemaking. Both are small and very cold. Saale-Unstrat is the most northerly in Germany. Its 1,200 acres (486 hectares) are planted primarily to lesser white varietals. The chalky soil adds some body to the light and pleasant wines. Sachsen (Saxony in English) is further east, with the old city of Dresden, left in ruins by World War II air raids, at its center. The 750 acres (304 hectares) of vineyards located along the River Elbe are planted to various white varietals, mostly Müller-Thurgau and Weissburgunder (Pinot Blanc). As investment in replanting the vineyards and in upgrading the winemaking facilities continues, the wine business in these two regions is expected to grow and improve.

Summary

It is somewhat ironic that one of the world's oldest and most traditional wine-producing countries is in a state of flux, holding on tenaciously to its heritage while looking to the future in its adjustments to the demands of a modern and international wine market. There is no doubt, however, that Germany will continue to produce extraordinary wines that complement a wide variety of cuisines, while the new, younger, highly educated and savvy generation of vintners, marketing executives, and government researchers move to increase international awareness, and acceptance, of these fine wines.

Review Questions

1. Describe the differences between *Einzellagen* and *Grosslagen.*
2. List the six *Prädikat* designations at the QmP level in Germany's wine classification system. By what criterion is a specific wine assigned to one of these tiers?
3. Is the Rheinhessen similar to the Rheingau in topography and soil content? If not, what are the differences?
4. Name the organization of winemakers in the Mosel that was formed in 1908. What is the stated purpose of this group?
5. What is the meaning of *Qualitätswein bestimmter Anbaugebiete?*
6. Recently the terms *trocken* and *halbtrocken* were officially added to German wine law. Define the terms, and describe their significance to the future of the German wine trade, domestically and internationally.
7. A small percentage of Germany's wine production is red wine. What grapes are used for these wines?

CHAPTER 9

Other European Regions and the Mediterranean

After reading this chapter, you should be able to

- identify the lesser-known but increasingly important wine regions of Europe and the Mediterranean.
- demonstrate an understanding of the winemaking heritage, the terroir, and the major varietals of these regions.
- describe the style of wine made in each region.

KEY TERMS

amphorae
Blauburgunder
Chasselas
Egri Bikavér
Furmint
Grüner Veltliner
Kadarka
Robola
Tokaji Aszú

When a wine consumer thinks of Europe, the countries that most likely come to mind are France, Italy, Spain, Germany, and Portugal. These five countries are, indeed, the most important wine producers on that continent. They are all famous for the variety and quality of their wines, and all have visible presence in the international wine market. However, these five are not the whole picture. There are other countries, in Central Europe (Austria and Switzerland), southern Europe (Greece), and in Eastern Europe (Hungary, Romania, Bulgaria), where grape growing and wine production have just as long a history, and where consumption of wine is still an inherent part of the culture. In this chapter, we will take a close look at these lesser-known countries and their histories. However, to fully understand European wine, one must go outside that continent and well back in history to the part of the world where production of wine began: the Middle East. Therefore, this chapter will also explore those countries along the eastern edge of the Mediterranean in which wine is still produced today (Israel and Jordan).

Central Europe

Unlike their neighbors in Eastern Europe, both wine-producing countries of Central Europe have been politically stable for centuries and are well-established members of the developed, industrialized community of Western nations. They make high quality wine and employ modern techniques while honoring age-old customs.

Austria

Austria's longstanding and proud wine business was dealt a nearly fatal blow in 1985 when a small number of unethical (and very shortsighted) wine merchants tried to improve their exports by adding to their supply of dessert wines a chemical meant to make the wines taste sweeter and feel more syrupy. The international press jumped on this scandal and greatly worsened the situation by reporting that it was a far more dangerous chemical that had been used. The public came to believe that antifreeze had been added to Austria's wines. The international reaction was immediate: In the year following the adulteration, exports from Austria were less than one-fifth of what they had been the year before (Robinson, 1994).

Vintners across the country worked closely with the Austrian government to reverse the bad image left by the "antifreeze" scandal. In late 1985 the laws pertaining to wine production were made stricter. The emphasis throughout the wine community turned from quantity to quality. That credo remains today. The sure sign that Austria has recovered from the setback of 1985 is that Austrian wines are now finding places on wine lists and retail shelves throughout the United States and Canada.

Austrian Wine—Historical Perspective

It is believed that the Celts who controlled what is now Austria grew grapes there as long as 1,000 years BC. Historians are not sure if the Celts made those grapes into wine, but it is known that under the domination of the Romans over ensuing centuries, viticulture was expanded. When Charlemagne became King of the Franks and Emperor of the Holy Roman Empire in approximately 770 AD, wine was definitely being produced in Austria, and he took a strong interest in improving that wine. The new king oversaw the passage of rudimentary grape classifications and wine laws. Many vineyards at the time were owned by monasteries, and monks were important wine producers throughout the Middle Ages. It was monks who brought Pinot Noir grapes from Burgundy and Riesling from Germany. By the end of the Middle Ages, acreage under vine was approximately 10 times what it is today (Robinson, 1994), and Austria was exporting large quantities of wine.

After the Napoleonic Wars, many church-owned vineyards reverted to ownership of farming families. In the nineteenth century, after losing many vineyards to hard frosts and later to phylloxera, Austrian botanists did considerable experimentation with grape varietals, bringing in more vines, including Chardonnay, from elsewhere in Europe to plant alongside their indigenous grapes. One of Europe's oldest schools of enology was established in Klosterneuburg, Austria, in 1860. At this school some varietals that are now the foundation of Austrian wine production were developed, primarily **Grüner Veltliner.**

The Wine Laws of Austria were first enacted in 1972, based essentially on the laws of neighboring Germany. Since the revamping of these laws in 1985, Austria's laws are among the most comprehensive and strict in Europe. The name of an Austrian wine indicates varietal, geographic place of origin, and sugar content. Even the wines from the lower rungs of classification, *Tafelwein* and *Landwein,* must indicate varietal and must, or sugar content. The varietal mentioned on the label, at all levels of classification, must compose at least 85 percent of the bottle's content. The same is true for vintage.

At the level of *Qualitatswein,* the quality level is indicated by ripeness, as in Germany. However, *Kabinett* is not a *Prädikat* in Austria, but rather its own quality subcategory. The *Prädikatswein* category includes seven classifications, from *Spätlese* through *Trockenbeerenauslese.*

Viticulture

The climate of Austria is definitely continental, with harsher winters and hotter, dryer summers than is the case in most of Western Europe. All viticulture, indeed most agriculture, is located in the eastern half of the country, where the climate and soil are more conducive to growing vines than would be the case closer to the towering Alps. There is some variation in mesoclimates among Austria's wine regions, but the majority of grapes are of the cold-resistant, early ripening varietals. Austria has 136,000 acres (55,000 hectares) under vine, and the most widely planted is the indigenous white grape Grüner Veltiner with 37 percent of the total. Second most prevalent is Müller-Thurgau, a lesser, but hardy, white developed in Switzerland. More acreage is being devoted to the noble Riesling, which produces some of Austria's greatest wines, in every style from very dry to lusciously sweet. As of 2004, only 3 percent of acreage, or just over 4,000 acres (1,619 hectares), was planted to Riesling. The two most prevalent red varietals are Portugieser and Blaurer Zweigelt. There is also some Pinot Noir, called Spätburgunder. Red wines make up less than 10 percent of Austria's production.

Most of Austria's wine (60 percent) comes from the five subdistricts within the large region of Niederösterreich (Lower Austria). Clustered along the banks of the famous Danube River in the western part of the region are the three districts of Kamptal, Kremstal, and Wachau. With their continental climate moderated somewhat by the Danube, and with their shale-filled rocky soil, these three small districts produce some of Austria's very best Rieslings. Legend has it that Europe's first Riesling vineyard was located in Wachau.

Incredibly, there are over 1,600 vineyards within the city limits of Austria's largest and most historic city, Vienna. This area is accorded status as a wine region, rather than as a subdistrict of Neiderösterreich.

The second largest wine region is Burgenland, which borders Hungary (Figure 9-1). It is somewhat warmer than the subdistricts of Niederösterreich and produces primarily soft, fruity reds and sweet dessert wines. The third major region is Stelermark (or Styria). It is located west of Burgenland, up against the foothills of the southern Alps. Grape growing is difficult in this mountainous area, and in a typical vintage, Styria's three districts combined will produce less than 5 percent of the country's wine.

Switzerland

Although wine has been produced in this alpine country since the time of the Romans, the high altitudes make grape cultivation difficult. The very cool temperatures and short growing season mean that each year it is a challenge to harvest fully ripe grapes (Figure 9–2). As early as the seventeenth century, Swiss winemakers were feeling the pinch as wine started to be imported from warmer climes, such as France's Rhône Valley. But winemaking has persisted, and today wine is made from local grapes in all 24 Swiss cantons (equivalent of a province or state). For most of history, Switzerland has consumed most of the wine produced in the country. In 1988 the country instigated a series of *Appellation Contrôlée* laws. At that time, the government started working with private companies to promote Swiss wines abroad. However, the Swiss people are enthusiastic wine consumers and are known to keep their most interesting wines for domestic consumption. For these reasons, the percentage of Swiss wines exported has climbed only to 1 percent of the total production.

FIGURE 9–1 The vineyards in Rust, Austria, in the wine region of Burgenland, are known for their sweet white wines, as well as elegant reds.

FIGURE 9–2 The snow-covered Alps beyond the vines at Château d'Aigle in Switzerland are indicative of the cool temperatures under which grapes must grow.

Wine Regions

In this small, mountainous country, land suitable for viticulture is scarce, so vineyards are planted everywhere, from deep valleys to alpine meadows to steep hillsides in the foothills of the Alps (Figure 9–3). There is a wide variety of microclimates throughout the many regions. Accordingly, Switzerland grows more than 30 different varietals. The most important are indigenous grapes, such as:

- ***Chasselas:*** Fully 37 percent of Switzerland's acreage is planted to this indigenous grape. It is an early ripener and produces a pleasant, dry white.
- *Sylvaner:* Second most widely planted varietal, the Sylvaner produces attractive whites with more body and more acidity that Chasselas. It is often sold as Johannisberg.
- *Müller-Thurgau:* Named for the botanist, Dr. Müller, who developed it in his home canton of Thurgau in the German-speaking part of the country, this cross of the noble Riesling with the conveniently early ripening Sylvaner, is well suited to cool climates. Sadly, it is woefully short of Riesling character.
- *Pinot Noir:* Called **Blauburgunder** in German-speaking cantons, this is Switzerland's most widely planted red varietal. It is sometimes blended with Gamay to produce Switzerland's best-known red wine, Dôle.
- *Merlot:* Found mostly in the southeastern, Italian-speaking section of Ticino.

There are also several other crosses that have been developed locally to withstand Switzerland's demanding growing conditions.

The geographic location of origin must show on Swiss wines. The most common appellations on exported wines are

- Valais, which is the southern section and accounts for one-third of the country's vineyards, produces a light, crisp white called Fendant (made with Chasselas grapes). The red wine Dôle is also made in Valais.
- Vaud, located in the southwest section, is planted almost entirely to Chasselas.

FIGURE 9–3 Vines growing on a hillside along the coast of Lake Geneva.

- Neuchâtel is north of Vaud on the shores of the lake of the same name. Of the 5,300 acres (2,145 hectares), most are planted to Blauburgunder, which is able to ripen adequately because of the warming influence of the lake.
- Ticino, very near Italy, and heavily influenced by Italian culture, produces primarily Merlots in the lighter, more straightforward Italian style.

Summary—Central Europe

Although the wines of both Austria and Switzerland are slowly becoming more available in North America, they are both still difficult to find, and can be quite expensive (in the case of Switzerland, very expensive). Switzerland may gradually relinquish more of its wines to export, but seems in no hurry to do so. Austria, on the other hand, seems poised to move forward, through promotional activities and consumer education, coupled with assertive marketing strategies, and claim a respectable share of the wine market in the United States and Canada.

Eastern Europe

History in Eastern Europe is replete with disruptions, mostly man-made. Although wine consumption was an inherent part of the culture of many countries in this region, the production of wine (or any alcoholic beverage) all but disappeared during the sixteenth and seventeenth centuries when many nations were under the rule of the Ottoman Empire. The Ottomans were Muslim, and that religion forbids the consumption of alcohol. The countries that did obtain freedom from the Ottomans replanted their vineyards, only to lose them to the deadly phylloxera invasion of the late nineteenth century. The twentieth century brought the devastation of two world wars, followed by Communist rule. Under the Communists, most vineyards and farms were reorganized into large collectives, whose purpose was to produce food and wine for the Communist bloc of nations.

The end of the Cold War and subsequent independence for many Eastern European nations did not bring immediate restoration of the heritage of wine production. Instead, political unrest continued, with various parties vying for power in each country. Also the style of wine that had been preferred by the Soviets was slightly sweet and/or fortified, a style that is not popular on the international market. Moreover, standards of cleanliness under the Communists had not been high, leading to frequent incidences of infected or spoiled wines. With support from their respective governments, and with funding from international agencies and the expertise of consulting enologists, Eastern European vintners are addressing these problems and looking to carve out a niche for themselves in the world market for wine.

Hungary

Since its liberation in 1989 when the Soviet bloc collapsed, Hungary has made great progress in restoring its long, proud heritage of wine production, which dates back to Roman times. As early as 1641, even before the country's emergence from Muslim rule, a set of Vine Laws had been drawn up, primarily to protect the legitimacy of the rich dessert wine **Tokaji Aszú,** for which demand was very high. Once the Turks were defeated in 1686 and Hungary became part of the Hapsburg Empire, vineyard plantings were increased across the country. After the phylloxera invasion, Hungarians began planting on resistant rootstock. Plantings were particularly wide in the Great Plain in the south-central portion of the country where the sandy soil was thought to be particularly inhospitable to the root-boring aphid. By the early 1990s, there were just over 272,000 acres (110,120 hectares) devoted to grapevines. During the 1990s there was an influx of foreign investment into the Hungarian wine business, attracted perhaps by the continuing fame of, and high demand for, the Tokaji dessert wines (Figure 9–4). Hungary has also benefited from the advice and assistance of visiting winemakers, most notably Englishman Hugh Ryman, whose father, Henry "Nick" Ryman, had, in 1973, purchased a dilapidated château in the Bergerac region east of Bordeaux and renovated it into a successful wine-producing estate. Using his father's experience as a model, Ryman has helped modernize and upgrade facilities across Hungary, usually encouraging the planting of varietals for which market demand exists.

FIGURE 9–4 Tokaji wines are aged for years in cellars similar to this one at the Oremus Winery in Hungary.

WINE REGIONS

Hungary lies east of Austria and south of Slovakia. It extends to Russia's western border. Hungary is landlocked but contains Europe's largest lake, Balaton. The Danube River (called Duna in Hungarian) runs through the country from north to south, dividing it almost exactly in half. To the west of the river is Transdanubia, and immediately to its east is the Great Plain. The northeast section, called the Northern Massif, contains the volcanic hills whose south-facing slopes are particularly well suited to viticulture. In the extreme northeast of the massif, right up against the border with Slovakia, is the region of Tokaj-Hegyalja, home of the famous dessert wine.

The climate throughout Hungary is entirely continental, with predictably cold winters and sunny, hot summers. Soils are varied, with sand predominating in the Great Plain, and basalt rock lying atop sandstone being typical all around Lake Balaton. The Northern Massif and Tokaj-Hegyalja are covered, not surprisingly, with volcanic rock covered with decaying lava.

The official Hungarian wine laws, which are overseen by the National Wine Qualifying Institute, have established four levels of quality for wine: table wine; regional wine; quality wine (*minöségi bor);* and extra-quality wine (*különleges minöségi bor).* The latter two levels are the ones found in export markets. The quality control laws are still being amended and improved, with an effort being made to bring them more in line with the laws of the European Union. Wines from Hungary are usually varietally labeled, which shows how deeply the Hungarians are aware of varietal differences. Unfortunately many of the indigenous varietals disappeared after phylloxera. Many Western European varietals are now planted in Hungary. Chardonnay, Riesling, Sauvignon Blanc, and Austria's Grüner Veltliner and Müller-Thurgau are planted along with the few indigenous vines, such as **Furmint** and Ezerjo. For red wines, there are Cabernet Sauvignon and Cabernet Franc, as well as Merlot and Pinot Noir growing along with Hungary's own **Kadarka.**

The existing wine laws divide the country into twenty official regions, spread among the three geographic sections mentioned above.

Transdanubia: This western part of Hungary contains 13 of the 20 designated wine regions. The very large lake, Balaton, has a moderating effect on the mesoclimate of the lake region, allowing grapes to fully ripen. The result is fairly full-bodied and aggressively flavored wines, mostly whites. The westernmost region, Sopron, is contiguous to Austria's Burgenland, which is famous for its sweet white wines. However, Sopron is known for its big, full reds, which are made from French varietals such as Cabernet Sauvignon and Merlot.

The Great Plain: This huge area accounts for over half the country's vineyard acreage. There are three official wine regions here.

The Northern Massif: In the northeast are three important wine regions: the Mátra Foothills (or Mátraalja), Eger, and Tokaj-Hegyalja. Most of the wine made in Mátraalja is dry white, made from two indigenous varietals, usually blended with Muscat, Chardonnay, or Sauvignon Blanc. Just to the east of the Mátra Foothills is the wine region named for the historic village of Eger. Because of its altitude in the Bükk mountains, spring often comes late, and rainfall is scarce. These conditions are conducive to producing the full-bodied, age-worthy red wine for which the region is famous, **Egri Bikavér,** or Bull's Blood. Once made primarily from the native Kadara grape, Bull's Blood is now mostly Cabernet, Merlot, and another, easier-to-handle native varietal, Kékfrankos.

Moving even farther east and north, one comes to Tokaj-Hegyalja, home of the only Hungarian wine that is more famous than Egri Bikavér. Once Europe's most popular sweet wine, and dubbed the Wine of Kings and the King of Wine, Tokaji has a long history. Hungarian folklore claims that the noble rot, and its ability to produce delicious balanced sweet wines, was discovered here in 1650, over 100 years before the first botrytized wines were made in Germany. The topography of the Tokaj-Hegyalja region, sheltered from drying icy winds on two sides by the Carpathian Mountains and at the confluence of two large rivers, creates the humidity and long warm autumns that are hospitable to the development of the fungus, here called *aszú*. The wine is still made from the native Furmint, plus one other indigenous white grape, and recently, small quantities of a member of the Muscat family has also been allowed. Demand for Hungary's lush dessert wine remains high, and production may increase as foreign investors, with their capital and expertise, continue to take interest in this small, remote corner of Hungary.

Eastern Europe: Other Countries

Four other countries in Eastern Europe produce wine. All are members of the former Soviet bloc: Russia, Bulgaria, Moldova, and the Ukraine. After the Second World War, all these countries saw their farms and vineyards organized into collectives under Soviet rule. Some effort was made to match varietals to the mesoclimates of various regions, but the emphasis was clearly on quantity, not quality, of wine. Total vineyard acreage increased considerably in the years between 1953 and the mid-1980s, when an effort by Soviet President Mikhail Gorbachev to reduce alcoholism forced reductions in the production of wine, as well as of the more likely culprit, vodka. Vines were pulled out for the 3 years of the Gorbachev initiative, reducing vineyard acreage by a third. Despite the loss of these vineyards, by 1989, when the U.S.S.R. disintegrated, its former republics were producing approximately 5 percent of the world's wines.

Since the ending of the Soviet regime, efforts by new governments to improve the quality of wine produced have had mixed results. There has been extensive research into viticulture and vinification at the Magaratch, a wine research institute on the Crimean Peninsula in the southern portion of the Ukraine. For instance, methods of effective winter frost protection have been developed for vineyards in Russia and the Ukraine, where winters are extremely harsh. Moreover, visiting consultants from the European Union have advised countries in Eastern Europe on matters of sanitation in the wineries and on the health of vineyards. Some foreign corporations have showed enough confidence in the future of wine production in this region to make major investments in wine-producing facilities. The most notable foreign investment in Eastern European wines was that of the Australian company, Penfold, in the country of Moldova. Penfold officials recognized that Moldova (formerly Moldavia), with its undulating hills and relatively mild continental climate, has excellent potential for the production of quality wines.

Summary—Eastern Europe

Only time will tell if Hungary, Moldova, and the other countries of Eastern Europe will succeed in their efforts to evolve from socialistic cooperative production of wine into a capitalistic system of privately held vineyards and wine-making facilities that can respond to the demands of the international marketplace and produce wines that will find a following. In a very crowded international market, it may behoove the Eastern European countries to stop the trend of abandoning their own indigenous grapes and age-old traditions in their quest for popular wines, and distinguish themselves from the competition by returning to the fine, unique wines they were making pre-U.S.S.R. As Tom Schmeisser, a highly-knowledgeable and well-respected New England retailer has put it, "As soon as more capital is crunched into Eastern Europe, their wines will have impact. The vineyards and varieties are there, but need a lot of nurturing" (Bouchard, 2005, p. 12).

Eastern Mediterranean Countries

The eastern Mediterranean is the true birthplace of viticulture and winemaking. Records reveal abundant production of wine in the ancient kingdom of Canaan as long ago as 1200 years. This coastal area, comprising a small part of what is now Turkey, and the coasts of modern Syria, Lebanon, and Israel, provided wine (and food) for the armies of the Pharaohs of Ancient Egypt. It is also well documented that wine was an inherent part of Greek civilization from the earliest of recorded times. Researchers have discovered evidence of wine production and consumption in what is now Greece from as far back as the second millennium before Christ. The modern countries of the Eastern Mediterranean continue their proud tradition of wine production to this day.

Greece

Greece is rightly called the cradle of civilization. Although it cannot be called the actual birthplace of viticulture and winemaking (that honor goes to the ancient countries at the eastern end of the Mediterranean), Greece is surely the cradle of wine. For this is where viticulture and enology have been studied, researched, and perfected for over 4,000 years. In modern times, Greece continues to make interesting, food-compatible, unique wines that are barely known outside that country.

GREEK WINE—HISTORICAL PERSPECTIVE

In Ancient Greece's agrarian economy, the ability to produce fine wine was a central component of economic stability. The ability to savor and appreciate fine wine was the sign of a cultured, civilized person. During Greece's Classical Period (approximately 6500 to about 200), wine was produced throughout the Greek mainland, on the island of Crete, and on the Aegean Islands. Through colonization, the Greeks spread viticulture to Sicily and the Italian mainland, the south of modern-day France, the Iberian Peninsula, and coastal regions around the Black Sea (present day Bulgaria, Russia, and Ukraine.) Although production was primarily on a small scale, with small farmers producing the majority of the wine, trade in wine was an important factor in the economies of many of the city-states. Wine was stored in large clay containers called **amphorae** (Figure 9–5). The amphorae were sealed with pine pitch (resin) and shipped by wagon to domestic markets or by boat to destinations around the Mediterranean. Athens, the largest and most populous of the city-states, was the most important market, but Greece's wine was transported throughout the Mediterranean world, especially to Egypt and Etruria (now Tuscany). Eventually, some colonies began to produce their own wine for export, but Greece remained the hub of the trade in wines even as the Roman Empire expanded. Many of the grapes now planted in Italy, such as Greco, Grechetto, and Aglianico, had their origins in Greek varietals.

During medieval times, Greece was part of the Byzantine Empire, formed around 300 AD and named for the great city that was its seat of power, Byzantium (now Istanbul in modern

FIGURE 9–5 Amphorae were very important in transporting wine throughout the Mediterranean.

Turkey). During this period, which extended into the mid-1500s, the Byzantines controlled the Mediterranean militarily, culturally, and commercially. In medieval Greece, most wine was produced by small property owners, or by monasteries. By the ninth century, monasteries were among the largest landowners, as many wealthy Greeks, perhaps in hopes of currying favor in the afterlife, bequeathed considerable land holdings to the monasteries. During this period, the Greeks forsook their amphorae for wooden casks, emulating their former colonists in Western Europe. Despite price competition from the wine-making monks at monasteries, who had to pay fewer taxes, the Greeks of the Byzantine era continued to make plentiful wine for domestic consumption and export. The wine recognized as the finest came from the Aegean Islands.

Beginning in 1300, the Ottomans, named for their leader, Osman I, began to acquire power and influence throughout the region. When they captured Byzantium in 1453, control of the Eastern Mediterranean fell into their hands, and they continued to build their power and influence through the sixteenth and seventeenth centuries. For the duration of the Ottoman Empire, viticulture and winemaking in Greece were set way back. The Muslim Ottomans did not forbid the Christian population of their colonies from producing wine, despite their religion's prohibition against alcoholic beverages. However, during the centuries of Ottoman rule, Greece's communications system and transportation infrastructure was fragmented, leading to a disruption in the trade in wine (and other agricultural products). The amount of wine produced fell off precipitously, and what was made was consumed locally. In the nineteenth and early twentieth centuries, while other European countries, most notably France, were perfecting their viticultural practices and vinification methods and establishing international demand for their wines, Greece regressed into vinous Middle Ages.

Even with the overthrow of the Turks in 1913, after a lengthy and costly battle for independence, Greece was not able to get immediately back on its feet. Set back severely, both in manpower and in revenues, by involvement in two world wars, the impoverished Greek state was set back even further by a prolonged civil war. Wine technology was essentially primitive. It was not until well into the 1960s that the Greek government was stable enough to begin devoting some effort to the modernization of the domestic wine industry. With admittance into the European Union in 1981, Greece marked the beginning of its modern wine era.

Wine Laws

During the 1970s, in preparation for its entry into the European Economic Community (EEC) which later became the European Union, Greece drafted quality control laws harmonious with those of member countries. Like France, Greece divides its wines into four levels of quality. (Greek law even uses some of the terms found in the French laws.) Quality wines are either of the sweet, dessert style, in which case they are labeled OPE (Controlled Appellation of Origin), or they are dry table wines and are labeled OPAP (Appellation of Superior Quality). The term Réserve on quality wines indicates additional aging in oak casks. About 12 percent of Greece's wine production qualifies for quality wine designation. The *vins de pays* designation applies to a variety of larger regions, where both indigenous and Western European varietals are allowed. The huge table wine category includes several successful branded wines and some nice wines made outside the parameters of the appellation laws.

Wine is made all over Greece, and its wine laws divide the country into 26 quality wine regions, each with a different mesoclimate. Among the regions that have established a reputation are Macedonia, Thrace, Zitsa, Peleponnese, Santorini, Cephalonia, and Crete.

Macedonia and Thrace in northern Greece are known for full-bodied red wines, mostly from native varietals. The better vineyards in these two regions are being studied for possible *grand cru* status.

In central Greece, Zitsa produces a pleasant light sparkling wine made from a native white varietal.

The southern peninsula of Peleponnese has a warm and hospitable climate. It produces fully 25 percent of national production, including a popular dry white wine that is fresh and spicy, a lush dessert wine from the Muscat grape, and a simple dry rosé from a native varietal. There are also several table wines and *vins de pays* made.

Many of the islands off Greece produce wine. Among the best known island appellations are Santorini, whose chalky subsoil, dry climate, and cooling winds produce a dry, crisp white. On the opposite coast, in the Ionian Sea, the island of Cephalonia is famous for the lemon-scented, aggressively flavored white made from the native grape, **Robola,** which botanists believe is the same grape as Italy's Ribolla. The southern island of Crete, in the Mediterranean Sea, produces a variety of wines, most notably big intense reds, all from native varietals.

Along with Greece's efforts to improve the health of vineyards and the technological modernization of wine-making facilities, there is a strong trend among the current generation of wine professionals to preserve indigenous varietals. This is done not only out of a sentimental attachment to their country's ancient heritage, but with an eye to modern marketing strategies. Young winemakers such as Yiannis Boutaris, whose family owns one of Greece's largest wine exporting companies, and Yannis Voyatzis, who makes wines in a rugged corner of Macedonia, and Yiannis Paraskevopoulos who owns a winery on the island of Santorini, believe strongly that if Greece falls prey to the temptation to produce wines only from immediately recognizable international varietals, they will lose out in the long run to countries already known for these wines. Greece must distinguish itself by being the source of fine wines made from indigenous varietals found nowhere else, wines that are part of an ancient tradition reaching back thousands of years to the dawn of civilization. This strategy seems to be reaping results in the United States. In 2004 almost 200,000 cases of Greek wine were imported, up 25 percent from 2000 imports (Walker, 2005).

Israel

Wine is an integral part of Jewish tradition and is mentioned frequently in the Talmud and the Old Testament. Archeological evidence shows that wine was produced in Palestine from Biblical times to the time of the Muslim conquest in 636 AD, at which time vineyards were destroyed. The Crusades, which occurred between 1100 and 1300 AD, brought Muslim rule to an end and allowed the reintroduction of viticulture under the Christians. However, with the subsequent exile of the Jews, wine production again ended. Although Jews, in their centuries-long banishment, continued to incorporate wine into their religious observances, it was not until their return to the Holy Land in the late nineteenth century that Jews were again able to produce their own wines in their own state. At the new Zionist settlements, the Jewish farmers were inexperienced with viticulture, and the resulting wine was of low quality. An enormously generous donation by France's Baron Edmond de Rothschild in 1882 allowed for the importation of French varietals to several settlements, and for the sponsorship of consulting enologists who worked with the settlers to perfect their viticultural and wine-making skills. The wine-making communes grew and thrived and began exporting kosher wine to Jewish communities around the world. (To be kosher, a wine must be produced under rabbinical supervision and cannot be touched by a Gentile or even a non-observant Jew once the grapes have been brought to the winery and crushed.) The kosher wines made were in a sweet, dense, alcoholic style, and were intended primarily for religious ceremonies. After the 1948 War of Independence, Israel continued to produce simple kosher wines in the same style. However, since the 1980s, the trend has been towards making dry, multidimensional wines that, although still kosher, can be enjoyed as part of a meal.

The varietals brought in by the French in the nineteenth century, mostly from the Rhône and Midi, such as Carignan, Grenache, Semillon, and Chenin Blanc, continue to thrive in Israel's dry, sunny climate. Recently the Bordelaise varietals Cabernet Sauvignon, Merlot, and Sauvignon Blanc have been widely planted. Irrigation is necessary because there is minimal rain from April through October. Viticulturally, Israel is divided into five regions.

Galilee in northern Israel is the most prized wine region, especially its comparatively cool Golan Heights, where the altitude and volcanic soil create the ideal mesoclimate for fine Cabernet Sauvignon and Merlot.

In upper central Israel, Haifa, including Mt. Carmel, is the site of one of the first Zionist wineries sponsored by the Baron de Rothschild's donation.

The country's largest wine region, Samaria (Shomron) is bounded by the Carmel Mountains and the Mediterranean. Most wine made here is consumed domestically.

The Judean Hills includes the vineyards planted on hills outside Jerusalem and on the West Bank.

In the southern, desert-like section of Israel, the relatively new region of Negev shows promise.

Lebanon

The Bekaa Valley in northern Lebanon has the ideal conditions for viticulture: altitude over 3,200 feet (975 meters), volcanic soil, adequate rainfall, and cool nights. The Jesuits founded Lebanon's first winery here in 1857. Despite the fact that Lebanon has been in a state of conflict and open warfare for forty years, wine is still made at that winery, Ksara.

The Bekaa Valley also holds the vineyards of one of the Eastern Mediterranean's best-known wineries. Château Mugar was founded in 1935 by a Christian, Gaston Hochar, who planted the French varietals Cabernet Sauvignon, Cinsault, and Carignan. From them he made a big, smooth, very complex, and age-worthy red wine. Today his son Serge, who trained in Bordeaux, continues the tradition. Château Mugar exports 95 percent of its production to markets in the United States, the United Kingdom, and even France. Both the red wine and the dry white made from local varietals are widely recognized as truly fine wines.

Summary—Eastern Mediterranean

The wines of the Eastern Mediterranean area do not yet have an important presence in North American markets. However, Greece seems to be undergoing a renaissance and may well become more evident. In Israel, as the trend away from sweet, highly alcoholic wines toward dry attractive wines continues, there is no doubt that Jews around the world will turn to these fine kosher wines, not just for religious observance, but as part of their daily cuisine. As for the wines from other countries in the region—Lebanon, Egypt, Turkey, and Cyprus—it is unlikely that any of them (other than Château Mugar) will find a following any time soon.

REVIEW QUESTIONS

1. As a reaction to what scandal did the Austrian government tighten up the country's wine production laws?
2. What is Grüner Veltliner? Where was it originally developed?
3. What is Müller-Thurgau? Where was it originally developed?
4. What is the most widely planted grape varietal in Switzerland?
5. For what two wines is Hungary most famous?
6. In Greece's wine laws, what does "OPAP" stand for?
7. When were French varietals first introduced into Israel's vineyards, and whose large gift made this planting possible? (Hint: the donor was a Frenchman).
8. The wine of one winery in Lebanon is well known and highly respected in Europe and the United States. What is the name of that winery, and what style of wine does it produce?

SECTION III

Wine Regions of North America

Wine Regions of North America are covered in three chapters: California; Washington and Oregon; and New York, Canada, and Other North American Regions. Domestic wines come in as many forms as their European counterparts and make up the majority of wine consumed in the United States. The wine regions of North America are covered in much the same way as the European wine regions in Section II, with sections on local grape growing and winemaking techniques.

CHAPTER 10

California

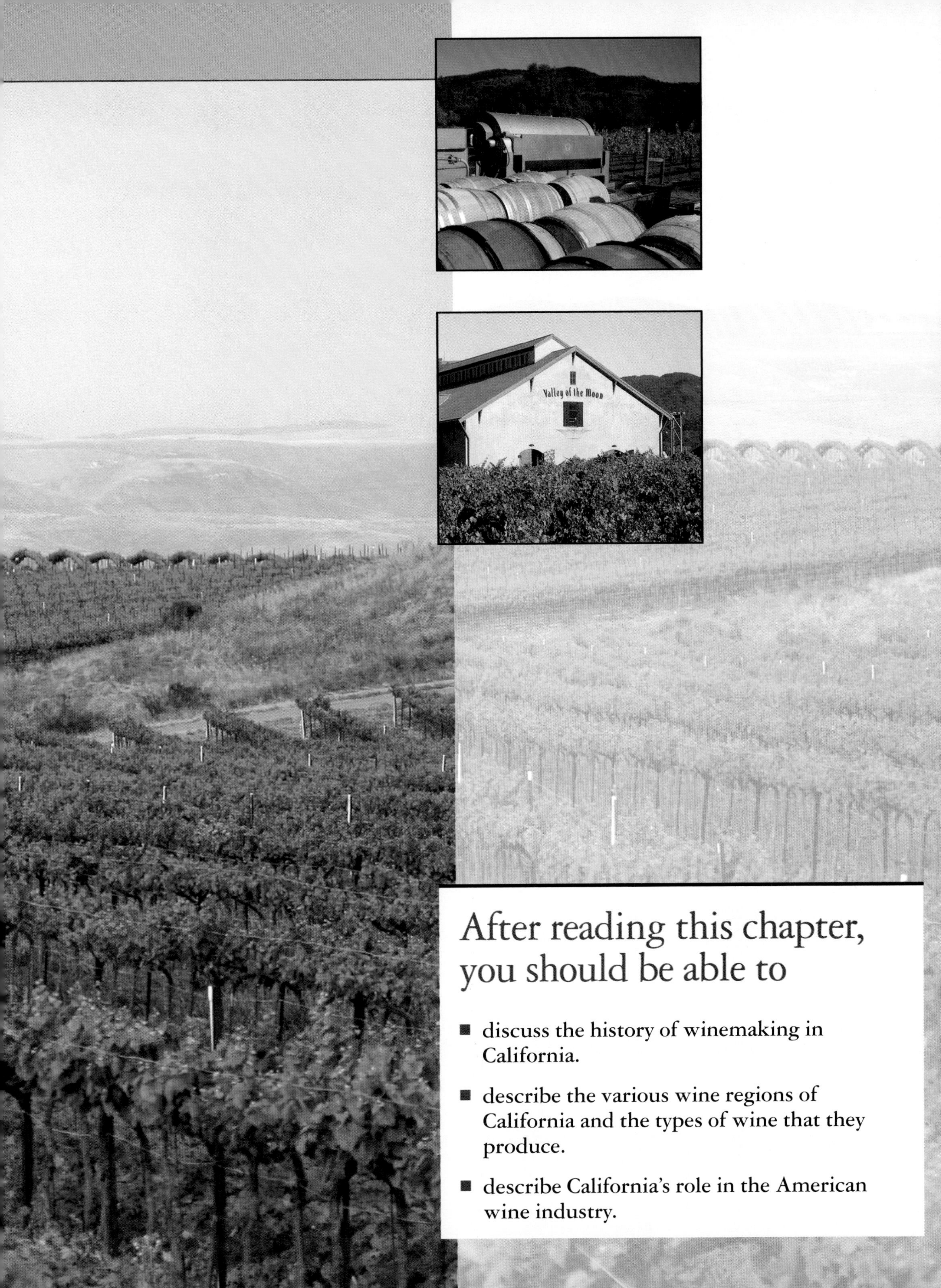

After reading this chapter, you should be able to

- discuss the history of winemaking in California.
- describe the various wine regions of California and the types of wine that they produce.
- describe California's role in the American wine industry.

KEY TERMS

American Viticultural Area (AVA)
appellation d'origine contrôlée
appellations
Bureau of Alcohol, Tobacco & Firearms (BATF)
bench land
boom and bust cycle
California Wine Association (CWA)
degree days
glassy-winged sharpshooter
heat summation
jug wines
Mission grape
mission period
Paris tasting
Pierce's disease
Serra, Junípero
Tax and Trade Bureau (TTB)
UC Davis
Vallejo, Mariano
Wine Institute

California is unquestionably the most important wine-producing region in North America (Figure 10–1). Containing more than 350,000 acres (142,000 hectares) of wine grape vineyards and more than 840 wineries, it is responsible for more than 90 percent of the wine that is produced in America (Wine Institute, 2002). Extending more than 700 miles (1,100 kilometers) north to south, California is known for both its mild Mediterranean climate as well as its beautiful landscape. Its temperate weather is characterized by wet winters followed by warm, dry summers. This cycle of wet and dry seasons makes it an ideal region to grow grapes as well as many other agricultural crops. Although vines require adequate water to grow, high humidity and rain during the summer when the fruit is on the vine can promote rot and mildew. The winter rains in California provide enough water, either by natural soil moisture or by irrigation, to mature the crop without relying on summer rain. Additionally, the summer weather also consistently provides enough warmth and sunlight to bring the grapes to their full ripeness.

FIGURE 10–1 Sonoma Valley vineyards.

California is also important from the perspective of wine consumption. In addition to being the most populous state in the union, it also has one of the highest per capita consumption of wine. These two factors combined result in Californians drinking 17 percent of the wine that is consumed in the United States. Because so much wine is both produced and consumed in the state, it has become home to many of the businesses that support the wine industry and plays a leading role in the United States wine market.

The development of winemaking in California reflects the melting pot of cultures that settlers to the state brought with them. European immigrants from different countries with diverse methods of viticulture and winemaking brought their knowledge and experience to the industry. Since the state did not have established winemaking traditions of its own, these Old World techniques flourished and were adapted in new ways that were suited to the conditions in California. In addition to this there was a great deal of innovation and modernization applied to the winemaking methods immigrants brought with them. This blend of cultures began in the 1800s when the state was first being settled and continues to this day with large multinational wine companies investing in California.

California Wine—Historical Perspective

Winemaking came to California with the Spanish missionaries, who were among the first Europeans to settle in California. Spain had occupied Mexico for more than 250 years when it expanded its territory, establishing 21 missions up the coast of California. The goals of the missions were threefold: to expand the territory of Spain and secure it for colonists, to convert the Native Americans in California to Christianity, and to develop the resources and send the proceeds back to Spain. The missionaries were led by a Franciscan friar, Father **Junípero Serra,** who established the first mission in San Diego in 1769 (Figure 10–2). Wine was essential to the new settlers who used it for sacramental purposes as well as for a beverage to be consumed with their meals. Soon after Father Serra arrived, he wrote in letters about the difficulty and cost of obtaining wine from Mexico, so the missionaries were very interested in producing their own wine. Although there were native grapevines growing throughout the state, they were unsuitable for winemaking, having smaller berries that were much less sweet than the European *vinifera* varieties. To alleviate this situation the friars imported *vinifera* cuttings to grow their own grapes to make wine. Although there is some academic dispute on exactly when and where the first vineyard was established, most historians believe it was in San Juan Capistrano in 1778 with the first vintage coming 4 years later in 1882 (Sullivan, 1998). Eventually grapes were planted at all but two of the missions, with the climates of San Francisco and Santa Cruz being considered too cold and foggy to ripen grapes. The Los Angeles Mission with its temperate climate and fertile soil had the most extensive vineyards of the missions.

While the exact year of California's first vintage may be open to discussion, it is known that the first variety grown was the **Mission grape**. Closely related to the varieties País and Criolla in South America, Mission is a red grape that is a prolific producer and adapts to a number of growing conditions. The friars used it to make a variety of wine styles, including white, red, dessert,

FIGURE 10–2
Franciscan friar, Father Junípero Serra, head of the California missions. During the mission period from 1769 to 1833, the friars introduced winemaking to California.

and brandy. The Mission grape produces a rather flavorless wine but this was probably of little consequence because the winemaking techniques were rudimentary even by the standards of the time. Grapes were trodden upon wooden platforms lined with animal skins with the juice collected in skin bags or wooden vats for fermentation and storage (Teiser & Harroun, 1983). In 1823 the last mission was established in the town of Sonoma, north of San Francisco Bay. Just 10 years later, in 1833, the now independent Mexican government ordered the secularization of the mission properties, and the missions and their lands went from church to government control. Although the winemaking methods of the mission may not have been advanced, their success inspired European immigrants who settled in the pueblos or towns that grew up around the missions.

Commercialization

By the time the **mission period** was coming to a close, private citizens were beginning grape-growing and winemaking operations in the state. The first large-scale commercial vintner was the appropriately named Jean-Luis Vignes (*vignes*, French for vines) in Los Angeles. Called Don Luis by the Californians, he came from Bordeaux, and unlike the friars he had extensive knowledge of winemaking practices. He also looked beyond the Mission grape, importing more premium varieties from France. He eventually produced 1000 barrels a year from his 100-acre (40-hectare) vineyard located along the Los Angeles River (Johnson, 1989).

In Northern California, Lieutenant **Mariano G. Vallejo** was sent to take over operations at the Sonoma Mission and pueblo after secularization. He quickly established a military barracks and began the process of laying out a town. He also restored the mission winery and its vineyard

FIGURE 10–3 The Vintage in California—At Work at the Wine-Presses. This well-known image illustrates how grapes were crushed and pressed in California during the middle of the nineteenth century. From the October 5, 1878 issue of *Harper's Weekly.*

that had fallen into disrepair. His success in civic duties soon earned him the rank of general. Vallejo also was known for his skill in grape growing and winemaking as well as cattle ranching and other agricultural operations. He inspired many other early settlers to come to the North Coast and aided them by using his power as a government administrator to grant them tracks of land to develop. One of the most notable settlers was George C. Yount, the namesake of Yountville, who was the first person to plant grapes and make wine in the Napa Valley. Vallejo remained active in civic affairs even after California became a part of the United States, helping to draft the state's constitution and serving as a senator in the first state legislature.

The success of Vignes and Vallejo as well as other early vintners was aided by California's expanding population. More and more settlers were arriving from the East Coast to take advantage of California's excellent climate for growing crops, and with the Gold Rush of 1849 there was an expanding base of consumers for their products. During the second half of the nineteenth century, commercial winemaking operations grew both in number and in size (Figure 10–3). With the completion of the transcontinental railroad in 1869, California wines became increasingly available on the East Coast.

The expansion of the industry was not without its ups and downs, and during this time grape growing and winemaking exhibited the classic **boom and bust cycle** that is common to many agricultural products. During boom times increased demand for wine leads to high prices for grapes which eventually results in over planting and excess production, ultimately lowering grape prices. When the price of grapes becomes too low, overall production stagnates or declines until the demand for wine increases, beginning the cycle over again. The boom and bust economic cycle was exacerbated by the destruction caused by the root louse phylloxera, discovered outside of Sonoma in 1873, and the economic depression of the late 1880s. Wine and grapes were often sold for less than it cost to produce them. Although these events were devastating, they did serve to weed out poor and inefficient producers and to replace the ubiquitous Mission grape with varieties that are more suited to winemaking. During this time some of California's most famous wineries were established: Buena Vista in 1857, Charles Krug in 1861, Beringer in 1876, Inglenook in 1879, and Korbel in 1882 (Laube, 1999) (Figure 10-4). All these wineries remain in operation today, albeit with different ownership.

FIGURE 10–4 Unloading grapes at an early California winery. Grapes are brought to the top level of the winery and then crushed into fermentation tanks in the cellar below.

In an effort to control production and to stabilize the wine market, some of the state's largest wineries joined together in 1894 to form the **California Wine Association (CWA).** Eventually the California Wine Association grew to include more than 52 wineries throughout the state, producing 80 percent of California's wine (Teiser & Harroun, 1983). As evidence of its size, the California Wine Association's bottling plant in San Francisco lost more than 10 million gallons of wine in the 1906 earthquake. Later that year a large shipping and bottling plant was built on the shores of San Francisco Bay at Point Richmond, and for a time it was the world's largest winery. The California Wine Association flourished, producing wine of good value, if of mediocre quality, until the onset of Prohibition in 1920.

Prohibition

The Eighteenth Amendment, also known as the Volstead Act, established national Prohibition in the United States and outlawed the manufacture and sale of alcoholic beverages from January 16, 1920, until December 5, 1933 (Figure 10–5). The anti-alcohol movement had been growing in the United States for more than 100 years and a number of individual states and communities passed their own "dry" laws during this time. During World War I, patriotic sentiment to preserve foodstuffs for the war effort rather than use them for alcohol production combined with the temperance movement to gain majority support for the Eighteenth Amendment.

The passing of Prohibition devastated the winemaking industry in California, closing all but the few wineries that were allowed to make wine for use in food flavoring or sacramental purposes. However, due to a clause in the Volstead Act that allowed the home production of up to 200 gallons of "non-intoxicating cider or fruit juice," which was interpreted to include homemade wine, the business of growing wine grapes in California actually flourished. Before Prohibition the price of grapes grown in the Central Valley averaged $10 to $20 per ton; with the harvest of 1920, the best varieties were fetching up to $125 per ton (Teiser & Harroun, 1983). Growers who were accustomed to barely making a profit were producing as much as they could for shipping out to eastern markets for home winemakers. The varieties that had thick

FIGURE 10–5 In New York during Prohibition, workers under police supervision dump contraband wine into the sewer.

skins and the most color were considered the most valuable because they could withstand the rail journey to the East Coast, and their deep color allowed home winemakers to stretch production by adding water and sugar to get more wine per pound of fruit. This meant that the classic wine varieties that replaced the Mission grape in the late 1800s were themselves replaced with varieties more suited for shipping. White wine grapes in particular were replaced with dark, thick-skinned grapes such as Petite Sirah and Alicante Bouschet, the latter variety being particularly prized because it is one of the few grapes that has dark red juice as soon as it is crushed or pressed, whereas most red grapes require skin contact after pressing to turn their juice red.

At the end of Prohibition in 1933, there was much anticipation by vintners that with its end would come a rapid resurgence of the wine industry. However, several factors prevented this from happening:

- During their long closure, wineries had fallen into disrepair, and there were few skilled winemakers available.
- Prohibition ended during the middle of the Depression, and there was little demand for wine and few resources available for rebuilding.
- People's tastes had changed and consumers had gotten used to poor quality homemade wine that was often sweetened or fortified with distilled alcohol to cover up the flaws.

In the first year after Prohibition there were 804 wineries in California; by 1940 one-third of them had failed. To help alleviate this situation, the **Wine Institute** was formed in 1934 as a trade organization to promote California wines and to lobby for regulations that were more favorable for wineries. Another organization that was instrumental in improving the quality of wine after Prohibition was the **University of California at Davis (UC Davis).** A pilot winery was built on campus, and an academic program was established to train students as winemakers and grape growers. One of its most significant accomplishments was categorizing the various grape-growing regions of the state by their climates and determining which varieties grew best in a given region. Unlike Europe, there was less of a tradition developed by years of trial and error to determine the proper growing regions for grape varieties. In a method called **heat summation,** or **degree days,** a vineyard is classified by the summation of its temperature above 50°F over the course of the growing season. In this system Region I is the coolest with 2,500 degrees summation or less, and then the regions have 500 degree increments to Region V, which is warmest with a summation of greater than 4,000 degrees. At best this is a very rough estimate of a vineyard's terroir and the grape varieties to which it is suited. However, even though the recommendations were not always followed, it proved to be useful in establishing what varieties might be suited for a given vineyard.

Through the work of these and other organizations, the quality of California wine improved, and consequently, sales increased. The wines produced during this time were generally inexpensive and of good, if not great, quality. Few people thought the wines matched the quality of imported wines. This trend continued until the 1960s.

The Wine Revolution

The late 1960s began a period of great expansion in the California wine business. From 1971 to 1980, the per capita consumption of wine in the United States doubled from 1.2 to 2.4 gallons per year, and the number of wineries in the state went from 227 to 470 (Lapsley, 1996). This coincided with an influx of new wineries that were established in California's coastal valleys (Figure 10–6). Many of them were started by wine enthusiasts who wanted to make wines like those they had tasted from Europe. Winemakers invested more effort in obtaining better grapes, and improved their methods of production such as using stainless steel tanks and French oak barrels. Sales of wine increased dramatically during this time, particularly for the premium end of the market. At the same time as consumers were buying more, their tastes were also changing. In 1968 table wines (dry wines) outsold dessert wines for the first time since before Prohibition. A few years later in 1976 white wines outsold red for the first time.

This was also a watershed year in terms of the reputation of Californian wines. In May of 1976 in Paris, a local wine merchant held an event where wines from some of California's best vintners were matched in a blind tasting against some of France's best producers. The judges were all French and included some of the country's best known wine experts. To the surprise of nearly all of those present, the California wines took the top honors for both red and white wine and six of the top 11 places overall (Conaway, 1990). Known as the **Paris tasting,** it helped to dispel the notion to wine consumers at home and abroad that California made second class wines.

FIGURE 10–6 Rodney Strong Winery in Sonoma County, built in 1970 it was one of the many new wineries that were started during the wine boom of the 1960s and 1970s.

American Viticultural Areas

For a wine to be labeled "California" the grapes that go into producing the wine must be grown entirely within the boundaries of the state. However, within the state there is a broad diversity of growing conditions that make its many regions suitable for producing a variety of grapes and wines. One method of identifying these grape growing regions, or **appellations**, is by political boundaries where the county of origin is used to describe the source of the grape. To be labeled "Monterey County Chardonnay" a minimum of 75 percent of the grapes used to produce the wine must be grown in California's Monterey County. Using the county of origin is not always adequate to describe a grape-growing area because political boundaries do not always follow the border between different grape-growing climates or terroirs.

In an effort to better identify unique growing regions, in 1980 the **Bureau of Alcohol, Tobacco & Firearms (BATF)** allowed for the creation of **American Viticultural Areas** more commonly known as **AVAs.** Vintners and growers could petition the BATF to form an AVA in a specific geographical area with a common climate, soil type, and history of winemaking. In 2002 the BATF was reorganized and wine is now regulated by the **Alcohol and Tobacco Tax and Trade Bureau (TTB).** The first AVA established was Augusta, Missouri, in 1980. As of January 2006 there were a total of 170 AVAs in the United States, 96 of which are located in California. A complete list of AVAs can be found in Appendix B.

The largest AVA in California is the Central Coast Appellation, which has over 5.4 million acres (2.2 million hectares) of land with nearly 100,000 acres (40,000 hectares) of vineyards. The smallest AVA is Cole Ranch with only 150 acres (60 hectares). Occasionally AVAs are initiated by growers or wineries to include the vineyards at their own estates and little else. In these situations often the only wine that is bottled and labeled with the AVA is what the winery produces.

Appellations can overlap political boundaries such as state and county lines as well as other AVAs. The Carneros district, for example, extends over the southern portion of both the Napa Valley and the Sonoma Valley appellations. A single vineyard in the Sonoma County portion of the Carneros district may also be part of the Sonoma Valley, Sonoma Coast, and the North Coast AVAs. To be labeled with an AVA, at least 85 percent of the grapes used to make a wine must be from that region. If wines are blended from several areas of the state that do not have a common political boundary or AVA, they are labeled as California. If the grapes are grown, produced as wine, and then bottled all on winery property they can be labeled "Estate Bottled." Frequently, wineries will acquire grapes from outside the appellation that the winery resides in. In this case, the winery can bottle the wine as an AVA designate but cannot call it "Estate Bottled" even if it owns the vineyard. To be labeled with both an "Estate" and an AVA designation the winery and the vineyard must be in the same AVA.

Unlike the French system of ***appellation d'origine contrôlée,*** American Viticultural Areas only govern the geographical origin of the grapes and do not dictate what varieties can be grown or winemaking techniques within the area. Furthermore, the TTB goes out of its way to state that the government sanction of the boundaries of an AVA makes no endorsement of quality of the grapes and wine that it produces. While having an AVA on the label does not make any official statement about the quality of a wine in the bottle, wines that are made from grape varieties that the appellation is renowned for do benefit from its reputation.

One complaint against the American system of viticultural areas is that to avoid litigation the government has been too free in allowing the establishment of new AVAs in areas that do not have a history of grape growing or a common terroir. Another criticism of the system is that large AVAs often include a number of climates and soil types that make many different types of grapes and wine. When the output of an AVA is too diverse, consumers have trouble knowing what to expect from the wine it

(continues)

American Viticultural Areas (continued)

produces. For example, the Central Coast AVA is very large and within its boundaries the climate ranges from very cool, where Chardonnay would be appropriate, to warm, where Zinfandel would do better. By simply labeling a wine "Central Coast" the consumer does not know which area, warm or cool, the grapes for the wine were grown in and whether he or she should purchase a Chardonnay or Zinfandel. Conversely, if the consumer sees a Chardonnay from the Edna Valley AVA, which is much smaller and has a cool climate, his or her choice would be clearer. It is also important to remember that differences in vineyard and winemaking practices at various wineries can eclipse the similarities of the wines that they make from grapes grown in the same viticultural area.

Throughout the 1980s and 1990s the California wine industry continued to evolve and improve its product. Although the growth in America's per capita wine consumption slowed and stabilized during this period, the market for premium wines continued to be strong. As the wine industry matured so did its consumers, and their tastes became more sophisticated. Wine producers referred to this phenomenon as "drinking less but drinking better" and it allowed the fine wine market to grow as sales of inexpensive **jug wines** slowed. During this time established wineries grew in size, and some consolidated, forming large companies with several brands of wine.

The industry's successes continued to inspire individuals to enter the business and start small boutique wineries that specialized in a particular wine. Because of the increase in the value of land in wine growing regions, this required considerably more investment than it did at the beginning of the wine revolution in the 1960s. The high cost of vineyard land in Napa and Sonoma Counties helped to spur development of premium wineries in other areas of the state such as Santa Barbara and Lake Counties.

Today at the turn of the new millennium, the cycle of boom and bust that has always been a part of the California wine industry continues. Increased sales of premium wine in the 1990s encouraged wineries to plant new vineyards and vintners to increase the amount of wine they produced. This new production came on line at the same time as there was decreased demand due to a sluggish economy. This in turn led to prices being lowered on many grape varieties grown throughout the state. Many premium wineries—not wanting to lower their price too much—sold off their excess production in bulk to other wineries that would blend the wine from multiple sources and then bottle it under their own name. The wine produced was less expensive and often of very good quality, which helped to spur consumer interest in wine.

Another factor affecting the economic cycles of the California wine industry is globalization. Consumers in the United States are becoming more familiar with wines from the Southern Hemisphere and Europe that compete directly with wines from California. These imported wines are often produced specifically for export to the United States and are frequently brought in by multinational companies that have winery holdings both in California and overseas. This global competition puts extra pressure on California winemakers to make a high quality product and keep their prices affordable.

The Wine Regions of California

California's large size and varied topography give it a wide range of growing conditions that are suitable for many different grape varieties and styles of wine. Grapes can be grown in most of the state except the northwest coast, which is too wet, and areas that are too high in elevation and

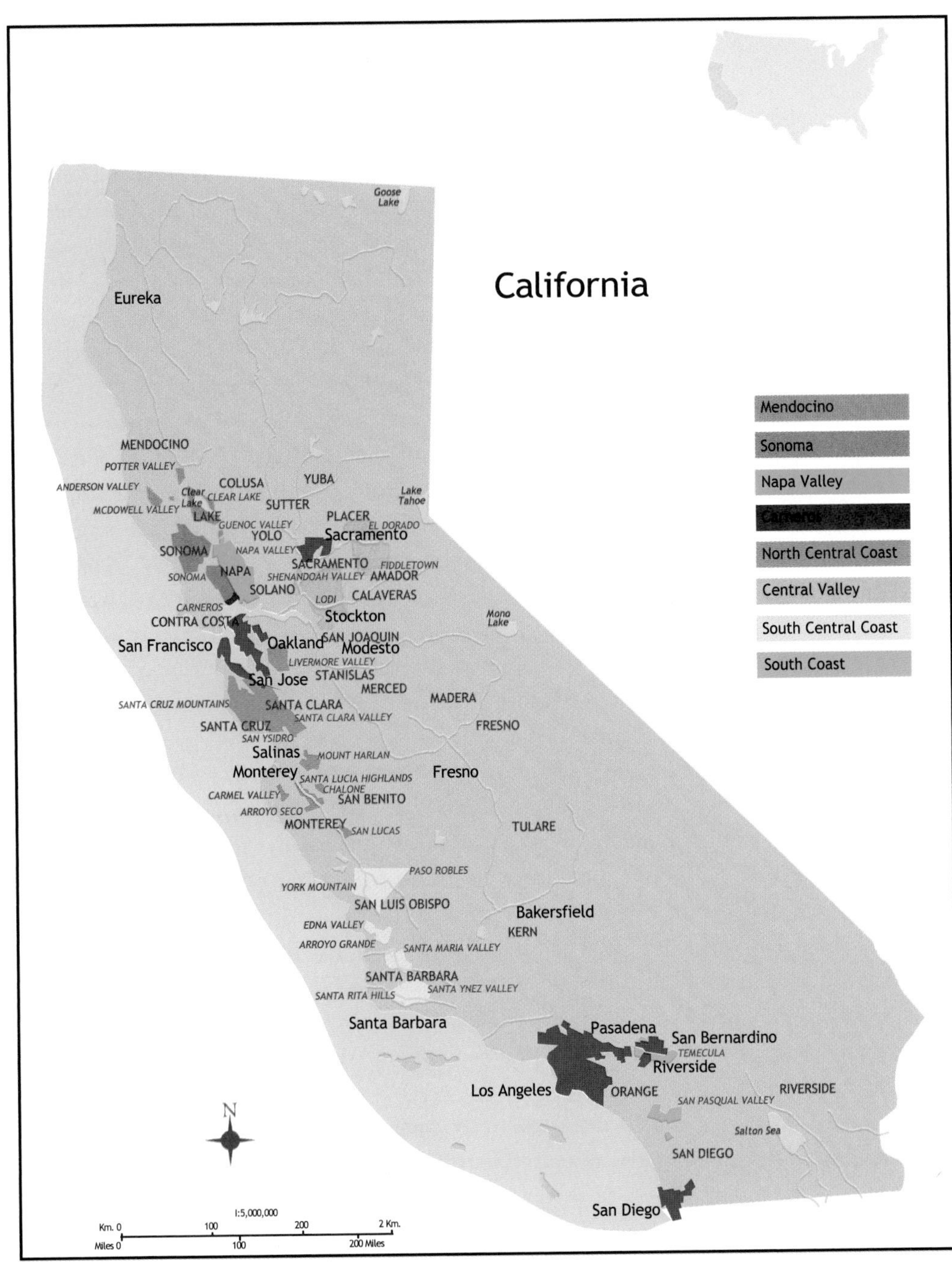

The wine regions of California.

therefore too cold. With irrigation even the desert areas in the southeast of the state can also support vineyards and are the home of much of California's table and raisin grape production. Most of the grapes that are used for premium wine production are grown in the state's coastal valleys. Here the Pacific Ocean has a moderating effect on the climate, keeping it cooler than vineyards that are located more inland. During the summer months, warm daytime temperatures in the interior of the state create rising air that draws cool breezes from the Pacific Ocean into the coastal valleys. These breezes often carry a layer of fog with them that arrives in the late afternoon or evening and is burned off by the sun the following morning. This cooler weather allows the grapes to retain more of their varietal character and natural acidity, making the wines they produce more intensely flavored. Inland, in the Central Valley region, the weather is warmer, and soils are more fertile, vineyards here yield more tons per acre, and the grapes that are grown often are used for less expensive wines. The Central Valley covers a large area, is the home to most of the state's agricultural output, and has the majority of California's vineyard acreage.

The Napa Valley

The Napa Valley lies in the temperate zone between Northern California's cool coast and warm interior. Beginning in the town of Napa, just north of San Pablo Bay (the northern section of San Francisco Bay), it runs in a gentle arc for 35 miles (56 kilometers) to the northwest until reaching the slopes of Mount St. Helena. The appellation covers the vast majority of the land in the county and includes most of the watershed for the Napa River as well as part of Pope Valley to the east. It ranges in elevation from near sea level on the valley floor to 2,700 feet (820 meters) along the ridges of the mountains. Grape growing is by far the dominant agriculture in the Napa Valley AVA with more than 35,000 acres (14,000 hectares) being planted to wine grapes.

Historically the valley has been producing wine since the 1830s, and by the 1880s it had built a reputation for producing fine wine. However, Prohibition decimated its wine industry and it recovered very slowly after repeal. Napa Valley regained its reputation for producing great wines only when it became the epicenter for the wine boom of the 1960s and 1970s. During this period, there was a great deal of investment coming into the valley, and the number of wineries grew from less than 30 in 1965 to more than 200 today. The Napa Valley is the most widely recognized of California's AVAs and is considered by many to be its premiere wine-producing region. This opinion is reflected in the fact that Napa Valley grapes and wines routinely command the highest prices in the state.

As the reputation of Napa Valley wine grew, more vintners came to the valley to establish vineyards and wineries. Some were wine enthusiasts who had been successful in another field of work and came to the valley to indulge in their passion for winemaking. Others were from large multinational corporations looking to diversify their business and take advantage of America's growing taste for wine. As the wineries grew in size and number, tourists began to flock to the area. This growth in tourism was aided by Napa's proximity to the San Francisco Bay Area. Visitors to San Francisco, as well as local wine enthusiasts, could visit the Napa Valley on an easy day trip from the city. In the 1970s, tourism became the county's largest employer after grape and wine production. By 2002, more than 5 million visitors a year were coming to the Napa Valley for wine tasting, causing traffic jams on the roads as well as crowding in the tasting rooms. To deal with these crowds, wineries routinely charge for sampling the wines in their tasting rooms, a practice that is still rare in other California wine regions.

As more of the area developed, there was greater concern that prime vineyard land was being replaced by subdivisions and that unmitigated growth was affecting the quality of life in the valley. In 1968, the first in a series of slow growth regulations was established in the form of an agricultural preserve. It was very controversial at the time because many growers and wineries did not want any constraints on how they could use or develop their property. Although land use regulations make it very difficult to start or expand wineries in the Napa Valley, they have controlled urban sprawl and preserved the valley's rural character. Today Napa has the smallest population of the 10 San Francisco Bay Area counties.

Regions of the Napa Valley

The Napa Valley is a large AVA with a wide variety of soil types and climates. The climate of the southern end of the valley is dominated by the cooling influences of San Pablo Bay. Here, particularly during the warm summer afternoons, fog will creep in, lowering the temperatures during the warmest part of the day. Cool climate varieties such as Pinot Noir and Chardonnay do best in this area of the valley. As you move up the valley toward the town of Calistoga, the bay has less

Reading a United States Wine Label

As the wine industry has advanced in recent decades, wine labels have become much more elaborate. While in the 1960s wine was usually bottled with relatively plain labels with a minimum of information, today they are designed to be more eye-catching with multiple colors, embossing, gold leaf, and distinctive shapes. This is to try to get potential customers to notice the wine on the shelf; also, expensive labels are used to try to convey an impression of quality for the wine in the bottle. The importance of packaging to wineries is evidenced by the industry saying, "you sell your first bottle of wine to a customer with the outside of the bottle, and the second bottle of wine to the customer with what is inside." In addition to their importance in selling wine, labels must also provide information to the consumer. Some of the information on wine labels is provided by the winery to describe what the wine tastes like and how it was made. The federal government also mandates what information must be placed on a bottle of wine to accurately describe what the wine is, including standards of composition that must be met before certain claims can be made on the label. In recent years the amount of information required by law has expanded to include warnings about health and whether there are sulfites present in the wine.

The next three pages outline the basic requirements for wine labels in the United Sates. Different countries have different standards and terms that they use to describe their wines. Information on the labels of imported wines is outlined in their respective chapters of this book.

Front Label

Winery Name: Most names are acceptable as long as they are not offensive or misleading.

Vintage Date: 85 percent of the grapes used to make the wine must be harvested in the year listed as the vintage. If an AVA is listed as the appellation, then a minimum of 95 percent of the grapes must be from the vintage stated. If a vintage date is used, the appellation must be stated as well.

Appellation: The district the grapes were grown in. If listed as a political region such as county or state, at least 75 percent must be from the region listed. Some states have laws requiring that for a state to be listed appellation, 100 percent of the grapes must be grown in that state.

Alcohol Content: The percentage of alcohol by volume must be listed; if it is 7 to 14 percent, the label may also state "Table Wine" or "Light Wine."

Variety: The varietal of the grapes used to make the wine. The variety listed must be at least 75 percent of the grapes used to make the wine. Some states have higher percentage requirements and *Vitis labrusca* varieties like Concord need to have only 51 percent. If a varietal is designated on the label, an appellation must also be listed.

Reading a United States Wine Label

Back Label

Wine Notes: Not required, but many wineries add them to describe how the wine is made and what it tastes like.

Web Address: Also not required but an excellent source of more information about the wine.

Production Statement: "Produced & Bottled By" means a minimum of 75 percent of the wine was fermented at the winery where it was bottled.

Sulfite Declaration: If the wine contains more than 10 PPM sulfur dioxide, "Contains Sulfites" must be included on the label.

KENWOOD

Sauvignon Blanc continues to be Kenwood's most popular wine. Partially aged in small oak barrels, this refreshing wine is superb as an apéritif and excellent with seafood, pasta and poultry dishes.

http://www.kenwoodvineyards.com

PRODUCED & BOTTLED BY KENWOOD VINEYARDS, KENWOOD, SONOMA COUNTY, CA

CONTAINS SULFITES

PRODUCT OF USA

GOVERNMENT WARNING: (1) ACCORDING TO THE SURGEON GENERAL, WOMEN SHOULD NOT DRINK ALCOHOLIC BEVERAGES DURING PREGNANCY BECAUSE OF THE RISK OF BIRTH DEFECTS. (2) CONSUMPTION OF ALCOHOLIC BEVERAGES IMPAIRS YOUR ABILITY TO DRIVE A CAR OR OPERATE MACHINERY, AND MAY CAUSE HEALTH PROBLEMS.

0 10986 00602 6

Winery Identification: The name of the winery and the city it is located in must be listed on the label.

Government Warning: Required on all beverages that contain more than 0.5 percent alcohol.

UPC Code: Not required by law but present on almost all wines sold at retail.

Country of Origin: Required on all imported wines and used on U.S. wines that may be exported. Wines from outside of the United States must be labeled "Imported by" instead of "Bottled by."

Other Terms Found on Wine Labels

Estate Bottled: For an estate designation, at least 95 percent of the grapes used to make the wine must be grown on the winery's own vineyards. Additionally, when the term "Estate Bottled" is used, both the vineyard and the winery must reside in the same appellation listed on the bottle.

European Names: It is permissible to use European place names such as Burgundy, Chablis, and Champagne as generic terms to describe domestic wines. However they must clearly state the place of origin such as "California Chablis."

Proprietary Names: Wineries are allowed to give their products a proprietary name. This is frequently done instead of varietal naming when the wine is a blend of several types of grapes with no one variety being more than 75 percent.

Reserve: Like the term "Old Vine," this has no legal definition. Although most wineries use "Reserve" for only a small amount of their best wines, some larger wineries use it on all their products to improve the wines' image to consumers.

More Items Found on Labels

Vintage Date: Since an AVA is listed as the appellation, 95 percent of the grapes must be harvested in the year listed as the vintage.

Vineyard designation: If a vineyard is stated on the label, 95 percent of the grapes used to make the wine must be harvested from that vineyard.

Appellation: If an American Viticultural Area (AVA) is listed as the appellation of origin on the wine, a minimum of 85 percent of the grapes used to make the wine must be from that appellation.

Old Vine: Legally "Old Vine" has no definition, but is commonly used to describe wines made from head-trained vines that are more than 50 years old.

Alcohol: If more than 14 percent, the term "Table Wine" cannot be used, and the percent of alcohol by volume must be stated.

LAKE SONOMA WINERY
Lake Sonoma crafts wines of uncompromising quality that are true expressions of individual appellation character. This "best of the best" philosophy is reflected in our Dry Creek Valley Old Vine Zinfandel from Saini Farms. This wine, crafted from vines planted in the 1930's, exhibits rich black cherry and berry fruit flavors over a solid oak framework.
VINTED AND BOTTLED BY LAKE SONOMA CELLARS
GLEN ELLEN, SONOMA COUNTY, CALIFORNIA
CONTAINS SULFITES
PRODUCT OF U.S.A.
GOVERNMENT WARNING: (1) ACCORDING TO THE SURGEON GENERAL, WOMEN SHOULD NOT DRINK ALCOHOLIC BEVERAGES DURING PREGNANCY BECAUSE OF THE RISK OF BIRTH DEFECTS. (2) CONSUMPTION OF ALCOHOLIC BEVERAGES IMPAIRS YOUR ABILITY TO DRIVE A CAR OR OPERATE MACHINERY, AND MAY CAUSE HEALTH PROBLEMS.

Production Statement: "Vinted and Bottled by" or "Blended and Bottled by" are used for wines when less than 75 percent of the wine was fermented at the winery where it was bottled.

Volume: Every wine bottle must state the amount of wine it contains in metric units. In most cases the volume in ml is molded into the glass at the base of the bottle.

of an influence and the climate becomes progressively warmer, making the area more suited to grapes that prefer warmer temperatures, such as Cabernet Sauvignon.

Elevation also affects the temperature. In much of the world, higher elevations are associated with cooler temperatures, but in coastal valleys like Napa the maritime influence on the environment makes the relationship between elevation and temperature more complicated. Since the fog often stays low as it moves up the valley in the evening, vineyards located above the fog layer will not benefit from its cooling effects. It is not unusual for hillside vineyards to have lower temperatures than the valley floor during the day and warmer conditions at night. These hillside vineyards typically experience more rainfall in the winter, and additionally are at lower risk of spring frost because cool night air will drain down from the hills and settle on the valley floor.

In addition to the variation in climate, there is also a complex array of different types of soils found throughout the valley. Like much of California, the Napa Valley is very seismically active, and the numerous fault lines that are found in the appellation contribute to the diversity of its soils. The volcanic origin of the mountains is evidenced by the presence of hot springs and geysers found in the northern part of the valley. On the valley floor, the soils are a mix of sedimentary layers of old sea beds with the material that has washed down from the surrounding hills. Here the soils range from being heavy with clay that retains water to light with lots of sand and gravel that provides good drainage. Alluvial soils that are a mix of silt, sand, and gravel are found on the lands that flank the Napa River and its tributaries. Some of the best soil is considered to be the **bench land** that lies above the floodplain of the Napa River. In the middle of the valley, the "Rutherford Bench" is particularly known for producing fine Cabernet Sauvignon. The surrounding hillsides are often made up of thinner, rocky soils that are composed of sedimentary layers uplifted by geological faults or are derived from volcanic activity. These hillside vineyards are typically low in vigor, and if managed properly produce some of the most intensely flavored wines.

The large size of the Napa Valley AVA combined with its wide diversity of soils and microclimates make it very difficult to generalize about it as a whole. Not long after the region was established as an AVA, smaller AVAs within the Napa Valley's borders started to be established. Although these smaller or sub-appellations do not have the notoriety of the Napa Valley AVA, they typically have more uniform terroir and are more suited to particular grape varieties. Some of Napa County's AVAs are very small, with few producers using the appellation that limits their relevance in the marketplace. As of 2004, Napa County was home to 14 different AVAs as well as being part of the North Coast appellation. The valley's numerous terroirs allow for the production of many different varieties of grapes, but economics and market demand make Cabernet Sauvignon and Chardonnay the most widely planted varieties (Table 10–1).

THE LOS CARNEROS AVA

The southernmost area of Napa County's viticultural regions is the Los Carneros appellation, commonly referred to as Carneros. In Spanish, the name means "the rams." It is made up of the flatlands and low hills at the base of the Sonoma and Napa Valleys. Being that it extends into both counties, wines made from the region can be labeled either Napa or Sonoma Carneros, or if the grapes are grown on both sides of the county line, it can be labeled as simply Carneros without a county designation. Being the closest of the smaller appellations of Napa County to the maritime influences of San Pablo Bay, the region is known for its cool and breezy weather. This climate makes it ideal for cool climate grape varieties such as Pinot Noir and Chardonnay, as well as grapes grown for sparkling wine production.

Table 10-1 Appellations within the Napa Valley AVA

APPELLATION	BEST KNOWN VARIETIES
Atlas Peak	Sangiovese
Chiles Valley	Sauvignon Blanc, Zinfandel
Diamond Mountain	Cabernet Sauvignon
Howell Mountain	Cabernet Sauvignon
Los Carneros (shares with Sonoma County)	Pinot Noir, Chardonnay, Merlot
Mt. Veeder	Chardonnay, Cabernet Sauvignon
Oak Knoll District	Chardonnay, Merlot
Oakville	Cabernet Sauvignon
Rutherford	Cabernet Sauvignon, Merlot
St. Helena	Cabernet Sauvignon, Zinfandel
Spring Mountain District	Cabernet Sauvignon
Stags Leap District	Cabernet Sauvignon
Yountville	Chardonnay, Merlot, Sauvignon Blanc
Wild Horse Valley (shares with Solano County)	Not known for any one variety

The Oak Knoll, Yountville, and Stags Leap AVAs

Moving up the valley from the Carneros region, one first encounters the Oak Knoll appellation just outside the city of Napa and then the Yountville AVA that is situated around the small town of the same name. These two appellations begin in the hills of the Mayacamas range on the west and extend across the valley floor to the base of the of the Vaca mountains on the east. Being relatively close to the cooling influences of the bay, these appellations encompass a moderate zone between the cooler climate to the south and the warmer appellations further up the valley. As in Carneros, Chardonnay and Pinot Noir are popular as well as Merlot and Sauvignon Blanc.

The Stags Leap district lies to the east of the Yountville AVA, straddling the Silverado Trail, one of the valley's major thoroughfares. It takes its name from the distinctive volcanic rock outcroppings of the Vaca range that overlook the appellation. Being on the east side of the valley, Stags Leap receives more afternoon sun then the west side of the valley and is slightly warmer. Cabernet Sauvignon is the variety for which the AVA is best known. The district's Cabernet was first made famous when the 1973 Cabernet Sauvignon made by Stags' Leap Wine Cellars won the influential 1976 Paris Tasting over notable Bordeaux producers. The district was established in 1989 after several years of legal arguments by local wineries on where exactly the borders should be drawn.

The Oakville, Rutherford, and St. Helena AVAs

Moving up the valley, the Oakville, Rutherford, and St. Helena AVAs lie in succession, each crossing from the Mayacamas Mountains on the west to the foothills of the Vaca Mountain range on the east (Figure 10-7). These three appellations lie in the heart of the valley and were all producing grapes when the Napa Valley first gained its notoriety for winemaking in the second half of the 1800s. The three AVAs are home to many wineries and have some of the Napa Valley's best-known and historic operations (Figure 10-8). These regions cover a number of soil types and

FIGURE 10–7 The Oakville region of Napa Valley appellation with the Vaca Mountains in the background.

microclimates that are suitable to many varieties of grapes, including Sauvignon Blanc, Zinfandel, and red Bordeaux varieties such as Cabernet Sauvignon, Cabernet Franc, and Merlot. This being said, the section of the Rutherford and Oakville appellations that runs from the western side of the Napa River to the base of the Mayacamas mountains is particularly well known for growing high quality Cabernet Sauvignon.

The Mount Veeder, Spring Mountain, Diamond Mountain, and Howell Mountain AVAs

The first three AVAs lie above the valley floor in the Mayacamas Mountain range on the west side of the Napa Valley, and Howell Mountain lies opposite them on the east side of the valley in the Vaca Mountain range. The Mt. Veeder appellation is the furthest south and extends from the foothills to the west of the city of Napa and north of the Carneros region for nearly 13 miles (21 kilometers) to the northwest through the mountains. It is bordered on the east by the Oak Knoll and Yountville appellations and by the ridgeline that marks the Sonoma County line on the west. Named for the pronounced 2,677-foot (816-meter) peak on its western edge, it has many different soil types, which combined with the variation in altitude and rugged terrain create a number of terroirs. This variation means the appellation is suitable for many varieties of grapes including Chardonnay, Zinfandel, and Bordeaux reds.

The Spring Mountain AVA begins several miles north of the Mt. Veeder appellation and directly west of the St. Helena AVA. It is warmer than the Mt. Veeder appellation but also has a varied mix of sedimentary and volcanic soils. There is no particular peak named "Spring Mountain"; instead the region takes its name from the numerous fresh water springs that dot the hillsides. The district was home to some of the first hillside grapes grown in the Napa Valley in the 1870s and is well known for its red wines.

Moving up the Mayacamas range the next appellation is the Diamond Mountain AVA. It is contiguous to the northern border of Spring Mountain and directly west above the town of Calistoga. It is slightly warmer than the Spring Mountain region and is more likely to receive

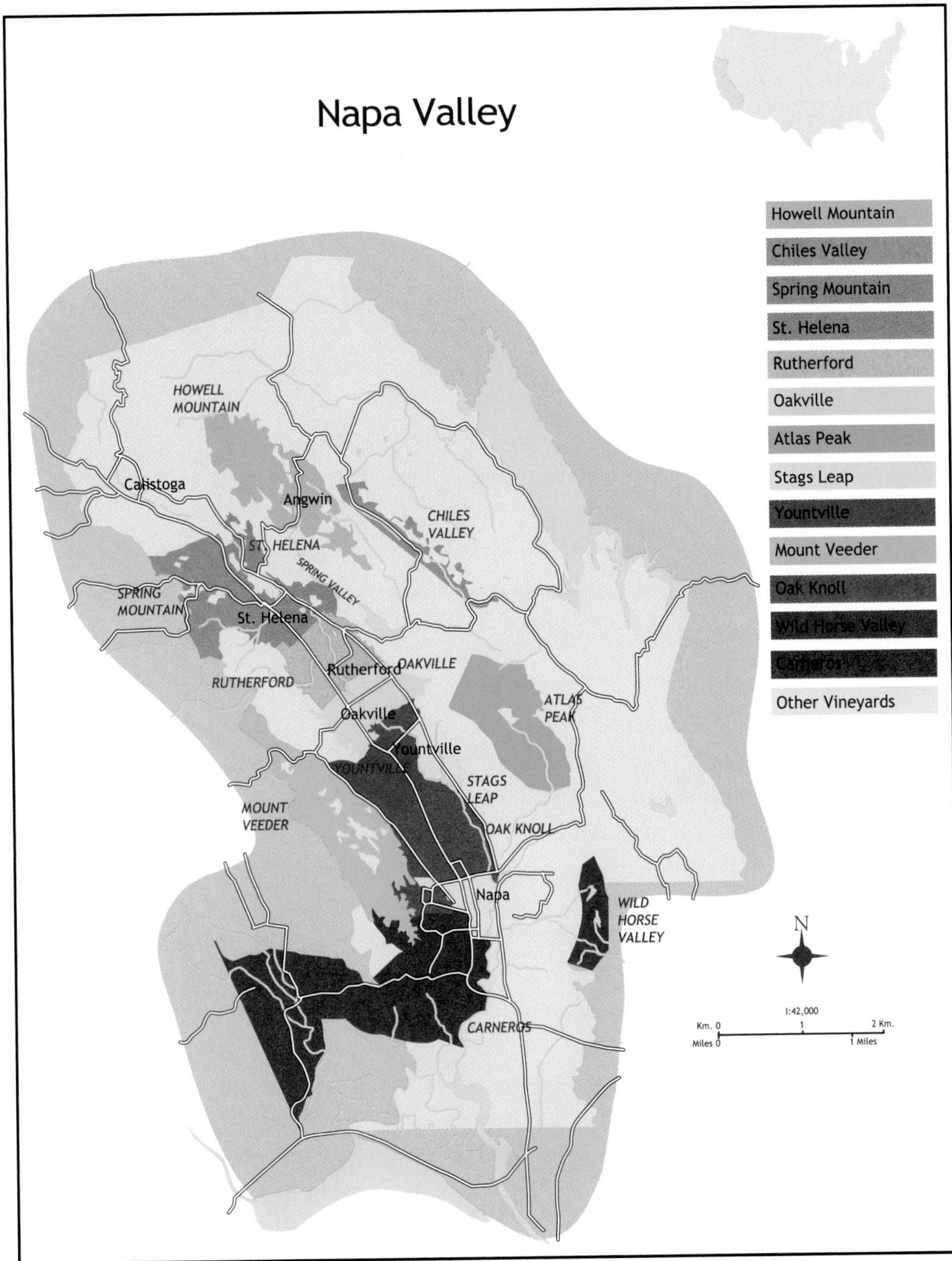

The appellations of Napa Valley.

cooling ocean breezes from the north through Sonoma's Knights Valley than from San Francisco Bay to the south. Its soils are more uniform than many of the mountain AVAs and are primarily of volcanic origin. It is best known for its Cabernet Sauvignon.

On the other side of the valley above the St. Helena AVA is the Howell Mountain appellation. It is a high region that begins at 1,400 feet (425 meters) and is well above the cool evening breezes and fog that comes into the valley in the summertime. During the day the elevation keeps the vineyards cooler than its counterparts on the valley floor, giving the region an overall moderate growing climate. It is one of the older sub-appellations established in the Napa Valley, and the district does well with Zinfandel and Cabernet Sauvignon.

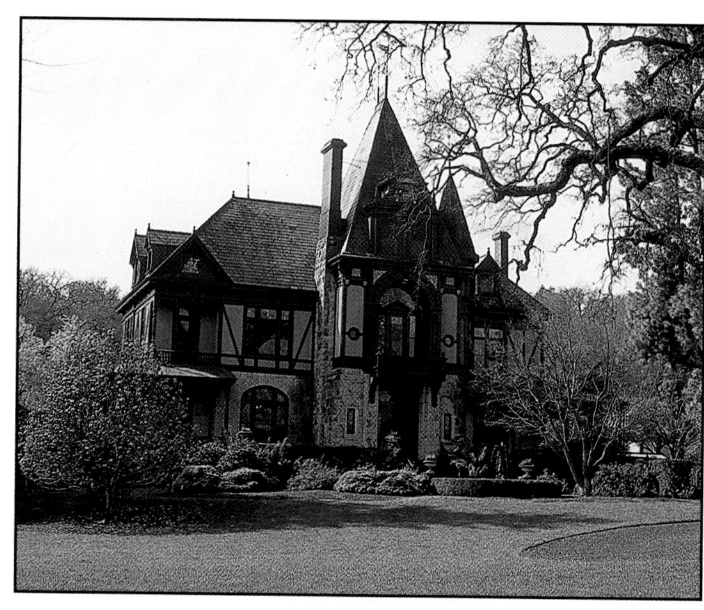

FIGURE 10–8 The Rhine House at Beringer Winery in St. Helena; built in 1883, it was modeled after the founder's home in Germany.

OTHER NAPA APPELLATIONS

These final three Napa County appellations are more limited in production than the other Napa appellations and are not as widely used by wineries. The Atlas Peak appellation is located in the Vaca range high above and to the east of the Stags Leap district. The appellation is slightly dryer than the other hillside regions of the Napa Valley, and most of the vineyards are owned by Atlas Peak Winery and planted to Sangiovese and Cabernet Sauvignon. The Chiles Valley AVA is in a narrow valley located in the Vaca range to the east of the Napa Valley. Being farther inland, it is slightly warmer than the Napa Valley, and Zinfandel is its best known grape. The Wild Horse Valley is located in the hills due east from the town of Napa and has limited vineyard acreage.

Sonoma County

Sonoma County lies directly to the west of Napa County. Its eastern border is the ridge of the Mayacamas range, and it extends westward to the Pacific Ocean. On the south it is bordered by San Pablo Bay and Marin County, and to the north it is bordered by Lake and Mendocino counties, both wine producing regions in their own right. Sonoma leads all other coastal counties in production of premium wine grapes with more than 59,000 acres producing over 180,000 tons of fruit, greater than 40 percent more wine grapes than Napa County. Its history goes back to the time of the missions, and its early successes helped to establish the wine industry in Northern California. Sonoma played its part in the rebirth of California's premium wine industry in the 1960s and 1970s when a number of small, quality-oriented wineries were founded. In particular, one pioneer was Hanzell Vineyards, located in the hills above the town of Sonoma. In the 1950s it was the first winery in the state to use stainless steel tanks and French oak barrels, both staples of modern premium wine production. Sonoma County wines are among the most famous in the state and rival those of the Napa Valley in quality if not quite in price.

The first vineyards were planted in the county in the 1820s at the site of the Sonoma Mission. Sonoma County's fertile soils and temperate climate made it a natural setting for many diverse types of agriculture, and the region soon gained a reputation for both the variety and quality of agricultural products it produced. Hops, apples, prunes, dairy, poultry, and vegetables all played a significant role along with grapes in the economy of the region. In recent years,

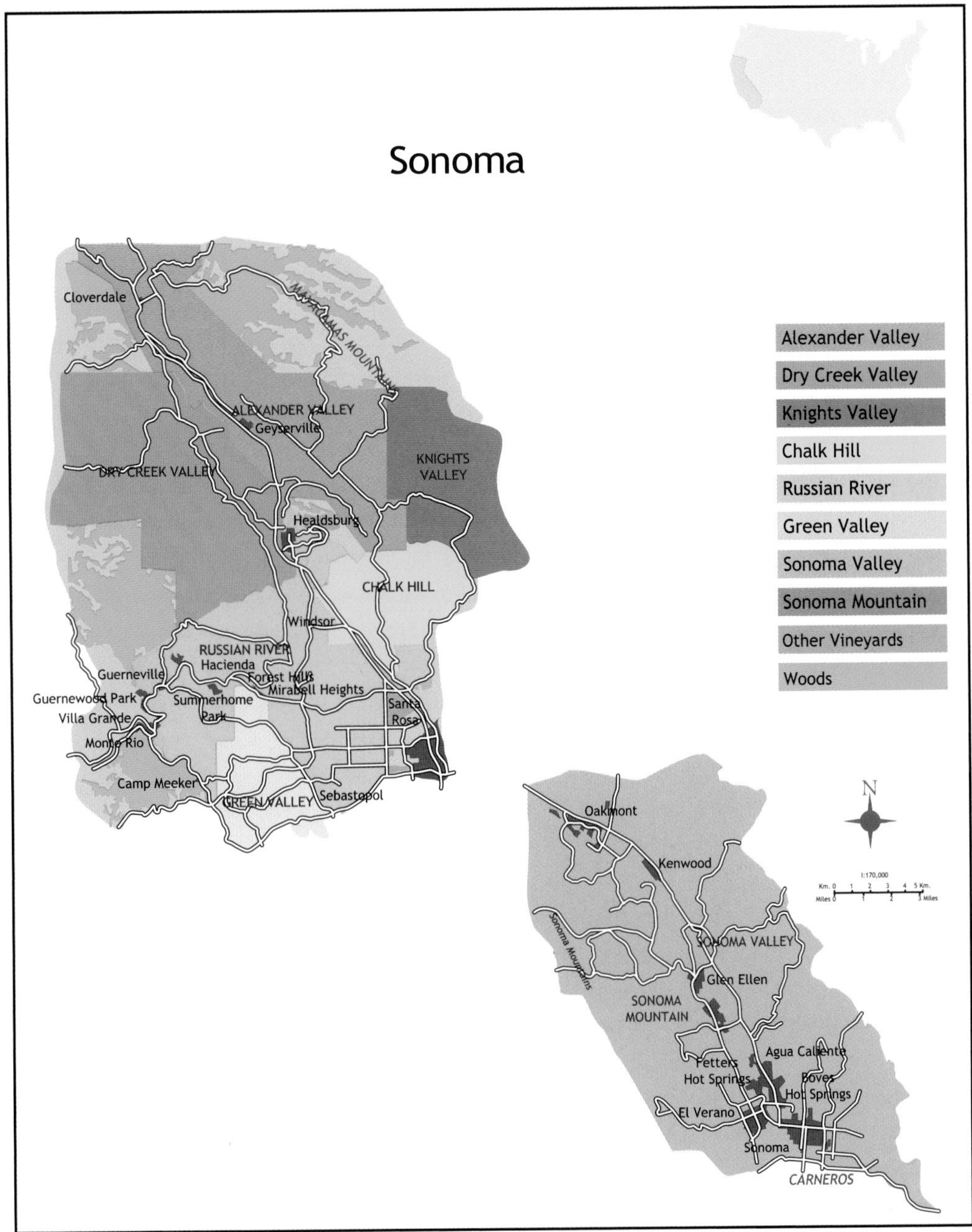

The appellations of Sonoma County.

California's wine boom has made grapes such a valuable commodity that vineyards have replaced most other types of crops in the county.

Invariably the similarities and proximity of Napa and Sonoma invite comparisons. Both appellations have comparable histories and played parallel roles in the development of California's premium wine industry. Additionally they are both large regions with a number of diverse terroirs that are well suited to a wide variety of grapes and wine. However, there are differences one notices when traveling between the two. Sonoma wine country, being generally in less of a limelight than Napa, is often less crowded and more rustic in character. Whereas the majority of Napa's vineyards and wineries are located in a chain as you move up the valley, Sonoma's are more spread out into grape-growing regions that are separated by either urban or undeveloped areas. Sonoma County has over three times the population as Napa County and has a much more diverse economy than Napa's, which is primarily based on the tourist and wine industries.

REGIONS OF SONOMA COUNTY

The most significant factor influencing the climate of Sonoma County is its proximity to the Pacific Ocean. In Sonoma County, like all of California's coastal wine regions, the ocean has a moderating effect, keeping the winter warmer and the summer cooler. Particularly during the warm season, the weather is dominated by the interaction between the cool maritime air at the coast and the warmer air inland. It receives coastal influences from both the San Pablo Bay to the south and the Pacific Ocean to the west. The cooling effects of the ocean and the bay are channeled through valleys and broken up by the county's numerous mountain ridges. This varied topography makes for a number of different grape-growing regions, each with its unique microclimate. In the more inland areas, the weather is warm and appropriate for grape varieties such as Cabernet Sauvignon. Near the coast and in the river valleys the climate is very cool and vineyards are better suited for Pinot Noir and Chardonnay grapes.

As in Napa County, the seismic activity that created the mountains has had a significant effect on the soil. Over millions of years sedimentary, alluvial, and volcanic soils were deposited and then often moved by the region's geological faults to new locations. The diversity is such that it is very common to have a single small vineyard containing several different types of soil. This variation can be challenging to grape growers, but with the proper rootstock and vineyard management, these difficulties can be overcome. In an effort to express its wide diversity of terroir, Sonoma County has been divided into a patchwork of 13 AVAs, including the western section of the Los Carneros district that was described under the regions of Napa Valley. Many of these appellations have gained superior reputations for producing particular varieties of grapes (Table 10–2).

THE SONOMA VALLEY AND SONOMA MOUNTAIN AVAS

Established in 1982, Sonoma Valley is the county's oldest AVA. It reaches from the shores of San Pablo Bay on the south, 23 miles (37 kilometers) to the northwest, and the small city of Santa Rosa. The appellation is framed by the Mayacamas range and Napa County to the east and to the west by Sonoma Mountain; it includes the entire Sonoma section of the Los Carneros AVA and the historic small town of Sonoma. Although there are many vineyards on the valley floor, the appellation also includes the mountainous terrain on either side of the valley. Fog and cool ocean breezes can come into the valley from the south, where there is a small gap at the southern end of

Table 10-2 Appellations of Sonoma County

APPELLATION	BEST KNOWN VARIETIES
Alexander Valley	Cabernet Sauvignon, Zinfandel
Bennett Valley	Merlot, Sauvignon Blanc
Chalk Hill	Chardonnay
Dry Creek Valley	Zinfandel, Merlot, Sauvignon Blanc
Knights Valley	Cabernet Sauvignon
Los Carneros (shares with Napa County)	Pinot Noir, Chardonnay
Northern Sonoma	Produces a number of varieties
Rockpile	Zinfandel
Russian River Valley	Pinot Noir, Chardonnay
Sonoma Coast (shares with Marin County)	Chardonnay
Sonoma County Green Valley	Pinot Noir, Chardonnay
Sonoma Mountain	Cabernet Sauvignon
Sonoma Valley	Merlot, Zinfandel, Cabernet Sauvignon

Sonoma Mountain, or from the north through Santa Rosa. This makes the midsection of the valley around the small town of Glen Ellen slightly warmer than its upper and lower sections. The Sonoma Valley is perhaps best known for its Zinfandel, but its diverse conditions allow for the production of a number of grape varieties. On the southern end, the cooler climate is ideal for Pinot Noir and Chardonnay, and in the midsections of the valley, Sauvignon Blanc and Merlot (Figure 10-9). Cabernet Sauvignon is well suited to hillside vineyards that are above the fog layer.

Within the borders of the Sonoma Valley AVA are the smaller regions of the Sonoma Mountain and Bennett Valley AVAs. The Sonoma Mountain AVA is above the valley floor on the slopes of Sonoma Mountain from 400 to 1,200 feet (120 to 360 meters). The soils are primarily of volcanic nature and are very well drained. Many of the vineyards have an eastern exposure that allows them to receive the full morning sun, which warms the vines quickly, and the afternoon sun on an oblique angle to moderate the temperature during the warmest part of the day. The appellation is home to the Jack London Vineyard, known for its Cabernet Sauvignon; it was the home of the famous author during his lifetime (Figure 10–10). The Bennett Valley region lays on the western edge of the Sonoma Valley between Sonoma Mountain to the south and Bennett peak to the north. The ocean breezes often pass through the valley on summer afternoons keeping the temperature moderate and making it well adapted for Sauvignon Blanc and Merlot.

The Russian River, Chalk Hill, and Sonoma Green Valley AVAs

Named for the Russian fur trappers who settled on Sonoma County's coast in the early 1800s, the Russian River enters the county from Mendocino County to the north and continues south until it is just above Santa Rosa. Here the river turns to the west, traveling through a gap in the coastal range of mountains to the Pacific Ocean. The Russian River AVA covers much of the lower drainage of the Russian River and begins 6 miles (9.7 kilometers) inland near the town of Guerneville, continuing up the river to the town of Healdsburg. The gap that is the

FIGURE 10–9 A Sauvignon Blanc vineyard in Sonoma Valley.

FIGURE 10–10 The Jack London Vineyard on Sonoma Mountain; the ranch was owned by the author during his lifetime and is best known for its Cabernet Sauvignon. Before the author's time, the ranch was also home to a large stone winery built by Kohler and Frohling, one of California's largest nineteenth century producers.

pathway for the river also is a corridor for cool ocean air that travels into the interior of the county from the coast. Particularly during the summer months, vineyards located in the Russian River Valley are covered with a cooling layer of fog brought inland by sea breezes in the late afternoon. The fog persists throughout the night and burns off from the heat of the day the following morning.

The closer to the coast that a vineyard is located, the earlier in the afternoon it receives fog and the later it will last in the morning. These conditions make the western portion of the Russian River AVA one of the coolest in the North Coast counties and excellent for producing grapes such as Pinot Noir, Chardonnay, and Pinot Gris. Inland the climate becomes slightly warmer, and varieties such as Sauvignon Blanc and Zinfandel do well. Within the Russian River AVA are two sub-appellations, Chalk Hill to the east and Sonoma Green Valley to the west. The Chalk Hill AVA is bordered by Knights Valley to the east and Alexander Valley to the north. The soils are primarily of volcanic origin and Chardonnay is the most widely planted variety. On the western side of the Russian River AVA is the Sonoma Green Valley AVA, so named to avoid confusion with the Green Valley AVA in California's Solano County. It is a very cool region and has a number of exceptional Pinot Noir vineyards.

THE DRY CREEK VALLEY AND ROCKPILE AVAS

The Dry Creek Valley AVA lies between the Russian River AVA to the south and the Alexander Valley AVA to the northeast. The Dry Creek Valley begins in the town of Healdsburg and follows the path of Dry Creek, a tributary of the Russian River, 15 miles (24 kilometers) to the northwest to Lake Sonoma at the head of the valley. The valley is somewhat narrow, being about 2 miles (3.2 kilometers) across at its widest point. The Dry Creek Valley AVA includes much of the mountainous areas that border the valley itself. In the Dry Creek region cool, winds from the coast are diverted by mountains, and the area is somewhat warmer than the Russian River AVA. Grapes were first planted in the 1860s and the potential for viticulture was quickly recognized. However the region suffered after Prohibition and prunes became the dominant crop until the 1970s. In the lowlands along the banks of Dry Creek the soils are more alluvial in nature, and in the bench land above the flood plain the soils are more volcanic in origin. The appellation is best known for its Zinfandel but Cabernet Sauvignon, Chardonnay, and Sauvignon Blanc are also widely planted.

The Rockpile AVA begins above Lake Sonoma in the western section of the Dry Creek AVA and extends to the Mendocino County border. Established in 2002 it is the county's newest appellation and is about one-fifth the size of the Dry Creek Appellation. The area is very rugged, and the grapes grow predominantly in hillside vineyards. Even though the AVA is very young, it has already gained a reputation for producing quality Zinfandel.

The Alexander Valley and Knights Valley AVAs

The Alexander Valley AVA is named for Cyrus Alexander, an early settler who received a land grant from General Vallejo. It is a wide valley created by the upper section of the Russian River as it passes through northern Sonoma County. It extends from just north of the town of Healdsburg 20 miles (32 kilometers) north to the Mendocino County line. It is bordered by the Dry Creek Valley to the west and the prominent mountain Geyser Peak to the east, whose slopes are home to a large geothermal power plant. Being more inland, it is slightly warmer than the Dry Creek Valley and typically has deep, fertile, sandy loam soils. These conditions can make the grapevines very vigorous and growers must manage their vineyards to control excess growth. The appellation is perhaps best known for its Cabernet Sauvignon, but there are also extensive plantings of Sauvignon Blanc and Chardonnay. In the Geyserville area located in the middle of the valley, Zinfandel does particularly well.

The Knights Valley AVA is positioned between Alexander Valley to the west and the upper Napa Valley to the east. Being further inland it is one of the warmest of Sonoma County's AVAs and is well known for its Cabernet Sauvignon as well as Sauvignon Blanc. The appellation's location on the eastern edge of Sonoma County is removed from urban areas, making it one of the least developed grape-growing regions in the county. Although it is located in Sonoma County, some think of the appellation as an extension of the Napa Valley. One reason for this could be that the largest grower in the Knights Valley is Beringer Vineyards of St. Helena.

Other Sonoma County Appellations

The two largest appellations in Sonoma County are the Northern Sonoma and Sonoma Coast AVAs, at 349,837 acres (141,574 hectares) and 516,409 acres (208,983 hectares), respectively. While they are the largest AVAs they are seldom used by wine producers and were primarily created by individual wineries that wanted to estate bottle grapes made from their widely scattered vineyards. The Northern Sonoma AVA covers the northern half of Sonoma County and includes all the territory of the Alexander Valley, Dry Creek Valley, Russian River Valley, Chalk Hill, Rockpile, and Knights Valley appellations. This large and diverse area covers a multitude of climates and soil types that are suitable to a wide selection of grape varieties. The Sonoma Coast appellation is even larger but has a more uniform cool climate. It extends along the western half of the county from San Pablo Bay on the south to Mendocino County in the north.

Lake and Mendocino Counties

The counties of Lake and Mendocino are part of the North Coast AVA and lie directly to the north of Napa and Sonoma Counties. As in Napa and Sonoma, the majority of the vineyards are located in the valleys between the mountains of the coast range. With the exception of Anderson Valley, the vineyards of Mendocino and Lake Counties are further inland, and the mountains that separate them from the ocean are more rugged and higher in elevation. These factors result in less of a coastal influence than in Napa and Sonoma and, consequently, a warmer climate. This warmer climate, along with lower land costs, makes the grapes less expensive than those of Napa and Sonoma. In addition to the wineries in the area producing wines under a Lake or Mendocino appellation, wineries from other nearby regions look to the area as an economical source of quality grapes for their own brands.

Like Napa and Sonoma, there is a long history of agriculture with the production of hops, tree fruits, and nuts all being an important part of the local economy. Similar to what happened in Napa and Sonoma, these crops are giving way to grapes because of their greater market value. The counties maintain much of their rural character and to this day are less developed, with a smaller population, than Napa and Sonoma. These qualities make Lake and Mendocino wine counties popular with tourists looking to avoid the crowds and high prices of their more famous neighbors to the south. Since the grape growing in Lake and Mendocino counties is dispersed, the AVAs of the region are spread out as well. Many of the 13 AVAs located within these counties do not overlap or have contiguous borders.

THE REGIONS OF LAKE COUNTY

Lake County is named for Clear Lake, a large natural body of water situated in the middle of the county. Originally part of Napa County, it was split off in the 1860s. Located inland and at a higher overall elevation than other North Coast counties there is little marine influence on the climate. These conditions make the days very warm and the nights cool. Much of the soil is of volcanic origin, produced from eruptions of the now dormant volcano Mt. Konocti that rises on the southern shores of the lake. The county has a long history of viticulture but was better known for its pears until the 1970s when large vineyards of Sauvignon Blanc and Chardonnay were planted in the rich valley soil that borders Clear Lake. In recent years vineyard development has accelerated due to growers moving in from Napa and Sonoma, where land costs have become prohibitively expensive. Many of the new vineyards are located in the hills that border the lake.

The Clear Lake AVA covers the land that surrounds the lake and includes most of the county's vineyards. The vineyards on the valley floor are best known for their Sauvignon Blanc with the hillside vineyards having more of a reputation for red varieties. There are two smaller sub-appellations located in the hills above the lake, the High Valley AVA to the north and the Red Hills AVA to the south. Here the warm climate—and well-drained soils in particular—produce Cabernet Sauvignon of good quality and value (Figure 10–11). Lake County also has two small appellations—the Guenoc and Benmore Valleys. The Guenoc Valley is located near the Napa County line and is only used by one winery. The Benmore Valley AVA lies in the hills on the western edge of Lake County near its border with Mendocino. The high elevation makes the region very cool, and Chardonnay is the predominate grape grown.

THE REGIONS OF MENDOCINO COUNTY

Mendocino County is situated between Lake County to the east and the Pacific Ocean to the west, and a large portion of the grape growing areas of the county was registered as the Mendocino AVA in 1984 (Table 10–3). The northern half of Mendocino County is heavily forested and better known for redwood trees than grapes, with the viticulture taking place in the southern portion of the county. Organic viticulture is popular in the county, with one-quarter of the vineyards certified organic. Of the seven AVAs located within Mendocino County, the Anderson Valley AVA is the closest to the coast and the coolest. Starting just 10 miles (16 kilometers) from the coast, it is a long narrow valley that travels for 20 miles (32 kilometers) inland along the route of the Navarro River and Highway 128 to the town of Booneville. The appellation grows slightly warmer as one moves inland or higher in elevation. The appellation has a number of small winer-

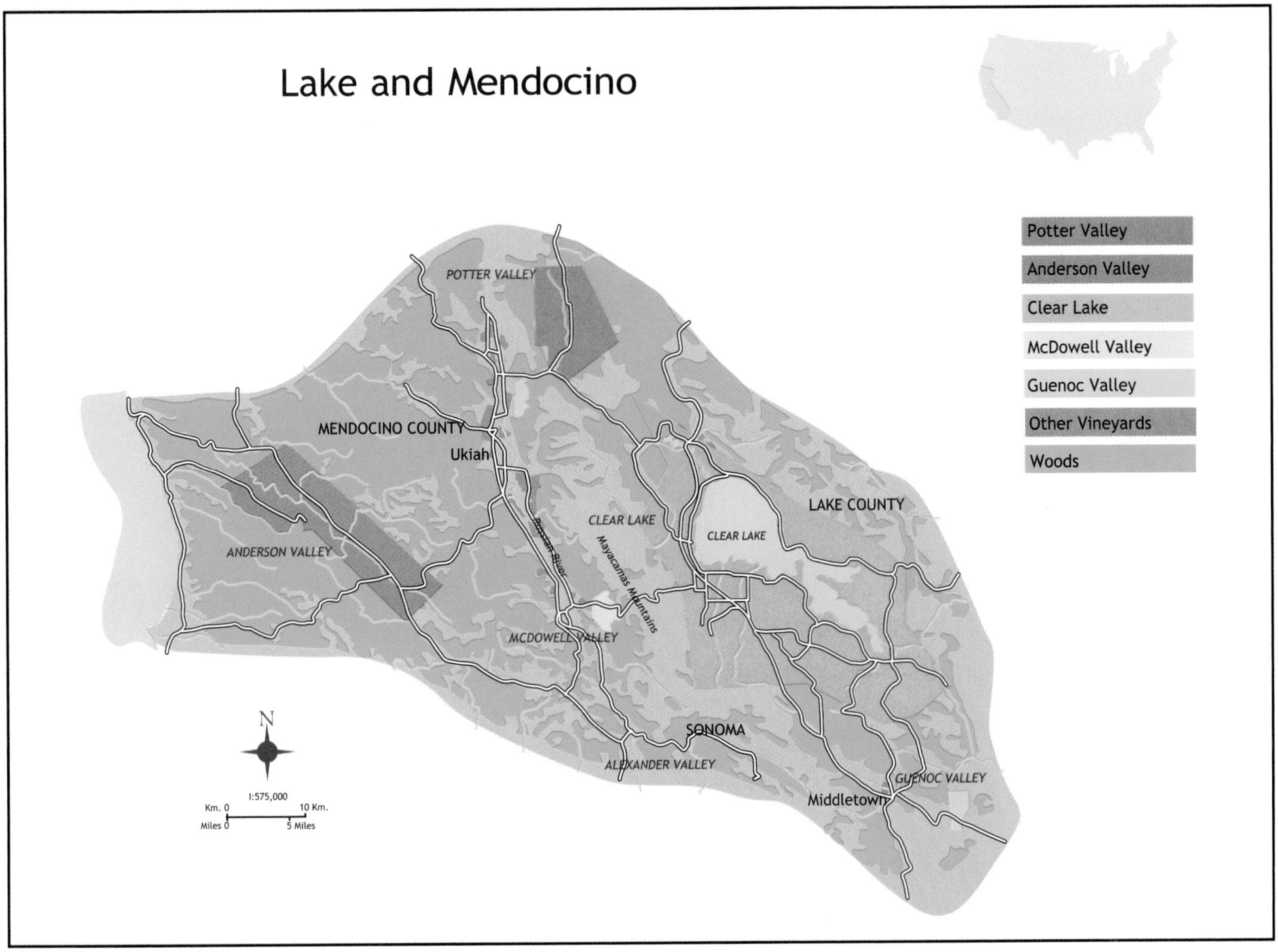

Lake and Mendocino

Potter Valley
Anderson Valley
Clear Lake
McDowell Valley
Guenoc Valley
Other Vineyards
Woods

POTTER VALLEY
MENDOCINO COUNTY
Ukiah
ANDERSON VALLEY
CLEAR LAKE
Russian River
Mayacamas Mountains
CLEAR LAKE
LAKE COUNTY
MCDOWELL VALLEY
SONOMA
ALEXANDER VALLEY
GUENOC VALLEY
Middletown
N
1:575,000
Km. 0 10 Km.
Miles 0 5 Miles

The appellations of Lake and Mendocino Counties

FIGURE 10–11 A Cabernet Sauvignon vineyard in the High Valley appellation of Lake County.

Table 10-3 Appellations of Lake and Mendocino Counties

Lake County

APPELLATION	BEST KNOWN VARIETIES
Benmore Valley	Chardonnay
Clear Lake	Sauvignon Blanc, Chardonnay
Guenoc Valley	Not known for any one variety
High Valley	Cabernet Sauvignon, Zinfandel
Red Hills	Cabernet Sauvignon, Merlot

Mendocino County

APPELLATION	BEST KNOWN VARIETIES
Anderson Valley	Chardonnay, Pinot Noir, Riesling, Gewürztraminer
Cole Ranch	Not known for any one variety
McDowell Valley	Syrah, Zinfandel
Mendocino	Produces a number of varieties
Mendocino Ridge	Zinfandel
Potter Valley	Sauvignon Blanc, Chardonnay
Redwood Valley	Red varietals
Yorkville Highlands	Cabernet Sauvignon, Sauvignon Blanc

ies that specialize in cool climate varieties such as Pinot Noir, Chardonnay, Riesling, and Gewürztraminer. The Anderson Valley is also known for its sparkling wine production.

On the eastern edge of the Anderson Valley begins the Yorkville Highlands AVA. It continues along the route of Highway 128 until it reaches Alexander Valley AVA at the Sonoma County border. Because it is warmer and higher than the Anderson Valley, Sauvignon Blanc and Cabernet Sauvignon do well here. In the southwestern corner of Mendocino County is the Mendocino Ridge AVA, which covers the large area between the Anderson Valley and the Sonoma County line. What makes this appellation unique is that it is noncontiguous, only including the land above 1,200 feet (366 meters) in elevation. By limiting the definition to the land above the fog line, the vineyards in the large AVA have a more uniform terroir.

The majority of Mendocino County's grapes are grown in the valleys that flank the Russian River as it travels southward through Mendocino County. North of the county seat of Ukiah are two AVAs, both with extensive planting of grapes. The Redwood Valley AVA is slightly larger and situated to the west, and the Potter Valley on the east tracks the east fork of the Russian River and is at a slightly higher elevation. Being far inland from the coast, they are warm regions that do not benefit from the cooling effects of the ocean. Traveling south along the river from the Redwood Valley there are two widely planted grape-growing regions that have not yet been granted AVA status. First is the Ukiah Valley that encircles the county seat. Here the valley broadens with the river plain and the weather becomes cooler. Below the Ukiah Valley is the Sanel Valley, which surrounds the town of Hopland. Mendocino County also has two small AVAs, the Cole Ranch and McDowell Valley AVAs, both created for use by individual wineries.

The Central Coast

This huge viticultural area extends from the east side of San Francisco Bay, all the way to Santa Barbara in Southern California. It includes the grape-growing regions of Alameda, Contra Costa, Monterey, San Benito, San Francisco, San Luis Obispo, San Mateo, Santa Barbara, Santa Clara, and Santa Cruz Counties. At nearly 5½ million acres (2.22 million hectares) the Central Coast is by far the largest AVA in California. Its size makes for such a great degree of variation in terroirs that the designation means little except to distinguish the cooler Central Coast vineyards from those located in the warmer regions of California's interior. Its climate ranges from being very warm in the inland valleys such as Paso Robles to very cool near the coast like the Edna Valley. Overall, it has slightly milder winters in the southern portion of the appellation than the north. Much like the large North Coast AVA, it is divided into a number of sub-appellations to help differentiate its varied growing conditions (Table 10-4). Wines from the Central Coast are frequently bottled using the county of origin as designation instead of using an AVA classification. The winemaking history of the district dates back to the time of the missions that stretched in a chain along the California coast, and the region has seen much new development in recent decades.

The San Francisco Bay Area and Santa Cruz County

Although the San Francisco Bay AVA is technically part of the Central Coast AVA, few people associate it with the more prolific areas of the appellation to the south. While the southern and central portions of the Central Coast appellation had little vineyard and winery development between the time of the missions and the 1960s, the San Francisco Bay area has been home to vineyards and wineries since the early 1800s. In the time since World War II, the majority of the vineyards in the Bay Area have been replaced by subdivisions. The Santa Clara Valley in particular had a reputation for being one of the finest grape-growing areas in the state in the 1880s. Today, despite the fact the Santa Clara Valley has its own AVA, the region has little agriculture and is better known as Silicon Valley. There are still some vineyards located in the southern end of the appellation near the town of Gilroy, but the historic grape-growing areas in the north of the appellation have been displaced by subdivisions and the high-tech industry. The Livermore Valley AVA located east of the bay also has a long history of winemaking and still has many vineyards, but this area is also threatened by urban development. Contra Costa County to the north of the Livermore Valley is another grape-growing region that has been decimated by housing and development. Prior to Prohibition there were more than 25 wineries and 6,000 acres (2,400 hectares) of grapes (Brook, 1999); now only a few remain.

The Santa Cruz Mountains AVA is located in the hills of the coastal range in Santa Cruz, San Mateo, and Santa Clara Counties. The appellation was one of the first in California, being formed in 1982, and is defined by the land that is above 400 feet (122 meters) in elevation. When the Central Coast AVA was approved in 1985, the Santa Cruz Mountains AVA was excluded so the appellations were contiguous but do not overlap. The terrain is rugged and the soil can be very stony, resulting in low vigor and yields. Although it is not nearly as developed as the appellations north of San Francisco Bay, it too has a long history that was renewed during the wine revolution in the 1960s and 1970s. Many wine enthusiasts with an independent viewpoint who were looking to start their ventures away from the more established regions of Napa and Sonoma were drawn to the isolation of the Santa Cruz Mountains. The vineyards and wineries are generally small and dispersed throughout the area, with many vintners acquiring grapes from outside the appellation to supplement their production.

Monterey and San Benito Counties

While the San Francisco Bay area to the north had some degree of viticulture and winemaking from the time the region was settled to the present day, Monterey and San Benito Counties had little viticultural activity from the mission period until the 1960s when a few vineyards were planted. Although there were few vineyards in the region, there was a great deal of agriculture.

The appellations of the North Central Coast region.

In Monterey County, the Salinas Valley in particular has ideal soils and climate for vegetable crops, but most growers considered it too cool and windy for wine grapes. In the 1970s wineries from outside the county, looking to expand their production, established large vineyard operations, and production rapidly grew (Figure 10-12). During the rapid development, varieties such as Cabernet Sauvignon were planted that were ill suited to Monterey's cool weather, and the results were far from perfect. The cool conditions gave the Cabernet Sauvignon that was produced a distinctly vegetative or "bell pepper" aroma that Monterey became known for. The unpopular flavors of the wines from Monterey slowed development, and vineyard acreage decreased in the early 1980s. In the 1990s high costs of vineyard land elsewhere in the state led vintners and growers to reconsider Monterey County, this time with a better understanding of what the marketplace desired and the appellation's terroir. Cool weather varieties like Chardonnay and Pinot Noir were planted in the northern part of the county, which was closer to the ocean and much

Table 10-4 Appellations of the Central Coast

The Central Coast AVA includes Alameda, Contra Costa, Monterey, San Benito, San Francisco, San Luis Obispo, San Mateo, Santa Barbara, Santa Clara, and Santa Cruz Counties

The San Francisco Bay Area AVA includes Alameda, Contra Costa, San Francisco, San Mateo, and Santa Clara Counties

SUB-APPELLATION BY COUNTY	BEST KNOWN VARIETIES
Alameda County	
Livermore Valley	Chardonnay, Cabernet Sauvignon, Mourvèdre
Monterey County	
Arroyo Seco	Chardonnay
Carmel Valley	Cabernet Sauvignon
Chalone	Pinot Noir, Chardonnay
Hames Valley	Cabernet Sauvignon, Merlot
Monterey	Chardonnay is best known of many varieties
San Bernabe	Produces a number of varieties
San Lucas	Chardonnay, Cabernet Sauvignon, Merlot
Santa Lucia Highlands	Pinot Noir, Chardonnay
San Benito County	
Cienega Valley	Produces a number of varieties
Lime Kiln Valley	Produces a number of varieties
Mt. Harlan	Pinot Noir
Paicines	Produces a number of varieties
San Benito	Produces a number of varieties
San Luis Obispo County	
Arroyo Grande Valley	Chardonnay
Edna Valley	Chardonnay, Pinot Noir
Paso Robles	Syrah, Zinfandel
Wild Horse Valley	Pinot Noir
York Mountain	Zinfandel
Santa Barbara County	
Santa Maria Valley	Chardonnay, Pinot Noir
Santa Rita Hills	Pinot Noir
Santa Ynez Valley	Chardonnay, Sauvignon Blanc, Syrah
Santa Clara County	
Pacheco Pass	Not known for any one variety
Santa Clara Valley	Chardonnay, Cabernet Sauvignon
San Ysidro District	Chardonnay
Santa Cruz County	
Ben Lomond Mountain	Not known for any one variety
Santa Cruz Mountains	Chardonnay, Pinot Noir, Cabernet Sauvignon

cooler, and warm weather varieties such as Merlot and Zinfandel were planted inland where it was warmer.

Today there are more than 40,000 acres (16,200 hectares) of grapes in Monterey County. The great majority of the vineyards are located in the large Monterey AVA that runs from Monterey Bay down Salinas Valley to the county border. Two-thirds of the grapevines in

FIGURE 10–12 A Chardonnay vineyard in the Salinas Valley of Monterey County.

appellation are white wine varieties and most of the crop goes to large wineries that use it as a source of inexpensive, high quality, cool-climate grapes for their California or "Coastal" blends. While the county's vineyards and wineries are predominately large operations, Monterey is also home to a number of small producers and sub-appellations. Moving down the Salinas Valley from the coast are the Santa Lucia Highlands, Arroyo Seco, San Bernabe, San Lucas, and Hames Valley AVAs, all located within the Monterey AVA. Generally, they have well-drained soils, and the appellations grow warmer the farther they are from the bay. The Santa Lucia Highlands on the western slopes of the valley is the closest to the coast; it is the coolest and has recently gained an excellent reputation for Chardonnay and Pinot Noir. The San Bernabe AVA, created in 2004, is the newest of the group. Like several other appellations in the state, it is home to only one vineyard and was created to benefit a single winery. However, unlike the others it is a very large vineyard with more than 5,000 acres (2,000 hectares) planted. Positioned just east of the Santa Lucia Highlands is the Carmel Valley AVA. It is much narrower than the Salinas Valley and protected by mountains from the maritime influences of the Pacific, making the appellation warmer and better suited to red varieties.

To the east of Monterey County is San Benito County. Smaller in both size and population, its vineyard development paralleled that in Monterey County. Like Monterey, the region has a mix of large vineyards and wineries as well as smaller producers. Additionally, in the north of the county on the valley floor, the climate is similar to that of the Monterey AVA, though to the south in the mountains of the Gavilan range this comparison ends and the climate becomes warmer. Similar to that found in the Burgundy region, the soil here has a great deal of limestone, which is rare in California. This drew vintners to the area for Chardonnay and Pinot Noir. There are several sub-appellations: Cienega Valley, Lime Kiln Valley, Paicines, Mt. Harlan, and Chalone AVAs. The last two are used only by the wineries that reside in them. The majority of the Chalone AVA is in Monterey County; however, since the climate is more similar to the other AVAs of San Benito, it is included here.

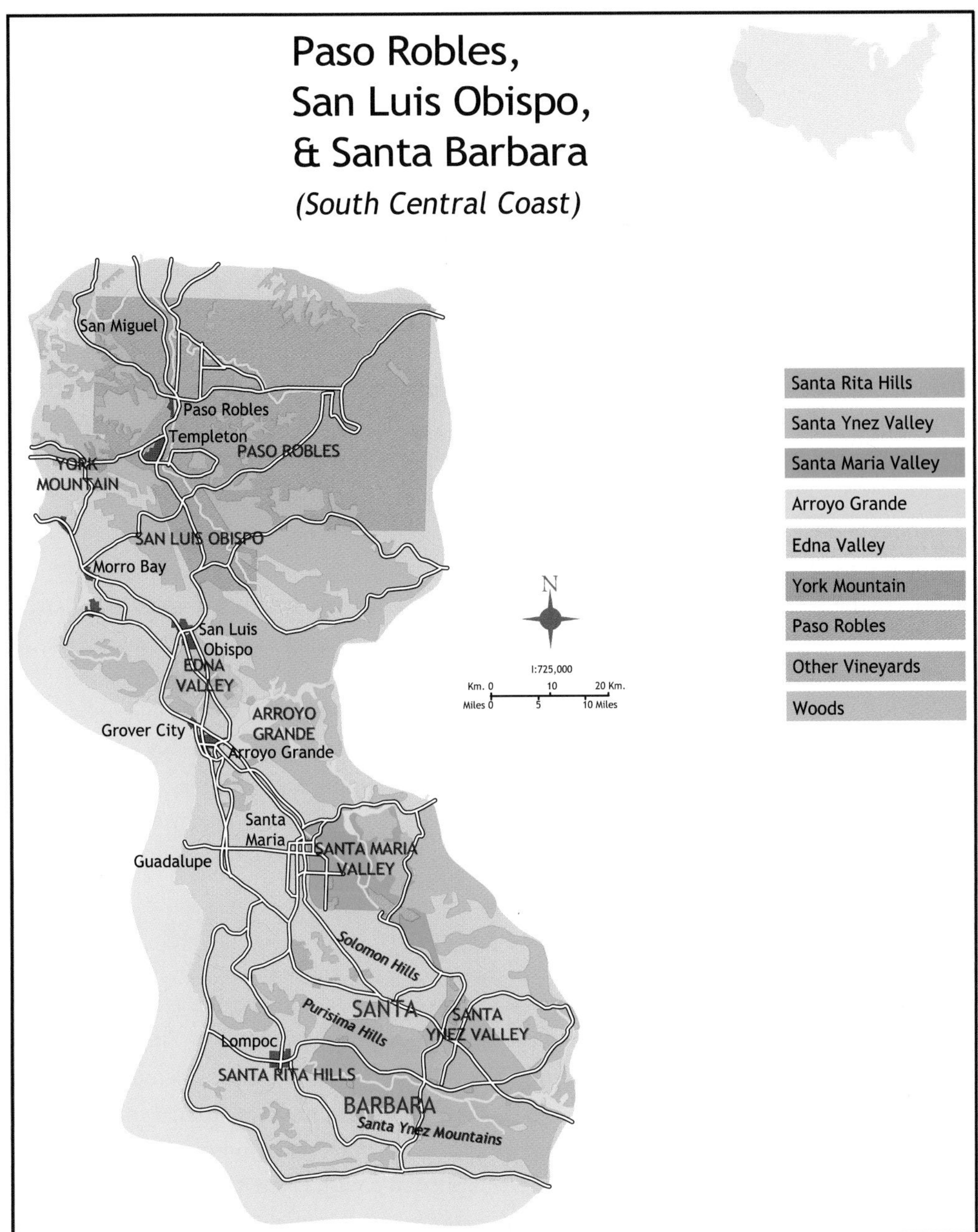

The appellations of the South Central Coast region.

San Luis Obispo County

San Luis Obispo County is located directly south of Monterey County. Much like Monterey, it has a long history of grape growing, dating back to the time of the missions, that was reborn during the wine boom, and the county has a dichotomy of large and small producers. The largest AVA in the county is the Paso Robles appellation with more than 609,000 acres (246,000 hectares). Its northern border is the Monterey County line and the western edge of the AVA begins about 10 miles (16 kilometers) in from the coast in the mountains of the southern portion of the Santa Lucia range. The appellation has experienced rapid growth in the last 20 years and currently has more than 20,000 acres (8,090 hectares) planted to wine grapes. Here the costal mountains block much of the cooling influence from the coast, and the region is generally warmer than Monterey AVA to the north.

The Paso Robles AVA has two distinctly different terroirs located on either side of the Salinas River. To the west, the terrain is hilly, and soils that are composed of ancient seabeds and limestone are common. It is slightly cooler here than on the eastern side, with more rainfall in the winter. Just south of the town of Paso Robles there is a break in the mountains called the Templeton Gap that creates a path for ocean breezes, and vineyards in this area have cooler weather conditions than other parts of the appellation.

The western Paso Robles AVA is home to numerous small vineyards and wineries. To the east of the Salinas River the land becomes a broad plane characterized by sandy loam soils. The conditions are drier and warmer than they are in Paso Robles' western portion and the elevation ranges from 600 to 1,000 feet (182 to 300 meters). The elevation helps to keep the nighttime temperatures cool during the summer months, and frequently temperatures swing 45°F (25°C) between day and night. The open terrain in this part of the appellation allows for plantings that are more expansive, and larger vineyards and wineries are more common. Similarly to Monterey County, wineries from outside the area have invested in vineyard land to augment their production. The region is best known for reds such as Zinfandel, Cabernet Sauvignon, and Syrah, but there is also a considerable amount of Chardonnay planted.

South of the Paso Robles appellation the mountains diminish and the weather has a greater maritime influence. Here lies the Edna Valley AVA, which is much smaller and considerably cooler than Paso Robles. Most of the vineyards are planted on the valley floor where the soils are composed primarily of sedimentary material derived from ancient seabeds. The appellation is best known for Chardonnay and Pinot Noir. Edna Valley's southern location in the state makes for an early spring budbreak. This, combined with the cool summers, results in a long growing season. There are two other appellations in the county worth mentioning, York Mountain and Arroyo Grande. York Mountain is located in the mountains on the western edge of the Paso Robles AVA. It is slightly cooler than the land to the east and is the site of the first commercial vineyards in the region in the 1880s. The Arroyo Grande AVA is just south of the Edna Valley AVA and has a similar terroir.

SANTA BARBARA COUNTY

The last of California's major viticultural areas before reaching the vast urban areas of Los Angeles, Santa Barbara County had some of the earliest vineyards in California. The friars of the mission period were the first people to plant grapes in the 1780s, but there was almost no further vineyard development until the 1970s. The climate here is comparable to that in the southern portion of San Luis Obispo County where winters are mild and have limited rainfall. Spring comes early to the appellation and budbreak occurs about a month before the cooler grape growing regions of California's North Coast. The summers are cool and often foggy which gives the grapes high natural acidity but also creates a risk of botrytis. White grapes outnumber reds by a ratio of nearly 3 to 1 and the county is home to more than 40 wineries. The proximity to the Los Angeles urban area brings in many visitors to Santa Barbara's wine county, and business from tourists has helped to fuel recent growth.

The two largest AVAs in the county are the Santa Maria Valley and the Santa Ynez Valley. The Santa Maria Valley AVA begins about 10 miles (16 kilometers) inland on Santa Barbara County's border with San Luis Obispo County just east of the town of Santa Maria and runs for 15 miles (24 kilometers) in a southeasterly direction. It is slightly cooler than the Santa Ynez Valley to the south, and Chardonnay and Pinot Noir are popular varieties. The Chardonnay from Santa Maria has a reputation for crisp acidity and tropical flavors. To the south is the Santa Ynez AVA and here, unlike most of the California coast, the mountains and valleys run in an east-west direction. Overall the appellation is a little warmer than the Santa Maria region but the vineyards near the coast are cooler than those that are more inland. Popular grapes include Chardonnay, Sauvignon Blanc, Riesling, as well as some Syrah.

Two smaller growing regions in the county are the Santa Rita Hills AVA and the Los Alamos Valley. The Santa Rita Hills AVA is carved out of the western section of the Santa Ynez Valley AVA; it is a cool area widely planted to Pinot Noir. Lying between the Santa Rita Hills and the Santa Maria AVA is Los Alamos Valley. Although it has more acres of wine grapes than Santa Ynez Valley, it does not currently have its own AVA. One reason for this is Los Alamos Valley has many large vineyards that produce grapes for big wineries that blend the wines they make with that from other appellations. Since wines made from Los Alamos Valley grapes are rarely kept separate, there is little incentive to give the region its own AVA.

The Central Valley

California's great Central Valley extends for 450 miles (720 kilometers) throughout the middle of the state. It drains the waters of the Sacramento River to the north and the San Joaquin River to the south through the Delta region into San Francisco Bay. Growing a diverse array of crops, its combination of fertile soils, warm weather, and irrigation make it the nation's most prolific agricultural area. It is also the state's most productive region for grapes, producing 73 percent of the wine grape harvest and 99 percent of the state's table grape and raisin production (California Department of Food and Agriculture, 2004). There are many large vineyards and the conditions allow for much higher croploads than the state's grape-growing regions near the coast. These vineyards generally produce a wine that has a more neutral flavor than coastal vineyards. The quality is also reflected by the price, with Central Valley grapes selling at an average of $300 per ton while Napa Valley grapes bring in more than $1,900 per ton. All the state's largest wineries are located in the Central Valley. Their huge outdoor stainless steel tanks give them a very different appearance than the small wineries of the coast (Figure 10–13).

The majority of the Central Valley's 200,000 acres (81,000 hectares) of vineyards lie mainly in its southern half, which reaches from the capital city of Sacramento down to Bakersfield. The Central Valley, despite its importance to the wine industry, does not have its own AVA, and the wines produced here are usually labeled "California." Although the climate makes the region better

FIGURE 10–13 An outdoor tank farm at Heck Cellars in the southern San Joaquin Valley. Large Central Valley wineries such as these produce the majority of the wine made in California.

suited for warm weather grapes, there are large plantings of cool climate varieties such as Chardonnay. During the 1980s, as consumers became more familiar with varietal names for their wine, growers began planting Chardonnay because it was popular with wine drinkers. Even though it does better in cool regions, it was worth more than white wine varieties traditional to the Central Valley such as French Colombard and Chenin Blanc, which were often simply labeled as "white table wine." The entire valley is warm, but the area just south of Sacramento where the Sacramento and San Joaquin Rivers join does receive some cooling influences from San Francisco Bay. This area, which is known as the Delta region, contains the Central Valley's two most prominent AVAs of Clarksburg and Lodi. The quality of the grapes grown in these AVAs is considered superior to those grown further south; consequently, they command higher prices.

Other Grape Growing Regions of California

The regions discussed so far contain the vast majority of commercial viticulture in the state of California. However, there are two additional appellations with smaller wine industries; they are the Sierra Foothills and Temecula. The Sierra Foothills AVA runs along the western flank of the Sierra Nevada mountain range above the Central Valley. It is a large appellation covering more than 170 miles (274 kilometers) from Yuba County in the north to Mariposa County in the south. Being in the hills, the terrain is generally rugged with volcanic soils, and the climate of a particular vineyard varies depending on the elevation and exposure. Grape growing has been going on in the area since it was first settled during the time of the Gold Rush, and there were a number of vineyards and wineries prior to Prohibition. Despite its long history and size, it only has about 4,000 acres (1,600 hectares) planted to grapes. The great majority of its vineyards are planted with red varieties and it is most well known for its Zinfandel, some of which grows on quite old vines. The appellation can have very warm afternoons during the summer months, but the elevation aids in keeping the nights relatively cool. It also contains the sub-appellations of Fiddletown, El Dorado, North Yuba, and California Shenandoah Valley.

Like the Sierra Foothills, the Temecula AVA, in Southern California, also has a long history of winemaking, going back to the time of the missions. It is a coastal valley that is situated in

FIGURE 10–14 Valley of the Moon Winery in Sonoma Valley. The winery, first established in the 1860s, has recently modernized its winemaking operations.

the southwestern corner of Riverside County about 20 miles (32 kilometers) from the ocean. The coastal influences keep the valley cool in spite of its location 70 miles (113 kilometers) from the Mexican border. Its cool growing conditions make it well suited for white grapes, and it is predominately planted to Chardonnay. The appellation contains about 1,800 acres (730 hectares) of grapes and recently has had severe problems with **Pierce's disease,** which is fatal to grapevines. Pierce's disease is caused by a bacterial infection and is spread from vine to vine by insects. Pierce's disease has always been present in the state, and there have been several serious outbreaks over the years. The outbreak in Temecula was spread by the recently introduced insect pest from the southern United States, the **glassy-winged sharpshooter.** Currently the state is investing a great deal of resources to control the pest and prevent it from expanding its range in California.

Summary

California's history and climate have made it the natural center of wine production in the United States. Its diversity of growing conditions allow for the creation of a wide range of wines from the inexpensive everyday table wines produced in the Central Valley to the unique high end wines of the North Coast (Figure 10-14). In recent decades, the increasing popularity of wine has resulted in the growth of new winemaking regions outside of California as well as the resurgence of the industry in areas where it has historic roots. In 2005, all 50 states had bonded wineries and there were 74 AVAs outside of California registered with the federal government. Despite the growth of these new wine regions, California remains the nation's most important wine producer, setting trends in winemaking much as it does for many other consumer items.

Review Questions

1. Who were the first people to make wine in California?
2. How did Prohibition affect the wine industry in California?
3. How did Prohibition affect the tastes of wine drinkers in the United States?
4. What was the "Paris tasting" and what role did it have in building the reputation of California wine?
5. What effect does a wine region's proximity to the coast have on the grapes that it produces?
6. What factors do vineyards within an AVA have in common?
7. What are the major wine regions of the state and the types of wine they are known for?
8. What is the importance of California in the United States wine industry?

CHAPTER 11

Washington and Oregon

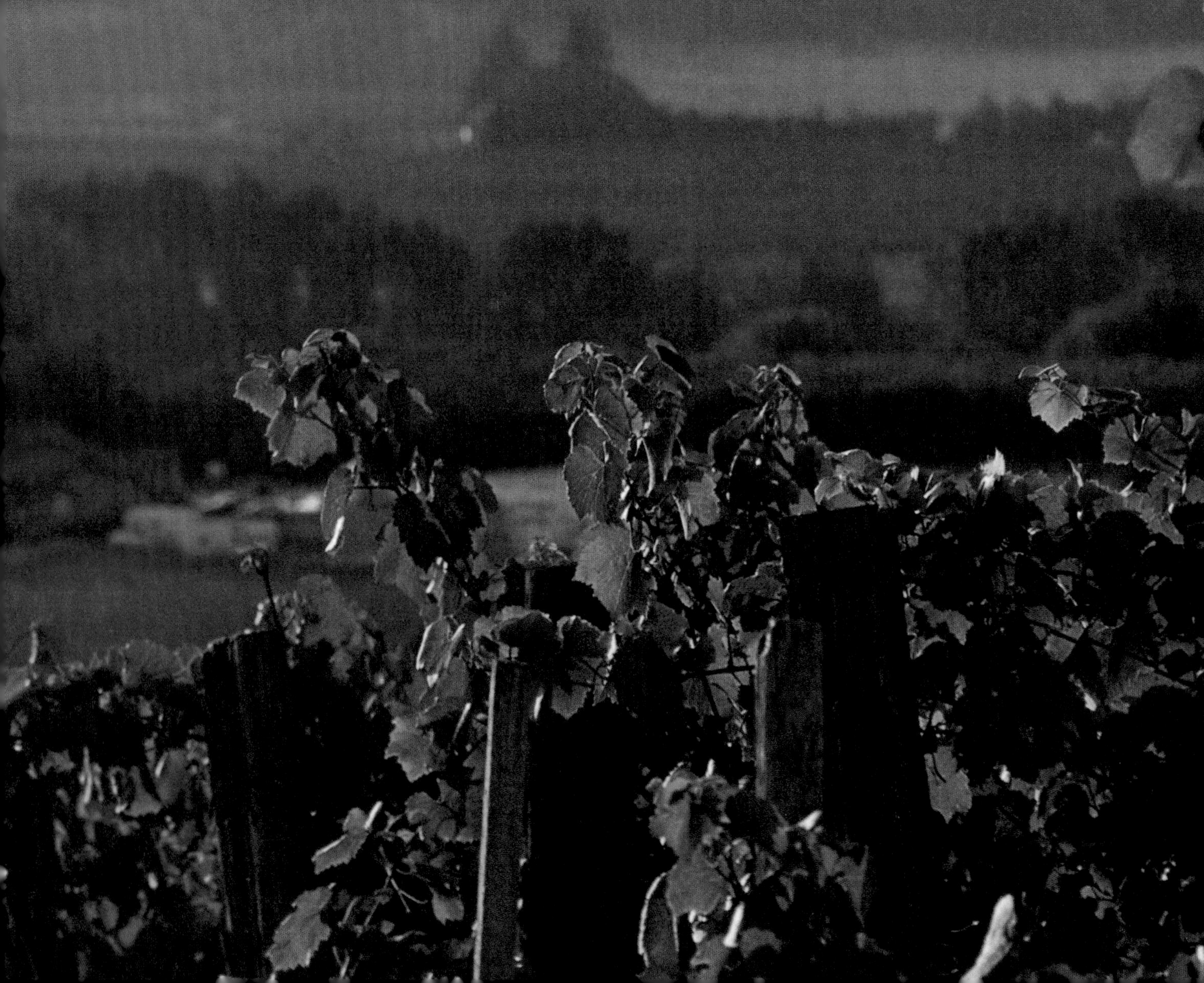

After reading this chapter, you should be able to

- display an understanding of the development of the wine industry in Washington and Oregon.
- describe the differences between the grape-growing regions of Washington and Oregon and the types of wine that they produce.
- compare and contrast the risks from weather in Washington and Oregon.

KEY TERMS

Concord
dry-farmed
Fort Vancouver
Isabella
own-rooted
rain shadow
Stimson Lane Vineyards & Estates
Washington Wine Quality Alliance (WWQA)
winterkill

Second only to California, the Pacific Northwest is one of the nation's most important wine producing regions. Oregon and Washington's wine-growing regions range from 400 to 600 miles (650 to 960 kilometers) north of California's Napa and Sonoma appellations. This places them at approximately the same latitude as the Bordeaux and Burgundy regions of France. Oregon and Washington combined account for about 4 percent of the United States production of wine. With comparable geography, history, and culture, the two states are similar in many respects. However, in spite of these similarities the wine industries of the two have developed differently, giving each its own unique identity. Washington's vineyards are predominately in the dryer eastern part of the state, and several large producers make up most of the wine (Figure 11–1). The Oregon vineyards, in contrast, are located primarily in the state's western half and small wineries and vineyards predominate (Figure 11–2). Washington and Oregon have three American Viticultural Areas (AVAs), the Walla Walla, Columbia Valley, and Columbia Gorge, which cross the boundary between the two states. The majority of the vineyard land in the Walla Walla and Columbia Valley AVAs lies on the Washington side of the border, while the vineyards of the Columbia Gorge appellation are more evenly distributed.

FIGURE 11–1 An expansive vineyard located in eastern Washington's Columbia Valley Appellation.

FIGURE 11–2 Vineyards growing in the low, forested hills of the Willamette Valley Appellation in Western Oregon.

Washington State

After California, Washington State is the nation's second largest producer of premium table wine, making nearly seven times as much wine as Oregon does. In recent decades there has been dramatic change in the state's wine industry; it has undergone rapid growth, with wine grape acreage increasing more than 250 percent and the number of wineries increasing 300 percent from 1993 to 2003. Chardonnay is the most widely planted variety but the state is better known for its red wines, Merlot and Cabernet Sauvignon (Figure 11–3).

FIGURE 11–3 The Red Willow Vineyard in the Yakima Valley was one of the first vineyards to grow Cabernet Sauvignon in 1973 and was the first vineyard to produce Syrah in 1988.

Washington State Wine—Historical Perspective

Compared to California, the history of winemaking in Washington is not nearly as extensive. The state's first grapes were planted by members of the Hudson's Bay Company at **Fort Vancouver** in 1825. The grapes were planted from seed instead of from cuttings, and the variety is unknown. However, since the seeds came from Europe they were undoubtedly a cultivar of *Vitis vinifera.* Although this is not much later than the winemaking began at the California missions; the commercial winemaking industry of Washington developed at a much slower pace. Fort Vancouver was a fur trading outpost on the north side of the Columbia River near Portland. Early settlers coming to the area for homesteading would stop at the fort for provisions, and some early immigrants acquired vine cuttings to take with them to their new farms. Here the grapes were cultivated, and the settlers would make small quantities of wine for personal use. In eastern Washington the first grapes were planted in the Walla Walla Valley around 1859. The settlers found the Walla Walla Valley to be an excellent growing region for many fruits and vegetables, and by the 1860s several plant nurseries were started and supplied local farmers with fruit trees and grapevines. Grapes grew particularly well in eastern Washington because it was much dryer during the growing season than the coastal areas were. The dry weather helped to prevent rot and mildew on the grape clusters as they developed.

Just northwest of Walla Walla in the Yakima Valley, grape growing began when a French immigrant named Charles Schanno planted grapes in 1869 and soon after began making wine. Schanno had brought vine cuttings with him when he moved from The Dalles, Oregon, on the Columbia River, protecting them during the journey by wrapping them in wet straw. While living in The Dalles, he had founded Oregon's first brewery and was very familiar with fermentation. Within a decade, grape growing had expanded in the Yakima area and traveled northward to the town of Wenatchee. Here John Galler, a Dutch trapper, and Phillip Miller from Germany had each established their own homesteads with fruit orchards as well as grapevines. By 1874, each operation was annually producing 1,500 gallons (5,700 liters) (Irvine & Clore, 1998).

By the end of the century, there were many small farming communities throughout the eastern half of Washington State. The region was found to be particularly well suited for tree crops and viticulture because of the dry growing season and plentiful water for irrigation from rivers. Despite the growth of agriculture in the region, winemaking remained mainly an amateur activity. Many settlers of European descent would grow grapes to make wine for their own families, and it was not a major source of income. At the turn of the century, the commercial industry began to develop in the Yakima Valley. In 1905, a Seattle attorney named Elvert Blaine founded Stone House Winery near the town of Grandview in the middle of the Yakima Valley. He hired a French Canadian winemaker named Paul Charvet, and soon they were using both traditional *vinifera* winemaking varieties as well as **Concord** grapes to produce wine. The Concord variety is native to the East Coast of the United States and is very cold tolerant as well as being a prodigious producer. Today it is more known for making grape juice and jelly than it is for making wine, although it is still used for that purpose in the Northeast United States. Because of their tolerance to the Northwest's cold winters, native American grapes from the East Coast like Concord and *vinifera*-native American hybrids such as **Isabella** became popular. The importance of grape growing increased as agriculture in Eastern Washington developed and by 1911 the first Columbia River Valley Grape Carnival was held in Kennewick, Washington, where over 40 different varieties of grapes were entered.

Prohibition and Rebirth

Prohibition had much the same effect on Washington's wine and grape business as it had in California. Although wineries were devastated by the law, vineyards flourished, producing grapes for juice as well as for home winemakers. At the beginning of Prohibition in 1920, the state's vineyards were producing 1,800 tons of fruit annually, and by 1929 they were producing 6,200 tons, the majority being the Concord variety (Irvine & Clore, 1998). After Prohibition was repealed at the end of 1933, the state set up the Washington State Liquor Control Board to regulate the consumption of alcohol. It did this by controlling the distribution and sale of wine and spirits through state-owned stores. Although the goal of the state stores was to limit consumption by being the only source for consumers to purchase alcohol, they also wanted to encourage the wine business in the state. They did this by giving a significant tax break on the wines that were produced from grapes grown in Washington, putting wines from California and Europe at a competitive disadvantage. The first winery started after appeal was St. Charles Winery, Washington Bonded Winery No. 1, on Stretch Island in Puget Sound. By 1937 there were 42 wineries operating in the state, but just five years later in 1942, wartime rationing and lack of demand had lowered the number to 26. Even though the number of Washington wineries declined, the protectionist tax code increased their market share, which reached a high of 65.7 percent in 1941 (Irvine & Clore, 1998).

The production of Washington wine declined in the decades following World War II, and consumption declined as well. However, grape growing did increase, with the extra tons going into Concord grape juice production. Many wines were made from Concord grapes picked at low sugar and then fortified to a higher alcohol. Non-grape wines were also made, taking advantage of the Northwest's production of fruits and berries. The quality of Washington wines during this time was generally substandard, and there was little incentive to improve production methods because the protectionist laws limited competition from out of state. This trend first began to change in the early 1960s; Washington State University (WSU) began conducting research on winemaking and growing premium wine grapes in eastern Washington. At WSU, Dr. Walter Clore and Dr. Charles Nagle established test vineyards of *vinifera* wine varieties and produced wines that demonstrated the region's potential for making fine wines.

In 1962 a group of amateur winemakers incorporated and formed the Associated Vintners. The group soon owned a winery and vineyard land and concentrated on producing premium wine from *vinifera* grapes. Associated Vintners would be renamed Columbia Winery in 1983. In 1967 one of the state's largest wineries, American Wine Growers or AWG, hired renowned California Winemaker André Tchelistcheff as a consultant and renamed their brand Ste. Michelle Vintners. Ste. Michelle was purchased by US Tobacco in 1974 and renamed Chateau Ste. Michelle. The new ownership invested a huge amount of capital and opened a number of wineries under the corporate name **Stimson Lane Vineyards & Estates.** By 2003 they were producing almost two-thirds of the wine bottled in the state.

In 1969 the Washington State legislature removed its tax on wine imported from out of the state. Although this slowed development for a time, the increased competition from out of state wineries, particularly those in California, forced producers to improve their product. In the late 1960s there was still greater than 20 times more vineyard acreage planted to Concord than there was with European wine varieties. Using knowledge gained from the research done at WSU, growers and vintners shifted efforts towards making wines from *vinifera* grapes that were more acceptable to consumers. Growth was steady but modest throughout the 1970s and there were 19 wineries in the state at the end of the decade (Figure 11-4).

As the consumption of table wines grew in America and Washington State wineries started producing better wines, the pace of development dramatically increased. In the early 1980s, two trade organizations were formed to help promote the wine industry in the state. In 1981, the

FIGURE 11–4 A mechanical harvester at Columbia Crest Vineyards, with the Columbia River in the background. On the left side of the photograph, a conveyer moves the picked grapes into a bin being towed alongside the harvester in the next row.

Washington Wine Institute was formed to promote the business interests of the wineries and was financed by winery dues. The Washington Wine Commission was created in 1983 with help from the Department of Agriculture. It focused on the marketing and promotion of Washington wine and was funded by a small tax on grape and wine production. These two organizations helped to support the wineries and vineyards of the state during the rapid growth that was to come.

The success of large wineries, like Columbia and Ste. Michelle, inspired many wine enthusiasts and grape growers in the 1980s and 1990s to establish wineries of their own. In 1999 another voluntary trade organization was formed, the **Washington Wine Quality Alliance (WWQA)**. The goal of the WWQA goes beyond simply promoting the state's wine industry, also setting quality standards for the wines its members produce. The group has three basic regulations that its members must adhere to:

- Reserve wines are limited to 3,000 cases or 10 percent of the winery's production.
- The grapes must be 100 percent from the State of Washington or a Washington AVA, or the percentage of each source should be listed on the label.
- Generic terms of European appellations such as Champagne and Burgundy cannot be used.

The group currently has a membership of more than one hundred wineries and growers. By the year 2002 there were more than 28,000 acres (11,300 hectares) of *vinifera* wine grapes and more than 240 bonded wineries in the state. Today Washington State is the second largest producer of fine wine after California, with 57 percent of the wine being made from red varieties and 43 percent from white (Washington Agricultural Statistics Service, 2002). During the rapid growth, the cost of vineyard land in Eastern Washington was much less than it was in California. This allowed Washington wineries to sell their product at a reasonable price and quickly gave the state a reputation for producing wines of good value. In recent years, as the industry has matured, it has also become known for making high quality wines.

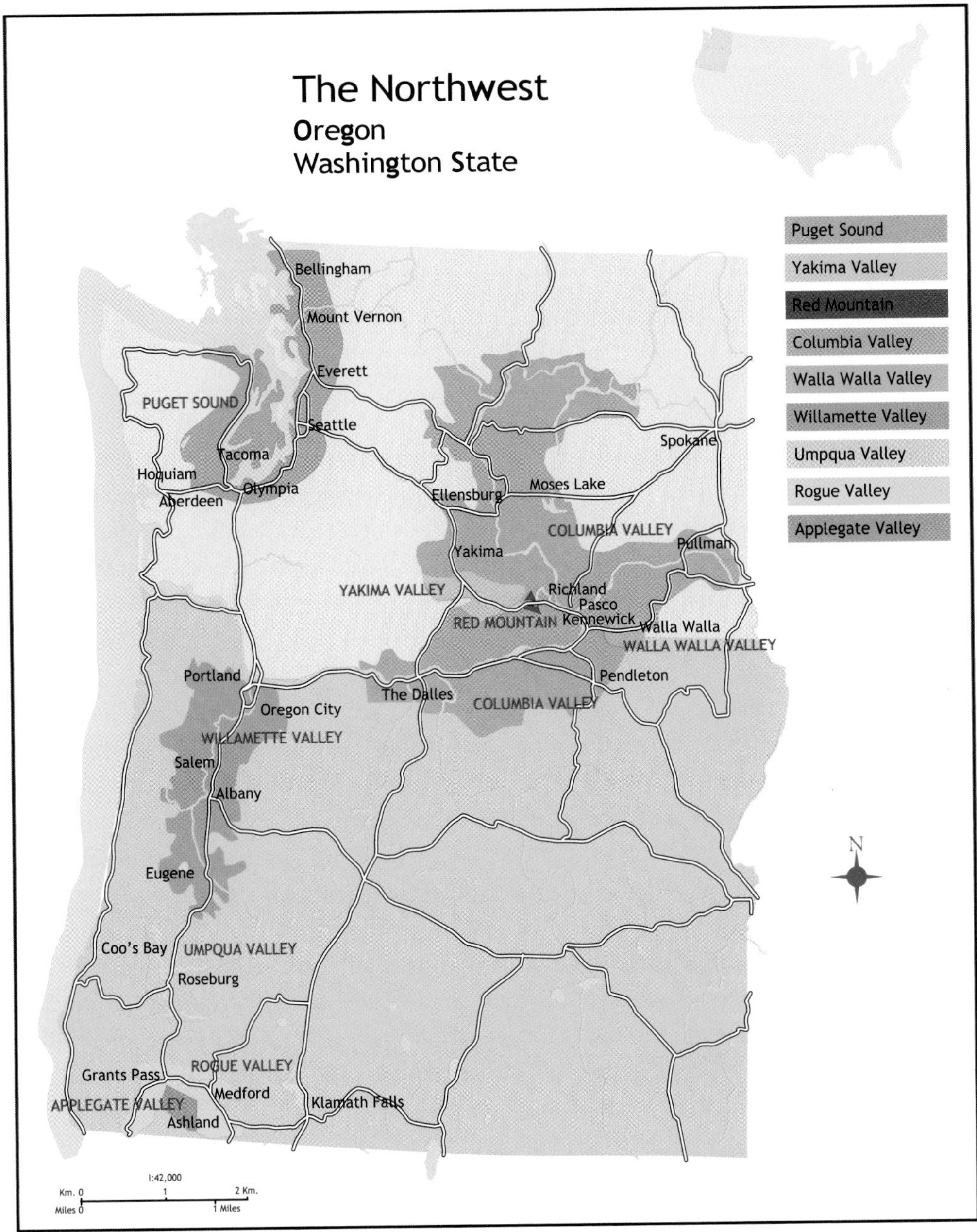

Map of Washington and Oregon State appellations.

The Wine Regions of Washington

The state of Washington is split into its eastern and western halves by the Cascade Mountain range. The western half is more urbanized and has the rainy weather and thick forests that most people associate with the state. Those who are unfamiliar with Washington's wine grape appellations wonder how it could be possible to grow wine grapes under such conditions. While there are some vineyards in the western half of the state, 99 percent of the wine grapes in Washington are grown east of the Cascades (Table 11-1). In contrast to the west side of the state, the eastern half is sparsely populated, with a dry climate and an agriculturally based economy. The dry, sunny, weather is due to the **rain shadow** that is caused by the Cascades. Storms coming in from the Pacific Ocean are blocked by the high mountains and release their rain on the western side before moving east. The rainfall averages about 8 inches (20 centimeters) per year, one-fifth of

Table 3-1 Appellations of Washington State

APPELLATION	BEST KNOWN VARIETIES
Columbia Valley (shared with OR)	Produces a number of varieties
Red Mountain	Cabernet Sauvignon, Merlot, Syrah
Walla Walla Valley (shared with OR)	Cabernet Sauvignon, Merlot, Syrah, Sangiovese
Yakima Valley	Chardonnay, Merlot, Cabernet Sauvignon
Columbia Gorge (shared with OR)	Produces a number of varieties
Puget Sound	Not known for any one variety

the rainfall on the coast. This amount of rain is not enough to support vines; therefore, the majority of the state's vineyards are irrigated. There is adequate water available for irrigation from wells and the rivers carrying snowmelt down from the mountains. The ideal climate during the growing season, combined with irrigation, makes Eastern Washington one of the country's most productive agricultural regions for many crops besides grapes, as well as what is perhaps its most famous product, apples.

Eastern Washington is also marked by much cooler winters than viticultural areas to the south (Figure 11–5). There is often subzero weather during the winter, and bud break is delayed, beginning about three weeks after it does in the Napa Valley. Despite the late start, harvest takes place about the same time as in Napa. This is due to the more northern latitude having longer days during the growing season, giving the vines a chance to catch up. Being east of the mountains, the area has more of a continental climate that is not as temperate as coastal grape growing regions. The cold winters are also responsible for **winterkill,** the biggest headache for Washington grape growers. If the temperature falls below 5°F (−15°C), it can cause damage to the dormant buds, and if it gets cold enough, it can actually kill the grapevine down to the ground level. Winterkill conditions are different from the spring frosts described in Chapter 2. During spring frosts, tender young shoots are susceptible to subfreezing conditions; winterkill damages dormant vines before budbreak occurs. This weather does not happen every year, and

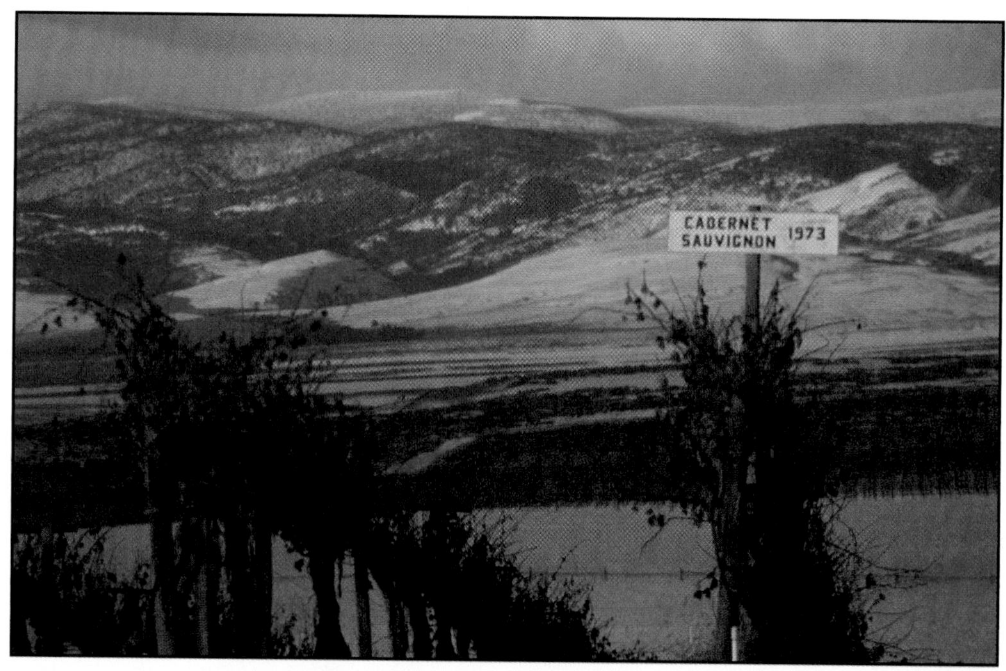

FIGURE 11–5 Mid-winter in a Cabernet Sauvignon vineyard in Eastern Washington. The inland location of Eastern Washington vineyards gives them a more continental climate than vineyards located on the coast and winter snow is common.

when it does, it has a greater effect on yield than quality. In the winter of 1996, a hard freeze reduced that vintage's Merlot production by 60 percent. During these winter cold snaps, cold air settles in low-lying areas just as it does during spring frosts, so a vineyard's slope and orientation are very important to its survivability.

The dry climate of Eastern Washington does have advantages; since bunch rot and mildew are encouraged by high humidity, they are less of a problem here. Phylloxera as well is less of a concern; although it was discovered in the state in 1943, it has not increased its range. It is not certain exactly why it never spread to other vineyards, but it is believed it is due to a combination of Washington's dry climate, sandy soils, and cold winters. Today almost all of the state's vineyards are **own-rooted,** that is, the vines are grown on their own roots without the use of a rootstock. This is important because if cold winter temperatures kill a vine to the ground level it is possible to train a new shoot up from the roots and establish a new vine more quickly than if it had to be regrafted.

The Columbia Valley

The Columbia Valley AVA is made up of the drainage of the Columbia River from the eastern half of Washington State as well as a small portion of north central Oregon. Established in 1987 it is one of the largest AVAs in the country, encompassing nearly 11 million acres (4.4 million hectares) with 17,000 acres (6,880 hectares) planted to grapes. Although the overall climate is similar throughout the region, there are many types of soils and microclimates that account for a number of unique terroirs. Only a small fraction of the appellation is planted to vineyards and there is much room for expansion, but not all of the land is suitable for growing grapes. The best vineyard sites must have the right soils, proper slope, and adequate water for irrigation (Figure 11-6). The Columbia Valley has three AVAs as well as a number of sub-appellations that are not official AVAs located within its boundaries. These appellations combined produce more than 98 percent of the state's grapes. Excluding these appellations from the figures, the Columbia Valley produces 60 percent of Washington's grapes with the most popular varieties being Cabernet Sauvignon, Merlot, Chardonnay, White Riesling, and Syrah in order. Syrah acreage in particular has been expanding very rapidly and shows promise to be one of the region's best varieties. The appellation includes the Tri-Cities of Richland, Kennewick, and Pasco at the confluence of the Columbia, Snake, and Yakima Rivers. While it is not an official AVA, there are a number of vineyards and wineries located in the Tri-Cities region.

FIGURE 11–6 The Seven Hills Vineyard in the Walla Walla Valley region of Washington's Columbia Valley Appellation.

YAKIMA VALLEY

The Yakima Valley is a broad valley formed by the Yakima River as it flows down from the Cascade Mountains. It runs for 60 miles (96 kilometers) from the southern edge of the town of Yakima to where the Yakima joins the Columbia River. Established in 1983, this appellation has a long history of winemaking and was Washington's first AVA. It contains just over one-third of Washington's vineyard acreage and a number of both small and large wineries. Chardonnay is the most widely planted variety followed by Merlot, Cabernet Sauvignon, and White Riesling. The valley has gentle rolling hills and most of the soil is made up of a gravelly silt-loam with a lot of gravel that drains very well; higher elevations in the valley tend to have soils with more clay content. Many crops other than grapes are grown in the valley including apples, cherries, hops, and alfalfa. Because of the cold winters, the topography of the vineyard site is very important, and vineyards compete with other crops for the best land. Near the southern end of the valley is the small community of Prosser, the home of several wineries as well as the Washington State University Research Station where much of the research on growing *vinifera* in the state was done. As the wine industry in the area grows, Prosser is becoming a destination for many people touring the Eastern Washington wine county.

RED MOUNTAIN

Red Mountain is Washington's smallest AVA and one of the most recent, being established in 2001, yet it already has a reputation for producing some of the state's best wines. Located about 20 miles east of Prosser, just outside the small town of Benton City, where the Yakima River turns to the north before it joins the Columbia, the appellation covers 4,000 acres (1,600 hectares), and its boundaries lay within both the larger Yakima and Columbia Valley AVAs. It lies above the valley floor on the western face of Red Mountain and is best known for its red wines; however, the mountain takes its name from the light red color of the cheat grass that grows on its slopes each summer (Figure 11–7). The first vines were planted in 1975 and currently there are 700 acres (280 hectares) of the appellation planted, making room for much more. However, expansion is limited by restrictions on developing new water sources for irrigation. The soil is light sandy loam and well drained, with the slopes of the mountain providing excellent air drainage during hard winter freezes. Red Mountain's weather is slightly warmer than most of the Yakima Valley to the west, and the grapes have little difficulty attaining ripe sugars at harvest. Cabernet Sauvignon and Merlot are the most widely planted varieties, but Syrah and Sangiovese are being established and show great promise. The grapes are used by a

FIGURE 11–7 Hedges Cellars in the Red Mountain Appellation; the region is the smallest of Washington's AVAs and is best known for its red wines.

number of Washington State wineries as well as several small premium producers located within the appellation.

THE WALLA WALLA VALLEY

The Walla Walla Valley lies at the southeastern edge of the Columbia Valley Appellation and the southern portion of the Walla Walla AVA extends over the border into Oregon. Walla Walla translates to "many waters," and the valley is marked by several tributaries that flow down from the Blue Mountains east of the valley into the Walla Walla River. It has a slightly warmer climate than most of the Columbia Valley AVA and receives more rainfall as well, 19 inches (48 centimeters) annually. This allows for some of the vineyards to be **dry-farmed,** grown without irrigation, a rarity in the desert-like climate of Eastern Washington. The valley has a rich history of agriculture and grape growing dating back to the 1860s. The region is still home to a number of vegetable crops including peas, strawberries, and onions that all do well in Walla Walla's rich volcanic soil. Beginning in the late 1970s small, premium wineries were established in the valley. Woodward Canyon Winery and Leonetti Cellars were both founded during this time, and remain two of the state's most highly regarded producers (Figure 11-8). In recent years, the Walla Walla Valley, like the rest of the state, has experienced dramatic growth, with vineyard acreage going from 450 acres (180 hectares) in 1999 to 1200 acres (485 hectares) in 2001 (Hall, 2001). Today it is the home of more than 40 small wineries that have an emphasis on producing premium wines, more than double the amount it had in 1999.

The Puget Sound Region

The Puget Sound Appellation contrasts in almost every way to Eastern Washington. Lying between the Pacific Ocean and the Cascade Mountains, it has a maritime climate and experiences much more rain and less prolonged freezes than the interior of the state (Figure 11–9). It is much more populated and home of the state's largest cities, Seattle and Tacoma. The wet weather makes it difficult to grow wine grapes and there are only about 80 acres (32 hectares) of *vinifera* varieties planted in the appellation (Washington Agricultural Statistics Service, 2002). Despite the fact there are few vineyards, Puget Sound has more than 30 wineries,

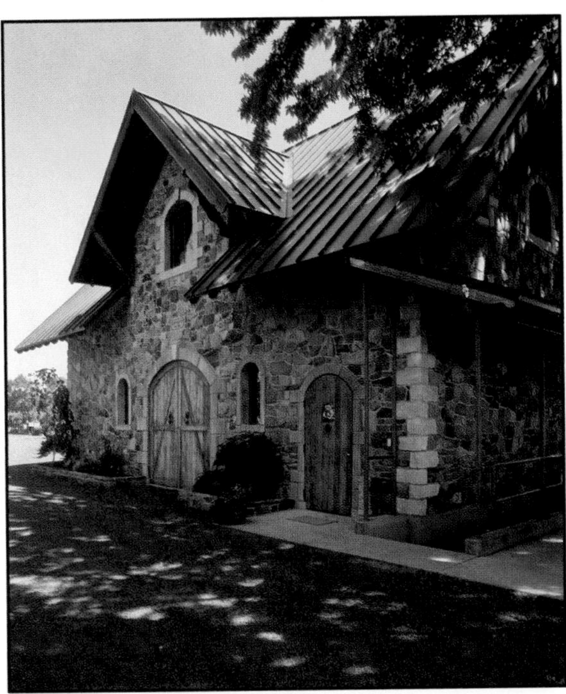

FIGURE 11–8 Leonetti Winery in Washington's Walla Walla Valley. One of the early premium wineries in the Walla Walla Appellation, it produces some of the state's most sought-after wines.

FIGURE 11–9 A fall vineyard in Washington's Puget Sound region.

including some of the state's largest producers. Smaller wineries often produce wine using hybrid grape varieties and other fruits and berries that do well in the appellation, as was once popular throughout the state. Most of the wineries source their grapes from east of the mountains to make their wines. Some Puget Sound wineries import the grapes directly to the winery for crushing; others have crush plants located in the Columbia Valley to be closer to the vineyards. After fermentation, the young wines are loaded into tanker trucks to be shipped to the Seattle area for blending and bottling. North of Seattle, the town of Woodinville has become a major wine-tasting destination with a number of both small and large wineries. Its two biggest are the headquarters of Château Ste. Michelle and Columbia Wineries, located across the street from one another (Figure 11-10).

FIGURE 11–10 The stainless steel tank cellar at Château Ste. Michelle Winery, one of Washington State's largest producers.

The Columbia Gorge Appellation

Established in 2004, the Columbia Gorge AVA is Washington's newest appellation. Its boundaries extend roughly 8 miles (13 kilometers) on either side of the Columbia River into Oregon and Washington, and it spans the two climates on either side of the Cascade Mountains. It begins just west of the southwest edge of the Columbia Valley AVA and extends westward through the gorge that is carved out by the river, ending 60 miles (96 kilometers) east of Portland. The soils are generally silty-loam and there is a great difference between the eastern and western portion of the appellation. The west is cool and wet with a climate that is similar to Oregon's Willamette Valley, and the east side is warm in dry more like the Columbia Valley. In between the two zones the temperature and rainfall is more moderate; however, the conditions are very windy. There are approximately 300 acres (121 hectares) of grapes in the region split between the Oregon and Washington sides.

Oregon State

Set in between California and Washington, Oregon is the country's fourth largest producer of wine (Figure 11–11). While the size of Oregon's wine business is overwhelmed by its neighbors, its reputation for producing fine wine, in particular Pinot Noir, is not. Pinot Noir makes up over one-third of the vineyard acreage and is the state's most important variety. Like Washington State, the industry has had tremendous expansion in recent years. Despite the growth, the majority of the producers remain very small, with more than half making less than 5,000 cases per year. This is evidenced by the fact that although the state bottles only a fraction of the wine that

FIGURE 11–11 Open-top stainless steel fermentation tanks at an Oregon winery. With open-top tanks, punching the cap down into the fermenting must is the most common method of extracting flavor and color from the skins.

Washington does, it has even more wineries, with more than 300 (Oregon Wine Board, 2005). The limited production of the wineries also means that Oregon wines usually command higher prices than those of Washington State.

Oregon State Wine—Historical Perspective

In the mid 1800s, settlers were attracted to Oregon's rich agricultural land and many came to settle the area by crossing the Oregon Trail. One of them, Henderson Luelling, was a horticulturist and planted Oregon's first grapevines in the Willamette Valley in 1847. Luelling, working with his son-in-law, William Meek, made wine, and in 1859 they won a medal at the California State Fair for their wine made from the hybrid grape Isabella (Hall, 2001). By the 1850s viticulture was also being developed in southern Oregon, in the Rogue River valley. Here Peter Britt grew grapes and established a winery named Valley View Vineyard. The winery and vineyards were restored beginning in early 1972 and continue to produce wine today. In the 1880s, two brothers named Edward and John Von Pessls came up from California to the Umpqua Valley region of southern Oregon. They planted cuttings they had brought with them of Zinfandel and other *vinifera* varieties obtained in the Napa Valley. To the north in the Willamette Valley, Earnest Reuter was making wine out of white *vinifera* grapes. Like his predecessor, Henderson Luelling, he won praise outside of Oregon with a gold medal at the 1904 world's fair in St. Louis.

Despite the fact these early vintners enjoyed some successes, viticulture in Oregon never developed to the extent it did to the south in California. Although the coastal river valleys in southern Oregon have a climate similar to California's North Coast AVA, most of the state's agriculture takes place further north in the Willamette Valley, which growers considered too damp and cool for production of wine grapes. What little industry existed was wiped out by Prohibition in 1920. After its repeal, the industry did not benefit from protectionist laws that Washington State had, and it could not compete with the more established vineyards in California, which enjoyed a more consistent climate. After Prohibition a handful of small wineries did exist; however, most produced fruit wines from berries and other crops that were grown on their own farms. Most of these wines were consumed close to home and rarely left the state. In 1938, there were 28 bonded wineries in Oregon; just 20 years later in 1958, nearly all were closed.

The Beginning of an Industry

The lack of development after the repeal of Prohibition lasted until the 1960s, when a new generation of winemakers began to attempt to make table wine from traditional wine varieties. Two of the early producers were expatriates from California. In southern Oregon, Richard Sommer began Hill Crest Vineyard outside the town of Roseburg. He had been trained at The University of California, Davis where he was warned that Oregon would be too wet and chilly to successfully grow *vinifera* grapes. In the Willamette Valley, Charles Coury began growing Alsatian varieties such as Pinot Blanc, as well as some Pinot Noir, on Wine Hill where Ernest Reuter had grown grapes in the 1880s. The year 1966 was an important year in Oregon winemaking, when David Lett of the Eyrie Vineyard Winery planted the first Pinot Noir vines in the Dundee Hills region of the Willamette Valley. Having spent time in France, Lett was convinced that the climate of Oregon more closely approximated that of Burgundy than California's did. After the vines were mature, he used traditional Burgundian production methods to make his wines. At the end of the decade, the wine industry in the state was still very small with only five bonded wineries producing wine.

Over the next two decades, Pinot Noir would become Oregon's most notable wine and raise the reputation of winemaking in the state. The wineries that were established in the 1960s and 1970s were small and often built by an owner/winemaker in contrast to the large capital-intensive wineries that were being built in California and Washington. Pinot Noir is a difficult grape to grow, and its delicate flavors are often lost during processing at the winery. The variety seemed well suited to Oregon's capricious weather and the handmade, labor-intensive techniques used at its small wineries (Figure 11–12). Oregon winemakers were also among the first in the country to pay close attention to the aspects of clone selection with Pinot Noir production. As described in Chapter 2, Pinot Noir has a number of different clones that growers can select from when planting their vineyards. In 1975, Oregon State University in Corvallis began working closely with growers to import Burgundy Pinot Noir clones best suited for table wine production and make them available to the public.

The quality of Oregon Pinot Noir gained worldwide attention in 1979 when David Lett entered a 1975 Pinot Noir from Eyrie Vineyard into an international Pinot Noir competition in Paris where it placed tenth, ahead of many of Burgundy's best producers. These results sent shock waves around the wine world in much the same way as the Paris tasting of 1975 did for California winemakers. During this time Oregon also formed some of the strictest labeling and composition laws in the United States. In 1977, the Oregon Liquor Control Commission enacted rules that state:

- A wine labeled "Estate" must be grown within 5 miles (8 kilometers) of the winery.
- The composition of a wine must be at least 90 percent of the varietal listed on the label.
- Generic terms of European appellations such as Champagne and Burgundy cannot be used.

The 90 percent minimum of a grape variety is considerably higher than the 75 percent required in California and Washington. By 1980 there would be 34 wineries and 1,100 acres (445 hectares) of wine grapes in the state.

FIGURE 11–12 Chardonnay grapes being loaded into a tank press for whole cluster pressing. Whole cluster pressing is gentler than pressing fruit that has been crushed and destemmed before pressing.

During the 1990s, the growing popularity of Oregon wine attracted new investment, and a few larger showcase wineries were built. During the 10 years from 1992 to 2002, the number of wineries in Oregon grew from 78 to 250 and it ranked fourth in production after New York State (Oregon Wine Board, 2005). Despite its successes, Oregon's wine industry has retained its modest character. Small independent producers are common, and the largest winery bottles only 125,000 cases a year. Oregon also has gained a reputation for white wines, most notably Pinot Gris and Chardonnay; however, Pinot Noir remains Oregon's most popular grape, representing 54 percent of the planted acres and one-third of the state's wine production (Oregon Agricultural Statistics Service, 2005). Phylloxera has been present in the state many years and is found in a number of vineyards, but it has spread slowly and remained somewhat isolated. Consequently half of the state's vines still grow on their own roots, although most new plantings are on rootstocks.

The Wine Regions of Oregon

Oregon has eleven AVAs. Eight are located west of the Cascade Mountains, two are in Eastern Oregon, and one spans the region between the east and the west along the Columbia River Gorge. Of the eight western appellations, half were recently established—in 2004 and 2005. The western appellations have more of a maritime influence on their climate, and all have boundaries entirely within the state. East of the mountains, the two appellations are the Columbia River and Walla Walla Valley Appellations. Both are shared with Washington State to the north and have drier climates than the western appellations. The Columbia Gorge AVA also shares a border with Washington; on its western edge the climate is more cool and wet, and on its eastern side it is dryer and warmer (Table 11–2).

The Willamette Valley

The Willamette Valley Appellation lies in the northwest of the state and is Oregon's most prolific region for agriculture. It is Oregon's largest and oldest AVA, established in 1984; it contains the majority of the state's vineyards and wineries producing nearly 75 percent of the grapes harvested in the state. The boundaries of the appellation are approximately formed by the watershed

Table 11-2 Appellations of Oregon State

APPELLATIONS BY GEOGRAPHY	BEST KNOWN VARIETIES
AVAs of Willamette Valley	
Willamette Valley	Pinot Noir, Chardonnay, Pinot Gris
Dundee Hills	Pinot Noir, Chardonnay, Pinot Gris
Yamhill-Carlton District	Pinot Noir, Chardonnay, Pinot Gris
AVAs of Southern Oregon	
Applegate Valley	Cabernet Sauvignon, Chardonnay, Zinfandel
Rogue Valley	Produces a number of varieties
Umpqua Valley	Cabernet Sauvignon, Merlot, Syrah
AVAs Shared with Washington State	
Columbia Gorge	Produces a number of varieties
Columbia Valley	Cabernet Sauvignon, Merlot, Syrah
Walla Walla Valley	Cabernet Sauvignon, Merlot, Syrah

of the Willamette River and extends from south of the city of Eugene to Portland, 125 miles (200 kilometers) to the north. The climate is generally cooler and wetter than Napa and Sonoma with a similar pattern of wet winters followed by dryer weather during the summers. This weather makes it an excellent region for the cool climate varieties Chardonnay and Pinot Gris as well as the Oregon's most popular variety, Pinot Noir (Figure 11–13).

Although there are vineyards and wineries located throughout the appellation, many are concentrated just to the southwest of Portland. If there is an epicenter of Oregon wine country it lies here in Yamhill County, particularly between the small towns of Newberg and McMinnville. Yamhill County alone has one-third of the state's vineyards and grows 45 percent of Oregon's Pinot Noir. To those who have visited California's Sonoma and Napa valleys, the area seems familiar in many respects. Like the wine country of the North Coast in California, the area is home to a number of vineyards and wineries. Here, however, there is much less development and the wineries are less crowded and more rustic in nature. This area has more than 50 wineries including some of the state's most well-known producers.

Many of the best vineyards in this part of the appellation are planted in the small ranges of hills that are laced throughout the countryside. Two of the most famous ranges are the Dundee and Eola Hills. The vines do well in their red volcanic soil, and the sloping hillsides provide good drainage for both water and cold air during the winter. If the vineyards are oriented to the south, they have better exposure to the sun, which is an advantage whenever growing grapes in a cool region. The northern part of the Willamette Valley is home to three of Oregon's newest subappellations, the Dundee Hills, Yamhill-Carlton, and the McMinnville AVAs. At the southern end of the valley there are fewer vineyards, and the soils have more clay content. Although the southern Willamette has fewer vineyards and wineries, it is home to the state's largest winery, King Estate.

Oregon is known for its rainfall, and it can be a major headache for vintners. The summers are usually dry, but storms often linger into the late spring, affecting bloom, or can come early in the fall during harvest. These conditions mean that mildew and bunch rot are always a concern, and there is not always enough warm weather for the grapes to ripen fully. The early fall rains are always a risk in the Willamette Valley, meaning that in some vintages the grapes never attain full maturity. For this reason the wines of the Willamette Valley experience more variation from

FIGURE 11–13 Bins of Pinot Noir grapes ready for transport to the winery for crushing.

year to year than those of most areas of California and Washington typically do. Oregon vintners are always quick to point out, however, that these conditions are similar to those of Pinot Noir's home in Burgundy. One advantage of Oregon's climate is that hard freezes that cause winterkill are much rarer than they are in Washington.

The Umpqua, Rogue, and Applegate Valleys

Established in 1984, the Umpqua Valley is centered on the town of Roseburg and was carved out by the Umpqua River and its tributaries. Located south of the Willamette Valley, it is a smaller appellation, only about a quarter the size of its neighbor to the north covering an area 70 x 35 miles (112 x 56 kilometers). It lies about the same distance inland as the Willamette Valley, but because the coastal mountains are higher in this part of the state, the climate of the region experiences less of a moderating influence from the Pacific Ocean than the Willamette Valley does. This fact, coupled with its more southerly location, allows the Umpqua to have a warmer, drier climate than the Willamette Valley. This dryer terroir means that there is less concern of early rains affecting the harvest, and warmer grape varieties such as Cabernet Sauvignon, Merlot, and Syrah do well in the Umpqua Valley (Figure 11-14).

The Rogue Valley Appellation encompasses the valley formed by the Rogue River as it travels along Interstate 5 just north of the California border. Here in southern Oregon, agriculture is less common than it is in the Willamette Valley, and the timber and forest products industry is more prominent. The Rogue Valley sits at a higher elevation than the rest of Oregon's appellations, with most of its vineyards lying between 1,000 and 2,000 feet (300 and 600 meters). It has a diversity of terroirs, with the areas to the west having a more coastal climate that is cooler, and has more rainfall than those areas that are further inland. The diversity of growing conditions allows for many different types of grapes and wines to be produced, and the appellation does not have a reputation for any one variety in particular. It was established in 1991 and in the year 2001 the sub-appellation of the Applegate Valley was formed within the borders of the Rogue Valley AVA.

In 2004, the Southern Oregon AVA was created, containing all of the Umpqua Valley, Rogue Valley, and Applegate Valley AVAs. Together these appellations grow roughly a fifth of Oregon's wine grapes. Although this large AVA has a number of microclimates, the entire region is generally much warmer and drier than the Willamette Valley. The Southern Oregon AVA has

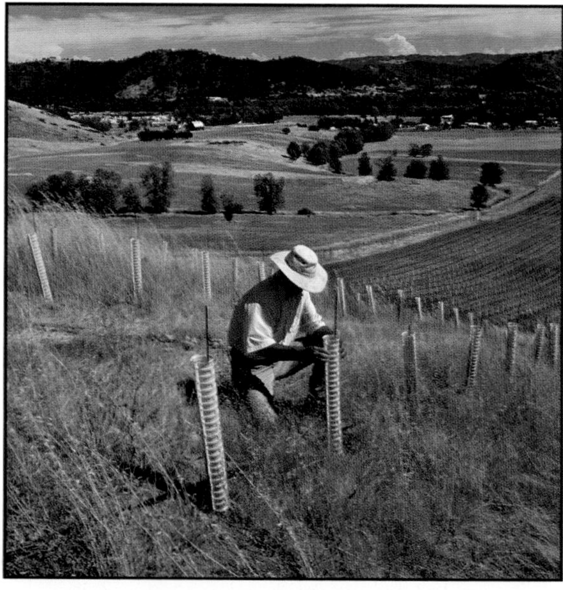

FIGURE 11–14 A vineyard manager examining a young Syrah vineyard in Southern Oregon's Umpqua Valley AVA.

most of the state's Cabernet Sauvignon, Merlot, and Syrah, varieties that do not always ripen in the Willamette Valley.

The Columbia and Walla Walla Valleys

These two appellations lie in the eastern half of the state along Oregon's northern border with Washington. The appellations together have about 1,000 acres (405 hectares) of vineyards, roughly 4 percent of Oregon's total acreage. The majority of both the Columbia Valley and the Walla Walla Valley appellations are in Washington State, and has much more in common with the growing conditions in Eastern Washington than with the appellations in Western Oregon. Although it can be confusing for consumers and producers alike to have AVAs cross over state boundaries, it makes perfect sense considering an AVA should encompass a similar terroir regardless of the political boundaries it crosses.

Summary

Although the wines of the Northwest are often eclipsed by the volume of wine made in California, they represent a significant and growing segment of American wine production. With Washington State wines, consumers have come to expect good Cabernet Sauvignon and Merlot presented at affordable prices. With Oregon wines, customers look for distinctive examples of Pinot Noir and Pinot Gris from small wineries with limited production. The wineries of the Northwest also enjoy a great deal of support in their home states and are becoming better known throughout the rest of the country. With both states, there is still plenty of opportunity for expansion in production, given increased demand from the marketplace.

Review Questions

1. What are the most popular grape varieties for Oregon and Washington?
2. How do the size of Oregon and Washington wineries compare to one another?
3. What is the most significant threat to vineyards in Eastern Washington?
4. Who introduced Pinot Noir to Oregon and when was it done?
5. How does the size of the wine industry in Oregon and Washington compare to California?
6. What is the most famous region in Oregon for growing Pinot Noir?
7. What is Washington's newest AVA and what varieties is it known for?

CHAPTER 12

New York, Canada, and Other North American Regions

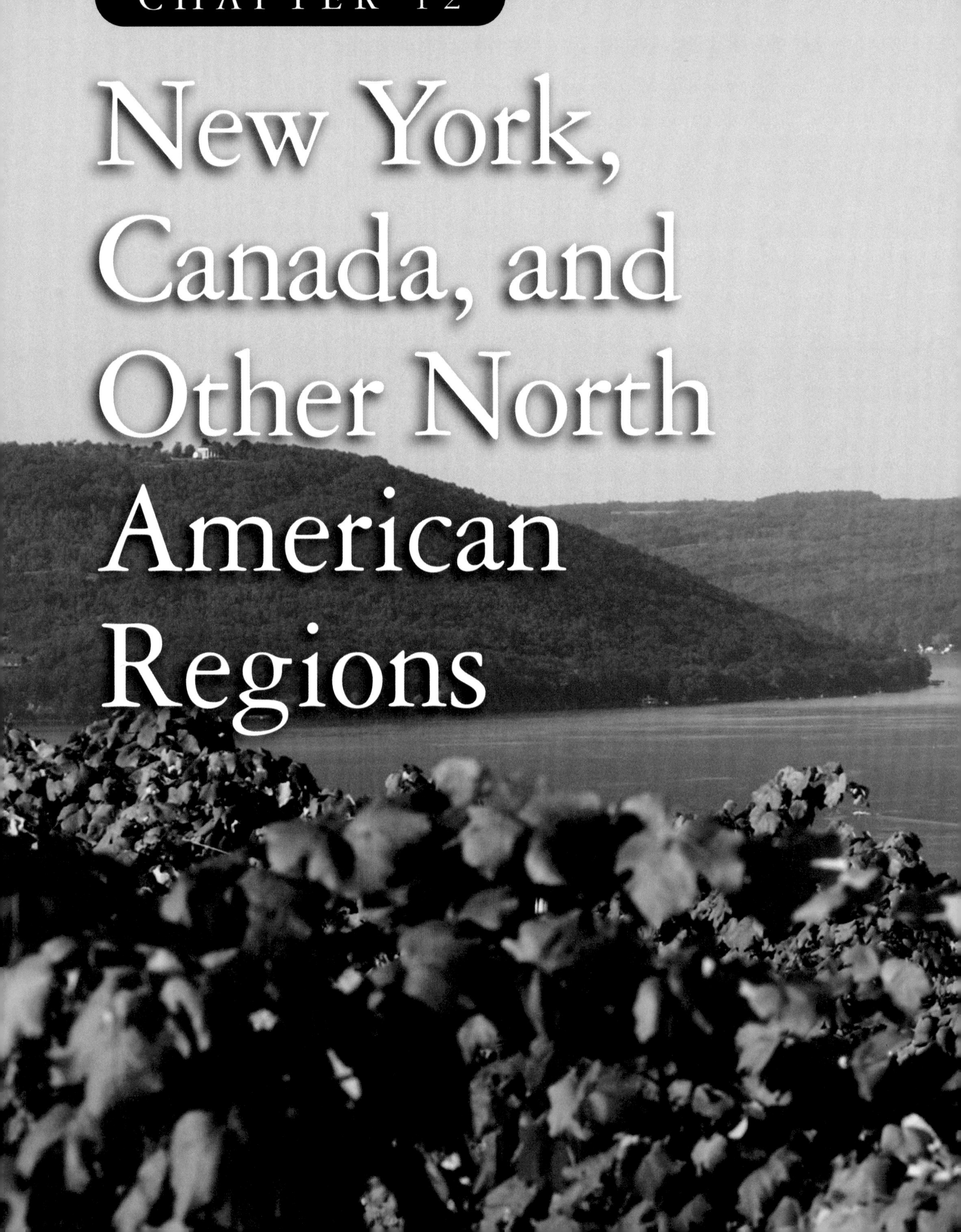

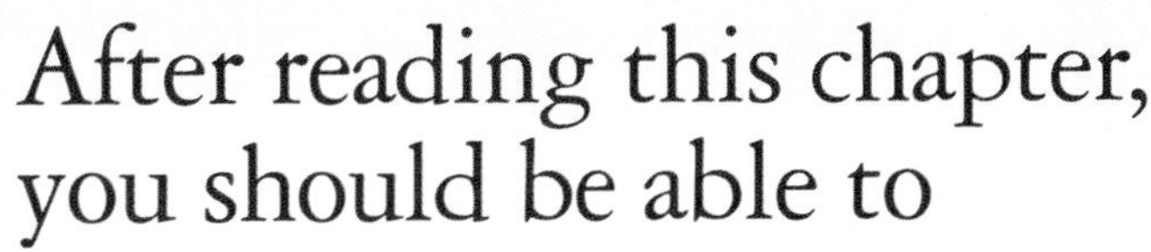

After reading this chapter, you should be able to

- trace the history of wine production in New York State, as well as that state's effect on the development of the American wine industry.
- identify the major wine regions, their respective climatic conditions, and the styles of wines produced.
- describe the unique climatic and geological characteristics of Canada's major wine regions and the styles of wines produced in each.
- understand Canada's role in the international wine market.
- outline the history of wine production in the Eastern, Southwestern, and Mountain regions of the United States.
- describe the types of wines made in these states, and the role of these states in the American wine industry.

KEY TERMS

Chambourcin
crosses
designated viticultural area (DVA)
hybrid
Maréchal Foch
phylloxera
Seyval Blanc
Vidal Blanc
Vitis labrusca
Vitis riparia
Vintners Quality Alliance (VQA)

There are currently federally bonded wineries in all 50 of the United States, as well in four provinces of Canada. These regional wines, from areas outside of America's Pacific West Coast, are of increasing importance in the North American wine industry. This chapter will look closely at wine regions of Canada, New York State, and the American Northeast, South, Southwest, and Mountain regions.

New York State

One could say that the American wine business started in New York State. The country's first commercial winery, Jacques Brothers, was established in 1839 in New York's Hudson River Valley. In 1885 the winery was renamed Brotherhood Winery, and in 2005 was recognized as America's longest continually operating winery. However, grape growing and winemaking had a foothold in New York well before the Jacques brothers began their winery. Dutch colonists

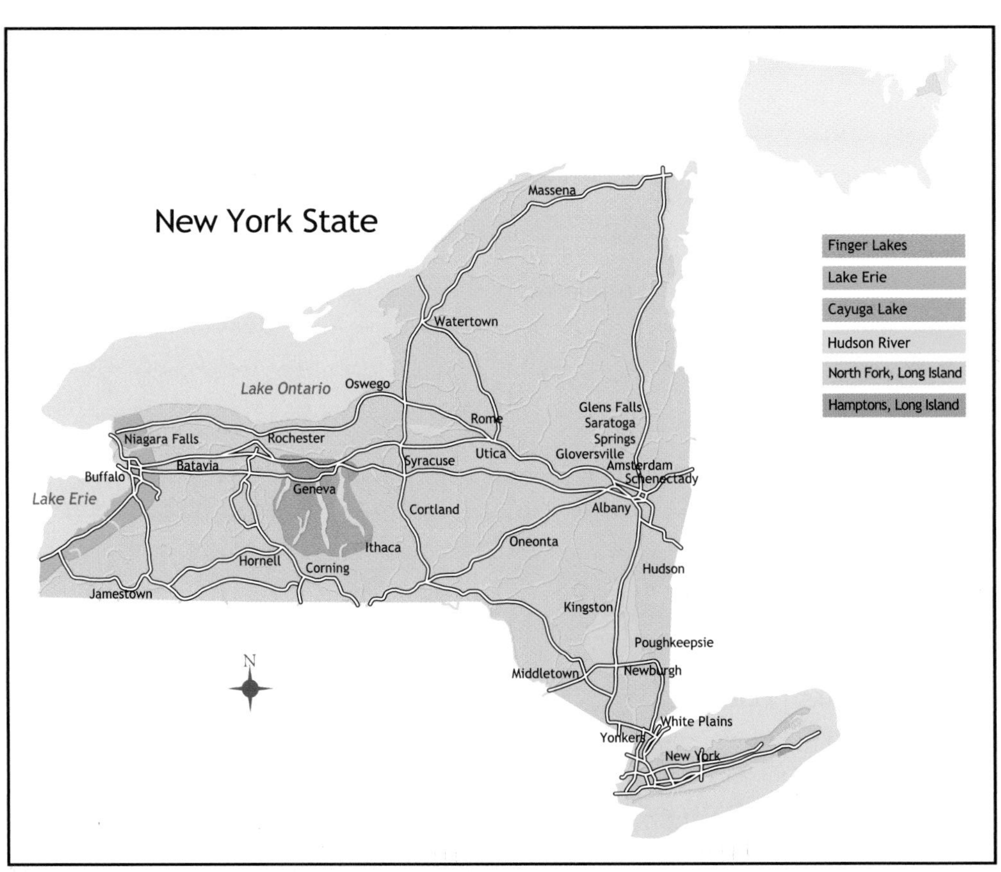

The wine regions of New York State.

settled in the region before the Mayflower brought English settlers to the Massachusetts Bay Colony. The Dutch were followed by French Huguenots, who, in 1677, began making wine near the town of New Pfalz. New York today is a vibrant wine region, producing 2,242,000 cases of fine wine per year. There are currently 163 wineries in the state (compared to only 15 as recently as 1960), and over 33,000 acres (13,360 hectares) planted to wine grapes. However, the history of wine production in this state has been a spotted and difficult, with several major setbacks.

New York State Wines—Historical Perspective

The first wines made in New York by French emigrants in the late seventeenth century were made with local indigenous grapes of the botanical species ***Vitis labrusca*** found growing wild. The results were less than impressive, so the French settlers imported *Vitis vinifera* vines from Europe. These transplanted varietals failed because of the extreme cold of winters, and later, **phylloxera,** the local root louse which killed *vinifera* vines by burrowing into the roots and sapping the plant's energy. (*Labrusca* vines were immune to the phylloxera because of genetic mutations over the many generations the two species had coexisted.) After the failed efforts with *vinifera* grapes, small production of wine continued using the Northeast's native *labrusca* grapes.

In the early and mid-1800s, viticulturists discovered that chance pollination of European *vinifera* by American species had produced some **hybrid** offspring. These were more capable of surviving the harsh climatic conditions. (It is from one of these hybrids, the Alexander, that the first commercial winery made its wines.) Later other chance-produced hybrids were defined, including Catawba and Isabella. By the early nineteenth century, botanists succeeded in creating **crosses** (that is, new species made by breeding two different native grapes) of *vinifera* with *labrusca* vines (as well as some of the native varietals from further south on the North American continent, like the *Vitis aestivalis,* and the ***Vitis riparia,*** a vigorous species found growing in damp sections along rivers and streams from Canada to the Gulf of Mexico.) From this careful, controlled breeding of hybrids emerged many new species suitable to wine production. These new varietals were widely planted along the Hudson River and around the Finger Lakes by new waves of French, German, and Swiss emigrants, starting in the 1850s. Further west, along the shores of Lake Erie, many acres were planted to these new hybrids starting after the Civil War, alongside the much larger plantings of Concord grapes that went into grape juice and jelly. Commercial winemaking took off across New York, and a large wine industry emerged, centered on the town of Hammondsport in the Finger Lakes.

Prohibition dealt a severe blow to the nascent wine industry of New York. Many of its wineries discontinued production of wine, turning instead to growing fruit grapes, or simply going out of business. When Prohibition was repealed in 1933, wine production re-emerged but maintained its emphasis on the native and hybrid grapes, producing heavy, slightly sweet wine, with a bouquet often described as "foxy" for its resemblance to the scent of animal fur. Most growers sold their harvested grapes to the two or three large companies that controlled the state's wine industry. A specialized section of the New York wine trade concentrated on the production of sweet kosher wines from the Concord grape. In the late 1930s some pioneers, such as Frenchman Charles Fournier from the Champagne region, began to plant some *vinifera* vines in the Finger Lakes district. But it was not until the arrival in New York in the 1950s of Dr. Konstantine Frank, a Ukrainian vintner and expert in *vinifera* grapes, that modern wine production started in New York. He did extensive research and identified areas where cool climate *vinifera* varietals like Riesling and Chardonnay could thrive.

Another boost to New York's wine industry was the Farm Winery Act of 1976, which was passed due to the growing interest in fine wine that was becoming evident across the United States. The act reduced fees for commercial wineries, increased tax benefits for small wineries, and allowed direct sales to consumers and restaurants. These critical changes made it economically feasible to own and operate a small winery (defined in the act as one that produces less than 50,000 gallons per year.) The result was a considerable increase in the number of small and medium-sized wineries in New York. Today most of New York's wineries are of this size, the type of boutique winery at which the emphasis is on quality and innovation.

Wine Regions of New York

The growing season in this large state varies from 180 days in northern, inland areas to up to 230 days in more moderate sections. The microclimates are strongly influenced by contiguous bodies of water. In upstate New York, many lakes were formed during the Ice Age as melting glaciers cut deep formations that gradually filled with water and became Lake Erie (one of the Great Lakes), the Finger Lakes, and the Hudson River (Figure 12-1). The glaciers also left rich soil behind. Closer to the Atlantic, it is the ocean that has a moderating influence on climate. Here the soils are sandier. The microclimate is very different from what one finds upstate. Among all the various growing conditions throughout New York, four regions have evolved as her premier wine-producing areas.

THE FINGER LAKES

Named for the lakes' resemblance to a hand's fingers, this region, although not New York's largest, produces 90 percent of the state's wine. There are 63 bonded wineries around the lakes, and 10,000 acres (4,050 hectares) of vineyards. The Finger Lakes AVA (American Viticultural Area) was established in 1982. There are 11 lakes in total. The important lakes for wine production are Canandaigua, Seneca, Keuka, and Cayuga. These are among the deepest lakes in North America, and all are large enough to have moderating influences on the climate. Cayuga was granted its own AVA in 1988 when local vintners were able to prove that its lower elevation and deeper depth provided mesoclimates suitable for the recently planted *vinifera* varietals.

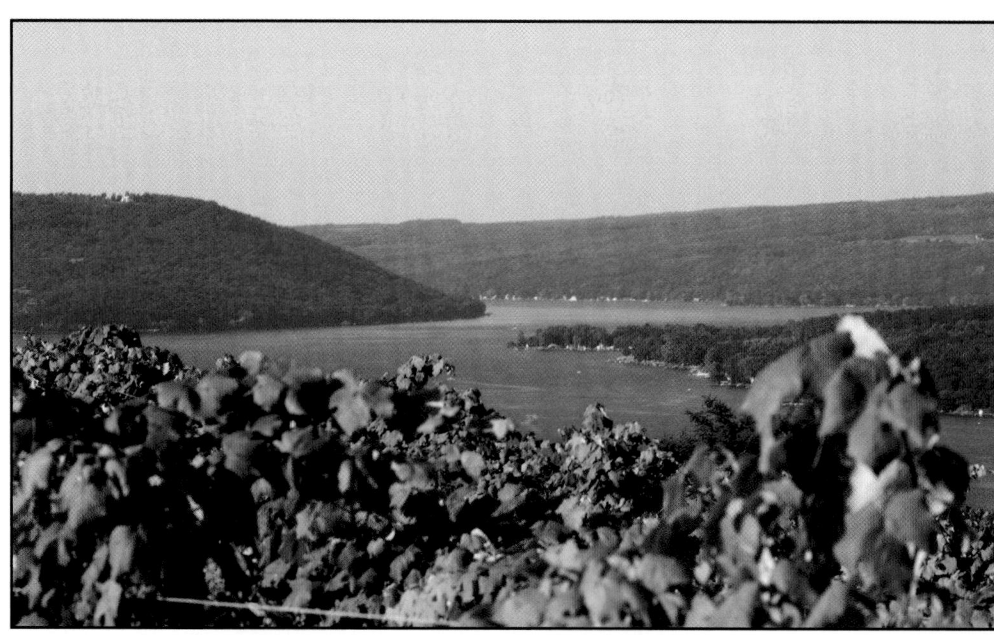

FIGURE 12–1 The sloping hillside vineyards in New York's Finger Lakes allow the ripening grapes to absorb sunlight, while the warmth reflecting back off the lake's waters further enhances the grapes' ripening process.

The vineyards of the Finger Lakes are planted to a variety of grapes—hybrids, crosses, and *vinifera.* George Fournier first introduced *vinifera* vines here in the 1930s, and hired Konstantin Frank as a consulting viticulturist. As Dr. Frank had predicted, Riesling does extremely well here, but its plantings are still small—only 500 acres (202 hectares). Also planted is Chardonnay, often made into lively fresh sparkling wine, by producers such as Glenora Winery. However, *vinifera* is still in the minority around the Finger Lakes, most of whose wines are still made from hybrids and native grapes, such as Niagara and Catawba. Concord is the most widely planted red varietal, and is made into a soft fruity wine. There have been promising reds made from the Bordelais varietal, Cabernet Franc.

Lake Erie

Lake Erie provides more climate-moderating influences than the other Great Lakes because it is lower in altitude and is downwind from Arctic air masses that prevail over Lakes Superior and Huron. Moreover, the huge Allegheny Plateau, 3 miles wide, acts to further trap the warmer air that radiates off the lake, thus increasing protection of the vineyards. The growing season averages about 185 days, from late April to early October. Precipitation is about 30 to 40 inches (76 to 102 cm) a year, and is evenly spread out over the year, allowing for adequate sunlight during the growing season. The Lake Erie AVA, established in 1983, extends into three states—New York, Pennsylvania, and Ohio. It has the largest amount of acreage in New York planted to grapevines (2,200 acres/890 hectares), but the majority of those grapes are destined for juice or jelly, or are eaten as table grapes. There only eight wineries in the AVA, producing dry table wines, sparkling wines, and a few select dessert wines, mostly from French-American hybrids such as **Seyval Blanc.**

Hudson River Valley

Besides its distinction as America's oldest wine region, where wine has been produced continually for over 300 years, the Hudson River Valley, just a short drive from New York City, is also one of North America's loveliest wine regions. The valley, with its historic small villages full of charming houses, antique shops, and cafes, provides views of stately green hills sloping toward the river. Many of the hills are covered with vineyards that are sheltered by the ridges and rock cliffs of the Catskills Mountains to the west. The Hudson River with its steep palisaded valley acts as a conduit for maritime air and weather patterns coming off the Atlantic. The mesoclimate here is, therefore, milder than is the case farther upstate, allowing several *vinifera* varietals to thrive, primarily Riesling, Chardonnay, and Cabernet Franc, in addition to the hybrids that the region has grown since before Prohibition. The Hudson River Valley was officially recognized as an AVA in 1982 (Figure 12–2). The emphasis now is strongly on delicately styled, *vinifera*-based table wines.

Long Island

This is the newest wine region in New York. It is also the most exciting and the most promising. The wine country is at the extreme easternmost section of the island, where the surrounding waters of Long Island Sound to the north, Peconic Bay to the south, and the Atlantic Ocean to the east provide a mild enough climate for *vinifera* varietals such as Merlot, Chardonnay, and Cabernet Franc to thrive. The growing season averages between 204 and 233 days, perfect for varietals that need long hang-time to fully evolve their flavors. The soil, for centuries planted primarily to vegetables, especially the lowly potato, is rich in minerals and drains well. The region has 1,950 acres (789 hectares) planted to wine grapes, spread among three AVAs. The Long Island AVA was defined and approved only in 2001, preceded by the original growing areas of the Hamptons (approved 1985) and the North Fork AVA (1986). Because of a thriving tourism

FIGURE 12–2 The Palisades, the 550-foot high cliffs along the western bank of the Hudson River, act as a natural funnel, pulling the moderating ocean air upriver, thus helping to ripen the grapes in vineyards along the river.

business (New York City is only 85 miles/137 km away) and considerable foreign investment (mostly by Europeans who recognize this as a promising grape-growing region), Long Island is the fastest growing wine region in the Eastern United States, producing some truly impressive wines. There are now 21 wineries in the North Fork and two in the Hamptons (Figure 12-3). Among the leading producers are Wölffer Estate, founded by a wealthy German businessman; Castello di Borghese, formerly Hargrave Vineyard (one of the first quality wineries here, first planted in 1973); and Pindar, another pioneer. In 1989, the Long Island Wine Council was founded with the objective of promoting Long Island as a producer of world class wines.

New York has played an extremely important role historically in the evolution of the wine industry of the United States. Now, led by exciting developments on Long Island's North Fork, and by increasing emphasis upstate on *vinifera* varietals, especially Riesling, New York seems poised to claim its place again as one of this country's finest regions for wine production.

COLLINA 48
NORTH FORK OF LONG ISLAND
MERLOT
ALC. 12.5% BY VOL.

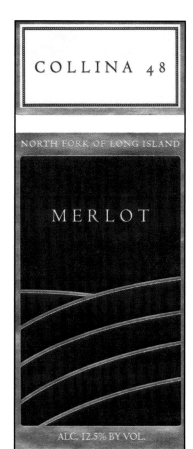

FIGURE 12–3 The North Fork of Long Island is proving to be a promising area for Merlot.

Other Wine Regions in the Eastern United States

Besides New York, other Northeastern states producing promising wines are Connecticut, Rhode Island, and the south-central coast of Massachusetts. Portions of these three states are included in the regional AVA, Southeastern New England. The boundaries of the AVA extend from just south of Boston down along the Rhode Island and Connecticut coasts to just north of New London. The boundaries never extend more than 15 miles (24 km) inland, assuring a truly coastal climate. The ocean moderates temperature extremes, so that the average temperature in January is 30°F and 70°F in July. Overall the climate is perfect for cold-hardy *vinifera* varietals such as Chardonnay, Pinot Noir, and Riesling, as well as certain French hybrids such as **Vidal Blanc.** Among the area's leading producers are Sakonnet Vineyards of Rhode Island, which makes a delightful rosé and lovely Gewürztraminer, and Westport Rivers of Massachusetts, whose sparkling wine made from Chardonnay is delicious (Figures 12–4 and 12–5).

FIGURE 12–4 Bill Russell, winemaker at Massachusetts' Westport Rivers Winery.

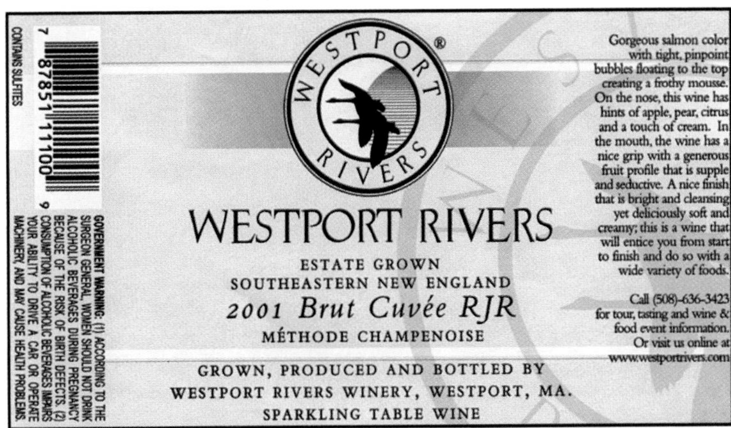

FIGURE 12–5 Westport Rivers' sparkling wine, made in the *méthode champenoise,* is 73 percent Pinot Noir and 27 percent Chardonnay. All grapes are grown in the winery's vineyards near Westport, Massachusetts, a few miles inland from the Atlantic Ocean.

Virginia

Winemaking in this Southern state dates back to the Jamestown settlement, where the earliest European settlers made themselves small batches from the indigenous wild grape, Scuppernong. The results were apparently quite unpalatable, as the colonists soon appealed for help. In 1609 the sponsoring company sent French vine cuttings to Jamestown. Better wine may have been produced, but the vines all died of fungus and other local diseases. Despite repeated attempts, Virginians were not able to protect their *vinifera* vines. Thomas Jefferson, a Virginian and a wine lover, was determined to make his state into a wine region. He traveled widely through Europe and brought back many cuttings. He planted several acres of his estate to these cuttings and to native grapes. There is no record of Jefferson's having ever made wine from his vineyard, as his vines probably also succumbed to disease and uncertain climate. But his legacy lives on. After the setbacks of phylloxera and other vine diseases, Virginia's farmers planted the hybrids that were developed by French and American botanists in the late nineteenth century. At the same time, crosses were also perfected. These hybrids and crosses, especially the crosses of the native red grape, the Norton, were hardy enough to resist the infections and pests that felled *vinifera* vines, as well as colder temperatures. Especially successful in Virginia were the hybrids of Seyval Blanc and Vidal Blanc, both whites, and **Maréchal Foch** and **Chambourcin,** both reds.

Prohibition, of course, caused the collapse of Virginia's wine industry, which did not revive until well into the 1970s, when new wineries were established that made interesting wines from the same French-American hybrids. With increased dedication to quality, Virginia vintners expanded their plantings of *vinifera* grapes in the 1980s. They now make very impressive Rieslings, Chardonnays, and Cabernet Francs, along with the more traditional Seyval Blanc, Vidal Blanc, and Norton. There are 70 wineries in Virginia, most small and family-owned. (The exception is the large, regional winery, Barboursville, owned by Zonin, a major Italian wine company). Virginia's wineries are helped by a well-established tourist industry. There are six AVAs in the state. Many experts consider Virginia to be one of the most exciting wine regions on the East Coast (MacNeil, 2001).

The Western United States

When people think of the Western states, they are more likely to picture cattle spreads and sheep ranches than wineries and vineyards. But several regions in the West, especially the Southwest and the Mountain states, are emerging as promising wine-producing regions. Among the more important are Texas, New Mexico, and Colorado.

Texas

The huge state of Texas is now the seventh-largest wine-producing state in the country. It traces its wine history to the mid-seventeenth century when Spanish missionaries first planted grapevines to make wine for the sacrament. (The grape they planted, the Criolla, has come to be known as the Mission grape.) Wine continued to be produced in Texas on through the eighteenth and nineteenth centuries, as new settlers brought vines with them. In hopes of replicating the wines of their native lands, they planted European *vinifera* grapes alongside the Mission. However, because of indigenous pests and fungal diseases and very cold winters, the only vines to survive were those of the Mission grape. With Prohibition, all efforts to produce quality wine in Texas came to a close for the next four decades.

In the 1970s, the increasing interest in fine wine across the United States inspired several Texas natives to give winemaking a try. Several grape-growers' associations were formed to do research and to support efforts at improving local viticulture. A professor at Texas Tech university, Bob Reed, began in 1976 to conduct experiments in viticulture. Soon his experiment had become the Llana Estacado Winery, now the state's largest winery. In the early 1980s, the University of Texas began planting experimental vineyards on land it owned. The research conducted by the university helped immeasurably to determine what varietals to plant on Texas' mesas and steep hillsides with their granite-limestone soil.

There are seven AVAs within Texas, stretching from the Texas High Plains AVA in the Texas Panhandle, to Texas Hill Country near Austin (the largest AVA in the United States), to Davis Mountains, near the border with Mexico. The prime determinants on what grapes to plant are climate and elevation. Higher elevations, above 3,000 feet (914 m), tend to produce the highest quality wines because of sunny days, very cool evenings, and lower humidity. At these elevations, Chardonnay is the principal white grape, along with Chenin Blanc, Gewürztraminer, Muscat Canelli, and Riesling. In these high vineyards, the reds are Cabernet Sauvignon and Merlot. At lower elevations, where temperatures tend to be higher, vineyards are more likely to be planted to Zinfandel, Sangiovese, Syrah, and Viognier, the popular white varietal from the Rhône. Some vintners, like Jim Johnson of Alamosa Cellars in the Texas Hill Country, are experimenting with the Tempranillo and Garnacha grapes of Spain.

There are now 106 wineries in Texas (*Wine Business Monthly,* Sept. 2005, p. 15). Many of them are small, family-owned operations, but Texas' largest winery, Ste. Genevieve, produces over a half a million cases a year. Despite the looming threat of Pierce's disease (see Chapter 10) in some parts of the state, and an incomplete process of viticultural winnowing (the determination of which grape varietal will grow best in which region), Texas has the potential to become a steady producer of high quality *vinifera*-based wines.

New Mexico

It is safe to say that without the French, there would be no wine industry in New Mexico. It was Spanish missionary monks who first planted wine grapes in the region, back in 1629, in order to provide themselves with sacramental wine for the daily Mass. From this humble beginning, New Mexico's production of wine grew, until by the late 1800s, the state was fifth in the nation in volume (Chittim, 2005). However, because of various calamities, some man-made (Prohibition) and some imposed by nature (flooding of the Rio Grande), New Mexico's wine industry ceased to exist by the 1940s. But the French brought it back.

In 1980, a contingent of French vintners visited New Mexico and recognized the tremendous potential for viticulture in the central part of the state, where broad plateaus at an average elevation of 4,000 feet (1,209 m) provided the ideal combination of sunny days and very cool evenings. The French vintners planted 3,000 acres (1,214 hectares) of *vinifera* vines in three locations. They hoped to eventually plant up to 21,000 acres (8,500 hectares). However, the project was allowed to peter out when market response was less enthusiastic than hoped. That failure could have been the end of winemaking in New Mexico, but the French arrived again in the form of one family from Champagne. In 1984, the Gruet family was vacationing in the Southwest when they visited the vineyards planted by the French vintners. After talking with the vintners and touring the area, Gilbert Gruet, whose family had been making Champagne for decades, realized his countrymen had been right. He bought a few acres, planted Chardonnay and Pinot Noir, and built a winery.

Today, Gruet is one of New Mexico's largest wine producers, run by Gilbert Gruet's son, daughter, and son-in-law. They turn out 50,000 to 60,000 cases a year of sparkling wine made in the *méthode champenoise.* Gruet's NV Brut, Blanc de Noirs, Demi-Sec, and Grand Rosé are widely recognized as among the best sparkling wines made in North America.

The New Mexico wine industry continues to grow. There are now three AVAs (one established in 1988 and the other two in 1985) and over 500 acres (202 hectares) planted. There are about 40 wineries, most very small. The only large enterprise besides Gruet is also French-owned: St. Clair Winery, owned by the Lescombes family, produces 50,000 cases a year of varietally named table wine. The state government has always been a strong supporter of the wine industry, and that support, plus a solid tourist trade, should assure the stability and continued growth of New Mexico's wine production.

Colorado

Winemakers whose vineyards and winery are at very high Rocky Mountain elevations face challenges that any winemaker at sea level cannot even imagine—everything from a winter snow cover of more than 12 feet (3.6 m), to spring frosts, to summer drought and elk munching on vine buds. Yet Colorado's Terror Creek Winery, elevation 6,417 feet (1,956 m), does turn out 850 cases a year of *vinifera* wines, including dry Alsace-style Gewürztraminer and Riesling.

As of early 2005, there were 62 licensed wineries in Colorado, most of them on the western slope of the Rockies. Total acreage is 750 acres (304 hectares) of vineyards. These numbers reflect a rapid growth rate: In 1990, there were no producing wineries in the state, and by 1999, there were only 300 acres (121 hectares) planted. Many wineries, especially in higher elevations like Terror Creek, concentrate on cool-climate *vinifera* varietals. Many vintners are now saying that Riesling is the most promising grape for their state. However, Merlot is still the most widely planted, taking up 28.9 percent of acreage, primarily in the warmer microclimates of the valleys. Presently, most of the production from Colorado's wineries is sold through their own tasting rooms, which are crowded four seasons a year with tourists. Whether Colorado's wines will eventually take their place in the national market remains to be seen.

Canada

Canadians are rightfully proud of the wines produced in their country. There are four provinces that produce wines, but the major wine regions are in two provinces: Ontario and British Columbia. Because these two provinces are literally a continent apart, the wines they make are very different. Ontario's vineyards lie primarily on the Niagara Peninsula, north of New York State. The climate here is very similar to that of New York's Finger Lakes. The specialty is ice wine, as well as German-style Rieslings. British Columbia is on Canada's West Coast, and its terroir closely resembles that of Washington State's Columbia Valley, and like the Columbia Valley, British Columbia is building a reputation for muscular reds, especially Merlot and Cabernet Franc. Its dry *vinifera*-based whites, like Chardonnay, can also be very good (Figure 12–6).

Canadian Wine—Historical Perspective

Winemaking in Canada can be traced to the early 1800s, when a German emigrant named Johann Schiller, recently retired from a military career, planted a small vineyard to native *labrusca* vines along a river west of Toronto. By 1890, there were 41 wineries throughout Canada. However, commercial winemaking did not have much time to develop, as Prohibition began in Canada in 1916, four years earlier than in the United States. The Great Experiment lasted only a few years in Canada, and upon its repeal in 1927, the provinces began granting licenses to new wineries. By the end of that first year after Repeal, 57 licenses to make wine were granted in the province of Ontario alone. In the ensuing decades, most wine made in Canada was of the slightly sweet, highly alcoholic style, and were made from *labrusca* grapes.

Canada's modern wine industry was born in 1975 when the small winery Inniskillen near Niagara Falls, was granted the first commercial license. At this time the taste in wines across North America was switching away from sweet and fortified wines to dry, balanced table wines. As the demand for this style of wine increased, so did the determination of Canadian vintners to make better, more sophisticated wines. As the number of commercial (albeit small) wineries increased, the Canadian government saw the need to institute some control over the production of wine and the use of names for the grape-growing regions. In 1988, a country-wide appellation system, **Vintners Quality Alliance (VQA),** was introduced, originally in Ontario and shortly thereafter in British Columbia. The VQA seal on a bottle's label signifies that the wine has been tested and meets a series of standards set by a board of local vintners, grape growers, and wine experts. The VQA also controls the use of appellations or **DVAs (designated viticultural areas).** The law stipulates that wines carrying the seal are made at least 95 percent from locally grown

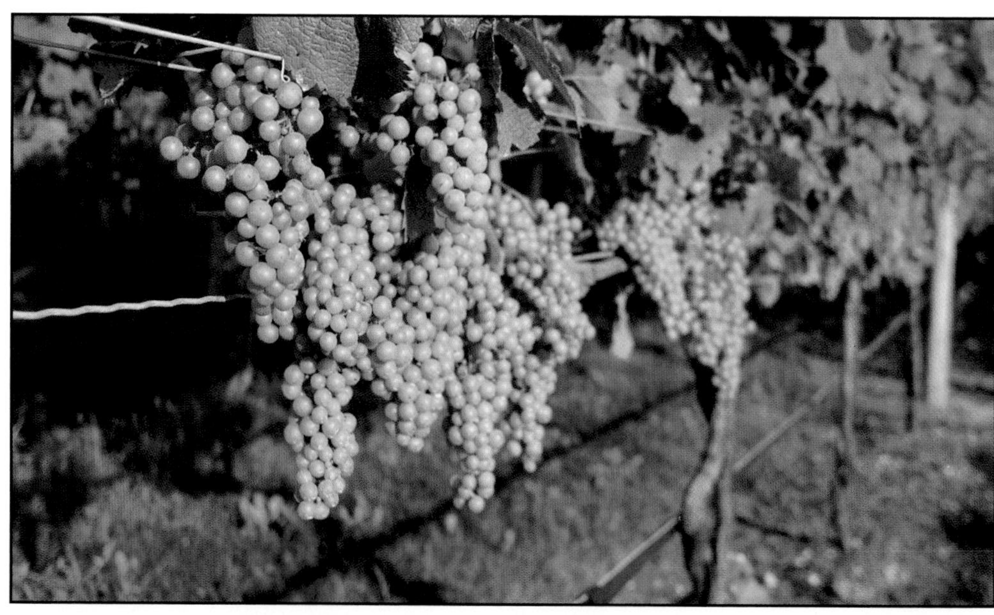

FIGURE 12–6 Chardonnay grapes can ripen in the northern sections of North America if temperatures are moderate and sunshine is plentiful.

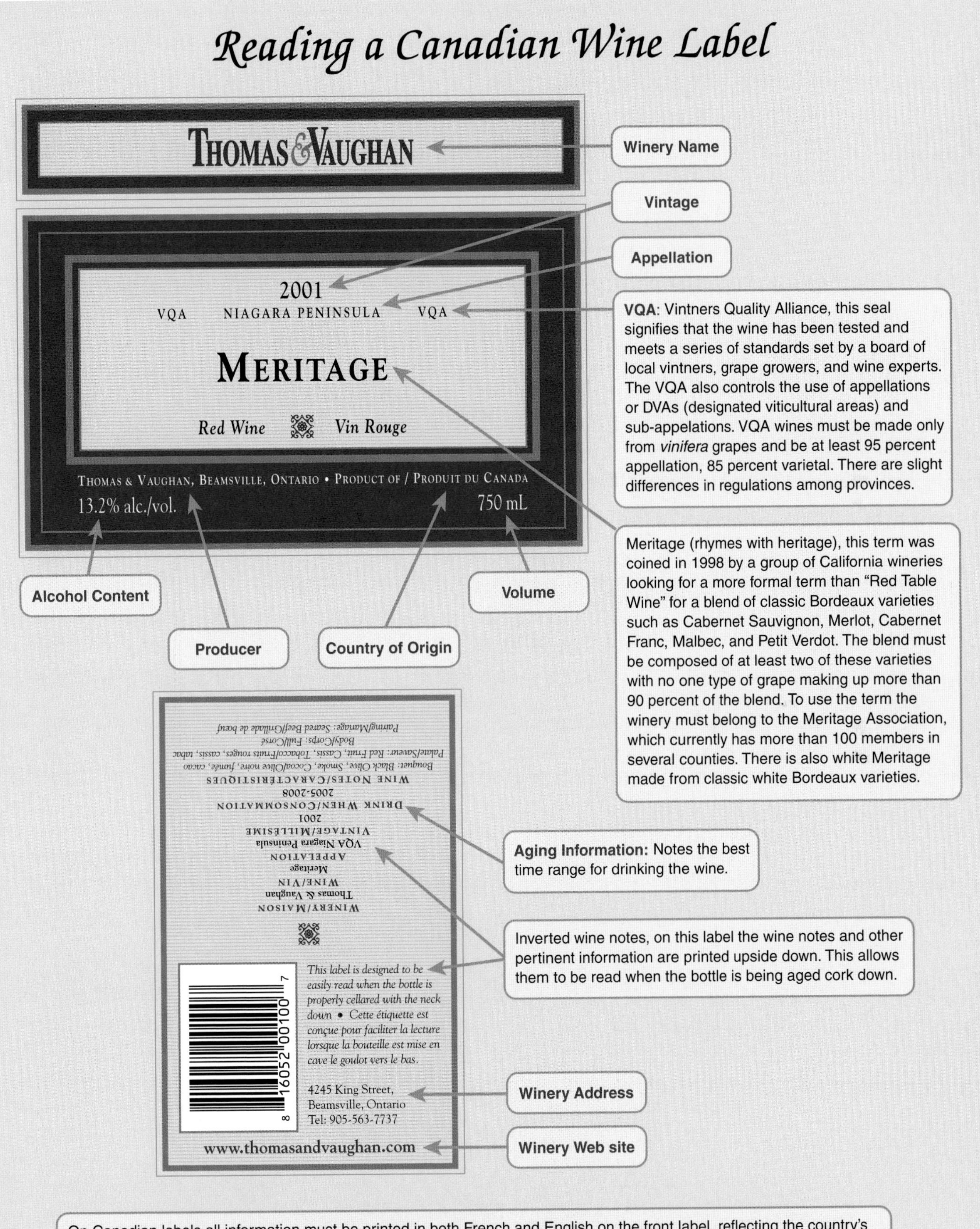

Reading a Canadian Wine Label
THOMAS & VAUGHAN
Winery Name
Vintage
Appellation
2001
VQA NIAGARA PENINSULA VQA
MERITAGE
Red Wine Vin Rouge
THOMAS & VAUGHAN, BEAMSVILLE, ONTARIO • PRODUCT OF / PRODUIT DU CANADA
13.2% alc./vol.
750 mL
VQA: Vintners Quality Alliance, this seal signifies that the wine has been tested and meets a series of standards set by a board of local vintners, grape growers, and wine experts. The VQA also controls the use of appellations or DVAs (designated viticultural areas) and sub-appelations. VQA wines must be made only from vinifera grapes and be at least 95 percent appellation, 85 percent varietal. There are slight differences in regulations among provinces.
Meritage (rhymes with heritage), this term was coined in 1998 by a group of California wineries looking for a more formal term than "Red Table Wine" for a blend of classic Bordeaux varieties such as Cabernet Sauvignon, Merlot, Cabernet Franc, Malbec, and Petit Verdot. The blend must be composed of at least two of these varieties with no one type of grape making up more than 90 percent of the blend. To use the term the winery must belong to the Meritage Association, which currently has more than 100 members in several counties. There is also white Meritage made from classic white Bordeaux varieties.
Alcohol Content
Volume
Producer
Country of Origin
WINERY/MAISON
Thomas & Vaughan
WINE/VIN
Meritage
APPELLATION
VQA Niagara Peninsula
VINTAGE/MILLÉSIME
2001
DRINK WHEN/CONSOMMATION
2005-2008
WINE NOTES/CARACTÉRISTIQUES
Bouquet: Black Olive, Smoke, Cocoa/Olive noire, fumée, cacao
Palate/Saveur: Red Fruit, Cassis, Tobacco/Fruits rouges, cassis, tabac
Body/Corps: Full/Corsé
Pairing/Mariage: Seared Beef/Grillade de boeuf
Aging Information: Notes the best time range for drinking the wine.
Inverted wine notes, on this label the wine notes and other pertinent information are printed upside down. This allows them to be read when the bottle is being aged cork down.
This label is designed to be easily read when the bottle is properly cellared with the neck down • Cette étiquette est conçue pour faciliter la lecture lorsque la bouteille est mise en cave le goulot vers le bas.
8 16052 00100 7
4245 King Street,
Beamsville, Ontario
Tel: 905-563-7737
Winery Address
www.thomasandvaughan.com
Winery Web site

On Canadian labels all information must be printed in both French and English on the front label, reflecting the country's bilingual status. On the back label having both French and English is optional. By treaty with the European Union, generic use of European names like "Chablis" is not allowed. Many Canadian wines are blends of both domestic and imported wines. These wines cannot be bottled under the VQA standards and have the statement "Cellared in Canada."

grapes. Only *vinifera* grapes are allowed. Furthermore, if a varietal name is given to the wine, 85 percent of the wine must be made from that varietal.

Wine Regions

Wine is produced in four Canadian provinces: Nova Scotia, Quebec, British Columbia, and Ontario. The amounts made in the first two are not commercially significant. In Quebec, wineries are making small amounts of hybrid-based wines, most of which is sold at the wineries' tasting rooms to tourists. Nova Scotia's three wineries are likewise dependent on tourism.

BRITISH COLUMBIA

British Columbia currently has 5,462 acres (2,211 hectares) planted to grapevines, with plenty of room for expansion, as the province has more land area than France and Germany combined. Most of that acreage is devoted to *vinifera* grapes, because in the late 1980s the provincial government offered a financial incentive to growers, encouraging them to tear out *labrusca* and

British Columbia's Okanagan Valley

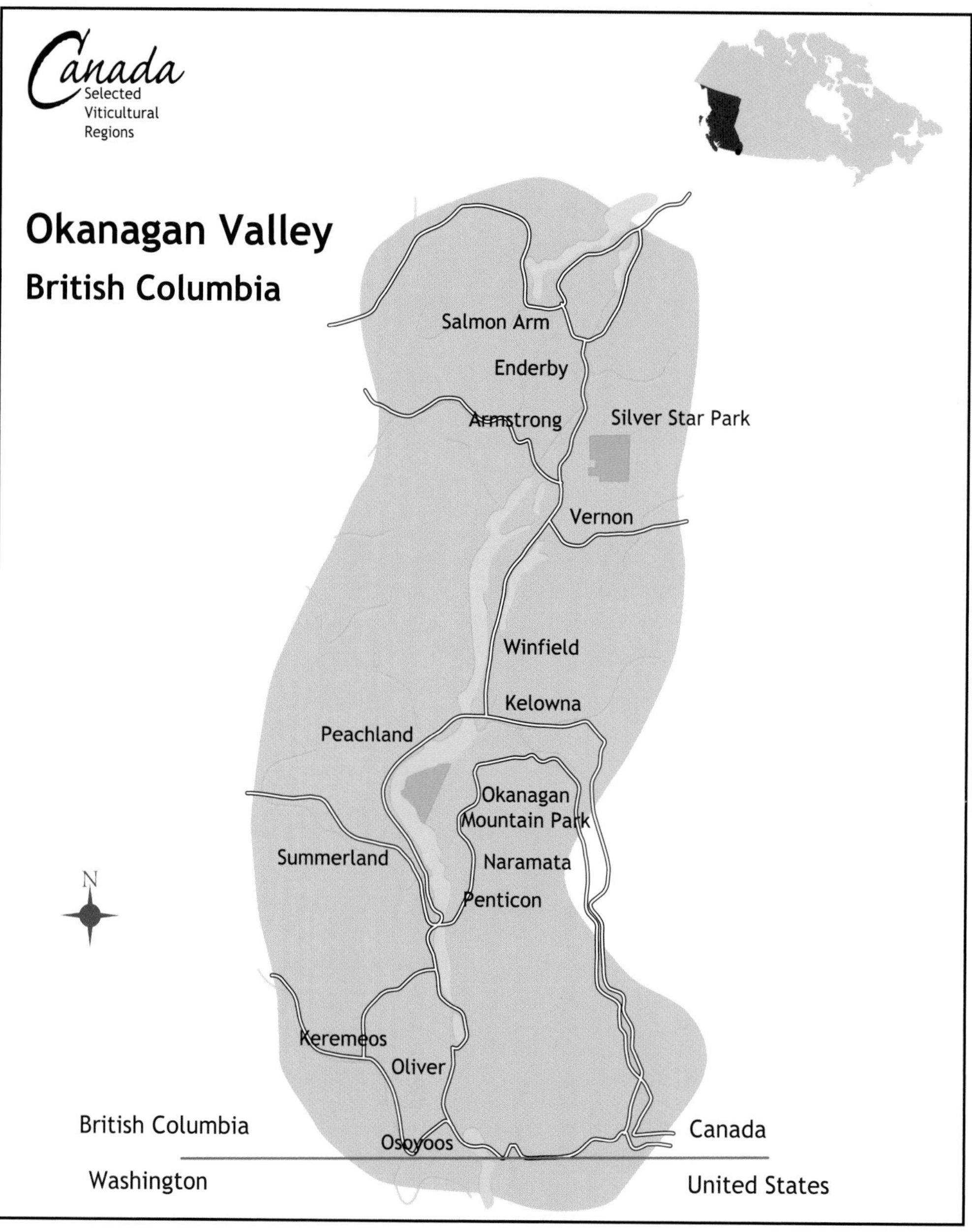

hybrid vines and replace them with *vinifera.* The most widely planted varietals are Merlot, Chardonnay, Cabernet Franc, Pinot Gris, Cabernet Sauvignon, and Riesling. The wine industry is spread among four distinct regions: the Okanagan Valley, Similkameen Valley, Vancouver Island, and the Fraser Valley. Most of British Columbia's wine comes from the Okanagan Valley, which extends for 100 miles (161 km) in the south-central part of the province. The valley receives minimal rainfall (the southernmost section is Canada's only classified desert.) Summers are hot, with July and August temperatures often reaching over 100°F. The northern latitude means days are long. Irrigation is necessary in such a hot, sunlit region, and several large nearby lakes provide plenty of water for the vineyards. Long sunny days, carefully controlled water content, and very cool nights, along with loamy-sandy soil, make for a grape grower's dream. Vintners are able to bring their crops to perfect ripeness, with acidity still intact. Although British Columbia is known for crisp clean whites such as Alsace-style Pinot Gris and vibrant Sauvignon Blanc, impressive progress has also been made with reds. British Columbia also makes sparkling wines from white *vinifera* varietals. It is also a reliable producer of Canada's most famous wine—ice wine.

Although British Columbia's wines are now sold in several European countries and a dozen U.S. states, fully 90 percent of its production is still sold within the province. Owners of many of the province's 97 wineries are working closely with the provincial government to create international market demand for their products.

Ontario

Being far inland, Ontario has a continental climate. Arctic air assures a short growing season and very cold winters. Without the moderating influences of the two Great Lakes that are contiguous with its borders, Ontario and Erie, the province would be too cold for any viticulture. Fortunately, the deep lakes warm the region enough to make the growing of certain cold-hardy

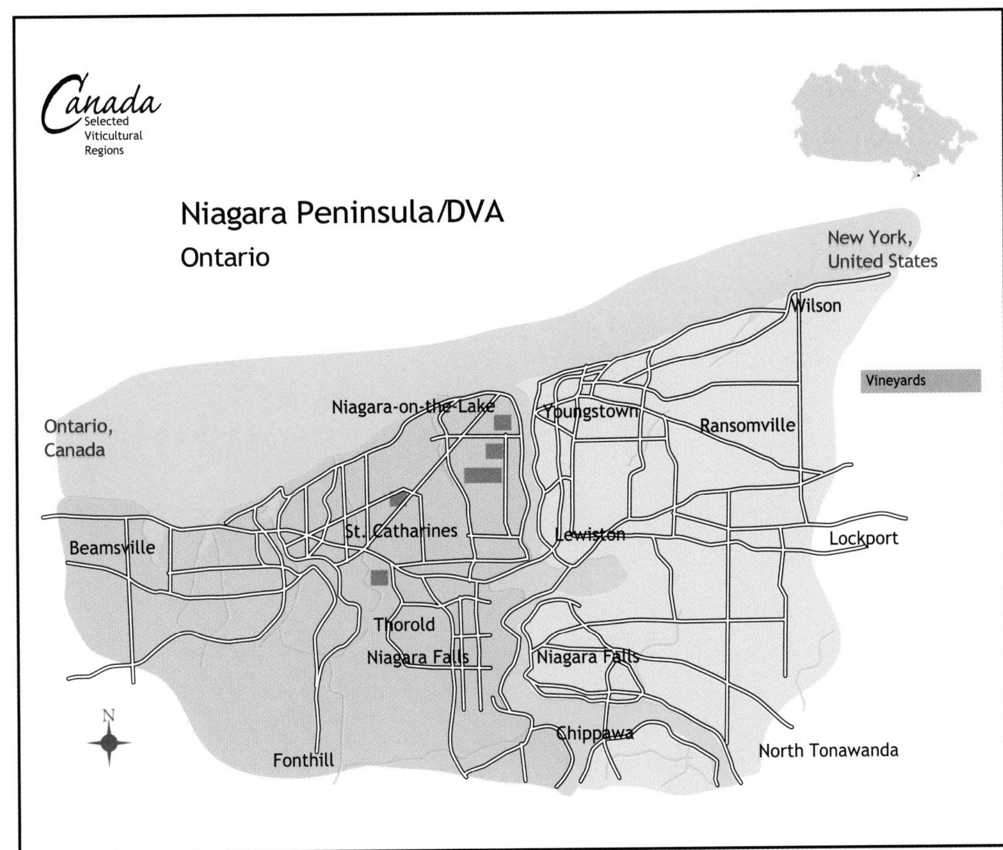

Ontario's Niagara Peninsula

varieties possible. Most of Ontario's vineyards are clustered along the Niagara Peninsula, where a huge escarpment, once the towering rocky shore of an Ice Age lake, provides additional protection from icy winds. The glaciers that carved out that ancient lake deposited a variety of deep, well-drained soils, also conducive to successful viticulture.

Ontario is Canada's largest producer of wine, accounting for about 75 percent of the country's wine. The VQA board in Ontario has approved three DVAs: Niagara Shore, Lake Erie North Shore, and Pelee Island. Although some table wines are made from *vinifera* grapes, for instance, Chablis-style Chardonnays and some off-dry Rieslings, many vineyards in all three DVAs are still planted to hybrids, predominately Vidal Blanc and Maréchal Foch.

Despite the DVAs' progress with table wines, the undisputed star in Ontario is ice wine, made from grapes (usually Vidal Blanc) that have been allowed to continue ripening on the vine until frozen by a sudden deep frost (Figure 12–7). The frozen water of the grape is separated out before fermentation. The result is a richly honeyed but cleanly balanced dessert wine of incredible complexity, with lovely nuances of ripe peaches or apricots. So serious are the Canadians about protecting the integrity of their ice wine that the VQAs of both Ontario and British Columbia have joined in an international agreement with Germany and Austria, pledging to use only the traditional, risky, and labor-intensive method of making ice wine from naturally frozen grapes. Producers vow never to sink to the shortcut sometimes used in certain New World regions of picking ripe grapes and placing them in large industrial freezers. A consumer can know, when buying a Canadian ice wine, that the product he receives will be a delicious, genuine example of this ambrosial wine for which Canada is famous.

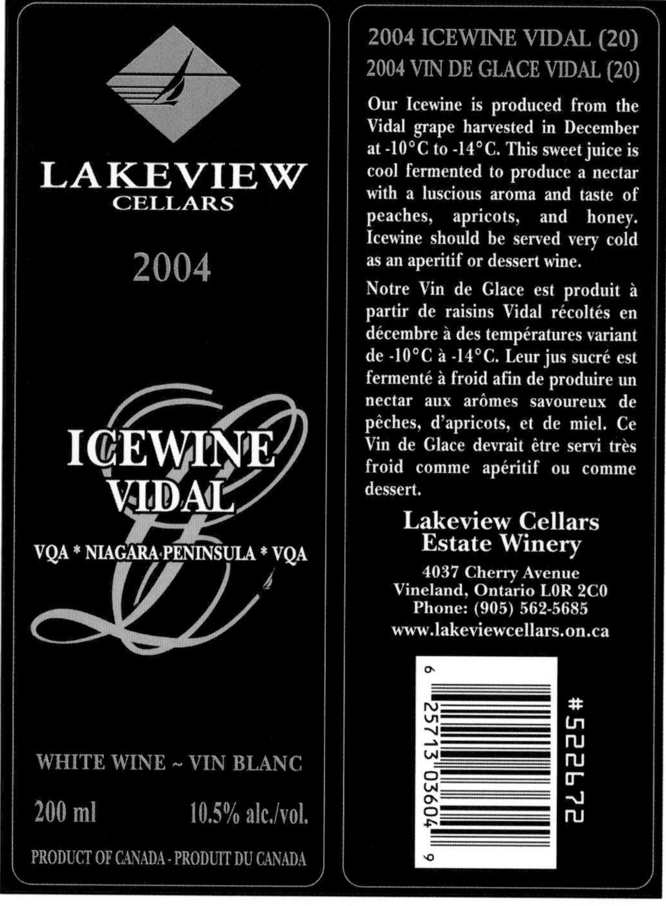

FIGURE 12–7 Canadians are rightfully proud of the many fine ice wines made in their country, such as this one from the Niagara Peninsula.

Summary

Producers of regional wines in North America are at a critical crossroads. Many wineries in lesser-known regions such as New York State and New England, or the American South and Southwest, or in Canada's British Columbia and Ontario, have achieved success in selling their wines locally, usually within a tourism-oriented economy, to which many wineries contribute through tasting rooms, B&B's, and special functions. As grape growing and winemaking in these areas continues to improve, the question is whether regional wines will move beyond being tourism curiosities and be able to find acceptance in the international wine market. Success will depend on a variety of factors. Finding the right varietal for the region is crucial, being sure first that the grape can thrive under the natural conditions of the region and second that adequate demand for that varietal exists. For instance, many experts feel the demand for Riesling is increasing in North America and that the Finger Lakes of New York should concentrate on that varietal, as it has proven to thrive in that mesoclimate. Similarly, the very hot and sunny parts of Texas may need to abandon efforts with certain *vinifera* grapes to concentrate on truly hot climate varietals, such as Spain's Tempranillo or Tuscany's Sangiovese. Niche marketing is also important—that is, creating a quality product that is unique to one's region. Successful examples include ice wine from the Niagara Peninsula, Gruet's sparkling wine from New Mexico, and Sakonnet's Rhode Island Red. Aggressive promotional efforts done in conjunction with other winery owners and quasi-governmental agencies, such as the Finger Lakes Wine Alliance or the British Columbia Wine Institute, could greatly heighten consumer awareness of regional wines. Over the next few decades, it appears that more of North America's regional wines will complete the processes of viticultural winnowing and strategic marketing, and then move on to take their rightful places in the international business of wine.

REVIEW QUESTIONS

1. What is the term used in Canada to indicate an officially approved viticultural area?
2. What changes were mandated by the Farm Winery Act of 1976, and how did those changes affect the New York wine industry?
3. The people of what European country had a particularly strong influence on the evolution of the modern wine industry in New Mexico? Name one important winery in New Mexico founded by a family from that country.
4. In what state was America's first commercial winery located?
5. There are four provinces of Canada that produce wine, but two are particularly important. Name those two provinces, and the style(s) of wine for which each is best known.
6. In what state is the United States' largest AVA located?
7. Portions of what three states are within the parameters of the Southeastern New England AVA?
8. What are the predominant red grapes planted in Virginia? What about white grapes?
9. What *vinifera* grapes do well in New York's Finger Lakes region?
10. What does the term "foxy" mean when used to describe wines made from *labrusca* grapes, or from hybrids of *labrusca*?

SECTION IV

Wine Regions of the Southern Hemisphere

This section consists of three chapters discussing the **Wine Regions of the Southern Hemisphere.** This is one of the fastest growing and innovative regions of winemaking in the world today. Chapters devoted to Australia and New Zealand, Chile and Argentina, and South Africa include their winemaking history, grape growing, and wine production methods.

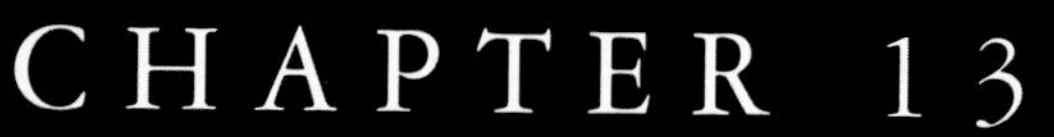
CHAPTER 13

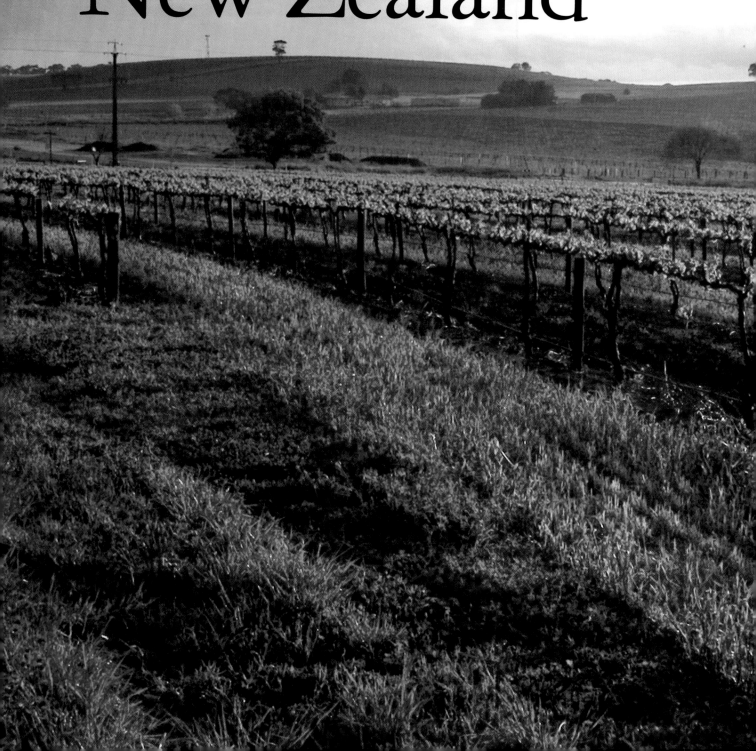

Australia and New Zealand

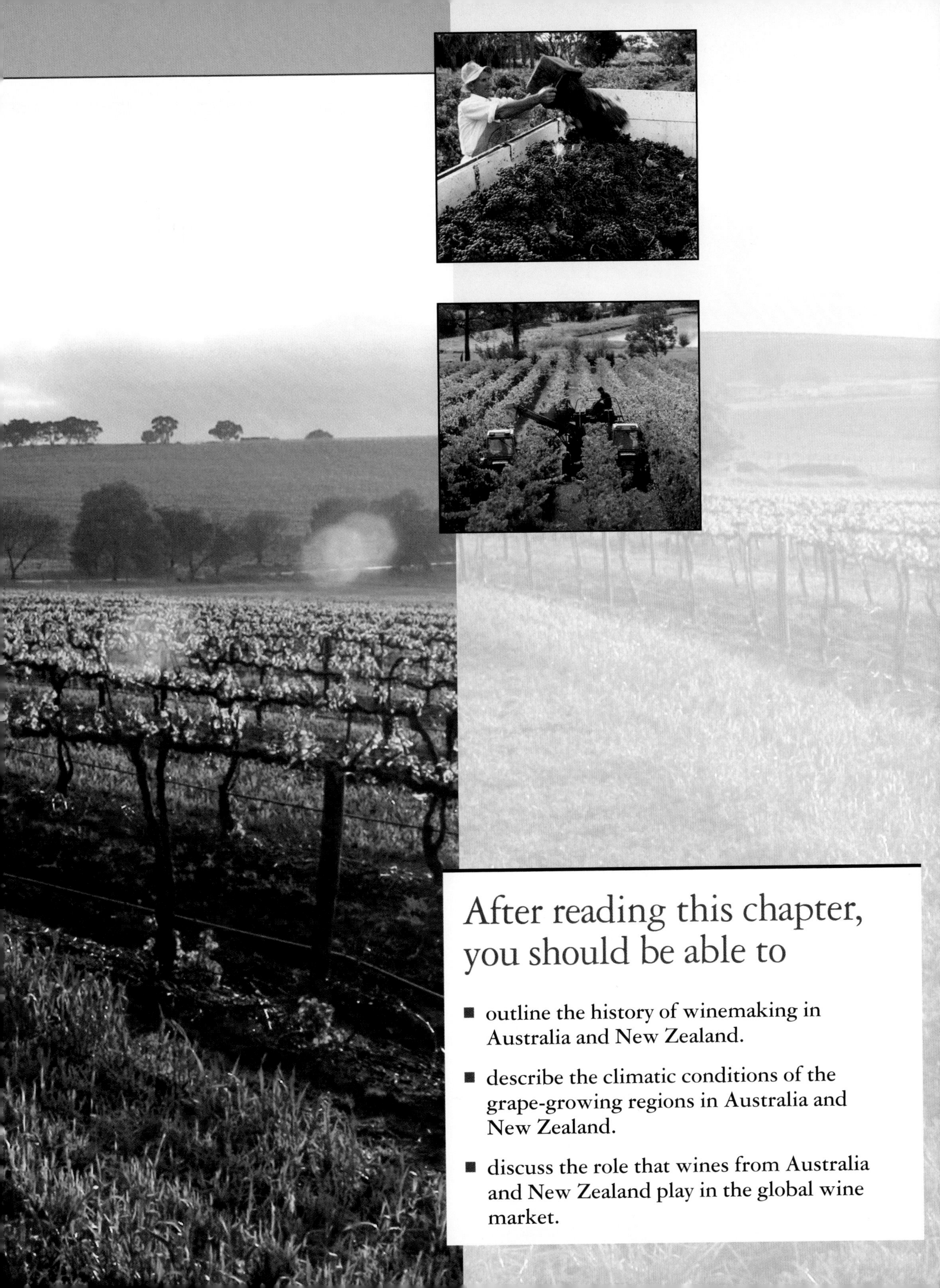

After reading this chapter, you should be able to

- outline the history of winemaking in Australia and New Zealand.
- describe the climatic conditions of the grape-growing regions in Australia and New Zealand.
- discuss the role that wines from Australia and New Zealand play in the global wine market.

KEY TERMS

Australian Wine and Brandy Corporation (AWBC)
Geographic Indications (GIs)
Hermitage
Hunter Valley Riesling
Rhine Riesling
Shiraz

Australia and New Zealand are considered by many wine consumers to be the preeminent wine-producing regions of the Southern Hemisphere. The two nations have much in common; both are located at the western edge of the South Pacific, isolated by water, and share a similar heritage. However, their viticulture and winemaking practices are different and both are uniquely suited to their terroirs and the types of wine that they produce. Being in the Southern Hemisphere, the seasons are opposite those in the United States and Europe, with harvest occurring in the months from February to April. This allows them to get young wines, particularly whites, from a given vintage to the market 6 months earlier than those made in the Northern Hemisphere. Australia and New Zealand enjoy a temperate climate that is well suited to grape growing, and in contrast to the United States, the climate becomes progressively cooler as one goes south. Australia is such a large island it is considered a continent, and it generally has warmer and dryer weather (Figure 13-1). In contrast, New Zealand, located to the southeast of Australia, is much smaller and has a cooler climate (Figure 13-2). The export market is important to both countries with Australia exporting 46 percent (Australian Wine and Brandy Corporation, 2004) of the wine it produces, and New Zealand exporting 49 percent (Keesing, 2003).

FIGURE 13–1 Vineyards in Australia's Barossa Valley.

FIGURE 13–2 Vineyards in Hawke's Bay in New Zealand.

Australia

Australia is a large country with a landmass nearly as large as the continental United States, but with a population that is only two-thirds that of California. The vineyards, along with the population, are concentrated in Australia's southeast. The nation has a long history of viticulture and winemaking, but like the much of the rest of the New World, Australia has undergone significant growth in the last 30 years. Since domestic consumption has been stable in recent years, much of this growth has been fueled by exports, with the United States and Great Britain being the biggest markets. Today Australia is the world's seventh largest producer and home to more than 1,600 wineries bottling 110 million cases of wine a year, making its industry about half the size of the United States (Australian Wine and Brandy Corporation, 2004). The majority of these wineries have been established in the last 20 years and are very small, with limited production of high-end wines. However, these wineries make only a small fraction of Australia's wine. The five largest companies, Hardy, Southcorp, McGuigan-Simeon, Orlando-Wyndham, and Foster's Wine Estates (formally known as Beringer-Blass) account for over 70 percent of the country's wine production. All these large producers are parent companies that own a number of wine brands that are more familiar to the public and, except for McGuigan-Simeon, are multinational corporations with winery holdings around the world.

The Australian wine industry has long had a reputation of being technologically innovative (Figure 13–3). The large multinational companies that are responsible for the majority of the wine production also have funded much of the innovation. Being far removed from the traditions of the Old World, many wineries have developed methods of grape growing and winemaking that are suited to their terroir and the tastes of Australian customers. As Australia's export market developed, consumers overseas also found a taste for these wines. Although it is difficult to generalize about the wine styles of a country as large and diverse as Australia, their exports have a reputation for being full-bodied and fruity wines that are reasonably priced (Table 13–1).

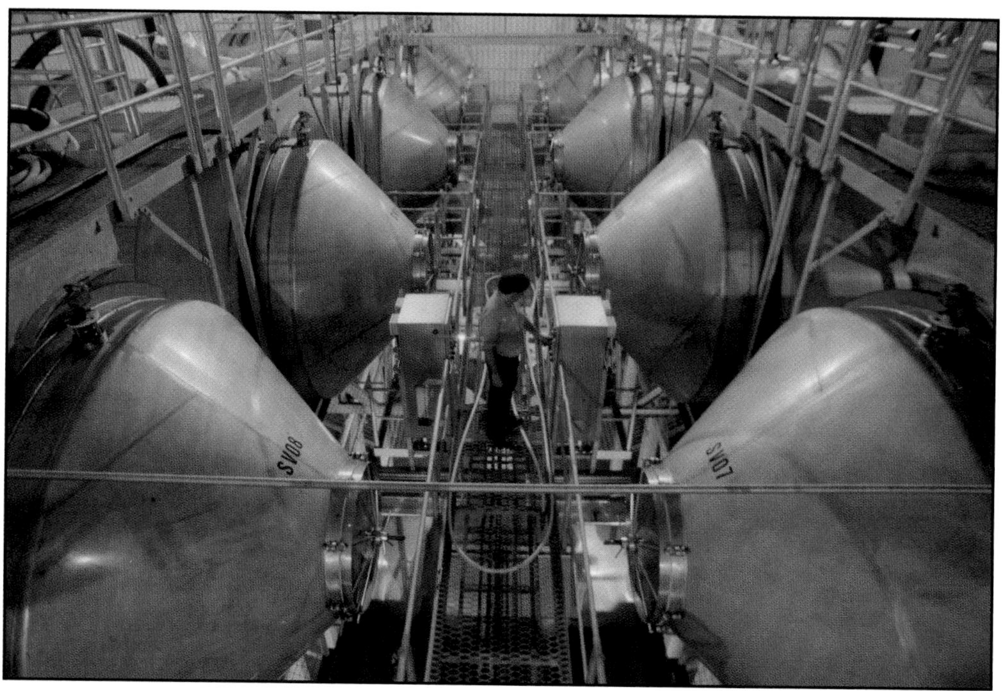

FIGURE 13–3 Rotary fermentation tanks in a large Australian winery. As described in Chapter 3, rotary fermentation tanks extract flavor and color from grape skins during red wine fermentation by revolving to mix the cap (skins) and juice together.

Table 13-1 Wine Brands of Major Australian Producers

Hardy
Hardys, Banrock Station, Leasingham, Houghton, Stonehaven, Barossa Valley Estate; part of the multinational company Constellation Brands that owns a number of U.S. and European producers.
Southcorp
Penfolds, Rosemount Estate, Lindemans, Wynns Coonawarra Estate, Seppelt, Coldstream Hills, Devil's Lair, Leo Buring, Rouge Homme, Edwards & Chaffey, Tollana, Secret Stone, Herrick, Talomas, Riccadonna, Lanson, Seaview, Killawarra, Queen Adelaide, Matthew Lang, Kaiser Stuhl, the Little Penguin, Glass Mountain, Blues Point, Minchinbury, as well as interests in Europe
McGuigan-Simeon
Bin Range, Black Label Range, Duck's Flat, Earth's Portrait, Genus 4, GTS, Harvest Range, Julian's, Lyndoch Valley, Superior Range, Yardstick, The Drainings, Miranda, Firefly, Sandra's 2LP, Tempus Two, Personal Reserve, Crocodile Rock
Orlando-Wyndham
Orlando, Jacob's Creek, Wyndham Estate, Poet's Corner, Carrington, Coolabah, Maison, Steingarten, Jacaranda Ridge, Morris Wines, Richmond Grove, Gramps Montrose, Craigmoor, Wickham Hill, as well as European producers
Foster's Wine Estates
Wolf Blass, Black Opal, Jamiesons Run, Yellowglen, Annie's Lane, Matua Valley, The Rothbury Estate, Greg Norman Estate, Saltram, Ingoldby, Yarra Ridge, St. Huberts, Mt. Ida, Mt. Tanglefoot, Pepperjack, Robertson's Well, Half Mile Creek, Maglieri of McLaren Vale, Metala, Mildara, Andrew Garrett, Baileys of Glenrowan, Shadowood, as well as numerous European and California producers

Australian Wine—Historical Perspective

In 1770, Captain James Cook reached Botany Bay on the southeast coast of Australia and sailed northward to Cape York, claiming the coast for Great Britain. In 1788, a group of soldiers, settlers, and convicts arrived to form a penal colony at Port Jackson where Sydney, in the state of

New South Wales, now stands. The colony was very isolated and needed to become self-sufficient as quickly as possible. Grapes were planted along with other food crops, but they did not do well in the humid climate of Sydney Harbor. By 1791 Arthur Phillip, the governor of the settlement, had established a small 3-acre vineyard 12 miles inland at the Parramatta River. Here the weather was dryer than Sydney Harbor, and the vines were more successful.

One of the first commercial grape growers in Australia was John Macarthur. Macarthur arrived in Sydney in 1790 and perhaps is more famously known for being the first person to import Merino sheep to Australia, which would become the mainstay of the nation's wool industry and Australia's first major agricultural export. In 1805, he was granted 2,000 acres (810 hectares) of grazing land outside of Sydney (Johnson, 1989). Ten years later in 1815, he began an 18-month journey through Europe with his two sons, James and William, to learn the craft of winemaking and to obtain grape cuttings to bring back to Australia. By 1820, he had established a 20-acre vineyard outside of Sydney, and within the decade they were producing more than 20,000 gallons of wine a year. His sons remained active in developing the Australian wine industry after their father's death in 1834.

Another early settler and vintner, Gregory Blaxland, is perhaps better known to Australians for being the first pioneer to cross the Blue Mountain range east of Sydney in 1813. He established a vineyard in 1818 on the 450 acres (182 hectares) he purchased in the Parramatta Valley. Here he experimented with a number of different grape varieties as well as other crops and was the first person to send wine from Australia back to Britain in 1822. The wines were fortified with brandy to protect them from spoilage on the long trip across the equator. In London his wines were awarded a silver medal in 1823 and a gold in 1828. His success raised the attention of others in Britain as well as in Australia.

James Busby is probably one of the best-known founders of Australian winemaking; born in Edinburgh in 1801, he immigrated to Australia in 1824. Convinced that the wine industry in his future home had potential, Busby traveled to France to study grape growing and winemaking. When he arrived in Australia he received a grant of 2,000 acres (810 hectares) in the Hunter River Valley north of Sydney and established vineyards in what was eventually to become one of Australia's most important wine regions (Evans, 1973). He named his estate Kirkton after his Scottish birthplace. Several years later, he traveled to Europe to obtain more information on winemaking and cuttings of more than 500 varieties of grapes. When he returned to Australia, Busby donated the majority of cuttings to the government for the purpose of establishing an experimental garden at Sydney, and the remaining cuttings were planted at Kirkton. In addition to importing the grapevines, Busby also promoted the industry by writing a number of books about winemaking and grape growing in Australia, as well as on his travels through the wine regions of Europe. His association with the Australian wine industry ended in 1833 when he immigrated to New Zealand. Although he had departed the country, his books on winemaking and his work importing grape varieties were to influence the development of winemaking in Australia for years to come.

By the 1830s, Great Britain had colonized the entire continent and settlements were established in many locations. Vineyard development was progressing rapidly in the Hunter Valley, and vines were also planted in what are now Victoria, South Australia, and Western Australia (Figure 13–4). In addition to new arrivals from Great Britain, Australia also became a destination for immigrants from other countries in Europe. Throughout the 1830s and 1840s, settlers arrived from many countries and were responsible for spreading agriculture across the continent. Many of the colonists from countries such as Germany, Italy, and France had knowledge of wine production that they had gained in their homelands. These settlers helped to expand the Australian wine industry, planting vineyards in desirable locations and bringing in new techniques. It was during this time that many of Australia's best viticultural regions were first estab-

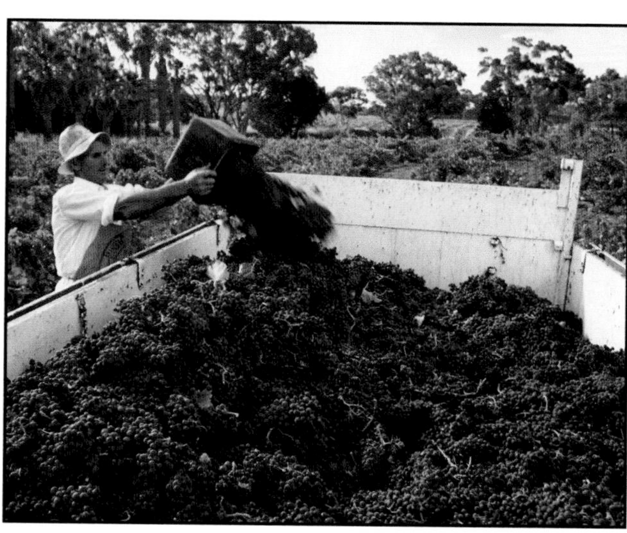

FIGURE 13–4 Harvesting grapes in South Australia.

lished. Two of the country's most famous wineries, Lindemans and Penfolds, were founded in the 1840s.

The discovery of gold in eastern Australia in 1852 had a significant effect on the country's fledgling wine business. Initially it slowed development because there was a shortage of labor as vineyard workers left to find their fortune in the mines. However, the gold rush that followed the strike brought in many new immigrants, and the population doubled to just over one million between 1850 and 1860. After the gold played out in the mines, these newcomers provided a source of labor as well as a market for Australia's wineries. Over the second half of the nineteenth century, the wine business slowly expanded as new vineyard land was developed. The pace of expansion was slowed by international and domestic tariffs that discouraged the shipping of wine out of its home state. The temperance movement was also active at the time but was unable to pass nationwide prohibition laws as it had in the United States. As duties on exported wines were relaxed, international trade increased, and by 1900, half a million gallons (1.9 million liters) were exported to Great Britain (Walsh, 1979). In 1901, the federation of the states lowered trade barriers within Australia and dramatically increased domestic consumption. In 1877, phylloxera was discovered in the Geelong area of the state of Victoria. While it destroyed nearly all the vineyards of the Geelong and Bendigo regions, the infestation was slow to spread elsewhere. Consequently, most of Australia was saved from the devastation that was experienced in California and Europe. Today phylloxera is absent from the states of South Australia, Western Australia, Tasmania, and most of New South Wales, and there are very strict quarantines in place to prevent it from spreading to new areas.

In the period between 1900 and the Second World War, the industry continued to grow but also experienced setbacks from droughts, economic depression, and regional outbreaks of phylloxera, thus exhibiting the boom and bust cycle of development that is common to the wine business. The Second World War severely cut exports, but when it was over a new wave of immigrants came to the country, providing a new market for wine. The situation in the 20 years following the war was similar to that in the United States; there was slow growth, but it was limited by lack of consumer interest and wines of unexceptional quality. This began to change in the 1960s when interest in wine began to grow and producers improved their product. It was not until 1968 that table wines outsold dessert wines in Australia, and by the mid 1970s exports were still only 2 percent of production.

During the 1980s and 1990s, production rapidly increased and Australia began to reach its potential for both quality and quantity of wine. Domestic consumption expanded during this time but not nearly at a rate to absorb the added production. There was, however, a rapidly growing market for Australian wines overseas. The modernization of the industry in the 1970s and 1980s allowed exporters to make a high quality product, and favorable exchange rates allowed them to keep their prices low.

The Wine Regions of Australia

Although Australia is a very large country, most of the continent is unsuitable for viticulture. The northern part of the country has a subtropical climate that is too warm for growing wine grapes, and the interior is too hot and dry. There are many areas where the soil and the climate are appropriate, but there is inadequate water available for irrigation. Grape-growing regions are concentrated in the areas with a temperate climate located in the valleys along the country's southeastern coast between Sydney and Adelaide. The moderate climate of this area is one of the reasons it is also where the majority of Australia's population lives. There are also viticultural districts along the coast of Western Australia near Perth, as well as on the island of Tasmania to the south.

The wines of Australia are labeled by the state they were grown in or by appellation of origin in a system called **Geographic Indications, or GIs.** This method of classification subdivides the territory of each state into a series of Zones, Regions, and Subregions (Table 13–2). These definitions are set by the **Australian Wine and Brandy Corporation,** or **AWBC.** The AWBC is a governmental organization that is responsible for the regulation and promotion of the Australian wine industry. This system of GIs is similar to the AVA system of viticultural areas used in the United States. Neither system of classification attempts to make any statement about a wine's quality, nor do they dictate which varieties can be grown in the region or what production methods can be used. Both GIs and AVAs only refer to the geographic origin of the

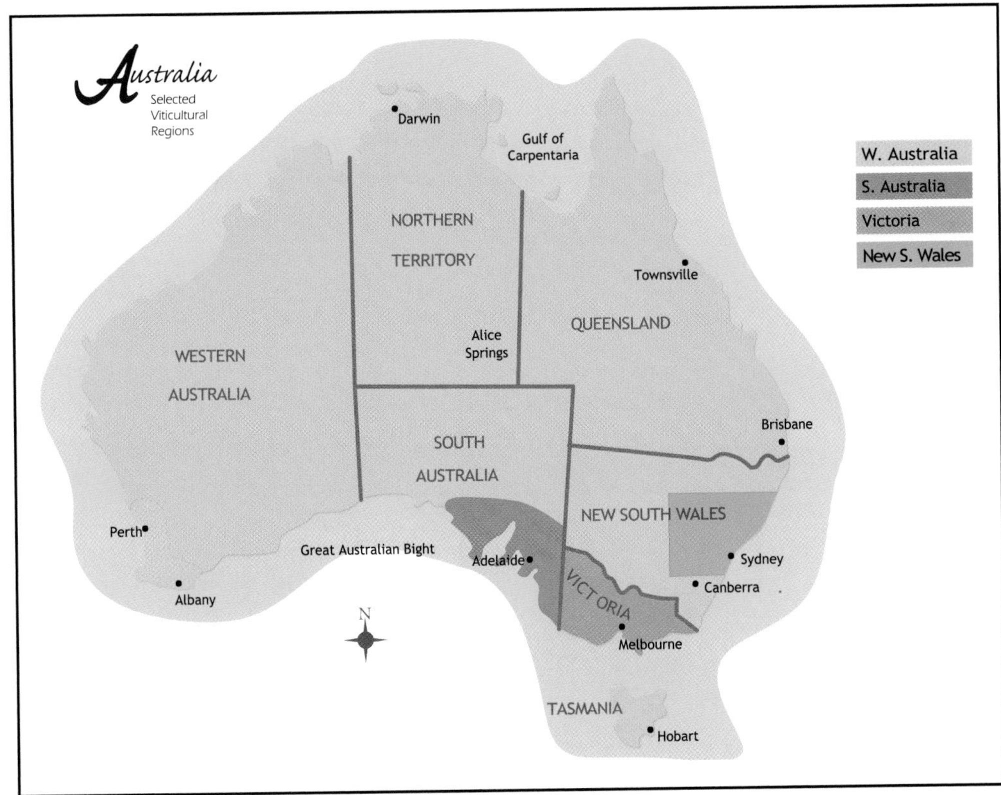

The wine regions of Australia.

Table 13-2 Australian Geographic Indications (Wine Regions)

AUSTRALIAN STATE OR ZONE	REGION	SUBREGION
South Eastern Australia includes the states of New South Wales, Victoria, Tasmania, and part of South Australia and Queensland		
South Australia		
Adelaide (Super Zone includes Mount Lofty ranges, Fleurieu, and Barossa)		
Barossa	Barossa Valley	
	Eden Valley	High Eden
Far North	Southern Flinders ranges	
Fleurieu	Currency Creek Kangaroo Island Langhorne Creek McLaren Vale Southern Fleurieu	
Limestone Coast	Coonawarra Mount Benson Padthaway	
Lower Murray	Riverland	
Mount Lofty ranges	Adelaide Hills	Lenswood Piccadilly Valley
	Adelaide Plains Clare Valley	
The Peninsulas		
New South Wales (NSW)		
Big rivers	Murray Darling (shared with Victoria) Perricoota Riverina Swan Hill (shared with Victoria)	
Central ranges	Cowra Mudgee Orange	
Hunter Valley	Hunter	Broke Fordwich
Northern rivers	Hastings River	
Northern slopes		
South Coast	Shoalhaven Coast Southern Highlands	
Southern New South Wales	Canberra District Gundagai Hilltops Tumbarumba	
Western Plains		
Western Australia		
Central Western Australia		
Eastern plains, inland and north of Western Australia		
Greater Perth	Peel Perth Hills	
	Swan District	Swan Valley

Table 13-2 Australian Geographic Indications (Wine Regions) (continued)

AUSTRALIAN STATE OR ZONE	REGION	SUBREGION
Western Australia		
South West Australia	Blackwood Valley Geographe	
	Great Southern	Albany Denmark Frankland River Mount Barker Porongurup
	Margaret River	
West Australian South East Coastal		
Queensland		
Queensland	Granite Belt South Burnett	
Victoria		
Central Victoria	Bendigo	
	Goulburn Valley	Nagambie Lakes
	Heathcote Strathbogie ranges Upper Goulburn	
Gippsland		
North East Victoria	Alpine valleys Beechworth Glenrowan Rutherglen	
North West Victoria	Murray Darling (shared with NSW) Swan Hill (shared with NSW)	
Port Phillip	Geelong Macedon ranges Mornington Peninsula Sunbury Yarra Valley	
Western Victoria	Grampians Henty Pyrenees	
Tasmania (no regions or subregions)		

grapes used to make the wine. As of 2004, there were 55 regions and 11 subregions of viticulture in Australia. There is also one all-encompassing appellation called South Eastern Australia that includes all of the grape-growing regions of the states of Victoria, New South Wales, South Australia, and Queensland. This appellation is primarily used by large wineries producing inexpensive wines blended from many parts of the country.

For a wine to be labeled as from a particular region, at least 85 percent of the grapes used to make the wine must be from that region. For varietal labeling, the wine must be at least 85 percent of the grape variety listed on the label and for vintage labeling the requirement is 85 percent. As in California, it is illegal to add sugar to raise the °Brix of musts before fermentation but permissible to add acid to lower the pH. This is not surprising given the similarities in the climate of California and Australia; grapes almost always reach full maturity at harvest but frequently have low acid.

New South Wales

As was previously stated, the first vineyards in Australia were planted in New South Wales. The state has 15 different wine regions and produces about one-quarter of Australia's wine. The wine industry of the state is centered about 90 miles (145 kilometers) north of Sydney in the Hunter Valley in the foothills of the Brokenback range. The wine industry in this region produces a number of red varieties, with Cabernet Sauvignon and **Shiraz** being the most prominent. The Shiraz grape, which is also called **Hermitage** in Australia, is known as Syrah to much of the rest of the world. Indigenous to the northern Rhône Valley of France, Shiraz is the most planted wine grape in Australia. Its popularity diminished in the 1960s and 1970s as the growers in New South Wales concentrated on Cabernet Sauvignon, but in recent years there has been a renewed interest in the variety. The weather in the Hunter Valley is always a concern for growers, hot and humid in the summer time with a perennial risk of fall rains. Additionally, much of the soil of the appellation is heavy in clay and drains poorly. Despite these difficulties, the Hunter Valley is one of Australia's best-known wine regions and produces a number of fine wines (Figure 13–5).

The principal white wines of the region are Semillon and Chardonnay. The Semillon from Hunter Valley is sometimes referred to as **Hunter Valley Riesling.** Historically, the predominant white wine produced in the region was Semillon, with Chardonnay not becoming popular until the 1960s and 1970s. During that time, the introduction of new technologies and techniques, such as cold fermentation and gentle pressing, resulted in more flavorful white wines. In 1973 Tyrrell, one of the area's wineries, began producing an award-winning Chardonnay called Vat 47. Because of the Hunter Valley's close proximity to Sydney, tourism has played an important role in the development of the area's wine business. Fledgling wineries were supported by sightseers coming up from the city to see the wine country and purchase wine. Wineries also put special effort into designing attractive buildings and landscaping to aid in drawing in tourists.

West of the Hunter Valley, on the opposite side of the Great Dividing Range of mountains, Mudgee is best known for its Chardonnay and Cabernet Sauvignon. Surrounded on three sides by mountain ranges, its elevation varies from 1,600 to 3,000 feet (500 to 1,000 meters). The higher

FIGURE 13–5 Early morning fog in the Hunter Valley region of New South Wales. Hunter Valley lies about 90 miles (145 km) north of Sydney and is the best known appellation in New South Wales.

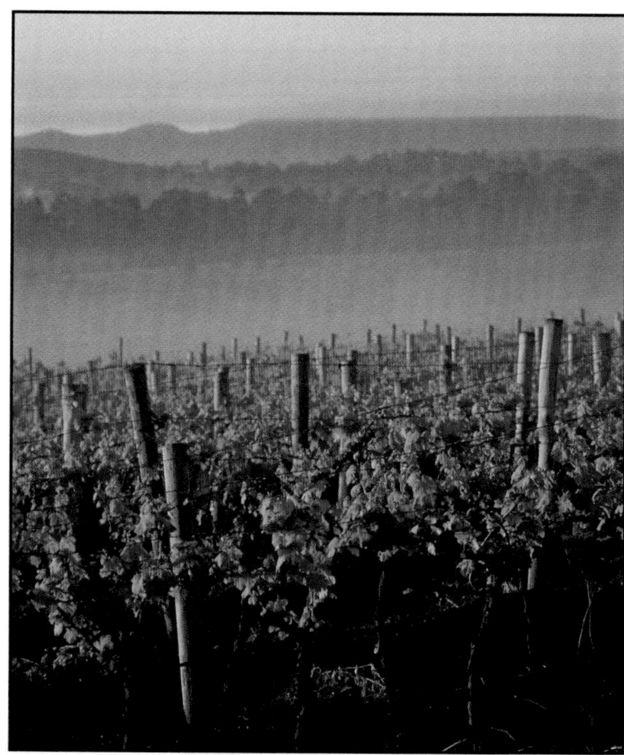

elevation gives the region a mild climate with cool nights. Settled in the mid 1800s, the early wine industry was shaped by German emigrants. Mudgee's vineyards grew extensively in the 1970s when many smaller wineries were established. Today many of the grapes it grows are shipped to Hunter Valley wineries for processing. Further inland, about 200 miles (320 kilometers) is the Riverina region; this large district has a number of expansive vineyards that harvest more grapes than the rest of the appellations in New South Wales combined. Most of these grapes are used by large wineries producing inexpensive blends for export; however, there are some small production lots of high-end dessert wines that are bottled from the region as well.

Victoria

The wine industry of Victoria was begun by Swiss immigrants in the middle of the nineteenth century. By 1900, vineyards had spread across the entire state, and Victoria was producing the majority of Australia's wine. Unfortunately, as in Europe and California, the phylloxera epidemic wiped out a huge number of vineyards. In recent decades, it has been one of the fastest growing wine regions, with a more than 10-fold increase in production from 1965 to 2000. Today its vineyard acreage is similar to that of New South Wales, growing about one-quarter of Australia's wine grapes. There are more than 500 wineries located throughout the state in 6 different wine

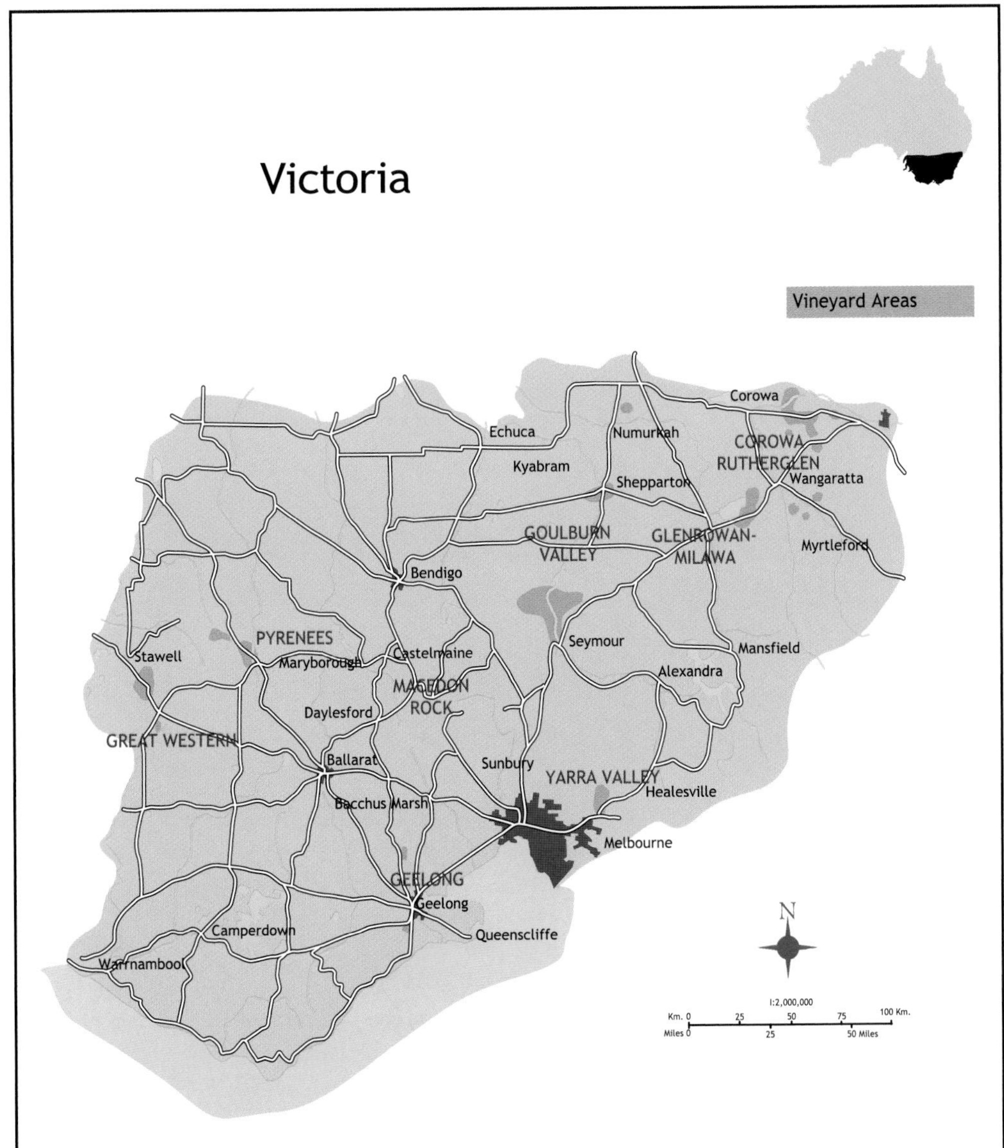

The wine regions of Victoria.

zones and 20 wine regions. There are several large wineries, particularly in the Murray Darling Region in northwestern Victoria, but most are small producers making limited amounts of high-end wine. Among the most famous areas are the Goulburn Valley in the northeastern part of the state, Yarra Valley just to the west of Melbourne, and Western Victoria where the Grampians and Pyrenees regions lie. The far southwest section of the state produces some fine Rieslings, while the northeast produces complex Muscats and Tokays. Two other regions, the small Mornington Peninsula and the much larger Gippsland, are on the southeastern edge of the state. Their location on southern tip of the continent gives them a cool climate, and Chardonnay and Pinot Noir do well here.

The Goulburn Valley is in the northeast section of Victoria near the border with New South Wales. The valley is formed by the Goulburn River, a tributary of the Murray River to the north. The region is best known for its Cabernet Sauvignon and Shiraz, as well as the white Rhône variety Marsanne. The Goulburn Valley is home to Chateau Tahbilk, which also uses Shiraz and Cabernet for its red wines. Established in 1860, it is one of the most historic wineries in Australia. The Yarra Valley region lies about 125 miles (200 kilometers) west of the city of Melbourne and also has a long history, dating back to the mid-nineteenth century. However, by the 1920s, economic depression and competition from wineries in other states had closed all the region's wineries. Beginning in the 1960s, a new generation of wineries were founded and since then the area has grown significantly (Figure 13–6). The Yarra Valley is particularly well known for its Pinot Noir, Chardonnay, and Cabernet Sauvignon.

Western Victoria is home to the Grampians region; previously know as the Great Western region, which is famous for its natural beauty, national parks, and rich history of winemaking. One of the best-known wines produced in the region is a *méthode champenoise* sparkling wine called Great Western. Named after a town on the Western Highway, it is made at the Great Western Estate winery established in the 1860s by Joseph Best and now owned by Seppelts. The winery has an extensive network of wine storage caves, or "drives," that were dug in the 1860s by ex-miners after the gold rush. Area wineries also produce red wines such as Pinot Noir,

FIGURE 13–6 Mechanical grape harvester at work in the Yarra Valley region in Victoria.

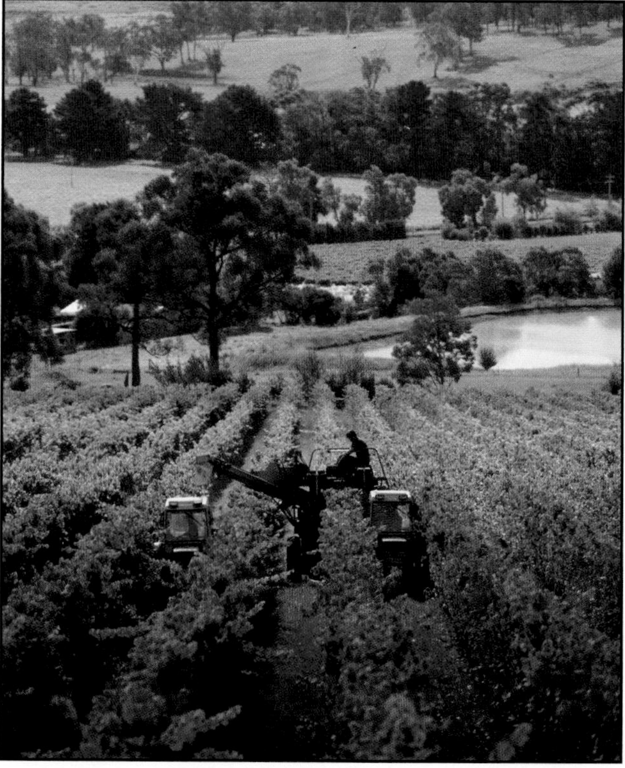

Cabernet, and Shiraz-Cabernet blends as well as a uniquely Australian product, sparkling Shiraz. Blends combining Cabernet with Merlot or Shiraz are particularly common in Australia. It should be noted that Australian wines are required to list the major component first; that is, a Shiraz-Cabernet contains more Shiraz than Cabernet. In the higher elevations of the district, Sauvignon Blanc is widely planted.

South Australia

The state of South Australia lies between Victoria and Western Australia and contains 7 wine zones and 15 regions. The state has more land in vineyards than any other in Australia and produces more than 46 percent of the nation's wine. While Victoria to the east is known for having primarily small wineries, South Australia by contrast is dominated by large producers. It is not uncommon for these big wineries to bring in grapes and juice from the large vineyards over the border in New South Wales and Victoria for fermentation and bottling. In addition to these big wineries, South Australia is also home to some of the country's preeminent grape-growing regions, making some of Australia's most expensive and sought after wines.

The majority of South Australia's vineyards lie primarily in its southeast corner. The appellations of the Barossa Valley, Adelaide Hills, McLaren Vale, and Clare Valley are clustered around the city of Adelaide. The area is also home to Roseworthy Agricultural College. Now part of the University of Adelaide, it is Australia's premiere institution for the study of viticulture and enology (Figure 13–7). The Coonawarra and Padthaway regions are located 200 miles (320 kilometers) to the southeast of Adelaide near the border with Victoria. Riverland is a large growing region directly north of Coonawarra on the Murray River near the border with New South Wales and Victoria. It is much farther inland than the other appellations, and here the vineyards are quite large and have higher yields, producing inexpensive wines. As previously mentioned, phylloxera has never come to the state and consequently most vines are own-rooted.

The Barossa Valley is about 40 miles (65 kilometers) to the northwest of the city of Adelaide and is one of Australia's oldest and most famous grape-growing regions. First developed in the

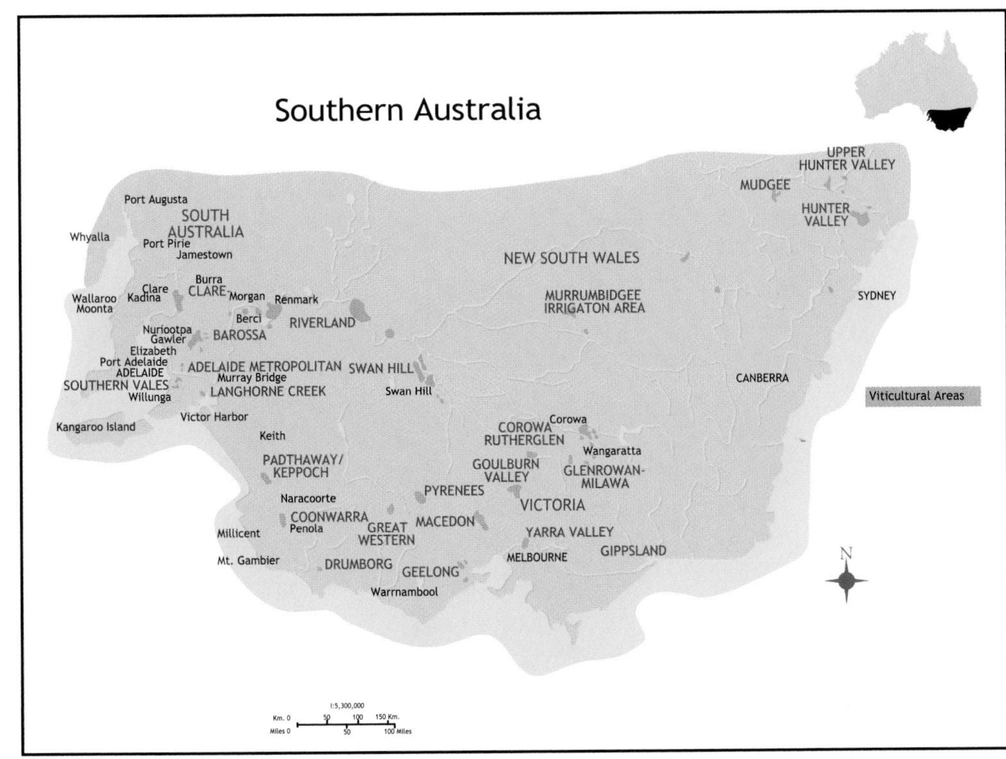

The wine regions of South Australia.

FIGURE 13–7 A professor examines vines at the Australian Wine Research Institute at the University of Adelaide in Adelaide, Australia. Nets are used to protect the ripe grapes from birds.

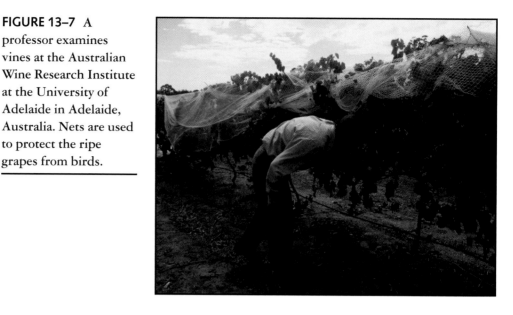

1840s by German immigrants, the Barossa Valley is home to the headquarters of many of Australia's largest producers. The vineyards spread throughout the lower part of the region, following the path of the North Para River. The climate has warm summers and cool winters, and Chardonnay and Semillon are both popular. In addition to the white varieties, the region also produces a number of red wines, including Cabernet Sauvignon and Shiraz. The principal winery for Cabernet is Seppeltsfield, owned by the same company that produces Great Western sparkling wine in neighboring Victoria. The best known Shiraz in the Barossa Valley is the Grange from Penfolds Winery, which has been made since 1951 (Figure 13–8). Made from dry-farmed old vines of Shiraz, it is aged in new American oak and occasionally has a small amount of Cabernet Sauvignon blended in for complexity. Grange is perhaps the most highly regarded of all of Australia's wines, and its intense flavors allow it to age for as long as several decades.

FIGURE 13–8 Penfolds Winery in the Barossa Valley region of South Australia. The winery is home to Penfolds Grange, a Syrah that is one of Australia's most famous wines.

Directly to the east of the Barossa Valley is the Eden Valley region, here the ground gradually climbs into the Barossa range of mountains and the elevation reaches 1,500 to 2,000 feet (450 to 600 meters). The soil is less fertile here than in the valley below, and the climate is cooler. The cool conditions make the region ideal for white grape varieties such as Chardonnay and **Rhine Riesling.** Rhine Riesling is an Australian name for the varietal White Riesling. To the south of the Barossa and Eden Valleys is the Adelaide Hills district, a series of gentle rolling hills about 9 miles (15 kilometers) to the west of the city of Adelaide. Here the terroir is even cooler than the Eden Valley, and spring frost can be a problem in low-lying areas. Varieties such as Pinot Noir and Sauvignon Blanc are planted along with the Chardonnay and Shiraz. To the south of the Adelaide Hills is the McLaren Vale region. The elevation is lower here, starting at the coast and increasing to 1,100 feet (350 meters) on the east. The appellation has a number of small wineries that produce a variety of red and white wines. The Clare Valley lies 75 miles (125 kilometers) to the North of Adelaide. Being further inland, it receives less of a moderating influence from the ocean and is warmer than the vineyard lands that surround Adelaide. Cabernet and Shiraz are popular, as well as Chardonnay and Riesling.

While Barossa Valley is recognized for its white wine, Coonawarra is regarded as one of the best regions for red wines in Australia. It lies between Naracoonte and Millicent and is the southernmost region in South Australia. The location results in much cooler temperatures and is the only region in Australia where there is a danger of frost. Coonawarra is particularly known for its soil, terra rosa, a vivid red topsoil that is found in thin bands that overlay soft limestone (Figure 13–9). The primary wine of the region is the Cabernet Sauvignon, which is deep in color with spicy fruit flavoring. Other grape varieties produced in the region include Shiraz, Rhine Riesling, Chardonnay, Pinot Noir, and Merlot. Although red wines predominate, the amount of land planted to Riesling is almost as large as that of Cabernet Sauvignon.

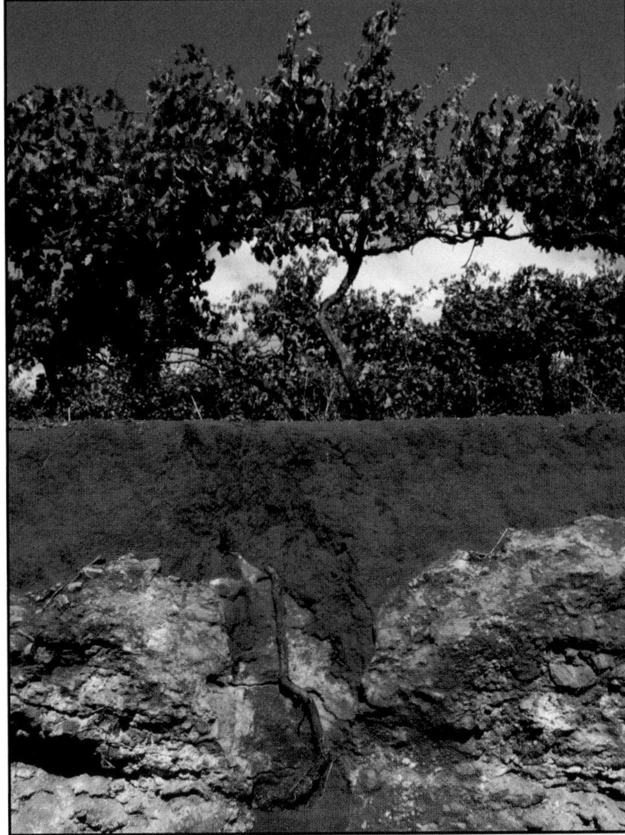

FIGURE 13–9 An excavation showing a soil profile of the root zone of a Shiraz vine grown in South Australia's Coonawarra region, illustrating the terra rossa for which the appellation is famous.

A new region Padthaway, just to the north of Coonawarra, is coming into prominence. Padthaway is actually two subregions, Padthaway and Keppoch. The climate in Padthaway is slightly warmer than Coonawarra, and although the frost is less of a threat here, in 1988 it succeeded in wiping out almost all of Padthaway's crop for the year. The soil in Padthaway is primarily a sandy soil similar to that found in the Barossa Valley. However, some of Coonawarra's terra rosa can be found in isolated areas. The primary grape varieties are similar to Coonawarra: Rhine Riesling, Cabernet Sauvignon, Chardonnay, Shiraz, Pinot Noir, Merlot, and Sauvignon Blanc.

Western Australia

Far from the urban centers and major grape growing regions to the east, there is a small but growing wine industry in the state of Western Australia. Western Australia is home to seven wine regions and produces approximately 6 percent of Australia's wine. Isolated by desert from

Wine regions of Western Australia.

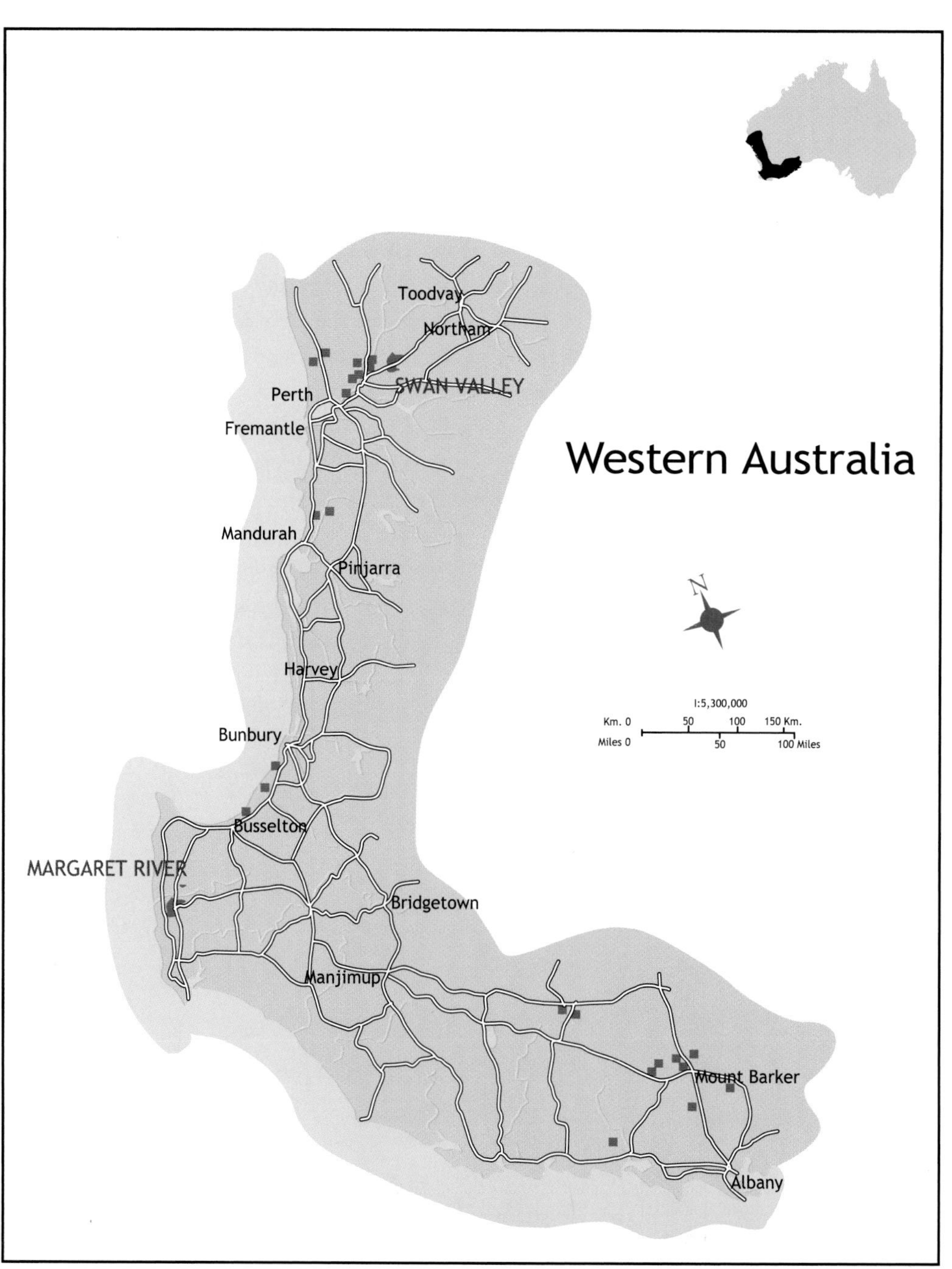

the rest of Australia's viticultural areas on the continent's southwest corner, the state has some of the oldest grape-growing regions in the country, as well as some of the newest. Grapevines were first planted outside of Perth as early as 1829 in what is now the Swan District region and the appellation is home to Olive Farm, the state's longest operating winery. The region is one of the warmest in Australia, and summertime temperatures can reach over 110°F (43°C). The Swan District and neighboring Perth Hills produce a number of red and white wines, but are best known for their whites, particularly Chenin Blanc.

On the southwestern corner of the continent, 250 miles (402 kilometers) to the south of Perth, there is another group of viticultural districts that include the Margaret River, Pemberton, and Great Southern. This part of the state has a much shorter history of winemaking than the region around Perth does and did not begin to develop until the 1970s. During the 1980s, there was much growth and foreign investment, but today most of the wineries are small operations, and the area is still not home to any of Australia's largest producers. The area is best known for its red wines, such as Cabernet Sauvignon, Shiraz, and Pinot Noir, that are grown in the Margaret River region and Mount Barker (a subregion of the Great Southern district). White wines are also produced, and the Pemberton region is famous for its Chardonnay. These regions experience a cooling influence from their proximity to the Indian Ocean and have a much more temperate climate than the vineyards around Perth.

Tasmania

Tasmania is a large island south of the state of Victoria. Although there are records of wine being produced in Tasmania in the early 1800s, winemaking had all but disappeared by the 1860s. The industry was not reestablished until the 1950s, when the Lalla Vineyard was planted on the Tamar River and Moorilla Estate was established in the Derwent Valley, north of Hobart, the capital city. These areas together with the Pipers Brook region, northeast of the Tamar River, and a small area on the east coast near the resort of Bichens continue to be the principal grape-growing regions. It is Australia's coolest area for grape growing and produces only about 1 percent of its wine. In 2002, there were 48 wineries on record, but only a few are of notable size. By far the biggest is Piper's Brook Vineyard, established in 1974. Another winery of note is Heemskerk, which produces sparkling wines in association with Louis Roederer of France. Because of the climate, the wine varieties that are grown here are early ripening. Of particular note is the Chardonnay and Pinot Noir, both of which are well suited to cool weather. Sauvignon Blanc also shows great promise, producing intensely grassy wines that are similar in style to the Sauvignon Blancs of New Zealand. Other varieties are Semillon, Merlot, and Gewürztraminer. This is the one area of Australia where little Shiraz is produced.

New Zealand

New Zealand lies 1,300 miles (2,100 kilometers) off the eastern coast of Australia across the Tasman Sea at roughly the same latitude as Tasmania. Its southerly location and maritime weather pattern gives the island a moderate but cool climate. Grapes have been grown in New Zealand nearly as long as they have been in Australia, but there was little development until the 1970s, and today the county still only produces about 6 percent of what Australia does. This was due to a number of factors including anti-alcohol regulations, more difficult growing conditions, and poor winemaking practices. Today, for the most part, these issues have been dealt with, and New Zealand is gaining a reputation for superior wines (Figure 13–10).

FIGURE 13–10 A vineyard in the Central Otago region on New Zealand's South Island. The region is one of the southernmost grape-growing areas in the world and the most inland of New Zealand's appellations.

New Zealand—Historical Perspective

The first *vinifera* vines were planted in 1819 by Samuel Marsden, near Kerikeri on New Zealand's North Island. Seventeen years later the famous Australian vintner James Busby settled nearby in Waitangi, planted grapes, and made New Zealand's first wine in 1836. Although New Zealand's commercial wine industry dates back to 1863, it was very slow to develop. Winemaking was widespread across the North Island but at a very small scale, and very little was exported. After phylloxera and powdery mildew arrived at the end of the nineteenth century, New Zealand growers imported native American grape varieties. Unlike the rest of the world, most growers did not bother to use them as rootstock, instead using them to produce wine grapes directly, along with French-American hybrids. It was also allowable to add both sugar and water to make up for under-ripe grapes and to increase yield. These vines and cellar practices produced wines of poor quality that were often fortified with alcohol to cover up their inadequateness. In 1960, the most widely planted grape variety in New Zealand was the American variety Isabella and only 12 percent of the wine that was produced was table wine (Halliday, 1991). Hardy and tolerant to cold conditions, Isabella was also popular in early vineyards in Oregon and Washington states.

In addition to inadequate vineyard and winemaking practices, New Zealand also had tariffs and legal restrictions that inhibited growth. While nationwide prohibition was narrowly defeated in 1919, there were a number of laws enacted that were designed to discourage the consumption of wine and alcohol products. Wine could not be sold by the bottle in shops until 1955, restaurants could not sell wine until 1960, and supermarkets could not sell wine until 1990. Most of what little wine was consumed during the first 70 years of the century was imported, usually from Australia (Robinson, 1999).

This less than ideal situation began to change in the 1970s when the wine industry started to put a greater emphasis on quality. Increasing local interest in fine wines gave producers incentive to change their ways and concentrate on producing a better product. Vineyards were replanted with superior-tasting classic *vinifera* wine grapes, and winemakers improved their techniques as well. The New Zealand Wine Institute was formed in 1975 to promote the industry, and the government also acted to help improve the quality of wine rather than solely trying to control its sale. In 1982, regulations were passed that limited the amount of water that could be added to wine, and in the mid 1980s, during a time of grape surpluses, the government helped to subsidize growers to pull out unpopular varieties. From 1973 to 1983, wine production grew more than 350 percent, and a healthy export market developed. The rapid growth continued through the 1980s and 1990s, and at times there were periods when supply outpaced demand. Today New Zealand's wines are popular throughout the world and often command high prices. It is also known for producing some of the world's most intensely flavored Sauvignon Blanc.

New Zealand's climate makes it particularly well suited to early ripening varieties that are more flavorful when grown in cool growing conditions such as Chardonnay, Sauvignon Blanc, and Pinot Noir. These three varieties now account for 70 percent of New Zealand's vineyard acreage. The cool conditions and occasional summer rains can cause problems for growers in terms of mold and mildew, particularly with Sauvignon Blanc. Sauvignon Blanc's thin-skinned grape berries are sensitive to rot, and its thick canopy of leaves inhibits air circulation around the clusters, keeping them from drying out. To deal with this problem, viticulturists in New Zealand have developed a number of innovative trellising systems to control vine vigor and to improve quality. These trellising systems have become popular in a number of wine regions around the world.

The Wine Regions of New Zealand

The two major islands of New Zealand form a long north-south chain, and no part of the country is more than 90 miles (145 kilometers) from the coast. For this reason, the weather all over the country is dominated by the influence of the ocean. There is still a great deal of variation in growing conditions, however, due to the varied terrain and long length of the country. New Zealand is more than 900 miles (1,500 kilometers) long, stretching from 35° south to 47° south in latitude. This compares roughly in distance and latitude as from Los Angeles to Portland, Oregon, in the Northern Hemisphere. Chardonnay vines grown in the cooler southern regions are picked 6 to 8 weeks after the beginning of harvest in the north. On the South Island, the vineyards of the Otago region are the southernmost vineyards in the world. The grape-growing appellations are divided into 10 growing regions, six on the North Island and four on the South Island (Table 13–3). The three largest regions—Marlborough, Hawke's Bay, and Gisborne—account for 80 percent of production (Keesing, 2003). Historically, vineyards had been concentrated on the North Island, but in recent decades extensive planting has resulted in the South Island having more vineyard acreage.

Gisborne

The Gisborne region lies on the eastern edge of the North Island and produces about 12 percent of New Zealand's wine. It is a cool region that is best known for growing white varieties, which make up 90 percent of the vineyards. There is ample rainfall throughout the growing season, which increases the likelihood of rot and can cause problems during harvest. The vineyards are primarily grown on the deep, alluvial soils that exist along the river plains. Abundant water and fertile soils allow for higher yield, and the grapes grown in Gisborne are usually crushed by larger wineries to make value-oriented wine. In recent years, smaller wineries have been moving into Gisborne and concentrating on producing a higher quality product. Chardonnay is the most popular grape, accounting for over half of the region's production.

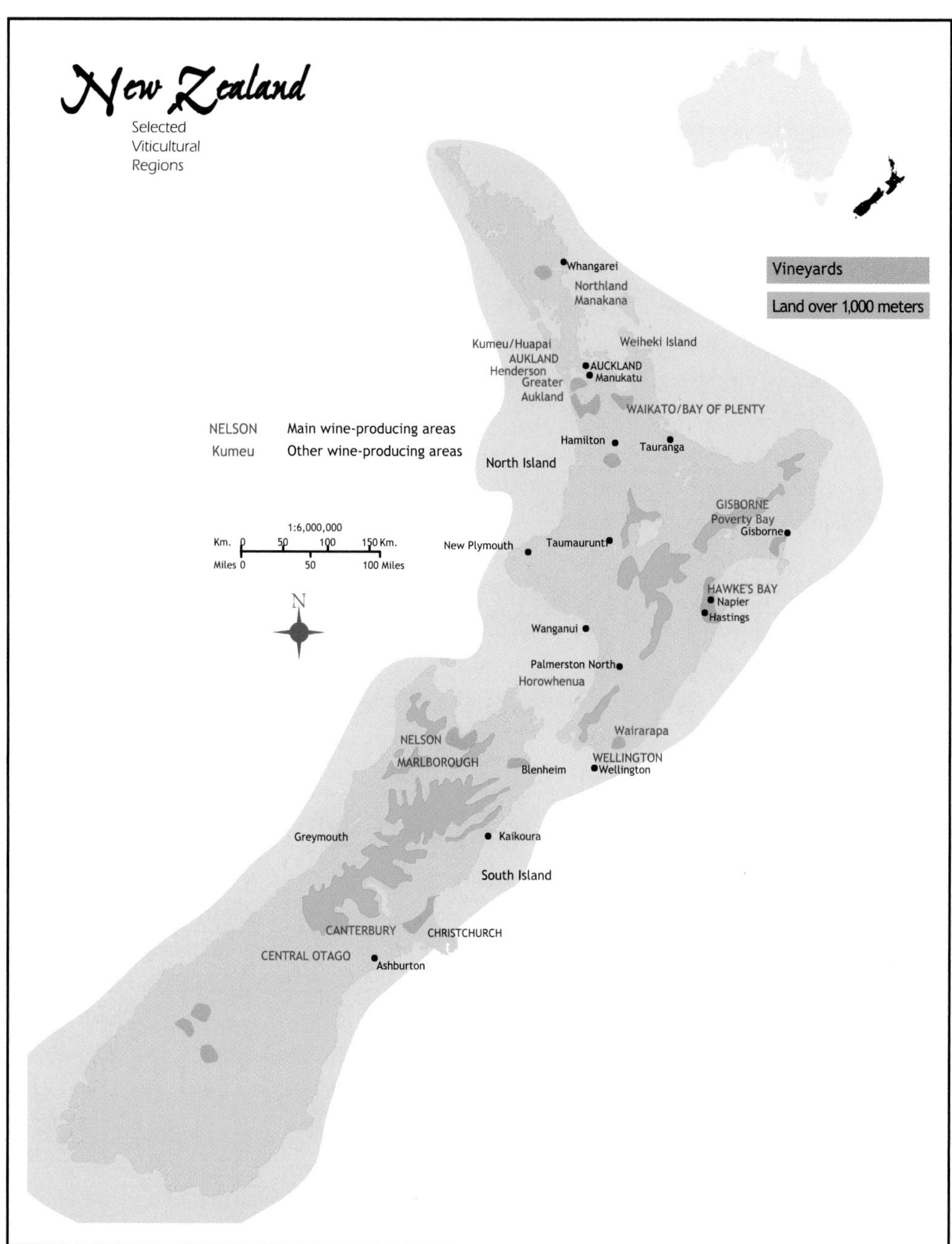

The wine regions of New Zealand.

Table 13-3 Wine Regions of New Zealand

APPELLATION	BEST-KNOWN VARIETIES
Auckland/Northern	Chardonnay, Merlot
Canterbury	Pinot Noir, Chardonnay, Riesling
Central Otago	Pinot Noir
Gisborne	Chardonnay
Hawke's Bay	Chardonnay, Merlot, Cabernet Sauvignon
Marlborough	Sauvignon Blanc, Pinot Noir, Chardonnay
Nelson	Sauvignon Blanc, Pinot Noir, Chardonnay
Waikato/Bay of Plenty	Produces a number of varieties
Waipara	Pinot Noir, Riesling
Wairarapa/Wellington	Pinot Noir, Sauvignon Blanc

Reading a New Zealand Wine Label

Domestic (New Zealand)

Front Label

Winery Name: Most names are acceptable as long as they are not offensive or misleading.

Variety: The varietal of the grapes used to make the wine. The variety listed must be at least 75 percent of the grapes used to make the wine. For wine labeled for export, the requirement is 85 percent.

Appellation or geographical indication: The district the grapes were grown in. Eighty-five percent of the grapes used to produce the wine must be grown in the region listed.

Vintage Date: 85 percent of the grapes used to make the wine must be grown the year listed as the vintage.

Single Vineyard: Designates that the grapes were grown in the Te Muna Road Vineyard in Martinborough.

Wine labels for Australia and New Zealand contain much of the same information that is found on United States wine labels. This label of Craggy Range Sauvignon Blanc is produced for the domestic New Zealand market and has several items, such as the allergen statement and servings per bottle, that are not required for export. Since this wine is not destined for the United States, the government health warning is also absent. If the wine were to be exported, it would have to be labeled with the label information required by the destination country.

Reading a New Zealand Wine Label

Domestic (New Zealand)

Back Label

Wine Notes: Not required but many wineries add them to describe how the wine is made and what it tastes like. They must not be misleading.

Alcohol Content and Container Volume: The percentage of alcohol by volume must be listed. If it is under 15 percent the label may also state "Table Wine." The volume of wine must also be listed in metric units.

Allergen Statement: Since 2002 both Australia and New Zealand have required that potential allergens, such as egg or milk products, be listed when they have been used (typically for fining) in the production of a wine.

Sulfite Declaration: If the wine contains any added sulfur dioxide it must be listed on the label. Here sulfur dioxide is referred to as "preservative 220."

Servings per Bottle: The number of "standard drinks" or servings must be listed. A standard drink is equal to the amount of the wine that contains 10 grams of alcohol.

TE MUNA ROAD VINEYARD

Craggy Range is a family owned winery specialising in the production of expressive Single Vineyard wines. Our Te Muna Road Vineyard is in the famous Martinborough Appellation. This wine is sourced exclusively from several parcels of vines growing on a stony, limestone influenced soil adjacent to the Huangarua River. The wine has rich ripe flavours of limes, peaches and apples alongside the characteristic herbaceous components and a unique dry grainy texture on the palate. As with all our Sauvignon Blanc wines intervention in the cellar is minimal out of care and respect for characters of the vineyard. Enjoy within three years of vintage, lightly chilled, wherever there is sun, surf or seafood.

SAUVIGNON BLANC
MARTINBOROUGH 2005

Alc 13% by vol 750ml

This wine was fined with traditional fining agents based on milk and fish products.
Contains preservative 220
Contains approx. 7.7 standard drinks

PRODUCED AND BOTTLED BY CRAGGY RANGE VINEYARDS LIMITED, WAIMARAMA RD, HAVELOCK NORTH, N.Z. WWW.CRAGGYRANGE.COM
WINE OF NEW ZEALAND

Production Statement: The name and address of the winery that produced the wine.

Country of Origin: Required on all New Zealand wines.

Hawkes Bay

Hawkes Bay is on the southeast coast of the North Island just below the Gisborne region. It is the second largest region in terms of production, growing nearly one-quarter of New Zealand's grapes. It has a long history of winemaking, with the first commercial vineyards being established in the 1890s. It has less rainfall and humidity than Gisborne and more sunny weather during the growing season. There is a great variety of soil types and terrain, making Hawkes Bay ideal for a number of grape varieties. As in the Gisborne, Chardonnay is the most widely planted grape; however, Hawkes Bay's warmer climate and added sunshine also allows red varieties such as Cabernet Sauvignon and Pinot Noir to do well. Sauvignon Blanc is also widely planted and makes a wine that has less of the varietal's grassy/herbaceous character than the grapes produced in the Marlborough region to the south. Hawkes Bay is widely considered one of New Zealand's best wine regions and is home to more than 60 wineries.

Marlborough

The Marlborough region lies on the northeast edge of New Zealand's South Island where the Wairau River empties into Cloudy Bay. Viticulture did not begin in the region until 1973 when the large winery Montana established vineyards there. In the three decades since then, the Marlborough appellation has grown to produce 44 percent of the country's wine (Keesing, 2003) and has become its most well-known appellation. The region is cooler than the grape-growing areas on the North Island, but it is also dryer with abundant sunshine during the growing season. These conditions combine for a long growing season that allows the flavors of cool climate varieties to fully mature. Marlborough has the reputation for producing some of the most strongly flavored Sauvignon Blancs in the world, with strong notes of grass and gooseberries (Figure 13–11). The soil is alluvial and is very rocky in areas, which help the vineyards to have adequate drainage. Its most popular variety is Sauvignon Blanc and it accounts for 55 percent of production followed by Chardonnay and Pinot Noir. The region is now home to more than 80

FIGURE 13–11 Vineyards in Brancott Valley, Marlborough, New Zealand. The region is best known for its intensely flavored Sauvignon Blancs.

wineries and in addition to table wine, there is a growing production of *méthode champenoise* sparkling wines.

Other New Zealand Wine Regions

The remaining wine regions of New Zealand contain only 20 percent of the country's vineyard land. These areas are growing rapidly, however, with many smaller premium wineries being established. This is evidenced by the fact that although these minor regions have one-fifth of the vineyards, they contain nearly two-thirds of New Zealand's 431 wineries (Keesing, 2003).

On the North Island the appellations are:

- Auckland/Northland, two small regions that run from the city of Auckland to the tip of the North Island. The area is warm and can be rainy during the growing season with a variety of soil types. Principal varieties are Chardonnay, Cabernet Sauvignon, and Merlot.
- Waikato/Bay of Plenty, an area to the north and east of the Hawkes Bay appellation. It has only a handful of wineries and about 350 acres (142 hectares) of grapes. It has one of New Zealand's warmer climates, with soils that are generally heavy clay loam. Principal varieties are Chardonnay and Cabernet Sauvignon.
- Wairarapa/Wellington, on the southern tip of the North Island just across the water from the Marlborough region. With a similar climate to Marlborough, it has a number of small producers, and Sauvignon Blanc, Pinot Noir, and Chardonnay are popular.

On the South Island:

- Nelson, on the northern tip of the South Island east of the Marlborough region. It has a varied topography with a number of microclimates and soil types. Principal varieties are Sauvignon Blanc, Chardonnay, and Pinot Noir.
- Canterbury/Waipara, on the eastern side of the South Island. Vineyards are located both around the city of Christchurch and about an hour north in the Waipara subregion. Canterbury has a cool and dry climate with alluvial soils. Waipara has chalky loam soils and is slightly warmer, although spring and fall frosts are a threat in both areas. Principal varieties are Pinot Noir, Riesling, Chardonnay, and Sauvignon Blanc.
- Central Otago, the coolest and most southern of New Zealand's wine regions. The appellation has less of a maritime influence. Most of the vineyards are planted on hillsides for better sun exposure and frost protection. Pinot Noir is by far the most popular variety with 75 percent of the plantings.

Summary

Australia and New Zealand are often grouped together in the minds of American consumers, but their diversity in terroir allows them to produce a number of varieties of wine in a wide range of styles. Australia is better known for its Shiraz, Cabernet Sauvignon, and Chardonnay that have full body and ripe flavors; while New Zealand is recognized for its Sauvignon Blancs and Pinot Noirs that have firm structure, good acid, and intense varietal character. Both Australia and New Zealand have a growing wine industry and a reputation for quality and value on the worldwide market. The two countries also have strong domestic markets but are dependent on exports to sell almost half of the wine that they make. The United Kingdom and the United States are the two biggest customers for New Zealand and Australian wines, and New Zealand and Australia also export a significant amount of wine to each other. This export market is aided by a relatively low cost of production when compared to the United States and Europe, as well as a favorable exchange rate. Australian and New Zealand wines have also had an influence on the American wine industry. In recent years the introduction of their wines into the United States market has kept pressure on domestic producers to keep their prices reasonable and, in some cases, to emulate the popular styles of New Zealand Sauvignon Blanc and Australian Shiraz.

Review Questions

1. What are the risks and benefits to the New Zealand and Australian wine industry by having a large percent of their production exported?
2. What types of varieties are suited to New Zealand's climate?
3. In what areas of the Australian continent are vineyards concentrated and what factors make them suitable for wine grapes?
4. How does the presence of Australian and New Zealand wines affect the wine market in the United States?
5. What grape varieties is Australia best known for, and how do their names differ from those used in the United States?

CHAPTER 14

Chile and Argentina

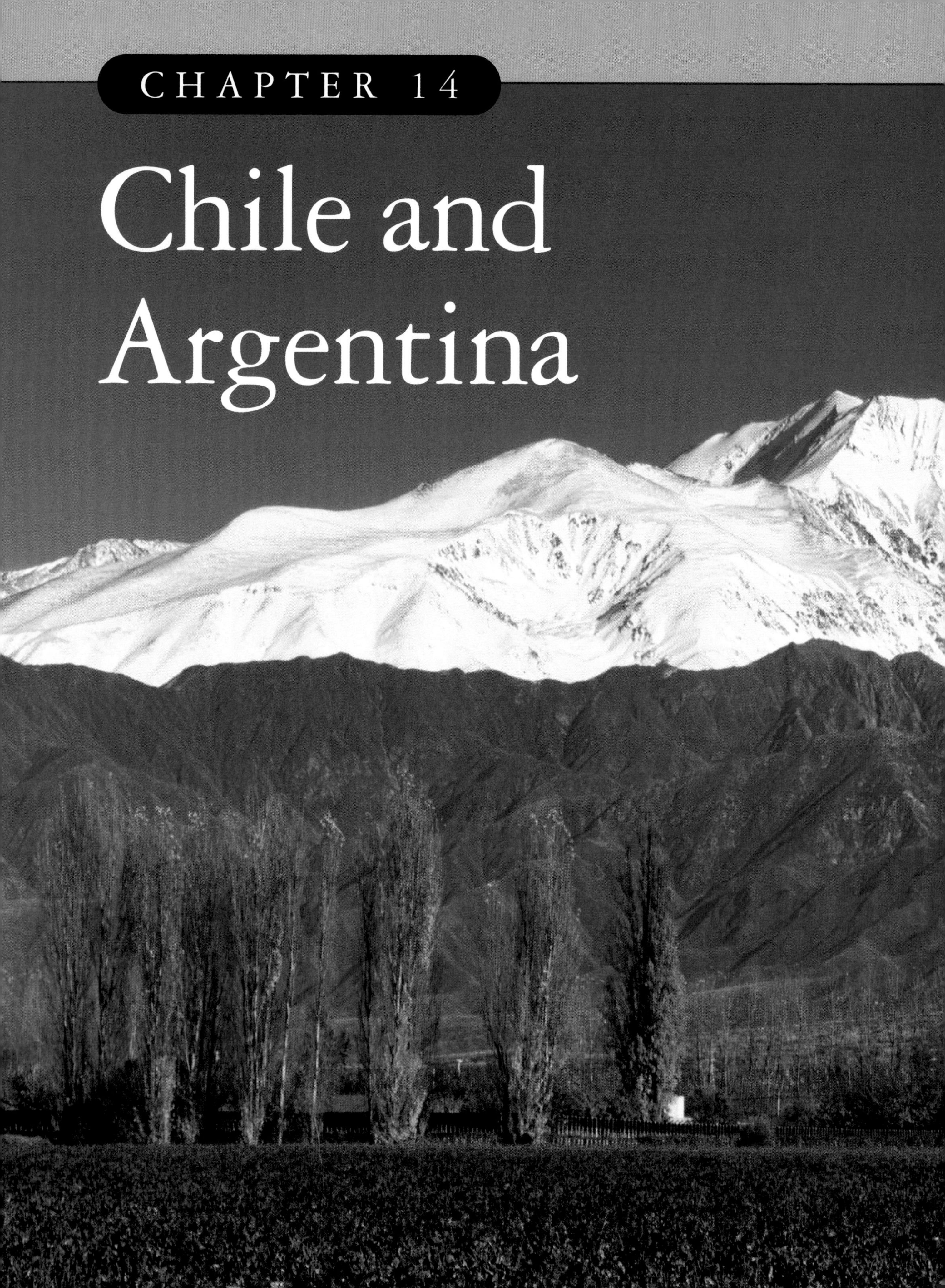

Chile

Chile is the world's eleventh largest producer of wine, bottling about one-fourth as much as the United States. Possessing the proper terroir to produce fine wines, it has grown dramatically in recent years to meet the demand brought on by the forces of globalization and the increasing world market for wine. The acreage of vineyard land in Chile expanded by more than 40 percent in the years from 1997 to 2002, primarily with new plantings of classic French varieties such as Cabernet Sauvignon, Merlot, and Chardonnay (Figure 14-3). These grapes from new vineyards have been going to more modernized wineries, producing wines targeted specifically for the export market in Europe and North America. The wine boom taking place in Chile today is reminiscent in many ways of those that previously took place in California and Australia. The low cost of land and labor allows Chilean vintners to keep their prices low, and inexpensive wines are their primary export. As the industry is maturing, reserve style wines are becoming more common as Chile attempts to move into the high-end wine market as it did in the $10-and-under category.

Chilean Wine—Historical Perspective

The European *vinifera* grapevine was brought to Chile by the Spanish conquistadors, who also introduced it to the rest of Latin America. As was the case in California, wine was important to the Spanish colonizers as both a beverage and for sacramental purposes, and it was much easier to produce it locally than to try to import wine from Spain. During the 1540s, former conquistador Francisco de Aguirre established the first Chilean vineyard in La Serena; shortly thereafter, Diego de Oro planted grapes outside of Santiago. The most popular grape variety that was cultivated during this period was the **País** variety. Similar to the Mission or Criolla variety, it is a red grape that produces well under a variety of growing conditions; however, it makes a rather flavorless wine. Although these varieties are descended from the European grapevine, *Vitis vinifera,* they are considered "native" grapes because they were imported as seeds and then bred in the new world instead of being brought in as cuttings. País and other "native" varieties continue to be widely

FIGURE 14–3 Sampling casks of wine at Viña Caliterra Winery located in Chile's Curicó Valley.

planted in Chile, and as late as 1997, País made up one-quarter of the grape harvest. These grapes are primarily used to make wines for domestic consumption and are being rapidly replanted with more familiar European grape varieties.

Over the next three hundred years, winemaking in Chile developed very slowly in spite of the excellent growing conditions. Two main factors contributed to this slow progress. The first was that it was difficult to expand vineyards beyond the area around Santiago into outlying areas due to the frequent raids and attacks against the colonialists by indigenous Indian tribes. The second was the lack of a market due to the small domestic population and the difficulties in exporting wine. One of the early efforts in exporting wine to Peru in 1578 was met with failure when Sir Francis Drake, the English explorer and privateer, captured 1,700 wineskins en route to Peru. During the sixteenth and seventeenth centuries, vineyard land expanded, and Chile began exporting inexpensive wine and brandy. Although Spain originally encouraged the export of wine as a revenue source, Spanish winemakers became upset when Chilean exports became a threat to their business. During the 1600s, the planting of new vineyards was banned for a period, and in 1774 the Spanish king forbade the export of any wine from Chile to other Spanish colonies. These laws were difficult to enforce, and in 1822 Chile won its independence, ushering in a new era in winemaking.

In 1830, the young country established an agricultural station called Quinta Normal, with the goals of teaching students and performing research on how to improve grape growing and wine production. Italian and French grape varieties were established at Quinta Normal, and French trained enologists came to aid in the work (Ureta & Pszczókowski, 1995). In 1851, Don Sylvestre Ochagavia imported classic French varieties for his vineyard, the first major commercial planting of traditional *vinifera* grapevines in Chile. The vines did very well, and his success inspired many others. In the second half of the nineteenth century, there was rapid expansion of vineyards planted to European varieties as well as improvement in winemaking techniques; this began the modern era of winemaking in Chile. During the period from 1850 to 1890, some of Chile's most famous wineries were established (Figure 14–4). Errázuriz, Concha y Torro, Cousiño Macul, and Santa Rita, among others, were all founded during this period and remain active today (Mathäss, 1997).

Production continued to expand primarily for the export market. In the 1880s, while phylloxera decimated the vineyards of Europe, wine from Chile was used to fill the demand. The root

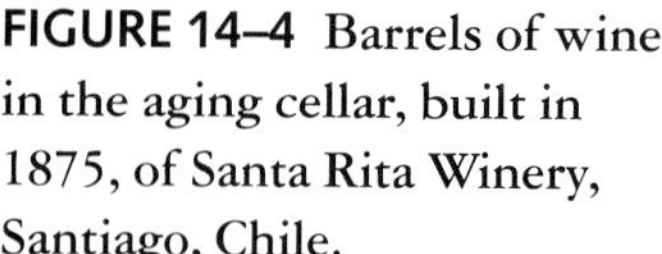

FIGURE 14–4 Barrels of wine in the aging cellar, built in 1875, of Santa Rita Winery, Santiago, Chile.

louse was never introduced to Chile, and today it is one of the few grape-growing regions where vines can be grown on their own roots. By the 1930s, European demand had slowed down, and Chile enacted prohibitionist laws that forbade the planting of new vineyards in an effort to restrict the production of wine. Soon after this World War II began, making shipping more difficult and isolating Chile from its European markets. After the war, the export market continued to decline, and the restrictive laws remained, which held production steady until they were repealed in the 1970s. During this time there was little advancement in wine quality, and nearly all the wine made was consumed domestically. When growers could expand their vineyards again, production soared at the same time per capita consumption was shrinking. This resulted in overproduction, exacerbated by the fact that the wines being made were not of a quality that could compete on the global market.

At the beginning of the 1980s, Chile was plagued by overproduction and poor quality wines. The wineries of Chile dealt with the situation by concentrating on improving their product through modernization and increasing their emphasis on the export market once again. This course of action was successful because Chile possessed the necessary conditions for success—a good terrior for growing wine grapes and a low cost of labor, allowing the wines to be attractively priced on the world market. During this time, Chile began to attract international investment. Some of the world's most famous wineries including Torres from Spain, Lafite-Rothschild from France, and Mondavi from the United States, all began operations in Chile. The development of the Chilean industry was also aided by winemakers from around the globe who came to take advantage of the opportunities and brought with them their experience and expertise. During the 1990s, the value of Chilean wine exports grew from $30 million to $600 million (Sparks Companies Inc., 2002). The number of wineries that exported also rose; however, both the export and domestic markets remain dominated by a few large wineries. Although the overall quality of Chilean wine has greatly improved, there are two distinct types of wine. One inexpensive and ordinary, made primarily for the domestic market, and a superior one made from classic European varietals, used mainly for the export market. Today wines from Chile have gained a reputation throughout Europe and North America for both quality and value.

The Wine Regions of Chile

Chile exists along the narrow strip of land between the Pacific Ocean and the Andes mountain range and is more than 2,800 miles (4,500 kilometers) long but averages only about 115 miles (180 kilometers) wide. Running north to south it has a great range of climates, from some of the world's driest deserts in the north of the country, to cool rain forests in the south. In the middle of the country lies a temperate zone with a Mediterranean climate that is ideal for the production of wine grapes. Most of the vineyards producing grapes for table wine are located in an area that begins about 100 miles (160 kilometers) north of the capital Santiago, and runs for 400 miles (640 kilometers) south. This area is also home to the majority of the nation's population as well. The wide variety of soils and microclimates in Chile's viticultural zone provide a number of terroirs that are suitable for many different varieties of both wine and table grapes. The more northern regions are generally much dryer than those to the south, yet have a mild climate. The southern vineyard regions are typically cooler and have more rainfall. The Pacific Ocean influences a vineyard's terroir, depending on how far inland the vineyard site is located. The coastal winds that bring this maritime influence inland are affected by the height of the mountains that lie between the vineyard and the ocean.

Chile's wine country is divided into five major grape growing regions. From north to south they are the Atacama, Coquimbo, Aconcagua, Central Valley, and the Southern Valley. These major

The major table wine producing subregions of Chile.

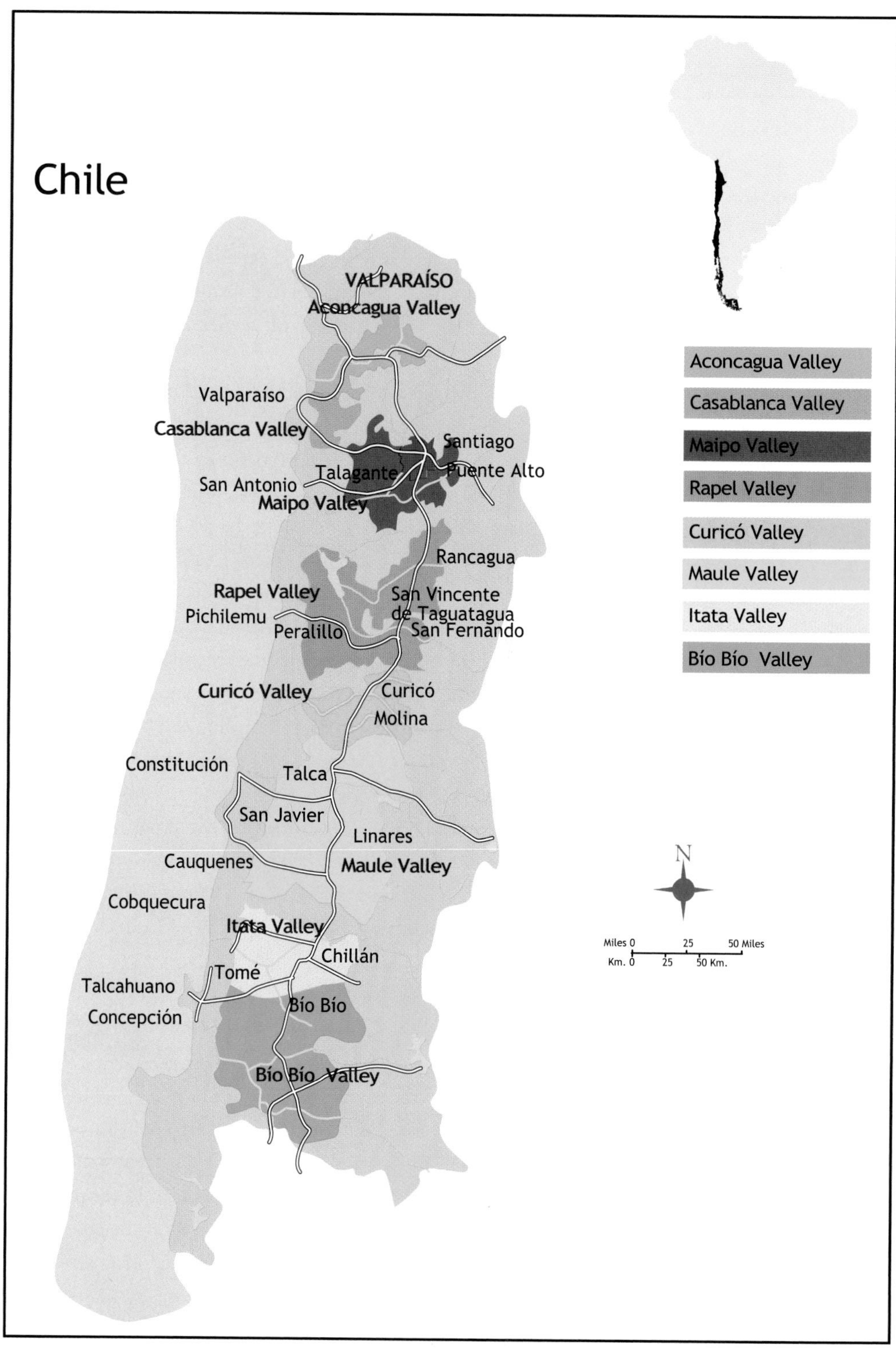

regions are separated by a number of river valleys that run from east to west, which drain the runoff from the Andean mountain range. These river valleys form eleven subregions that are further broken up into a number of zones and areas. This is similar to the Australian system of appellations except that in the Chilean method of organization, zones represent the smallest geographic area instead of the largest (Table 14–1). Chilean law states that for a wine to be listed with a specific geographical area it must be made using at least 75 percent of grapes grown in that area. The 75 percent requirement also applies to the vintage and variety listed on the label. However, wine made for export follows the European standard of 85 percent for variety, region, and year.

Table 14–1 The Grape-Growing Regions of Chile from North to South

REGION	SUBREGION	ZONE
Atacama	Valle de (Valley of) Copiapó	
	Valle del Huaco	
Coquimbo	Valle del Elqui	
	Valle del Limarí	
	Valle del Choapa	
Aconcagua	Valle de Aconcagua	
	Valle de Casablanca	
Valle Central (Central Valley)	Valle del Maipo	
	Valle del Rapel	Valle de Cachapoal
		Valle de Colchagua
	Valle de Curicó	Valle del Teno
		Valle del Lontué
	Valle del Maule	Valle del Claro
		Valle del Loncomilla
		Valle del Tutuvén
Valle del Sur (Southern Valley)	Valle del Itata	
	Valle del Bío-Bío	

The Atacama and Coquimbo Regions

The two northernmost regions, Atacama and Coquimbo, are hot and dry and are considered too warm for the production of grapes for fine table wine. This being said, there are many vineyards in these areas that have a history of winemaking and viticulture that goes back to the sixteenth century (Figure 14–5). Although there are still several wineries in the Coquimbo region, the

FIGURE 14–5 Vineyards in the northern grape-growing Elqui Valley region of Chile. The grapes produced in this desert region are primarily used for the production of Pisco.

Pisco

Pisco is considered by many to be the national drink of Chile. It is a light-colored brandy and is enjoyed either by itself or mixed in a cocktail such as a Pisco Sour or Pisco and Cola. It is very popular in Chile and little known outside of South America. Of the more than 5 million cases a year that are produced, less than 25,000 cases are exported (Duijker, 1999). Pisco is also made in Peru but uses different grape varieties and has a different flavor than Chilean Pisco. The production of Pisco is strictly regulated by the government with laws that were first established in 1931 and revised in 1985. The grapes that are grown for Pisco can only come from the Atacama and the Coquimbo wine regions, and the fermentation and distillation must also take place in these appellations as well. There are 13 varieties of grapes that are allowed to be used for Pisco, the most common being various clones of Muscat and the varieties Pedro Jiménez and Torontel (called Torrontés in Argentina). The high percentage of Muscat in the base wine that is used for distillation gives the resulting brandy a distinct floral/fruity character.

FIGURE 14–6 A pot still used for the production of Pisco. This is similar to the type of still that is used for the production of Cognac in France.

The winemaking for Pisco is basic. The grapes are crushed and then given a small amount of skin contact before pressing to give the juice more flavor. After the juice has been fermented into wine it has about 13 percent alcohol. It is then distilled into brandy at 60 percent alcohol in copper **alembic stills,** which are the same type as those used for making Cognac (Figure 14–6). Traditionally the Pisco was then aged in casks of **raulí** wood, which is a type of South American beech. Today most producers use stainless steel or concrete vats for the majority of their production, saving the use of raulí casks for their reserve lots. After aging, the Pisco is graded, blended, and bottled at 30 to 40 percent alcohol.

vineyards mainly produce table grapes or grapes that are used for the production of **Pisco,** a type of brandy.

The Aconcagua Region

The Aconcagua Region is the northernmost of Chile's table wine regions; it is made up of two subregions, the Aconcagua and Casablanca Valleys (Figure 14–7). The Aconcagua River valley lies about 60 miles (100 kilometers) north of Santiago, just to the west of Mount Aconcagua, the Western Hemisphere's tallest mountain at an elevation of almost 23,000 feet (7,000 meters). The region has alluvial soils and a warm climate with a long growing season. It is best known for

FIGURE 14–7 A vineyard in Aconcagua Valley, the northernmost grape-growing region used for the production of table wines.

growing red grapes such as Cabernet Sauvignon. The first wine grapes were established by Maximiano Errázuriz in 1870, and today the Errázuriz Winery remains active in the region and is one of Chile's largest producers (Figures 14–8 and 14–9).

The Casablanca Valley lies along the coast about 50 miles (80 kilometers) west of Santiago. Viticulture is relatively new to the area; the first vineyards were established in 1982. Another of Chile's major wineries, Concha y Torro, was one of the first wineries to use grapes from Casablanca, and their success inspired many others to start vineyards here. Although the appellation is very young, it has had rapid growth in new vineyards and now has almost four times as much acreage as is used in the Aconcagua Valley. Because it is near the ocean, there is a strong maritime influence, and the climate is very cool. Morning fog is common during the growing season, and in springtime there is always a risk of frost. This cool weather makes the Casablanca Valley ideal for growing white wine grapes, particularly Sauvignon Blanc and Chardonnay, as well as cool-weather reds such as Pinot Noir.

The Central Valley Region

The Central Valley is Chile's largest viticultural region and has by far the most acres in grapes. It is made up of four subregions. From north to south they are the Maipo, Rapel, Curicó, and the Maule. In the Central Valley, the rainfall gradually increases as you move south. The Maipo River valley is just south of Santiago, and its proximity to the capital has given prominence to the region from the very beginning. It is known for producing superior wines, particularly Cabernet Sauvignon. The Maipo was also one of the first viticultural areas in Chile established in the sixteenth century, and it was the first area to grow classic European varietals in the nineteenth century. The Maipo is a diverse region with a number of soil types and a wide range of elevation and rainfall. The growth of Santiago has put pressure from urbanization on many of the vineyards close to the city, and newer plantings are being set up further into the countryside. Just south of the Maipo region is the Rapel Valley, formed by the Cachapoal and the Tinguiririca Rivers. The elevation of the coastal range of mountains in this part of the country is low, and

FIGURE 14–8 The Errázuriz Winery in the Aconcagua Valley, Chile. The winery is one of the most historic in the region.

FIGURE 14–9 Robert Mondavi and a delegation from California visit the Errázuriz Winery in 1996 to taste wines being made as part of a joint venture between the two wineries.

they do not block cooling ocean breezes. This gives the area a maritime climate in spite of its somewhat inland location. Like Casablanca, the Rapel Valley has experienced considerable growth in recent years, primarily in red grapes. The Carmenère that is grown in the Rapel Valley is considered one of the best that is produced in Chile.

Carmenère is a red grape that is similar in taste and appearance to Merlot and is rarely cultivated outside of Chile. Also known as Grand Vidure, prior to the outbreak of phylloxera it was popular in the Bordeaux region of France, but due to Carmenère's low yield, when the vineyards were replanted it was widely replaced with other varieties. Merlot and Carmenère were often confused, and during the 1990s most of what was thought of as Merlot in Chile was determined to be misidentified Carmenère. It produces a full-bodied deeply colored wine that is becoming popular as a varietal wine that is unique to Chile.

The Curicó subregion is centered on the town of Curicó, 120 miles (200 kilometers) south of Santiago. This is the widest part of the Central Valley and the terrain is made up of plateaus with small hills. This broad expanse allows vineyard operations to spread out and many of them are quite large (Figure 14–10). The soils are fairly uniform and generally made up of clay loam and decomposed volcanic material. The region has slightly less area in vineyards than the Rapel region to the north. Red and white varieties are grown here in roughly equal proportions. The Maule River valley lies just to the south of the Curicó subregion and is the southernmost appellation in the Central Valley. The valley has the greatest acreage of any area in Chile, more than 41,000 acres (17,000 hectares) planted to wine grapes (Duijker, 1999). The area is mainly planted to white grapes, in particular Chardonnay and Sauvignon Blanc. Merlot acreage is also significant, and there is still a great deal of the País variety grown for the domestic wine market.

Reading a Chilean Wine Label

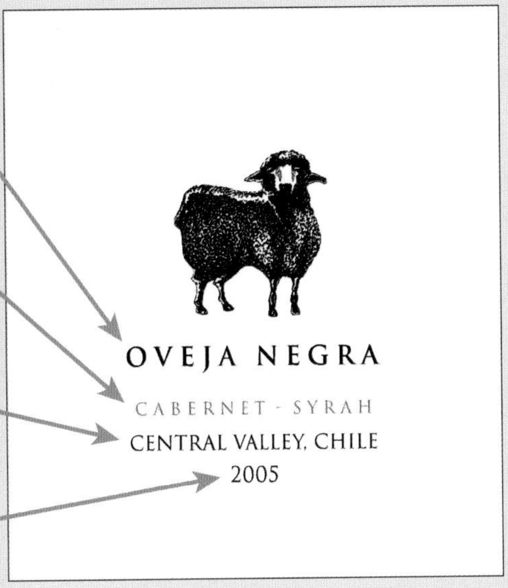

Winery Name: Most names are acceptable as long as they are not offensive or misleading.

Variety: For export, when more than one grape variety is listed the total percentage of the varieties listed on the label must equal 100 percent. If there is only one variety listed on the label, the wine must be at least 75 percent of that variety for domestic (Chilean) use and 85 percent for export.

Appellation: When listed, 75 percent of the grapes used to produce the wine must be from that appellation. For export, the standard is 85 percent.

Vintage Date: Like variety and appellation, the standard for domestic sales is 75 percent, for exported wines it is 85 percent.

Front

OVEJA NEGRA
CABERNET - SYRAH
CENTRAL VALLEY, CHILE
2005

Composition: The percentage of the varietals used to make the wine.

Back

OVEJA NEGRA
CABERNET 60% - SYRAH 40%
CENTRAL VALLEY
WE ALL KNOW A BLACK SHEEP.
THAT SOMEONE SPECIAL. OUT OF THE ORDINARY.
SOMEONE WHO CALLS YOUR ATTENTION.
WHO INTRIGUES YOU.
THIS WINE IS OUR BLACK SHEEP.
A WINE WITH CHARACTER AND PERSONALITY.
INTRIGUING.
IT WILL NOT PASS UNNOTICED ON YOUR TABLE.
WINE OF CHILE
www.viawines.com
GOVERNMENT WARNING:
(1) ACCORDING TO THE SURGEON GENERAL, WOMEN SHOULD NOT DRINK ALCOHOLIC BEVERAGES DURING PREGNANCY BECAUSE OF THE RISK OF BIRTH DEFECTS. (2) CONSUMPTION OF ALCOHOLIC BEVERAGES IMPAIRS YOUR ABILITY TO DRIVE A CAR OR OPERATE MACHINERY, AND MAY CAUSE HEALTH PROBLEMS.
CONTAINS SULFITES.
8 58472 00049 6
750 ml 13.5% alc. by vol.
Produced and bottled by VIA S.A., Chile.

Wines Notes: Not required, but many wineries add them to describe how the wine is made and what it tastes like. They must not be misleading.

Web Address: Not required but an excellent source of more information about the wine.

Country of Origin: Required on all imported wines.

Government Warning: Required on all beverages that contain more than 0.5 percent alcohol sold in the United States.

Alcohol Content: The percentage of alcohol by volume in the wine must be listed.

Sulfite Declaration: If the wine contains more than 10 ppm sulfur dioxide "Contains Sulfites" must be listed on the label to be sold in the United States.

Production Statement: The name of the winery that produced the wine.

Volume: Every wine bottle must state the amount of wine it contains in metric units. In most cases the volume in ml is molded into the glass at the base of the bottle.

As in the United States, the term **Reserva** (reserve), does not indicated any government sanctioned designation. However, when the terms **Reserva, Gran Reserva,** and **Reserva Especial** are used, the wine must have its appellation listed.

FIGURE 14–10 A large vineyard in Chile's Valle Central (Central Valley) region.

The Southern Region

The Southern Region is made up of the Itata and the Bío-Bío river valleys. Being at the southern end of Chile's viticultural regions, it is the coolest and wettest appellation. The Itata River valley has been growing grapes since the time of the Spanish colonization and accounts for about 17 percent of Chile's wine grapes. It still has large areas of País and other less desirable wine grapes; however, lately there have been efforts to replant them with better varieties. To the south of Itata is the Bío-Bío River valley. At the southernmost and coldest part of Chile's wine country, spring frost can be a problem, and cool climate varieties such as Pinot Noir, Chardonnay, and Gewürztraminer do best here. This subregion is also very wet. Some areas receive more than 45 inches (115 centimeters) of rainfall per year.

Argentina

Argentina has over a half million acres in grapes, nearly twice as much as Chile. By comparison, Argentina's wine industry is about half that of the United States. Investment in vineyards and wineries in Argentina has come more slowly than it did in Chile, being hindered by the unstable economic situation that has beleaguered the country. Recently, despite the economic meltdown that occurred in 2001, the situation in Argentina is beginning to improve. Like Chile in the 1990s, investors are beginning to be drawn to the country to take advantage of the low cost of vineyard land and labor. Although older plantings of more traditional Argentine varieties are still common, acreage is expanding in the classic European varieties in Argentina, and wineries are being updated (Figure 14–11). However, this modernization is coming slowly and the majority of the country's output is destined for the domestic market for inexpensive wine.

Argentine Wine—Historical Perspective

The first grapevines of European origin were introduced to Argentina from Chile in 1556. They were planted in the colonial settlement of Salta, in the north of the country. The early settlers discovered that some of the best locations for growing vines were in the foothills of the Andes.

FIGURE 14–11 French oak casks in a Mendoza winery in Argentina.

Vineyards were soon established to the south of Salta around the town of Mendoza. Here the native variety **Criolla,** which is similar to the Chilean País, was widely planted. The production of wine grew slowly for the next 250 years, with what was made being split between the small domestic market and the limited export trade. After independence from Spain in 1816 all this began to change. A new wave of emigration from the winemaking countries of Europe brought in new residents, forming a healthy domestic market for wine as well as expertise on how to make it. Many of the new residents came from Italy and Spain, giving more Spanish and Italian influence on the varieties of grapes grown and wines produced in Argentina.

In 1853, Argentina established a school of agriculture in the Mendoza province. The director of the school was French-born Miguel Pouget, who trained his students in modern French winemaking techniques and imported classic French wine grape varieties to Argentina for the first time. Many irrigation projects were completed during this time, which allowed the expansion of vineyards and many other forms of agriculture. The industry continued to grow and have a strong export market until the 1920s, when it experienced setbacks. The worldwide depression of the late 1920s, affected the demand for both the domestic and export markets. Additionally phylloxera was introduced during this time. Argentine growers, familiar with what phylloxera had done to the vineyards of Europe decades earlier, responded quickly with the widespread use of grafting and rootstocks. Unfortunately, this also had the effect of spreading viruses and other diseases. Phylloxera still exists in Argentina today, but because it is not widespread and causes little damage, most vines are grown on their own roots.

Since the 1920s, Argentina has experienced severe economic and political instability, which has adversely affected the development of the wine industry. There was little foreign investment and modernization, and so the quality of the wine suffered. In spite of this, the national wine consumption remained strong, and a great deal of mediocre wine was produced for the domestic market. This situation began to change in 1970 when the per capita wine consumption began to decline from a high of over 23 gallons (90 liters) per year to less than half that amount in 1997. Consequently during the 1980s there was a severe overproduction problem and 36 percent of Argentina's vineyards, primarily the lower quality varietals, were taken out of production (Lapsley, 2001). Although vineyard acreage has declined, Argentina is the fifth largest producer of wine after France, Italy, Spain, and the United States (Wine Institute, 2004).

Today Argentina has realized that its wine future lies in the export market. Moreover, much like Chile, Argentina possesses the proper environment for growing grapes and a low cost of land and labor. This situation has spurred new development and has attracted international investors despite the unstable economic situation. One advantage Argentina has over Chile is that there is much more land available for planting. Currently Argentina remains in the shadow of its neighbor on the international wine market, exporting only 7 percent of its production compared to Chile's 54 percent. However, given the proper resources and development, Argentina has the potential to equal and perhaps surpass Chile's export wine production (Figure 14–12).

The Wine Regions of Argentina

Much of Argentina's viticultural regions lie on the eastern slopes of the Andes Mountain range. Being in the rain shadow of the mountains, the climate is generally dryer and warmer than what is found in the vineyards of Chile. To deal with the dry conditions, most vineyards are irrigated with runoff from the mountains and are planted at higher elevations, which helps to keep the vineyards cool. Argentina also has more of a continental climate and less maritime influence than Chile does. Without the moderating influence of the Pacific Ocean, Argentina's vineyards experience more difficulties with spring frosts, hail, and heat waves than the vineyards in Chile do. Chile also receives more of its rainfall in the winter, while Argentina has more rain during the growing season. Argentina is a large country and there is a great deal of suitable land for viticulture (Table 14–2). Most vineyards are concentrated in the northwest of the country, but grapes are grown from the northern border with Bolivia to as far south as the border of the Patagonia region, two-thirds of the length of the country.

The Mendoza and San Juan Regions

The Mendoza region is the largest appellation in Argentina at 58,000 square miles (150,000 square kilometers) and has about 75 percent of the country's vines. The appellation is centered on the city of Mendoza, which is only about 150 miles (240 kilometers) east of Santiago, but it is

FIGURE 14–12 The modern wine cellars of Bodegas Salentein, Uco Valley, Mendoza, Argentina. The winery is an example of the modernization of the wine industry that is currently taking place in Argentina.

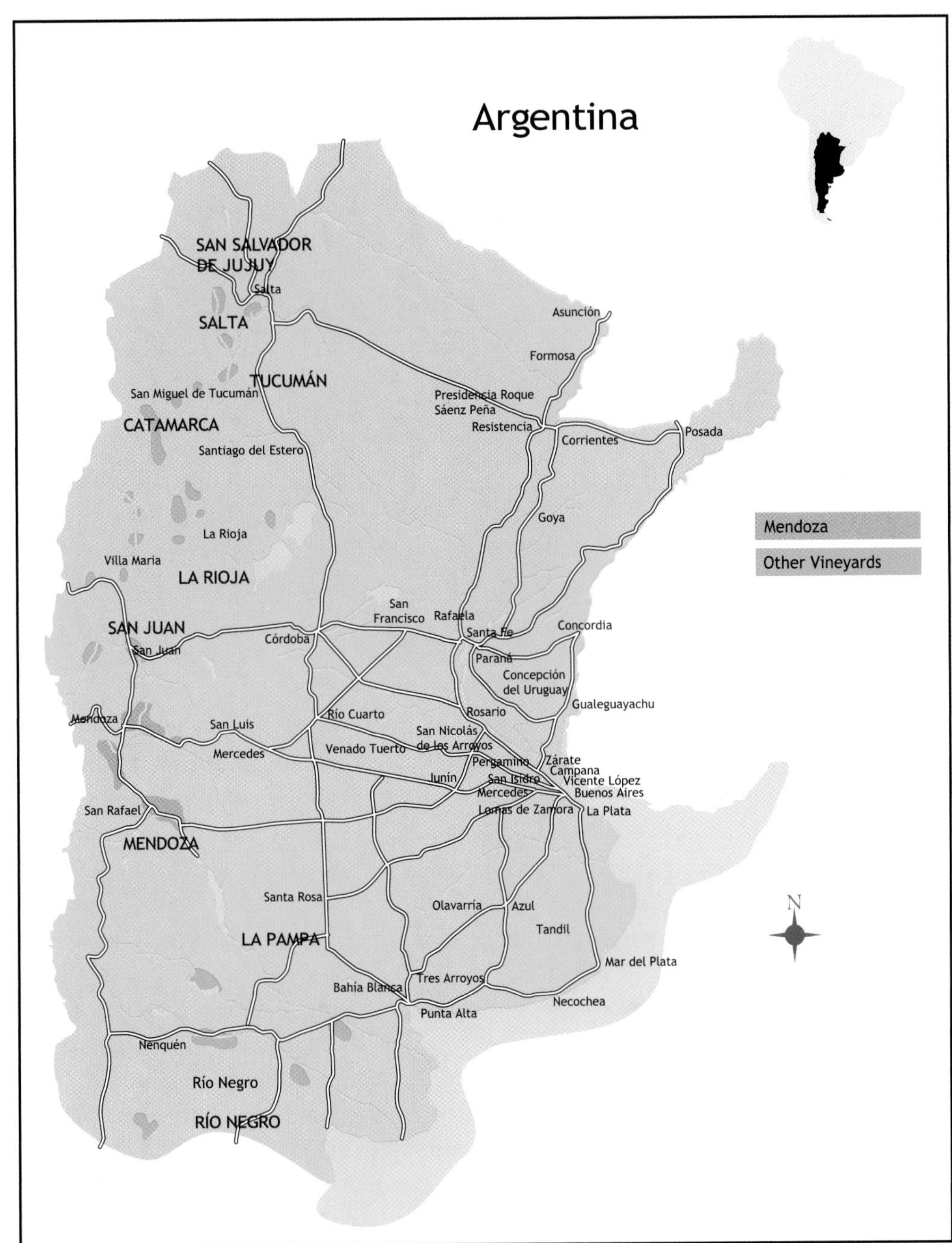

The wine regions of Argentina.

Table 14–2 The Grape-Growing Regions of Argentina from North to South

REGION	BEST-KNOWN VARIETIES
Jujuy	Torrontés, other white varieties
Salta	Torrontés, other white varieties
Tucumán	Torrontés, many other varieties
Catamarca	Not known for any one variety
La Rioja	Malbec, white varieties
San Juan	Average quality table wines and dessert wines
Mendoza	Torrontés, Malbec, many other varieties
Rio Negro	Sauvignon Blanc, Chardonnay, Pinot Noir

FIGURE 14–13 Fall vineyards with the Andes Mountains in the background, Luján de Cuyo, Mendoza, Argentina.

more than 600 miles (950 kilometers) west of Buenos Aries. The soil is generally sandy and the climate is desertlike, but because of irrigation there is a great deal of agriculture in the region. The elevation ranges from 2,000 to 3,000 feet (600 to 900 meters), and this altitude helps keep the vineyards cool. Mendoza is considered one of Argentina's best appellations for quality, and there are more than 1,000 wineries in the district. Due to Mendoza's size and variety of growing conditions, it has been broken into several subregions including the Agrelo, Luján de Cuyo, San Rafael, and Tupungato (Figure 14–13).

The Mendoza region has many vineyards of both classic European cultivars as well as native varieties like Criolla but it is best known for two other varieties, **Malbec** and **Torrontés.** Malbec is a grape that is native to the Bordeaux region of France, where it is called **Côt.** Considered to be a lesser grape in its home region, Malbec flourishes in its adopted home of Argentina, making full-bodied wine that is inky in color with a strong fruity character. Called Torontel in Argentina, Torrontés is a white grape native to the Rioja region of Spain that, like Malbec, has become better known in the Mendoza region than it is in its native country. It produces light-bodied, crisp wines that have a fragrant aroma similar to Muscat or Gewürztraminer. There are three varieties of Torrontés: Torrontés Riojano, which is the most common cultivar, Torrontés Sanjuanino, and Torrontés Mendocino. It is considered by many in Argentina to be the country's best white wine; however, Chardonnay commands higher prices and is more frequently exported.

The San Juan region lies north of Mendoza and is Argentina's second largest grape appellation, producing nearly 20 percent of the country's grapes. Being further north and closer to the equator, it is warmer than Mendoza and is best known for making average quality wines for domestic consumption and for growing table grapes. The San Juan region contains about 250 wineries, and the warm climate is ideal for making dessert wines and brandy. Collectively, the Mendoza and the San Juan regions are referred to as the Cuyo region.

The La Rioja and Salta Regions

The La Rioja region lies between the San Juan region on the south and the Salta region to the north. The climate is warm and dry, but the wines are generally of better quality than those in

FIGURE 14–14 The vineyards in the Calchaquí Valley of Argentina's Salta region. Here the vineyards reach between 7,200 feet (2,200 meters) and 9,800 feet (3,000 meters) above sea level. The high elevation helps to moderate the temperatures of the region's warm latitude.

the San Juan region. Recently there has been much planting of new vineyards, particularly to white varieties. The Salta is a small appellation in the far north of the country. It is located at a latitude of about 25 degrees south, roughly equivalent to the latitude of Miami in the Northern Hemisphere (Figure 14–14). The Salta region is about 600 miles (960 kilometers) north of Mendoza. At low elevations the climate is subtropical, but at higher elevations grapes for table wine can be grown. Most of the area's vineyards are planted between 5,000 and 6,800 feet (1,500 and 2,000 meters), making them some of the highest vineyards in the world. This high altitude makes the terroir very cool, and the Torrontés grape does quite well here, with the chilly climate preserving its fruity character. Nearby are the small appellations of Jujuy and Tucumán, which have similar growing conditions to those found in the Salta.

The Rio Negro Region

Located in the far south of Argentina's wine country, Rio Negro is a cool region that shows great promise. It has a long growing season with plenty of sunny days in which cool-climate varieties such as Sauvignon Blanc and Chardonnay thrive. Pinot Noir also does well here and is grown for both table wine production and as a base for sparkling wine. The soils are more chalky than those in the other wine regions of Argentina. Compared to the vineyard lands of the Cuyo region, there are relatively few acres in production. However, vineyard land is expanding, and the region has the potential to become one of the country's best appellations.

Summary

Chile and Argentina have the terroir to match some of the world's best wine-growing regions. In the past their potential has not been realized due to restrictive laws, unfavorable economic conditions, and a lack of modernization. Chile was the first of the two to look outward to the export market to deal with its expanding production and declining national consumption (Figure 14–15). Chile's efforts to make better wine that can compete on the world market have been met

FIGURE 14–15 A winery worker inspects bottles on a bottling line at Santa Rita vineyard, south of Santiago, Chile. Santa Rita is Chile's largest wine producer.

with success, and much more of Chile's wines are exported even though it produces less than Argentina. This being said, Argentina is attempting to modernize its vineyards and wineries to make better wine for export. Both countries have come far in recent decades and have achieved a reputation for producing quality wine at a good value.

Review Questions

1. Compare how the export markets for wine developed in both Chile and Argentina.
2. How do the wines that are made for the domestic market in Chile compare with those made for export?
3. What are the most popular grape varieties grown in Argentina and what are their origins?
4. Outline the conditions that limit production in Chile.
5. Compare how Spanish colonialists developed winemaking in South America to how it was done in California (Chapter 10).

CHAPTER 15

South Africa

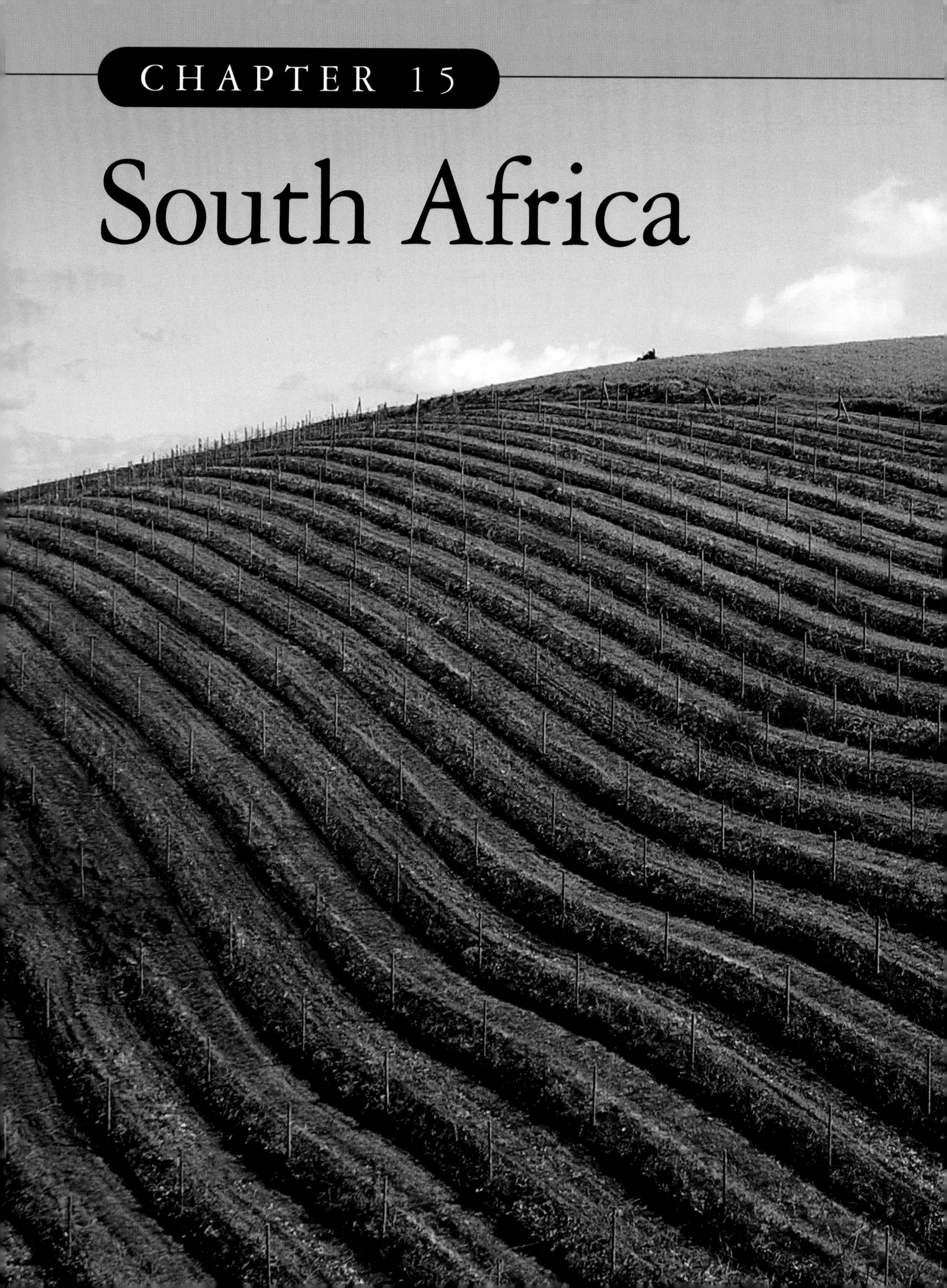

After reading this chapter, you should be able to

- outline the history of winemaking in South Africa.
- describe the climatic conditions of the grape growing regions in South Africa.
- discuss the competition that winegrowing has in South Africa.
- discuss the role that South African wines play in the global wine market.

KEY TERMS

aspect
Biodiversity and Wine Initiative
Black Association of the Wine and Spirit Industry (BAWSI)
Cape Doctor
Cape Floral Kingdom
Chenin Blanc
cooperatives
Hanepoot
Huguenots
Integrated Production of Wine
jerepigo
Méthode Cap Classique
Muscadel
oidium
Pinotage
potstill brandy
Wine of Origin Scheme

Until the demise of the apartheid regime and lifting of international sanctions in the early 1990s, South African wine was banned and its citizens unwelcome in many countries. During this repressive regime, South African producers had to look to the local market for sales, and their knowledge of winemaking trends was severely restricted, as travel opportunities were limited. Once democracy was established, with the first democratic elections held in 1994, momentum for South Africa's re-entry into the international arena grew rapidly; it was marked by many far-reaching changes in the wine industry. These run in tandem with the

Map of South Africa

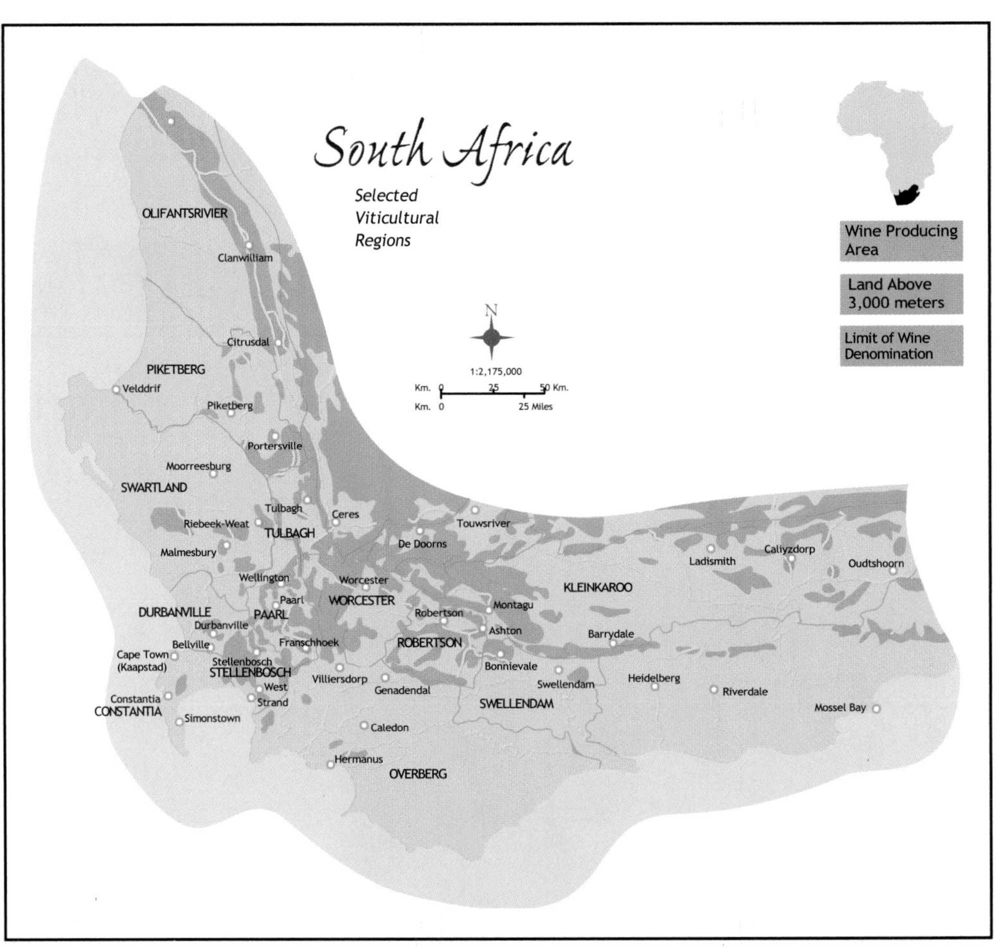

surge of interest in wine worldwide and are, perhaps most immediately, measured through the growth in exports. In 1991, 5.7 million gallons (23 million liters) of South African wine left the country. By August 2005 that annual figure had risen to 70.752 million gallons (283,007 million liters) (SA Wine Industry Information Systems [SAWIS], 2005, no. 29, p. 25). While the UK remains the most important volume market, the U.S. is gaining ground and now fills sixth place (SAWIS 2005, no. 29, p. 28).

Behind those bare figures lie many reasons for this remarkable growth, reasons that go beyond the demise of apartheid and lifting of sanctions: the interest in wine worldwide; big shifts in vineyard makeup with quality international varieties being planted, many in areas where wine grapes had never before been grown; improved winemaking skills and South Africa's historical links with European countries; many South Africans with family or ties in the UK, Holland, Germany and other countries, which not only provides a ready market for the wines, but encourages the growth in tourism, another benefit for the sales of South African wine.

Thanks to this international interest and increased sales, the Cape (as the area where the South African winelands are located is generally known) has seen an explosion of private cellars whose owners are focused on producing quality individual wines that speak of their South African origin, thus differing from the big brand, commercial ranges. These individuals have helped build the popularity of South African wine by showing its many faces. There have also been big shifts in vineyard makeup and opening up of areas where wine grapes had never before been grown. Prior to 1992, when the quota system, which had restricted where vines could be planted, the authorities then having an eye more on quantity than quality, was dropped, a trickle of new private cellars opened; the leap from 141 in 1991 to 477 in 2004 (SAWIS, 2005, no. 29, p. 7) reflects the attraction of the wine lifestyle, in the scenically beautiful Cape with its reliable harvest weather, good soils and many mesoclimates, suitable for a wide variety of styles, if not of the ability of such new ventures to soak up capital! It's not only South Africans who have been attracted by the thought of owning a winery. Many high-profile foreigners, individuals rather than large corporations, have also recognized the Cape's wine potential. Notable investors include Californian viticulturist-husband-and-winemaker-wife team, Phil Freese and Zelma Long; also Mme May-Eliane de Lencquesaing, Alain Moueix, Bruno Prats, and Hubert de Boüard, all well-known names in Bordeaux. There are many more and the number keeps growing. For the French, there is also the freedom from the restrictive regulations such as those which apply in Bordeaux and, for the Europeans in general there is the advantage that the Cape is virtually in the same time zone and a convenient 11 hours' flight away.

Despite this rush of new wineries, 80 percent of the annual crush is still produced by South Africa's 66 **cooperatives** (SAWIS, 2005, no. 29, p. 16), wineries jointly owned by the grape farmer members, the advantage being the pooling of winemaking and marketing costs; the crop is vinified at one cellar and while small quantities of the wine may be bottled under that cellar's own label, the majority is usually sold in bulk to one or more of the big merchants for blending into their branded wines.

Increase in demand has also seen vineyard area increase; from 240,429 acres (99,763 hectares) in 1991 to 300,645 acres (124,749 hectares) in 2004 (SAWIS, 2005, no. 29, p. 9). (These figures include table grapes and varieties used for drying, both of which are sometimes used for wine.) By 2002, South Africa was the world's ninth largest producer but filled only fifteenth place in area under vine (SAWIS, 2005, no. 29, p. 35).

Vineyard expansion appears limited compared with countries such as Australia or Chile because many vines were uprooted either because they were lesser varieties, such as the white

Palomino, Kanaan, or Cape Riesling, or because they were planted under the wrong conditions. Replanting on better sites and with internationally recognized grapes such as Cabernet Sauvignon, Chardonnay, Sauvignon Blanc, and Shiraz as well as with other Bordeaux and Rhône varieties, has brought the country into line with international trends. Such changes also had the effect of shifting the imbalance between white and red varieties from 84.1 percent in favor of whites in 1991 to a more equitable 54 percent by 2004 (SAWIS, 2005, no. 29, p. 12). However, only in the districts of Paarl and Stellenbosch do red varieties outnumber whites. Because the upsurge in plantings gathered momentum only since the early 1990s, the majority of the Cape's vineyards are less than 10 years old.

It is also significant that around 90 percent of Cape vineyards are planted within the **Cape Floral Kingdom,** one of the smallest and richest plant kingdoms on earth and internationally recognized as a World Heritage Site; the Cape Floral Kingdoms is an area in which there are many different plant species indigenous to this area (Figure 15–1). Conservation is critical, yet the re-alignment of vineyards to sites designated for quality rather than quantity has created concerns that some of the most vulnerable natural habitat such as the renosterveld and fynbos will be targeted for vineyard expansion.

Partners in the **Biodiversity and Wine Initiative** (environmental bodies and the wine industry), a project set up in 2004, aim to minimize the further loss of threatened natural habitat and to contribute to sustainable wine producing practices, through the adoption of biodiversity guidelines by the South African wine industry. Guidelines and minimum standards for environmentally friendly practices, from vineyard to cellar to packaging, had already been introduced through the **Integrated Production of Wine,** a scheme introduced via an Act of Parliament in 1998.

If exports have shown a healthy increase, sales at home have been less promising; per capita consumption has fallen from 2.31 gallons (9.06 liters) per annum in 1991 to 1.95 gallons (7.65 liters) per annum in 2004 (SAWIS, 2005, no. 29, p. 32). Only a small sector of South Africa's 44 million people drink wine; for the vast majority of the black South African population, wine is

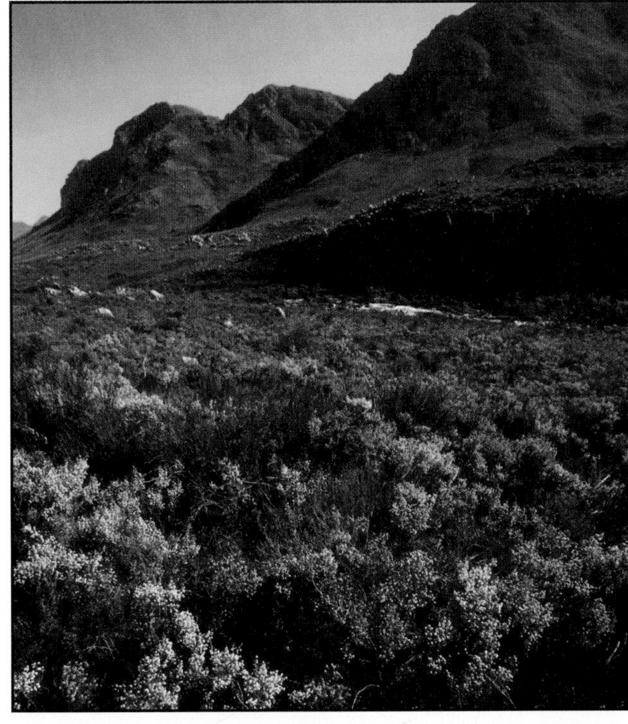

FIGURE 15–1 Vines and the Cape's indigenous vegetation are learning to live side by side, as vineyards extend into cooler areas and up the hillsides that are the natural habitat of this diverse floral kingdom.

not yet a beverage of choice, although interest is growing. This isn't so surprising when one realizes there is still no racial equality in the wine industry. The **Black Association of the Wine and Spirit Industry (BAWSI)** is driving the change in the status quo. This body not only aims to empower its members to enable them to become owners in the industry but also help with the social upliftment of people on farms and implementation of fair labor practices. Individuals are also playing a role, helping workers to buy their own houses and set up their own winemaking operations. From the hands-on point of view, the first few young black South African viticulture and enology graduates from Stellenbosch University are now working in various Cape cellars; many more are still completing their degrees. Bursaries and scholarships offered by various companies and other bodies assist them with their studies.

South African Wine—Historical Perspective

South Africa is unusual in that it is able to trace the origins of its wine industry back to day one. Such precise information is thanks to Jan van Riebeeck, the first Governor of the Cape, who was meticulous in keeping a diary during his tenure as the representative of his Dutch masters at the tip of Africa.

The region known today as the Western Cape was, for two reasons, a logical place to start winegrowing in South Africa. First, it has a Mediterranean climate; the cold, wet winters and warm, dry summers provide ideal conditions for producing quality wine grapes. With the majority of South Africa's present day vineyards still grown in the Western Cape province, the terms Cape and South Africa are interchangeable when talking about the country's wines.

Second, the Cape of Good Hope was also conveniently halfway between Holland and the Dutch East Indies, a route plied by the Dutch East India Company in their profitable spice business. This convenient location, as well as the verdant growth noted by those who had landed there, encouraged the Lords Seventeen, as the Company bosses were known, to start a settlement in what is today Cape Town at the foot of Table Mountain (Figure 15–2). This stopover would be

FIGURE 15–2 This is the same view of Table Mountain, apart from the buildings of modern Cape Town, that Jan van Riebeeck and his colleagues would have had as they sailed into Table Bay in 1652. He planted the first vines at the foot of Table Mountain, in present day central Cape Town.

used as a repair station for their ships and as a place for growing corn and vegetables to augment the natural vegetation, all of which would help to prevent scurvy among the crews.

When the decision to go ahead was finally taken, Jan van Riebeeck, previously a ship's surgeon, was chosen to head the expedition and establish the settlement. While the Company considered wine an important part of a ship's rations, partly to help fend off scurvy but also because the water carried on board soon became undrinkable, there is no evidence to show vines were part of any agricultural plan for the new settlement.

Commander van Riebeeck and the other members of the expedition arrived in Table Bay on April 6, 1652. At van Riebeeck's request, the first batch of vine cuttings arrived in 1654; none survived. A second delivery, planted alongside the vegetables in the Company gardens, a plot today located in the heart of Cape Town, provided more positive results. On Sunday, February 2, 1659, van Riebeeck recorded in his diary: "Today—God be praised—wine pressed for the first time from the Cape grapes, and from the virgin must, fresh from the coop, a sample taken . . . pressed from the three young vines that have been growing here for two years . . . yielding 12 mengles must from French or Muscadel grapes, the Hanepoot Spanish not yet ripe." (Louis, 2004) (*Note:* Twelve mengles equals 3.8 gallons or 15 liters)

What was that first wine like? Regrettably, the commander, so meticulous in keeping other records, failed to comment. What his diary entry does hint at is the type of grape that featured among the first plantings, although there are no records to verify any conjecture. Logically, the vines came from French vineyards, which were the nearest to Holland. Synonyms for the variety known in South Africa as **Muscadel** are Muscat de Frontignan or Muscat Blancs à Petits Grains, a vine widely grown in the south of France. It is thought the Spanish Hanepoot may be Muscat d'Alexandrie, **Hanepoot** being a common South African synonym for this grape today.

Encouraged by the success of this maiden vintage, the first vineyard that bore any resemblance to such a title was planted in an area appropriately named Wynberg (Wine Mountain) on the slopes of the Peninsula Mountain chain to the south of Cape Town. Today, this now vineless land is the home of the Archbishop of Cape Town.

If Jan van Riebeeck was the Cape's first *vigneron,* another governor of the Cape, Simon van der Stel, is regarded as the father of winegrowing in South Africa. He was also the creator of what was to become the country's first internationally recognized wine. Van der Stel became Governor in 1679, the same year he founded the town of Stellenbosch, some 28 miles (45 kilometers) east of Cape Town. Today, Stellenbosch, both the town and the area, are regarded as the hub of the Cape wine industry (Figure 15–3). It boasts the greatest number of wineries of any of the Cape's wine regions and has an enviable reputation for producing many of South Africa's best wines, reds especially.

But Simon van der Stel left another legacy that was to bring the Cape international acclaim during and long after his lifetime. In 1685 he was granted land which he called Constantia. This farm, lying on the slopes of the Peninsula Mountain chain, further south than van Riebeeck's Wynberg, proved ideal for growing vines. The soils are deep and well drained; winter rainfall is plentiful, and in summer, the southeasterly wind blowing off nearby False Bay keeps things cool. Among the wines van der Stel made was a sweet liqueur type in white and red versions; Muscadel, the variety mentioned by van Riebeeck, is again suggested as one of the grapes used. This fragrant, unfortified sweet wine, recognized in its own right, was the forerunner of the Constantia dessert wine that, in the hands of Hendrik Cloete in the latter part of the eighteenth century, was to find fame throughout the world. Among its celebrated admirers were Frederick the Great of Prussia, Bismarck, British royalty, and perhaps most notably, Napoleon. Literary greats Charles Dickens, Jane Austen, and Baudelaire wrote in its praise.

The Dutch, not primarily known for their knowledge of winegrowing, might have introduced vines to the Cape but a greater depth of experience became available with the arrival of the

FIGURE 15–3 Stellenbosch town and district is renowned for its beautiful old Cape Dutch architecture, mountainous scenery, vines, and excellent wines.

French **Huguenots** in 1688. The protestant Huguenots were forced to flee from their homeland after the revocation of the Edict of Nantes, which had guaranteed them religious tolerance. Many of these refugees came from the south of France, where some had been actively involved with the production of wine; even those who had no firsthand knowledge were familiar with the culture of wine. The area known as Franschhoek, or French Quarter, is indelibly associated with these French immigrants, and many of their descendants are still well known names in the wine industry.

The industry grew and reached a pinnacle of prosperity toward the beginning of the 1800s. By then, the English had taken over at the Cape and the Napoleonic Wars cut off French exports to England, opening the way for Cape wine; this was a boom time for the Cape farmers. It wasn't to last long. By the 1860s the English had patched up their relationship with the French, and reduction in tariffs on French wines saw imports of Cape wines dramatically slump.

Not long after, nature joined man in edging the Cape wine industry into difficult times. First reports of the fungus **oidium,** or powdery mildew, were made around 1819; by the mid 1800s, it was widespread. Fortunately, it was soon discovered that by dusting the vines with sulfur the fungus could be kept at bay. By the 1880s, phylloxera was devastating Cape vineyards with the same ruthlessness the louse had shown in France. Even Groot Constantia, the home of Cloete's famed dessert wine, lost its prestige after being sold to the government in 1885.

The start of the twentieth century was marked by the problem of over-production; a wine farmer's success was then measured by high yields. By 1918 there were almost 87 million vines producing some 14.3 million gallons (56 million liters) of wine, most of which could find no buyers (Hughes, Hands, and Kench, 1988).

This situation, with the government's backing, led to the establishment in 1918 of the giant ruling cooperative, known as Ko-operatiewe Wijnbouwers Vereniging (KWV). Its objective was to stabilize the price of wine and brandy produced by members and to promote the sale of their products. This body became both player and industry regulator. In 1957 a quota system was introduced which determined where vines could be planted; quantity rather than quality was an important factor. The KWV also controlled importation and quarantine of vine material,

FIGURE 15–4 Nelson Mandela, South Africa's first democratically elected President, and his African National Congress colleagues toast his 1993 Nobel Peace prize award with South African wine. International acceptance and soaring exports followed this endorsement.

a hindrance to any producer intent on making quality wine from recognized international varieties.

If that made life inconceivably difficult for the few independent wineries, the industry as a whole suffered with the introduction of international trade sanctions in the 1980s in protest against the apartheid regime. It was somehow appropriate that the date then-President F. W. de Klerk made his momentous speech unbanning the African National Congress and announcing Nelson Mandela would be released from prison was February 2, 1990, exactly 331 years after the first Cape wine was made.

South African wines were officially welcomed back into the international fold when Mr. Mandela, the country's first democratically elected president, endorsed them and toasted his 1993 Nobel Peace Prize with Cape wine (Figure 15–4).

The Wine Regions of South Africa

The Cape's vineyards lie roughly between 31° S and 35° S, stretching some 250 miles (400 kilometers) along the coastal belt north of Cape Town and a similar distance along the coast to the east. Further north, around 500 miles (800 kilometers) from Cape Town, vines are also grown along the banks of the Orange River. While wine and grape concentrate (for fruit juices) are produced in this summer rainfall region, the major variety, Sultana, is used for drying. Grapes in both areas are harvested any time between January and the end of March or early April. These are the main wine-growing areas, but there are vineyards beyond their borders, some of which are receiving official recognition from the authorities.

Within the Western Cape's basically Mediterranean climate zone, there is considerable variation in rainfall, temperature, and wind, as well as in soils, **aspect** (the direction in which the site faces), and topography. Although generally a winter rainfall area (a season lasting from approximately May to September), annual precipitation can vary from as much as 4.93 feet (1.5 meters)

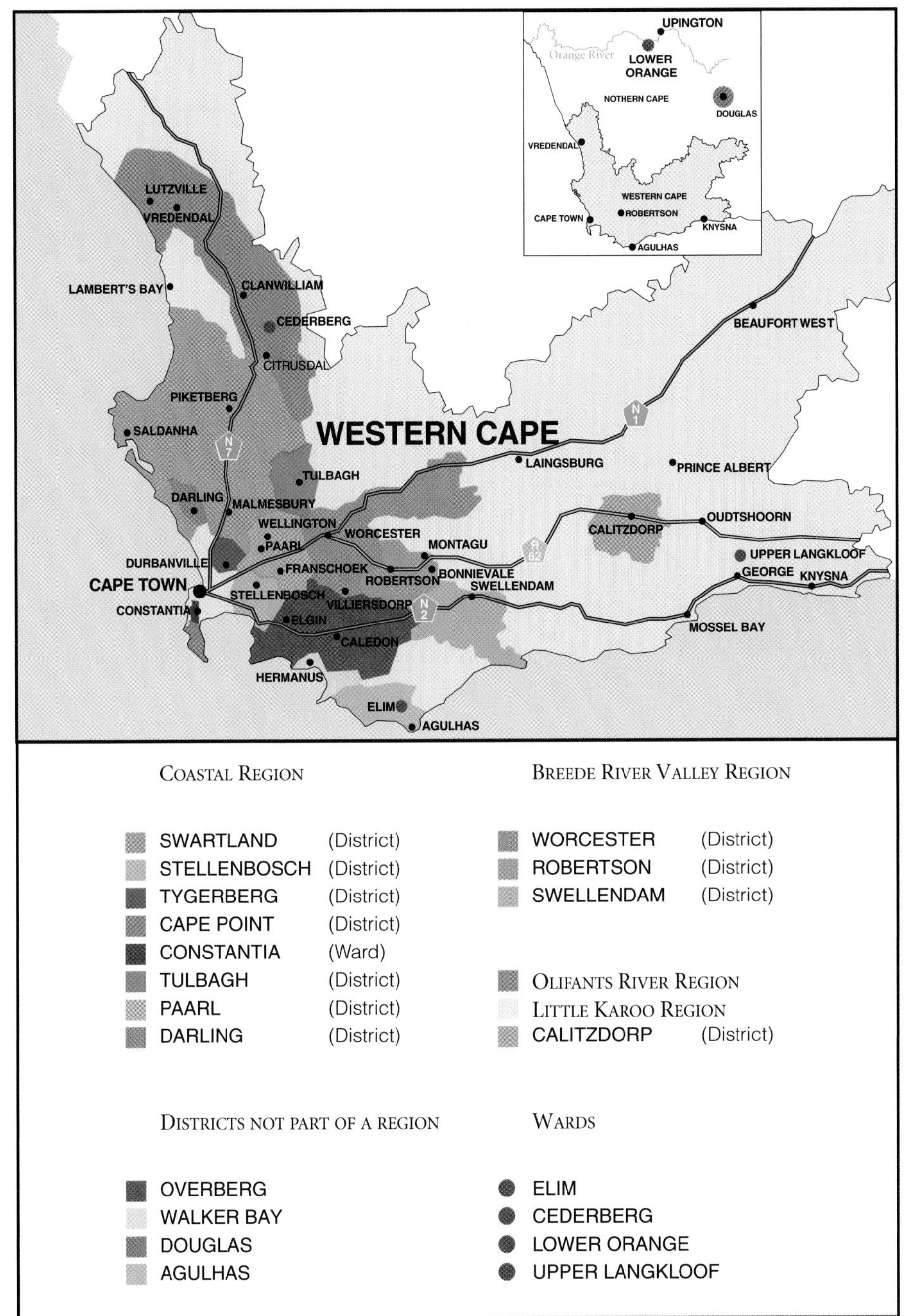

Wine Producing Regions of the Western Cape

to less than 7.8 inches (20 centimeters) in different parts of the winelands. There are also periods, sometimes lasting 2 or 3 years, of drought. The need for irrigation, or at least its availability, in the production of quality grapes is now widely recognized.

Inland areas tend to be warmer and drier, while on-shore summer breezes cool vineyards close to both the south and west coasts. Because South Africa has limited scope for finding cooler growing conditions through heading south—the Atlantic Ocean soon gets in the way—higher ground in the mountains spread around the Western Cape is being explored with good success. Slowly, it is becoming apparent that within each region, there are many microclimates suited to a wide spectrum of grape varieties.

The **Wine of Origin Scheme** officially demarcated South Africa's wine regions in the early 1970s and was first implemented in 1973. This scheme came about when the KWV

Reading a South African Wine Label

Front Label

"Kumkani"–Brand Name: English, Afrikaans, and African names are all used on South African wines. Provided they are not already a registered trademark, most names are acceptable.

"Lanner Hill"–Single Vineyard: Single vineyards have to be registered, not more than 14.5 acres (6 hectares), and from one grape variety only.

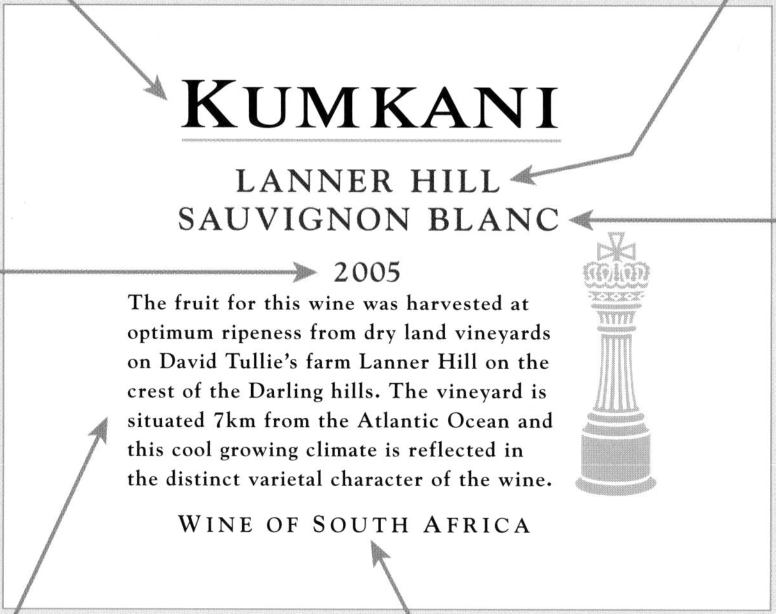

KUMKANI

LANNER HILL
SAUVIGNON BLANC

2005

The fruit for this wine was harvested at optimum ripeness from dry land vineyards on David Tullie's farm Lanner Hill on the crest of the Darling hills. The vineyard is situated 7km from the Atlantic Ocean and this cool growing climate is reflected in the distinct varietal character of the wine.

WINE OF SOUTH AFRICA

Wine Notes: Not obligatory, but the information in this case must be factual and be approved by the Wine & Spirit Board.

Sauvignon Blanc Grape Variety: A single varietal wine must contain a minimum of 85 percent of that grape. Only certain grape varieties are permitted under the Wine of Origin Scheme.

Vintage: The year in which the grapes were harvested. A minimum of 85 percent of the contents must come from the year declared on the label.

Wine of South Africa: Country of origin. Here not a compulsory item, as it appears without the other compulsory information.

Certification Sticker: Certification confirms all the legal requirements under the Wine of Origin Scheme have been met and the wine has been approved by the Wine and Spirit Board. The numbers enable the wine to be tracked right back to the vineyard. This is the small rectangular white label with numbers on it, stuck on the neck of the bottle.

Reading a South African Wine Label

European Export Back Label

Wine Notes: Explanations of the meaning of foreign names helps the consumer have a better understanding of why the wine was given this particular name.

Wine of Origin: Appellation, the demarcated area, in this case a ward Groenekloof, where the wine was grown.

Product of South Africa*: Compulsory Country of Origin. Words with the same meaning may be substituted.

Contains Sulfites*: Compulsory; must be indicated on all wine filled/bottled or sold after November 25, 2005. The spelling must be in the language stipulated by the destination market. Only required if the total sulfur level is more than 10 milligrams per liter.

KUMKANI

SINGLE VINEYARD

LANNER HILL SAUVIGNON BLANC

Derived from the Xhosa word meaning King – Kumkani is a leader amongst South African wines. A wine with stature & pedigree, it epitomizes the quality inherent in the rich diversity of the ancient soils, unique climates and winemaking heritage of South Africa.

The grapes for the Single Vineyard range of wines are grown in partnership with private, family owned farms – each wine reflecting a specific terroir.

This expressive, classic Sauvignon blanc reflects its cool climate origins. Ripe gooseberry flavours, flinty and full-bodied with a crisp lingering finish.

WINE OF ORIGIN GROENEKLOOF
WINE OF SOUTH AFRICA
CONTAINS SULPHITES, ENTHÄLT SULFITE
INNEHÅLLER SULFITER

Alc 14.5% A581 750ml

IMPORTED BY: OMNIA WINES, MOORBRIDGE COURT,
MOORBRIDGE ROAD,MAIDENHEAD, SL6 8LT, UK

ALC–14.5% by Vol*: Compulsory; in the United States, the given figure may differ by not more than 1.5 percent if the alcohol is less than 14 percent, over 14 percent a difference of 1 percent is permitted; in the EU a difference of 0.5 percent is permitted; and locally, in South Africa, the alcohol level on the label may vary by 1 percent in the actual wine.

Produced and Bottled: Omnia Wines, Stellenbosch, South Africa*, Winery identification. Either the A number or name and address is required for the local market. At present these may be replaced by the importer's details on wine for export, but from 2007 onward, the same rules that apply to South Africa will apply to exports.

A581: The A code numbers enable identification of the producer of the wine, if this information isn't provided on the label. The list of A numbers is held by the Liquor Products Division of the Department of Agriculture.

Compulsory items as indicated by an asterisk above have to appear "within the same field of vision on one or more labels of a container," that is, either front or back labels, or, if it's a wrap-around label, on one end of that label.

Reading a South African Wine Label

United States Export Back Label

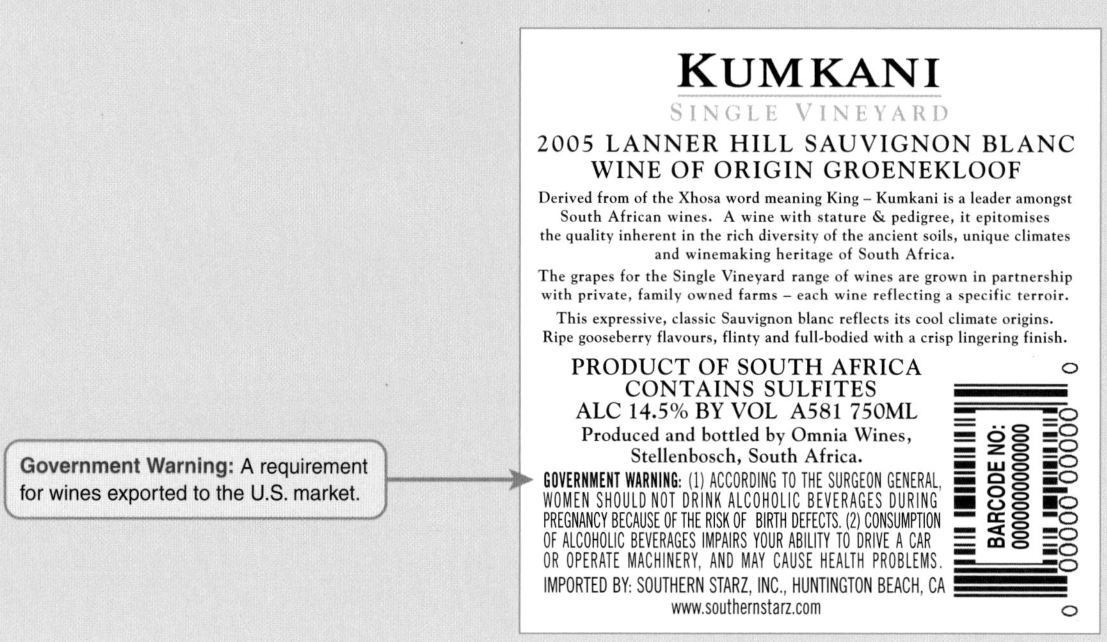

KUMKANI
SINGLE VINEYARD
2005 LANNER HILL SAUVIGNON BLANC
WINE OF ORIGIN GROENEKLOOF

Derived from of the Xhosa word meaning King – Kumkani is a leader amongst South African wines. A wine with stature & pedigree, it epitomises the quality inherent in the rich diversity of the ancient soils, unique climates and winemaking heritage of South Africa.

The grapes for the Single Vineyard range of wines are grown in partnership with private, family owned farms – each wine reflecting a specific terroir.

This expressive, classic Sauvignon blanc reflects its cool climate origins. Ripe gooseberry flavours, flinty and full-bodied with a crisp lingering finish.

PRODUCT OF SOUTH AFRICA
CONTAINS SULFITES
ALC 14.5% BY VOL A581 750ML
Produced and bottled by Omnia Wines,
Stellenbosch, South Africa.

GOVERNMENT WARNING: (1) ACCORDING TO THE SURGEON GENERAL, WOMEN SHOULD NOT DRINK ALCOHOLIC BEVERAGES DURING PREGNANCY BECAUSE OF THE RISK OF BIRTH DEFECTS. (2) CONSUMPTION OF ALCOHOLIC BEVERAGES IMPAIRS YOUR ABILITY TO DRIVE A CAR OR OPERATE MACHINERY, AND MAY CAUSE HEALTH PROBLEMS.
IMPORTED BY: SOUTHERN STARZ, INC., HUNTINGTON BEACH, CA
www.southernstarz.com

BARCODE NO:
000000000000

Government Warning: A requirement for wines exported to the U.S. market.

realized that to export to Europe, it was necessary to fall in line with the European Economic Community (now European Union) appellation requirements. The winelands were broadly divided into three levels of decreasing size: regions, districts, and wards. The Western Cape (embracing all these demarcations) and Northern Cape (covering all the vineyards around the Orange River) are also now designated Wine of Origin units, as is the province of KwaZulu Natal, a tropical area on the country's east coast. The latter is very new and the few vines planted there have yet to prove the worth of the venture.

Smaller, individual origins are the estate wine and, most recently, the single vineyard. Where applicable, these will be indicated on the label, the Wine of Origin designation being indicated by the letters WO as well as the words Wine of Origin.

Geographical features defined each of the four regions, which took the name by which the area as a whole is known: Breede River Valley, Klein Karoo, Olifants River, and Coastal.

The district is a smaller area of origin; these were originally drawn up along geopolitical boundaries and, among the current 17, include such well-known places as Stellenbosch, Paarl, and Robertson. Delimitation of the ward, smaller than either the region or district, takes into account more physical and climatic similarities, a sense of terroir, as well as the sense of community among the people within the defined area. With the interest in more closely defining the Cape's wines, the 52 wards presently registered are set to increase. Among the more familiar ward names are Constantia, Franschhoek Valley, and Durbanville.

While each of the regions contains districts and wards, not all districts and wards fall within a region; much of this is due to areas opening up where vines have never before been grown.

An estate wine has to be grown, made, and bottled on a single production unit (farm with vineyards and cellar) registered for this purpose. Likewise, single vineyard wines have to come from a registered unit, limited to a maximum of 14.5 acres (6 hectares) and from one grape variety only.

Recent legislation has opened the way for wines made from more than one official Origin to state the various sources on the label. Claims for vintage or variety, whether locally or for export, require a wine to contain a minimum of 85 percent of both.

The Olifants River Region

This area is home to the most northern vineyards in the Western Cape. Inland, it is hot and dry, but with irrigation water readily available from the Olifants River, high yields are easily achievable. This led to the region to be associated with poor quality, bulk wines. With the opening up of export markets, better quality varieties have been planted; vines have also moved up from the fertile valley floors to mountain slopes. Viticultural practices have been upgraded, resulting in farmers being paid on fruit quality rather than simply on sugar levels. These improvements have encouraged lower yields and better quality grapes. The climate favors organic viticulture and several farmers are now taking this approach. In the cellars, winemakers have learned to craft consumer-friendly, early-drinking wines. Chenin Blanc and Colombard still hold sway but there are also considerable plantings of Shiraz, Cabernet Sauvignon, Merlot, Chardonnay, and Sauvignon Blanc.

Sauvignon Blanc might seem an unlikely variety introduction in such a hot region, but with the quota system a thing of the past, winemakers' adventurous spirit has led them to explore the length and breadth of the winelands; thus they are discovering small enclaves within their own, individual mesoclimates. For instance, in the Olifants River, the young ward of Bamboes Bay fronts directly on to the Atlantic. The vineyards, lying virtually on the beach, are cooled by sea mist that curls around them early morning and by breezes brought up the coast by the cold Benguela Current.

To the south and about 50 miles (80 kilometers) inland, the vineyards in the Cederberg Ward (a stand-alone ward, just outside the Olifants River boundary) perch 3,614 feet (1,100 meters) up in the Cederberg Mountains. Both of these diverse spots have the ability to produce crystalline or pure Sauvignon Blanc; reds also look set to perform really well, although the vines are very young. Further wards within the region continue to be identified and demarcated.

Klein Karoo (Little Karoo) Region

This region, the furthest east from Cape Town, is also the dryest (Figure 15-5). Generally speaking, it is semi-desert, ideal for ostrich breeding, for which the region is renowned, but making irrigation essential for growing vines and limiting their spread. Apart from limited water, birds, of the smaller, flying kind, are the main problem Calitzdorp farmers have to contend with. The grapes conveniently ripen when there is little else in the food line for the birds; the damage they can cause has led some farmers to net their vines, as happens in New Zealand. Nevertheless, a wide range of grape varieties are grown here, most on an experimental basis.

Where the region has made its mark is with fortified wines, Port-styles in particular. These are wines made in the same way as Portuguese Port but South Africa's trade agreement with the European Union means the word "Port" will be phased out on export markets by 2007 and locally by 2014.

The District and Karoo town of Calitzdorp is affectionately known as the Port capital of South Africa (though it would be wrong to say it's the only place where this style excels; excellent examples are also made in Stellenbosch and Paarl). The area has climatic similarities with Portugal's Douro Valley, and most of the wineries make one or more Port-styles, which are described as Cape Ruby, Cape Vintage, and so on, on the label, that is, the Portuguese terms plus

FIGURE 15–5 The Klein Karoo is one of South Africa's dryest vine-growing regions. It has a climate similar to Portugal's Douro Valley, and most wineries produce Port-style fortified wines, many well respected by their Portuguese counterparts.

the Cape identifying origin. With such a concentration of top notch Port-style producers plus the biennial Port Festival held in Calitzdorp, the title appears appropriate.

It is no secret that the Portuguese believe South African Port-styles to be closest to their own in quality. In the early days Tinta Barocca, as well as nontraditional Port varieties such as Cinsault, Cabernet Sauvignon, and Shiraz, were used for these wines; today they have been vastly improved through the addition of other traditional Port varieties, particularly Touriga Nacional, regarded as one of the finest. With the formation of the South African Port Producers' Association, guidelines were drawn up regarding styles and their labelling; these broadly follow their Portuguese counterparts.

As in other regions, there are exceptions to the general climate rule. The ward of Tradouw and its 253 acres (105 hectares) of vineyard (SAWIS, Vines in the Wine of Origin Areas, 2004) sit at the top of the eponymous pass, looking straight out to the Indian Ocean, some 28 miles (45 kilometers) to the south. The vines thus enjoy an unhindered and beneficial cooling influence, allowing for the production of elegant Bordeaux-style blends and Chardonnay. Other cooler localities are also being explored.

Breede River Valley

An inland region, where the two main districts, Worcester and Robertson, track the Breede River (Wide River) part of the way along its course from the Witzenberg Mountains down to the Indian Ocean. When quantity was esteemed over quality, the warm Worcester vineyards were mainly planted on the fertile, alluvial soils next to the river. The trend today is toward less vigorous hillside sites, though irrigation remains essential. With 21 percent of the national vineyard, Worcester is the most densely planted of any area (SAWIS, 2005, no. 29, p. 9).

Chenin Blanc and Colombard for brandy making still head the varietal list but Chardonnay, Sauvignon Blanc, Sémillon, Cabernet Sauvignon, Shiraz, and Merlot are gaining ground. Smatterings of other Bordeaux reds and the more unusual Sangiovese, Barbera, Tannat, and Viognier are of a more experimental nature. Apart from brandy, Worcester's main output from the many cooperatives is bulk wine for the merchant trade.

Downstream, Robertson is making a name for Chardonnay, where the variety has the biggest percentage share of vineyard of any area where it is grown (Figure 15–6). Initially, the grape was recognized for producing a distinctive table wine but, increasingly, it is being transformed into smart sparkling wines made in the traditional *méthode champenoise.*

Chenin Blanc and Pinotage

Of the many wine varieties growing in the Cape's vineyards, Chenin Blanc and Pinotage are of particular significance to South African winemakers.

Chenin Blanc, the same grape that produces wines of renown in the Loire, has for many decades been the most planted variety in South Africa. Even today, it heads the list, accounting for 19.1 percent of vineyard area, although this is a dramatic decrease over its 31.2 percent share back in the early 1990s (SAWIS, 2005, no. 29 p. 12).

Wine producers' appreciation of Chenin Blanc originally stemmed from the grape's versatility—it can be turned into anything from a quality sparkling wine to a fortified **jerepigo** (very sweet, unfermented grape juice fortified with grape spirit) and everything in between—good yield, and its suitability for making brandy. Brandy has always been an important grape-based product and, as with table wine, the quality has been improving over recent years.

In 2004, over 21.7 million gallons (85 million liters) of wine, out of a total wine crop of just over 259,510,581 gallons (1 billion liters), was earmarked for **potstill brandy.** Such wine will be batch distilled twice in a potstill before undergoing a minimum of three years' aging in small oak casks. South Africa also boasts the world's largest potstill distillery; the KWV's Worcester premises house 120 potstills under one roof. Interest in brandy isn't exclusive to the large spirits companies; there are also around 20 smaller and private cellars that produce pure potstill brandy. Nevertheless, most commercial, big brand brandies are a blend of distillate from the continuous still system with a legal minimum of 30 percent potstill product (SAWIS, 2005, no. 29, p. 15).

Brandy apart, Chenin Blanc devotees' main focus is to elevate the grape to the same noble status and quality it enjoys in the Loire. Decent, everyday drinking wines with good fruit and fresh acid balance have always been produced, but they tend to be overnight wonders and lose their fruity appeal before the following harvest. Through the founding of the Chenin Blanc Association in 1999, the members intended to identify the best vineyards, some of them very old by South African standards, and ensure they were retained. The Association also categorized Chenin Blanc into six different styles from fresh and fruity to rich and wooded, with the goal of giving the variety better focus and improving its image in consumers' eyes. Thanks to greater visibility in the market place, consumers are showing more interest in Chenin Blanc, although styles are still being refined and maturation potential assessed.

The same can be said about South Africa's own variety, **Pinotage.** Professor Abraham Perold bred this red variety, a cross between Pinot Noir and Cinsault, in 1924. At that time Cinsault was, somewhat strangely, also known as Hermitage, hence the name Pinotage. At one stage, when he moved house, the professor's new variety was nearly lost, but the four seedlings were rescued. Some 35 years later, this unlikely crossing provided the Champion red wine at the Cape Wine show, and in 1961 the first Pinotage appeared on the market. Like Chenin Blanc, it is a very versatile performer, slipping as easily into the role of sparkling wine made in the traditional *méthode champenoise,* as it does into fortified dessert wines. Naturally, it held huge novelty value for foreign consumers when South Africa reentered the global market in the 1990s; everyone wanted to try wine from this typically South African grape. To build on this interest the Pinotage Association was formed, its agenda being to research, improve wine quality, and promote Pinotage.

The enthusiasm with which Pinotage was planted, especially after lifting of sanctions, failed to take into account that it is a niche variety, both relatively unknown and eliciting diverse opinions. Whereas demand for international big guns Cabernet Sauvignon and Shiraz is still increasing, Pinotage has yet to find a comfortable level of production; since 2001, its vineyard share has been decreasing. The best wines, whether rich and ripe or more elegantly

(continues)

Chenin Blanc and Pinotage (continued)

reflective of their Pinot Noir parentage, are on a par with those from any classic variety. But research continues in an effort to eliminate a sometimes worrisome residual bitterness, and it takes a practiced winemaker to manage Pinotage's inherent astringent tannins. Another challenge many Pinotage makers have set themselves is to use the grape in a blend to create a new, uniquely Cape style. The Cape blend, as they would like it known, remains very much a work in progress, and there is vigorous debate as to whether Pinotage should be a mandatory ingredient.

Although South Africa remains the variety's most important proponent, worthy examples are also made in New Zealand and California. On a smaller scale, Pinotage is found in New York, Canada, Brazil, Israel, and Zimbabwe.

South African producers have adopted the name **Méthode Cap Classique** for these wines, since the term Champagne is not permitted. *Méthode* and *Classique* neatly convey traditional production methods, while *Cap* signifies the South African origin. Underpinning this affinity with Chardonnay and sparkling wines are the limestone soils, unusual in the Cape. Vines don't have these chalky soils all to themselves; Robertson is also recognized as a great racehorse breeding area, and visits to the town in early summer showcase the beauty of the massed beds of roses.

Like its upstream neighbor, Robertson early on proved ideal brandy making country; Chenin Blanc and Colombard with Chardonnay, today covering nearly three-quarters of the area's vineyards, are a legacy of this history. Nevertheless, within the past 10 years, reds—Shiraz and Cabernet Sauvignon in particular—have proved they can produce quality wines with medium-term aging potential.

Summer is a long, hot season with occasional flash downpours. Afternoon breezes beating up the Breede River from the sea temper the heat; this, together with the limestone soils, helps maintain good acid levels in the grapes. With its low average annual rainfall of around 10.6

FIGURE 15–6 Robertson is known for its limestone soils, which produce distinctive Chardonnay and, increasingly, Méthode Cap Classique sparkling wines.

inches (27 centimeters), irrigation is also essential to Robertson's success. This comes not from the Breede River but via a canal from the Brandvlei Dam near Worcester.

If these wines are the modern face of the Breede River, the fortified dessert wines reflect the area's much-loved traditional offerings. There are two types: Muscadel, both white and red, from Muscat à Petits Grains; and Muscat d'Alexandrie, or Hanepoot. These wines, made from unfermented, very sweet grape juice fortified with neutral grape spirit and generally unoaked, are usually enjoyed young, but the best can age impressively.

Coastal Region

The Coastal Region, as such, will rarely be mentioned, simply because of the wealth of well-known, quality districts and wards within its boundaries. Names such as Stellenbosch, Paarl, Franschhoek, Constantia, and even Durbanville and Swartland have much greater recognition factor than Coastal.

Stellenbosch

Stellenbosch is considered the hub of the South African wine industry (Figure 15–7). The University of Stellenbosch and Elsenburg Agricultural College, both offering courses in viticulture and enology, are based here, as is the agricultural research center, Nietvoorbij. The town is also home to Distell, South Africa's largest wine and spirits producer. Vineyards and cellars add to this concentration of wine-related bodies; the greatest number of private cellars found anywhere in the Cape are located in this prime wine-growing area. Taking into account cooperatives and wholesalers, the total is around 104 but never stands still long enough for an accurate count. With quality always the focus of the majority producers, a goal assisted by the temperate climatic conditions, it's not surprising many of the international investors have chosen Stellenbosch as the base for their South African ventures; Mme May-Eliane de Lencquesaing, Alain Moueix, Bruno Prats, and Hubert de Boüard among them.

Stellenbosch, with just over 17 percent of the national vineyard (SAWIS, 2005, no. 29, p. 9), covers an area surrounded by an amphitheater of mountains and hills with important frontage on False Bay. In summer, the southeasterly wind blows off the sea, cooling the vines and helping prevent disease. This wind is known as the **Cape Doctor.** Soils, aspects, and elevations are varied, allowing for a wide spectrum of varieties to thrive and be equally successful with a whole range of wine styles, anything from Cap Classique sparkling to Vintage Port style.

FIGURE 15–7 Although a new Stellenbosch winery, Tokara employs scientific methods to match site to grape variety and pinpoint where grapes are ripening; the human input from viticulturist and winemaker is of equal importance. The excellence of the maiden vintage is testimony to these carefully tended vineyards.

This diversity is illustrated through the district's five wards. Simonsberg-Stellenbosch, farthest from the sea, is known for its full, rich reds. Bottelary, with an unusual majority of north- and west-facing slopes, also favors fuller-bodied red wines. Shiraz is a variety of note here. On the top of the Bottelary Hills, with their unhindered exposure to False Bay and its cooling influence, Sauvignon Blanc is making its mark. In Devon Valley, the mix of aspects allows for both Chardonnay and Cabernet Sauvignon to shine. Cabernet Sauvignon also performs well in the Jonkershoek Valley Ward. There are fewer producers in the Papegaaiberg (Parrot Mountain) Ward, so among the mix of white and red varieties, no one stands out.

Although the area's reputation has been built on red wines, with greater viticultural knowledge, even site-sensitive varieties such as Sauvignon Blanc now produce excellent results. Cabernet Sauvignon, Merlot, Pinotage, and Shiraz dominate red plantings; Chenin Blanc, Sauvignon Blanc, and way behind, Chardonnay account for the majority of white varieties. The land might be some of the Cape's most expensive real estate, but it doesn't deter adventurous winemakers from experimenting with recently introduced or more unusual varieties. Among those finding a foothold in this densely planted area are Viognier, Grenache, Mourvèdre, Barbera, Nebbiolo, Sangiovese, Tempranillo, and Zinfandel.

PAARL

Paarl (Afrikaans for Pearl) District borders on Stellenbosch to the north and, although part of the Coastal Region, is landlocked and generally warm. It is second only to Worcester in vine density, accounting for 18 percent of the national vineyard (SAWIS, 2005, no. 29, p. 9). Running roughly southeast to northwest, mountains are again a defining feature in this district (Figure 15–8). At the southern end, the Franschhoek Peaks tower above the long, narrow Franschhoek Valley, one of the earliest and best-known wards. The Berg River, which begins in these mountains, runs along the valley floor, through the town of Paarl itself, and eventually out into the Atlantic Ocean; en route it offers irrigation facilities to those vineyards still planted along its banks. In Franschhoek these include some century-old Semillon vines, but as in other areas, new vineyards are being established up the mountain slopes, where the vines grow less vigorously and it is cooler.

Depending on slope and aspect, anything from fine Cap Classique sparklers to robust reds can perform well. There are even spots where Pinot Noir, in the right hands, produces enjoyable if not great wines. If midsummer temperatures reaching well into the 90s °F (30s °C) make this seem unlikely, the southeasterly wind again comes to the rescue as it pours over the mountain

FIGURE 15–8
Boschendal was established in 1685, one of the earliest farms in the area. Today, its 494 acres (200 hectares) of vineyard cover many different mesoclimates and produce a wide range of quality wines. These, together with its excellent restaurants, old manor house, and dramatic scenery, make it a favorite stopover for wine lovers.

pass above the town, beneficially cooling the higher vineyards. Traditionally, white varieties have dominated plantings in an area where vines have always had strong competition from fruit orchards. Today red and white varieties are more evenly matched in this pretty valley, which is renowned for having some of the finest restaurants in the winelands.

Further north, the valley opens up, though is still tracked by mountains to east and west, the most notable western peak being the north face of Simonsberg, now demarcated as the Ward of Simonsberg-Paarl. The gentle slopes are well exposed and characterized by deep soils, ideal for the red varieties planted on the lower ground, the whites higher up to catch any cool breeze. At the western edge, the land flattens out before it once again rises as Paarl Mountain. The soil here, a mix of gravel and clay, is less fertile, but that is precisely what attracted Californians Phil Freese and Zelma Long when they were looking for land to start their Vilafonté project—the name reflects the main soil type of their vineyards.

Around the town of Paarl and further north there is little relief from the summer heat and rainfall decreases. The wines from Paarl Mountain, behind the town, the Ward of Wellington, downstream and centred on the Groenberg and Hawekwaberg Mountains and the Ward of Voor Paardeberg, reflect their still, sunny origins. These are mainly reds, Shiraz with its Rhône cohorts, Mourvèdre and Grenache, Cabernet Sauvignon, Merlot, and Pinotage, popular both as varietal wines and blends. Whites, able to hold their own in a richer style, also do well, Viognier, Chardonnay, and Chenin Blanc among them. In fact, Chenin Blanc remains the most planted variety in Wellington, primarily on the flatter land around the Berg River. This is a relic of its suitability for sherry making, a wine style still produced by the KWV and Monis, a division of Distell. Both still have soleras, although these are much smaller than even a decade ago.

Its mix of wineries matches the variation in Paarl's soil, slope, aspect, and climate. At one end of the scale is the vast KWV company, at the other garagistes producing a few hundred cases; in between there is a big mix of cooperatives, merchant and private cellars.

Tulbagh

Spilling over from Paarl's northwestern boundary, Tulbagh is encircled by towering peaks (Figure 15–9), the narrow valley exit at its southeastern end following the Breede River on its

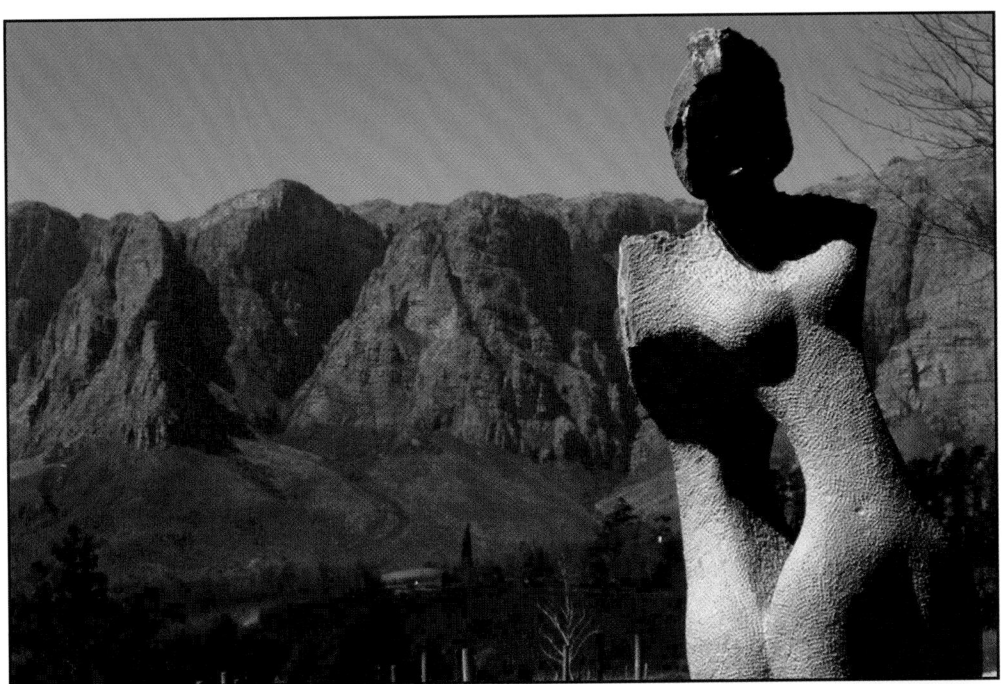

FIGURE 15–9 Wine and art make natural partners. Visitors to many of the Cape's wineries will find art works on display either inside or outside the cellar; they also frequently appear on wine labels.

route to Worcester, Robertson, and the sea. This district is very warm in summer, but snow is a common occurrence on the higher peaks in winter. Soils here feature many large boulders among the river sand. Very much the Cinderella of wine areas until recently, over the past 5 to 8 years, Tulbagh has seen new, private wineries start up all over the valley. Many of these are already producing award-winning wines, both red and white. This is shaking up the area's somewhat lackluster image and the old theory that Tulbagh was good for producing everyday drinking white wines only. White varieties still account for over half of the 1,376 acres (571 hectares) of vines (SAWIS, Vines in the Wine of Origin Areas, 2004), but the potential for quality reds is already evident.

SWARTLAND

Swartland, the "blackland," is named for the indigenous dark, bush-like renosterveld vegetation that dots the tops of the sweeping hills characterizing this district (Figure 15–10). As part of the major grain growing area of the Western Cape, golden wheat stretching to the horizon is the Swartland's very visible trademark with vines clustered close to the hilltops. This is all changing with the veritable hurricane of vine planting, especially in the southern parts. The soils here are red and very deep, with good water retention properties, thus little allowance was made for irrigation in the past; today, it is a quality option being installed alongside the new vines. Individual farm dams, replenished by the winter rains, are the main source of water for irrigation. Anticipate big, rich wines from this hot, sunny area. Cabernet Sauvignon might be the dominant red variety, but the Swartland is fast getting a reputation for being prime Shiraz country, with other Rhône varieties, such as Grenache, Mourvèdre, and Viognier as complexing partners. Chenin Blanc outstrips all other white grapes.

DARLING

A recently delimited district at the southern end of the Swartland, closer to the Atlantic Ocean, Darling enjoys basically the same soil types and similar inland, hot conditions. But on its western edge, it is exposed to the Atlantic coastline, making it a good deal cooler. The Groenekloof Ward is making a strong bid for one of the leading places for excellent Sauvignon Blanc, although a wide variety of both white and red grapes are grown.

FIGURE 15–10 Both wheat and vines share the deep, rich soils of the warm Swartland. Wheat greens the area in winter and spring, while vines take over the green mantle from late spring through summer.

Wineries around Swartland and Darling used to be composed mainly of cooperatives. Today, there are many private players starting cellars, merchants with their own vineyards, and contract growers who supply wineries outside the region.

Tygerberg

Competition for the ward of Durbanville, better known than the Tygerberg District within which it falls, extends beyond the realm of grapes. This ward borders the outer reaches of Cape Town's northern suburbs; thus its 3,374 acres (1,400 hectares) of vineyard (SAWIS, Vines in the Wine of Origin Areas, 2004) are always under pressure from property developers. Fortunately, the handful of producers, mainly private with one big but very good merchant, is as one intent on quality. The soft folds of the Tygerberg Hills, allowing for both exposure and shelter, assist them, as do the cooling effects of the breezes and mists which roll in off the nearby Atlantic in late summer afternoons. With unhindered exposure to False Bay, some distance away to the southeast, Durbanville receives some beneficial cooling effect from the famous Cape Doctor. Over the past 8 to 10 years, Sauvignon Blanc has grown in stature and quality and is by far the most planted white variety. Although there is more Cabernet Sauvignon than Merlot, locals believe the latter variety, which is earlier ripening and stresses more easily, produces the better wines.

Constantia

Constantia has the same problem as Durbanville. Urban Constantia is a very upmarket southern suburb of Cape Town; property developers are always on the lookout for land on which to develop new housing estates. Fortunately, the mountain slopes Simon van der Stel recognized 320 years ago as being ideal for growing great wine continue to justify his faith, although the taxes for such prime real estate do place a burden on the farmers. Until recently there was a mere handful of wineries, three of them forming part of van der Stel's original land grant. Now a few more are opening up, their vineyards scaling even greater heights along the mountain chain. All are privately owned, small, and quality driven. Modern-day Constantia's forte is white varieties in general, Sauvignon Blanc in particular, but also Semillon and Riesling; for the reds, Merlot is showing promise on the lower, warmer slopes.

But if South Africa has any wine that could be described as an icon, it is the reincarnation of Cloete's Constantia dessert wine. After meticulous research into the original to ensure authenticity of every detail from vine to wine, the Jooste family of Klein Constantia re-created the renowned wine in 1986, the same year that they harvested their maiden vintage on this restored property. It is made from low-yielding, very ripe Muscat de Frontignan vines, aged 2 years in oak barrels and packaged in a replica of the nineteenth-century bottle. Like its predecessor, Vin de Constance has been applauded worldwide.

If Constantia represents the earliest face of South African wine, the most up-to-date is found on the other, Atlantic, side of the Peninsula Mountain chain. The district of Cape Point currently boasts one winery and 75 acres (31 hectares) of vines (SAWIS, Vines in the Wine of Origin Areas, 2004). Sauvignon Blanc is again the main player and is doing as well as its counterpart on the other side of the mountain. Reds have yet to make an appearance.

Overberg, Walker Bay, and Cape Agulhas

These three districts, following the coast east from Cape Town, cover some of the Cape's newest wine territory, developed from virgin ground since the fall of the quota system in 1992. Their importance far exceeds the tiny area under vine, around 3,374 acres (1,400 hectares) between the three districts, although more vines are being planted every year (SAWIS, Vines in the Wine of Origin Areas, 2004).

All are in sight of the sea—in some cases, a wave or two away from the beach—so it might appear strange none fall into the Coastal Region (in fact there is no all-embracing region), but this is one of the idiosyncrasies of the Wine of Origin Scheme, possibly because the majority of vineyards have been developed since the WO Scheme was implemented.

The Overberg is best known for its one ward, Elgin, internationally recognized for its apple orchards. As the fruit market fortunes have fluctuated, so vines have gained a small foothold. Results from the first winery, Paul Cluver, bottling since 1997, and a few others who have sold grapes to producers outside the region, have proved so promising, that more vineyards are being established by some of the Cape's most highly regarded winegrowers. As in other cool areas, Sauvignon Blanc is the focus of attention, but other cool climate varieties such as Gewürztraminer, Riesling, and Pinot Noir along with Cabernet Sauvignon, Merlot, Shiraz, and Chardonnay are all sharing in Elgin's newfound popularity. The whole area lies on a plateau, roughly 986 feet (300 meters) above sea level, although some vineyards are now close to the 2,957-foot (900-meter) mark. Soils feature the shale, granite, and sandstone that are found in many other areas. What sets Elgin apart is the effect of the southeaster. This summer wind that moderates temperatures around Stellenbosch at sea level does the same on this plateau but also blows in a cloud cover that carries with it the likelihood of rain. With rot an ever-present threat, meticulous viticultural practices are vital. This climatic feature also applies to the other two districts.

Walker Bay, to the south and east of Overberg and down at sea level, has a more temperate climate but is still exposed to the onslaught of unwanted rain the summer southeasterly wind can bring. It is here, in the Hemel en Aarde (Heaven and Earth) Valley that in the mid 1970s, Burgundy lover, Tim Hamilton Russell found what he thought was the best spot to plant Pinot Noir and Chardonnay. Over the years his faith has been vindicated, especially with the fickle Pinot. Today, the six or so small cellars in the valley (not yet a ward but this is under consideration) all produce respectable to very good examples of Pinot Noir. Within the greater spread of the district, Pinot's stepchild Pinotage proves it enjoys both warm and cool conditions, producing aromatic wines. Shiraz is received as enthusiastically as it is elsewhere; with Merlot and Cabernet Sauvignon, it makes up the major red presence in the district. Sauvignon Blanc, Chardonnay, and Chenin Blanc are leading white varieties.

Cape Agulhas is a new district at the southernmost point in Africa. Vines were first established here as recently as the end of the 1990s but growth has been exponential. From about 120 acres (50 hectares) in 2003, the ward of Elim shot up to 270 acres (112 hectares) by the end of the following year. In total, Cape Agulhas has 648 acres (269 hectares) under vine (SAWIS, Vines in the Wine of Origin Areas, 2004). Sauvignon Blanc and Shiraz are again the favored varieties; the former has already demonstrated it has found yet another spot in the Cape where it's happy to put down its roots and produce potentially benchmark wines.

The area around Cape Agulhas is very exposed and flat; the vines are thus at risk from wind, which can break the young shoots, from rot from the cool, sometimes damp, summer conditions, and from birds. At present most grapes are vinified outside the area but it is possible that as confidence in the quality grows, more cellars will be put up there.

Until recently this was the farthest coastal point east with an official demarcation; in January 2006, a new district is destined to come into being. The town of Plettenberg Bay is roughly 187 miles (300 kilometers) east from Cape Agulhas and primarily known as a holiday resort; until now, vines have not been part of the attractions. But today there are 12 acres (5 hectares) of vineyard on the chalky slopes just outside the town, with the first wine, a Cap Classique bubbly, due for release early 2006. It is too early to predict whether more vineyards will be planted, but according to farmer Peter Thorpe the positive points are the good soils and cool summer temperatures. Rain falls in both winter and summer, so spraying is necessary; but more than mildew, birds are the big problem. Nevertheless, this does indicate there are perhaps more possibilities outside South Africa's traditional wine-growing areas than have yet been explored.

Summary

Since democratization and the dropping of sanctions, South Africa has adapted to global requirements remarkably quickly. With a weak currency, exports soared; the return of a stronger Rand affected many companies. This spurred them to greater efficiency, with the result that exports continue to be strong foreign currency earners for the country. Now a broader spectrum of the local market needs to be persuaded to drink wine.

But even after 13 years, the industry remains in a revolutionary phase. New areas and varieties are still being explored and well over half the vineyards are less than 10 years old. Better plant material and yet more quality varieties are required to allow South Africa to show its potential through its many and diverse mesoclimates. The young generation of viticulturists, winemakers, marketers, and researchers, who have not known the restrictive practices of the previous era, are well traveled and able to help the country realize this potential in the future.

REVIEW QUESTIONS

1. What are some of the main changes in the South African wine industry since the dismantling of the apartheid regime?
2. Why is Chenin Blanc considered such an important variety?
3. Why can such a range of grape varieties be grown across the Cape vineyards?
4. Discuss the influence the French have had in the South African wine industry.
5. Outline the conditions that limit production in South Africa.

SECTION V

The Business of Wine

The Business of Wine and how it is presented and managed in food service are covered in the three chapters of this section. Topics include selling and serving wine; developing and managing a wine list; and buying and cellaring wines. Important subjects include identifying goals of a wine program, understanding major considerations in choosing wine for a business, tableside wine service, and staff training.

CHAPTER 16

Selling and Serving Wines

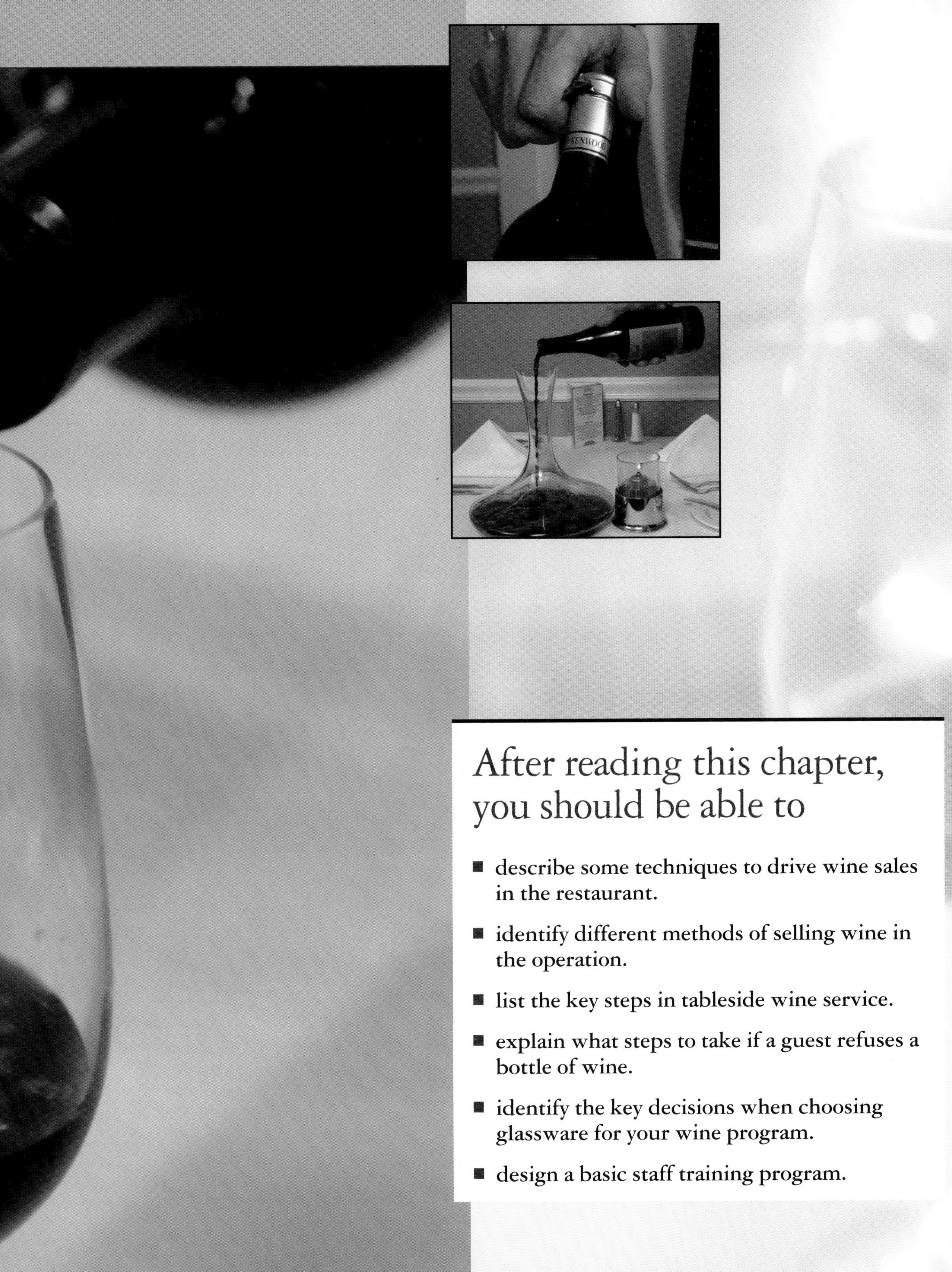

After reading this chapter, you should be able to

- describe some techniques to drive wine sales in the restaurant.
- identify different methods of selling wine in the operation.
- list the key steps in tableside wine service.
- explain what steps to take if a guest refuses a bottle of wine.
- identify the key decisions when choosing glassware for your wine program.
- design a basic staff training program.

KEY TERMS

3 v's (vintner, varietal, vintage)
aeration
capsule
corkage fee
cork taint
decanter
decanting
half bottle
large format
magnum
sediment
sommelier
table tent
trichloroanisole (TCA)
twist top
unique selling point (USP)
wine flight
wine key
worm

The business of selling and serving wine is distinctly different from enjoying wine at home or with friends. When a guest chooses to purchase a glass or bottle of wine in a restaurant she makes a conscious decision to do so for a variety of reasons. The process of making this decision is greatly influenced by the management of the restaurant and the manner in which they present their wine program to the guests. The sale is further supported and executed by the service staff, and all pieces need to be in place to ensure a helpful and seamless delivery of product to the customers. Proper execution of the sales and service of wine in the restaurant is no accident; to the contrary, it is the culmination of many hours of preparation and planning, staff training, and a sizeable investment on the part of the operation. It is not difficult to start a concise and inclusive wine program in a restaurant setting, it simply takes a plan and some effort on the part of the management and service team. Depending on the size of the operation, the initial investment can be small or large and can expand as time and the budget allow. The first step is to commit to providing a service to the guests, and then to formulate a plan of action. This chapter and the following two chapters focus on the planning and execution of a successful restaurant wine program and outline some essential steps for success.

Wine Service and the Role of the Sommelier

Throughout history, the role of promoting wine sales and ensuring appropriate wine service has often been the job of the **sommelier.** Historically the sommelier was responsible for stocking and maintaining the provisions and ensuring that products being served were sound and had not spoiled (MacNeil, 2001). This role has changed considerably through the years, and the modern view of the sommelier focuses on supporting the guest in the restaurant wine experience. This can involve design and execution of the wine list, acquiring and maintaining inventory, and selling and serving the wine tableside, as well as suggesting appropriate wine and food pairings.

Today this role is occasionally undertaken by a full-time, well-trained staff member whose focus is solely on the wine program, but not every operation can afford this level of investment. In the vast majority of operations, the duties of a sommelier may be divided between the manager, service staff, and bar manager. Several people may help access and purchase wine, maintain inventory, develop the wine list, and sell the wine tableside. All staff must be prepared to discuss wine with, and sell it to, the guests in order to fully support the wine program.

Once guests have chosen to purchase wine, steps must be taken to ensure prompt delivery of the wines to the table, proper presentation of the chosen bottle, and ongoing support of the guest experience. There are several things that we can do as dining room professionals to make certain that the guests receive maximum enjoyment from their wine selection. Servers and sommeliers have the most direct impact on this process, but everyone from the bartender to the dining room manager can play a positive role. The proper execution of wine service further reinforces an operation's commitment to the guest experience and can easily build strong customer loyalty.

Start Early

A potential wine sale in the restaurant begins before the guest even walks through the door. For example, many restaurants have a signboard or kiosk outside the door where their food and wine menus are posted to make selections clear to potential guests (Figure 16–1). A restaurant that has won an award for its wine list may choose to post this accolade on the menu board as an attempt to sway undecided diners. For serious wine aficionados, a well-organized and diverse wine list may be as important (if not more important) than the selection of food offered by the chef. By marketing your wine list with your menu board, you have the chance to convince diners that you are serious about your wine program before they even enter your door.

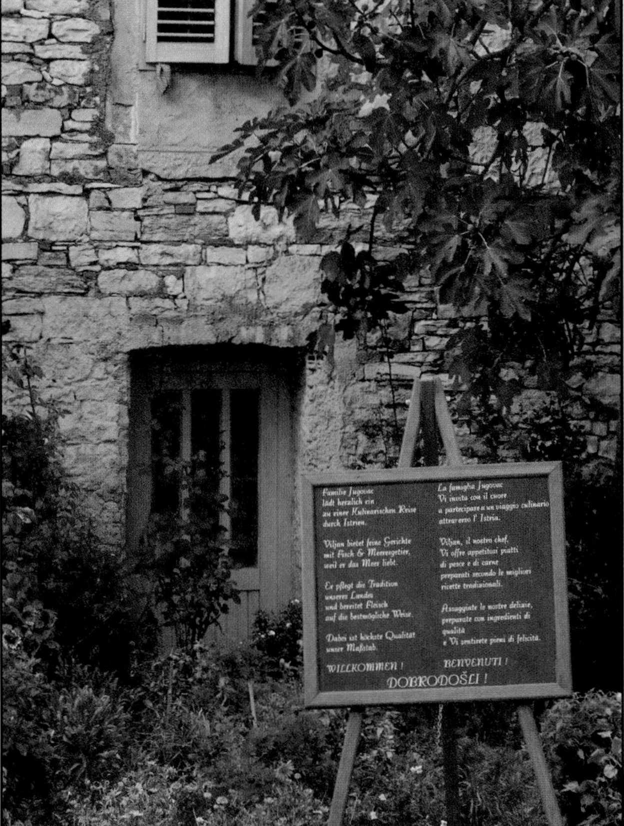

FIGURE 16–1 Placing the wine list on a sign board outside the restaurant can help the operator to encourage wine sales. Here the menu board advertises the wines that are being offered.

Guests should be regularly reminded that wine is an enjoyable aspect of dining, and we plan to provide this service for them. The more we remind them of this fact, the more we encourage wine sales in the operation. Using a variety of the following techniques, you can design and implement a successful and profitable wine program that also focuses on guest service and satisfaction.

Promoting Successful Wine Sales

Why on earth would guests who visit a restaurant choose to spend twice or even three times as much money for a bottle of wine in our operation than they would spend for the very same bottle at a retail store? The answer to this question lies in the level of service provided and the ability to deliver added value to their purchase. The concept of added value relates to the fact that a guest may pay more for a product, but the execution and delivery of that product enables them to get more as well. Restaurant wine sales are an excellent example of how added value can benefit both the guests *and* the operation.

Adding value to the guest experience takes different forms in each operation. For example, the chef has worked to prepare and to offer distinct, flavorful, and unique dishes that the guests can't easily replicate in their own homes. Even if a chef gives out recipes, the techniques employed in the kitchen, the inviting décor, and the fine level of service simply can't be matched in the home. For the wine drinker, added value is measured differently. Value is added for these customers by having appropriate, polished glassware; by having attentive, knowledgeable servers who will open and pour the wine; and by having a professional level of service supporting the sale of wine through to the end of a meal.

From the time the wine is chosen to the time the bottle is empty, the guests should be supported in their enjoyment and appreciation of the product by the service staff. This allows the guests a complete experience that cannot be replicated at home. The guests have the opportunity to be entertained and educated by the staff, and they are able to pair a delicious wine of their choice with the entrée of their choice, prepared by a skilled chef. Both the wine and the food can benefit from this pairing. If the service supports this potential, the guests will be truly impressed and will feel like the money was well spent; if any piece is not satisfactory, the guests are not likely to return. If they do choose to return, they are unlikely to want to make the same level of investment because the value isn't there.

The job of the restaurateur is as much to build value into the guest experience as it is to meet the guests' needs. The more the restaurateur works to build value into the guest experience, the more likely the guests are to enjoy their time and to return. Loyal, repeat business is the cornerstone of our operation, and we need to work to build that whenever we can. Wine aficionados can be among the most loyal guests, and they are likely to buy more than a simple entrée every time they visit. It is the rest of the dining public that might need to be convinced (or reminded) that having wine with their meal is a good idea. We do this in many ways, but the more consistently and more thoroughly we do it, the greater the likelihood that the guests will decide to invest in wine.

Promoting wine sales and service is a critical aspect of success in the restaurant industry. Aside from the obvious benefits of increased sales and profitability, tableside wine service gives the server or sommelier the opportunity to add value to the guest experience by demonstrating professionalism and attention to detail. Proper tableside wine service is an art in itself and improper service is unlikely to ever be forgotten. In many cases, guests view purchasing a bottle of wine as an investment in their experience at the restaurant, and service staff should treat it similarly, whether it is a bottle of house Merlot or the finest offering from the wine list. When the staff does so, it demonstrates to the guests that the management cares about their experience, regardless of the amount of money that is being spent.

Every restaurant should work to include education as a key part of their wine program—this includes education for the customer as well as for the staff. If a well-trained server is able to pass on some wisdom to the guest, the guest experience is positively affected and the guests are encouraged to return for another visit. As the service staff continues to pass on their knowledge, they become more comfortable and confident tableside and are further encouraged to improve their knowledge.

Driving On-Premise Sales

Wine sales that take place on the site of the restaurant are called on-premise sales. Once the guest has entered the premises, wine sales take on several new dimensions. It is not uncommon to encounter a prominent display of bottles of wine that are served by the glass, usually on a table near the host stand. The main goal of these displays is to get your guests to think about purchasing wine with their meal before they even sit down. By creating a display of the options available, you remind the guests that wine should be a part of their meal and entice them to try one of your offerings by allowing them to make a visual connection. This is a relatively passive approach compared with other methods, but it is an easy and important component of a larger program designed to stimulate wine sales. Guests often need several reminders before they commit to a wine purchase.

Many restaurants make use of promotional bottles donated by a winery or a wholesaler. These bottles—sometimes called "dummy bottles" because they are not filled with wine but are sealed and empty—are usually large format bottles designed to draw attention and to reinforce the restaurant's commitment to its wine program. Wineries that donate these display bottles benefit from the visibility and exposure that they get.

Featured Wines

Once the guest is seated, the bulk of the wine selling can really take place. One option that is commonly employed is to have a featured selection of wine on the table. Usually this is an unopened bottle of the house's featured wine, sitting on the table along with the flatware and glassware. This serves partly as table decoration and partly as suggestive selling tool, but either way is sure to be noticed. Once the server greets the table and relays the specials to the group, it is easy to transition to wine sales by drawing the guests' attention to the featured bottle. The server will usually say something such as, "I would like to point out our featured wine this evening . . . " or "If you were considering having wine with your meal . . . " and then go on to explain some key facts about the wine, offer some tasting notes, and suggest appropriate menu pairings. This allows the guests to learn about the wine without having to ask, and also helps to guide them in their selections. If the wine is appropriately priced and the server has worked to help the guests understand it, increased sales of the wine are virtually assured.

This approach does have some drawbacks, however. For one thing, the restaurant must have enough inventory of the featured wine to be able to display a full bottle on *every* table. In a small operation, this is not a big concern. In an operation that may have 300 seats, however, this could involve having six or eight cases of the wine on hand, just to display on the tables. If the featured wine is in the middle to high price bracket, this can be a very costly endeavor. A second possible drawback to this approach is that targeting a specific wine as a featured special could increase sales of that wine, but also deter guests from ordering other wines on the list. If the featured wine is inexpensive, it may prompt guests to spend less than they otherwise would have done. This is a relatively shortsighted concern, however, because any wine sale is a good thing, regardless of the price point. If the server can also make a personal connection with the guests while creating a focused and informative dialogue about the wine, then it truly is a win-win situation. It is a safe bet that the added value the guests experience will encourage them to repeat their visit and will help to build customer loyalty.

Setup for Sales

Another effective way to promote a wine sale is by including a wine glass in the standard setup of the tables. This sends the message to the guests that "we assume you will be enjoying wine with your meal, so we have planned ahead." This also makes the server's job a little easier because they don't have to set glassware if a party orders a bottle of wine. If they order wine by the glass, the server could either remove the preset glass and return with the requested glass of wine, or could bring the open bottle of house wine to the table to pour tableside. This technique includes added challenges for the management. First, there must be enough clean, polished wine stems to allow setting one at *each* spot on *every* table. In many operations, this would tie up most of the clean wine glasses available and leave very little additional glassware if it were needed. It is also possible that this practice could increase breakage, especially if servers are not careful with the stems. The standard procedure for this type of setup should include the servers removing stemware from any setting where the guest has chosen not to order wine. This will allow the clean glasses to be available for guests who *do* order wine, and will also help to reduce breakage from having a cluttered table.

The biggest challenge with pouring house wines by the glass at the table (as opposed to at the bar) is controlling the portion size. It is easy to overpour wine in a guest's glass, especially if the server is in a rush or is improperly trained about the proper amount to pour. Overpouring will quickly result in an increased beverage cost for the operation, effectively counteracting any benefit from increased wine sales at the tables. All service staff must be trained to monitor the proper pour size, not just the bar staff.

Another common way to promote featured wines is by using a **table tent** or other tabletop display (Figure 16–2). Although this type of table display does not fit with every restaurant concept, it is a highly visible and very effective method of focusing guest attention on specific products. This is usually a simple A-frame device that stands up on the tabletop and is included as

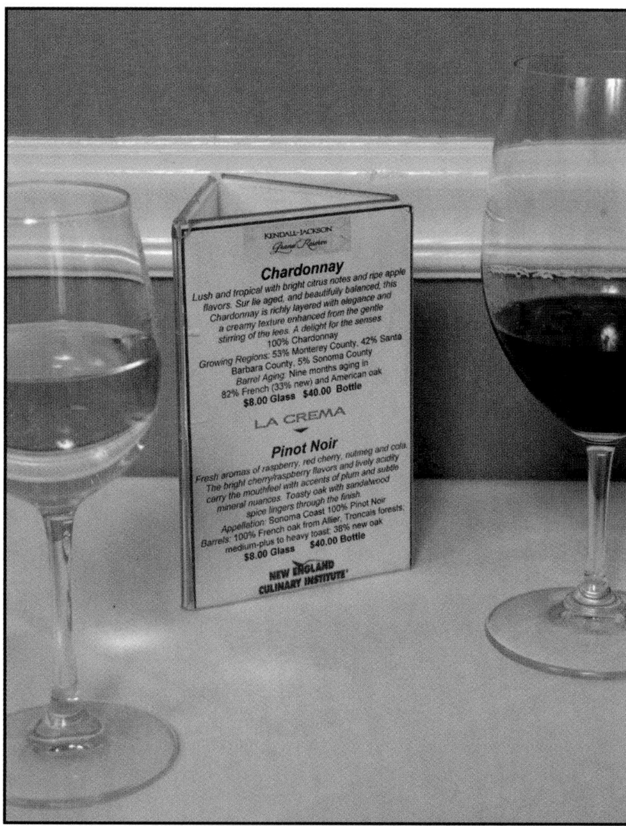

FIGURE 16–2 This table tent is a useful sales tool when combined with a strong wine program. It is often used to advertise featured wines and can also give the customer useful information about the wines.

part of the standard table setting. This allows the guest to see product information, prices, tasting notes, and pairing suggestions at their leisure throughout the meal. Servers may simply call the guests' attention to the table tent as part of their introduction or recitation of specials—for example, "And our salmon special this evening will pair particularly well with our featured wine, described on the table tent." This effort ties the special entrée with the special wine and helps to build guest interest in both products.

Server Suggestions

The impact that servers have on promoting wine sales should not be underestimated. The servers have the most personal contact with the guests, and once a rapport has been created this trust can be parlayed into a meaningful wine experience for the guest and a profitable one for the restaurant. Proper training of staff is a critical precursor to successful tableside wine sales, but once the staff is comfortable talking about wine with guests, the real results can be realized—happy, well-supported guests and increased wine sales. One easy way to get servers involved in wine sales is to conduct regular and focused tastings of selections from the wine list. Once servers have found a wine that they truly enjoy and understand, selling that wine becomes easy for them. They use their enjoyment of the wine as a selling point and offer a firsthand view of the product in question. As a rapport is established, the guests look to the server for suggestions and guidance in their dining experience. Offering enjoyable wine suggestions is an obvious extension of this bond between the server and guests.

As part of preservice lineup, the management team or chefs should suggest an appropriate wine pairing for each special and as the server shares the specials with the table, the wine should be mentioned (Figure 16–3). It is possible to offer a stock pairing suggestion for each entrée item on the menu and have each server suggest the same wines for each dish. This takes the guesswork

Today's Specials

appetizer

herb goat cheese tart
beet confit, balsamic syrup
Paired w/ *Charles de Fere* Brut Sparkling Wine

salad

local greens salad
grapefruit & Campari vinaigrette
Paired w/ *Matanzas Creek* Sauvignon Blanc

entrees

pecan crusted chicken breast
mole sauce, summer vegetable hash

seared flat iron steak
sauce poivrade, pommes frites
Paired w/ *St. Francis* Cabernet Sauvignon

dessert

lemon angel food cake
wild berries
Paired w/ *Kiona* Late Harvest Gewürztraminer

FIGURE 16–3 This menu offers suggested wine pairings for each course to help guide the guests and encourage wine sales.

out of the experience because management will have chosen appropriate wines ahead of time, which helps prevent unsuccessful pairings that might decrease the guests' enjoyment of the meal or wine.

This focus on predetermined pairing suggestions can limit creativity on the part of the server who looks to offer each table a unique experience. Also, it can be uncomfortable for a server who suggests the standard Sauvignon Blanc with the goat cheese tart and is promptly informed that the guest dislikes Sauvignon Blanc. This server is caught off guard and is hard put to suggest an alternative, especially if management has not properly trained the staff in the basics of pairing wine with food but instead simply relied on predetermined pairings.

Menu Suggestions

One of the easiest ways to encourage wine sales is to suggest a wine pairing for every item listed on the menu. For each item offered, whether appetizers, entrées, or desserts, consider adding a wine pairing suggestion directly below the listing of the item. This is an easy way to reinforce the idea that wine is suitable for every course during the meal and to get the guests to think about wine at the same time they are thinking about food. Most people know that food and wine go hand in hand, but sometimes guests need a gentle reminder. These predetermined pairings allow management to suggest wines that will work well with the food items without putting pressure on the guests. As mentioned above, it also reduces the chances that guests will experience an unpleasant pairing that might mar their experience.

Visible Storage and Special Seating

When designing restaurants, many operators choose a highly visible spot for their wine racking, making it an integral part of the floor plan. This design incorporates the wine bottles and racks as part of the décor and helps to create a wine ambience. This may manifest itself as a designated wine room or as racking that is part of the overall design of the dining room. In either case, visibility of the wine is paramount, and the theory is that this visibility helps encourage wine sales (Figure 16–4). Either situation has its challenges, however, and these will be addressed in this section.

To start, the practice of storing wine at cool temperatures has a specific purpose (see Chapter 18 for more on this). Because the vast majority of dining rooms are much warmer than the ideal temperature for wine storage, storing the bulk of the wine inventory in the dining room guarantees that it will be kept too warm unless a temperature control system is used. This not only risks the longevity and quality of the wine, but also means that most of the wine may be served too warm. Wine served too warm does not have the quality or character that is demonstrated when it is served at the proper temperature. Although some may argue that this really only matters with the best wines, the reality is that *every* guest should have the best wine experience possible, regardless of what wine they choose to purchase, from the most basic glass up to the most expensive bottle. If even one guest experience is overlooked, loss of return visits could result.

Another challenge inherent in dining room wine racking is access by servers or sommeliers. If the wine is not easily accessible by the service staff, it can cause a myriad of problems. Delivery of wine to a table could be delayed because the server has to spend extra time trying to locate the bottle ordered. If the bottle is near another table, that table might be inconvenienced when the service staff works around them to retrieve the selected bottle. This negatively affects one party for the sake of another, a practice that should be avoided as much as possible.

On the other hand, visible wine racking serves the same purpose as a featured bottle of wine or a table tent announcing a special wine: high visibility and suggestive selling. A well-designed and appealing wine racking system or a designated wine room is sure to help stimulate wine sales. Guests may actually ask to sit in a special wine room because of the enjoyable atmosphere.

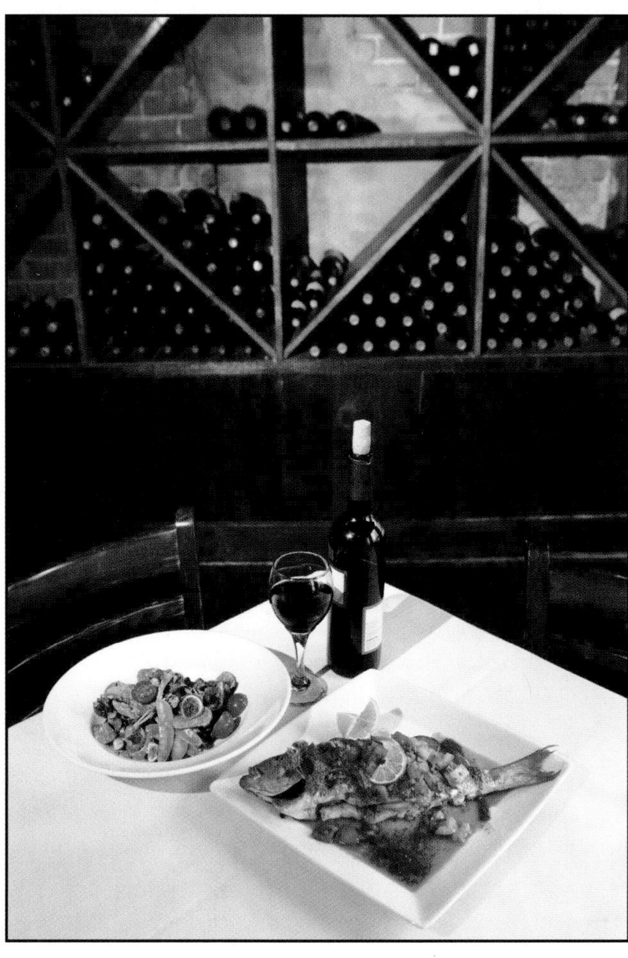

FIGURE 16–4 This dining space uses wine racks and bottles to create a special experience for the guests. Rooms such as this are both functional and decorative and can create a unique atmosphere when incorporated into restaurant designs.

A wine room can also be used and promoted as a venue for special events. For example, management may work with the chef to organize wine and food pairing events, such as dinners, that take advantage of the special space in the wine room. Weekly or monthly wine dinners bring in revenue and offer diners a unique event outside the ordinary offerings of the restaurant. The atmosphere of the wine room helps reinforce the special nature of the event. Guests that enjoy these events will no doubt develop into regular patrons based on their experience. And, if the wine dinners are offered on slower nights, they can help generate sales on otherwise quiet evenings with little effect on regular seating and kitchen performance.

Offer Options

Part of the reason that guests dine out is that they like choices, and the wine list is no exception. By offering guests a broad selection of wines to choose from, we are able to appeal to many different taste preferences. But the quest to offer options should not stop with representing different varietals and regions on the wine list; it should also include different pour sizes and possibly even different bottle sizes. This topic will be covered in more depth in Chapter 17, but parts of it will be touched on here.

Offer Flights

A **wine flight** is a selection of wines offered together as a package. The general goal of offering a flight of wines is to allow guests to compare the wines and discern differences between them. A flight might include three Merlots from three countries, encouraging a comparative tasting of the same grape from three distinct climates or regions. A typical white wine flight might consist of three Chardonnays, one produced with no oak influence, one produced with a little oak, and one that employs lots of oak in the profile. By serving these three wines together, guests can analyze the effect that oak has on the wine and decide which one they like best. By selling the flight, revenue is generated; by allowing guests to decide which wine they like best, we encourage further sales of that wine. Even if guests don't order more wine on that same visit, chances are good that their experience will encourage them to return. And when they do, it is likely that they will try more of the wine that they liked best or try a different flight of wines. The experience is enjoyable and informative, allowing guests to formulate their own opinions as they experience and learn about the differences between the wines.

In general, flights are offered in smaller portion sizes than regular wines by the glass. This is practical; serving a guest three full glasses of wine at once would be irresponsible and could easily result in the guest becoming intoxicated. Additionally, by the time the guest has finished one glass of white wine, the other two glasses may be too warm to be fully enjoyed (obviously not a concern with red wines). Usually the pour size for a flight of wines is 2 to 3 ounces of each wine. This is enough to allow a good taste of each, but the total amount is little more than a full glass of wine. Pouring in this amount also allows guests to follow up by ordering an additional glass of the wine they enjoyed the most. Some operators offer flights in various sizes, including 1, 2, or 3 ounces of each wine, and then price the pours according to the amount being ordered. This method encourages guests to try a taste of a wine that they might not want a full glass of, but would rather just sample. For example, if an expensive wine was priced at $10 per ounce, a guest could afford a sample of a wine that he or she might not normally get to try without spending $250 to buy the bottle. Offering wine flights should be seen as a starting point for wine sales, not an ending point; flights offer customers an analytical comparison of various wines and allow the operator to spark a deeper interest in wine for the guest.

Most restaurant operations will offer guests samples of wine, but the methods for delivering this service to the guests vary. Some operations will simply offer a half ounce to 1 ounce taste of a specific wine that the guest has questions about, at no charge. The idea is that the guest is going to purchase a glass of wine but wants to try it before buying it. It could be argued that this isn't the best practice financially because every ounce that is given away is going to increase the operation's beverage cost. The other side of the coin is that this provides service for guests; if allowing them to try a free sample of the wine ensures a happy guest and a wine sale, why not do it? The amount of wine being offered is usually quite small, and the ultimate outcome for the restaurant is positive. Guests who are offered a sample of a wine have an increased likelihood of purchasing a glass, whether it is the wine they sampled or another, and they will feel good about their experience. This will result in happy, repeat customers who form the cornerstone of the business.

One way to address the guests' desire to sample wine is to offer various pour sizes for *all* house wines, not just wines offered in flights. If guests are allowed to order a 1-, 3-, or 5-ounce glass, they can select the amount of each wine that they would like. It is almost like a "choose your own flight" option, where guests can custom-tailor their own selections according to their interests. The management simply prices the wines by the ounce; this allows the guests to sample the wines that they desire and allows the operation to generate sales from every ounce of wine that is poured.

Bottle Sizes

Another way to offer the guests choices is to offer wine in several different bottle sizes. One common bottle size is the **half bottle** (375 ml; 12.7 oz)(sometimes also called a split), which is half the volume of a traditional bottle (750 ml; 25.4 oz) and contains enough wine for about two glasses. Half bottles encourage guests who would not normally order a whole bottle of wine to order a bottle rather than a glass, and the reduced size can accommodate a couple who wishes to have a glass of wine with their meal but does not want to invest in (or consume) a full bottle. For some people, having an entire bottle of one wine is too much so they may choose to try two different half bottles. The volume is the same as a full bottle, but they have the added benefit of trying two different wines. Because half bottles are usually priced at half the cost of a full bottle, some diners may elect to try a more expensive wine but only assume half the cost. This allows them to experience a nicer bottle without the significant investment of a full bottle.

The traditional bottle size is 750 ml (25.4 oz.), which contains enough wine for four to five glasses. This is the most common size of bottle and the one that our guests have come to expect. These size bottles make up the bulk of most wine lists and are usually priced at many different levels to appeal to a wide range of diners. (See Chapter 17 for more information.)

A **magnum** (1.5 liters; 50.7 oz.) is a good choice for parties of four or more because it contains the same volume of wine as two regular size bottles. These are the smallest bottles in the **large format** category. Large format bottles are those that are larger than the traditional 750 ml (25.4 oz.) bottles. Magnums contain eight to ten glasses of wine, so they are best sold to larger parties who would otherwise order two bottles of the same wine. Offering magnums demonstrates a diverse and complete wine list and can appeal to larger parties looking to purchase wine. On an operational level, magnums command extra attention in the dining room. Just like a bottle of wine on the table can be a focal point for the dining experience, a large format bottle can convey a special occasion for the diners and an additional investment in the meal. The sheer size and presence that a large format bottle has makes it an immediate conversation piece, and the fact that they are often much rarer than normal size bottles adds to the allure. An added benefit of large format bottles is that the wines contained in them age at a slower pace than smaller bottles. With all other factors being equal, the smaller the bottle size the faster the wine inside ages. (This topic is covered in more detail in Chapter 18.)

Even larger bottle sizes are also available, although they are somewhat rarer, especially in most restaurant settings (Figure 16–5). Very large bottles tend to be impractical in most restau-

FIGURE 16–5 These large bottles are great for drawing attention to a restaurant's wine program when put out for display. Large format bottles can also be appealing to larger parties looking to purchase several bottles of one wine.

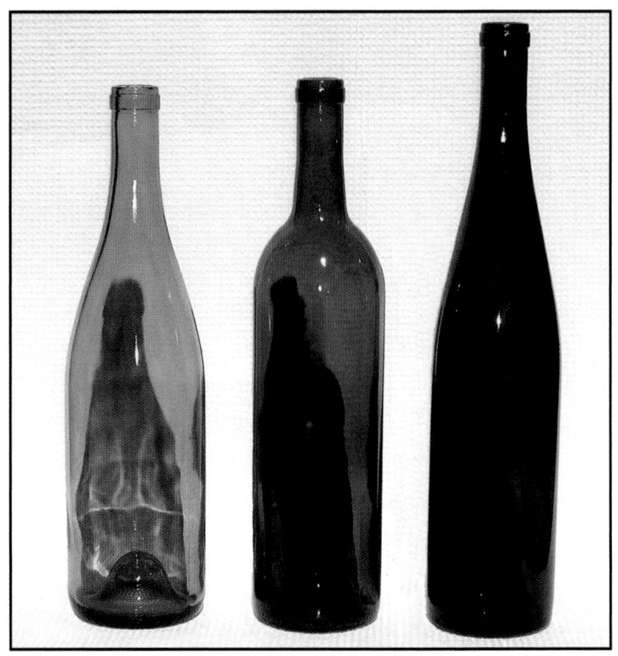

FIGURE 16–6 The three most common bottle shapes, from left to right: Burgundy, Bordeaux, and Hock. The Burgundy bottle with sloped neck is usually used for traditional Burgundian varieties such as Pinot Noir and Chardonnay. The Bordeaux bottle, also called Claret, has more prominent shoulders and is used for traditional Bordeaux varieties such as Cabernet Sauvignon. Finally, the tall and slender Hock bottle is traditionally used for varietals native to Germany and Alsace such as Riesling and Gewürztraminer.

rant settings, because of both the challenges of storing them and of serving them at a table. These bottles can be unwieldy for servers to handle tableside, so special considerations should be taken into account when serving large format wines. (See the section on Tableside Wine Service for more detail.) In most cases, they are suitable only for quite large parties and therefore tend to sell infrequently. However, due to their larger size, they can command some hefty prices on the wine list.

In addition to the wine bottles shown in Figure 16-5, there are also less common large format bottle sizes. The various sizes have names that differ slightly depending on whether the bottle is a Claret shape (Figure 16-6), usually used for red wines, or whether it is used for sparkling wine (Table 16-1).

Bringing Their Own

Some restaurants allow guests to bring their own bottle of wine from home to enjoy with their meal. Many operations allow this because they do not have a license that allows them to sell liquor on their premises. Other restaurants that do offer a wine list may allow guests to bring their own

Table 16–1 Standard Bottle Sizes

NUMBER OF STANDARD SIZED BOTTLES	VOLUME	NAME FOR BORDEAUX SHAPED BOTTLE	NAME FOR SPARKLING WINE BOTTLE
1/2	375 ml	Half Bottle or "tenth"	Same
1	750 ml	Bottle or "fifth"	Same
2	1.5 liters	Magnum	Same
4	3 liters	Double Magnum	Jéroboam
6	4.5 liters	Jéroboam	Rehoboam
8	6 liters	Impériale	Methuselah
12	9 liters	Usually used for sparkling wine	Salmanazar
16	12 liters	Usually used for sparkling wine	Balthazar
20	15 liters	Usually used for sparkling wine	Nebuchadnezzar

wine but charge the guests a fee for doing so, called a **corkage fee.** Corkage fees offset the loss of a sale from the restaurant's own wine list, help cover the costs associated with the server's efforts, and also cover the cost of the glassware. This is not allowed by all states, however, and each operation should be sure to check with its state liquor board to be certain that it is in compliance with state law before allowing guests to bring their own bottles. Regardless of state law, some operations choose not to allow corkage, in an effort to sell their own inventory of wine.

Tableside Wine Service

The presentation of the wine to guests, from wineglasses to pouring, is important to the guests' perception of value.

Glasses First

The first consideration when preparing to serve wine to a table is the glassware that the wine will be served in. There are almost as many different types of wineglasses as there are types of wine. Any wineglass should be large enough to accommodate the amount of wine being poured without the glass being filled to the rim, to allow the aroma to develop within the bulb of the glass. In any size glass, the amount of wine should never be more than about two-thirds of the space in the glass, although an ideal amount would probably be half of the volume of the glass, As guests swirl the wine in their glasses the surface area is increased and the aromatic components of the wine are released. Without releasing these volatile constituents, the wine will smell less pungent and will not show very much of its character.

An exception to this standard fill level would be with those glasses used for a restaurant's house wines served by the glass. In this case, perceived value to the guest is almost as important as the ability to appreciate the wine's aromas. Many times, wines offered by the glass arrive at the table in a small- to medium-sized glass that is filled almost to the rim. Although this makes the guest feel that he or she is are getting good value for the dollar, the ability to swirl the wine and fully appreciate the aroma is greatly diminished. The restaurant needs to find the balance between perceived value and full enjoyment of the wine. For example, a 5-ounce portion of wine in a 15-ounce glass will allow the wine's aroma to benefit from the extra air space, but may leave the guests wondering why the amount of wine in their glass is so small, despite the fact that 5 ounces is the typical pour size offered in the industry (Figure 16–7).

The size and shape of the bowl on the wineglass is as much a matter of preference as it is practicality. Although there are specific types of glasses for almost every type of wine, few restaurants actually have the money to invest in more than a couple styles of glass. It is widely accepted that most wines smell and taste best in the appropriate wine glass; there is a marked difference between a Pinot Noir that is tasted in a glass with a tall, narrow bowl and the same Pinot tasted in the bulbous glass actually designed for the varietal.

In some cases, restaurants may simply have one multipurpose glass that is used for every situation. This is less than ideal, but it is a situation often encountered in the industry. Many operators are hesitant to invest in expensive glassware that is prone to breakage. But paying lots of money for nicely styled glassware is not necessary; it is the size and shape of the glass that has the most effect on the sensory perceptions of the wine, not the price of the glass itself. A wine glass with a bowl that tapers in at the top will be more effective at trapping a wine's aromas than one that flares outward at the top. Similarly, the shape of the bowl affects where on the palate the wine hits as it is sipped; a broad top to the bowl allows the wine to contact the sides of the palate as it enters the mouth, which can increase the perception of a wine's acidity.

FIGURE 16–7 Each of these three glasses contains the standard 5-ounce pour of wine, although the perception of value changes based on the glass size. The largest glass on the right may best accentuate the wine's aromatics, while the small glass on the left appears fullest and therefore may appeal to consumers who seek value.

At the very least, the operation should have two different glass styles, chosen to complement the wines offered on the wine list. For example, having a large glass for red wines and a slightly smaller glass for whites would suffice. It is not uncommon to encounter one glass that is used for the restaurant's wines sold by the glass and another, perhaps more elegant, glass used for serving bottled wine tableside. This reflects the fact that wines poured by the glass should convey value to the guest through the fill level, while guests purchasing bottles for their table are more concerned about having an appropriate glass to enjoy their wine selection.

Many operations will have two different levels of glassware; those glasses that are used for glass sales and bottles priced below $40 (for example), and a nicer set of glasses that are used when selling bottles of wine priced over $40. This system seems to imply that more expensive bottles are more worthy of nice glassware and therefore does a disservice to those guests purchasing less expensive wines. The truth, however, is that very nice glasses represent an added expense for the operation and function as an added value to the customer who elects to spend more for a bottle of wine. A guest who requests a nicer or larger glass for wine, however, should be immediately accommodated, regardless of the price or style of wine that he or she is enjoying.

One more important factor in selecting restaurant glassware is the durability of the glasses. The treatment of a wineglass in a commercial restaurant setting is very different from that of a glass in a private home. Glasses are put into dish racks, they often have to endure dozens of washings in a commercial washer, and they are frequently banged around by service staff in the heat of a busy service period. Durability, therefore, should be one of the factors that is considered when purchasing wineglasses. The finest glasses should always be washed by hand at the bar and not put through the washing machine, where they are prone to break. Some breakage is a reality in the industry but it should be monitored in order to ensure it is kept at appropriate levels. It does not take long for the adequate supply of nice wine glasses to be reduced to barely enough as they periodically break. The restaurant's glassware budget, therefore, should not just include investment for the initial supply of glasses, but a periodic restocking to ensure adequate numbers of glasses.

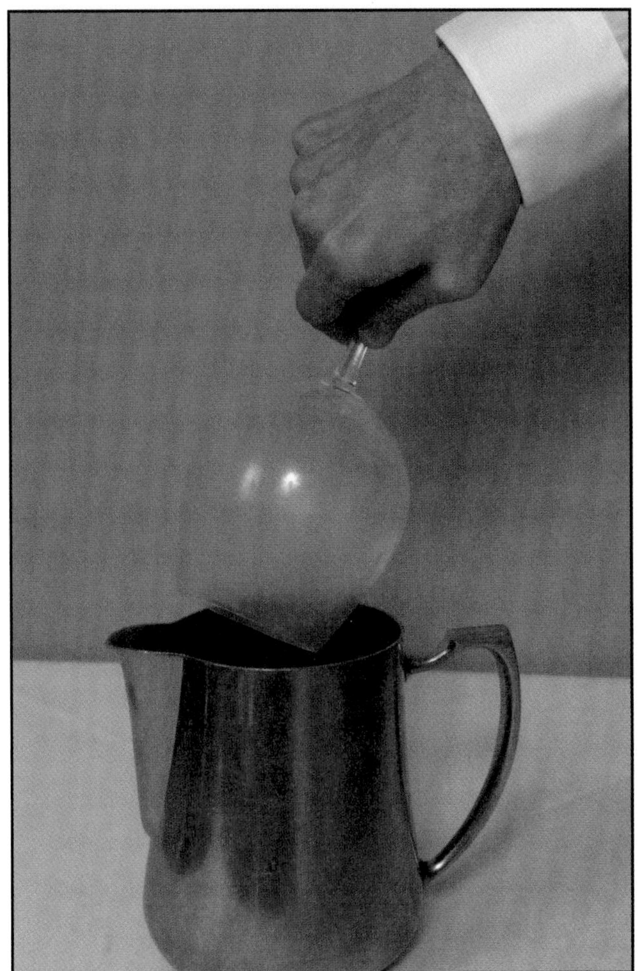

FIGURE 16–8A Using a pitcher filled with hot water, the server steams the wineglass prior to polishing it.

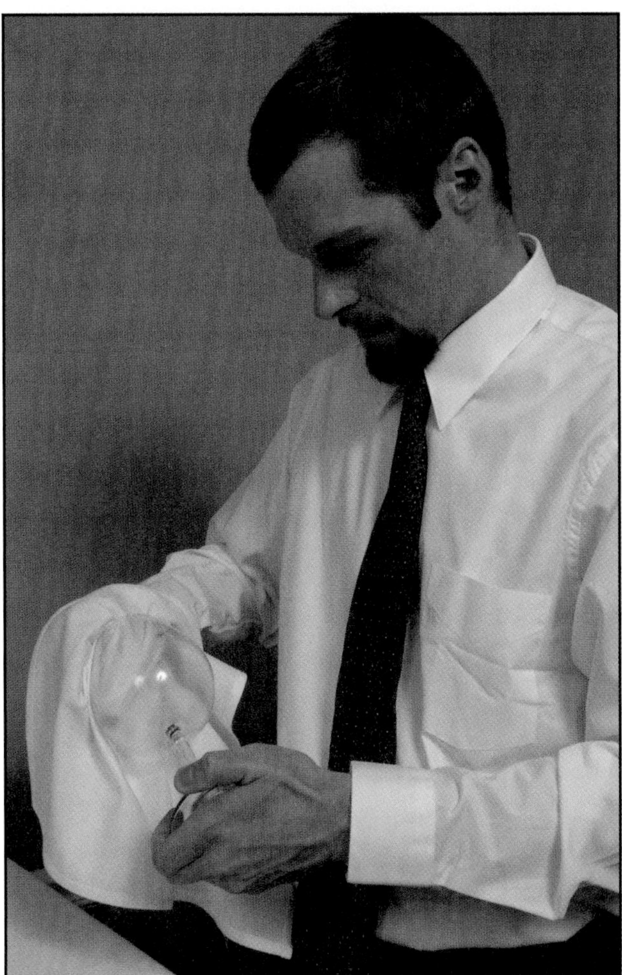

FIGURE 16–8B After the glass is steamed, it can be easily polished using a clean, dry service linen. All glassware should be polished in this manner prior to being presented to the guests.

Regardless of the style and size of the glasses, they must be served clean and polished. Any type of soap or detergent residue that remains on the glasses will distort the aromas of the wine and can greatly affect the perception of the wine's profile. Glassware with water spots is unsightly and a sure sign of lack of attention to detail. Wineglasses should be dried with a lint-free linen prior to being shelved at the bar or server station. Glasses that are not used often can accumulate dust as they sit on the shelf and should be stored upside down to prevent dust and foreign matter from settling in the bowl. Polishing glassware is an easy activity, although completing this task in the heat of a busy service period can be a challenge. Service staff should be trained how to properly polish glassware, and they should understand the importance of doing so. Training them to be gentle with the glass is critical to reducing breakage.

The simplest way to polish glassware is to use steam and a clean, lint-free linen. Most operations have silver coffee or tea pots that can also be used for polishing glasses. A large coffee mug may also work, depending on the size of the wine glass. The process is simple: Fill a pot or cup with hot water, then hold the wineglass by the stem upside down over the container. The steam rising from the container will condense inside the glass, which should then be polished with the clean linen (Figures 16-8 A and B). This process will remove the water spots and leave the glassware shiny and polished. Servers often use the same process to polish silverware, although often they add a little vinegar to the water. Vinegar should not be used with wine glasses because the vinegar aromas might linger in the glass and negatively affect the aroma of the wine.

Once the glassware is clean and polished, it can be set on the table. Wineglasses may be brought to the table in the server's hand, held upside down by the base of the stem. This allows a server to carry many glasses at the same time without getting the bowl of the glass dirty. Another option would be to deliver the glasses to the table using a service tray. Either way is acceptable. The wineglasses are placed to the guests' right side, next to the water goblet, although this may vary depending on the operation and table setup. Placing glasses to the right allows the service staff to pour wine from the right without having to reach over the guest. Presetting the glassware allows the server to focus on properly presenting and pouring the wine.

Proper Serving Temperatures

(This section deals with the *serving* temperatures of wine. Chapter 18 will address the proper temperatures for *storing* wine.) Any wine is at its best when it is served at the appropriate temperature. This can be challenging in a restaurant setting because of storage. Traditionally, wine has been served at room temperature. This is historical and predates the era of modern climate systems. During much of history, actual ambient temperature in most rooms was quite a bit cooler than what is commonly thought of today as room temperature. If wine is served at modern room temperature, it is often served too warm to be fully enjoyed. When wines are served at the incorrect temperature, their profile suffers and they do not fulfill their potential for enjoyment. However, differences in serving temperatures can also be used to alter the perception of a wine in a positive way.

The commonly accepted serving temperature for various wines appears in Chapter 4, Tasting Wines. Serving wines at the appropriate temperatures is a little more challenging in a restaurant setting, however, and care should be taken to deliver the wine to the table at the appropriate temperature whenever possible. White wines are often stored in a refrigerated area in the restaurant setting, and this can result in the wines being served too cold and thus seeming less aromatic and fruity. When served too warm, the perception of alcohol in a wine is heightened, and the wine often tastes hot. At lower temperatures, the alcohol is less noticeable, and the wine tastes more balanced. These points should be reinforced with service staff through demonstrations so that they fully understand the importance of proper serving temperature.

Opening the Wine

The selected wine should arrive at the table as promptly as possible after it has been ordered. This seems simple, although in many larger operations with extensive wine lists, accessing reserve bottles of wine can be a time-consuming process. If a wine that is selected to accompany the entrée course still has not arrived by the time the guests get their entrées, by the time the server has finished presenting and pouring the wine, the guests may be well into their meal. This is embarrassing for the establishment and the guests may no longer want the wine because of the delay. A clear, organized system of storage for bottles is critical to prompt delivery of wines to the table, and the wine list must be kept accurate to reflect the current inventory. This is covered in more detail in Chapter 17.

Once the correct bottle is located, it should be wiped free of any dust or residue that has accumulated on the outside or on the label. Older bottles should be handled delicately to avoid disturbing any sediment that may have formed. The server should approach the host of the table or whomever ordered the bottle of wine, show the bottle to the guest, and confirm that it is the correct bottle, while also allowing the guest to see the label and inspect the bottle if desired. Many embarrassing moments occur when servers assume they have the correct bottle, only to be told that it is not what was ordered. To verify, the server should show the label to the guests, telling them the **3 v's (vintner, varietal, vintage):** The vintner or producer of the wine, the varietal (the grape or the blend of grapes used to make the wine), and the vintage. For many Old World wines, the varietal is not listed, in which case it is important that the server say the region (and/or vineyard) from which the wine originates. At this point, the guest's job is to confirm the bottle being presented is indeed the one ordered.

It is important that the guest confirm the vintage is correct for several reasons. First, in most wine regions in the world different vintages produce different characteristics and qualities in the grapes and therefore in the wines made from them. One year may produce exceptional fruit and the next merely mediocre. This will impact a guest's decisions to choose one wine over another. Commonly, when one vintage is sold out, wholesale suppliers will replace it with the newer vintage. If the newer vintage is of lesser quality than the older, the guest may not want that wine and request a substitute. Keeping proper vintages on the wine list is a critical aspect of any wine program, and the list should be updated whenever needed.

Wine Keys

The tool most commonly used to open a bottle of wine is called a **wine key,** or corkscrew (Figures 16-9A and B). Regardless of the name, this tool is used to cut the foil and efficiently remove the cork from the bottle. When selecting an appropriate wine key, there are many styles to choose from and a few important features to look for. The traditional styles have been largely improved upon through new designs, and the wine keys of today are far superior to and more efficient than those from years ago. Although the choice of style is partly a matter of personal preference, there are several items to keep in mind when selecting one for use in a restaurant setting:

- *The blade:* This is the knife used to cut through the **capsule** or foil topping that covers the cork and the bottle top. Aside from the size of the knife, the main difference is whether it is a serrated or straight blade. Serrated blades tend to dull more slowly, so they are effective longer. Some wine bottles still use a foil-style capsule and the repeated use of a straight blade will cause it to dull.

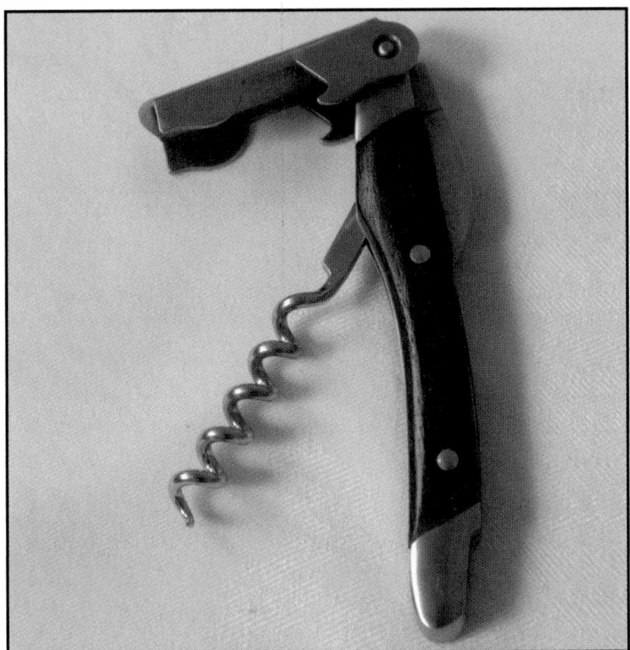

FIGURE 16–9A A waiter's-style wine key or corkscrew is preferred in a restaurant setting, as it makes opening a bottle of wine much easier tableside.

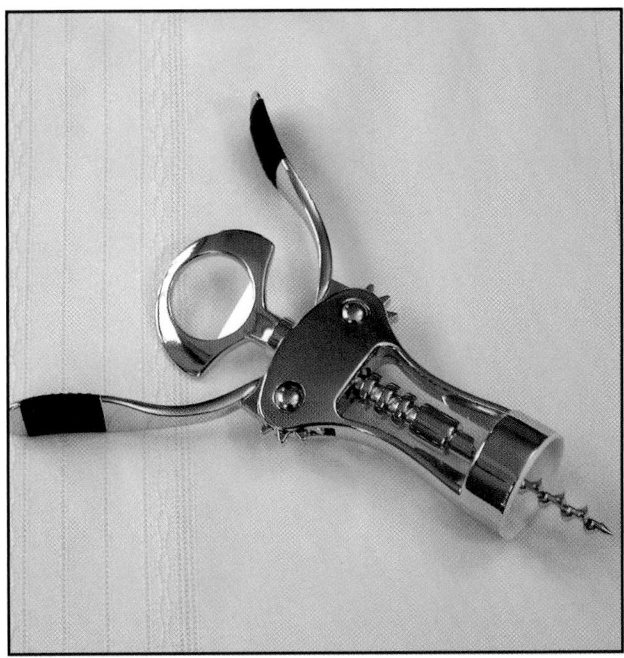

FIGURE 16–9B The butterfly-style wine key is larger than the waiter's style, so it isn't as easy to carry.

- *The worm:* The spiral screw, or **worm,** is the piece that is twisted into the cork, allowing it to be removed. Most modern worms have five or six turns on their spirals, which is enough gripping power to facilitate cork removal, although inexpensive and older styles often have shorter worms with only three or four turns. These should be avoided. Many worms are now coated with black Teflon, which eases insertion and removal of the cork from the spiral.
- *The lever:* This is the part of the wine key that applies leverage to the bottle to allow the cork to be removed. The biggest innovation in wine keys has come from the creation of a double-hinged, or two-step, lever. This style has two points of contact or *steps* and allows the cork to be eased out in two stages. This reduces the instances of corks breaking in half as they are removed because there is less pressure applied to the cork at any given time.

Regardless of the style of wine key chosen, all service staff and bartenders should carry one as a matter of course. No one should have to scramble around trying to find a wine key for tableside presentation at the height of service. Servers should also have access to service linens that are used during the process of wine presentation. The linen should be clean and folded, draped over the server's arm as they approach the table. A white wine service should use a white linen and a red wine a red linen; having red wine spots staining a white linen is unsightly, and a red linen does not showcase the color of the white wine. The linen is used to wipe the neck of the bottle and to prevent wine from dripping onto the table or the guests as the server pours for the party.

After the guest has confirmed that the wine being presented is the proper wine, the server should proceed to open it for the guests. This may happen in different ways. The first method is for the server to open the bottle of wine while holding it, never setting it down on the table. Holding the bottle, opening the wine key, cutting the foil, and removing the cork can be a challenge. Thus, some operations choose to have the service staff open bottles of wine on a wine coaster, or small plate, placed on the table near the host. This makes it easier for servers to open the bottle of wine because they have less to juggle in their hands. The wine coaster also catches any drips that may come from the neck of the bottle and prevents them from staining tablecloths and other service linens. These coasters add a little extra touch to the process of wine presentation.

Regardless of the method in which the wine is presented, the process of opening should be consistent between staff members. Service staff should all be trained on the proper method of opening wine tableside, and they should be allowed to practice until they feel comfortable with the techniques. Servers should always open bottles of wine in front of the table; bottles should *never* be opened in the back station and then brought to the table already opened (Figure 16-10). This may make the guest suspicious about when the bottle was opened and why it was not presented to the table first. Additionally, if the server has mistakenly grabbed the wrong bottle, there is no chance for the guest to ask for the correct bottle because the wrong one has already been opened. This will clearly increase the overall beverage cost for the operation.

Using the knife on the wine key, the server should proceed to cut the foil on the lip of the bottle (Figure 16-11). There are several techniques used to cut the foil, depending on the style of service desired. Some wine opener sets come with a foil cutter, a crescent-shaped device with rollers inside, that is rotated around the neck of the bottle. The rollers make a clean cut on the foil and allow it to be easily removed, but they also require the server to carry one more piece of equipment. The blade on the wine key works fine, and it is very effective at removing the foil. The server can cut either above or below the lip of the bottle. Cutting below the lip is preferred because it helps prevent pieces of the foil from falling into the guests' glass of wine and because securing the knife under the lip gives better control, allowing a clean cut of the foil. A sure sign of a novice is to destroy the foil in an attempt to remove it. Some operations may choose to have their servers remove the entire foil, although this is sometimes seen as a less formal method and should be avoided if possible. After the foil is removed it should be placed in the server's pocket, not presented to the guest.

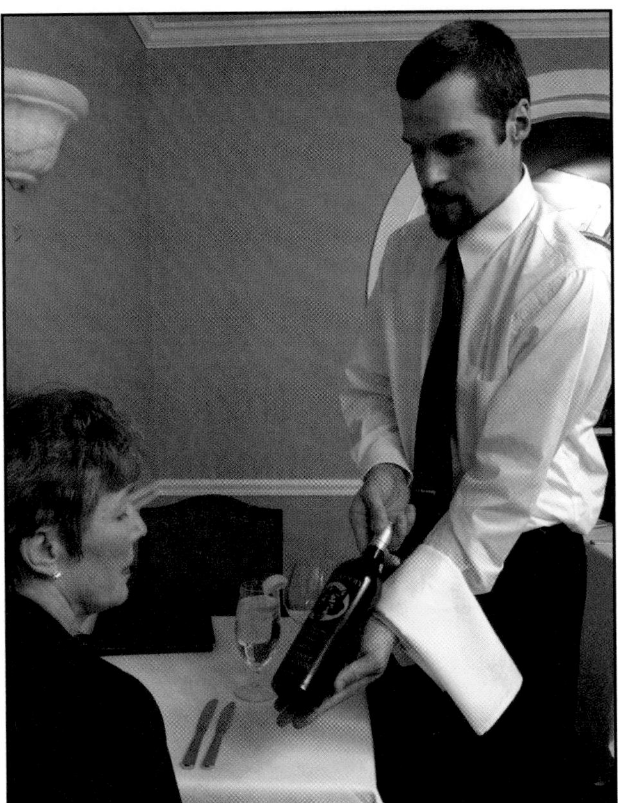

FIGURE 16–10 The process of presenting the wine to the guest allows her to confirm that it is the correct bottle and vintage. The server should identify the vintner, varietal (or blend), and vintage as the bottle is presented to the guest.

FIGURE 16–11 Using the blade on the wine key, the foil should be cleanly cut below the lip of the bottle and removed to allow access to the cork.

Many bottles of wine are now sealed with a wax-based coating instead of the traditional foil capsule (Figure 16–12). This decision may be made by the winery because it sets their product apart on crowded store shelves or because they feel that it adds an extra level of appeal to their bottle. This wax coating may take the form of a small disc of wax that simply covers the top of the cork, or it could be more elaborate and involve a significant area of wax at the top of the bottle. Either way, servers are often tempted to cut the wax and attempt to remove it; this usually ends up creating a waxy mess on the bottle and leaves chipped wax on the table. Instead, service staff should be trained to ignore the wax and simply insert the worm of their wine key into the wax as if it isn't even there. As the cork is removed from the bottle, the wax is removed as well. This process is usually less messy than trying to remove the wax. Servers should then use their service linen to wipe the remaining wax from around the lip of the bottle prior to pouring.

Once the foil or wax is removed, the spiral or "worm" of the corkscrew can be used to remove the cork. Placing the point of the worm in the center of the cork ensures that it will enter the cork in the middle and allow easier extraction of the cork (Figure 16-13). The worm is twisted into the cork until about half a turn remains (Figure 16-14). It is important to insert the worm far enough to guarantee proper leverage to extract the cork. It is also important, however, not to have the worm go all the way through the bottom of the cork because it can dislodge pieces of cork that get into the wine and float on the surface as it is poured. Particularly

FIGURE 16–12 Some producers use wax blends instead of foil to seal their bottles. This bottle has a handy pull-tab to allow removal of the wax top.

FIGURE 16–13 The point of the worm should be inserted into the middle of the cork. It is important to insert the worm straight to allow clean and easy removal of the cork.

old or brittle corks can cause a film of floating cork that is unsightly and may be offensive to the guests. At the very least, this can convey a lack of experience or focus on the part of the server.

After the worm is used to safely remove the cork from the bottle, the cork should be removed from the worm and presented on the guest's right side, either on a bread and butter plate brought for the occasion or simply on the table (Figure 16–15). There should be no assumption made that the guest needs to do anything with the cork. Do not wait for them to smell it or suggest that they need to, since this may make the guest uncomfortable. There is a historical perspective for the process of presenting the cork as well as a ritualistic one; the historical piece is that by inspecting the cork the guest can be sure that the cork removed is actually from the bottle being poured. In the past, unscrupulous restaurateurs may have partaken in fraud by removing the contents of a particularly desirable bottle of wine and replacing it with an inferior product, then recorking the bottle and selling it as the original product, thereby deceiving the guest. By inspecting the cork, the guest is verifying that the vintage and producer stamps match the bottle being poured.

This type of fraud is a virtual nonoccurrence in the modern age, however, and the inspection process should instead focus on a more relevant issue: Is the cork sound and in good shape? If not, then there is an increased likelihood that the defective cork has caused the wine to spoil, probably due to the presence of too much oxygen contacting the wine. When inspecting the

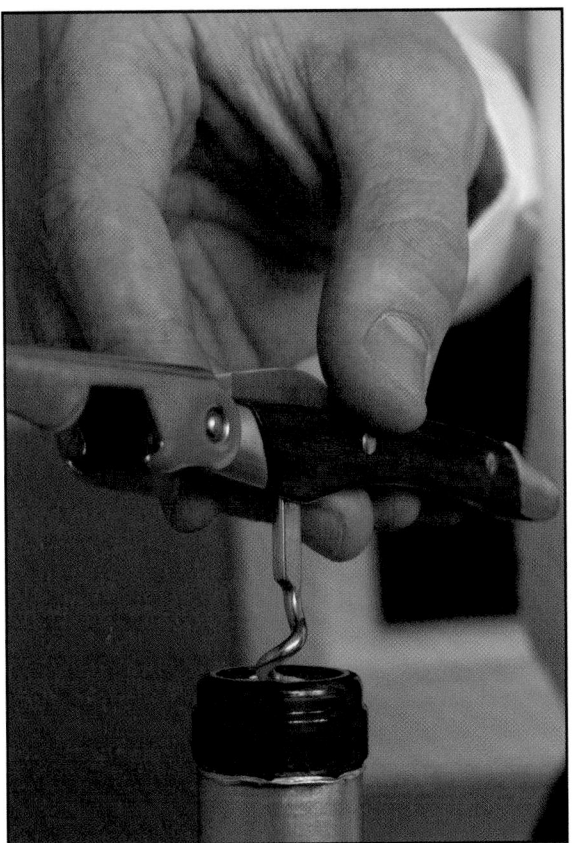

FIGURE 16–14 The worm is inserted almost all of the way into the cork, until only about half a turn remains. This will ensure that the server has enough leverage to fully remove the cork from the neck of the bottle.

FIGURE 16–15 After removal, the cork is presented to the right of the host for inspection.

cork, it should be firm but pliable and squeezable (not too dry), and the end that was closest to the wine should be moist, which indicates that the wine was properly stored on its side. If the sides of the cork show streaking that extends to the top end of the cork, this could indicate that the cork did not seal the bottle securely. If the wine could leak out the side of the cork, oxygen could leak in and increase the chance of the wine spoiling.

Inspecting the cork is more important than smelling it, although many people still feel that it is imperative to smell the cork. This practice was instituted with the belief that a taster can detect "off odors" in the wine by smelling the cork, but this is not as common a practice today. Try it yourself: smell a fresh cork, just removed from a bottle of wine; what do you smell? Chances are very good that you will smell *cork,* with perhaps a hint of aroma from the wine, but not much. Since one of the off-odors that can be present in wine is the smell of cork, what sense does it make to smell the cork in order to detect off-odors? A much more effective method is to smell the wine for off-odors, not the cork. By inspecting the cork, however, the server or guest can get an idea of how the cork has held up during storage, giving a preview of the possible condition of the wine. When encountering a funky, decrepit cork, be immediately skeptical of the condition of the wine. It is not always spoiled, but when the cork is in terrible shape, the wine has a lower chance of being sound. The cork is the main barrier between the detrimental effects of oxygen exposure, and as the barrier breaks down the chances of spoilage increase dramatically.

Corks and Cork Taint

To many consumers a bottle that is sealed with a natural cork is the mark of a quality wine. Since they were introduced for use in wine bottles in the seventeenth century corks made from the bark of the cork oak tree, *Quercus suber,* are the closure of choice for the majority of the world's best wines. This being said, today the most common flaw found in bottled wines is spoilage due to bad corks. Because corks are made from a naturally occurring material, tree bark, there is a great deal of variation in density and porosity from cork to cork. The denser and less porous the bark used for the cork the better the seal it will make with the bottle. If the moisture content of the cork is too low when inserted into the bottle, it may not be flexible enough to make a proper seal. Cork manufacturers grade their corks by appearance and quality to try to limit this variation, but it is difficult to obtain perfect quality control.

Cork taint is a musty smell that is the most common problem associated with natural corks, affecting approximately 3 to 5 percent of the wines in which they are bottled. It occurs when the cork has been exposed, either in the forest or during processing and storage, to mold growth, and the corks absorb a compound called **trichloroanisole (TCA)** (proper term 2,4,6-trichloroanisole) from the mold. TCA is the most prominent of several compounds that are produced by mold that have a strong, rank smell often described as being similar to the smell of wet newspapers or a damp basement and that can be detected in the aroma of a wine in levels as low as several parts per trillion. TCA can also be produced when empty wooden wine containers such as barrels are improperly stored damp, and mold growth occurs in or near the barrels. The presence of chlorine can exacerbate the production of TCA; therefore, cleaning agents that contain chlorine should never be used around barrels.

When a server at a restaurant pours a small taste of wine to be sampled before it is served to the rest of the table, it is primarily to determine if the wine has cork taint or is "corked," and if so the bottle is refused and another bottle can be ordered. Although cork taint is the most common flaw found in commercial wines, unfortunately the majority of consumers are unfamiliar with the aroma and do not recognize it in the glass. Those who do notice the smell often assume it is a result of poorly made wine and not from a bad cork. Additionally, subthreshold TCA levels can mask the fruity aroma of a wine giving it a dull aroma even if the concentration is not strong enough to produce a noticeable musty character. Servers should be trained to identify TCA contamination in order to help support the guest when they decide to refuse a bottle of wine because it is faulty.

To combat this problem, cork manufacturers are changing their processing methods to reduce the chance

(continued)

Types of wine bottle closures, in the front row from left to right: Screw cap (Stelvin), two types of extruded synthetic corks, two types of synthetic corks cast from a mold, an agglomerate cork, an agglomerate cork with natural cork disks on the ends. In the back row: two natural corks being cut from a block of cork bark, and two natural corks separated from the bark.

Corks and Cork Taint (continued)

for mold growth during processing and are performing more analysis with sophisticated laboratory equipment to monitor corks for TCA levels. An increasing number of wineries are also using synthetic closures such as corks made of plastic or screw caps to seal their bottles. Screw cap or **twist top** closures are made of aluminum and are similar in appearance to the capsule that goes over the top of a cork finished bottle. The caps have a seal that breaks when the top is removed and have a food grade liner under the cap so the wine in the bottle does not come into contact with the aluminum of the cap. Stelvin is the most common producer of this style and all twist top closures are often mistakenly referred to as "Stelvins," despite the fact that many producers now make this style of closure.

Serving Wine from a Bottle with a Twist Top

The process of opening a bottle of wine with a synthetic cork is the same as for a bottle with a traditional cork seal, but twist tops are a little different. The principal difference that is noticed with a twist top closure is that a wine key is not needed at any point in the opening process. The twist top bottle can be presented in the same manner, and the guests may notice the twist top closure and comment on it. Servers should be trained to understand and explain the reasons behind a producer using twist top closures, and this can function as an interesting point of discussion with the table as the wine is opened. Because the servers have a linen over their arm during wine presentation, they may simply use the linen to grasp the twist top and subtly remove it. The twist top should not be presented to the guests unless they request to see it, but should instead be placed in the server's apron pocket. Although wines from bottles with synthetic corks or screw caps will not have cork taint, other types of spoilage are possible and the wine should always be tasted by the host before it is served to the rest of the table.

Some parts of the wine-producing world have embraced the use of screw caps more readily than others, although their ability to greatly reduce the risk of TCA contamination is widely accepted. New Zealand, Australia, and some parts of the United States have employed twist tops widely, and other countries are following their lead. They are quickly gaining acceptance in many areas of the world, but are most commonly used as a closure for white wines. Many producers traditionally used twist tops on their lowest category of wine, commonly called jug wines. Some consumers still regard twist top sealed wines as inexpensive, inferior products, simply because they are sealed with a twist top. Although this may have been true 20 years ago, the trend has changed considerably in recent years; some producers use twist top closures on all tiers of wine, from the cheapest to the most expensive.

The move away from natural corks is gradual because of tradition, and many wineries are concerned with consumer reaction. Although these modern closures do not have cork taint, a wine that they are bottled with may not age in the same way that a wine bottled with natural cork does. No closure of any kind is perfect, natural and synthetic corks, as well as screw caps, have their limitations. However, an increasing number of winemakers are moving away from natural corks made from tree bark because they feel that the problems associated with them outweigh their benefits.

Taste Test

After removing and presenting the cork, the server should pour a sample of the wine for the host to evaluate (Figure 16–16). There are some restaurants where the server (or preferably the sommelier) will smell and possibly taste the wine prior to the guest doing so, in an attempt to prevent the guest from encountering a spoiled wine. Obviously this should only be done in a

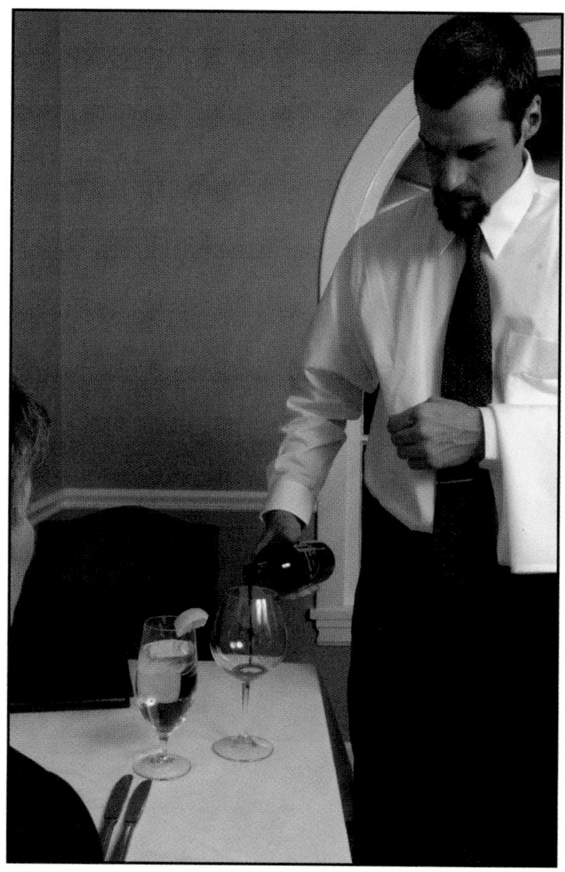

FIGURE 16–16 The server should pour a small (1- to 2- ounce) sample for the host to evaluate and check for off-aromas or flaws.

situation where the staff is properly trained to detect a flawed wine. The amount being poured should be about 1 ounce, just enough to taste the wine and evaluate the aromas for flaws, but not much more than a taste. The server should then wait patiently for the guest to approve of the wine before offering to pour for the other guests.

In a situation where a party has ordered both a white and a red wine, the server should be sure to bring two glasses for the host to use for tasting, so that the residue from the first wine doesn't interfere with the second wine. If a party orders a second bottle of the same wine, it is appropriate for the server to bring a new glass for the host to sample the fresh bottle. If the wine is the same as the first bottle, it is not necessary to replace the guests' glasses with new ones. If the party switches wines, however, all used glassware should be replaced with clean glasses before the new wine is poured.

The guest smells and tastes the wine sample to determine if the wine is sound or if it is flawed in some way. Some aroma flaws are readily apparent as soon as the bottle is opened, but others may develop in intensity as the wine is exposed to air in the glass. Oxygen in the air can also be a source of spoilage while the wine is still in the bottle. If the cork has not made an adequate seal, the bottle may leak allowing excess air to come in contact with the wine. While a sound cork does allow a small degree of exposure to the air helping the wine to age, excess oxygen will cause the wine to take on a Sherry-like, or oxidized character. Older wines will often have a slight degree of oxidized character even if they have had a sound cork. Another common flaw encountered is cork taint which is caused by the cork itself.

Sending Back the Wine

In some cases, the guest may choose to send the wine back; this situation should be taken seriously. Prior to removing the bottle from the table, the server should find out why the guest is dissatisfied with the wine. There are many reasons why a guest may choose to refuse the wine selection, and each one has a different solution. For example, a situation where the guest simply doesn't find the *style* enjoyable will need to be resolved very differently from one where a guest feels that a particular wine is flawed, although both cases should result in replacement of the bottle. If the guest is disappointed by the style of wine, a knowledgeable staff member should make a concerted attempt to help the guest choose a replacement bottle to their liking. This should be undertaken by a manager or sommelier preferably, or at least someone who has enough knowledge to properly guide the guest through the many options available on the list. If the guest feels the wine is faulty, however, then the bar or service manager should evaluate the faulty bottle and offer a replacement bottle immediately.

Resolving guest dissatisfaction with a selected bottle of wine should be taken seriously and addressed as promptly as any other issue which arises during the service period. In most cases, guests view purchasing a bottle of wine as an investment in their meal, and most bottles cost far more than the average food item. Prompt and proper resolution of the complaint can be very effective at showing guests that we care and that every aspect of their experience matters. If the wine truly *is* flawed, it can be used to educate staff in the detection of flawed wine or it could be sent back to the distributor for credit, so there is no excuse for not taking back a bottle that guests find unsatisfactory. It should be noted, however, that it can be in bad taste for guests to refuse a bottle simply because it is "not what they expected." If guests have a question about a style of wine, they should feel comfortable requesting information or guidance from the service staff. Opening several bottles of wine to try to find one that the guests like is not effective on a business level and is not convenient during a busy service period.

In the case of an extremely rare or expensive bottle of wine that is sold "as is," returning the bottle may not be allowed by the operation. If a special bottle of wine is virtually irreplaceable, the guest may be expected to pay for it, regardless of the condition of the bottle once it is opened. Obviously this is a relatively rare situation, but the situation may arise. With properly trained staff who work to help the guest make appropriate selections, however, the chances of a guest refusing a bottle of wine can be greatly lessened.

The guest will usually smell and taste the sample portion before granting permission for the server to pour for the other guests at the table. Once the host has approved, the server should proceed to pour the wine for the guests, making sure to pour for ladies first, then the gentlemen, and ending with the host. A typical pouring may involve serving the ladies first in a clockwise direction, then circling back around the table in a counterclockwise pattern while pouring for the gentlemen, and ending with the host, leaving the bottle at this position on the table. The server should pour from the guests' right side unless otherwise restricted. Depending on the size of the party, more than one bottle of wine may be needed. Each 750 ml (25.4 oz) bottle of wine contains from four to six glasses of wine, depending on the size of the pour. If the pour size is 4 ounces, there are six glasses to a bottle, but if the pour size is 6 ounces, then there are only four glasses per bottle.

With larger sized groups (six or more), the server should be sure to pour enough so each guest receives an appropriate amount, but not so much that the bottle is empty by the time the host is served. If the pour size is controlled properly, all guests can be served some wine without the immediate need for another bottle. A classic example of bad etiquette is to not have enough left for the host and then to ask if he or she would like to order another bottle. Not only does it show a lack of focus, it sends the message to the guests that they are being coerced into spending more money. This is a bad precedent to set and should be avoided at all times.

If the guests order two different bottles, the server should inquire from the host as to the order of preference of delivery. Whites are usually brought before reds, but if two whites or two reds are ordered, the server should prompt the guest for guidance on their preference of delivery.

If a white and a red are both to be brought at the same time, be sure to have the host sample the white first so the palate is kept fresh to taste the red (see Chapter 4 for more information on the proper tasting order). It might be possible for another server to assist in the presentation of two bottles simultaneously, to be sure that the guests receive both bottles in a timely fashion. A savvy server will have already asked each guest whether they will be having white or red wine, and so can set appropriate glassware. This display of teamwork positively affects the guest experience and helps maintain the flow of the dining experience.

A useful tool for service staff who frequently pour bottled wine tableside is a laminated Mylar disc that is rolled up to form a spout and inserted into the neck of the bottle. There are several different brand names associated with these products such as Pour Discs (Figure 16-17), and they can be ordered with the operation's logo or name imprinted on the side. These discs are very effective at eliminating the drips from pouring wine from the bottle. They also add an extra touch of polish to the service, showing the guest that the operation has thought about the details. Since the discs are inexpensive and disposable (although limited reuse is possible), it may also function as a souvenir that the guest is able to take home after the bottle is empty.

Part of the service of wine in the restaurant setting involves the service staff repouring wine for the guests as necessary. This can be a delicate issue with some guests; when in doubt, ask the host what should be done about refilling guests' glasses. Traditional protocol involves the server pouring wine for guests who have emptied their glasses, but no assumption should be made that every guest would like more wine. A simple and effective technique is to approach the guest with the bottle and, prior to pouring into the glass, pause for a second or two. This simple pause allows guests to see that the server is going to pour wine and enables them to decline if they don't want more wine. A server who does not allow the guest a moment to decline may pour

FIGURE 16–17 This laminated disc is inserted into the neck of the bottle to enable drip-free pouring of the wine.

wine for guests who no longer wish to drink, thereby leaving less wine available for those who do. Some parties may desire to pour the wine themselves and may ask the server not to pour once the initial pouring is done. The wishes of the guests should be honored at all times.

Large format bottles require an extra level of attention and service, both because they represent an extra level of investment on the part of the guest and because their sheer size can make them unwieldy tableside. Magnums are still manageable tableside in the usual manner, but any bottle larger than a magnum requires extra steps. Large bottles that are the equivalent of four or more regular bottles may need to be handled by two service staff members, if only for safety. These size bottles are large and heavy and dropping one would guarantee not just a large mess but also a significant loss of product for the operation.

Pouring for each guest from a huge bottle is dramatic but impractical. A better solution is to have service staff open the bottle in the usual way, being sure to place it on the tabletop rather than trying to hold and open it at the same time. The servers should then pour a good amount into a decanter and use this to pour for the host and then the guests. This allows the guests to enjoy the presence of the jumbo bottle and allows the servers to easily pour the wine for the guests. As the decanter is emptied it is refilled from the larger bottle until the bottle is emptied. The wine benefits as well because it receives a good amount of air as it is consumed and re-poured.

Sparkling Wine

The process of opening Champagne and other sparkling wines tableside follows the same protocol as still wines, although there are a couple of key differences. The first main difference is that a wine key is not necessary during the process. The foil covering with sparkling wines usually has a perforation that allows the server to manually remove the foil capsule. The blade of a wine key could be used if necessary, but is usually not needed. When the foil is removed the wire cage that holds the cork in place is exposed (Figure 16–18A). On the cage is a wire tab that is untwisted to loosen the cage and allow access to the cork. When the cage is loosened, great care must be taken with the cork because there is significant pressure inside the bottle. The neck of the bottle should never be aimed at a guest or another person because the cork is capable of causing injury if it flies out accidentally. A service linen is typically placed over the exposed cork to allow the server to maintain a proper grip on the cork. The cork should be twisted gently in one direction while the bottle is twisted in the other direction. The cork has a tendency to fly out of the bottle, and it should be carefully controlled. The server should gently ease the cork out of the bottle, taking care to avoid the all-too-common popping noise that is often heard. A gentle sigh is more appropriate and professional, and demonstrates skill and confidence on the part of the server. The wine is poured for the guests in much the same fashion as for still wines, except that the glasses should be partly filled and the foam should be allowed to settle slightly before each glass is topped off (Figure 16–18B).

Decanting the Wine

In some cases the guests may ask to have the bottle of wine decanted. **Decanting** a wine involves separating the wine from the bottle sediment through careful pouring of the wine into a glass container. The process of decanting is relatively straightforward but can greatly affect the overall impression and enjoyment of the wine. The vessel used for this process is a type of glass carafe that is often called a **decanter.** The decanter is then used to serve the wine after it has been separated from the sediment in the bottle. As wines age, their composition changes and they often precipitate **sediment** in the bottle. Although this will be addressed more in Chapter 18, the basic premise is that sediment is unsightly and does not taste good. Having a glass of wine with a lot of sediment is like having sand in your clams—not dangerous or unhealthy, simply

FIGURE 16–18A The cork used to seal a bottle of Champagne has a characteristic mushroom shape. The cork is held in place with a wire cage to ensure that the pressure in the bottle does not force the cork out of the neck.

FIGURE 16–18B A small amount of Champagne should be poured in the glass and then topped off once the bubbles subside to avoid overflow.

unpleasant and showing a lack of focus and attention to detail. With wines, the same premise applies. Older wines may have accumulated a significant amount of sediment as they aged, and it is appropriate to separate the wine from this gritty deposit. To do so is relatively easy but demands a couple of easily acquired tools and some time on the part of the server.

Service staff should have access to a decanter that can be used for this process. Although there are dozens of styles to choose from, the basic design involves a bulbous bottom and a fluted top (Figure 16–19). The purpose of the bulb is to maximize the surface area of the wine in an effort to achieve the optimum level of aeration and release the esters to enhance the aroma profile of the wine. In an ideal situation, a wine that is in need of decanting would have been stood up straight for at least a couple hours and preferably a couple of days to allow the sediment to settle to the bottom of the bottle. In most restaurant settings, the wines that are most in need of decanting have been stored lying down where the sediment has settled along one side of the bottle. Once the bottle is picked up, the sediment is distributed throughout the wine once again, and

FIGURE 16–19 Various sizes and shapes of decanters are available on the market today. These are some of the most common shapes and are used to aerate the wine to increase aromatics.

decanting becomes more difficult. It is critical, therefore, that a server or sommelier who is retrieving an older bottle of wine for a guest handle the bottle as gently as possible to avoid stirring up the sediment.

The goal of decanting is to pour the wine off of this sediment and leave the gritty residue in the bottle. A server decanting tableside should remove the entire foil capsule from the bottle of wine to allow a full view of the neck. A light source should be available for the server; historically this has been a candle, although a small flashlight may be easier to use, depending on the context. Restaurants that include candles as part of the table setting could easily employ them for decanting when needed. The light source is placed on the table and the wine bottle is held above the light as the server pours the wine into the decanter (Figures 16–20 A and B). As the wine is poured, the server should watch carefully. As the sediment approaches the neck of the bottle, the server stops pouring so that the sediment does not enter the decanter. This process leaves the wine in the decanter free of sediment and usually results in several ounces of wine being left in the bottle, but this remaining wine is full of gritty sediment.

In some cases a restaurant may choose to invest in a decanting funnel which often includes a screen that nests inside the funnel to catch any sediment which would otherwise enter the decanter. These funnels span the range from inexpensive to very costly, but the basic function of the different styles is the same; depending on the style of service and the investment from the operation, servers may or may not have access to these devices. They are particularly handy at filtering bottles which may have had the sediment stirred up on the way to the table, and are generally a good investment in a service setting for this reason.

An alternate (although related) use of a decanter would be when a guest requests the wine be aerated. The process of **aeration** is similar to decanting, although sediment is not a concern. Most wines, regardless of age, will benefit from some exposure to air to allow the release of the aromatic esters. Some wines, particularly younger, more robust wines, really benefit from air exposure. These wines might be described as "tight" or "closed," indicating that they are not as aromatic as might be expected. This subdued nose may be because of the wine being recently bottled, or because not enough aromatic esters have formed yet for the wine to be readily aromatic. Either way, the lack of significant aroma is sure to detract from the overall impression of the wine, but this situation can be helped. To encourage these wines to release their aromatics

FIGURE 16–20A A candle is traditionally used as a light source for decanting.

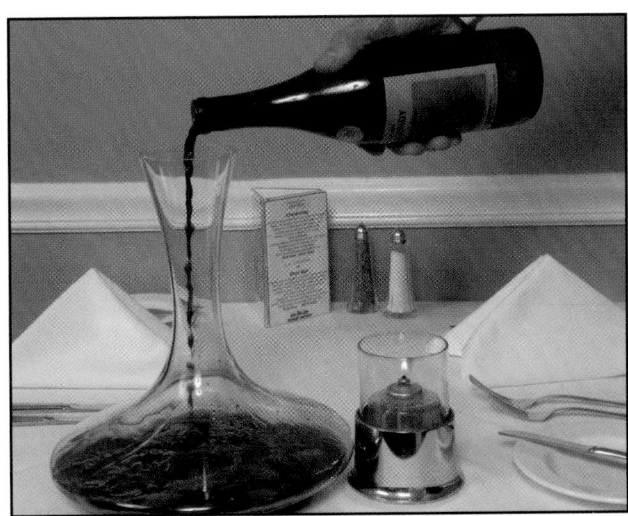

FIGURE 16–20B The process of decanting to allow the server to see the wine's sediment as it approaches the neck of the bottle.

and "open up," they should be aerated. Because sediment is not an issue, the use of a candle is not required, and the foil capsule can be cut as with traditional wine presentation. The server simply adds the extra step of pouring the wine into the decanter to oxygenate it prior to pouring for the guests. The simple act of pouring the wine into the decanter pumps it full of oxygen and encourages the release of aromatic elements. Most wines become increasingly aromatic as they are exposed to air, and this simply helps to speed up the process. Although many wines would benefit from this simple process, most guests will not request it unless they have a specific knowledge of the process or another reason to do so. Wines that are especially old or fragile to start with should not be aerated because they could easily lose character when exposed to excessive oxygen.

Staff Training

One of the critical success factors in any restaurant's wine program is the service staff that executes and supports the sales of wine at the guests' table. A basic staff training program can easily be set up to allow the staff to increase their knowledge and understanding of wine, and also to help ensure that wine-drinking guests are getting the best level of service possible. There are several important steps to take when training staff, and details are included in this section.

Management must first decide on the timing and the method of delivering staff training. Perhaps the easiest way to begin the process is during a staff meal before service. Many restaurants have a "family meal" or staff meal before commencing service for the evening, and this is a logical and comfortable time to introduce wine training. Often this time is used for team development, menu tasting and education, and addressing comments and concerns from the chef. It is easy to introduce wine education in this setting because the staff is already focusing on the impending service and expects to embark on some type of training, whether about the specials, the new menu, or the chef's input. Wine training that begins in this format often expands to include additional training sessions and helps to spark an increased staff interest in wine outside of the context of service.

Once the training program has begun in earnest, it becomes easier to incorporate it into the daily routine of the staff. Shorter training sessions can then focus on new labels that are added to the list, including a tasting of the wine and suggested menu pairings. Getting the staff comfortable with the basics of wine is the first critical step and forms a solid foundation for ongoing training.

Many wine producers or local distributors are willing to visit the establishment and help with (or lead) tastings and training events. All industry professionals, regardless of their role, understand the importance of proper training, especially in the restaurant setting. Taking advantage of resources in this respect will help guarantee a solid and diverse training program. Sparking a passion for wine in the service staff is sure to carry over to their lives outside of work and will continue to push them to learn more. This will benefit the operation because staff who are passionate about wine can share that passion with their guests. Increased sales and satisfied customers are the end results.

Training During Staff Meetings

One of the key benefits of training staff during staff meetings is that most of the service staff is present and can be addressed as a group. A specific wine might be chosen to be discussed because it is the featured wine of the evening or because it will work particularly well with the evening's special. Either way, the tasting should be organized and meaningful, allowing freedom to share information and opinions openly without fear of judgment. Building this supportive and education-based culture around wine training can take a little practice, but once the staff realizes the rewards of their efforts, the momentum will continue to grow.

Staff training is not simply an investment from the staff but one that management makes as well. Some of the costs include the actual cost of wine used for staff tastings, which can easily total hundreds, if not thousands, of dollars over the course of a year. Although some distributors may be willing to donate bottles of wine for daily tastings, this is not always allowed by state liquor laws, so the management should research the state laws regarding the donation of wines to be used for tastings. There is also the cost of labor because staff members should be paid for the time they spend in training. But considering that the additional wine sales generated by a strong, well-trained service staff could total thousands of dollars per month, the investment is well worth the cost.

When training staff, it is critical to start simple and to give them meaningful pieces of information that are useful tableside and that help them understand wine on many levels. Chances are good that servers will encounter some guests who know at least something about wine, if not quite a bit, so if their knowledge isn't up to speed it will be immediately noticed. The staff should have a good grasp of what wine is and how it is produced; they should also know the essential aspects of grape production and climatic factors that influence grape production. Then they should have the opportunity to taste wines that demonstrate the concepts being taught. Scientific data and confusing terms should be kept to a minimum, at least to start, with the information becoming increasingly technical as the staff demonstrates understanding.

It would be ideal, for example, to do the first training on "What is Wine?" and focus on the basics of its production. The next training could cover the basics of terroir and include the key environmental factors that influence grape quality and production. Subsequent trainings could focus on winemaking and the production steps that affect the character and quality of the finished wine. Throughout these trainings, specific examples of wines from the wine list could be referenced, or ideally tasted, to reinforce the material. For example, once the staff gains a good understanding of when, how, and why malolactic fermentation takes place, they should be guided through a tasting of Chardonnays that demonstrate the effects that malolactic fermentation has on the character of the wine.

As the staff gains an understanding of the basic principles of wine, they can then apply the material to the wines offered on the wine list. When the learning is tied to the wine list, it creates a point of reference for the staff and helps to make them more comfortable tableside. Staff training that is undertaken outside of the context of driving sales in the restaurant and providing proper tableside service is only partially successful and is missing the opportunity to positively affect the operation's bottom line and the guest experience.

With the staff's increasing comfort with the basics of what wine is and how it comes to be, trainings can focus on more intricate details of the wine world. After discussing and becoming familiar with terroir, the staff might undertake a tasting of different Pinot Noirs from the list that showcase the effect that terroir can have on the wine—perhaps a comparative tasting of red Burgundy, or a comparison between the Pinot Noir from Burgundy and those from California and Oregon that focuses on the differences that climate makes on a wine's profile. This would allow the staff to try key wines from the list and to compare them to other similar offerings. The differences in the wines are as important as the similarities, and both should be discussed. This helps to increase staff comfort tableside, as they are then able to compare and contrast different Pinots for guests who are interested in ordering one of these wines.

Another approach for staff training could include focusing on a specific varietal, and using that as the basis for the day's training. For instance, a discussion about the origins and different styles of Chardonnay could be followed by a tasting of various Chardonnays from the wine list. This would give the staff some background about the grape and allow them to apply that information to the wines offered on the wine list. The staff could then speak generally about the character of Chardonnay as well as specifically about the Chardonnays on the list. They could then accurately discuss the similarities and differences between various Chardonnays offered tableside.

Regardless of the method of training, some type of assessment should be employed to ensure that staff is understanding and retaining the information and is ultimately held accountable for the material. This may be a written exam that the staff needs to pass or an oral exam that mirrors an exchange likely to occur with a guest tableside. The focus should be on developing staff knowledge and increasing the comfort level in talking and teaching about wine. Some distributors will offer incentives to the service team or management for increased sales of specific wines. These incentives can take many forms, including promotional bottles or wine related accessories, although the state liquor laws may not allow this. Again, it is crucial that management research the state laws before taking any incentives from a distributor.

Focus on the USP

As the servers open a bottle of wine tableside, it can sometimes be uncomfortable for them as the table watches what they do and analyzes their every move. Some servers are comfortable with this role and enjoy being on the spot while others may get intimidated. It takes a strong server to ignore these analytical stares, and a creative one to avoid them altogether. It helps to give the service staff some tips on engaging the table as they open and pour wine for the guests so that they feel comfortable during this process. This can be as simple as asking the guests if they have ever tried the selected wine and perhaps sharing some insight about it.

Every wine has uniqueness, a specific selling point that makes it appealing to the consumer. The **unique selling point (USP)** of the wine is part of what draws a guest to a specific bottle, the aspects that intrigue and encourage experimentation with various products. Restaurants have USPs, and so does each and every bottle of wine. It may be the specific region of origin, a story about the winemaker or the winery, or simply some information about the label design. The USP might be more geared toward the restaurant, about why the restaurant chose that bottle for the list, the server's personal experience with the bottle, or how previous guests have enjoyed the wine. The culture of wine is steeped in history, and most wine regions have very rich histories of their own, full of unique selling points that can be captured and shared with guests as we prepare to serve them their wine. It takes the focus off of the server and directs it to the bottle, creating another point of connection between the guests and the wine. Knowing even a little history can make for interesting discussion and add further value to the guest experience. And best of all, it gives the guests something to take home besides a cork or an empty bottle.

An easy entry point to understanding a wine's history or USP is to identify some of the best-selling wines on the list and assign a small research project to the service staff. Once a week, a specific server is asked to research a given wine and report to the team the following week during staff meal. The server should research the history of the winery, the vineyard climate, perhaps tasting notes on the wine, and any other relevant information, focusing on finding the uniqueness of the wine, whether it is the special vineyard site, the storied history of the winery, a unique grape used in the blend, or the splendor of an excellent vintage. With all the information on the Internet and in reference texts, this is a simple endeavor. This information is reported to the group, and the servers then discuss how to sell the wine tableside.

Every wine tasted should be evaluated for the key selling points, in relation to what makes the wine appealing to guests. This helps reinforce the idea that each wine is unique and builds an understanding of the differences in production and terroir. If this technique is employed regularly, the team will quickly gain a whole repertoire of stories and selling points tied to the wines on the list. Aside from building team competency, this shared knowledge benefits the guests and we will have added value to their experience. Chances are good that they will share the story with others, and will return for another visit.

Some operations build the USP of the wines into the layout of their wine list by including a simple sidebar with interesting bits of information related to specific wines on the list. This technique gives guests some guidance as they peruse the list and helps to focus their attention on specific wines. These could be wines that are exceptional examples of their style or varietal, or they might simply be wines that we enjoy selling due to their quality. Either way, calling guests' attention to specific areas of the list will help to encourage sales of those products. More on this in Chapter 17.

Summary

The business of wine sales and service relies on a commitment from the service staff and management of a restaurant operation. A properly designed and executed sales plan coupled with well-trained service staff is the cornerstone to successful restaurant wine sales. The operators need to decide on appropriate methods for encouraging wine sales during every stage of the dining experience and need to support those sales with a professional level of service. All members of the service staff should be comfortable recommending wines from the wine list, discussing wines, and opening wine tableside. A lack of focus on any aspect of the wine sales procedure may result in disappointed guests and can deter them from returning to visit the establishment. Focused and professional wine service is necessary to maintain a solid and well-rounded restaurant operation.

Review Questions

1. What are some methods that restaurant operators use to encourage wine sales in the restaurant setting?
2. What are the proper steps to follow when opening a bottle of wine tableside?
3. Why would a server present the cork to the guest when opening wine at the table?
4. List some strategies used to organize successful staff trainings. What are some considerations when developing a training plan?
5. Compare the processes of decanting and aerating. What are the similarities? What is different about the two procedures? When and why would each be used?
6. What are some factors to consider when purchasing glassware for the wine program?

CHAPTER 17

Developing and Managing a Wine List

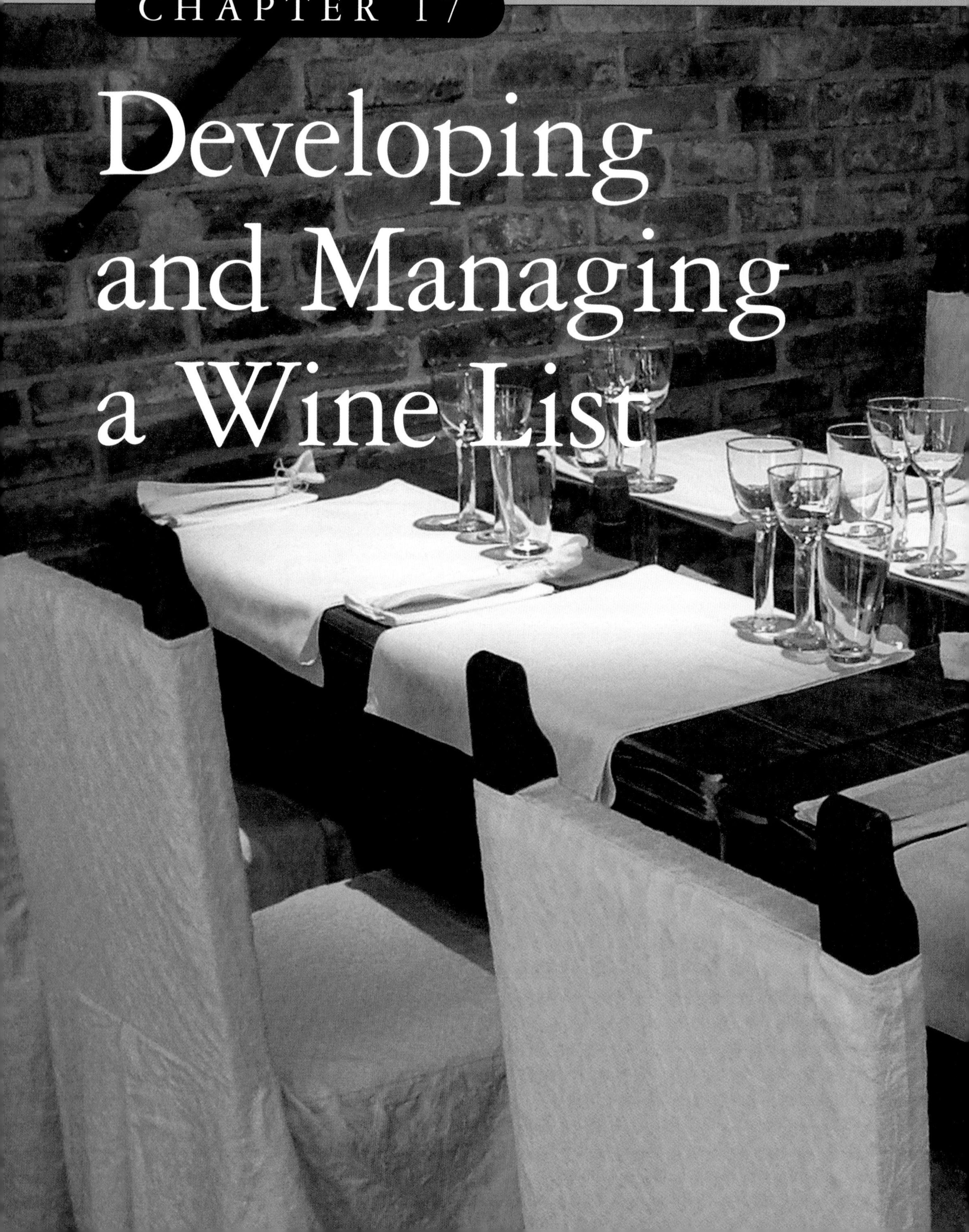

After reading this chapter, you should be able to

- identify the key considerations when creating a wine program.
- organize the layout of a wine list according to the goals of the operation.
- develop appropriate pricing strategies for wines by the glass and the bottle.
- explain the major considerations when selecting wines to sell by the glass.
- be able to choose wines for a wine list.

KEY TERMS

by the glass (BTG)
cost percent
fixed markup
house wines
inert gas
markup
post off
sit-down tasting
sliding scale
three-tiered system of distribution
trade tasting
walk-around tasting
wholesaler

The first step in creating or updating a restaurant wine list is to identify the goals for the wine program. The number one goal for the wine list in a restaurant setting should be to encourage wine sales. Wine lists that encourage guests to browse and purchase wine through a clear layout and fair pricing strategies will be more successful than those that do not. Depending on the style and concept of the operation, the specific method of laying out the list and pricing of the wine will vary, but the operator's goals should be identified before starting this process. For restaurants that focus on educating the guest and offering a unique experience, the wine list may be an extension of this educational philosophy and may include educational features in the design. Value-oriented restaurants should similarly focus on offering value in the wine list. For restaurants that already have a solid wine program in place, but look to increase overall wine sales, a simple restructuring of the design may be all that is needed. By starting with a clear set of goals the operator will be able to focus on achieving desired outcomes as he or she builds and adjusts the wine program.

Choosing Which Wines to Sell

Thanks in large part to Prohibition that spanned from 1920 through 1933, the United States uses a **three-tiered system of distribution** for alcohol (Figure 17–1). In most states, alcohol must go from the *producer* to a *distributor* to a *retailer* or *restaurant* before it can be purchased by the consumer, although some states do allow direct shipping from producer to consumer. This system affects both the general consumer and the restaurateurs who are seeking to purchase wine. One noticeable effect is that each tier of product handling adds costs to the product. Each tier must make a profit, and these additional costs affect the final price of the product. The typical **markup** for the wholesaler is about 35 to 40 percent and for the retailer it is from 25 to 50 percent, depending on the product and the retailer. Many independent retailers may use a higher markup than chain operations because chain operations tend to sell a higher volume of wine. Restaurants are able to purchase wine at wholesale prices, which are typically 30 percent below retail cost. For example, a wine that is in a store for $15 would cost a restaurant about $10.

Another major effect that Prohibition had on the distribution of alcohol is that each state is in charge of creating its own laws regulating the transportation and sale of alcohol. This has led to a muddy and sometimes difficult system of wine availability, especially between states (For more on US wine laws, consult Appendix A at the back of the book). Depending on the state where the restaurant operates, the ability to select and purchase wine could be markedly affected.

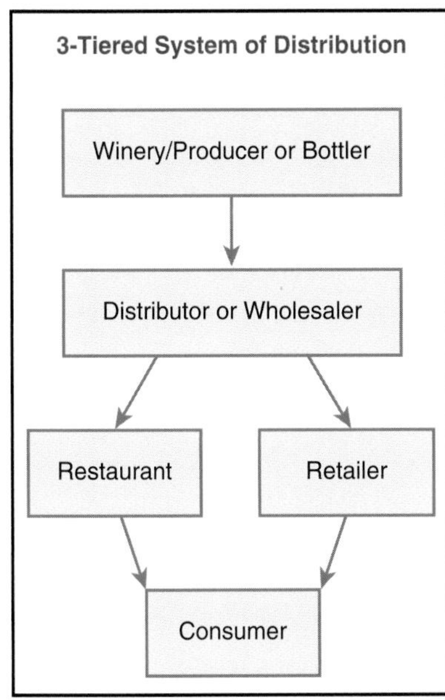

FIGURE 17–1 This chart outlines the three-tiered distribution system that still exists in many states. Wines (and other alcohol) must be sent from the producer to a licensed wholesaler who then sells it to a retailer or restaurant. From there it can be sold to a consumer.

Most wineries and producers will sell their wines to **wholesalers** (also called distributors) who then distribute to retailers. Wholesalers are operations that procure and redistribute the wines to retail and restaurant operations. Most wines purchased by restaurant operations are selected and purchased from the offerings of their local wholesalers. When an operation is looking to add labels to expand the list, they turn to their local wholesaler representatives for guidance and recommendations about which products to offer. A restaurant operator *must* be familiar with the local and state laws regarding the acquisition and sale of alcohol, both to ensure compliance and to understand the best manner in which to obtain certain wines.

The system is structured in this way for a variety of reasons. It ensures that minors do not have access to alcohol, and it ensures that states are able to collect appropriate taxes on the sale of alcohol.

The three-tiered system of distribution can affect a restaurant's ability to source and purchase specific wines. It also mandates that restaurants purchase their wines from wholesale distributors. Some regions of the country are affected by this and others are not. For example, if a restaurant wanted to find a specific Pinot Noir that was not currently available from a local distributor, it would need to have the distributor import the wines, then purchase the wines from the distributor. Although the Supreme Court is currently debating the legality of this system, it pervades the industry as of now.

The first challenge involved in selecting wines for a restaurant wine list is to decide which wines to choose. With thousands of wines to choose from, where to begin? The best and most obvious place to start is with wines that reflect the goals of the restaurant operation and meet the demands of the customers who dine there. Any wine chosen should be enjoyable, free from production or storage flaws, offer a good value-to-price ratio for the guests, and ideally be reflective of the grape used, the terroir, and the region of origin. Wine representatives often visit the restaurants where they have accounts and taste new wines with the bar team, management, or service staff. They bring samples of the wines and discuss them with the team, hoping that the operator will commit to making a purchase. This type of "try before you buy" approach is invaluable in selecting wine for the wine list. Sales reps understand that their restaurant accounts can be very lucrative when they sell large volumes of wine, and they constantly work to maintain good relations with the management team. Establishing a good rapport with account reps can help the operator get good pricing and may allow the restaurant access to small production wines or limited allocations of product.

Special Pricing

Some distributors offer periodic discounts on certain wines called post offs. A **post off** is a discount applied to the price of bottles or cases of wine. These discounts are usually offered to the restaurant operator for wines sold by the glass. This benefits the wholesaler, who sells a higher volume of these wines to the restaurant, and the restaurant, which receives lower prices on these high volume wines. A savvy operator can take advantage of post offs by purchasing when the desired wine is on special rather than paying full price. If a wine that is offered by the glass posts off every other month, then the operator can purchase this wine when on special and use the savings per bottle to help improve profitability. Strategic buying can affect the operation's bottom line, but it also requires the operation to purchase enough wine for several months' worth of sales by the glass. Thus, the restaurant has a higher amount of money tied up in its wine inventory, but the savings per bottle can sometimes make it worth the effort.

Clearly the operator needs to predict the amount of wine needed for a 2-month supply and purchase accordingly and also needs enough storage space to accommodate the larger volume purchasing. If the selected wine is discounted $1 or $2 per bottle and the wine is bought several cases at a time, the restaurant can save hundreds of dollars on one wine alone. If discounted purchase prices are used for several of the house wines, it can greatly affect the operation's profit. Wines priced to maintain an appropriate beverage cost at regular prices are even more profitable when the discount is considered. Selecting wines that post off regularly for glass sales can help to ensure an adequate level of profitability from the wine program.

Trade Tastings

In many areas of the country distributors are able to host wine tastings for those in the industry and occasionally the general public. These tastings are often referred to as **trade tastings** because they are open to members of the retail wine trade and restaurants. There are different styles of tastings that restaurateurs may attend. A **walk-around tasting** is a casual affair, where tables are set up with wine displays and samples are poured for the attendees (Figure 17–2). The attendees can survey the wines available and request a sample of any wine they would like to try. These tastings often include the presence of winemakers or winery representatives who are there to discuss and represent their wines or local purveyors who are familiar with the products. A **sit-down tasting** is usually a more formal training seminar that includes a lecture-style format and involves a guided tasting led by a selected panel of experts. These types of tastings are often very informative and more focused than a simple walk around tasting. These events should not be missed, because of both the professional networking opportunities and the opportunity to taste and evaluate the newest wines on the market before committing to making a purchase for the wine list. By approaching the tasting with a clear plan and an idea of the goals for the event, an operator is able to maximize the outcome of the event.

For example, if a restaurant manager is searching for a new Merlot to serve by the glass, then a simple plan might involve identifying the Merlots that are in the appropriate price point and tasting them to choose one for the list. If the restaurant's wine list is missing some mid-priced Riesling, then the tasting plan should involve trying several appropriate wines in an attempt to choose one for the list.

In any case, during the tasting event it is important to keep the palate fresh and the mind clear. The easiest way to do this is to be sure to spit out the wines as they are tasted. Although the amount of wine being poured is small (1 to 2 ounces), the sheer volume of wine available for tasting during the event could easily result in intoxication. It is important to draw a distinction

FIGURE 17–2 A walk-around tasting event.

between *tasting* wine and *drinking* wine. *Tasting* wine is usually undertaken for purposes of analysis and implies that the participant is evaluating the wine; *drinking* wine is focused on the act of enjoying wine through actually consuming it. Aside from being unprofessional, consuming the alcohol at a tasting event clouds the judgment and will affect the ability to make conscious, informed decisions about the wines. As described in Chapter 4, professional tasters make a habit of spitting regularly to ensure objective analysis and evaluation and to demonstrate professional responsibility. Regardless of the attendee's goals for the tasting event, it is important that participants pace themselves, drink plenty of water, eat any available food, and spit out all the wine tasted. Doing so will ensure that the event is as informative and fun as possible.

Selling Wines by the Glass

For many guests, their introduction to purchasing wine comes from exploring a restaurant's **by the glass (BTG)** offerings. These wines used to be commonly referred to as **house wines.** This term has fallen out of favor in restaurant contexts because it can imply a very basic wine to the guests. Many operations now offer two tiers of wines by the glass—an entry level, value-priced tier, and a second tier of slightly more expensive wines designed to encourage incremental sales. As guests develop their knowledge and interest in wine, they can begin to explore the more expensive wines (reserve tier) available by the glass (Figure 17–3).

Buying a single glass of wine is a manageable investment and allows the guest to experiment with different wines without spending too much time or money doing so. The "risk" associated with it is minimal as well because a guest who does not enjoy a particular wine has only invested in the price of one glass. For the restaurant, selling wines by the glass is the most profitable way to sell wine and it represents the greatest volume of wine that the operation will sell. Many restaurants attribute 60 to 75 percent of their wine sales to by the glass sales. Obviously, certain establishments (such as fine dining) may sell a higher number of *bottles* of wine rather than glasses, but they are an exception. The average profit on wines sold by the glass is relatively

FIGURE 17–3 Most restaurants will offer a selection of wines available by the glass. These have traditionally been called "house wines," although this term is gradually falling out of favor.

House Wines by the Glass

Whites

Columbia Crest Chardonnay	$5.50
Campanile Pinot Grigio	$5.00
Hugel Riesling	$5.75

Reds

Castle Rock Cabernet Sauvignon	$5.50
Mark West Pinot Noir	$5.75
Cline Zinfandel	$5.75

Reserve Wines by the Glass

Whites

Brancott Sauvignon Blanc	$7.00
St. Urbans Hof Riesling	$7.25
Ferari Carano Chardonnay	$8.00

Reds

Kenwood Cabernet Sauvignon	$8.00
Patricia Green Pinot Noir	$8.75
Rosenblum Zinfandel	$8.50

high. Although this can help to offset the lower profits of other items, it also helps to offset the spoilage that can occur with serving wine by the glass (especially for red wines).

A simple selection of wines sold by the glass will ensure that diners are able to find wines that are appealing and affordable, having a minimum of six to eight of these wines will satisfy most guests. The wines offered by the glass are usually typical examples of their variety or style. For example, when most guests order a glass of Chardonnay, they expect one that has a fair amount of oak in the profile. If an operation only offered a crisp, lean, and steely Chablis as their house Chardonnay, this would not meet the expectations of most guests who order the wine. Although there are plenty of fans of Chablis, it is a distinctly different style of Chardonnay than that produced in California, which many Americans tend to expect.

Choosing Wines to Sell by the Glass

The available selection of house wines is the driving force in many restaurants' wine programs and can make a marked difference in both the guest experience and the overall profitability of the program. Careful selection can ensure happy guests and a profitable program. When selecting wines to pour by the glass, there are a few key points to remember: availability, pour size, price point, and style.

AVAILABILITY

One strategy employed by many operations is to carefully choose a house wine and to stick with it for an extended length of time. A wine that is chosen to be a long-term offering by the glass should be produced in sufficient quantities as to be readily available for the given period of time. It would be disappointing to have strong sales of a certain wine by the glass and then to find it no longer available when the manager attempts to reorder. Obviously, only a limited amount of each wine is made each year, but many value wines are produced in increasingly large volumes. Wines chosen for sales by the glass usually fall into this category.

An alternative way to approach selecting house wines is to commit to frequently changing the available selections. This method is better because it keeps the list fluid and the guests' interest in the wine program high. This strategy involves regularly choosing new offerings that meet the requirements of the wine program but still offer the guests variety and value. The popularity of personal computers and printers has made the process of updating and changing wine selections more manageable on a frequent basis. Most restaurants now print their own wine lists in house, and this allows the operator to adjust the selections and update the list as needed. Historically, changing the list was a more difficult process to undertake, and consequently was done far less often. Frequent changes also allow the operator to take advantage of special pricing because a wine that is discounted can be added to the house wine selections. If that wine runs out or increases in price, it is simply replaced by a more appropriate selection, and the list is reprinted. The operator has the opportunity to take advantage of pricing while offering customers the variety and value that they seek.

Pour Size

The standard pour size for wines served by the glass is 5 ounces. With the standard wine bottle containing about 25 ounces, this means that each bottle contains five glasses of wine. If the operation were to offer a smaller, 4-ounce pour, then there are six glasses per bottle and if the restaurant offers a larger, 6-ounce pour then there are four glasses per bottle. The size of the house pour can depend on many factors, but the 5-ounce pour is the most common. If the wine is used for a catering event, it may possibly be poured in a smaller size and be offered at a lower price to the guests. If the operation offers a larger 6-ounce pour then their prices may seem high when compared to other restaurants offering the same wine. Whichever pour size is chosen will affect the pricing of the wines and this should be taken into account when deciding on the appropriate pour. Restaurants that offer smaller, tasting portions or flights of wine (as discussed in Chapter 16) can use this to help determine the cost per ounce of wine, which will then be useful in deciding on appropriate pricing for various pour sizes.

Guests who purchase wine by the glass seek a certain perception of value. If a 5-ounce pour is served in a 15-ounce glass, it does not seem like a value to the guest, even if the amount of wine is the standard pour. Most guests are seeking perceived value, and a glass that is only one-third full does not "feel" like a value. A restaurant must find the balance between their glass size and their pour size in order to present the perception of value to the customer.

Price Point

Wines chosen to be poured by the glass often fall into the value end of the spectrum. Guests who want to purchase a glass of wine generally expect it to be affordable and reasonably priced, and when chosen appropriately, the house wines can offer a good value-to-price ratio for the guest. By choosing appropriately priced bottles, an operation can ensure profitability with sales of these wines (Figure 17–4). A general rule of thumb would be to look to wines that are priced at $10 or less per bottle for sale as a house wine. Bottles in this price range allow the operator to price each glass in the $4 to $8 range, which is where most house wines are priced. Wines that cost more per bottle will need to be priced higher per glass or they may limit potential profitability for the restaurant.

Wine programs typically run a **cost percent** between 28 and 35 percent. The cost percent is the ratio of money spent to money earned for the operation. Having a cost percent of 30 percent means that 30 percent of the money earned from a wine sale was spent on purchasing the wine.

FIGURE 17–4
An example of estimated markups.

Cost per Bottle	Estimated Markup
$8 and under	Three times cost
$8 to $15	Two and a half times cost
$15 to $20	Two times cost
$20 to $40	One and a half times cost
$40 and up	Cost plus $20

It is very common for house wines to have a lower cost percent than reserve bottles, but the overall average usually falls within this range. For a bottle that has a cost of $10, with five glasses per bottle, the cost per glass is $2. If the operation sells each glass of wine for $6, then the cost percent per glass is 33 percent ($2/$6 = .33). If the operation decided to sell that same glass for $8 instead, then their cost percent would be 25 percent ($2/$8 = .25).

Spoilage of wines served by the glass is an issue that needs to be addressed by every restaurant operation. In cases of more expensive or less popular wines, a restaurant may sell only one or two glasses every couple days. This situation can lead to the remaining half bottle being spoiled from oxygen exposure before the operator has a chance to sell it. Profits can quickly erode because of spoiled products. For example, if a restaurant pays $10 for a bottle of wine but only sells one glass for $6 and the rest of the bottle spoils, there is a loss of $4 on the bottle instead of the potential gain from selling five glasses at $6 each. For this reason, proper preservation of open bottles should be a priority (this is addressed in detail in a later section).

Deciding on an appropriate pricing strategy for wines sold by the glass involves determining the operation's ideal cost of goods as well as the general sensitivity of the customers to prices. Although it can be difficult to decide on an appropriate pricing strategy, careful analysis of the customers, the competition, and the restaurant concept can help an operator determine the most effective range of prices. Through careful analysis of sales and periodic adjustments, the restaurant can continue to price its offerings to please the guests and promote the bottom line.

STYLE

Wines offered by the glass are almost always consumed with food from the menu, so the wine should be chosen so that the style and characteristics complement the menu as much as possible. Wines that are high in alcohol or have a heavy oak profile are not very versatile with food and do not make the best selections for house wines. Some people prefer heavily oaked Chardonnay, and this style should be available for these guests, but the house Chardonnay should have a lower level of oak to ensure that it works well with the menu.

Restaurants that focus on a certain style of cuisine should make extra efforts to have the wine list reflect their concept. Italian cuisine tends to work very well with Italian wines, and the regional connection between local food and wine cannot be overemphasized.

What to Have

Depending on the type of restaurant, the clientele, and the location, several styles of wine and varietals should be available as house wines. These are the wines that customers ask for by name and expect to be available. For white wines, these include Chardonnay, Pinot Grigio (Pinot Gris), Riesling, and Sauvignon Blanc (Fumé Blanc). Most guests who are interested in a glass of white wine will be expecting at least these to be offered. For reds, be sure to have Pinot Noir, Cabernet Sauvignon, Merlot, Zinfandel, and Syrah (Shiraz). It is clearly in an operator's best interest to offer good examples of these wines by the glass to please the guests, but the available offerings should not stop there. In addition to these classic varietals, offering wines from unusual grapes or regions exposes the guests to a unique experience and helps to generate interest in the wine program.

It is entirely possible to develop a comprehensive program of house wines that meets the expectations of the average guest and exceeds the expectations of many others. It would seem very unusual to a guest if Chardonnay were not available by the glass, even though you may stock a perfectly acceptable wine that would compare in profile to a Chardonnay. A comprehensive wine program would include the Chardonnay that the guest expects but also provide an appropriate alternative that the guest could also enjoy. It would not be too hard to convince that same guest to try a Pinot Blanc, for example, which offers many of the characteristics that are enjoyed in the Chardonnay but may be unfamiliar to the guest. The challenge faced by industry professionals is to meet the guests' expectations by providing the Chardonnay they seek while also educating them about similar wines that are not Chardonnay, which they may come to enjoy as much as their favorite Chardonnay. By doing so, we can open our guests' eyes to the diverse world of wine, hopefully making a loyal repeat guest in the process.

The wines poured as the restaurant's house wines should be consistent, quality examples of the varietal or blends offered. Despite these wines usually being among the most affordable selections on the list, care should be taken to ensure that acceptable wines are offered. Most restaurant wine sales come from selling wine by the glass, and more guests are exposed to these wines than any other wine on the list. If the operator chooses to pour a wine simply because it is cheap, chances are good that the guest will not enjoy it and won't purchase any more in the future. Instead of increasing profits by selling cheap wine at high markups, the operation will lose sales due to guests not wanting to purchase the wines. The real challenge in creating a profitable wine list is to select quality wines at affordable prices that offer a great price-to-quality relationship for both the operator *and* the customer.

Responding to Industry Trends

American taste preferences change from year to year and even from month to month. This greatly affects the products that Americans are interested in buying and should directly affect the wines that a restaurant offers to the guests. Staying current with consumer trends will help to ensure that the restaurant is offering what people seeking. As buying patterns change, the savvy operator is monitoring industry trends and regularly adapting the list, phasing out those wines that are becoming less popular and adding those wines that are gaining in popularity.

A clear example of changing tastes would be the popularity of California Merlot during the mid to late 1980s. The demand and interest in these wines (and Merlots in general) experienced a huge surge in popularity during this time period and restaurant wine lists reflected this demand. Then in 2004, along came the movie *Sideways,* where Pinot Noir was touted and Merlot was knocked, and consumer purchasing trends swung like a pendulum. In the months following the release of *Sideways,* sales of Pinot Noir jumped 15 percent.

The fickle nature of the American consumer can seem daunting to restaurant operators, but it need not be. Simply because a mainstream movie bashes Merlot and lauds Pinot Noir, everyone does not stop drinking Merlot altogether. But a wine list that does not contain many Pinot Noirs in the wake of a movie like *Sideways* is certainly missing an opportunity to appeal to those guests who want to try these wines. A simple adjustment would be to add a couple more Pinot labels and perhaps reduce the size of the Merlot section according to guest preferences. Both wines have a place on any well-rounded wine list, but consumer demand can fluctuate significantly.

Preserving Open Bottles of Wine

Because the heart and soul of any restaurant wine program is the by the glass program, a critical factor to consider is preserving the opened bottles of wine. Once a bottle of wine is opened, it begins a steady, often rapid decline due to the action of oxygen (oxidation), which can cause the wine to develop a stale or Sherry-like character. Because restaurants have to open a whole bottle of wine to sell a single glass, some system must be in place to ensure that the remaining portion of wine in the bottle remains in good quality until the next glass is sold. In some cases, the whole bottle is sold relatively quickly, and this is not of major concern. Wines sold less frequently, however, are likely to deteriorate before the entire bottle is sold.

Despite the relatively good markup on wines sold by the glass, wasting the unsold portions of the bottle after selling a glass will quickly erode a restaurant's profit margins. The full revenue from selling wines by the glass can only be realized if an operation sells the entire bottle. If half the bottle is wasted due to spoilage, the product cost percent doubles. The operator can ensure that the entire bottle of wine remains saleable by employing some form of preservation for the open bottles of house wine. Some of the most common systems of preservation will be discussed in this section.

INERT GAS

Perhaps the simplest and most affordable method of preserving open bottles of wine is using one of the canisters of **inert gas** available on the market. Inert gas (or nonreactive gas) is a gas that does not interact with the substances that it contacts. These canisters are relatively affordable, often costing less than $10 and capable of preserving up to several dozen bottles of wine, depending on the canister size and the amount of wine left in the bottle. These canisters usually contain a blend of inert gases (such as nitrogen and carbon dioxide) which form a "blanket" over the wine and displace the oxygen in the bottle. The layer of gas prevents the oxygen from directly contacting the wine, and helps to protect the wine from the oxidation that would otherwise occur. When used properly, these gas canisters are capable of preserving opened bottles of wine for several days (Figure 17–5). When using this technique, operators should taste their opened wines on a regular basis to ensure that the wines are holding up sufficiently before they are sold to guests. Serving oxidized wines is a sure way to stop a successful wine program in its tracks.

FIGURE 17–5 Canisters of inert gas can be used to preserve open bottles of wine and help prevent oxidation. To use, simply insert the plastic tube into the bottle and spray several short bursts inside. Quickly re-insert the cork and the wine is then protected by a blanket of inert gas and will be less susceptible to damage from oxygen.

The key benefits of this system are obvious: The gas is inexpensive to purchase and easy to use with very little training or time invested. Additionally, it is more effective than using no system at all and will help to protect the opened wine at least until the next day's service period. For house wines that sell frequently, this is usually all that is needed with the rapid level of turnover. These wines are sold often enough that the entire contents of the bottle have been sold before the gas may become ineffective. For these higher volume wines, the gas system works fine. It is an easy and affordable way for an operation to preserve their open wines.

Cabinet Systems

Establishments that are serious about their wine programs may choose to invest in some form of cabinet system for preserving their open bottles of wine. Cabinet systems consist of an inert gas system coupled with a dispenser spigot and tubing to pour the wine. The cabinet and tubing provide a pressurized blanket of gas in the opened bottles and ensure minimal chance of oxidation to the open wines. These systems have advantages and drawbacks that need to be considered before an establishment commits to purchasing one. Depending on the size, the initial investment can be quite pricey, but these systems can help ensure longer term preservation and are therefore more useful with wines served by the glass that don't sell as frequently. Restaurants that invest in these systems are able to open and offer some more expensive bottles by the glass, without having to worry about the shelf life. This can further expand the options for guests by allowing the restaurant to provide some premium wines—usually only available by the bottle—as a glass offering.

There are several different producers of these cabinet style systems, most notably Cruvinet. The clear and obvious advantage to these systems is the high profile that they can play in an operation. A large, polished cabinet system displayed in the dining room or bar area sends a clear message to the guests: we are serious about our wine program. The cabinet functions as a centerpiece for the area and also acts as a sales tool: it intrigues the guests, makes them take notice, and hopefully, further entices them to purchase wine. The cabinet systems also give the operator peace of mind because these systems can preserve an opened bottle of wine for several weeks. If an operation does not sell the entire bottle of wine within that time frame, that bottle should probably not be offered by the glass.

The drawbacks to these systems are less obvious, although they should be considered before making the investment to purchase one. They tend to be quite costly depending on the model, features, and the size of the unit. Even the basic models can cost several thousand dollars, and the largest ones can run into the tens of thousands of dollars. A smaller, basic cabinet will cost around a thousand dollars, depending on the maker. Aside from the initial investment, these systems can have additional costs tied to upkeep. Because they rely on the presence of an inert gas to form a preservative blanket and prevent oxygen from contacting the wine, the tank used to supply this gas must be refilled periodically, varying with the volume of wine being preserved and the size of the system. The cost of tank refills should not be overlooked; in some cases it can be close to $100 per tank, depending on the size. The tank may last a month or more, so although it is not a daily expense, over the course of a year the costs can add up.

Another concern is that in some models, a small amount of wine is trapped in the tubing between the preservative gas and the serving spigot. This wine is not covered by the blanket of gas and is therefore prone to oxidation. This can result in having to "bleed," or pour off, this wine from the line before reaching the preserved wine in the bottle. The amount of wine poured off may not be much, but if the wine being stored in the cabinet cost $100 per bottle, the value of an ounce is about $4. Bleeding the line may be the equivalent of pouring $4 down the sink in order to preserve the remaining $96 worth of wine. Although some wine programs can usually absorb this expense with little trouble, every ounce of wine wasted can affect overall long-term beverage costs. One less-noticed drawback to using this system is that as the number of wines offered by the glass increases, there may not be space in the cabinet system because a limited number of spigots are available. Therefore, many operators may choose to preserve their higher-priced wines in the cabinet system and use a simple gas or vacuum system for all of their house pours. As the costs, benefits, and drawbacks are considered for this system, it becomes more obvious why they are only common in operations of a certain size and scale. The benefits can outweigh the drawbacks in certain situations, but each operation must examine its own goals and budgets and decide on which method will work best.

VACUUM SYSTEMS

Various vacuum systems are available, including a simple version for the home and industrial models useful for restaurant settings. A vacuum system consists of a simple rubber stopper with sealed holes in the top and a separate pump which is held on the top of the stopper and pumped several times to remove air from the bottle. The pump works by removing most of the air in the bottle and creating a vacuum seal which minimizes both the presence of oxygen and its destructive effects on the wine. With some of the smaller home systems, the pump removes most of the air from the bottle. With these simple handheld pumps, it is virtually impossible to remove *all* of the oxygen, so these systems are still used primarily for short-term storage in the home.

Commercial versions of these vacuum pump systems are available, and they are more efficient at removing more oxygen from the bottle. One such system consists of a wall-mounted electric pump and a set of rubber stoppers. A stopper is placed into the top of an opened bottle of wine, and the bottle is held up to the electric pump. In a matter of seconds, the pump removes the air from the bottle and creates a vacuum seal capable of preserving the wine. This system is simple and quick to use. The initial investment can still be significant, with some systems costing several thousand dollars to purchase, although the upkeep costs really only include the cost of replacing the rubber stoppers. It is easy to add additional wines by the glass, as well, because the operator simply has to have extra stoppers to seal the bottles. This system gives the management the most flexibility with changing wines served by the glass; changes can be made quickly and efficiently based on product availability and price changes.

If a restaurant operator is serious about recognizing the profits that can come from its wine program then the operator should be sure to employ *some* form of preservation system for open bottles of wine. Over the long term, the benefits and potential profits far outweigh the initial investment in a preservation system. Restaurateurs who elect to use no preservation system are risking dumping a portion of their profits down the drain in the form of spoiled wine.

Storage and Inventory Levels

When deciding on how many labels and bottles are appropriate for the operation, one of the key things to consider is how much room is available to commit to storing wine? If the answer is very little, then creating a large inventory of wine is not a good idea. If, on the other hand, the operation has reserved a fair amount of space to be used for wine storage, the wine inventory can grow accordingly. Wines that are poured by the glass generally are sold and replaced frequently, so a large amount of storage is less of an issue. Many reserve wines sold by the bottle will remain in inventory for a considerable length of time, and these wines will demand sufficient storage space.

Within the room for wine storage, some type of racking is necessary (Figure 17–6). Although this does not need to be expensive, it is critical to allow easy access to the wines as they are ordered by the guests. The racking often corresponds to bin numbers, which are an easy reference point for

FIGURE 17–6 A clean, well-organized system of racking is essential to maintaining inventory in a restaurant wine program. Various racking systems exist, but a key requirement is for service staff to be able to find bottles quickly and easily.

locating specific wines. As a guest orders a certain wine, it can be found in the numbered bin. Neck tags that can be labeled with the producer and vintage are also readily available and reasonably priced, and they allow service staff to select the appropriate wines quickly and accurately.

A safety factor with wine storage that must be considered is securing the wine inventory away from employee theft and misuse. Theft can occur in many situations, whether from an unscrupulous bartender deliberately misstating inventory or from wayward servers taking a bottle as an after-shift perk. Misuse can occur when the new line cook grabs a bottle of a '96 Chablis Grand Cru instead of the box of Chablis that is used for cooking in the kitchen. Although it may sound far-fetched, stranger things have happened. Any wine storage should be able to be secured with a lock, and access to the wine should be limited to management and bar staff.

Pricing the Wine List

Perhaps no other aspect of restaurant wine sales is as challenging to management as deciding on an appropriate pricing strategy. There are many different pricing strategies and viewpoints on what is considered an appropriate **markup** on wine. The markup is the difference between the bottle cost and the selling price. A wine that costs $10 and is sold for $30 has a markup of $20. This wine also has a cost percent of 33 percent ($10/$30 = .33). The factors which affect the markup and cost percent include the restaurant concept and service style, the menu prices for food, the selections offered on the list, and the prices charged by the competition. After analyzing these factors, the management will decide on an appropriate markup strategy and employ it to price the wine list.

On one side of the pricing debate are the proponents of high markups who justify their exorbitant pricing by citing the extremely low profits achieved by most restaurant operations. They also cite the difficulties with purchasing and storing many bottles of fine wine, many of which will be held in storage as "overhead" for months (or years) before they are eventually purchased by a guest. Add to this the hidden costs involved with the service and sale of products to guests and it is tempting to side with the proponents of high markups. These operators may charge two or three times the price of a bottle when it is sold in their restaurant. A bottle that cost $20 to buy may be sold for $60 or more on the list. This practice is sure to upset the vast majority of guests, many of whom are very aware of the retail prices of the wines on the list. If a guest knows that they can buy the same bottle of wine for $20 in the store, they will be unlikely to want to purchase it for $60 in the restaurant, no matter how good the food or service. This high markup mars the image of the restaurant and makes the guest feel like he is being taken advantage of by high prices. It also ends up producing less revenue in the long run as guests decide to purchase less wine because of the high prices.

The other pricing perspective focuses on maintaining fair pricing in an attempt to deliver value to the guests and increase sales with appealing prices. This method of pricing seeks to institute less of a markup in favor of moving a higher volume of wine. This approach takes less profit from an individual bottle but seeks to appeal to a greater number of guests with better prices. It is a classic example of higher volume with lower prices: If more people are able to afford a bottle of wine in the restaurant, more people will buy bottles of wine. Even though the operator may get less money for each bottle sold, over the long term the restaurant will sell more wine and therefore end up with higher wine sales and consequently end up with more loyal customers. Many operations do the same thing with food coupons and discounts: Guests may save 10 percent off of their meal, but if 10 percent more people come in to eat because of the promotion, then the restaurant ultimately earns more revenue.

In the modern era of wine, technology and a keen focus on quality have enabled producers to create better wines at better prices than at any point in history. Add to this the global wine surplus due to increasing production, and access to quality wine is at an all-time high. This has allowed producers to flood the market with value brands that deliver higher quality, even at the lower price points. Consumers can now select from a wide range of bottles priced in the $6 to $10 category, and they are quickly coming to expect a similar level of value in the restaurant setting. The good news is that restaurant operators have access to the same value wines that consumers have, and this can translate to a wine list with a wide range of wines, offering quality and value at *every* price point. It would be far better for a restaurant to set themselves apart by having fair, even low, prices on wine than to be known for having high prices. The guests will be thankful and respond by buying more wine which will, in turn, lead to increased sales overall. In this situation, everybody wins.

Sliding Scale Pricing

A **sliding scale** method of pricing takes advantage of the strengths of high markups and the benefits of lower pricing. A sliding scale is a pricing system that utilizes a variety of markups to balance profit for the operation with value for the consumer. In this situation, a restaurant might employ the highest markups on the lowest priced wines and assign a lower markup on higher priced wines. The target beverage cost percent can fluctuate in relation to the product price, so that an average of 30 percent could be attained by selective pricing. For example, if a restaurant sold one wine with a 25 percent cost, a second wine with a 30 percent cost and a third wine with a 35 percent cost, the average of these three is 30 percent. Despite the fact that only one of the three products actually has a 30 percent cost, the mix of cost percents ensures that they average out to the target percent.

Many operators have begun to focus on lowering their markups on higher priced wines in a distinct effort to increase sales of those wines. This increases sales of wine, and also encourages guests to invest in more expensive bottles. If an operation can offer relative value to the customer, even at the highest price points, it will help the guests feel that they are being treated fairly. Besides helping to increase wine sales overall, this will result in the operation making more money in the long run through sales of higher priced wines.

In the case of the wine list this same strategy could be employed on a larger scale. A restaurant operator may assign a target cost percent of 35 percent to the wine program. The operator would then assign the lowest tier of wines the highest markup, the middle tier an average markup, and the highest priced wines the lowest markup. The wines that cost between $6 and $15 per bottle may have a target cost percent of 25 percent. Wines that cost between $16 and $25 may have a target cost percent of 35 percent, and those wines that cost above $26 may have a target cost percent of 45 percent. The average of these three percentages is 35 percent. This method sets a fair price for both the operator and the guests. It also helps to generate higher-end bottle sales because the highest cost wines are marked up the least. To guests who are interested in buying these wines, they will seem like relative values in relation to other restaurants that don't employ a sliding scale. The reality with this system is that since more wines are likely to be sold in the lower to middle price points, the overall cost percent for the program could easily come in below the target of 35 percent.

For wines at the highest end of the price spectrum, an operator may simply assign a **fixed markup** to the cost of the bottle, instead of using the cost percent method. The fixed markup method assigns a set amount of money to the cost of the bottle to determine the sales price. For example, for wines that cost $100 or more, the additional markup may be fixed at $30. This allows the operator to offer reasonably priced wines to the consumer and ensures that the operator is receiving a fair markup on its wines. This system is occasionally employed at all price points on wine lists, but is more common at the higher price points.

Incentives

Just as a savvy operator will work to drive wine sales through the marketing of the wine program, restaurateurs who wish to increase the number of bottles of wine sold will create incentives to encourage guests to purchase bottles. One such incentive might be to offer 10 or 15 percent off of all bottles of wine from the list on a given day of the week. Many restaurants now take this to another level, and offer half-priced bottles from the list on certain nights of the week. The operator is willing to drastically reduce their profits from selling the bottle in order to get people in the door, and they hope to build customer loyalty every night of the week. These promotions can be done on slower days of the week, thereby increasing both the number of guests during these slower periods and wine sales overall. A promotion can even be tied to special menus, such as when running a special Italian dish, offering 10 percent off of Italian bottles of wine. This type of selective marketing can be very effective at increasing both wine sales and guest traffic.

Filling Out the Wine List

Once the storage space is identified and the goals of the wine program have been decided, it is time to select the labels for the list. The key to selecting appropriate wines is to choose those that complement both the style of food offered and the expectations of the clientele that will be purchasing the wine. Having dozens of great French wines does very little for guests choosing wine in an Italian restaurant. Similarly, having many high priced wines in a popularly priced restaurant might not be the best way to stock a wine list.

One common mistake is to choose and price wines out of the range of affordability of the customers. If the check average per person is about $15, then having most selections priced above $50 will not encourage guests to purchase bottles of wine. By tailoring pricing to the clientele and service level, a wine list can be priced affordably and will be more effective at generating sales. A good rule of thumb is to price the bulk of your wines near the average check for a party of two. This price bracket represents the "sweet spot" for wine sales. The sweet spot is the price range where the majority of the guests will find relative value and where the bulk of wines should be priced. The sweet spot fluctuates widely based on the style of restaurant, the menu offerings, and the purchasing patterns of the guests, but it should be a focal point in assembling the wine list.

For example, if the check average for a party of two is around $30, then pricing the bulk of the wine list around that price point will help customers feel like they are getting value. Knowing that most guests will not purchase the cheapest bottles on the list and that the most expensive ones will be purchased in even smaller quantities allows the operator to put the bulk of the wine offerings into the most popular price brackets. The more expensive bottles make the cheaper ones look like relative values. Although there is no specific formula that can be followed for every operation, having a good idea of the clientele and their spending habits will help ensure proper pricing for the list. If the restaurant is a startup operation, it will not have the luxury of knowing the customers' spending habits. In this scenario, trial and error will help to find the best pricing methods. Regularly evaluating wine sales will help the operator to find the right balance of price for the guest and profit for the operation.

To help understand this, imagine an example where the restaurant list will be set up with 100 selections (or labels) available. For an easy example, assume that 50 of these selections are red and 50 are white. Fifty of these wines would be in the sweet spot for the list, 40 would be just outside the sweet spot, and 10 selections would be in the extreme ends of the pricing. If the sweet spot for the list is in the $40 to $60 price range, then this would allow the operator to select 25 white and 25 red wines priced in this range. The operator would then source 20 white and 20 red wines that are outside of this range—perhaps 10 labels that are above $60 and 10 labels that are in the $25 to $40 range. The operator would then fill in the two extreme ends of the list by choosing 5 wines that are priced in the highest bracket and 5 wines that are at the value end of the spectrum.

An operator using this formula can easily choose wines that are appropriate for the setting and are priced to sell. This ensures that the bulk of the available wines are priced to appeal to diners, and it also allows a wide enough range of pricing to allow the list to appeal to diners at both ends of the spectrum. As the operator assembles the wine list and analyzes the spending habits of the clientele, selections should be adjusted accordingly, tailoring the list to meet guest demand and to maintain proper profit levels. Using this system, the restaurant is able to assemble a complete wine list that covers the basic price points and appeals to a range of diners.

Bottle Formats

One way to offer options to the guests and allow the operator to fill out the list is to source wines in various bottle formats. Many operations now offer a separate section that includes several half bottles. This appeals to some guests, both because half bottles are usually a more affordable option and also because it allows the guests to experiment with two different wines instead of a full bottle of just one wine. Half bottles are increasingly popular in the restaurant setting and would be a good consideration on any restaurant wine list.

Large format bottles should also be considered, although perhaps more selectively. These large bottles of wine are great for larger parties and certainly command attention when served tableside and displayed on the table. They reinforce the special nature of the guests' meal and allow a party to enjoy the wine of their choice without having to order several individual bottles of the wine. Unlike half bottles, however, they tend to sell infrequently, and they should assume a supporting role on the wine list. Inclusion of both these formats offers the customers a wide variety of choices and ensures that the wine list is well rounded and appealing to all guests.

Organizing the List

Even though they may not always ask, most guests would like some guidance in selecting wine for their meal. Staff members can offer this guidance, but the structure and layout of the list can also assist guests in finding the wine that they are looking for. The way in which a wine list is structured reflects the goals of the operation and directly affects the manner in which guests find wines that they are interested in purchasing. There are many different ways to structure and lay out a wine list, depending on the style of restaurant and the objectives of the list itself. Some of the most common methods for wine list layout will be discussed in this section. Although there is no *best* method to use, the layout should be carefully thought through to ensure it represents the goals of the operator and meets the demands of the customers. In most cases, the wine list layout will include several different aspects of these styles as the operator desires.

By Region of Origin

Wine lists that are organized strictly by region would include a separate section for each of the major world wine regions represented on the list (Figure 17–7). This type of list would include a section for French wines, another section for Italian wines, and so forth. Each section would be

Wine List Organized by Region

	France	**Prices**
Whites		
Burgundy	1999 Louis Jadot *Meursault*	$73
Bordeaux	2001 Lynch Bages *Pauillac*	$68
Alsace	2002 Trimbach *Riesling*	$74
Reds		
Bordeaux	1999 Châteaux Gloria	$48
Rhône	1998 Domaine Janasse *Châteauneuf-du-Pape*	$62
Burgundy	2002 Louis Latour *Gevrey Chambertain*	$152
Provence	1998 Domaine Tempier *Bandol*	$52
	Italy	
Whites		
Fruili	2004 Campanile *Pinot Grigio*	$25
Veneto	2002 Anselmi *Soave*	$30
Reds		
Tuscany	2001 Rocca Delle Macie *Chianti Classico*	$32
Piedmont	1999 Stefano Farina *Barolo*	$53
Veneto	1997 Tommasi *Amarone*	$105
	Australia	
Whites		
Margaret River	2001 Cape Mentelle *Chardonnay*	$47
Clare Valley	2002 Tim Adams *Riesling*	$36
Reds		
Barossa Valley	2003 Thorn-Clarke Shotfire Ridge *Shiraz*	$35
Coonawarra	2000 Wynn's *Cabernet Sauvignon*	$27
Margaret River	2001 Cape Mentelle *Merlot/Shiraz* Blend	$42

FIGURE 17–7 This wine list is organized according to the wine's country of origin. This method allows guests to choose wines based on their favorite regions.

further divided, such as Bordeaux, Burgundy, Rhône Valley, and Champagne for France. The benefits to this type of layout include the ability of guests to quickly and conveniently find wines from regions that they prefer. For example, a guest who is interested in having a French wine with dinner will know exactly where to look and will not need to waste time browsing through many selections that he or she is not interested in purchasing. It is easy enough to lay out a list by region and place appropriate wines in each category. One of the drawbacks is that this organization might limit the focus of the guest; that same guest who had intended to have a French wine might be equally happy if he were helped to select a California wine that used French grape varieties, for example. If the guest browsed only the French section of the list, he or she may miss out on an opportunity to try something a little different and to positively affect the dining experience.

By Varietal

Most wine consumers can name their favorite wines and they often choose their wines by the varietal used to make the wine (Figure 17–8). For these guests, the easiest way to structure a wine list is to organize by the main grape used in the wine. This format would have a section for each of the major varietals, including a Chardonnay section, a Sauvignon Blanc section, and so on. This allows guests who have a favorite grape to easily find wines that fit their interest. Under this format, those wines from the New World and the Old World would be listed together. For example, Chardonnay from Chablis, France, would be in the same category as Chardonnay from

FIGURE 17–8 Because most guests can identify their favorite varietals, many operators will organize their wine list by the varietal used. This helps guests to find their favorite varietal wines easily, but groups various countries' wines together in one category (French Chardonnay is in the same category as California Chardonnay).

Wine List Organized by Varietal

	White Wines	
Chardonnay		**Prices**
Burgundy, France	Louis Jadot *Meursault*	$75
Chablis, France	La Chabliesienne *Petit Chablis*	$32
Napa, California	Ferari Carano	$68
Riesling		
Mosel, Germany	St. Urbans Hof QbA	$29
Finger Lakes, New York	Red Newt Cellars	$45
McLaren Vale, Australia	Peter Lehmann	$36
	Red Wines	
Pinot Noir		
Burgundy, France	Domaine de Courcel *Pommard*	$82
Oregon	King Estate	$55
Tasmania, Australia	Ninth Island	$62
Cabernet Sauvignon		
Bordeaux, France	Château Gloria	$75
Sonoma, California	Kenwood Vineyards *Jack London*	$42
McLaren Vale, Australia	D'Arenberg *High Trellis*	$36
Zinfandel		
Paso Robles, California	Ridge	$58
Sonoma, California	Spencer Roloson	$44
Napa, California	Coppola Edizione Pennino	$70

California. Guests looking for these wines would have all of the offerings listed together and would not have to look very far to find a wine to their liking.

One of the challenges with this layout comes when a wine is not a single varietal, but a blend of two or more grapes. For example, Rhône wines can sometimes contain a dozen different grapes. Wines such as this would not fit neatly into any single varietal category and may be included in a section titled "blends." Another challenge with this layout is that wines that are from the same grape but distinctly different in style will fall into the same category.

By Style

Another common organizational method is by style—with a section titled "Light-Bodied, Crisp Whites," one titled "Medium-Bodied Whites," one titled "Rich, Full-Bodied Whites," and the same for reds (Figure 17–9). The benefits of this method of labeling are clear: guests can quickly and easily select wines based on their stylistic preferences. This also helps expose customers to wines that they may not be familiar with. For example, a guest looking for an Italian Pinot Grigio in the "Light-Bodied, Crisp Whites" section might stumble across a New Zealand Sauvignon Blanc that she wasn't familiar with. As both wines are in the selected category, the guest may be encouraged to try a new wine and fall in love with it. If these same wines had been listed solely by region of origin the guest may never have encountered the wine from New Zealand.

Wine List Organized by Style

White Wines		
Light-Bodied, Crisp Whites		**Prices**
Bordeaux, France	Château Haut Rian *Sauvignon Blanc*	$26
Marlborough, New Zealand	Brancott *Sauvignon Blanc*	$40
Friuli, Italy	Anselmi *Pinot Grigio*	$22
Medium-Bodied Whites		
Willamette Valley, Oregon	King Estate *Pinot Gris*	$30
Clare Valley, Australia	Peter Lehmann *Semillion*	$38
Napa Valley, California	Ferari Carano *Fumé Blanc*	$50
Full-Bodied, Rich Whites		
Sonoma, California	Sonoma Cutrer *Chardonnay*	$36
Mosel, Germany	JJ Prum *Riesling* Auslese	$74
Wachau, Austria	Hiedler *Gruner Veltliner*	$55
Red Wines		
Light-Bodied, Fruity Reds		
Burgundy, France	Louis Jadot Beaujolais *Gamay*	$25
Willamette Valley, Oregon	Adelsheim *Pinot Noir*	$48
Veneto, Italy	Bolla *Bardolino*	$20
Medium-Bodied Reds		
Rioja, Spain	Marques de Caceres *Tempranillo*	$30
Chianti, Italy	Rufina *Chianti Classico*	$52
McLaren Vale, Australia	d'Arenberg "Custodian" *Grenache*	$37
Full-Bodied, Rich Reds		
Napa Valley, California	Cakebread *Cabernet Sauvignon*	$84
Ribera del Duero, Spain	Casa L'Ermita *Tempranillo*	$38
Rhône Valley, France	Guigal *Châteauneuf-du-Pape*	$65

FIGURE 17–9 By grouping various wines according to their style and body, the guests can choose a wine that suits their preferences. This list groups together "light, crisp white wines" from various countries.

By Price Point

Some wine lists are organized according to price point, with the lowest priced wines being listed first and increasingly more expensive wines down the list. This method allows guests to easily access wines that they can afford, but it also encourages them to select their wines based on price. It also discourages them from browsing the entire list because the natural tendency is to identify the upper limit of the budget and to select a bottle from that point or below. This may seem like a minor issue, but if guests are made aware of the full breadth of selections on the list regardless of price, it may encourage them to return for the more expensive bottle in the future. If they never look beyond wines that fit neatly into their price bracket, they will never know what else is there to be enjoyed. If guests are encouraged to explore the list in an effort to find a wine that suits their value perspective, then they may find a special selection that they hadn't quite expected.

To Describe or Not to Describe?

One of the key factors when deciding on the layout and design of a wine list is whether or not to include tasting descriptions on the list. Obviously, if space is an issue with the list, then descriptors may not be an option. But if space is less of an issue, then the operator may choose to have some tasting notes or wine descriptors for each wine in order to help guide guests in their selection. Some people feel very strongly about whether or not to have wine descriptors on the list, but this is partly a decision based on the style of operation as well as the type of clientele that frequents the operation. For example, if the majority of the guests at the operation are new to the world of wine, then descriptors may be very effective at helping them make an appropriate selection. If the majority of the clientele are wine aficionados, however, then a large number of descriptions may not be necessary.

If an operator elects to have descriptors on the menu, several things should be considered as the descriptions are written. The descriptors should give a basic overview of the style of wine and a general commentary on its aroma and flavor profile. Descriptions should not be too elaborate or creative because this may not help the guest in making a decision, but rather deter or confuse them. Similarly, if a wine is described as "crisp and flinty with hints of melon and spice," a guest who does not know what is meant by "flinty" may decide that this wine is not for them despite the fact that they may have fully enjoyed the wine and not even recognized a hint of flint. The basic goal of a descriptor on a wine list should be to guide guests in making a selection that they will enjoy and not to completely summarize the profile of the wine.

Staff Favorites Section

One way to market wines on the wine list is to include a section that lists staff favorites or house favorites. This section should include wines priced in the sweet spot of the list and those wines that are excellent examples of their style or varietal. These could be wines chosen by the staff during tastings or those that offer exceptional price-to-quality ratios. Drawing attention to these wines in a special section makes them stand out and draws the guests' attention to them. And if the staff is involved in helping to select the wines for this section, they take some ownership of these wines and will be encouraged to sell them. This section will also help those guests who aren't interested in looking at the whole list, but simply want to choose from a limited number of selections.

Perhaps the best method for laying out the wine list employs a combination of these methods, with the addition of some descriptions designed to guide the guests in selecting their wine. An excellent, comprehensive layout could include a regional heading with varietal subheadings. In the white wine section of the list, this may include a heading for France, with subheadings for the main varietals, including Chardonnay, Sauvignon Blanc, Muscadet, and so forth. This allows the guests to focus on both the region of origin as well as the primary grapes used, and helps to guide those guests who desire Chardonnay but don't know that white Burgundy wines are made from Chardonnay. Accompanying the Burgundy listings could be a description to help inform the guest of the style of Chardonnay and to give key profile descriptors.

List Formats

Once the wines are selected and the pricing strategy decided, the actual format should be chosen. The most common method of presenting the wine list to the customer is in some form of booklet or menu holder. Menu holders often contain clear plastic sheets that display the typed list. The wine list is typed on sheets of paper and inserted into the menu presenter. This method has some key benefits—namely the ability to change and update the list at a moment's notice. Because the inventory of wine can change frequently, it is critical to be able to update the list regularly. Nothing is worse than a guest requesting a special bottle of wine only to be informed that it is out of stock. When the listings are kept current, what the guests see is what they can get.

Some restaurants present their wine list in a formal, bound-book style format. These lists have a nice look and feel to them, but the ability to change and update selections can be somewhat limited. There is even one restaurant in Washington, D.C., that has eliminated the actual list altogether; they simply display the bottles on a stand in the dining room and guests are encouraged to browse the table, looking at the labels and selecting a wine simply by seeing the bottles. Regardless of the manner in which you present your wine list to the guest, it should reflect the goals of the operation and the desires of the guest. Being creative allows the guests to feel that the experience is special and allows the operator to respond to changing labels and inventory levels.

Summary

Choosing wines for the list and deciding on appropriate pricing can be a daunting task. By starting with a good idea of the spending habits of the guests, the competition's prices, and the goals for the operation, management can create a complete and diverse list that offers potential profit for the restaurant and perceived value for the customers. There is no specific formula for creating a list; it can be highly individualized based on the specific operation. If management sets goals for the wine program before starting, they can make their decisions based on their desired outcomes. Regardless of the format and pricing, guests should feel that they are being offered value, while the operator focuses on selling more wine.

Review Questions

1. Describe the three-tiered system of distribution. How does this affect restaurant operators when creating a wine list?
2. List some of the key factors to consider before creating a restaurant wine list.
3. Identify several pricing methods for wines sold by the bottle. How is pricing by the bottle different from pricing wines sold by the glass?
4. What is meant by "house wines"? What are the benefits to offering house wines?
5. How would a restaurateur decide which wines to put on a wine list and which wines might *not* be good on the list?
6. How could an operator preserve open bottles of wine? What are the benefits and drawbacks to the various systems?

CHAPTER 18

Buying and Cellaring Wines

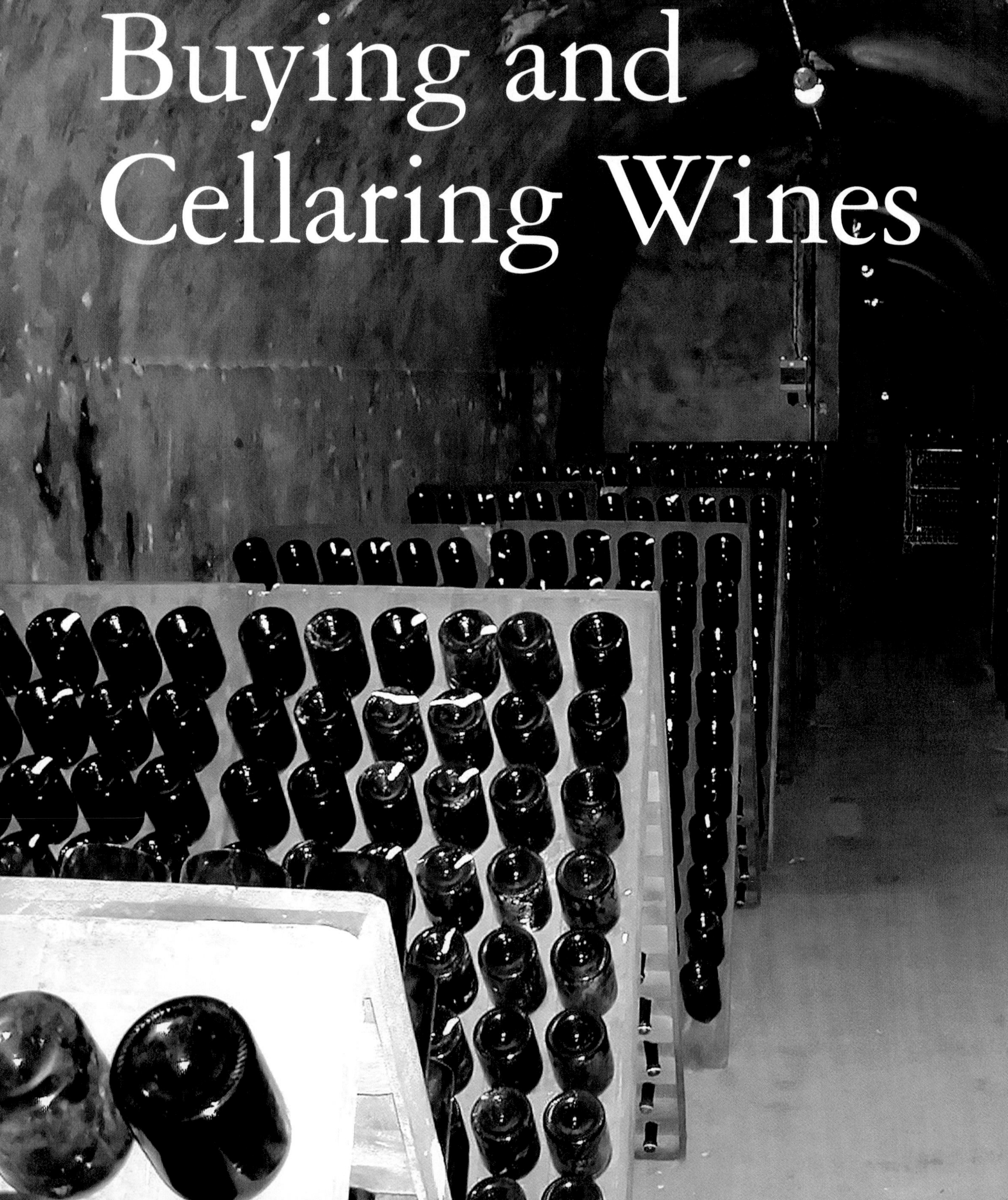

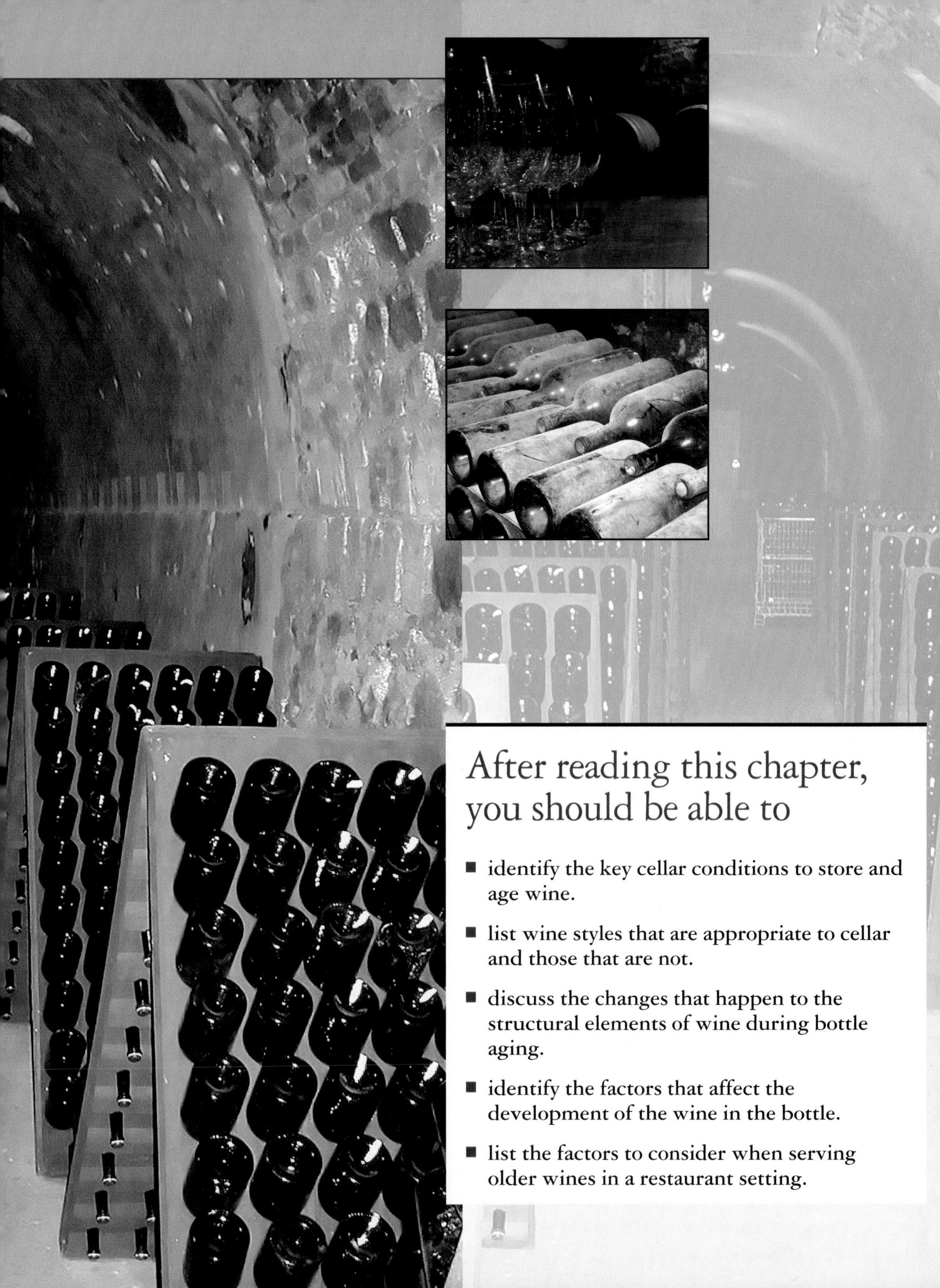

After reading this chapter, you should be able to

- identify the key cellar conditions to store and age wine.
- list wine styles that are appropriate to cellar and those that are not.
- discuss the changes that happen to the structural elements of wine during bottle aging.
- identify the factors that affect the development of the wine in the bottle.
- list the factors to consider when serving older wines in a restaurant setting.

KEY TERMS

- anthocyanins
- bin number
- bottle bouquet
- esters
- futures
- passive cellar
- sediment
- ullage

Most Americans do not purchase wine with the intention of aging it for months or even years; they simply purchase a bottle to have with dinner that night or later in the week. Despite that, a small percentage of the population does purchase wine to age, sometimes for a period of decades. This chapter will examine the reasons and methods behind aging wine, focusing on the factors that will influence which wines are capable of aging, as well as how this issue can be addressed in the restaurant setting. Having a wine cellar can be a fun, rewarding, and educational experience for the consumer, and for restaurants, having access to older wine can be a special service for the guests. Cellars don't need to be huge or very complicated as long as some basic storage criteria are met, and the rewards of owning one far outweigh the effort of creating an ideal situation for storage.

The main reason to age wine is to allow it to improve in the cellar area. As it ages, the wine undergoes a series of small changes that accumulate to have a large impact on the wine. These changes impact the wine's sensory profile, specifically the aromas and the palate impressions from the wine. The wine gains complexity in aromatics and the impression on the palate softens, creating a more enjoyable drinking experience. Some wines can start out with little to offer and gain incredible complexity with several years of bottle age.

Those wines that will truly benefit from long-term aging are a small minority of the wine world. The vast majority of wines being produced are enjoyable upon release and designed for immediate consumption. Winemakers work to produce these wines by minimizing tannin extraction during winemaking, producing softer, rounder, and more fruit-driven wines. Although some wines of this style can age, the majority are drinkable from a young age, and most will not benefit from extended aging.

A fair amount of guesswork comes into play when considering how a given wine will age. Winemakers and industry professionals have experience with many wines and their performance in aging, but certain factors can greatly impact the rate and the way in which any given wine develops. It is impossible to know *exactly* how a wine will age when considering how complex the individual chemical makeup of any given wine is, but generalizations can be helpful in determining an appropriate aging window.

If a wine is to age appropriately and gain complexity as it does so, it must start life with a balanced profile. Aging a wine that lacks balance will simply result in an older, unbalanced wine. If the wine has too much alcohol or too much oak when it is young, these elements will not become better integrated with bottle age. Ideally the wine will balance its acid levels, tannin, alcohol, fruit qualities, and its general chemical makeup to age appropriately.

It is important to remember that very few of the world's wines *need* to age to be pleasurable. The true collectors who assemble vast cellars of age-worthy wine are a small minority indeed, and they do not represent the consumer market targeted by most producers. Additionally, the best wines to age are mostly reds, since whites do not hold up nearly as well in the cellar.

Before aging any wine, it is important to focus on what the owner finds appealing about the wine, since this will impact the length of time the wine should age. For example, with Cabernet Sauvignon, astringency from tannin is a key component of the varietal and is especially noticeable in young wines. If the collector *enjoys* this tannic structure, long-term aging might not be the best option because aging diminishes the perception of tannin. If a taster enjoys the fresh, fruity nature of a young white wine, aging that wine is not the best idea because the fruit character will change significantly with bottle age. An exception to this strategy would be if a collector was cellaring the wine as a financial investment, in which case the personal palate preference is less important than the resale value of the aged wine. Thus, what the cellar owner is looking for in any given wine helps determine the appropriate aging time for the wine. It is safe to say, however, that it is better to consume a wine too early rather than too late, simply because wine that is past its peak has lost much of the character that makes it enjoyable.

What Happens as Wines Age?

As wines age, several key changes take place that greatly affect the sensory profile. In this section, these changes will be discussed, along with the factors that affect the rate at which the changes take place. There is still a concentrated focus on researching the exact nature of the molecular changes that take place in a bottle of wine as it ages, and these studies are sure to help winemakers and consumers alike. The following sections deal with the larger and more obvious changes that students of wine will encounter as they taste older wines.

Color

The most obvious thing that occurs as wines age is that the color changes dramatically. In general, white wines get *darker* with age while red wines get *lighter* in color. White wines change from being a bright yellow or light gold color to take on more of a golden or amber color. Red wines change from dark red or purple to take on a brick-red or tawny brown color. This color change is especially noticeable when examining the rim of the wine. In red wines, this color change is the result of changes in the **anthocyanins,** or color compounds, present in the wine. Anthocyanins begin life as blue or reddish pigments that affect the color of the wine. The makeup of the anthocyanins is directly connected to the variety and species of grape, as well as the pH level of the wine, and affect whether the wine will have a purple or reddish hue. In the presence of acid and alcohol molecules, these color compounds change their structure, binding together and gradually altering the color of the wine. These color compounds eventually become so large that they precipitate out to become a part of the **sediment** that accumulates in the bottle as the wine ages (Robinson, 2001). See the photo in Chapter 4 that demonstrates the effect that bottle age has on color.

Aromatics

One of the most desirable changes that occurs as wines age is the increase in aromatic qualities and complexity. Red wines in particular change from having varietal fruit characteristics and aromas to having more of a **bottle bouquet.** Bottle bouquet refers to the increasingly complex smells that develop in wine over time as a result of the interactions of the various compounds. These interactions occur primarily between the alcohol and acid compounds in the wine, but also include reactions with color compounds and other phenolic components. Odorless alcohol and odorless acid molecules connect to create aromatic compounds called **esters.** Esters are the primary molecules responsible for the bouquet of wine. The esters that are present in wine as a result of fermentation have primarily fruity aromas; more complex aromas that develop as a result of interactions in the bottle gradually replace the original esters so as wines age they become noticeably less fruity. As time passes, the rate at which these compounds connect and disconnect affects the wine's particular smell. Some of these molecules connect after a short time, some take a little longer, and some take *years* to connect. The specific rate at which these reactions occur directly impacts the bouquet of any given wine at any given time. This is why the bouquet of a wine evolves over time, and explains the key reason why anyone would choose to age wine. As the reactions occur, the wine's bouquet develops in complexity and depth and becomes more intriguing with each passing day.

The rate at which these changes take place is dependent on many factors, including the style of wine, the pH level and total acidity, the tannin structure, the temperature and humidity of the cellar or storage area, the chemical composition of the wine, and many other factors (Baldy, 1997). In some cases, a significant change in aromatics is noticeable in just 3 to 6 months, especially with young, recently bottled wines. In other cases, it may take several years to begin to notice the development of the bottle bouquet. Periodic tasting of wine at various stages of development is the best way to gauge the progress and decide upon the best time for consumption.

Aging and Tannins

In red wines, tannins act as a preservative. They help to prevent oxidation of the wine as it is being made; higher levels of tannins help to ensure a longer life as the wine ages. Wines with higher levels of tannin tend to be more age worthy than those with little tannic structure. Since red wines have much higher levels of tannins than whites, they are more ageable overall. Although there are white wines that *will* improve with age, they are a very small percentage of the market.

As wines age, the perception of tannin diminishes. Older wines are often described as having a "soft mouthfeel" because of the decreased astringency. When the wine is young, the tannin molecules are relatively short, and these short molecules are perceived as being quite astringent on the palate. As time passes, these smaller molecules link up to form longer molecules that are perceived as much less astringent. Eventually, these tannin molecules become so large that they fall out of solution, adding to the sediment that forms in the bottle. Wines that are intensely astringent when young can soften considerably with enough time in the bottle (Robinson, 2001). It is important to remember that as a wine softens with bottle age, it also loses much of its fruit character. Determining when a specific bottle is ready is partly a matter of striking a balance between softened tannins and enough remaining fruit to ensure the wine is enjoyable.

Acids

The pH level of wine affects both its color and its development in the cellar. Acid functions as a preservative, just as tannin does, and wines that are more acidic (lower pH level) tend to be more age worthy than those with less acid. The level of acid in wine also affects the rate at which the esters form, so a wine's acidity impacts the development of the bottle bouquet as much as the mouthfeel of the wine. Since acid molecules bind with alcohol molecules to create esters, the perception of the wine's acidity also changes with time in the cellar. Just as the perception of tannin diminishes with time, so does the perception of acidity. Acid molecules also precipitate out of solution and eventually add to the bottle sediment.

Effects of Temperature on Aging

One of the most important conditions in aging wine is the temperature of the cellar or storage space. The ideal temperature range is between 55° and 60°F (12.8° and 15.6°C), although a wider range of 50° to 65°F (10° to 18.3°C) would not necessarily be harmful. Below 50°F, the rate at which the wine ages is slowed; wines aged at very low temperatures do not show the same development as those aged in a warmer environment. Research has shown that the temperature of storage directly affects the rate at which changes take place in the wine bottle. A rough estimate would be that for every 18°F (7.8°C) increase in temperature, the rate at which the changes happen doubles. Two bottles of the same wine aged in different areas, 18°F apart will show distinctly different profiles at any given time (MacNeil, 2001).

Aside from storing wine within the appropriate temperature range, it is important that the temperature remain *constant.* Wide variations of temperature, even within the range of acceptability, can lead to wine spoilage. The more the temperature fluctuates, the greater the likelihood that the wine will be spoiled. Ensuring a *stable* temperature is as important as ensuring an *acceptable* temperature.

Sediment

As wines age, they often throw off silty sediment that accumulates on the sides or bottom of the bottle. The connections that help to develop a wine's bouquet also cause the sediment to form. This sediment is a combination of the smaller color, acid, and tannin compounds, which have bonded together to form larger molecules until they are so large that they can no longer stay dissolved in solution. They then precipitate out of solution and form the sediment in the bottle.

The amount of sediment does not always reflect just the age of the wine because many factors can affect the accumulation of sediment. For example, those wines that are darker and have richer color to start have more pigments to shed and can therefore create more sediment. A lightly colored wine such as Pinot Noir will create proportionately less sediment than a darker wine. Wines with higher levels of tannins will also create more sediment as those tannins bind together and settle out of solution. Tartaric acid also settles out of solution and becomes part of the bottle deposit. Depending on the composition of the young wine and the filtering practices by the winemaker, wines can throw varying amounts of sediment as they age. During tableside wine service, this sediment is removed during the process of decanting. See Chapter 16 for more on this.

Bottle Size

A general rule is that the larger the bottle, the slower it ages. For any development in color or aromatics to be noticed, it must have happened in a high enough proportion to affect the profile of the wine. Small changes gradually accumulate to make a large impact on the wine's profile. Because the volume of the wine in the half bottle is much less, the changes that affect a wine's development can happen more quickly.

Another factor in the different rates of aging based on bottle size has to do with the air space (or **ullage**) present in the bottle (Figure 18–1). Despite the fact that a magnum is twice the volume of a standard bottle, they have relatively similar amounts of air space between the end of the cork and the wine. The ratio of wine to air is higher in a large bottle (more wine per volume of air) than in a smaller bottle, and this greatly impacts the rate of aging. This is one reason why many producers will bottle their wines in different size bottles. Many collectors focus on acquiring large format bottles because their potential longevity is much greater. Older wines bottled in large formats are often highly desirable for this reason. Tasting the same older wine from a half bottle, a standard bottle. and a magnum bottle will reveal the differences in aging patterns attributable to the different bottle sizes.

The Effect of Vintage on Ageability

The composition of the grapes at harvest greatly impacts the style and profile of the finished wine and directly affects the ageability of the wine. For example, wines with lower pH values develop more slowly and have more potential ageability. Cooler regions and vintages tend to pro-

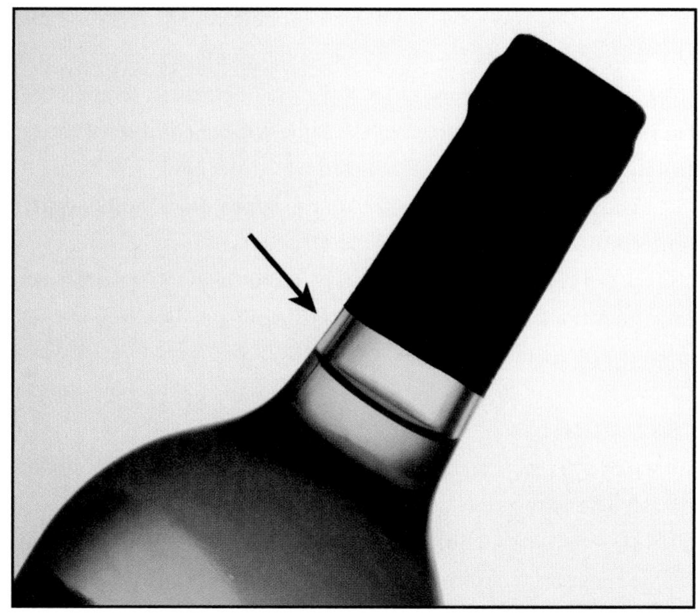

FIGURE 18–1 This photo shows the ullage, or air space, between the wine in the bottle and the cork. The amount of air in this space can impact how the wine develops through the aging process. Older wines often show a larger amount of ullage as the wine slowly evaporates through the aging process. Larger bottles have a lower ratio of air space to wine volume and therefore tend to age more slowly.

duce wines with lower pH levels, and this in turn helps to ensure that the wines will hold up better in the cellar. Those wines with the longest potential lives in the cellar need to begin life with the composition to age. This potential comes from the structural makeup of the wine, and higher acid wines have better potential life in the cellar. Vintages directly impact the character, balance, and composition of the grapes, so certain vintages in any given area are recognized as creating more ageable wines than other years. For details on this, consult any one of the many vintage charts that are currently available from wine publications or online.

Styles of Wine to Age

Wines can be cellared for the short term (0 to 1 years), the medium term (2 to 5 years), or long term (6 to 10+ years). Each age bracket will include different wines, depending on the results desired from aging, but most wines are highly enjoyable at a wide range of points in their development.

Certain wines have a good track record of long-term ageability and have proven the ability to develop increasing complexity over medium- to long-term cellaring. Typically, these wines are at the upper echelon of their categories, and they represent a small minority of the world's vast sea of wine. Those wines that would fit this category would include vintage Port, the top tier wines of Bordeaux, red (and some white) Burgundy, Cabernet Sauvignon, many Rhône Valley wines, and the upper echelon of dessert wines, specifically Tokaji from Hungary, Sauternes, and Vouvray in the Loire Valley of France. Although there are many other wines which could be deemed "age worthy," the individual character of the wine will help determine the aging window and the right time for consumption. There comes a point where the wine reaches a plateau of development, with no further improvement, from which it begins a slow, steady decline. The time at which this happens is highly variable, based on many of the same criteria that determine how the wine ages to begin with, as well as the particular style of wine. Although wines past their prime may be of interest to the taster due to their stage of development, they usually show a shadow of their former glory.

Which Wines Should *Not* Be Aged?

There are more wines that should not be aged than those that should. This is partly because of the current trend to produce fruit-forward wines that cater to the idea that wines should be enjoyable upon release. Most of the world's wines do not need to be aged, and many are best consumed as soon as possible.

The vast majority of white wines are best served sooner rather than later. Whites are appreciated for their up-front fruit character and simply do not have the character to develop with long-term aging. If these wines *were* aged, the fresh fruitiness would diminish significantly, and they could develop very different profiles that not everyone would find enjoyable. Many reds would suffer the same fate, especially those produced from lighter grapes, via carbonic maceration, or from grapes grown in warmer climates. Wines from less successful vintages often do not fare as well in the cellar and are better consumed when young. These are just some examples of wines that are best enjoyed in their youth, when the characteristics that define them are the most apparent. When in doubt, consult industry publications, vintage charts, or other resources to find guidance on recommended drinking windows for wines. It can also be an educational endeavor to age some wines that are regarded as not age worthy, in an attempt to understand what happens to these wines. This will give some perspective on *why* certain wines should not be aged and what they become when they are past their prime.

Ideal Cellar Conditions

Once the appropriate wines have been chosen to place in the cellar, an essential aspect of proper aging is providing the ideal conditions for storage. Because wine is a perishable product, care must be taken to store it in an area that will provide conditions for perfect development that will ensure the wine is in the optimum condition when it is ready to be opened. A relatively strict set of circumstances are required to properly age wine, and they are outlined in this section.

Temperature and Humidity Control

Proper control of the cellar temperature and humidity is of the utmost importance to the fine wine collector, but should also be high on the priority list of anyone who wishes to store wine for extended periods of time. As previously mentioned, the best range of temperatures to store wine is between 55° and 60°F (12.8° and 15.6°C). Wines stored outside these temperatures run the risk of less than ideal development, and in a worst-case scenario can spoil before being consumed. For collectors with extensive collections of prized wines, maintaining a proper temperature is critical, and a number of methods might be employed to maintain these temperatures.

Many wine aficionados begin their wine collection with a few bottles, simply stored on a rack in the kitchen or dining room. Often times, this slowly expands to include a couple cases of wine, and the passion for collecting is born. The storage space does not need to be large or complicated, but the collection of wine should *not* be stored in the warmest room of the house or a hot closet or pantry. Finding space in a cooler area is critical for wine storage, even with smaller collections. Most larger collections grow out of small investments.

One increasingly common situation is for collectors to have a custom-built cellar created to house their wine (Figure 18–2). These expansive (and expensive) cellars often employ a temperature control unit, similar to an air conditioner, which regulates the internal temperature and humidity of the cellar to keep it within the ideal range. These units often cost several hundred dollars, but they are relatively inexpensive compared to the money wasted when special bottles of wine spoil. These units can be installed in many different storage spaces, and some companies now specialize in the design of custom cellar spaces. Custom-designed cellars usually cost at least several thousand dollars, and in some restaurant settings, the investment is worth the return. In restaurants with these cellar spaces, the area can function as an event room and as a marketing

FIGURE 18–2 This custom wine cellar is beautifully designed and includes a small wine tasting area. It employs a temperature and humidity control unit to ensure optimum conditions to store and age wine.

tool for the restaurant's wine program. When the operation has invested tens or hundreds of thousands of dollars in wine inventory, proper storage conditions are a necessity.

Humidity control is also important to wine collectors because the corks that provide the airtight seal on wine bottles can dry out without enough humidity. The ideal level of humidity is relatively high, usually in the 70 to 80 percent range. The drying of the corks can cause oxygen to enter the bottle and spoil the wine. Higher levels of humidity can cause mold to form on the bottle labels; although this is unsightly, in most cases it does not affect the integrity of the wine in the bottle. Moldy labels would, however, affect the resale value of the wine in the event that the cellar owner is looking to return a profit on investment-grade wines. In custom-designed cellars, a vapor barrier is often applied between the walls to ensure proper retention of moisture in the storage space. A thermometer and hygrometer (humidity-monitoring device) should be kept in the storage space to regularly monitor the cellar environment.

Passive Cellars

For those who are lucky enough to have a basement to store their wine, this space can function as a great space to store the collection. A **passive cellar** is a storage area that does not employ any type of temperature or humidity control system to maintain consistent levels (Figure 18–3). The success of these cellars relies on the ambient temperature and humidity readings staying within an acceptable range to ensure that the wine ages safely. These cellars may experience fluctuations in both temperatures and humidity, depending on the season and even the time of day. These fluctuations should be slight enough to not impact the development of the wine, but in extreme situations the longevity of the wine can be at risk. Many châteaux in Europe often store both current and back vintages in deep, passive cellars near the winery and even these cellars experience *some* minor fluctuations. Passive cellars are beneficial because the owner does not need to purchase and maintain a temperature control unit, so the initial start-up expense is minimal. Because there is no unit to maintain or purchase, money can be used to purchase wine to put in the cellar. This is the most common way beginners start to assemble a wine collection: simply identifying an appropriate storage space and assembling a number of wines to be saved for future consumption.

FIGURE 18–3 This passive cellar may look less dramatic than the custom cellar in Figure 18–2, but it can be just as effective at providing the ideal storage conditions for aging wine.

The drawbacks to passive cellars can be numerous, but temperature and humidity fluctuations are the biggest concern. Big changes can jeopardize the quality of the wine and perhaps even spoil it before it is ready to be consumed. Even some experts disagree on exactly how much fluctuation is needed to notice a decline in the wine quality of the wine. If a collector is purchasing and aging wine as a financial investment to later sell at auction, storing wine in a passive cellar is not the best idea. Passively cellared wines attain much lower prices at auction than those wines that come from controlled cellars. To purchase investment-grade wines and store them in a passive cellar is not a sound financial investment. If a collector is looking to gain a profitable return from a wine investment, documentation of the storage helps to prove the provenance and can drastically affect the selling price.

Renting Storage Space

There are a growing number of operations that offer to rent storage space for fine wines. These spaces are set up to ensure proper storage of fine wine, including temperature and humidity control. These services are often very affordable, depending on the volume of wine stored there. Some restaurant operators may elect to store their wine collection in a rental space if the restaurant itself does not have proper storage. Where maintaining a top-tier inventory of fine and rare wine is of the utmost importance, these storage options often represent viable choices for management of the restaurants' inventory, but this added cost can add to the expense of maintaining these wines on the list. Obviously, the operator will have more limited access to the wines in their inventory, and this will affect their ability to put certain wines on the list to be made available to the guests.

Alternatives to Cellars

Not everyone who purchases wine for aging stores it in an actual cellar. The market is flooded with temperature-controlled storage cabinets that function much as a cellar does, but are designed to cater to people who do not have access to an ideal storage area. These temperature- and humidity-controlled cabinets (or cases) have racking to hold the bottles of wine and control panels that allow adjustment of the internal temperature. Some even offer two temperature control zones, one for storing white wines and another for reds. These cabinets range in size from countertop models that hold five or ten bottles all the way up to systems designed to hold upward of 1,000 bottles. Depending on the size and the desired features, these can be relatively affordable or quite expensive, but they are very effective for preserving a fine wine collection.

The benefits to these systems are obvious: They allow people without an ideal storage environment to store and age their wines without fear of spoilage. The cabinets often have roll-out shelves that allow easy access to the wine, and some have glass doors to allow viewing of the bottles. They can also act as a decorative and functional piece of furniture in the home of the wine lover. Some of the most common and available versions come from the manufacturers Eurocave and Vintage Keeper.

Purchasing Wine

Most wines are released to market in the spring and fall of each year. This is for practical reasons as much as tradition because shipping cases of wine to the far reaches of the globe in trucks, boats, and planes during the height of the summer or the dead of winter is not a good idea. Thus the spring and fall offer the most temperate climate to ship wine, ensuring that quality products are delivered intact to customers. The timetable of producing wine in the winery often includes a certain amount of time for the wine to age, whether in barrel or bottle, and the release dates for any given wine will vary accordingly. Wines that have finished fermenting by the spring following the harvest are clarified, stabilized, and bottled, and those meant for immediate consumption are ready to be shipped by the fall following the year of harvest. Release dates for specific wines depend on many factors, some of which are tied to marketing factors while others are tied to winery operations.

Knowing that these are the key times when wines are released makes it easy to say that spring and fall are the best times to purchase wine. However, this is only partly true. Since this is the time when *new* wines (or vintages) are released, it is a good time to buy current releases. But many retailers have not fully sold the previous vintage of the wine, and still have it on their shelves. Prior to the arrival of the newest vintage, many retailers may seek to move the remains of the previous vintage to make room for the new wines, and so they may actually offer special price discounts to the consumer. This would be a good time to purchase wines at a significant discount.

Regardless of how the wine is purchased, those people interested in discerning how a wine develops over time should buy multiple bottles of each wine. This may mean simply purchasing three or four of a wine instead of just one bottle, or it may mean buying wine a case or two at a time, depending on the budget and available storage space. By acquiring multiple bottles of each wine, the collector can then taste the wine periodically through its development, noticing both how the wine has changed and also making note of the best time to consume the remaining bottles. Taking tasting notes on its profile and comparing these notes at various points in the wine's life will allow a window into the changes that have taken place. This is the best way to begin to understand how wine develops with age, and should be undertaken whenever possible.

Racking for the Wine

There are many different racking options available on the market that cater to the demands of budget and storage space. The most common are metal or wood racks that allow the bottles to be laid down during storage (Figure 18–4). Storing bottles in this manner ensures that the corks

FIGURE 18–4 This photo shows another system of racking wine for a cellar setting.

stay moist and continue to provide a proper seal against oxygen entering the bottle. The size and style of racking depends largely on the aesthetic demands of the owner, as well as the budget available to invest in this aspect of the cellar. Racking can be simple or elaborate, so long as it stores wine on its side and allows the owner access to the collection.

For a private cellar or a cellar for the restaurant, the same set of storage and racking conditions apply. There are some key differences in the layout and design of a restaurant's cellar, however, and these differences mainly relate to the need to access wine at a moment's notice in the restaurant setting. Because of this, many restaurants employ systems to organize the storage area to facilitate locating specific wines. One of the most common methods of doing this is to use a **bin number** system (Figure 18–5). A bin number is simply a reference number that directs the service staff to the location of a specific wine in the storage area. These numbers are usually listed on the wine list as well, so guests who choose to do so may order according to the wine's bin number.

Buying Futures

Some wine producers will offer their wine as **futures.** Buying futures means that the wine is purchased before it is actually released to the market. This manner of purchasing is done for several reasons, but primarily for collectors to ensure they have access to the wines that they desire for their cellars. The customer can secure the quantities of the desired wines before the general public has a chance to buy the inventory. The wine is purchased and paid for before the wine is acquired. Sometimes, the wine is not delivered to the customer for several years because it may be bought while it is still in barrel (or before it is even harvested).

Many producers will offer discounts to customers when they purchase futures; this is advantageous for the winery because they are able to create capital to help fund winery operations. The practice also helps to ensure that the wine gets sold, instead of having excess inventory sitting in storage waiting for a buyer. This practice is most common in Bordeaux, although many other markets have recognized the benefits of selling wine as futures, and have begun to follow Bordeaux's lead. For highly allocated wines with limited productions, futures may be the only way to access the wines because little is available on the open market after the futures are purchased. For wines that are produced in high volumes and are not difficult to access, buying on futures is usually not recommended because they will be readily available in the market upon release.

FIGURE 18–5 These labeled bins allow easy access to stored wines. This method of storage allows service staff to find bottles quickly and easily when they are ordered during service. The bin numbers are sometimes included on restaurant wine lists.

Buying Wine at Auctions

The fast-paced and competitive nature of the auction suits some people just fine. These people enjoy the thrill of outbidding other collectors in order to acquire sought-after wines for their cellar. For others, however, the hectic nature of an auction can be overwhelming and not very fun. Each collector's feelings determines if this is a good method for purchasing wine. Because a group of people at the auction are all working to acquire the same wines, the prices that are paid sometimes exceed the estimated value of the wine. To the people who win the bidding, this may be of little consequence because of the scarcity of the wine or the enjoyment of the auction process itself. But those who are looking for value or are not prepared to stop bidding when the price gets too high can end up paying far too much for the wine they purchase in this setting.

Auctions are a way to gain access to scarce or older vintage wines when there is little of it available on the open market. Auctions offer the chance to buy wines which were sold out long ago, but have been offered as a lot to be sold to the highest bidder. Many of these wines come from the cellars of private collectors or are re-releases from the wineries themselves, and in many cases the storage has been pristine. Wines sold at auction from passive cellars often command much lower prices than those with guaranteed provenance.

Older Wine and Restaurants

Clearly, very few restaurant operations have the space to age wines for years, although the financial investment for doing so can be great. Not every restaurant will have customers who expect offerings of older fine wines, and the demand for these products should be evaluated before making a large investment for the list. For restaurateurs wishing to offer older wines to their guests, there are a couple key ways to make this happen and several things to keep in mind when selling and serving these wines to guests.

Accessing Older Wines

The primary way that restaurants have access to older wines is purchasing them at public auctions. The auction market in the United States has taken off in the last 20 years and has begun to rival the long-standing and very popular auctions that have taken place for years in Europe. The most famous auction houses include Christie's, Sotheby's, Sherry-Lehmann, and Zachy's. These houses obtain various wines from the cellars of private collectors, estate sales, and the winery's own stocks and offer them to the highest bidder. These events often have very high profiles, drawing bidders from all over the United States and often, the world. Some bidders use the phone to call in bids, and some auctions take place on the Internet. Regardless of the location and the wines offered, these events can be fun, dramatic, and challenging.

The most challenging aspect of these events is a limited number of any given wine with plenty of people competing for the same lots. This can quickly and dramatically inflate the purchase price beyond that of the current market. Once the buyer's premium is tacked on (the auction house's percentage) the price can be far above the current market value. By the same token, certain lots of wine may be underappreciated by the audience in attendance and could be snatched up below market value. Either way, the bidder must carefully consider his or her budget and be willing to stop bidding and walk away when necessary, regardless of the allure of the wine lot.

Traveling to these auctions is not always easy, and the Internet has fostered the growth of many online auction sites. These sites allow buyers to stay in the comfort of their own home (or restaurant) and bid on the wines they would like. The buyer's premium and potential shipping costs are still present, both of which can add 20 percent or more to the hammer price of the wines.

Auction events offer access to rare caches of wine that are otherwise not available to purchase. For a restaurant, buying these wines allows the operation to offer older vintages without having to age them for years. Restaurants get the luxury of offering older wine to their guests without assuming the overhead costs of holding a large inventory that will not be sold for years. This alone can make the higher auction prices well worth it.

Private Cellars

In some areas of the country, it is legal to offer wine from a private cellar for sale on the wine list. Depending on the individual state's laws, this may be a viable option for offering older wines to guests at the restaurant. This is a beneficial situation for both the restaurant and the cellar owner, since the restaurant has access to special wines without having to age them in-house or purchase them at auction. The cellar owner is able to sell wines from their collection without having to navigate the auction setting, and the guests have access to special wines through the restaurant's wine list. If the laws are conducive to this, it can be a truly beneficial situation.

Serving Older Wines in the Restaurant

When serving older wines tableside, there are several key factors to keep in mind in addition to those discussed in Chapter 16. Most older wines will contain some sediment, and this should be addressed through proper tableside decanting. Many very old wines are extremely fragile—too much oxygen will cause them to deteriorate quite rapidly. For these wines, decanting may do more harm than good. When in doubt as to how to handle the wine, consult with the guest to see how he or she would like the service delivered.

Because wines gain in complexity with age and have often softened considerably, they are often more dependent upon the food that they are served with. Wines that have developed intense complexity may be best served with simple dishes because intense flavors in the food may overwhelm the wine. Also, if a guest has chosen to purchase a costly and rare wine, then that wine is clearly the focal point of their meal. This is not to say that food is not important, but in this situation, the job of the food should be to complement the wine and maximize the enjoyment of it. Sometimes a simple dish will offer the guest the most enjoyment and will help ensure that the wine is able to showcase its best attributes.

Summary

The desire to cellar and age wine stems from the idea that certain wines improve with bottle age. Over time, the formation of the bottle bouquet offers a greater level of complexity and can allow the taster a more enjoyable experience. The key requirements for cellaring wine include proper selection of wines to cellar, storage conditions that are conducive to aging, and the desire to monitor a wine's development over time. Wine cellars need not be huge or expensive to create, but the limit is the budget and imagination of the collector. By choosing the appropriate wines to age and providing proper storage conditions, any collector can discover the rewards of older wine: more complex and intriguing profiles, and a greater understanding of how the age of wine can affect the perception of its character.

Review Questions

1. What are the ideal conditions for storing wine?
2. What criteria will make a wine age-worthy? Which wines should not be aged?
3. What happens to a wine's profile as it ages?
4. What is meant by the "bottle bouquet"?
5. What considerations should be made when serving older wines in a restaurant setting?

Wine Law in the United States

Introduction

The business of wine and alcohol is more heavily regulated than any other legal product in the United States with the possible exception of firearms. Multiple agencies at the federal, state, and local level all regulate the production, distribution, and sales of wine. There are numerous laws at every level of government that control wine and there is little uniformity of these statutes between state and local districts. These conditions make for a confusing array of wine laws across the United States. The laws are designed with two principle aims—first to collect taxes, and second to control and limit consumption.

The taxation of alcohol is considered a "sin tax" like those that are levied on cigarettes and gambling. A common justification for taxes on alcoholic products is that they help to compensate the government for dealing with the consequences of excessive consumption. All wines sold in the United States are subject to a federal tax, and states also apply additional excise taxes to wines sold inside their boundaries. State excise taxes range from $0.11 per gallon in Louisiana to $2.50 a gallon in Alaska. These taxes are in addition to the federal tax of $1.07 per gallon for wine containing 14 percent or less alcohol and $1.57 per gallon for wine over 14 percent, as well as any sales taxes that may be applicable. There are different tax levels for different types of alcoholic beverages; typically the greater the alcohol content of the product the higher the tax. However, there is no distinction made between moderate consumption of table wine with food and excessive consumption by alcoholics. There are also many laws by all levels of government designed to control how wine is distributed and sold. These laws govern who can sell what categories of wine, where and in what types of locations it can be sold, and what hours or days of the week it can be made available to consumers.

Historical Background

The origins of this complex multitude of laws that govern alcohol go back to the repeal of Prohibition in 1933. The era of Prohibition had profound effects on the country's demand for wine as well as what styles of wine became popular with consumers. Similarly, Prohibition's repeal had a direct effect on how wine was regulated by government agencies. As discussed in Chapter 10, the Eighteenth Amendment to the Constitution enacted nationwide Prohibition on the production, sale, or consumption of alcohol in 1920 and was repealed with the Twenty-First Amendment in 1933. One of the stipulations of the repeal amendment is that states were allowed to govern all aspects of alcohol within their borders. This resulted in each state forming a set of its own unique laws to govern wine. In turn, many states granted a great deal of authority to cities and counties to control alcohol under their own local jurisdiction. Since standards of what is acceptable regarding wine consumption vary greatly from one area to the next, there are still many counties, particularly in the South and Midwest, which continue to ban the sale of alcohol.

The ability of a state to regulate and limit the sale of wines produced in other states directly contradicts another part of the Constitution, the Interstate Commerce Clause. The Interstate

Commerce Clause generally prohibits states from having exclusionary tariffs or laws on products imported from other states. However, courts have decided that the Twenty-First Amendment supercedes the Interstate Commerce Clause and that the Interstate Commerce Clause does not cover wine and other alcoholic products. There are exceptions to this; in 1984, the Supreme Court decided that states cannot have lower taxes on wines made within their own borders than they levy on wines produced in other states (Lee, 2005). Additionally, in May 2005, the Supreme Court decided that states could not prohibit the direct shipping of wine from out of state wineries to consumers, yet allowed direct shipping from in state wineries. There have been a number of cases regarding wine regulation brought before courts at every level in the United States, and the laws governing wine continue to evolve. If one can make any generalization, it is that laws that are designed purely to be exclusionary to wines produced out of state are not allowed, whereas laws designed to limit or control consumption of all wines regardless of the state of origin are permissible.

Federal Wine Laws

The majority of federal laws that govern wine are usually concerned with either the collection of federal excise taxes or the regulation of wine labeling. The agency that oversees the enforcement of federal laws is the Alcohol and Tobacco Tax and Trade Bureau, or TTB, formerly known as the Bureau of Alcohol Tobacco & Firearms (BATF). Wineries, distributors, and retailers must all keep detailed records that ensure any bottle of wine that is produced or shipped is accounted for and its taxes are paid. Wineries must report to the TTB monthly to account for the amount of wine they produce and ship. Before it is sold, wine is stored in a bonded warehouse, a warehouse where the tax on the wine has not yet been paid. Once it leaves the bonded warehouse to be shipped to a distributor, the winery is responsible for paying the federal tax.

The TTB is also responsible for regulating wine labels. The TTB stipulates what information must be printed on all wine labels for wine sold in the United States, as outlined in Chapter 10. Wineries must submit their labels to the TTB to be approved before they are allowed to be used for a bottle of wine. The TTB examines potential labels to ensure that they have all of the required text and the text is printed at the size specified by law. The TTB must also approve of all graphics and names on the label as well as any statements the winery makes about its product. Wineries must be able to document all claims made on the label, such as the variety or appellation of the wine, are true and the alcohol content printed on the label is accurate.

Federal wine laws regarding labeling are currently evolving and becoming more stringent. Recently enacted federal bioterrorism laws require that producers be able to trace any product used in wine, such as yeast, fining agents, and corks to its origin. All producers of food products will be required to have this level of traceability; wine is ahead of most other food products because wineries have always been required to keep good records to prove statements about a wine's variety and appellation. In the future, ingredient labeling may be required, particularly for allergens such as egg whites, which are used for fining.

State Wine Laws

The state laws that govern wine and alcohol are so complex and vary so much from state to state it is difficult to make any sweeping statements that would apply to all states. This being said, state wine statutes generally fall under one of the following categories:

- Laws that govern the distribution of wine
- Laws that govern the selling of wine
- Laws that govern the state taxation of wine
- Laws that govern the shipping of wine by producers and consumers

State Distribution Laws

There are four entities that are involved in getting wine into consumers' hands.

- Suppliers (wineries) that produce the wine
- Brokers, companies, and individuals that represent the winery and sell its product for a commission
- Distributors, companies that purchase large amounts of wine in cases from suppliers and store it their own warehouses while they sell it to retailers by the case or by the bottle
- Retailers, either on-premise (restaurants and bars) or off-premise (wine shops and grocery stores), who sell wine directly to consumers

Every state has different laws that govern the actions of all of these groups; some states prohibit some of them from operating. States can be divided into two categories based on the degree of direct management they have over distribution channels. Thirty-two states have a competitive model of distribution, where the private sector distributes and sells the wine. These states include many of the large markets for wine consumption such as California, New York, and Florida. The 18 remaining states use the control model of distribution. Here, state agencies manage and control either distribution or sales and in some cases both operations. Major control states include Pennsylvania, New Hampshire, and Michigan. Because of the efficiencies inherent in private companies competing against each other, the competitive model usually offers consumers better prices and selection of wine. The control model offers states a lucrative source of revenue from their stake in the profits generated by wine sales. Even in states where the control model is not popular, it is difficult to repeal because if the revenue from alcohol sales is lost other taxes would have to be raised.

In many states, the three-tier system of distribution is required by law. In this system, producers are not allowed to sell directly to retailers; instead they must go through a middle distributor who buys the wine from the winery before it can be sold to the retailer. These laws exist to prevent wineries from having undue influence on retailers. Some states have franchise laws that do not allow wineries to have more than one distributor or to change to another distribution company if they are dissatisfied with the one that they are currently using. In a three-tier state, if a winery cannot find a distributor that wants to handle its product it will not be allowed to sell its wine even if there are wine shops or restaurants that wish to carry it. These laws are particularly difficult for small wineries with limited production where the small amount of sales possible in a particular state do not justify the amount of effort it takes to acquire a distributor.

State Laws Governing Wine Sales

As previously stated, state and local governments have great latitude in establishing laws regarding alcohol, including how, where, and when it can be sold. In some control states such as New Hampshire, Utah, and Pennsylvania, wine can only be purchased through state stores. By controlling the sales of all wine, the state is assured that no one underage buys wine and that it receives all of the state excise and sales tax; in addition the states receives profits from the sale as well. In an effort to limit consumption, some states such as New York do not allow wine to be sold in grocery stores. This requires consumers to travel to a wine shop to purchase their wine instead of simply buying it with their groceries. Other laws govern when wine and alcohol can be sold, either by banning it after a certain hour of the night or by not allowing the sale of wine on Sunday through the use of so-called "blue laws."

States also have Alcohol Beverage Control (ABC) agencies that monitor wine retailers and enforce the states' laws. These agencies are in charge of issuing liquor licenses that retailers need to sell wine and other alcohol products as well as monitoring them to insure all regulations are being followed. These agencies also control who is allowed to serve wine—in some states, waiters must be of legal age to drink alcohol before they are allowed to serve wine to patrons. In other states, it is illegal for a employee of a winery to pour their own brand of wine at a tasting; it must be poured by a third party.

STATE LAWS GOVERNING SHIPPING

Many states do not allow the shipping of wine into the state unless it passes through established distribution channels. This type of shipping usually occurs when a consumer orders wine directly from the winery via phone or the Internet. States generally do not like these types of purchases because the state taxes may not be paid on such a transaction and wine could potentially be sold to minors. Suppliers counter that they would be happy to pay state taxes if they were allowed to ship wine into the state and that an adult signature can be required upon delivery to prevent underage people from obtaining wine.

The prohibition on direct shipping may not only cover wine sales but also any shipping of wine across the state's borders. Some states ban any outside importation of wine by consumers even if they are carrying it over the states' border themselves for personal consumption. Wine being sent by a private citizen as a gift or as a donation to a charity is also banned in these states. In six states—Florida, Georgia, Indiana, Kentucky, North Carolina, and Tennessee—it is not only illegal but actually a felony to direct ship wine from outside the state to consumers. These laws are actively supported by many distributors that have a vested interest in keeping things the way they are.

Anti-direct shipping laws are seldomly enforced against private citizens returning from vacation with a bottle of wine or people who send a bottle of wine to a relative as a gift. These bans do, however, have an effect on wineries that are not allowed to ship wines directly to consumers. Small wineries in particular that do not have nationwide distribution are the most affected, while larger wineries whose product is available in every state are less affected. Different shipping companies such as UPS, FedEx, and DHL, all have varying policies on which states they will ship wine to, and the U.S. Postal Service does not ship wine under any circumstances.

Other states, particularly those with a significant wine industry, have more liberal laws governing the direct shipping of wine. In 1986, California began a campaign of reciprocity, where it would allow direct shipping of a reasonable amount of wine into the state by wineries and private citizens if the state it was being shipped from would do the same. As of June 2005, there were 13 reciprocal states as well as 16 that allow limited direct shipping under certain conditions.

Summary of State Laws

Since the laws of the different states are very complex and constantly changing, the most up to date information can be obtained from the visiting the Wine Institute or TTB Web sites at:

Wine Institute	http://www.wineinstitute.org
TTB	http://www.ttb.gov/alcohol/index.htm

Table A-1 Shipping Laws, Distribution Models, and Web Addresses of Alcohol Beverage Control Offices Listed by State

STATE	DIRECT SHIPPING	DISTRIBUTION MODEL	STATE ABC WEBSITE
Alabama	Prohibited	Control State	http://www.abcboard.state.al.us/
Alaska	Limited	Competitive	http://www.dps.state.ak.us/abc/
Arizona	Limited	Competitive	http://www.azll.com/
Arkansas	Prohibited	Competitive	http://www.state.ar.us/dfa/abcenforcement/index.html
California	Reciprocal	Competitive	http://www.abc.ca.gov/
Colorado	Reciprocal	Competitive	http://www.revenue.state.co.us/liquor_dir/toc.htm
Connecticut	Prohibited	Competitive	http://www.dcp.state.ct.us/licensing/liquor.htm
Delaware	Prohibited	Competitive	http://www.delawarepublicsafety.com/dabc.cfm
Florida	Prohibited	Competitive	http://www.state.fl.us/dbpr/abt/index.shtml
Georgia	Limited	Competitive	http://www.etax.dor.ga.gov/alcohol/index.shtml
Hawaii*	Reciprocal	Competitive	http://www.co.honolulu.hi.us/liq/
Idaho	Reciprocal	Control State	http://www.isp.state.id.us/abc/index.html
Illinois	Reciprocal	Competitive	http://www.state.il.us/lcc/
Indiana	Prohibited	Competitive	http://www.in.gov/atc
Iowa	Reciprocal	Control State	http://www.iowaabd.com/
Kansas	Prohibited	Competitive	http://www.ksrevenue.org/abc.htm
Kentucky	Prohibited	Competitive	http://www.abc.ky.gov/
Louisiana	Limited	Competitive	http://www.atc.rev.state.la.us/atcweb/home.htm
Maine	Prohibited	Control State	http://www.state.me.us/dps/liqr/homepage.htm
Maryland**	Prohibited	Competitive	http://www.comp.state.md.us/
Massachusetts	Prohibited	Competitive	http://www.state.ma.us/abcc
Michigan	Prohibited	Control State	http://www.michigan.gov/cis/0,1607,7-154-10570__,00.html
Minnesota	Reciprocal	Competitive	http://www.dps.state.mn.us/alcgamb/alcgamb.html
Mississippi	Prohibited	Control State	http://www.mstc.state.ms.us/abc/main.htm
Missouri	Reciprocal	Competitive	http://www.mdlc.state.mo.us/
Montana	Prohibited	Control State	http://www.mt.gov/revenue/forbusinesses/liquorlicensees/liquorlicensees.asp
Nebraska	Limited	Competitive	http://www.nol.org/home/NLCC/
Nevada	Limited	Competitive	http://tax.state.nv.us/
New Hampshire	Limited	Control State	http://webster.state.nh.us/liquor
New Jersey	Prohibited	Competitive	http://www.state.nj.us/lps/abc/index.html
New Mexico	Reciprocal	Competitive	http://www.rld.state.nm.us/AGD/index.htm
New York	Prohibited	Competitive	http://www.abc.state.ny.us/
North Carolina	Limited	Control State	http://www.ncabc.com/
North Dakota	Limited	Competitive	http://www.state.nd.us/taxdpt/alcohol/
Ohio	Prohibited	Control State	http://www.liquorcontrol.ohio.gov/liquor.htm
Oklahoma	Prohibited	Competitive	http://www.able.state.ok.us/
Oregon	Reciprocal	Control State	http://www.olcc.state.or.us/
Pennsylvania	Prohibited	Control State	http://www.lcb.state.pa.us/
Rhode Island	Limited	Competitive	http://www.dbr.state.ri.us/liquor_comp.html
South Carolina	Limited	Competitive	http://www.sctax.org/default.htm
South Dakota	Prohibited	Competitive	http://www.state.sd.us/drr2/revenue.html
Tennessee	Prohibited	Competitive	http://www.state.tn.us/abc
Texas	Limited	Competitive	http://www.tabc.state.tx.us/
Utah	Prohibited	Control State	http://www.alcbev.state.ut.us/
Vermont	Prohibited	Control State	http://www.state.vt.us/dlc
Virginia	Limited	Control State	http://www.abc.state.va.us/
Washington	Reciprocal	Control State	http://www.liq.wa.gov/
Washington, DC	Limited	Competitive	http://abra.dc.gov/abra/site/default.asp
West Virginia	Reciprocal	Control State	http://www.wvabca.com/
Wisconsin	Reciprocal	Competitive	http://www.dor.state.wi.us/
Wyoming	Limited	Control State	http://revenue.state.wy.us/

* Hawaii has different ABC offices for the different islands.
** Two counties in Maryland have control model distribution.

American Viticultural Areas

American Viticultural Areas by State (Excluding California)

STATE	AVA
Arizona	Sonoita
Arkansas	Altus
Arkansas	Arkansas Mountain
Arkansas	Ozark Mountain [AR, MO, OK]
Colorado	Grand Valley
Colorado	West Elks
Connecticut	Southeastern New England [CT, MA, RI]
Connecticut	Western Connecticut Highlands
Indiana	Ohio River Valley [IN, KY, OH, WV]
Kentucky	Ohio River Valley [IN, KY, OH, WV]
Louisiana	Mississippi Delta [LA, MS, TN]
Maryland	Catoctin
Maryland	Cumberland Valley [MD, PA]
Maryland	Linganore
Massachusetts	Martha's Vineyard
Massachusetts	Southeastern New England [CT, MA, RI]
Michigan	Fennville
Michigan	Lake Michigan Shore
Michigan	Leelanau Peninsula
Michigan	Old Mission Peninsula
Minnesota	Alexandria Lakes
Mississippi	Mississippi Delta [LA, MS, TN]
Missouri	Augusta
Missouri	Hermann
Missouri	Ozark Highlands
Missouri	Ozark Mountain [AR, MO, OK]
New Jersey	Central Delaware Valley [NJ, PA]
New Jersey	Warren Hills
New Mexico	Mesilla Valley [NM, TX]
New Mexico	Middle Rio Grande Valley
New Mexico	Mimbres Valley
New York	Cayuga Lake
New York	Finger Lakes

(continued)

STATE	AVA
New York	The Hamptons, Long Island
New York	Hudson River Region
New York	Lake Erie [NY, OH, PA]
New York	Long Island
New York	Niagara Escarpment
New York	North Fork of Long Island
New York	Seneca Lake
North Carolina	Yadkin Valley
Ohio	Grand River Valley
Ohio	Isle St. George
Ohio	Kanawha River Valley [OH, WV]
Ohio	Lake Erie [NY, OH, PA]
Ohio	Loramie Creek
Ohio	Ohio River Valley [IN, KY, OH, WV]
Oklahoma	Ozark Mountain [AR, MO, OK]
Oregon	Applegate Valley
Oregon	Columbia Gorge [OR, WA]
Oregon	Columbia Valley [OR, WA]
Oregon	Dundee Hills
Oregon	McMinnville
Oregon	Red Hill Douglas County
Oregon	Ribbon Ridge
Oregon	Rogue Valley
Oregon	Southern Oregon
Oregon	Umpqua Valley
Oregon	Walla Walla Valley [OR, WA]
Oregon	Willamette Valley
Oregon	Yamhill-Carlton District
Pennsylvania	Cumberland Valley [MD, PA]
Pennsylvania	Central Delaware Valley [NJ, PA]
Pennsylvania	Lake Erie [NY, OH, PA]
Pennsylvania	Lancaster Valley
Rhode Island	Southeastern New England [CT, MA, RI]
Tennessee	Mississippi Delta [LA, MS, TN]
Texas	Bell Mountain
Texas	Escondido Valley
Texas	Fredericksburg in the Texas Hill Country
Texas	Mesilla Valley [NM, TX]
Texas	Texas Davis Mountains
Texas	Texas High Plains
Texas	Texas Hill Country
Texas	Texoma
Virginia	Monticello
Virginia	North Fork of Roanoke

(continued)

STATE	AVA
Virginia	Northern Neck George Washington Birthplace
Virginia	Rocky Knob
Virginia	Shenandoah Valley [VA, WV]
Virginia	Virginia's Eastern Shore
Washington	Columbia Gorge [OR, WA]
Washington	Columbia Valley [OR, WA]
Washington	Horse Heaven Hills
Washington	Puget Sound
Washington	Red Mountain
Washington	Wahluke Slope
Washington	Walla Walla Valley [OR, WA]
Washington	Yakima Valley
West Virginia	Kanawha River Valley [OH, WV]
West Virginia	Ohio River Valley [IN, KY, OH, WV]
West Virginia	Shenandoah Valley [VA, WV]
Wisconsin	Lake Wisconsin

American Viticultural Areas in California with County Information

CALIFORNIA AVA	COUNTY
Alexander Valley	Sonoma
Anderson Valley	Mendocino
Arroyo Grande Valley	San Luis Obispo
Arroyo Seco	Monterey
Atlas Peak	Napa
Ben Lomond Mountain	Santa Cruz
Benmore Valley	Lake
Bennett Valley	Sonoma
California Shenandoah Valley	Amador, El Dorado
Capay Valley	Yolo
Carmel Valley	Monterey
Central Coast	Alameda, Contra Costa, Monterey, San Benito, San Francisco, San Luis Obispo, San Mateo, Santa Barbara, Santa Clara, Santa Cruz
Chalk Hill	Sonoma
Chalone	Monterey, San Benito
Chiles Valley	Napa
Cienega Valley	San Benito
Clarksburg	Sacramento, Solano, Yolo

(continued)

CALIFORNIA AVA	COUNTY
Clear Lake	Lake
Cole Ranch	Mendocino
Cucamonga Valley	Riverside, San Bernardino
Diablo Grande	Stanislaus
Diamond Mountain District	Napa
Dos Rios	Mendocino
Dry Creek Valley	Sonoma
Dunnigan Hills	Yolo
Edna Valley	San Luis Obispo
El Dorado	El Dorado
Fair Play	El Dorado
Fiddletown	Amador
Guenoc Valley	Lake
Hames Valley	Monterey
High Valley	Lake
Howell Mountain	Napa
Knights Valley	Sonoma
Lime Kiln Valley	San Benito
Livermore Valley	Alameda
Lodi	Sacramento, San Joaquin
Los Carneros	Napa, Sonoma
Madera	Fresno, Madera
Malibu-Newton Canyon	Los Angeles
McDowell Valley	Mendocino
Mendocino	Mendocino
Mendocino Ridge	Mendocino
Merritt Island	Yolo
Monterey	Monterey
Mt. Harlan	San Benito
Mt. Veeder	Napa
Napa Valley	Napa
North Coast	Lake, Marin, Mendocino, Napa, Solano, Sonoma
North Yuba	Yuba
Northern Sonoma	Sonoma
Oak Knoll District	Napa
Oakville	Napa
Pacheco Pass	San Benito
Paicines	San Benito
Paso Robles	San Luis Obispo
Potter Valley	Mendocino
Ramona Valley	San Diego
Red Hills Lake County	Lake
Redwood Valley	Mendocino

(continued)

CALIFORNIA AVA	COUNTY
River Junction	San Joaquin
Rockpile	Sonoma
Russian River Valley	Sonoma
Rutherford	Napa
Salado Creek	Napa
San Benito	Stanislaus
San Bernabe	Monterey
San Francisco Bay	San Benito, San Francisco, San Mateo, Santa Clara, Santa Cruz
San Lucas	Monterey
San Pasqual Valley	San Diego
San Ysidro District	Santa Clara
Santa Clara Valley	Santa Clara
Santa Cruz Mountains	San Mateo, Santa Clara, Santa Cruz
Santa Lucia Highlands	Monterey
Santa Maria Valley	San Luis Obispo, Santa Barbara
Santa Rita Hills	Santa Barbara
Santa Ynez Valley	Santa Barbara
Seiad Valley	Siskiyou
Sierra Foothills	Amador, Calaveras, El Dorado, Mariposa, Nevada, Placer, Tuolumne, Yuba
Solano County Green Valley	Solano
Sonoma Coast	Sonoma
Sonoma County Green Valley	Sonoma
Sonoma Mountain	Sonoma
Sonoma Valley	Sonoma
South Coast	Riverside, San Diego
Spring Mountain District	Napa
St. Helena	Napa
Stags Leap District	Napa
Suisun Valley	Solano
Temecula Valley	Riverside
Trinity Lakes	Trinity
Wild Horse Valley	Napa, Solano
Willow Creek	Humboldt, Trinity
York Mountain	San Luis Obispo
Yorkville Highlands	Mendocino
Yountville	Napa

FRANCE
Bordeaux: The 1855 Official Classification of the Médoc

CHÂTEAU	COMMUNE
First Growths	
Château Lafite-Rothschild	Pauillac
Château Latour	Pauillac
Château Margaux	Margaux
Château Haut-Brion	Pessac (Graves)
Château Mouton-Rothschild	Pauillac (Reclassified in 1973)
Second Growths	
Château Rausan-Ségla	Margaux
Château Rauzan-Gassies	Margaux
Château Léoville-Las Cases	Saint-Julien
Château Léoville-Poyferré	Saint-Julien
Château Léoville-Barton	Saint-Julien
Château Durfort-Vivens	Margaux
Château Gruaud-Larose	Saint-Julien
Château Lascombes	Margaux
Château Brane-Cantenac	Cantenac-Margaux
Château Pichon-Longueville-Baron	Pauillac
Château Pichon-Longueville	Pauillac
Château Pichon-Longueville-Comtesse de Lalande	Pauillac
Château Ducru-Beaucaillou	Saint-Julien
Château Cos d'Estournel	Saint-Estèphe
Château Montrose	Saint-Estèphe
Third Growths	
Château Kirwan-Cantenac	Margaux
Château d'Issan-Cantenac	Margaux
Château Lagrange	Saint-Julien
Château Langoa-Barton	Saint-Julien
Château Giscours-Labarde	Margaux
Château Malescot-Saint-Exupéry	Margaux
Château Cantenac-Brown	Cantenac-Margaux
Château Boyd-Cantenac	Margaux

(continued)

CHÂTEAU	COMMUNE
Château Palmer	Cantenac-Margaux
Château La Lagune	Ludon
Château Desmirail	Margaux
Château Calon-Ségur	Saint-Estèphe
Château Ferrière	Margaux
Château Marquis d'Alesme-Becker	Margaux
Fourth Growths	
Château Saint-Pierre	Saint-Julien
Château Talbot	Saint-Julien
Château Branaire-Ducru	Saint-Julien
Château Duhart-Milon-Rothschild	Pauillac
Château Pouget	Cantenac-Margaux
Château La Tour-Carnet	Saint-Laurent
Château Lafon-Rochet	Saint-Estèphe
Château Beychevelle	Saint-Julien
Château Prieuré-Lichine	Cantenac-Margaux
Château Marquis-de-Terme	Margaux
Fifth Growths	
Château Pontet-Canet	Pauillac
Château Batailley	Pauillac
Château Haut-Batailley	Pauillac
Château Grand-Puy-Lacoste	Pauillac
Château Grand-Puy-Ducasse	Pauillac
Château Lynch-Bages	Pauillac
Château Lynch-Moussas	Pauillac
Château Dauzac-Labarde	Margaux
Château d'Armailhac	Pauillac
Château du Tertre	Margaux
Château Haut-Bages-Liberal	Pauillac
Château Pédesclaux	Pauillac
Château Belgrave	Saint-Laurent
Château de Camensac	Saint-Laurent
Château Cos-Labory	Saint-Estephe
Château Clerc-Milon	Pauillac
Château Croizet-Bages	Pauillac
Château Cantemerle	Haut-Médoc

FRANCE The 1855 Official Classification of Sauternes and Barsac

CHÂTEAU	COMMUNE
Great First Growth (*Grand Premier Cru*)	
Château d'Yquem	Sauternes
First Growths	
Château La Tour-Blanche	Bommes
Château Lafaurie-Peyraguey	Bommes
Château Clos Haut-Peyraguey	Bommes
Château de Rayne-Vigneau	Bommes
Château Suduiraut	Preignac
Château Coutet	Barsac
Château Climens	Barsac
Château Guiraud	Sauternes
Château Rieussec	Fargues
Château Rabaud-Promis	Bommes
Château Sigalas-Rabaud	Bommes
Second Growths	
Château de Myrat	Barsac
Château Doisy-Daene	Barsac
Château Doisy-Dubroca	Barsac
Château Doisy-Vedrines	Barsac
Château D'Arche	Sauternes
Château Filhot	Sauternes
Château Broustet	Barsac
Château Nairac	Barsac
Château Caillou	Barsac
Château Suau	Barsac
Château de Malle	Preignac
Château Romer du Hayot	Fargues
Château Lamothe	Sauternes

FRANCE
The 1959 Official Classification of Graves

1959 the *INAO (Institut Nationale des Appellations d'Origine)* completed the classifications of Graves estates. There are two lists, one for producers of exceptional white wine, and one for producers of exceptional red wine. Some estates are on both lists.

CHÂTEAU	COMMUNE
Classification of Red Wines of Graves	
Château Bouscaut	Cadaujac
Château Haut-Bailly	Léognan
Château Carbonnieux	Léognan
Domaine de Chevalier	Léognan
Château de Fieuzal	Léognan
Château Olivier	Léognan
Château Malartic-Lagravière	Léognan
Château La Tour-Martillac	Martillac
Château Smith-Haut-Lafitte	Martillac
Château Haut-Brion	Pessac
Château La Mission-Haut-Brion	Talence
Château Pape-Clément	Pessac
Château La Tour-Haut-Brion	Talence
Classification of White Wines of Graves	
Château Bouscaut	Cadaujac
Château Carbonnieux	Léognan
Château Domaine de Chevalier	Léognan
Château Malartic-Lagravière	Léognan
Château Olivier	Léognan
Château La Tour-Martillac	Martillac
Château Laville-Haut-Brion	Talence
Château Couhins-Lurton	Villenave d'Ornon

FRANCE
The Revised Official Classification of Saint-Émilion, 1996

When Saint-Émilion was first classified in 1954, the *INAO* and local vintners agreed that the law should be written so as to allow a possible revision of the classification every 10 years. By this law, only *grand cru* estates are eligible to become *grand cru classé* or *Premier grand cru classé.* The most recent reclassification was in 1996.

First Great Classed Growths (Category A)	
Château Ausone	
Château Cheval-Blanc	
First Great Classed Growths (Category B)	
Château L'Angelus	Château Beau-Sejour-Becot
Château Beausejour	Château Belair
Château Canon	Château Figeac
Château La Gaffeliere	Château Magdelaine
Château Pavie	Château Trottevieille
Château Fourtet	
Great Classed Growths	
Château Balestard-La Tonnelle	Château Bellevue
Château Bergat	Château Berliquet
Château Cadet-Bon	Château Cadet-Piola
Château Canon-La Gaffelière	Château Cap De Mourlin
Château Chauvin	Clos des Jacobins
Château Corbin	Château Corbin-Michotte
Château Cure-Bon la Madeleine	Château Dassault
Château Faurie de Souchard	Château Fonplegade
Château Fonroque	Château Franc-Mayne
Château Grand Mayne	Château Grand Pontet
Château Guadet-St. Julien	Château Haut-Corbin
Château Haut-Sarpe	Château L'Arrosee
Château La Clotte	Château La Clusiere
Château La Couspaude	Château La Dominique
Château La Serre	Château La Tour Du Pin-Figeac
Château La Tour-Figeac	(Giraud-Belivier)
Château La Tour Du Pin-Figeac-Moueix	Château Lamarzelle
Château Laniote	Château Larcis-Ducasse

Château Larmande	Ch. Laroque St.-Chris.-Des Bardes
Château Laroze	Château Le Prieure
Château Les Grandes Murailles	Château Matras
Château Moulin Du Cadet	Château Pavie-Decesse
Château Pavie-Macquin-Ripeau	Château Petite Faurie De Soutard Château Château Saint-George Cote Pavie
Château Soutard	Château Tertre-Daugay
Château Troplong-Mondot	Château Villemaurine
Château Yon-Figeac	Clos de L'Oratoire
Clos Saint-Martin	Château Couvent des Jacobins

FRANCE
Pomerol (Not Classified)

Pomerol has never officially been rated. However, it is widely accepted that the following are the area's leading estates, with Château Pétrus in a special category of its own.

Château Pétrus	
Château Certan-de-May	Château la Conseillante
Château L'Évangile	Château La Fleur Pétrus
Château Lafleur	Château Latour à Pomerol
Château Petit Village	Château Trotanoy
Vieux Château Certan	Château Beauregard
Château Le Bon Pasteur	Château Certan Giraud
Château Clinet	Clos L'Église
Clos Rene	Château la Croix de Gay
Château L'Église Clinet	Château Gazin

Germany Communes

Communes of the following wine regions are listed below: Mosel-Saar-Ruwer, Mittel Mosel, Rheingau, Rheinhessen and Rheinpfalz.

Mosel-Saar-Ruwer		
Ayl	Eitelsbach	Kanzen
Kasel	Maximin Grunhaus	Oberemmel
Ockfen	Wiltingen	
Mittel Mosel		
Bernkastel	Brauneberg	Graach
Piesport	Trittenheim	Urzig
Wehlen	Zeltingen	
Rheingau		
Erbach	Eltville	Giesenheim
Hattenheim	Hochheim	Johannisberg
Kiedrich	Rauenthal	Rudesheim
Winkel		
Rheinhessen and Rheinpfalz		
Bingen	Bodenheim	Deidesheim
Forst	Kallstadt	Konigsbach
Nackenheim	Nierstein	Oppenheim
Ruppertsberg	Wachenheim	

Wine Organizations and Publications

Below are lists of some organizations and publications to further enhance your learning. Many other resources exist for researching wine, but these will help you get started.

ORGANIZATION	WEB ADDRESS
American Wine Society	www.americanwinesociety.org/web/welcome.htm
Appellation America	www.appellationamerica.com
Australian Wine & Brandy Corporation	www.awbc.com.au/Default.aspx?p=1
California State University Fresno Winery	www.fresnostatewinery.com
Canadian Vintners Association	www.canadianvintners.com
National Restaurant Association Education Foundation	www.nraef.org
New Zealand Winegrowers	www.nzwine.com/index.html
Professional Friends of Wine	www.winepros.org/index.htm
Rhone Rangers	www.rhonerangers.org
Society of Wine Educators	www.societyofwineeducators.org
The Oregon Wine Board	www.oregonwine.org/index.php
U.C. Davis Department of Viticulture and Enology	http://wineserver.ucdavis.edu
Washington Wine Commission	www.washingtonwine.org
Wine & Spirits Wholesalers of America	www.wswa.org/public
Wine Brats	www.winebrats.org
Wine Institute	www.wineinstitute.org
Wine Lovers Page	www.wineloverspage.com/index.phtml
Wines of Chile	www.winesofchile.org/index.htm
Zinfandel Advocates and Producers (ZAP)	www.zinfandel.org

PUBLICATION	WEB ADDRESS
California Grapevine	www.calgrapevine.com/home.html
Connoisseurs' Guide to California Wine	www.wineaccess.com/expert/connoisseurs/home.html
Decanter	www.decanter.com
Practical Winery & Vineyard	www.practicalwinery.com
Quarterly Review of Wines	www.qrw.com
Vineyard & Winery Management Magazine	www.myvwm.com
Wine Business Monthly	www.winebusiness.com
Wine Enthusiast	www.winemag.com/ME2/Default.asp
Wine Spectator	www.winespectator.com/Wine/Home
Wines & Vines	www.winesandvines.com

APPENDIX J

Wine Tasting Sheet

Wine #	Comments	Personal Rank	Group Rank

A

aeration (air-AY-shun) The process of using a decanter to oxygenate a wine.

Aglianico Widely planted red grape of southern Italy.

Airén The most widely planted grape in Spain, accounting for about one-third of total acreage. Especially prevalent in the dry central parts of the country, the grape is used in the production of brandy. It is also increasingly vinified into non-descript, but fresh, dry white wines.

Albariño A white grape grown in Galicia, Spain (also called *Alvarinho,* grown in the Vinho Verde region of Portugal). It produces complex, aromatic wines with bracing acidity.

alembic still A type of copper "pot" still used for making Cognac Brandy, as well as Pisco.

alluvial Soils created by the flooding along rivers and streams.

Alvarinho See *Albariño.*

***amabile* (ah-MAH-bee-lay)** Italian. Off-dry or semisweet.

American Viticultural Area (AVA) (VIHT-ih-kuhl-cher-uhl) A particular area of grape growing with specific boundaries sanctioned by the government.

amphora (AM-fuhr-uh) A jar with an oval body, narrow neck, and two handles used by the ancient Greeks to transport wine. Plural, *amphorae* or *amphoras*.

***Anbaugebiete* (AHN-bow-geh-beet)** German. A wine region. There are thirteen *Anbaugebiete* in Germany. A smaller, officially recognized wine-producing district within an *Anbaugebiet* is called a *Bereich*.

anthocyanins (an-tho-SIGH-uh-nins) The compounds responsible for the color of red wine.

***Appellation d'Origin Contrôlée (AOC)* (ah-pehl-lah-SYAWN daw-ree-JEEN kaw*n*-traw-LAY)** French. "Controlled Place of Origin." In the French system, the highest level of classification for wine regions. The designation applies to all wines made from grapes grown in that region if the wine is made according to all guidelines for that appellation, as outlined in the AOC laws.

appellation (ap-puh-LAY-shuhn) An area of origin of grapes.

aspect The direction in which a slope faces.

***Auslese* (OWS-lay-zuh)** German. Selected. Describes grapes picked at a high level of ripeness, thus with considerable sugars.

Australian Wine and Brandy Corporation (AWBC) A governmental organization responsible for the regulation and promotion of the Australian wine industry.

autolysis (aw-TAHL-uh-sihss) The breakdown of yeast cells.

AVA See *American Viticultural Area*

B

Bacchus (BAK-uhs) The Roman god of wine.

BATF Abbreviation for the Bureau of Alcohol, Tobacco & Firearms.

***Beerenauslese* (BAY-ruhn-OWSlay-zuh)** German. Literally, "selected berries," the term refers to grapes which are picked one-by-one depending on the level to which they have been affected by botryitis. The wine produced is one of the two levels of dessert wine within Germany's Prädikat system.

bench land A plateau that lies above the floor of a river valley.

***Bereich* (beh-RIKH)** German. A smaller, officially recognized wine-producing district within an *Anbaugebiet*.

bin number A number assigned to a wine that describes its area of storage in a restaurant's cellar area; bin numbers are used to facilitate ordering and locating specific wines when they are ordered by guests.

Biodiversity and Wine Initiative A project set up between environmental bodies and the South African wine industry with the aim of minimizing the loss of threatened natural habitat from the expansion of vineyards.

Black Association of the Wine and Spirit Industry (BAWSI) A South African association founded in 2002 with the aim of helping its black South African members play a more meaningful role in the wine industry.

blanc de blanc (BLAH*N* du BLAH*N*) A Champagne made from white grapes only.

blanc de noir (blah*n* duh NWAHR) A Champagne made from red grapes only.

Blauburgunder Swiss name for the Pinot Noir grape.

blind tasting A tasting where the identities of the wines being served are not known by the tasters.

bloom The period of flowering when pollination takes place.

bodega (boh-DAY-gah) Spanish. A winery.

boom and bust An economic cycle characterized by periods of overproduction and underproduction, typical for many agricultural products.

***Botrytis cinerea* (boh-TRI-this sihn-HER-ee-uh)** A type of mold that grows on grape clusters that can be used to make dessert wine.

bottle bouquet The term used to describe the increasingly complex aromatics that develop in wines as they age.

***Brettanomyces* (breht-tan-uh-MI-sees)** A yeast that can grow in wines while they are being aged that will produce a distinctive "barn yard" smell, considered by some to be a type of spoilage.

brilliant The appearance of a wine clear of any visual defects.

brix See *degrees brix*

brut (BROOT) The term used to describe a dry Champagne; also used by sparkling wine producers in other countries.

by the glass (BTG) Wines that are sold on an individual glass basis (as opposed to selling an entire bottle). This manner of selling wine usually represents the bulk of wine sales in the restaurant setting.

C

California Wine Association (CWA) A large cooperative winery that produced most of California's wine from 1894 to 1919.

canes The shoots of a grapevine on which the buds form.

cap The layer of skins that is formed on the surface of a fermenting container of red must.

Cape Doctor Summer southeasterly wind that moderates temperatures and helps prevent disease in the vines.

Cape Floral Kingdom A World Heritage site in the western Cape that contains the greatest number of indigenous plants in the smallest area on earth.

capsule (KAP-suhl) The foil or laminate covering that covers the cork and top of the wine bottle.

carbonic maceration (kar-BAHN-ihk mas-uh-RAY-shuhn) A form of fermentation in which whole bunches of grapes are placed in a large container which is then filled with carbon dioxide and sealed. The weight of the bunches above breaks the bunches on the bottom, thus releasing the juice and allowing fermentation to begin. While fermentation is occurring in the bottom of the tank, in the top levels fermentation proceeds inside the unbroken grapes because the lack of oxygen causes the skins to die, releasing an enzyme inside each grape that transforms sugars into alcohol. The process of carbonic maceration, widely used in Beaujolais, results in wine that is softer and fruitier than would be the case if the whole batch had been crushed, and then fermented.

Carmenère (car-men-HER) A red grape that is similar in taste and appearance to Merlot, widely planted in Chile but rarely grown elsewhere.

Cava (KAH-vah) Spanish sparkling wine made in the *méthode champenoise.* The term, from the Catalonian word for "cellar," was adopted in 1970.

Chambourcin (shahm-boor-SAN) A recent French hybrid; a very vigorous and productive red grape, planted frequently in Virginia.

Champagne (sham-PAYN) A sparkling wine from the Champagne region of France.

chaptalization (shap-tuh-luh-ZAY-shuhn) Addition of sugar after fermentation. Named for Jean Chaptal of France. Forbidden in all Qualitätsweins of Germany.

Charmat process (shar-MAHT) A method of sparkling wine production where the secondary fermentation takes place in tanks rather than in the bottle.

Charta (KAR-tah) German. A group of wine producers in the Rheingau dedicated to producing wines of higher quality than specified by German wine law. Founded 1984.

Chasselas (shas-suh-LAH) One of Switzerland's major white grapes.

Chenin Blanc (CHEN-ihn BLAHN) White grape variety producing wines of renown in the Loire and the dominant grape variety in South Africa.

***classico* (KLA-sih-koh)** Italian. A geographic designation for the highest quality zone within an Italian *DOC* region.

clone (KLOHN) A grapevine that is genetically identical to the parent grapevine that it is propagated from, retaining the same attributes.

***clos* (KLOH)** French. A walled-in vineyard. The term is often found in the names of Burgundy's vineyards.

***Colheita* (cuhl-YAY-tah)** The Portuguese word for "harvest," used to signify the vintage year of a wine. It is also the name of a style of Port made from grapes grown all in one year and aged at least seven years in barrel before bottling.

Concord (KAHN-kord) A native American grape variety used for grape juice production and some winemaking.

cooper A barrel maker.

cooperative A winery jointly owned by grape farmer members who deliver their crop to a central cellar, the majority of the wine usually sold in bulk to one or more merchants for blending into their branded wines.

coopers Barrel makers.

cordons The branches of a grapevine that the canes grow from.

corkage fee (CORK-ihj) A fee that a restaurant may charge guests who bring their own bottle of wine to enjoy with their meal.

cork taint A musty smell that is the most common problem associated with natural corks. It occurs when the cork has been exposed, either in the forest or during processing and storage, to mold growth and the corks absorb a compound called 2,4,6-trichloroanisole, more often called trichloroanisole or TCA, from the mold. TCA is the most prominent of several compounds that are produced by mold and can be detected in the aroma of a wine in levels as low as several parts per trillion.

Cortese (kohr-THE-zeh) Italian. Important white grape of Piedmont.

cost percent The cost percent is the percentage of the selling price that was spent to buy the product.

Côt A red grape variety native to the Bordeaux region of France; popular in Argentina, where it is called Malbec.

crianza (kree-AHN-zah) In Spain, the youngest classification for wood-aged wines.

Criolla A name for the Mission grape variety, similar to País that is grown in South America.

cross A new grape species resulting from the cross-pollination of two native varietals of the species *labrusca.*

***cru* (KROO)** French. Literally "growth," often used to signify a rated, or high quality vineyard.

***cuvée* (koo-VAY)** A blend of wine, a French word that translates literally to "tub full" or "vat full."

D

decanter A glass carafe used to serve wine, often in conjunction with decanting and aeration.

decanting (dee-KANT-ing) The process of using a decanter to separate a wine from its sediment.

degree days A method of using average daytime temperature to provide a rough estimate of a vineyard's terroir.

degrees Brix (°Brix) The percent sugar by weight of a liquid.

***demi-sec* (DEHM-ee-sehk)** French. Literally "half-dry," but used to describe a Champagne that is quite sweet.

***Denominação de Origem Controlada* (deh-naw-mee-nah-THYON deh aw-REE-hen con-traw-LAH-tah)** Portuguese. Literally, controlled denomination of origin, meaning the wine is from one of Portugal's protected appellations, and has been produced in compliance with the regulations of that region.

***Denominación de Origen* (deh-naw-mee-nah-THYON deh aw-REE-hen)** Spanish. A controlled appellation with specific regulations, the enforcement of which is overseen by a local branch of the national *INDO*.

***Denominazione di Origine Controllata (DOC)* laws (deh-NAW-mee-nah-TSYAW-neh dee oh-REE-jee-neh)** Italy's set of laws enacted in 1963 to regulate the production of wine, protect the defined wine zones, and guarantee authenticity and consistency of style.

***Denominazione di Origine Controllata e Garantita (DOCG)* (deh-NAW-mee-nah-TSYAW-neh dee oh-REE-jee-neh con-traw-LAH-tah eh gah-rahn-TEE-tah)** The highest designation in Italy's *DOC* laws, given only to the country's most prestigious wine zones.

depth The intensity of color of a wine.

descriptive analysis The process of describing and identifying the different smells present in a wines aroma.

designated viticultural area (DVA) In Canada, an officially approved appellation.

dessert wines Wines made with appreciable sugar.

disgorging The process of removing the yeast from the neck of the bottle in *méthode champenoise* production.

Dionysus (di-uh-NI-suhs) The Greek god of wine.

***dolce* (DOHL-chay)** Italian. Sweet.

Dom Pérignon (dom pay-ree-NYON) A blind Benedictine monk who was involved with the introduction of Champagne.

dosage (doh-SAHJ) The addition of a small amount of wine, usually sweetened, after disgorging.

***doux* (DOO)** French. "Sweet"; used to designate the sweetest style of Champagne, which is rarely imported into the United States.

dry-farmed Growing grape vines without supplemental irrigation.

dry wine A wine without perceptible sweetness.

dull The appearance of a wine that is turbid or cloudy.

E

Egri Bikavér (EH-g*r*ih BIH-kah-vah*r*) Hungary's most famous red wine. The name derives from a legendary battle against the Turks in the early 1500s, during which the residents of the town of Eger, known as the Magyars, were fiercely defending their town. During the battle, the Magyars drank large quantities of red wine, and, legend has it, the Turks, upon seeing the red-stained beards of their opponents, withdrew because they believed the Magyars obtained their great strength and ferocity by drinking the blood of bulls.

***Einzellage* (I'n-tsuh-lah-guh)** German. A single vineyard. There are 2,600 *Einzellagen* in Germany's wine regions.

***Eiswein* (ICE-vyn)** German. Ice wine. Made from nonbotryitised but very ripe grapes that have been frozen by a sudden drop in temperature. The grapes are picked from the vine while still frozen. The frozen water crystals are discarded during the crush, leaving a high concentration of natural sugars. The resulting sweet wine must have minimally the same sugar content as a *Beerenauslese*.

enologist Someone who studies wine, from the term *enology*.

enology (ee-NAHL-uh-jee) The study of winemaking.

***Erstes Gewächs* (AYR-sters GER-vehks)** German. "First Growth"—one of the terms proposed as vineyard ratings by the group of vintners in Germany calling themselves the "First Growth Committee"

esters (EHS-tuhrs) Aromatic compounds created by acid and alcohol molecules binding together during the course of fermentation and bottle age.

estufa From the Portuguese word for stove; the tanks used on the island of Madiera to heat wine, thus accelerating the maturation process.

ethanol Beverage alcohol that is produced by yeast, also called ethyl alcohol.

extended maceration (mas-uh-RAY-shun) Delaying the pressing of red must until several weeks after fermentation has stopped.

extra brut (BROOT) Champagne to which no *dosage* or additional sugar, is added before bottling.

extra dry A confusing term used in Champagne to indicate a wine that is slightly sweet.

F

fighting varietals (vuh-RI-ih-tuhl) Inexpensively priced varietal wines that were introduced in the early 1980s.

fining (FI-ning) The addition of a compound that effects the composition of a wine but does not stay in solution.

fino A light, dry style of Sherry.

fixed markup A method of pricing that assigns a standard markup value to all wines within a certain price point; used to offset the high prices that would result from assigning one markup strategy to wines at all price points.

flight A group of wine evaluated at the same time in a wine tasting.

flor yeast (FLAWR YEEST) A harmless, film-forming yeast that floats on the surface of Sherry while it ages in casks.

fortified wine A wine fortified through the addition of extra alcohol, usually in the form of neutral, that is, colorless and flavorless, brandy. The best-known fortified wines are Sherry, Port, Madiera, and Marsala.

Fort Vancouver An early settlement on the north shore of the Columbia river started by the Hudson's Bay Company.

free run The first fraction of juice that is released during pressing; usually it is the highest quality juice.

French paradox The paradox that although the French have higher saturated fat in their diets they have lower heart disease than Americans, suspected to be caused by the moderate consumption of red wine.

frizzante **(freet-TSAHN-teh)** Italian. Slightly sparkling wine, as opposed to a truly sparkling wine or *spumante.*

fungicide A compound that is applied to a vineyard to control rot.

Furmint (FOOR-mint) White varietal indigenous to Hungary.

futures A method of purchasing wine before it is released to the market, allowing buyers to secure the wines that they desire.

G

Garnacha A widely planted *vinifera* grape used in many of Spain's red wines. In France, the grape is called Grenache.

garrafeira (gah-rah-FAY-ruh) Used in Portugal to signify a wine that is from an exceptional year and has aged in cask for at least two years and, after bottling, is aged an additional year before being released.

Geographic Indications (GIs) A method of classification that subdivides the territory of each Australian state into a series of Zones, Regions, and Subregions.

glassy-winged sharpshooter A vineyard pest that spreads Pierce's disease.

grafting The process of taking a cutting, or scion, and affixing it to a rootstock to produce a single grapevine with the positive aspects of each.

gran reserva (grahn ray-SAYR-vah) Spanish. Wine from excellent vintage years and from fine vineyards that have received considerable aging before being released for sale. Gran Reservas, by law, have been aged in oak casks for a minimum of two years, and then aged further in bottle or tank for at least another three years.

Grosslage **(GROSS-lah-guh)** German. Literally "large site," the term describes a number of contiguous vineyards that have been grouped together under one popular name. About 150 *Grosslagen* were created by the German Wine Law of 1971. Some of these are so large (average size: 1,500 acres) as to be meaningless, geographically and viticulturally.

Grüner Veltliner (GROO-ner FELT-lih-ner) Austria's most important indigenous grape.

Gutsabfüllung **(GOOTS-ab-few-lung)** German. Estate-bottled. The wine so labeled must be made by the same entity that grew the grapes and made the wine. A more restrictive term than *Erzeugerabfüllung.*

H

half bottle This size bottle holds half of the volume of a standard bottle of wine (375 ml).

Hanepoot A white table grape and wine grape, also known as Muscat d'Alexandrie.

heat summation A method of using average daytime temperature to provide a rough estimate of a vineyard's terroir.

heavy soil A soil with a high percentage of clay that has a high capacity for holding water.

herbicide A compound that is applied to a vineyard to control unwanted plant growth.

Hermitage An Australian name for the grape variety Syrah, also known as Shiraz.

house wine A term commonly used to describe a restaurant's standard offerings by the glass. This term has taken on a negative connotation in recent years because it implies an entry level and basic wine.

hue The shade of color of a wine.

Huguenots French Protestants who fled their homeland after the revocation of the Edict of Nantes in 1688, many of whom travelled to South Africa and helped establish the Cape's wine industry.

Hunter Valley Riesling (REEZ-ling) An Australian name for the grape variety Semillon.

hybrid A new grape variety produced by breeding two varieties from different botanical species, most commonly a varietal from an American species, e.g., *labrusca,* with a French varietal from the *Vitis vinifera.*

I

imbottigliato dal produttore all'origine **(ihm-boh-tee-LYAH-toh)** Italian. Estate-bottled. A term that can appear on an Italian label; equivalent to France's *misen bouteilles au château or au domaine.*

Indicazione Geografica Tipica Italian regional wine that reflects the terroir of the geographic region which shows on the label; equivalent to France's *vin de pays.*

Instituto Nacional de Denominaciones de Origen (INDO) The government agency that oversees Spain's *DO* system; equivalent to France's *INAO.*

inert gas A gas that does not interact with the substances that it contacts.

Instituto da Vinha e Vinho (IVV) The government agency at the national level that oversees Portugal's systems of controlled appellations.

Integrated Production of Wine A scheme introduced in 1998 setting guidelines and minimum standards for environmentally friendly practices in South African vineyards, cellars, and packaging.

Isabella A hybrid grape once popular in the Northwest.

J

jerepigo Wine style consisting of very sweet, unfermented grape juice fortified with grape spirit.

jug wines Inexpensive table wines that are sold in large bottles.

K

Kabinett **(kah-bih-NEHT)** German. The driest *Prädikat* within the QmP rankings.

Kadarka (KAH-dahr-kah) Hungary's most characteristic red grape.

L

lagar (lah-GAHR) A shallow vat used to produce Port.

Landwein **(LAHNT-vyn)** German. Regional wine, a step above *Tafelwein* in quality in Germany's quality control laws.

large format bottle Any size bottle that is larger than the standard 750 ml format.

late-harvest Wines that are made from grapes picked at a much higher sugar level than those for table wines.

lees (LEEZ) The solids that settle out at the bottom of a vessel of wine.

light soil A soil with a high percentage of sand that has a low capacity for holding water.

loam A type of soil that is a mixture of clay, silt, sand and organic matter that is fertile and drains well.

M

Macabeo (mah-kah-BEH-oh) The dominant white grape of Spain, particularly in Rioja. Also called Viura.

macroclimate The broad weather conditions of a particular wine-growing region.

magnum The smallest of the large format bottles. It holds the same amount of wine as two standard bottles (1500 ml or 1.5 liters).

Malbec (mahl-BEHK) A red grape variety popular in Argentina, native to the Bordeaux region of France, where it is called Côt.

malolactic bacteria (ma-loh-LAK-tihk) The bacteria that convert malic acid to lactic acid.

Manzanilla A light style of Sherry, aged in the coastal region of Sanlúcar de Barrameda, Spain.

marc **(MAHR)** Grape skins after they are pressed.

Maréchal Foch (MAH-ray-shahl FOHSH) A French hybrid named for France's famous World War I general, widely planted in Canada and New York.

markup The difference between the restaurant's cost and the selling price of a product.

mesoclimate Local weather conditions that affect a vineyard or a portion of a vineyard, more commonly known as microclimate.

Méthode Cap Classique (MAY-tohd KLAH-seek) South African term for sparkling wine made in the same method as Champagne, which term isn't permitted.

méthode champenoise **(may-TOHD shahm-peh-NWAHZ)** The traditional method of producing sparkling wine by fermenting it in the bottle.

metodo classico/metodo tradizionale **(MEH-toh-doh CLAH-see-coh/MEH-toh-doh trah-dee tsyoh-NAH-lay)** Terms used to describe a sparkling wine made in the *méthode champenoise.*

microclimate Local weather conditions that affect a vineyard or a portion of a vineyard, less commonly used to refer to the climatic conditions around a single vine.

Mission grape A *vinifera* red grape variety grown by the missions. See also Criolla.

mission period From 1769 to 1833, when a series of missions were established along the coast of California by Spain and Mexico.

moelleux **(mwah-LEUH)** French. A term used to describe sweet wines, often used for dessert wines in the Loire Valley.

Moscato A white grape often used to make sparkling wines; Muscat.

mousseux **(moo-SEUHR)** French. Sparkling; used to describe sparkling wines made outside of Champagne.

mouthfeel The texture or body of a wine felt in the mouth.

Müller-Thurgau (MEW-luhr TOOR-gow) A lesser white grape indigenous variety developed in Switzerland in the late nineteenth century and now widely planted in Germany. The grape produces characterless, flabby wines, and is slowly being phased out.

Muscadel (mus-kuh-DEHL) A grape variety and wine style. The grape variety, both white and red versions, is a synonym of Muscat à Petits Grains. The wine style is the unfermented very sweet juice of the grape variety, fortified with grape spirit, in other words, a jerepigo.

Muscadinia rotundifolia **(MUHS-kuh-dihn roh-tuhn-dih-FOHL-ee-uh)** A grapevine that is native to the Eastern United States, also known as *Vitis rotundifolia.*

must Unfermented grape juice, either before or after it has been separated from the skins and seeds.

N

Nebbiolo (neh-b'YOH-loh) Italian. Principal red grape of Piedmont.

négoçiant **(nay-goh-SYAHN)** The middleman in wine production who purchases grapes from vineyard owners, makes the wine and markets it.

noble rot (***pourriture noble*** **in French;** ***Botrytis cinerea*** **in Latin)** A mold that attacks the skins of grapes while they are still on the vine, causing the skins to crack, thus allowing the watery juice of the grape to evaporate. This evaporation greatly elevates the concentration of sugar in the grapes, which can result in every sweet but balanced dessert wines.

nonvintage A wine that is made from grapes harvested in two or more years is a nonvintage wine. Champagne is often nonvintage (abbreviated "N.V."). Sherry is always nonvintage.

O

Oechsle (UHK-sluh) German. A measurement of the level of sugar in grapes as an indication of ripeness. Named for German scientist, Ferdinand Oechsle.

oidium Also known as powdery mildew, a fungus that attacks the vine but can be controlled by dusting the vines with sulfur.

olfactory bulb The organ above the sinus that receives signals from the olfactory epithelium before sending them on to the brain.

olfactory epithelium The membranes in the nasal septum that detect volatile compounds that are inhaled.

oloroso The heavier, richer, dessert style of Sherry, aged in casks without the protective flor yeast.

own-rooted When grape vines grow on their own roots instead of using a rootstock.

P

País A red grape variety native to Chile, similar to Mission or Criolla.

Palomino (pah-loh-MEE-noh) A white grape of the Jerez region of Spain, widely used in the production of Sherry.

Paris tasting A tasting that took place in Paris in 1976 in which California wines took top honors.

passive cellar A cellar that does not employ any mechanical method of temperature or humidity control.

Pedro Ximénez (PEH-droh hee-MEE-nihs) A white grape widely planted in the southern parts of Spain, and often used for Sherry.

pesticide A compound that is applied to a vineyard to control unwanted organisms.

phylloxera (fihl-LOX-er-uh) Root-eating aphid native to North America that devastated the vineyards of Europe and California in the late nineteenth century.

Pierce's disease A bacterial disease that can kill grapevines.

Pinotage (pee-noh-TAHJ) South African red grape variety, a cross between Pinot Noir and Cinsaut (then known as Hermitage), bred by professor Abraham Perold in 1924.

Pisco A lightly colored brandy that is very popular in Chile.

polymerize The process of smaller molecules joining together to form larger molecules.

pomace (PAH-muss) Grape skins after they are pressed.

Port A fortified red wine with about 10 percent sugar and 20 percent alcohol from the Douro Valley in Northern Portugal.

Port-style A wine made in the style of Port, but produced outside the Port region of Portugal.

post off A periodic discount offered on bottles and cases of wine, usually offered on wines sold by the glass in the restaurant operation.

potstill brandy Wine that has been batch distilled twice in a potstill (alembic) before undergoing a minimum of three years' aging in small oak casks.

Prädikat **(preh-dih-KAHT)** In German and Austrian wine laws, the ripeness of grapes at harvest time and thus the sweetness of the resulting wine.

press fraction The juice that is released in pressing after the free run, usually of a lower quality.

primary fermentation The first alcoholic fermentation, utilizing sugar from the grape juice.

primary lees (LEEZ) The lees that form in juice before fermentation.

Prohibition The period of time from 1920 to 1933 when the purchase or sale of alcohol was illegal in the United States.

propriétaire **(proh-pree-ay-TEHR)** French. Indicates that the entity that made a wine also owns the vineyard where the grapes were grown.

pumping over Irrigating the cap with juice during fermentation.

punching down Breaking up the cap during red wine fermentation by pushing it down into the juice.

Q

Qualitätswein bestimmter Anbaugebiete (QbA) **(kvah-lih-TAYTS-vine behr-SHTIHMT-tuhr ahn-BOW-geh-beet)** German. The third level of quality in Germany's wine laws, used to show that the grapes in a bottle were all grown within one of the thirteen wine regions *(Anbaugebeite).* There are few regulations or standards at this level. The majority of wines exported from Germany are at the QbA level.

Qualitätswein mit Prädikat (QmP) **(kvah-lih-TYATS-vine mitt PRAY-dee-kaht)** The highest level of quality in Germany's system of quality control laws. The term means literally "quality wine with designation." There are five Prädikats, or designations, based on the quality of the grapes used and their level of ripeness when picked.

quercetin A compound found in wine that has positive health effects.

quinta **(KEEN-tah)** A Portuguese name for a farm or vineyard, it also applies to an estate that produces wine.

R

racking The process of transferring clean wine or juice off the lees that form in a tank or barrel.

rain shadow The weather condition where Pacific storms lose their moisture on the west side of the Cascades, creating a dry climate in eastern Oregon and Washington.

raulí A type of South American beech tree used for making wine casks but being phased out with modernization.

Recioto (reh-CHAW-toh) Italian. The ripest grapes in a bunch, usually those on the upper sides, often further concentrated by being dried on special mats in ventilated rooms.

reserva (ray-ZEHR-vah) A Spanish term used to indicate additional aging. The law specifies a total of three years of aging either in cask or in bottle for reserva wines. In Portugal the term signifies a wine from a very good vintage.

resveratrol (rez-VEHR-ah-trawl) A compound found in wine that has positive health effects.

Rhine Riesling (RINE REEZ-ling) The Australian name for the grape variety White Riesling.

riddling (RIHD-ling) The process of moving the yeast to the neck of the bottle in *méthode champenoise* production.

***riserva* (ree-ZEHR-vah)** Indicates additional aging for Italian wines, usually partly in oak barrels.

Robola White grape native to Cephalonia in Ionian Sea off Greece's west coast.

rootstock A grapevine that is resistant to soil pests that is used as a root system for grafting wine grape varieties.

rosé A pink wine made in one of two methods. The first method is to use red grapes but allow only a very short maceration period so that minimal color is extracted from the skins. The second method is to add a small amount of white wine to a red wine after fermentation is complete.

rotary fermentor (fer-MEN-tor) A tank designed to mix the cap and juice during red wine fermentation automatically.

ruby Port The simplest, most straightforward style of Port (and the most affordable). Bottled after only two or three years in cask, it is deep ruby in color and has a vivid fruitiness.

S

Saccharomyces cerevisiae The species of yeast that is most often used for winemaking. The name is derived from the Latin terms for sugar-fungus and grain, the latter referring to its most common use in breadmaking.

Sangiovese (san-joh-VAY-zeh) Principal red grape of Tuscany.

scion (SI-uhn) A cutting from a cane of a grapevine used to propagate another grapevine that is genetically identical to the parent.

***sec* (SEHK)** French. Dry.

***secco* (SHE-koh)** Italian. Dry.

sediment The gritty accumulation that appears in the bottle after considerable age—a combination of tannin, acid, and color molecules that have precipitated out of solution as they became larger.

***Sekt* (ZEHKT)** German. A quality sparkling wine.

sensory evaluation The process of using the effect a wine makes on one's senses to review and describe a wine.

Serra, Junípero A Franciscan priest who founded the missions in California.

Seyval Blanc (say-vahl BLAHN) A French hybrid widely planted in Canada and parts of the Eastern United States.

shatter The process of unfertilized flowers falling off the grape cluster.

Sherry A wine native to Spain done in a wide variety of styles that have an oxidized character.

Shiraz An Australian name for the grape variety Syrah, also known as Hermitage.

sit-down tasting An industry wine event where the participants are invited to attend lecture-style tastings and educational events. These may include meeting the winemakers or vineyard managers in addition to tasting their wine offerings.

sliding scale A method of pricing that applies various cost percents to wines at different price points. This ensures that the guests are offered reasonable value at all levels of the wine list.

***solera* (soh-LEH-rah)** A sequential system of aging wine in barrels. Used primarily in Jerez, a *solera* allows wine from various vintages to be aged in the same system, thus evening out quality differences between vintages.

sommelier (saw-muh-LYAY) The person responsible for organizing and sustaining a wine program in a restaurant setting.

sparkling wine Wine with bubbles or effervescence.

Spätburgunder The German word for the Pinot Noir grape.

***Spätlese* (SHPAYT-lay-zuh)** German. The second designation in the QmP level. "Spätlese" means "late," and these wines are picked later in the fall, after the Kabinett grapes. The exact ripeness, or must level, is explicitly spelled out in German wine law for each region and each varietal. Spätlese wines are not sweet, but off-dry.

***spumante* (spoo-MAHN-tay)** Italian. Sparkling wine.

stemmer-crusher A machine used to separate grape berries from stems and break the berries open to release the juice.

still wine A wine without effervescence.

Stimson Lane The parent company of Chateau Ste. Michelle Winery, the largest in the state of Washington.

***superiore* (soo-payr-YOH-reh)** Italian. Designation for wine of better quality, with higher alcohol content and/or additional aging.

***sur lie* (soor LEE)** Aging your wine on the yeast lees after fermentation.

***sur pointe* (soor PWANT)** Bottles in the neck-down position.

T

table tent A small tabletop display that can be used to advertise specials or specific wines.

table wine A still wine, that is also dry, and with moderate alcohol content.

***Tafelwein* (TAH-fuh-vyn)** Table wine, the lowest level of quality for German wines. A very small percentage of Germany's wines fall into this category.

tannin buildup In red wines, repeated sips of the same wine will taste increasingly astringent.

tannins Polyphenolic compounds that contribute bitterness and astringency to wine and play an important role in a wine's flavor, texture, and ageing qualities. They are primarily derived from grape skins, but seeds, stems, and oak barrels can also contribute to a wine's tannins.

tawny Port A wine aged much longer in wooden casks than a Ruby, and which thus has lost its reddish color and taken on a golden-brown, or tawny, hue. Most Tawny Ports are bottled with an indication of age on the label.

Tempranillo (tem-prah-NEE-yoh) The most important grape of Spain, the backbone of many of its greatest red wines.

terroir (tehr-WAH*R*) A French term used to describe the unique character a wine exhibits due to the specific physical characteristics (i.e., terrain, soil composition, drainage, precipitation, prevailing winds, average temperatures, etc.) of the location where the grapes were grown.

***tête de cuvée* (koo-VAY)** French. Literally, "top batch," used in Champagne to designate the deluxe bottling of a producer.

three-tiered system of distribution The method of distribution of alcohol in the United States mandating that products are transferred from a producer to a wholesaler to a retailer or restaurant operation.

threshold The level at which a taster can detect a given flavor compound.

tirage (tee-RAHZH) The process of aging a bottle of sparkling wine on its fermentation lees.

Tocai Friulano A white grape widely planted in northeast Italy.

Tokaji Aszú Hungary's famous dessert wine, made from botrytized grapes from the Tokaj-Hegyalja region.

Torrontés A white grape variety with a fruity flavor, native to the Rioja region of Spain.

trade tasting An event sponsored by a winery or wholesaler to allow industry professionals to taste wines that are recently released.

trellis A support for grapevines.

trellising Using supports either man-made or natural, such as a tree or post, to elevate a grape vine off the ground.

triage The labor-intensive process of hand-sorting grapes by quality, and rejecting inferior ones, before winemaking begins.

trichloroanisole (TCA) A chemical compound that results from the interaction of chlorine and wood products, most commonly corks.

***Trockenbeerenauslese* (T*R*AWK-uhn-bay-ruhn-OWS-lay-zuh)** German. Literally, "selected dried berries," this is the fifth designation within the QmP level of quality. The grapes are picked late in the fall, when they are fully botrytized, and thus shriveled (dried). The resulting wines are wonderfully rich and honeyed, yet retain enough acidity to be beautifully balanced. Very rare, risky, and labor-intensive—and thus, expensive.

TTB Abbreviation for the Alcohol and Tobacco Tax and Trade Bureau.

twist top Also known as a screw cap. A twist top closure is a metal top that is used to seal a bottle instead of a traditional cork.

U

UC Davis A University of California campus with research and degree programs in viticulture and enology.

ullage (UHL-ihj) The head space between the wine and the container it is stored in. For example, in a wine bottle, the space between the wine and the end of the cork.

unique selling point (USP) The aspects which make any item unique and appealing to the consumer.

V

Vallejo, Mariano A general and rancher who helped to settle the Napa and Sonoma Valleys.

VDQS (*Vin Délimité de Qualité Supérieure*) In French law, the classification of wine regions just below *Appellation d'Origin Côntrolée.*

véraison (vay-ray-ZO*N*) The beginning of ripening of a grape berry.

Verband Deutscher Prädikatsweinguter (VDP) A group of German winemakers dedicated to traditional styles in the production of quality wine. Founded 1908 in the Mosel region.

Verdejo A white grape of considerable complexity and vibrant acidity that is widely planted in the Rueda region of Spain.

Verdicchio (vehr-DEEK-kyoh) White grape planted throughout southern Italy.

vertical flight A group of the same type of wine from consecutive vintages.

Vidal Blanc (vee-dahl BLAH*N*) A French hybrid, the basis for many of Canada's famous late-harvest dessert wines, as well as many ice wines.

***vin délimité de qualité supérieure* (VDQS) (va*n* deh-lee-mee-TAY duh kah-lee-TAY soo-pehr-YUR)** In French law, the classification of wine regions just below *Appellation d'Origin Côntrolée.*

***vin de pays* (va*n* doo pay-YEE)** "Country wine," the second level of classification in French wine laws.

***vin de table* (va*n* deu TAH-bl)** The lowest level of classification in French wine laws.

***vin doux naturel* (va*n* doo nah-tew-REHL)** Indicates a naturally sweet wine made by arresting fermentation early on through the addition of alcohol to kill the yeast, leaving high levels of unfermented sugar.

***vino da tavalo* (VEE-noh dah TAH-voh-lah)** Italian. Table wine; the lowest designation in Italy's *DOC* laws.

vintage (VIHN-tihj) The year in which the grapes for a wine were harvested.

Vintners Quality Alliance (VQA) Canada's quality control agency.

viticulture (VIHT-ih-kuhl-cher) Grape growing.

***Vitis labrusca* (VEE-tihs luh-BRUSH-kuh)** Species of the *Vitis* genus that is indigenous to North America, especially in the Northeastern regions.

***Vitis riparia* (VEE-tihs rih-PEHR-ee-uh)** A very hardy species of the *Vitis* genus, prevalent in the Southeastern United States, usually in damp areas along streams.

***Vitis vinifera* (VEE-tihs vihn-IHF-uh-ruh)** The species of grape that is most commonly used for producing wine.

Viura The dominant white grape of Spain, particularly in Rioja. Also called Macabeo.

volatile A compound that can evaporate and become airborne.

Volstead Act The 18th Amendment implementing the prohibition of the sale or consumption of alcoholic beverages.

3 v's (vintner, varietal, vintage) The vintner is the producer of the wine (who made it); the varietal is the grape that is used to make the wine; the vintage is the year that the grapes were harvested.

W

walk-around tasting An industry event where wines can be sampled and tasted. Usually less formal than a sit down tasting.

Washington State University A university in eastern Washington that did pioneering work on growing *vinifera* in the Northwest.

Washington Wine Quality Alliance (WWQA) A voluntary trade organization formed to set standards for Washington State wines.

wholesaler An operation that acts as a go-between for the wine producer and the wine retailer. This is the second tier of the three-tiered system of distribution.

wine flight A series of wines that are poured in smaller portions and served together or with accompanying courses.

Wine Institute A trade organization for California wineries.

wine key Also called a corkscrew. This is the tool that is used to open wine bottles. The key components of a wine key are the worm, the knife blade, and the lever.

Wine of Origin Scheme The Scheme drawn up to demarcate South Africa's wine regions, first implemented in 1973, which divided the winelands into three levels of decreasing size: regions, districts, and wards.

winterkill Damage or death of grapevines caused by severe cold during their dormant period.

worm The spiral piece of a wine key that is twisted into the cork, allowing it to be removed.

REFERENCES

Abel, E. L. (1995). An update on incidence of FAS: FAS is not an equal opportunity birth defect. *Neurotoxicology and Teratology 17*(4): 437–443.

Adams Beverage Group. (2004). *Adams wine handbook.* Norwalk, CT: Author.

Allen, L. (2003, November) Greece's Enduring Wine Heritage. *Wine Spectator,*

Australian Wine and Brandy Corporation. (2004). *AWBC Wine facts Information,* Adelaide, South Australia: Author.

Baldy, M. W. (1997). *The University Wine Course.* The Wine Appreciation Guild: San Francisco.

Banning, B. (June, 2004). Jean-Michel Deiss. *Wine International.*

Bouchard, F. (2003, May). Profile: Nik Weis. *Beverage Business,* p. 4.

Bouchard, F. (2005, September) Profile: Tom Schmeisser. *Beverage Business,* p. 12.

Boulton, R. B., Singleton, V. L., Bisson, L. F., & Kunkee, R. E. (1996). *Principles and Practices of Winemaking.* New York: Chapman & Hall.

Bowen, P. (2005, June) Sprinkler vs Drip Irrigation in the Okanagan Valley. *Wines & Vines,*

Bowers JE. Meredith CP. The Parentage Of A Classic Wine Grape, Cabernet Sauvignon. [Article] *Nature Genetics. 16(1):84-87, 1997 May*

Brook, S. (1999). The Wines of California. New York: Faber and Faber.

Burman, J. (1979). *Wine of Constantia.* Pretoria, South Africa: Human & Rousseau.

California Department of Food and Agriculture (2003) *Grape Crush Report 2002 Crop.* Sacramento, CA: The State of California.

California Department of Food and Agriculture (2004). *Grape Crush Report 2003 Crop.* Sacramento, CA: The State of California.

Chittim, C. (2005, October) New Mexico Wine-making: A Colorful History. *Wines & Vines,*

Coates, C. (2000). *The Wines of France.* San Francisco: The Wine Appreciation Guild.

Conaway, J. (1990). *Napa.* Boston: Houghton Mifflin Company.

Clarke, O. (2000). *Introducing Wine.* New York: Harcourt.

Cooke, J. (2002, November). Barolo's New Generation. *The Wine Spectator,* 78–81.

Cowham, S. & Hurn, A., (2001, April). French Pinot Noir clones—an Australian perspective. *The Australian Grapegrower & Winemaker. 447: 93–95.*

Duijker, H. (1999). *The Wines of Chile.* Utrecht, Netherlands: Spectrum.

du Plessis, C. (Ed.). (2005). *SA Wine Industry Directory 2005/2006.* Sulder Paarl, South Africa: Wineland Publications.

Elia, R. L. (1992, Spring). "60 minutes" The French Paradox. *Quarterly Review of Wines. 14*(2), 44–46.

Epstein, B. S. (2005, December). Campania: Modern Wine from Ancient Grapes. Retrieved from http://www.winepages.com

Evans, L. (1973). *Australia and New Zealand Complete Book of Wine.* Dee Why West, NSW Australia: Paul Hamlyn Pty. Ltd.

Federation of Dining Room Professionals. (1998). *The Professional Service Guide.* New Jersey: Author.

Gedeon, J. (2005, September) British Columbia Becomes a Contender. *Wine Business Monthly,*

Goldfinger, T. M. (2003, August). Beyond the French paradox: the impact of moderate beverage alcohol and the consumption in the prevention of cardiovascular disease. *Cardiology Clinics,* 21-3.

Hall, C. M., & Sharples, L. (Ed.). (2000). *Wine Tourism Around the World.* Oxford: Butterworth-Heinemann.

Hall, L. S. (November, 2001). Changing Trends in Alsace. *Wine Business Monthly.*

Hall, L. S. (2001). *Wines of the Pacific Northwest.* London: Octopus Publishing Group Limited.

Halliday, J. (1991). *Wine Atlas of Australia and New Zealand.* London: Harper Collins.

Heald, E., & Heald, R., (2004, Autumn). Spain's greatest wines. *Quarterly Review of Wines.*

Herbst, R. & Tyler, S. (1995). *Wine Lover's Companion.* New York: Barron's Educational Services, Inc.

Hughes, D., Hands, P., & Kench, J. (1988). *The Complete Book of South African Wine* (2nd ed.). Cape Town: C. Struik Publishers.

Irvine, R., & Clore, W.J. (1998). *The Wine Project: Washington State's Winemaking History.* Vashon, WA: Sketch Publications.

Jackson, R. S. (1994). *Wine science.* San Diego, CA: Academic Press.

Jackson, R. S. (2000). *Wine science.* San Diego, CA: Academic Press.

Jefford, A. (2002). *The New France: A Guide to Contemporary French Wine.* London: Mitchell Beazley.

Johnson, H. (1989). *Vintage: The Story of Wine.* New York: Simon and Schuster.

Johnson, H. (1994). *The World Atlas of Wine.* New York: Simon & Schuster.

Keesing, A. J. (2003). *Grape and Wine industry Statistical Annual 2003.* Hastings, New Zealand: Robert Kale & Associates Ltd.

Kladstrup, D., & Kladstrup, P. (2001). *Wine and War: The French, the Nazis and the Battle for France's Greatest Treasure.* New York: Random House.

Kolpan, S., Smith, B. H., & Weiss, M. A. (2002). *Exploring Wine.* New York: Wiley & Sons.

Kramer, M. (1989). *Making Sense of Wine.* New York: William Morrow & Co.

Lapsley, J. (2001, November). Argentina: A giant awakening – wine industry. *Wines and Vines, 82*(11), 114-120.

Lapsley, J. T. (1996). Bottled Poetry. Berkeley, CA: University of California Press.

Laube, J. (1999). *California Wine.* New York: Wine Spectator Press.

Leipoldt, C. L. (2004). *Food & Wine.* Stonewall: Cape Town.

Lord, T. (1988). *The New Wines of Spain.* San Francisco: The Wine Appreciation Guild.

MacNeil, K. (2001). *The Wine Bible.* New York: Workman Publishing.

Mathäss, J. (1997). *Wines from Chile.* Amsterdam: Qué Más.

Molesworth, J. (2002, November). Tasting America's Bounty. *The Wine Spectator,*

Molesworth, J. (2002, November). Wine Across America. *The Wine Spectator,*

Molesworth, J. (2005, December) Riesling. *Wine Spectator,*

National Center for Statistics and Analysis. (2001). *2001 FARS Annual Report File.* Washington DC: Author.

National Institutes of Health. (2001). *Alcohol Related Mortality United States, 1979-96.* Bethesda, MD: Author.

Ness, C. (2002, August 21). Organics take root: State-of-the-art wines quietly shed sprouts-in-a-bottle image. *The San Francisco Chronicle,* p. 1WB.

Noble, A. C., Arnold, R. A., Buechsenstein, J., Leach, E. J., Schmidt, J. O., & Stern, P. M., (1987). Modification of a Standardized System of Wine Aroma Terminology. *American Journal of Enology and Viticulture, 38* (2), 143-146.

Oregon Agricultural Statistics Service. (2005). *2004 Oregon Vineyard and Winery Report.* Portland, OR: Author.

Oregon Wine Board. (2005). *Oregon State Wine Facts.* Portland, OR: Author.

Osborne, L. (2004). *The Accidental Connoisseur: An Irreverent Journey through the Wine World.* New York: North Point Press.

Parker, R. (1991). *Bordeaux: A Comprehensive Guide to the Wines Produced from 1961-1990.* New York: Simon & Schuster.

Parker, R. (2003). *Bordeaux: A Comprehensive Guide to the World's Finest Wines.* New York: Simon & Schuster.

Peterson, J. (2002). *Sweet Wines: A Guide to the World's Best.* New York: Stewart, Tabori & Chang.

Phillips, R. (2000). *A Short History of Wine.* New York: HarperCollins.

Raabe, S. (2005, October) Colorado's Wine Industry. *Wines & Vines,*

Robinson, J. (1986). *Vines grapes wines.* New York: Alfred A. Knopf Inc.

Robinson, J. (Ed.). (1994). *The Oxford Companion to Wine.* Oxford: Oxford University Press.

Robinson, J. (Ed.). (1999). *The Oxford Companion to Wine* (2nd ed.). New York: Oxford University Press.

Sanderson, B. (2005, December) Great Grape: Riesling. *Wine Spectator,*

Scheme for the Integrated Production of Wine (IPW), Liquor Products Act, Act 60. (1998).

Schieldknecht, D. (2002 April) Slate Soul: The Mosel's Essential Rieslings where Soil and Style Converge. *Wine & Spirits Magazine.*

Schoenfeld, B. (2005, December). Tradition and Ambition. *The Wine Spectator,*

Shesgreen, S. (2003, March 7). Wet Dogs and Gushing Oranges: Winespeak for a New Millenium. *The Chronicle of Higher Education.*

Smith, D. V., & Margolskee, R.F. (2001, March). Making Sense of Taste. *Scientific American, 284*(3), 32-39.

Smith, R. (2002, July 10). Wine; Solved: The Great Zinfandel Mystery; The birthplace of California's signature grape turns out to be Croatia. *Los Angeles Times,* p. H.1.

Smart, R. (June, 2004). Terroir Unmasked. *Wine Business Monthly.*

Sparks Companies Inc., Gale group, (2002, April 29). Chilean Wine Industry contends with its own success. *Food and Drink Weekly,*

Spurrier, S., & Dovaz, M. (1983). *Academie du Vin Complete Wine Course.* New York: Putnam.

Standage, T. (2005). *A History of the World in 6 Glasses.* New York: Walker & Company.

Suckling, J. (2002, November). Piedmont's Silver Lining, *The Wine Spectator,* 70–77.

Sullivan, C. L. (1998). *A Companion to California Wine.* Berkeley, CA: University of California Press.

Teiser, R., & Harroun, C. (1983). *Winemaking in California.* New York: McGraw-Hill Book Company.

Tinney, M. (2005, September) Colorado's Passionate Wine Industry Reinvents Itself. *Wine Business Monthly,*

Ureta, F.C., & Pszczólkowski, P.T. (1995). *Chile Culture of Wine.* Santiago, Chile: Editorial Kactus.

Vine, R. (2000). *Wine Appreciation.* New York: John Wiley & Son.

Wagner, P. (October, 1974). Wines, Grapes, Vines and Climate. *Scientific American,* 107-115.

Walker, L. (2005, December) World Beat. *Wines & Vines,* 57.

Walsh, G. (1979). The Wine Industry of Australia 1788–1979. In *Wine Talk* Canberra: Australian National University.

Washington Agricultural Statistics Service. (2002). *2002 Wine Grape Acreage Survey.* Olympia, WA: Author.

Welch, E. (2005). *Shopping in the Renaissance: Consumer Culture in Italy, 1400–1600.* Yale University Press.

Wilson, J. E. (2002). *Terroir: The Role of Geology, Climate and Culture in the Making of Wine.* San Francisco: The Wine Appreciation Guild.

Wine Institute. (2001). *American Viticultural Areas.* San Francisco: Author.

Wine Institute. (2002). *2001 Statistical Highlights.* San Francisco: Author.

Wine Institute. (2004). World Wine Production by Country. San Francisco: Author.

Yair, M. (1997). *Wine chemistry.* San Francisco: The Wine Appreciation Guild.

Web Sites

www.chianticlassico.com, Chianti Classico

http://www.deutschweine.de, Website of German Wine Institute

www.GermanWineUSA.org, Website of German Wine Information Bureau.

www.italtrade.com, Italian Trade Commission

www.sawis.co.za, Website of SAWIS (SA Wine Industry Information and Systems).

http://www.sopexa.com

http://www.VDP.de, Website for Verband Deutscher Prädikatsweingüter

www.vintagenewyork.com

www.weingut-st-urbans-hof.de Website for St Urbans-hof

INDEX

Pages numbers followed by "t" indicate tables; page numbers in italics indicate figures.

A

Abruzzi, 195, 197t
Access to wine, 450, 458
Acetaldehyde, 70
Acid, 379
Acidity of wine, 193, 455
Acidity of grapes, 24, 47
Acids, 72, 502
A code numbers, 427
Aconcagua region, 401, 403t, 404–405
Aconcagua Valley, 403t, 404–405, *405*, *406*
Added value, 446
Adelaide Hills Region, 378t, 383, 385
Adelaide Plains Region, 378t
Adelaide Zone, 378t
Aegean Islands, 288
Aeration, 471–472
African National Congress, *424*
Ageability, 503–505
Aged Tawny Port, 242
Aging cave, *234*
Aging cellar, *124*, *400*, 505–508
Aging information, 363
Aging wines, 500–501
- Barolo, 192
- Château Gruard-Larose, 118
- humidity control, 506
- Mercurey, 143
- restaurants, 510–512

Aglianico, 213, 214t, 288
Aglianico del Vulture, 214, 214t
Agrelo subregion, 412
Ahr, 274
Airén, 229, 237
Alameda County appellations, 325t
Alamosa Cellars, 360
Alba, 193
Albana, 191t
Albana di Romagna, 190, 191t
Albany Subregion, 379t
Albarino, 229, 230–231
Albumin, 62
Alcohol
- effect on fermentation, 48
- perception of wine, 79, 458
- and tears, *80*

Alcohol and Tobacco Tax and Trade Bureau (TTB), 303, 514, 516
Alcohol Beverage Control (ABC) agencies, 515
Alcohol consumption
- mortality rates, *98*
- negative effects, 97–98
- positive effects, 98
- women, 98–99

Alcohol content, 47, 126, 156, 185, 188, 307, 309, 363, 392, 407, 427
Alcoholism, 98
Alembic stills, 404
Alentejo, 245
Alexander, Cyrus, 319
Alexander Valley AVA, 316t, 319, 520
Alexander VI (pope), 222
Alianca, 244
Alicante Bouschet, 301
Aligoté, 143
Allergen statement, 392
Alluvial soil, 21, 207, 273
- Alsace, 164
- Graves region, 120
- Northern Rhône, 149

Almonds, 138, 141, 150
Aloxe, 141
Aloxe-Corton, 131t, 133, 141
Alphonse, 223
Alpine valleys Region, 379t
Alsace, 163–167, 275
- appellations, 169–172, 170t
- terroir, 164

Altare, Elio, 192
Alto Adige, 207t, *212*, 212
Alvarinho (Albarino), 229, 239
Amabile, 188
Amarone, 207t, 210–*211*
American Viticultural Areas (AVAs), 303–304, 309, 518–522
American Wine Growers (AWG), 337
American Wine Society, 531
Amontillado, 238
Amphorae, 6, *182*, *288*
Anbaugebiete, 256, 257t
Andalucia, 237–239, *238*
Andeol Salavert, 153
Anderson Valley AVA, 319, 320–322, 322t, 520
Andes Mountains, *398*, *412*
Anjou, *171*
Anjou Mousseux, 171
Anjou/Saumur, 169, 170t, 171–172
Annee Rare R.D., 163
Anselmi, 209
Anthocyanins, 501
Antifreeze scandal, 280
Antinori, Niccolò, 200
Antinori, Piero, 186, 200
Antinori winery, *200*
AOC (Appellation d'Origine Contrôlée), 108–109, 159
Apartheid, 418
Appearance, 77
Appellation America, 531
Appellation Contrôlée (Swiss laws), 282
Appellation d'origine contrôlée (AOC), 108–109, 159
Appellation of Superior Quality (OPAP), 289
Appellations, 309
- Argentina, 410–413, 411t
- Australia, 378t–379t
- Burgundy, 131t
- California, 303–304
- Chile, 407
- French concentric circles, *130*
- Napa Valley, 311t
- New York State, 356–358
- New Zealand, 390t
- Oregon, 348t
- washington, 340t
- on wine labels, 307, 391

Applegate Valley AVA, 348t, 350
Apulia (Puglia), 213, 214, 214t
Aquilegia, 207
Aragonez (Tempranillo), 245
Argentina
- exports, 410
- history, 408–410
- map, *411*
- wine regions, 410–413

Arinto do Dao (Malvasia Fina), 244
Arizona AVA, 518
Arkansas AVAs, 518
Arneis, 190, 191t, 194
Arneis di Roero, 191t, 194
Aroma, 77–79
- evaluation by, 86–88
- lack of, 471
- standards, 93

Aromatics
- age, 501–502
- decanters, *470*

Arroyo Grande Valley AVA, 325t, 328, 520
Arroyo Seco AVA, 325t, 326, 520
Arsac, France, 117
Artificial carbonation, 67
"As is", 467
Aspect, 424
Aspersion, 160
Assemblage, 161
Associated Vintners, 337
Asti, 190, 191t, *194*–195
Asti zone, 193
Astringency, 79, 501
Aszú, 286. *See also* Botrytis
Atacama region, 401, 403–404, 403t
Atlas Peak AVA, 311t, 314, 520
Auckland/Northland region, 390t, 394
Auctions, 510, 511
Aude, 174
Augustus, 251
Auslese, *254*, 258
Ausonius, 124
Australia
- appellations, 377–387
- climate, 372
- exports, 394–395

Australia *(continued)*
  Geographic Indications (GIs), 378t–379t
  history, 374–377
  major producers, 374t
  map, *377*, *381*
  wine labels, 391
Australian Wine and Brandy Corporation (AWBC), 377, 531
Australian Wine Research Institute, *384*
Austria, 280–282
Autolysis, 66
Auxerrois, 164
AVA (American Viticultural Areas), 303–304, 309, 518–522
Avignon, 107, 148, 149
AWBC (Australian Wine and Brandy Corporation), 377, 531
AWG (American Wine Growers), 337

B

Bacchus, 7
Baden, 275
Bad Kreuznach, 275
Baga, 244
Balthazar, 454t
Bamboes Bay ward, 429
Bandol, 174
Banfi, *202*
Banyuls, 175, 176t
Barbaresco, 183, 187, 190, 191t, 192–193, 195
Barbera, 32, *34*, *188*, 190, 191t, 195, 434
  Breede River Valley region, 430
Barbera d'Alba, 191t
Barboursville winery, 359
Bardolino, 207t, 208–209, 209–*210*
Bardolino Superiore, 207t, 210
Barolo, 183, 187, 190, *191*–192, 191t, 195, 217
Barossa Valley Region, 378t, 383–384
Barossa Zone, 378t
Barraida, 244
Barrels, 58–61, *400*
Barriques, 202
Barsac, 122, 125t
  Classification of 1855, 115, 122, 525
Barsac, France (village), 122
Basalt rock, 286
Basement as passive cellar, 506–*507*
Basilicata, 213, 214, 214t
Basket presses, 54, *55*
Bas-Medoc. *See* Medoc
Basserman Jordan, 274
Batard-Montrachet, 133
BATF (Bureau of Alcohol, Tobacco & Firearms), 303, 514
Baumard family, 172
BAWSI (Black Association of the Wine and Spirit Industry), 421
Beaujolais, 131t, 134, 145–148
  compared to Dolcetto, 194
  map, *146*
Beaujolais Blanc, 144
Beaujolais Nouveau, 53, 145, *147*, 210, 213
Beaujolais Superieur, 145
Beaujolais-Villages, 131t, 145
Beaumes-de-Venise, 151t, 156
Beaune, 131t, *132*, 133, 141–142
Beechworth Region, 379t
Beerenauslese, *254*, 258
Beiras, 243
Bekaa Valley, 291
Belemnite chalk, 159
Bellet, 173
Bendigo Region, 379t
Benguela Current, 429
Ben Lomond Mountain AVA, 325t, 520
Benmore Valley AVA, 320, 322t, 520
Bennett Valley AVA, 316t, 317, 520
Bentonite, 62
Bereiche, 256, 257t
Bergerac, 176, 176t
Bergerac region, 285
Berg Schlossberg, *272*
Beringer-Blass, 373
Beringer Vineyards, 299, *314*, 319
  food and wine pairing experiment, 96–97
Bernard Morey et Fils, *132*
Bernkastel, *264*, *268*, 268–269, *269*
Bertani, 209
Bertrand the Goth, 107
Best, Joseph, 382
Beyer, Léon, 165
Beyer, Marc, 165–166
Beyer, Yann, 165
Bianco d'Alcamo, 214t, 215
Bianco di Custoza, 207t, 209
Bienvenues-Batard-Montrachet, 133
Big Rivers Zone, 378t
Billecart-Salmon, 163
Binge drinking, 97
Bin numbers, 488–489, 509, *510*
Bio-Bio Valley, 403t, 408
Biodiversity and Wine Initiative, 420
Biodynamic viticulture, *172*, 172
Biondi-Santi, Ferruccio, 201
Black Association of the Wine and Spirit Industry (BAWSI), 421
Blackwood Valley Region, 379t
Blade of wine key, 459, 460, *461*
Blaine, Elvert, 336
Blanc de blanc, 162
Blanc de Blancs, 163
Blanc de noir, 162
Blanchot, 135t
Blauburgunder (Pinot Noir) , 283, 284
Blauter Zweigelt, 281
Blaxland, Gregory, 375
Bleeding the line (cabinet systems), 487–488
Blending wine, *61*, 61–62, 161
  cuvée, 66
Blends, 383
  Bordeaux, 113
  Carmignano, 202–203
  Italy, 188
  Supertuscan, 186
  Valpolicella, 210
  wine lists, 495
Blind tastings, 84
Bloom, 28–29
Bockstein, 267
Bodega, *225*, 234
Bodega Montecillo, 234
Bodegas Salentein, *410*
Bolgheri, 197t, 200
Bollinger, 163
Bommes, France, 122
Bonnes Mares, 133, 138, 139
Bonnezeaux, 172
Boom and bust economy, 13, 299, 304
Borba, 245
Bordeaux, 111–127
  aging wines, 504
  appellations, 125t, 126
  vs. Burgundy, 148t
  climate, 113
  history, 111–113
  map, *116*
  trade with England, 112
  wine regions, 115–127
Bordeaux Superieur, 126
Borrado das Moscas, 244
Boschendal, *434*
Botrycine, 69
Botrytis, 122, 286
Botrytis cinerea, 68
Botrytized wines
  Beerenauslese, 258
  Coteaux du Layon, 172
  Monbazillac, 176
  Quarts de Chaume, 172
  Sauternes and Barsac, 122, 125t
  Trockenbeerenauslese (TBA), 258–259
Bottelary ward, 434
Bottle bouquet, 501–502
Bottle display, 447
Bottles ordered, 467–468
Bottle shapes, *454*
  Alsace, 166
  Bordeaux vs. Burgundy, 148t
  Franken, 276
Bottle shock, 64
Bottle sizes, 452–454, 454t
  aging wines, 503
  large format, 469, 492–493
  pricing, 492–493
Bottling, 63–64, *414*
Boucherottes, *132*
Bougros, 135t
Bourgueil, 170t, 171
Boutaris, Yiannis, 290
Brachetto, 191t
Brachetto d'Acqui, 191t
Brancott Valley, *393*

Brandy, 430, 431–432, 432
Brauneberg, 269
Breast cancer, 98
Breede River Valley region, 428, 430, 432–433
Bressandes, 142
Breuer, Bernard, 272
British Columbia, 364–365
  map, *364*
Britt, Peter, 346
Brix (degrees Brix), 46–47
Broke Fordwich Region, 378t
Brotherhood Winery, 354
Brouilly, 131t, 148
Brumont, Alain, 176
Brunello di Montalcino, 196, 197t, *201*–202, 217
Brut, 161
Bual, 246
Budbreak, 26–28, *27*, 340
Buena Vista winery, 299
Bulgaria, 287
Bull's Blood, 286
Bülönleges minöségi bor, 286
Bunch rot, 28
Bureau of Alcohol, Tobacco & Firearms (BATF), 303, 514
Burgenland, 282
Burgundy, *23*, 127–148
  aging wines, 504
  appellations, 131t
  vs. Bordeaux, 148t
  classification system, 130, 132–134
  history, 129–130
  maps, *128*, *134*, *137*, *144*, *146*
Bürklin-Wolf, 274
Busby, James, 375, 388
Buying futures, 509
BYOB, 454–455
By the glass wines, 448, 481–485
  price point, 483–484
Byzantine Empire, 204, 288–289

C

Cabernet d'Anjou, 171
Cabernet de Saumur, 171
Cabernet Franc, 32–33, *34*
  British Columbia, 365
  California, 311–312
  France, 113, 120, 123, 124, 125, 169, 170t, 171, 176, 176t
  Hungary, 286
  Italy, 200, 207, 207t, 208, 212, 213
  New York, 357
  Virginia, 359
Cabernet Sauvignon, *34*–35
  aging, 501, 504
  Australia, 380, 382, 383, 384, 385, 386, 387
  British Columbia, 365
  California, 311, 312–313, 314, 316, 317, 319, 320, *321*, 322, 324, 328
  Chile, 405
  effects of age, *78*
  France, 113, *117*, 120, 122, 124, 125, 125t, 171, 173, 176, 176t
  Hungary, 286
  Israel, 291
  Italy, 186, 197t, 198, 200–201, 203, 207, 208, 212
  Lebanon, 291
  New Zealand, 393, 394
  Oregon, 350
  Spain, 235, 237
  South Africa, 420, 429, 430, 431, 432, 434, 435, 437, 438
  Texas, 360
  Washington, 335, *340*, 341, 342–343
Cabinet systems, 487–488
Cachapoal Valley, 403t
Cahors, 112, 176t, 177
Calabria, 213
Calabria, Italy, 214
Calcareous soil, 140, 207
California
  appellations, 303–304, 311t, 316t, 325t, 520–522
  Central Coast AVA, 322–329, *324*, 325t, *326*
  Central Valley, 329–330
  climate, 304–306
  Gold Rush, 299
  map, *305*
  Mission period, 297–298
  Napa Valley, 306, 310–314, 311t, *313*
  Prohibition, 300–301
  Sonoma County, 314–319, *315*, 316t
  vineyard, *13*
  wine regions, 304–306, 310–331
California Grapevine, 531
California Shenandoah Valley AVA, 330, 520
California State University Fresno Winery, 531
California Wine Association (CWA), 300
Calitzdorp, 429
Campania, 213–214, 214t
Campo Viejo, 234
Canaan, 287
Canada
  history, 362
  maps, *364*, *365*
  wine labels, 363
Canadian Vintners Association, 531
Canaiolo, 196, 197t, 200
Canberra District Region, 378t
Cancer, increased risk, 98
Candle for decanting, *471*
Cane pruning, 31, *32*
Cannonau, 214t, 216
Cannonau di Sardegna, 214t, 215
Canon-Fronsac, 125
Cantenac, France, 117
Canterbury region, 390t, 394
Cap, 51, *52*
Capay Valley AVA, 520
Cape Agulhas, 437, 438–439
Cape blend, 432
Cape Doctor, 433, 437
Cape Floral Kingdom, 420
Cape Riesling, 420
Cape Ruby, 429
Cape Town, South Africa, *421*
Cape Vintage, 429
Cap management, 51–53
Carbon dioxide, 64
Carbon (for fining), 62
Carbonic maceration, 53, 145–146, 237
Cardiovascular health, 98
Carignan
  Israel, 291
  Lebanon, 291
  South of France, 173, 175, 176t
Carinena (Mazuelo), 236
Carmel Mountains, 291
Carmel Valley AVA, 325t, 327, 520
Carmenère, 113, 406
Carmignano, 196, 197t, 202–203
Carpathian Mountains, 286
Carthaginians, 221
Casablanca Valley, 403t, 404–405
Casks, *399*, *409*
Cassis, 174
Castelao Frances (Periquita), 229
Castello di Borghese, 358
Castello di Fonterutoli, 197
Castilla y León, 231–232
Cataluña, 235–237
Catamarca region, 411t
Catawba grapes, 355, 357
Catherine of Aragon, 222
Catholic Church, 8–9, 105, 129, 163, 182, 197
Cato, 181
Cava, 235–236
Cayuga AVA, 356
Cedar, 156
Cederberg ward, 429
Cellars, 505–508, 511
Celts, 281
Cencibel. *See* Tempranillo
Central Coast AVA, 322–329, 520
  appellations, 325t
  map, *324*, *326*
Central Europe, 280–284
Central Italy, 197t
Central Loire, 169
Central Otago region, *388*, 390t, 394
Central ranges Zone, 378t
Central Valley, California, 329–330
Central Valley region (Chile), 401, 403t, 405–406, *408*
Central Victoria Zone, 379t
Central Western Australia, 378t
Cephalonia, 289, 290
Certification sticker, 426

Châlone, France, 142–143
Chablis, 131t, 134, 135–136
  map, *134*
Chalk Hill AVA, 141, 316t, 318, 520
Chalone AVA, 520
Chalone AVA (California), 325t, 327
Chambertin, 133, 138
Chambolle, 138
Chambolle-Musigny, 131t, 133, 138–139
Chambourcin Virginia, 359
Champagne, 10, 157–163
  artificial carbonation, 67
  Charmat process, 67
  climate, 160
  exports, 160
  fraud, 106
  history, 158–159
  methode champenoise, 64–67
  pouring, 469, *470*
  styles, 161
  terroir, 159
  transfer method, 67
  on U.S. labels, 64
  viticulture, 159–160
Champagne flute, *65*, 83
Champagne houses, 163
Champagne method. *See Méthode champenoise*
Chante-Perdrix, 156
Chaptalization, 252, 257
Chardonnay, *35–36*, *362*
  Argentina, 412, 413
  Australia, 380, 382, 384, 385, 386, 387
  Austria, 281
  British Columbia, 365
  California, 310, 311, 312, 316, 317, 318, 319, 320, 322, 324, 326, *327*, 328, 330
  Cava, 236
  Champagne, 64, 159
  Chile, 405, 406, 408
  France, 127, *131*, 131t, 135, 136, 138, 141, 143, *145*, 164, 171
  Germany, 266
  Hungary, 286
  Italy, 190, 191t, 208, 212, 253
  New Mexico, 361
  New York, 356, 357
  New Zealand, 389, 394, 393
  Ontario, 366
  Oregon, 348
  South Africa, 420, 429, 430, 432, 434, 435, 438
  Southeastern New England AVA, 358
  Spain, 235
  sparkling wines, *359*
  Texas, 360
  Virginia, 359
  Washington, 335, 341, 342
Charlemagne, 9, 105, 163, 251, 281
Charles Heidsieck, 163
Charles Krug winery, 299
Charles VIII, 168
Charmat process flowchart, *67*
Charta wines, 262
Charvet, Paul, 336
Chassagne-Montrachet, 131t, *132*, 133
Chasselas, 164, 283
Château Ausone, 115, 124
Château Barat, 122
Château Cabannieux, 122
Château Calon-Segur, 119–120
Château Canon, 124
Château Cheval Blanc, 115, 124
Château d'Aigle, *283*
Château d'Archambeau, 122
Château de Beaucastel, 156
Château de Carrolle, 122
Château de Fuissé, 143
Château de la Jaubertie, 176
Château de la Roche-aux-Moines, 172
Château d'Yquem, 122
Château Figeac, 124
Château Finegrave, *117*
Château Fortia, 107, 149, 156
Château Grand-Puy-Lacoste, 119
Château-Grillet, 153
Château Gruard-Larose, 118
Château Haut-Brion, 115, 120–121, *121*
  Classification of 1855, 114
Château Haut-Marbuzet, 119
Château La Conseillante, 125
Château Lafite, 114
Château Lafite-Rothschild, 118, 200
Château La Gaffeliere, 124
Château La Mission Haut-Brion, 121
Château la Nerthe, 156
Château Latour, *118*
  Classification of 1855, 114
Château Le Pin, 125
Château Les Ormes-de-Pez, 119
Château Margaux, 117
  Classification of 1855, 114
Château Meyney, 119
Château Montus, 176
Château Mouton-Rothschild, 118
  Classification of 1855, 114
Château Mugar, 291
Châteauneuf-du-Pape, *107*, 149, 150, 151t, 156, *157*
Château Pavie, 124
Château Petrus, 125
Château Pichon-Longueville-Comtesse de Lalande, 118–119
Château Rahoul, 122
Château Roquetaillade La Grange, 122
Château Sansay, 122
Château Ste. Michelle, 337, *344*
Château Tahbilk, 382
Château Terre-Roteboeuf, 124
Château Trotanoy, 125
Château Woltner, 121
Chénas, 131t, 147
Chenin Blanc, *36*
  Australia, 387
  California, 330
  Charmat process, 67
  France, 169, 170t, 171, 172
  Israel, 291
  South Africa, 429, 430, 431–432, 434, 435, 436, 438
  Texas, 360
Chenin Blanc Association, 431
Chevalier-Montrachet, 133
Chianti, 183, 187, 196, 197, 197t, 198–200
Chianti Classico, *188*, 197t, 199–200
Chianti Colli Senesi, 197t, 198
Chianti Rufina, 197t, 198
Chile
  appellations, 407
  climate, 401
  exports, 413–414
  grape-growing regions, 403t
  history, 399–401
  map, *402*
  wine labels, 407
  wine regions, 401–408
Chiles Valley AVA, 311t, 314, 520
Chinon, 170t, 171
Chiroubles, 148
Chlorine, 464
Choapa Valley, 403t
Choosing wines, 479–480
  by the glass wines, 482–485
Chouacheux, 142
Christianity, 7–8, 221, 251
Christian Moreau, 136
Cienega Valley AVA, 325t, 327, 520
Cinsault, 431
  Klein Karoo region, 430
  Lebanon, 291
  Southern Rhône, 150, 151t, 156–157, 157
  South of France, 173, 174, 175, 176t
Ciron River, 122
Cirrhosis of the liver, 97
Cistercians, 129, *252*
Clairette, 173
Clare Valley Region, 378t, 383, 385
Clarification, 62
Clarity, 77, 86
Clarksburg AVA, 330, 520
Claro Valley, 403t
Classico, 188
Classification of 1855, 114
  Medoc, 523–524
  Sauternes and Barsac, 525
Classification of Graves, 526
Classification of St. Emilion, 527–528
Classification (on label), 126
Clear Lake AVA, 320, 322t, 521
Clement V (pope), 107
Climate, 21–24
Cloete, Hendrik, 422

Clones, 19t
  Pinot Noir, 39, 347
  Sangiovese, 201, 202
Clore, Walter, 337
Clos, *127*, 138
Clos de Beze, 133, 138
Clos de la Roche, 138
Clos de Mouche, 142
Clos de Vougeot, 133, 139
Clos du Papillon, 172
Clos de la Bousse d'Or, 142
Clos de Mesnil, 163
Clos Floridene, 122
Clos St. Denis, 138
Clos Ste.-Hune, 166
Closures, *464*
Clos Vougeot, *127*
Clovis, 158, 163
Coastal areas, 23
Coastal region (South Africa), 428, 433–439
Coates, Clive, 110
Codorníu, 235, 236
Colchagua Valley, 403t
Cold fermentation, 206
Cold stabilization, 62, *63*
Cold War, 285
Cole Ranch AVA, 322, 322t, 521
Colheita selecionada, 229
Colline Teramane, 197t
Collio, 207, 207t
Colli Orientali, 207, 207t, 208
Color, 77, 86
  age, 77, *78*, 501
  Chablis, 135
Colorado, 361, 518
Columbard, 113
  France, 125
  South Africa, 429, 430, 432
Columbia Gorge AVA, 334, 340t, 345, 348t
Columbia Valley AVA, *334*, 340t, 341, *341*, 348t, 351
Columbia Winery, 337, 344
Columbus, Christopher, 221, *222*
Columella, 8
Comércio e Tourismo de Portugal (ICEP), 230
Comité Interprofessionel du Vin de Champagne (CIVC), 157, 159
Commune appellation, 130, *132*, 133, 151
Communism, 284
Comte de Champagne, 163
Comtes d'Equisheim, *166*
Concentric circles, *130*
Concha y Torro, 400, 405
Concord grapes, 336, 337, 355, 357
Condrieu, 151t, 153
Conero, 197t
Connecticut AVAs, 518
Connoisseurs' Guide to California Wine, 531
Consejo reguladores, *225*, 226–227
Constantia, 422, 437
Controlled Appellation of Origin (OPE), 289
Cook, James (Captain), 374
Coonawarra Region, 378t, 383, *385*, 385
Cooperatives, 253, 276, 419, 437
Coopers, 7, 58
Copiapó Valley, 403t
Copper sulfate, 112
Coquimbo region, 401, 403–404, 403t
Corbières, *175*, 175, 176t
Cordon training, *32*
Cork, 10, 462–463
  Alentejo, 245
  inspection, 462–463
  presentation, *463*
  sparkling wines, 64, 469, *470*
  trade agreements, 223–224
Corkage fees, 455
Corkscrews, *459–463*
Cork taint, 88, 464–465
Cornas, 151t, 154–155
Corsica, 177
Cortese, 188, 190, 191t
Cortese di Gavi, 191t, 194, 195
Corton, 133, 141
Corton-Charlemagne, 133, 141
Corvina, 207t, 209, 210 211
Cos d'Estournel, *119*
Cost percent, 483–484
  sliding scale, 490–491
Côt, 412
Coteaux d'Aix-en-Provence, 173–174
Coteaux d'Aix-en-Provence-les-Baux, 173–174
Coteaux du Languedoc, 175–176
Coteaux du Layon, 170t, 172
Coteaux du Tricastin, 151t, 155
Coteaux Varois, 173
Côte Chalonnaise, 131t, 134, 142–143
Côte de Beaune, 131t, *132*, 133, 134, 137, 138, 141–148
Côte de Beaune-Villages, 141
Côte de Brouilly, 131t, 148
Côte de Nuits, 22, 131t, 133, 134, 137, 138–141
Côte d'Or, 133, 134, 137–142
  map, *137*
Côte Rôtie, 151, 151t, 153
Côtes de Blaye, 126
Côtes de Bourg, 126
Côtes de Provence, 173
Côtes des Blancs, 159, *160*
Côtes du Lubéron, 151t, 157
Côtes du Rhône, 148–157
  appellations, 150–151, 151t
  maps, *152*, *155*
  terroir, 149–150
Côtes du Rhône-Villages, 150
Côtes du Roussillon, 174–175
Côtes du Roussillon-Villages, 174–175
Côtes du Ventoux, 151t, 157
Country of origin, 392, 407, 426–427, 427
  wine lists, *493–494*
Coury, Charles, 346
Cousiño Macul, 400
Cowra Region, 378t
Crémant d'Alsace, 167
Crémant de Bourgogne, 143
Crete, 288, 289, 290
Criadera, *239*
Crianza, *226*, 227
Crimean Peninsula, 287
Criolla, 297, 360, 409, 412
Criots-Batard-Montrachet, 133
Crljenak Kastelanski (Zinfandel), *42*
Croft, 239, 241, 243
Cropload, 28–29, 32
Crozes, 153
Crozes-Hermitage, 151t, 153–154
Cru Beaujolais, 145, 146–148, *147*
Crusades, 10, 129, 154, 290
Cruvinet, 487
Cucamonga Valley AVA, 521
Curicó Valley, *399*, 403t, 405, 406
Currency Creek Region, 378t
Custom-designed cellars, 505–*506*
Cuvée, 66
Cuvée Columbus, 163
Cuvée Florens-Louis, 163
Cuvée Grand Millesieme, 163
Cuvée Winston Churchill, 163
Cuyo region, 412
CWA (California Wine Association), 300

D

The Dalles, Oregon, 336
Danube River, 282
Dão, 244
Dark Ages, 105, 182
Darling district, 436–437
Deacidification, 59
De agri cultura (Cato), 181
De Aguirre, Francisco, 399
De Boüard, Hubert, 419, 433
Decanter, 531
Decanters, *470*, 471–472
Decanting, 469–472, 511
  with large format bottles, 469
Decanting funnel, 471
Degree days, 302
Degrees Brix, 46–47
Deidesheim, 274
Dejuicing tanks, 56
De Klerk, F. W., 424
Delaforce, 243
De Lencquesaing, May-Eliane, 419, 433
Della Rocchetta, Mario Incisa, 200
Della Rocchetta family, 186
De Médicis, Catherine, 168
Demi-sec, 161, 171
De Mouchy, Duchesse, 121
Denmark Subregion, 379t
Denominacão de Origem Controlada (DOC), *225*, 228–229, 244

Denominacion de Origen Calificada (DOCa), *225*, 227
  Rioja, 232
Denominacion de Origen (DO), *225*, 226–227, 227
Denominazione d'Origine Controllata (DOC), 184–189
Denominazione d'Origine Controllata e Garantita (DOCG), 184, 185, 187–188
De Oro, Diego, 399
De Pombal, Marquis, 244
Depth of color, 77
De Riscal, Marques, 232
De Rothschild, Edmond, 290, 291
Descriptions on wine lists, 496
Designated viticultural areas (DVA), 362, 363, 366
Dessert wines, 68–71
  aging, 504
  Constantia, 422, 437
  German, 258–259
  label, *259*
  Madeira, 245–246
  Marsala, 215
  Muscadel, 433
  Muscat Beaumes-de-Venise, 156
  Muscat d'Alexandrie, 433
  Port, 241–243
  Quarts de Chaume, 172
  Sherry, 237–239
  South of France, 175
  Vin Santo, 203–204
Deuxieme cru (Classification of 1855), 114, 523
Deuxieme cru (Sauternes and Barsac Classification of 1855), 122, 525
Devon Valley, 434
Dewavrin-Woltner, Francoise and Francois, 121
Diablo Grande AVA, 521
Diacetyl, 59
Diamond Mountain AVA, 311t, 312–313
Diamond Mountain District AVA, 521
Dijon, France, 134, 138
Dillon family, 120–121
Dionysus, 6
Discussing wines, 89
Diseases, 20, *24*
  downy mildew, 112, 253
  France, 112
  Italy, 183
  powdery mildew, 183, 388, 423
Disgorging, 66–67, *161*
Disgorging machine, *66*
Displays
  bottles, 447
  cabinet systems, 487
  featured wines, 447
  in lieu of wine lists, 497
  table tents, 448
  wine lists, 445
Distell, 433, 435
Distribution laws, 515, 517
Distributors, 478, 479
Dôle, 283
DOC (Denominazione d'Origine Controllata), 184–189
DOCG (Denominazione d'Origine Controllata e Garantita), 184, 185, 187–188
Doktor, *269*
Doktor vineyard, *264*
Dolce, 188, *194*
Dolcetto, 190, 191t, 194, 195
Domaine de la Baume, 174
Domaine de la Romanee-Conti, 140
Domaine de Mont Redon, 156
Domaine Lamarche, 140
Domecq, 239
Dom Perignon, 163
Dopff, 163
Dordogne region, 112
Dormancy, *26*, 31–32
Dornfelder, 266
Dosage, 67, 161
Dos Rios AVA, 521
Double magnum, 454t
Douro region, 227, 240–241
Doux, 161
Dow, 241
Downy mildew, 112, 253
Dow, 241, 243
Drake, Francis, 400
Dresden, 276
Drip irrigation, 21
Driving under the influence, 97–98
Dry Creek Valley AVA, 316t, 318–319, 521
Dry farming, 29, 384
Drying grapes, 204
Dundee Hills AVA, 348t, 349
Dunnigan Hills AVA, 521
Durance River, 155, 157
Durif, 38
Dutch East India Company, 421
DVA (designated viticultural areas), 362, 363, 366

E

Eastern Europe, 284–287
Eastern Mediterranean, 287–291
Eastern plains Zone, 378t
Eastern Washington, 340–341
Echezeaux, 133, 139
Eden Valley Region, 378t, 385
Edict of Nantes, 423
Edna Valley AVA, 325t, 328, 521
Education, 447
Effects of age, *77*, *78*
Effervescence, 64, 159
  perception of wine, 79
Eger, 286
Egg whites (for fining), 62
Egon Muller winery, *268*
Egri Bikavér, 286
Egypt, *5*, 5–6
Eighteenth Amendment, 11, 300, 513
Einzellagen, 256, 257t
Eiswein, 69, *254*, 258, *259*
El Dorado AVA, 330, 521
Eleanor of Aquitane, 10, 105, 111
Elgin Ward, 438
Elim Ward, 438
Ellison, Curtis, 98
Elqui Valley, *403*, 403t
Elsenberg Agricultural College, 433
Emilia-Romagna, 190, 191t
Encruzado, 244
Enologist, 76
Enology, 9
Entre-Deux-Mers, 125, 125t
Epenots, 142
Épernay, 159
Errázuriz Winery, 400, 405, *406*
Erstes Gewächs, 261
Esslingen, *275*
Estate bottled wines, 126, 303, 307
Esters, 501–502
Estufas, 245
Ethanol, 71–72
Etienne Sauzet, *131*
Etruscans, 180–181, 197
Eurocave, 508
European Economic Community, 289, 428
European names (US wines), 308
European Union, 428
  denomination laws, 235, 255
  Greece, 289
  Hungary, 286
  Italy, 185
  Portugal, 224
  Southern Italy, 213
Evora, 245
Exports, 291
Exposition Universelle, 114
Extended maceration, 53
Extra brut, 161
Extra dry, 161
Eyrie Vineyard Winery, 346, 347
Ezerjo, 286

F

Fair Play AVA, 521
Falernian, 213
Family meal, 472
Fargues, France, 122
Farm Winery Act of 1976, 355–356
Far North Zone, 378t
Fatigue in tasting, 91
Featured wines, 447
  table tents, *448*
Ferdinand, *222*
Fermentation, 47–*48*
Fermentation cellar, *48*
Fermentation tanks, *48*, *51*, *345*, *374*
Fernandez, Alejandro, 231

Fetal alcohol syndrome (FAS), 99
Fiano, 214t
Fiano di Avellino, 214, 214t
Fiddletown AVA, 330, 521
Fifth, 454t
Fighting varietals, 13, 35
Figs, 156
Filling machine, *64*
Filtering, 62, *63*
Finger Lakes AVA, *356–357*
Fining, 62
Fino, 70–71, 237
First growth (Classification of 1855), 114, 115, 523
First Growth Committee, 261
First growths appellation, 133
Fixin, 131t, 133
Flagy-Echezeaux, 139
Flanders, 168
Flavor, 79
Fleurie, 148
Fleurieu, 131t, 378t
Flights, 84, 85, 451–452
Flint soil, 138
Florence, Italy, 203
Flor yeast, 70, 71, 237
Flower cluster, *28*
Foil capsule, *461*, *464*
  cutting, 460
  sparkling wines, 469
Fonseca, 241, 243, 245
Food and wine pairing, 94–95, 97t, *449*–450
  experiment in, 96–97
Fortant de France, 174
Fortified wines, 68
  Madeira, 245–246
  Marsala, 215
  Muscadel, 433
  Muscat d'Alexandrie, 433
  Port, 69, 241–243
  Sherry, 70
  trade with England, 223–224
Fort Vancouver, Washington, 336
Foster's Wine Estates, 373, 374t
Fournier, Charles, 355, 357
Foxy bouquet, 355
France
  Alsace, 163–167
  appellation hierarchy, *130*
  appellations, 125t, 131t, 151t, 170t, 176t
  Bordeaux, 111–127
  Bordeaux vs. Burgundy, 148t
  Burgundy, 127–148
  Champagne, 157–163
  history, 104–106, 111–113, 129–130, 158–159, 163–164, 168–169
  Loire Valley, 167–172
  maps, *110*, *116*, *128*, *134*, *137*, *144*, *146*, *152*, *155*, *168*
  South of France, 173–177
  trade, 105, 112, 168, 183
  wine labels, 126
  wine laws, 108–110
  wine regions, 115–127
Franciacorta Spumante, 190, 191t
Franco, Francisco, 223
Frank, Konstantine, 355–356, 357
Franken, 276
Frankland River Subregion, 379t
Franschhoek, 423
Franschhoek Valley Ward, 434
Frascati, 195, 197t
Fraser Valley, 365
Fraud
  antifreeze scandal, 280
  Burgundy, 129–130
  Champagne, 159
  cork, 462
  French names, 106
  production numbers, 107
Freese, Phil, 419, 435
Freixenet, 236
French Columbard, 67, 330
French paradox, 98
French Revolution, 106, 112
Fritz Haag, 270
Friuli, 206
Friuli Grave, 207t
Friuli-Venezia Giulia, 205–208, 207t
Frizzante, 195, 209, 210, 211
Fronsac, 125
Frost danger, 26–28, 160, 169
Fuissé, *145*
Full-bodied wines, food pairing, 95, 97t
Fumé Blanc. *See* Sauvignon Blanc
Fumigation, 24
Fungi, 24
Fungicides, 28
Furmint, 206, 286, 287
Future of wine, 13–14
Futures, 509
Fynbos, 420

G

G7, 224
Gaglioppo, 214
Gaja, Angelo, 192
Gaja, Giovanni, 192
Galen, 8
Galicia, 230–231
Galilee, 291
Galler, John, 336
Gamay
  carbonic maceration, 53
  France, 131t, 143, 145, 146, 171
  Switzerland, 283
Gard, 174
Garganega, 207t, 209
Garnacha, 215, 229, 231, 233–234, 235, 236, 360. *See also* Grenache
Garnacha Blanca, 234
Garonne River, 112, 120, 122, 125
Garrafeira, *225*, 229
Gascony, 112
Gattinara, 191t, 193–194
Gaul, 105, 182
Gavi/Gavi di Gavi. *See* Cortese di Gavi
Geelong Region, 379t
General appellation, 130, 133
Geographe Region, 379t
Geographic Indications (GIs), 377–387, 378t–379t
Geographic Region of Origin, 185
Georg Breuer, *259*, *262*, 272
Georges DuBoeuf, 143
Gerard Tremblay, 136
Germany
  Alsace, 163, 164
  Charta wines, 262
  Classic wines, 260–261
  climate, 263–264
  communes, 530
  First Growth Committee, 261
  influence on Italy, 205–206, 211
  land designations, 257t
  map, *263*
  Selection wines, 260–261
  trade organizations, 262
  trade with England, 251–252
  wine labels, *254*, 259–60
  wine laws, 255–262
  wine regions, 263–270, 272–276
  World War II, 112
Gevrey, 131t, 133, 138
Gevrey-Chambertin, 138, *139*
Gewurztraminer, *36*, 266
  Alsace, 163, 164, 166, 167
  California, 322
  Chile, 408
  Colorado, 361
  South Africa, 438
  Southeastern New England AVA, 358
  Spain, 235
  Tasmania, 387
  Texas, 360
Ghemme, 191t, 193–194
G. H. Mumm, 163
Giacosa, Bruno, 192
Gigondas, 151t, 156
Gippsland Zone, 379t, 382
Gironde, 111, 113
Gisborne region, 389, 390t
GIs (Geographic Indications), 377–387
Givry, 131t, 142, 143
Glassware. *See* Wineglasses.
Glassy-winged sharpshooter, 331
Glen Ellen, California, 317
Glenora Winery, 357
Glenrowan Region, 379t
Globalization, 304
Glycerol, 72
Golan Heights, 291

Gold Rush, 299
Goldtropfchen, *254*, *269*
Gonzalez-Byass, 243
Gorbachev, Mikhail, 287
Goria, Giovanni, 184
Gosset, 163
Goulburn Valley Region, 379t, 382
Gout de terroir, 106
Gouveio, 241
Government warning, 407, 428
Graben, *269*
Graciano, 234
Graft, *20*
Grampians Region, 379t, 382–383
Grand Cordon Rouge, 163
Grand cru, 108, 132, *133*, 135t, 138, 139, 140, 141, 142, 165, 166, 289
Grande Cuvée, 163
Grands Échezeaux, 133, *133*, 139
Grands marques, 160
Grand Vidure (Carmenère), 406
Granite Belt Region, 379t
Gran reserva, *225*, 227, 407
Grape press, *12*
Grauburgunder (Pinot Gris), 265
Grave del Friuli, 207, 208
Graves region, 120–122, 125t
  official classification, 115, 526
  soil, 120, *121*
Great Depression, 108
Greater Perth Zone, 378t
Great growths appellation, 133
Great Plain region, 286
Great Southern Region, 379t, 387
Great Western winery, 382–383
Grechetto, 288
Greco, 214, 214t, 288
Greco di Tufo, 214, 214t, 216
Greece, 6, 148, 180, 288–290
Gregory the Great (pope), 8
Grenache, *37*, 434
  France, 150, 151t, 155, 156, 157, 173, 174, 175, 176t
  Israel, 291
  South Africa, 435, 436
Grenouilles, 135t
Grèves, 142
Grillo, 214t
Groenekloof Ward, 436
Groot Constantia, 423
Grosslagen, 256, 257t
Gruet, Gilbert, 361
Gruet winery, 361
Grüner Veltliner, 281, 286
Guenoc Valley AVA, 320, 322t, 521
Guest dissatisfaction, 466–467
Guigal company, 151, *153*
Gundagai Region, 378t
Gutsabfullung, *254*, 269–270
Gyropalette, *236*

H

Haag, Wilhelm, 270
Hac, 150
Haifa, 291
Halbtrocken wines, 260
Half bottles, 452–453, 454t
  aging wines, 503
  pricing, 492
Hames Valley AVA, 325t, 326, 521
Hamptons, 357–358
Hanepoot, 422, 433
Hanzell Vineyards, 314
Hapsburg Empire, 183, 222, 285
Hardy, 174, 373, 374t
Hargrave Vineyard, 358
Harvest, 30–31, 49, *153*, *183*, *376*, *382*, *398*
Harvey, 239
Hastings River Region, 378t
Haut-Medoc, 115–120, 125t
Hawke's Bay region, *373*, 389, 390t, 393
Hazelnuts, 138
Head training, *25*, *26*, *32*
Health effects of wine, 97–99
Heart disease, 98
Heat retention, 118, 119, 159
Heat stabilization, 62
Heat summation, 302
Heck Cellars, *329*
Hedges Cellars, *342*
Heathcote Region, 379t
Heemskerk, 387
Heidelberg, 275
Hemel en Aarde Valley, 438
Henri II, 168
Henri III, 168
Henri IV, 167, 168, 169
Henri Laroche, 136
Henry II (Henri of Anjou), 10, 105, 111
Henry (prince), 245
Henry VIII, 222
Henty Region, 379t
Hermitage, 431. *See also* Shiraz (Syrah); Syrah
Hermitage (subregion of Côtes du Rhône), 151t, 154
Hessiche Bergstrasse, 275
High Eden Subregion, 378t
High Valley AVA, 320, *321*, 322t, 521
Hill Crest Vineyard, 346
Hilltops Region, 378t
Hochar, Gaston, 291
Hochar, Serge, 291
Hoffman, Rudolf, 271
Hohenmorgan, 274
Holy Roman Empire, 182
Hoppers, 49
House wines, 481–485
Howell Mountain AVA, 121, 311t, 312–313, 521
Huaco Valley, 403t
Hudson River Valley AVA, 357
Hudson's Bay Company, 336
Hue, 77
Hugel, 163
Huguenots, 423
Humidity control, 506, 507, 508
Hundred Years War, 10
Hungary, 285–287
Hunter Region, 378t
Hunter Valley Zone, 378t, *380*
Hybrids, 19–20, 355, 359, 388
Hygrometer, 506

I

Iberian peninsula
  climate, 229
  map, *220*
Ice wine (Canada), 365, 366
IGT (Indicazione Geografica Tipica), 186, 187, 201
Imbottigliato dal produttore all'origine, 189
Impériale, 454t
INAO (Institut National des Appellations d'Origine des Vins et Eaux-de-Vie), 108
Incentives, 474, 491
Indiana, 518
Indicazione Geografica Tipica (IGT), 186, 187
  Supertuscan, 201
INDO (*Instituto Nacional de Denominaciones de Origen)*, 226–227
Inert gas, 486–487
Inglenook, 299
Inniskillin, 362
Instituo do Vinho do Porto, 228
Institut National des Appellations d'Origine des Vins et Eaux-de-Vie (INAO), 108
Instituto da Vinha e Vinho (IVV), 228
Instituto do Vinho da Madeira (IVM), 246
Instituto do Vinho do Porto (IVP), 241
Instituto Nacional de Denominaciones de Origen (INDO), 226–227
Integrated Production of Wine, 420
Interstate Commerce Clause (U.S. Constitution), 513–514
Inventory, 447, 488–489
Ionian Sea, 290
Irrigation, 29, 291, 365, 409, 436
Isabella grapes, 336, 346, 355, 388
Isabella (queen), *222*
Isole e Olena winery, *204*
Isonzo, 207, 207t
Israel, 290–291
Istanbul, 288–289
Italy
  history, 180–184
  map, *181*, *189*, *196*, *205*
  Piedmont, 190–195
  Southern Italy, 213–215
  Tre Venezia, 204–213
  Tuscany, 195–204
  wine labels, 185

wine names, 187–189
wine regions, 189–216, 191t, 207t, 214t
Itata Valley, 403t, 408
IVM (Instituto do Vinho da Madeira), 246
IVP (Instituto do Vinho do Porto), 241, 242
IVV (Instituto da Vinha e Vinho), 228

J

Jack London Vineyard, 317, *318*
Jacques Brothers Winery, 354
Jamestown, Virginia, 359
Jefferson, Thomas, 114, 359
Jerepigo, 431
Jerez, 237–239
Jéroboam, 454t
Jerusalem, 291
Jesuitgarten, 274
Jesuits, 291
Jesus Christ, 7–8
Jewish culture, 7
Jews banned from Spain, 222
J.J. Prum, 270
Joan of Arc, 10
Johannisberg, 272, *273*
Johannisberg Riesling. *See* Riesling
Johnson, Hugh, 194
Johnson, Jim, 360
John XXII (pope), 107
Joly, Nicolas, 172
Jonkershoek Valley ward, 434
Jooste family, 437
Jo Pithon estate, *171*
Joseph Drouhin, 136, 142, 143
J.P. Vinhos, 245
Judean Hills, 291
Juffer, 269
Jujuy region, 411t, 413
Juliénas, 131t, 147
Julius Caesar, 213
Junta Nacional do Vinho, 224

K

Kabinett, *254*, 257, 281
Kadarka, 286
Kamptal, 282
Kanaan, 420
Kangaroo Island Region, 378t
Kékfrankos, 286
Kentucky, 518
Kenwood Vineyards, *61*
Keppoch Subregion, 386
Kirkton, 375
Klein Constantia, 437
Klein Karoo region, 428, 429–430
Klosterneuburg, Austria, 281
Knife of wine key, 459, 460, *461*
Knights Valley AVA, 316t, 319, 521
Koblenz, 270
Kobrand Corporation, 174
Kohler and Frohling, *318*
Ko-operatiewe Wijnbouwers Vereniging (KWV), 423–424, 425, 428, 431–432, 435
Korbel, 299
Kosher wines, 290
Kremstel, 282
Krug, 163
Ksara winery, 291
KwaZulu Natal, 428
KWV (Ko-operatiewe Wijnbouwers Vereniging), 423–424, 425, 428, 431–432, 435

L

Labarde, France, 117
Labels, reading
Australia, 391
Canada, 363
Chile, 407
France, 126
German, 254, 259–60
Italy, 185
New Zealand, 391–392
South Africa, 426–428
Spanish, 225
United States, 307–309
Laboratory analysis, 76, 90
La Chapelle Vineyard, *154*
La Coulée de Serrant, *172*
Lactic acid, 59
Lafite-Rothschild, 401
Lagar, 69
La Grande Dame, 162, 163
La Grande Rue, 140
Lagrein, 207t, 212
Lake Balaton, Hungary, 286
Lake County, California, 319–320
appellations, 322t
map, *321*
Lake Erie AVA, 357
Lake Erie North Shore DVA, 366
Lake Garda, 209
Lake Geneva, *284*
Lakeview Cellars, *366*
Lalande de Pomerol, 126
La Landonne, 151
Lalla Vineyard, 387
La Mancha, 237
Lambrusca, 191t, 207t
Laminated discs, *468*
La Mission, 121
La Mouline, 151
Landwein, 256, 281
Langhorne Creek Region, 378t
Languedoc-Roussillon, 174–176, 176t
Large format bottles, *453*–454, 469, 492–493, 503
La Rioja region, 411t, 412–413
La Romanee, 133, *140*
La Royale, 163
La Tache, 133, 140
Late-bottled vintage (LBV) Port, 243
Late-harvest wines, 68–69
Latisana, 207
Latitude, 21
Latium, 195, 197t
La Vielle Ferme, 157
LBV (late-bottled vintage) Port, 243
Lebanon, 291
Le Cloux, 140
Le Corton, 133, 141, 142
Lees, 55, *58*
Legs, *80*
Le Montrachet, 133
Lenswood Subregion, 378t
Leognan, 120
Léon Beyer, 165–166
Leonetti Cellars, *343*
LeRoy, Baron, 107, 149
Les Amoureuses, 139
Les Caillerets, 142
Les Charmes, 139
Les Clos, 135t
Lescombes family, 361
Les Pépinières, *171*
Lett, David, 346, 347
Lever of wine key, 460
L'Hérault, 174
Libournais, 122–123, 125t
Libourne, France, 122
Liebfraumilch, 253, 265, 273
Liger Belair family, 140
Liguria, 215–216
Limarí Valley, 403t
Lime Kiln Valley AVA, 325t, 327, 521
Limestone Coast Zone, 378t
Lindemans, 376
Liquer de triage, 161
Lison-Pramaggiore, 207
Listrac, 116
Little Karoo region, 428, 429–430
Livermore Valley AVA, 323, 325t, 521
Llana Estacado Winery, 360
Locorotondo, 214, 214t
Lodi AVA, 330, 521
Lodovico, Marchese, 186
Loire Valley, 167–172
history, 168–169
map, *168*
Lombardy, 190, 191t
Loncomilla Valley, 403t
Long, Zelma, 419, 435
Long Island AVA, 357–358
Long Island Wine Council, 358
Lontué Valley, 403t
Loosen, Ernst, 270
Lords Seventeen, 421
Los Alamos Valley, 329
Los Carneros AVA, 310, 311t, 316, 316t, 521
Louisiana, 518
Louis Jadot, *132*, 143, *147*

Louis Latour, 141
Louis Roederer, 387
Louis the German, 163
Lower Mosel, 270
Lower Murray Zone, 378t
Luelling, Henderson, 346
Luján de Cuyo, *412*
Lurton family, 122
Lustau, 239
Luxury item, wine as, 13
Lyon, 134, 149

M

Macabeo, 229
Macarthur, James, 375
Macarthur, John, 375
Macarthur, William, 375
Macedonia, 289, 290
Macedon ranges Region, 379t
Maceration, 53, 119
Mâcon, 143
Mâcon-Lugny, 143
Mâconnais, 131t, 134, 143–144
  map, *144*
Mâcon-Villages, 131t, 143
Macrobrunn, 272
Macroclimate, 21, 23
Madeira, 224, 245–246
Madera AVA, 521
Madiran, 176, 176t
Magaratch, 287
Magnello, Deidre, 216
Magnums, 453, 454t, 469
Maipo Valley, 403t, 405
Malbec
  Argentina, 412
  France, 113, 120, 176, 176t, 177
Malibu-Newton Canyon AVA, 521
Malic acid, 24, 59, 62
Malmsey, 246
Malolactic fermentation, 59, 161, 193, 473
Malvasia, 197t, 198, 200, *204*, 214, 234
Malvasia Fina, 241, 244
Mandela, Nelson, *424*
Manuel II, 224
Manzanilla, 70
Marc, 54
The Marches, 195, 197t
Marconnet, 141
Maréchal Foch, 359, 366
Maremmo region, 200
Margaret River Region, 379t, 387
Margaux, 116, 117, 125t
Margaux, France (village), 117
Markup, 478, *484*, 484, 489–490
  sliding scale, 490–491
Marl, 117, 142, 164
Marlborough region, 389, 390t, *393*, 393–394
Marques, 160
Marsala, 214t, *215*, 215
Marsannay, France, 138
Marsanne, 150, 151t, 153, 174
Marsden, Samuel, 388
Maryland AVAs, 518
Masi winery, 211
Massachusetts AVAs, 518
Mátra Foothills (Mátraalja), 286
Maule Valley, 403t, 405, 406
Maury, 175
Mazuelo (Carignan), 234
Mazzei family, 197
McDowell Valley AVA, 322, 322t, 521
McGuigan-Simeon, 373, 374t
McLaren Vale Region, 378t, 383, 385
McMinnville AVA, 349
Mechanical harvesters, 30, *31*, *382*
Medicinal qualities of wine, 8
Mediterranean climate, 149, 155, 174, 421, 424
Medoc, 111, 120, 125t
  Classification of 1855, 114, 523–524
Meek, William, 346
Melon de Bourgogne, 170t, 172
Mendocino AVA, 322t, 521
Mendocino County, California, 319–322
  appellations, 322t
  map, *321*
Mendocino Ridge AVA, 322, 322t, 521
Mendoza region, *398*, *409*, *410*, 410, 411t, 412, *412*
Menetou-Salon, 170, 170t
Menu board, *445*
Menu, food and wine pairing on, *449*, 450
Menu holders, 497
Mercurey, 131t, 142, 143
Meritage, 363
Merlot, *37*, *114*
  Australia, 385, 386, 387
  British Columbia, 365
  California, 311, 312, 317, 325
  Chile, 406
  France, 113, 117, 119, 120, 123, 124, 125t, 126, 176, 176t, 177
  Hungary, 286
  Israel, 291
  Italy, 207, 207t, 208, 212
  New York, 357, *358*
  New Zealand, 394
  Oregon, 350
  South Africa, 429, 430, 434, 435, 437, 438
  Spain, 235
  Switzerland, 283
  Texas, 360
  Washington, 335, 341, 342–343
Merlot Blanc, 113
Merritt Island AVA, 521
Mesoclimate, 23
Methode Cap Classique, 432, 434, 439
Méthode champenoise, 64–67, 211, 430, 431
  flowchart, *65*
Methuselah, 454t
Methyl bromide, 24
Metodo classico, 190, 211, 212
Metodo tradizionale, 190, 211, 212. *See also Méthode champenoise*
Meursault, 131t, 133
Mexico, 297
Michigan AVAs, 518
Microbes, 62
Microclimate, 23
Middle Ages, 8–9, 105, 129, 182, 204–205, 281
Middle East, 290–291
Middle Mosel, *267*
The Midi, 174–176, *175*, 176t
Miller, Phillip, 336
Minervois, 175, 176t
Minnesota, 518
Minöségi bor, 286
Mis en bouteilles au château, 126, 189
Mis en bouteilles au domaine, 189
Mission grape, 297–298, 301, 360
Mission period, 297–298
Mississippi, 518
Missouri AVAs, 518
Misuse, 489
Mittelrhein, *253*, 274
Moët et Chandon, 163
Moldova (Moldavia), 287
Molinara, 207t, 210
Molise, 195
Monasteries, 8–9, 105, 182, 251, *252*, 289
Monbazillac, 176
Mondavi winery, 401
Mondavi, Robert, *406*
Monferrato, 193
Monis, 435
Montagne de Marne, 159
Montagny, 131t, 142, 143
Montalcino, *202*
Montana winery, 393
Montélimar, 149, 155
Montepulciano, 197t
Montepulciano d'Abruzzi, 195, 197t
Monterey AVA, 325–326, 325t, 521
Monterey County, 323–327
  appellations, 325t
  map, *324*
Montravel, 177
Moorilla Estate, 387
Moors, 8, 221
Morey, 138
Morey-St. Denis, 131t, 133, 138
Morgon, 131t, 148
Mornington Peninsula Region, 379t, 382
Mortality rates, *98*
Moscatel de Setubal, 244–245
Moscatel (Muscat), 244–245
Moscato d'Asti, 191t, 195
Moscato (Muscat), 190, 191t
Mosel River, *266*
Mosel-Saar-Ruwer, 266–270
Moueix, Alain, 419, 433

Moulin-a-Vent, 131t, *147*–148
Moulis, 116
Mount Barker Subregion, 379t, 387
Mount Benson Region, 378t
Mount Lofty ranges Zone, 378t
Mourvèdre
 Franc, 150, 151t, 155, 156, 157e, 173, 174, 175, 176t
 South Africa, 434, 435, 436
Mouthfeel, 79–80, 502
Mt. Carmel, 291
Mt. Harlan AVA, 325t, 327, 521
Mt. Veeder AVA, 311t, 312–313, 521
Mucadelle, 113
Mudgee Region, 378t, 380–381
Müller, Hermann, 265, 283
Müller-Thurgau, 265, 273, 274, 275, 276
 Austria, 281
 Hungary, 286
 Italy, 207t, 212
 Switzerland, 283
Murray Darling Region, 378t, 379t, 382
Muscadel, 422, 433
Muscadelle, 122, 177
Muscadet, 170t, 172
Muscadinia rotundifolia, 19
Muscat, 190, 244–245, 404
 Australia, 382
 France, 151t, 156, 164, 166, 175
 Greece, 290
 Hungary, 286, 287
Muscat Beaumes-de-Venise, 156
Muscat Blanc, 37, *38*
Muscat Blancs à Petit Grains. *See* Muscadel
Muscat Canelli, 360
Muscat d'Alexandrie. *See* Hanepoot
Muscat de Frontignan (Muscadel) 175, 437
Muscat de Lunel, 175
Muscat de Mireval, 175
Musigny, 133, 139
Muslims, 284, 289, 290
Must, 51, *52*
Must pump, *54*

N

Nagambie Lakes Subregion, 379t
Nagle, Charles, 337
Nahe, 274–275
Nantes, 170t, 172
Napa Valley, 306, 310–314, 521
 appellations, 311t
 climate, 306, 310
 map, *313*
Napoleon, 106
Napoleonic Code, 106, 129
Napoleonic Wars, 129, 281, 423
Napoleon III, 114
National Restaurant Association Education Foundation, 531
National Wine Qualifying Institute (Hungary), 286
Navarra, 234–235
Nebbiolo, 190, 191–192, 191t, 192–193, 434
Nebuchadnezzar, 454t
Neckar River Valley, *22*
Negev, 291
Negociants, 107–108, 114, 127, 136, 141, 142, 143, 150, 154, 174, 192, 195
Negroamaro, 214
Nelson region, 390t, 394
Nematodes, 24
Neolithic Period, 180
Neuchatel (Switzerland), 284
New Jersey AVAs, 518
New Mexico, 360–361, 518
New South Wales, 378t, 380–381
New York State
 appellations, 356–358, 518–519
 history, 354–356
 map, *354*
New Zealand
 appellations, 390t
 climate, 372, 387, 389
 exports, 394–395
 history, 388–389
 map, *390*
 wine labels, 391–392
 wine regions, 390t
New Zealand Winegrowers, 531
Neyret-Gachet family, 153
Niagara grapes, 357
Niagara Peninsula, 366
 map, *365*
Niagara Shore DVA, 366
Niederösterreich, 282
Niersteiner, 273
Nietvoorbij (agricultural research center), 433
Noble, Ann, 92
Noble rot, 122, 286
Noble varietals, 104, 150, 164
Nonreactive gas, 486–487
Nonvintage Champagne, 162
Norman Conquest, 10
North Carolina, 519
North Coast AVA, 319–322, 521
North East Victoria Zone, 379t
Northern Cape, 428
Northern Massif, 286
Northern Rhône, 149–150
 appellations, 151–155, 151t
 map, *152*
Northern Rivers Zone, 378t
Northern Slopes Zone, 378t
Northern Sonoma AVA, 316t, 319, 521
North Fork AVA, 357–358
North West Victoria Zone, 379t
North Yuba AVA, 330, 521
Norton grapes, 359
Novella di Teroldego, 212–213
Novello, 212
Nuits-St. Georges, 131t, 133, 140–141

O

Oak, 59–60
 American vs. French, 234
 cork, 464
Oak Knoll AVA, 311, 311t, 521
Oakville AVA, 311–*312*, 311t, 521
Ochagavia, Sylvestre, 400
Ockfen, 267
Oechsle, Ferdinand, 252
Oechsle scale, 252
Oenotia, 180
Off-dry wine, 47, 97t
Office of Champagne USA, 159–160
Ohio AVAs, 519
Oidium (powdery mildew), 183, 388, 423
Okanagan Valley, *364*, 365
Oklahoma, Ozark Mountain AVA, 519
Older wines, restaurants, 510–512
Old Vine, *132*, 307, 309
Olfactory epithelium, 77–78
Olfactory pathway, *78*
Olifants River region, 428, 429
Olite, 235
Olive Farm, 387
Oloroso, 70–71, 238–239
Online auctions, 511
On-premise sales, 447
Ontario, 365–366
 map, *365*
OPAP (Appellation of Superior Quality), 289
OPE (Controlled Appellation of Origin), 289
Open bottles, preserving, 485–488
Opening wine at the table, 460
Oporto, 239, *242*
Orange Region, 378t
Oregon
 appellations, 348t, 519
 climate similar to France, 346
 latitude relative to France, 334
 map, *339*
Oregon Liquor Control Commission, 347
Oregon State University, 347
Oregon Wine Board, 531
Oremus Winery, *285*
Organic viticulture, 33, 320
Organic winemaking, 33
Orlando-Wyndham, 373, 374t
Ornellaia, 186
Orvieto, 195, 197t
Osborne, 243
Osman I, 289
Ottoman Empire, 284, 289
Overberg District, 437–438
Overcropping, 32
Overhead, 489–490
Overpouring, 448
Own-rooted vines, 383
Oxidation, 58, 61, 70, 466

P

Paarl, 420
Paarl District, 434–435
Pacheco Pass AVA, 325t, 521
Padthaway Region, 378t, 383, 386
Paicines AVA, 325t, 327, 521
País, 297, 399–400, 406, 408, 409
Palace Hotel at Bucaco, 244
Palacios, Alvaro, 237
Palestine, 290
Palette, 173
Palomino, 70, 229, 231, 232, 237, 420
Papegaaiberg ward, 434
Paraskevopoulos, Yiannis, 290
Parellada, 236
Paris tasting, 302, 311
Parker, Robert, 117, 118, 127
Paso Robles AVA, 325t, *326*, 327–328, 521
Passito, 204, 209
Passive cellars, 506–*507*
Pato, Luis, 244
Pauillac, 116, 118–119, 125t
Paul Cheneau, 236
Paul Cluver winery, 438
Paul Jauboulet, 154
Pebbly soil, *216*
Pécharmant, 176
Pedro Ximénez, 229, 237, 238–239, 404
Peel Region, 378t
Pelee Island DVA, 366
Peleponnese, 289, 290
Pemberton Region, 387
Penèdes, 235–236
Penfolds Winery, 287, 376, *384*
Peninsulas (Australia), 378t
Pennsylvania AVAs, 519
Pérignon, (Dom) Pierre, 10, *158*–159
Periquita, 229, 245
Perold, Abraham, 431
Perricoota Region, 378t
Persia, 5
Perth Hills Region, 378t, 387
Pesquera, 231
Pessac, 120
Pessac-Leognan, 120, 125t
Petite Sirah, *38*, 301
Petit Verdot, 113, 120
Pfalz, 273–274
Pfersigberg, 166
pH, 21, 379, 501, 501–502
Phenolic compounds, 501–502
Phillip, Arthur, 375
Phoenicians, 5, 177, 221
Phylloxera, 11, *20*, 20, 383
  Argentina, 409
  Australia, 376
  Chile, 400–401
  France, 106, 112, 129–130, 163
  Germany, 253
  Hungary, 285, 286
  Italy, 183
  New York State, 355
  New Zealand, 388
  Oregon, 348
  Portugal, 224
  South Africa, 423
  Spain, 223, 231, 232, 235
  Virginia, 359
Piccadilly Valley Subregion, 378t
Piedmont, 190–195, 191t
  map, *189*
Pierce's disease, 331, 360
Pieropan, 209
Piesport, *269*
Pindar, 358
Pinotage, 431–432, 435, 438
Pinotage Association, 431
Pinot Bianco (Pinot Blanc), 190, 206, 207t, 212
Pinot Blanc, 38, *39*, *265*
  Alsace, 164, 167, 194
  Champagne, 66
  Willamette Valley, 346
Pinot Grigio (Pinot Gris) , 206, 207t, 208, 212, 217
Pinot Gris, 38, *39*, *265*
  Alsace, 163, 164, 166, 167
  British Columbia, 365
  Oregon, 348
  Piedmont, 190
  Russian River Valley AVA, 318
Pinot Meunier, 64, 159
Pinot Noir, *18*, *39*–40, 265, 274, 275, *349*, 431, 432
  Argentina, 413
  Australia, 382, 385, 386, 387
  Austria, 281
  California, 310, 311, 316, 317, 318, 322, 324, 326, 327, 328, 329
  carbonic maceration, 53
  Champagne, 64, 159
  Chile, 405, 408
  clones, 347
  France, 127, 131t, 137–138, 140, 141, 143, 164, 167, 169
  Hungary, 286
  Italy, 190, 191t, 213
  New Mexico, 361
  New Zealand, 393, 394
  Oregon, 345, 346, 347, 348
  sediment, 503
  South Africa, 434, 438
  Southeastern New England AVA, 358
  sparkling wines, *359*
  Switzerland, 283
Piper Heidsieck, 163
Piper's Brook Vineyard, 387
Pisco, *403*, 404
Place settings for tastings, *82*
Plettenberg Bay, 439
Pliny the elder, 8
Pneumatic presses, 55
Pollination, 28
Pol Roger, 163
Polymerization of tannins, 53
Pomace, 54
Pomerol, 124–125, 125t
  listing, 115, 529
Pommard, 131t, 133, 142
Porongurup Subregion, 379t
Port Phillip Zone, 379t
Port-style wines, 69–70, 429–430
Portugal
  climate, 229
  map, *240*
  trade with England, 222, 223
  wine laws, 227–229
  wine regions, 239–246
Portugieser, 266, 273, 281
Port wine, 241–243
  styles, 70t, 241–243
  wine laws, 227
Post offs, 480
Potassium bitartrate, 62, *63*
Potstill, *404*
Potstill brandy, 431–432
Potter Valley AVA, 322, 322t, 521
Pouguet, Miguel, 409
Pouilly-Fuissé, 131t, 143–144, *145*
Pouilly-Fumé, 169–170, 170t
Pouilly-sur-Loire, 169
Pour Discs, 468
Pourriture noble. See noble rot
Pour sizes, 448, 455, 467
  flights, 452
  by the glass wines, 483
  tastings, 480
Powdery mildew, 183, 388, 423
Practical Winery & Vineyard, 531
Pradikät, *254*
Pradikätswein, 281
Prats, Bruno, 419, 433
Pregnancy, 99
Preignac, France, 122
Premier cru, 108, 132, 133, 135, 136t, 138, 139, 140–141, 142, 143
Premier cru (Classification of 1855), 114, 115, 523
Premier cru (Sauternes and Barsac Classification of 1855), 122, 525
Premieres Côtes de Bordeaux, 125
Premier grand cru classe A, 115, 124
Premier grand cru (Sauternes and Barsac Classification of 1855), 122, 525
Presentation to customer, 458–459, 460, *461*
Preserving wine, 485–488
Press fraction, 54
Pressing the skins, 53–55, *347*
Preuses, 135t
Price point
  by the glass wines, 483–484
  wine lists, 492, 496
Pricing, 480, 489–493

Primary fermentation, 59
Primary lees, 57
Primitivo, 214
Primitivo di Manduria, 214
Primogeniture, 106, 129
Priorato, 236–237
Private cellars and restaurants, 511
Problem drinkers, 98
Production statement, 308, 309
Professional Friends of Wine, 531
Prohibition, 11, *301*, 478
  Canada, 362
  Chile, 401
  New Zealand, 388
  United States, 300–301, 337, 346, 355, 359, 360–361, 513
Propriétaire labels, 128, 130, 136, 143, 153
Proprietary names, 308
Prosecco, 207t, 209, 211
Prosecco di Congliano, 211
Prosecco di Valdobbiadene, 211
Provence, 173–174, 176t
Prugnolo. *See* Sangiovese
Prum, Manfred, 270
Pruning, 31–32
Publications, 531
Puckery sensation, 79
Puget Sound AVA, 340t, 343–*344*
Puglia (Apulia), 213, 214, 214t
Puligny-Montrachet, 131t, 133
Pumping over, *52*
Punching down, 51, *52*
Purchasing wine, 508
Pyrenees Mountains climate, 229
Pyrénées Orientales, 174
Pyrenees Region (Australia), 379t

Q

Qualitätswein, 281
Qualitätswein bestimmter Anbaugebiete (QbA), 256
  label, *256*, *257*
Qualitätswein mit Pradikät (QmP), 256–259
Quality wines, Spain, 227
Quarterly Review of Wines, 531
Quarts de Chaume, 170t, 172
Queensland Zone, 379t
Quercetin, 98
Quercus (oak), 59–60, 464
Quincy, 170, 170t
Quinta Normal, 400
Quintas, 69, *225*
Quota systems, 423

R

Racking, 55, *58*, 62, 161
Racking (storage), 488–*489*, 508–*509*
Radda, 185
Raisining, 6, 204, 209
Ramandola, 207t, 208
Ramona Valley AVA, 521
Rapel Valley, 403t, 405–406
Raulí wood, 404
Raventós, José, 235
Recioto di Soave, 207t, 209
Recioto di Valpolicella Amarone, 211
Red and white wine, differences, 55–56
Red Hills AVA, 320, 322t, 521
Red Mountain AVA, 340t, 342–343
Redondo, 245
Red Willow Vineyard, *335*
Red wine, effects of age, *78*, 505
Red winemaking, *49*–55, 145–146
Redwood Valley AVA, 322, 322t, 521
Reed, Bob, 360
Refilling glasses, 468–469
Refosco, 207, 207t
Refrigeration, 458
Regional appellation, 130, 133
Region of origin, 185, *493*–*494*
Régnié, 131t, 148
Rehoboam, 454t
Renaissance, 10, 182–183
René Barbier, 237
René Dauvissat, 136
Renosterveld, 420
Rental space, 507
Reserva, *225*, 227, 229, 407
Reserva especial, 407
Réserve, 289
Reserve, 307
Réserve St. Martin, 174
Restaurants and older wines, 510–512
Resveratrol, 98
Retailers, 478–479
Reuilly, 170, 170t
Reuter, Earnest, 346
Rheingau, 270, 272
Rheinhessen, 273
Rheinpfalz. *See* Pfalz
Rhine Riesling, 385, 386. *See also* White Riesling; Riesling.
Rhode Island, 519
Rhône Rangers, 531
Rhône Valley
  aging wines, 504
  history, 149
Rías Baixas, 230–231
Ribeiro, 231
Ribera del Duero, *226*, 231
Ribolla, 290
Ribolla Gialla, 207, 208
Richard the Lion Hearted, 10
Richebourg, 133, 140
Riddling, 66, *161*, *236*
Riesling, *40*, 252, 265, *269*, 275
  Australia, 382, 385
  Austria, 281
  British Columbia, 365
  California, 322, 328
  Colorado, 361
  France, 163, 164, 166, 167
  Germany, 271, 272, 274, 282
  Hungary, 286
  New York State, 356, 357
  New Zealand, 394
  North Fork AVA, 358
  Ontario, 366
  South Africa, 437, 438
  Southeastern New England AVA, 358
  Spain, 235
  Switzerland, 283
  Texas, 360
  Virginia, 359
Right Bank (Libournais), 122–123, 125t
Rioja, *225*, 232–234
  map, *232*
Rio Negro region, 411t, 413
Riserva, 188
River Elbe, 276
Riverina Region, 378t, 381, 383
River Junction AVA, 522
Riverland Region, 378t
River Neckar, 275
Rivesaltes, 175
Robertson district, 430, *432*–*433*
Robola grape, 290
Rockpile AVA, 316t, 319, 522
Rodney Strong Winery, *302*
Roero, 191t
Rogue Valley AVA, 348t, 350
Romagna, 190
Romanée-Conti, 133, 140
Romanée-St. Vivant, 140
Roman Empire, 6–8, 105, 129, *140*, 149, 181–182, 197, 221, 251
Rondinella, 207t, 209, 210
Room temperature, 458
Rootstocks, 11, 20. *See also* phylloxera
Rosé, 155, 157, 162
Rosé d'Anjou, 171
Rosengarten, 272
Roseworthy Agricultural College, 383
Rosso di Montalcino, 197t, 202
Rotaliano, 212
Rotary fermentors, 52, *53*, *374*
Roussanne, 150, 153
  Southern Rhône, 156
Route du Vin, 115, 118
Ruby Port, 242
Rudesheim, *272*
Rueda, 232
Rugiens, 142
Rully, 131t, 142, 143
Russell, Tim Hamilton, 438
Russia, 287
Russian River Valley AVA, *23*, 316t, 317–318, 522
Rust, Austria, *282*
Rutherford AVA, 311–312, 311t, 522
Rutherglen Region, 379t
Ruwer, 268–269
Ryman, Henry "Nick", 176, 285
Ryman, Hugh, 285

S

Saale-Unstrat, 276
Saar, 267
Saccharomyces cerevisiae, *47*
Sachsen, 276
Sagrantino, 196, 197t
Sagrantino di Montefalco, 196, 197t
Sakonnet Vineyards, 358
Salado Creek AVA, 522
Salazar, Antonio de Oliveira, 224
Sales, on-premise, 447
Salice Salentino, 214, 214t, 216
Salmanazar, 454t
Salta region, 411t, 412–413
Samaria (Shomron), 291
Samples, 452
San Benito County, 323, 327
  appellations, 325t, 522
San Bernabe AVA, 325t, 326, 522
Sancerre, 170, 170t
Sandeman, 239
Sandrone, Luciano, 192
Sanel Valley, 322
San Francisco Bay AVA, 323, 522
San Gimignano, 203
Sangiovese, *40*, 197t, 434
  California, 314
  South Africa, 430
  Italy, 186, 196, 198, 200–202, 203
  Texas, 360
  Washington, 342–343
Sangiovese Grosso. *See* Brunello
San Joaquin Valley, *329*
San Juan Capistrano, 297
San Juan region, 411t, 412
Sanlúcar de Barrameda, 70
San Lucas AVA, 325t, 326, 522
San Luis Obispo County, *326*, 327–328
  appellations, 325t
San Pasqual Valley AVA, 522
San Rafael subregion, 412
Santa Barbara County, *326*, 328–329
  appellations, 325t
Santa Clara County
  appellations, 325t
  map, *324*
Santa Clara Valley AVA, 323, 325t, 522
Santa Cruz County
  appellations, 325t
  map, *324*
Santa Cruz Mountains AVA, 323, 325t, 522
Santa Lucia Highlands AVA, 325t, 326, 522
Santa Maddalena, 207t, *212*
Santa Maria Valley AVA, 325t, 328, 522
Santa Rita Hills AVA, 325t, 329, 522
Santa Rita Winery, *400*, *414*
Santa Ynez Valley AVA, 325t, 328, 522
Santenay, 142
Santorini, 289, 290
San Ysidro AVA, 325t, 522
Sardinia (Sardegna), 213, 214t, 215–216
Sassicaia, 186, 200–201
Saumur, 171
Saumur Mousseux (Saumur d'Origine), 171
Sauternes, 68, *108*, 122, 125t
  aging, 504
  Classification of 1855, 115, 122, 525
Sauternes, France (village), 122
Sauvignon Blanc, *25*, 40–*41*
  Argentina, 413
  Australia, 385, 386, 387
  British Columbia, 365
  botrytized, *68*
  California, 311–312, 317, 318, 319, 320, 322, 328
  Chile, 405, 406
  effects of age, 77
  France, 120, 122, 113, 125, 125t, 169, 170t, 171, 174, 176t, 177
  Hungary, 286
  Israel, 291
  Italy, 206, 207t, 208
  New Zealand, 393, 394
  Spain, 232, 235
  South Africa, 420, 429, 430, 434, 437, 438–439
Savennières, 170t, 171–*172*
Savigny-les-Beaune, 141
Scavino, Paolo, 192
Schanno, Charles, 336
Scharzberg, 267
Scharzhofberg, 267, *268*
Scheurebe, 273
Schiava, 207t, 212
Schieferton, 266
Schiller, Johann, 362
Schioppettino, 207
Schloss Johannisberg, 272, *273*
Schloss Vollrands, 272
Schmeisser, Tom, 287
Screw caps, *464*, 465
Scuppernong grapes, 359
Seasons and wine releases, 508
Secco, 188
Second growth (Classification of 1855), 114, 161, 523
Sediment, 469–471, 501, 503, 511
Seguin, Gerard, 116–117
Segura Viudas, 236
Seiad Valley AVA, 522
Sekt, 267, 273
Selbach, Johannes, 270
Selbach-Oter, 270
Sélection de Grains Noble (SGN), 165, 167
Selling points, 474–475
Semillon, *114*
  Australia, 380, 384, 387
  France, 113, 120, 122, 125t, 173, 176t, 177
  Israel, 291
  South Africa, 430, 434, 437
Sending back wine, 466–467
Sensory evaluation, 76–80
  difficulties in, 90–91
  and tastings, 81
  and wineglasses, 455
Seppelts, 382
Seppeltsfield, 384
Sercial, 246
Serra, Junípero, 297, *298*
Server suggestions, 449–450
Service, importance of, 446
Service linens, 460
Servings per bottle, 392
Serving temperatures, 458
Setúbal, 244–245
Seven Hills Vineyard, *341*
Sèvre et Maine district, 172
Seyval Blanc, 357, 359
Sforzato, 191t
Sforzato di Valtellina, 190, 191t
SGN (Sélection de Grains Noble), 165, 167
Shale, *267*
Shatter, 28
Sherry, 70–71, 71t, 222, 223, 237–239, 435
Shesgreen, Sean, 216
Shipping wine, 515
Shiraz (Syrah), 41, *385*
  Australia, 380, 382, 383, 384, 385, 386, 387
  South Africa, 420, 429, 430, 431, 432, 434, 435, 436, 438–439
Shoalhaven Coast Region, 378t
Shomron (Samaria), 291
Sicily, 213, 214t, 215, *216*
Sideways (movie), 485
Siena, Italy, 199
Sierra Foothills AVA, 330, 522
Sight, 77, 86
Signboard, *445*
Silicon Valley, 323
Silt, 21
Silvaner, 265, 275
  Tre Venezia, 207t
Similkameen Valley, 365
Simonsberg-Paarl Ward, 435
Simonsberg-Stellenbosch ward, 434
Sinkoff, Martin, 174
Skins, removal of, 53–55
Slate, 266–267, *267*, 272
Sliding scale pricing, 490–491
Smart, Richard, 116–117
Smell, 77–79
Soave, 207t, 208–209
Soave Classico Superiore, 209
Soave Superiore, 207t
Society of Wine Educators, 531
Sogrape, 244
Solaia, 186
Solano County Green Valley AVA, 522
Solera, 71, 237–238, *239*, 435
Sommelier, 444–445
Sommer, Richard, 346

Sonnenberg, 272
Sonnenuhr, 269, *270*
Sonoma Coast AVA, 316t, 319, 522
Sonoma County, 314–319
  appellations, 316t
  climate, 316
  map, *315*
Sonoma Green Valley AVA, 316t, 318, 522
Sonoma Mountain AVA, 316t, 317, *318*, 522
Sonoma Valley AVA, 316–317, 316t, 522
Sopron, Hungary, 286
Soussans, France, 117
South Africa
  climate, 424–425
  exports, 419
  history, 421–424
  map, *418*, *425*
  trade with England, 423
  wine labels, 426–428
  wine regions, 424–439
South African Port Producers' Association, 430
South Australia, *376*, 383–386
  map, *383*
South Burnett Region, 379t
South Coast AVA, 522
South Coast Zone, 378t
Southcorp, 373, 374t
South Eastern Australia, 378t
Southeastern New England AVA, 358
Southern Fleurieu Region, 378t
Southern Flinders ranges Region, 378t
Southern Highlands Region, 378t
Southern Italy, 213–215, 214t
Southern New South Wales Zone, 378t
Southern Oregon AVA, 350–351
Southern Rhône, 149–150, 155–157
  appellations, 151t
  map, *155*
Southern Valley region, 401, 403t, 408
South of France, 173–177
  appellations, 176t
South West Australia Zone, 379t
Southwest (of France), 176–177, 176t
Soviets, 285, 287
Spain
  climate, 229
  map, *230*
  trade with England, 222
  wine laws, 226–227
  wine label, 225, 226
  wine regions, 230–239
  wine trade, 222–223
Spanish Civil War, 223
Spanish Inquisition, 222
Spanna (Nebbiolo), 191t, 193–194
Sparkling winemaking, 64–67
Sparkling wines
  Argentina, 413
  Asti, 190, 194–195
  bottle sizes, 454t
  California, 322
  Cava, 235–236
  Champagne, 159–163
  France (non-Champagne), 143, 167, 171
  food pairing, 95
  Italy, 190, 195, 209, 210, 211, 212
  Methode Cap Classique, 432, 434, 439
  New Mexico, 361
  New Zealand, 394
  opening, 469
  South Africa, 430, 431, 432, 434, 439
  Australia, 383, 387
Spätburgunder (Pinot Noir), 265, 274, 275, 281
Spätlese, *254*, 257–258, 281
Special pricing, 480
Speranza winery, *215*
Spitting, 481
Splits, 452–453, 492
Spoilage, 62
Spring Mountain AVA, 311t, 312–313, 522
Sprinklers, *27*, 160
Spumante, 212
Spur pruning, *25*, *26*, 31, *32*
Stability, 62
Staff favorites on wine lists, 496–497
Staff meetings, 472–473
Staff training, 449–450, 472–475
Stags Leap AVA, 311, 311t, 522
St. Amour, 131t, 147
Standardization of wines, 13
State law, 473, 478, 514–517
  BYOB, 455
St. Aubin, 142
St. Charles Winery, 337
St. Chinian, 175
St. Clair Winery, 361
Ste. Geneviève winery, 360
Stelermark (Styria), 282
Stellenbosch, 420, 422, *423*, 433–434
Stelvins, *464*
Ste. Michelle Vintners. *See* Château Ste. Michelle
St. Émilion, 123–124, 125t
  classification, 115, 527–528
Stemmer-crushers, 49–51
St. Estèphe, 116, 119–120, 125t
St. Helena AVA, 311–312, 311t, 522
Still wine, 46
Stimson Lane Vineyards & Estates, 337
St. Joseph, 151t, 153
St. Julien, 116, 117, 125t
Stone Age, 180
Stone House Winery, 336
Storage cabinets, 507–508
Storing wine, 450–451, 458, 480, 488–489, 507
Strathbogie Ranges Region, 379t
Straw flasks, *199*
St. Urbans-hof, *254*, 270, 271
St. Véran, 131t, 144
Style of wine, *495*, 504–505
Styria (Stelermark), 282
Sugar
  adding, 379
  Alsatian wines, 165
  effect on fermentation, 48
  late-harvest wines, 68
  residual, 62
Suggestive selling, 447, 448, 450
  wine storage
Suisun Valley AVA, 522
Sulfite declaration, 308, 392, 407, 426–427
Sulfur, 28, 33, 423
Sulfur dioxide, 33
Sultana, 424
Sunbury Region, 379t
Superiore, 188
Supertuscan, 186, 200–201
Supreme Court (U.S.) , 479, 514
Sur lie, 58
Sur pointe, 66
Sustainable viticulture, 33
Swan District Region, 378t, 387
Swan Hill Region, 378t, 379t
Swan Valley Subregion, 378t
Swartland, 436, 437
Sweet spot for pricing, 492
Swirling wine, 86–87
Switzerland, 282–284
Sylvaner, 164, 283
Syrah, *41*
  California, 328
  France, 150, 151t, 153, 154, 155, 156, 157, 173, 175
  Oregon, *350*
  Spain, 237
  Texas, 360
  Washington, *335*, 341, 342–343

T

Table Mountain, *421*
Table setting, 448–449
Table tents, *448*
Table wine, 46
  U.S. vs European terminology, 258
Tafelwein, 255–256, 281
Taittinger, *161*, *162*, 163
Talmud, 290
Tank presses, *56*
Tannat, 176, 176t, 177, 430
Tannins, 13, 53
  aging wines, 192, 501, 502
  astringency, 79–80
  Barolo, 192
  and fining, 62
  Pinotage, 432
  St. Julien, 117
  Syrah, 150
Tarn River, 112
Tarragona, 236
Tartaric acid, 59, 62, 156
Tart wines, food pairing, 95, 97t

Tasmania, 379t, 387
Taste, 79
Taste test, 465–*466*
Tasting notebook, 89
Tastings
  analytical wine tasting, *82*
  blind, *84*
  evaluation, 86–89
  fatigue, 91
  industry, 480–481
  order of sampling, 85, 468
  place settings, 81–*82*
  public, *81*
  staff training, 449–450, 472–473
  swirling wine, 86–87
  walk-around, 480, *481*
Tasting sheets, 89, *90*, 531
Taurasi, 213–214, 214t, 216
Tavel, 151t, 156–157
Tawny Port, 242, *243*
Taxes, 513
Taylor Fladgate & Yeatman, 243
TCA (trichloroanisole), 464–465
Tchelistcheff, André, 337
Tears, *80*
Technology, 10–11, 13–14, 373
Temecula AVA, 330–331, 522
Temecula Valley AVA
Temperance, 376, 388. *See also* Prohibition.
Temperature
  aging, 502–503
  cellar, 505–506
  fermentation, 47, 58
  passive cellar, 507
  perception of wine, 79
  room temperature, 458
  serving, 458
  storage cabinets, 507–508
  white grapes, 56
  and wine storage, 450, 502–503, 505–506, 507–508
Tempranillo, 41, *42*, 229, *233*, 235, 237, 360, 434
Tennessee, 519
Teno Valley, 403t
Tenth, 454t
Teroldego, 207t
Teroldego Rotaliano, 207t, 212
Terra Montosa label, *262*
Terra rossa, *385*, 386
Terroir, 22, 106–107, 113
Terror Creek Winery, 361
Tête de cuvée, 162, 163
Texas, 360, 519
Texas High Plains AVA, 360, 519
Texas Hill Country AVA, 360, 519
Texas Tech, 360
Texture, 79–80
Theft, 489
Thermometer, 506
Thirty Years' War, 163, 252
Thomases, Daniel, 193
Thorpe, Peter, 439
Thrace, 289
Three-tiered system of distribution, 478, *479*
Three V's, 458–459, *461*
Threshold, 90–91
Ticino, Switzerland, 283, 284
Tight wines, 471
Tignanello, 186, 200–201
Tinta Barocca, 430
Tinta Cão, 69
Tinta Negra Mole, 246
Tinta Roriz (Tempranillo), 41, *42*, 69, 229, 231, 240–241
Tinto Fino. *See* Tempranillo
Tirage, 66
Tocai, 207t, 208
Tocai Friulano, 206–207, 209
Tokaj-Hegyalja, 286
Tokaji, *285*, 504
Tokaji Aszú, 206, 285
Tokara winery, *433*
Tokay, 382
Topography, 21, *22*
Torgiano Rosso Riserva, 195–196, 197t
Torontel (Torrontés), 404, 412, 413
Torres, Jaime, 235
Torres, Juan, 235
Torres, Miguel, 235
Torres, Miguel, Jr., 235
Torres winery, 401
Torrontés, 231, 404, 412, 413
Touraine, 169, 170–171, 170t
Touraine Mousseux, 170
Touriga Francesca, 69, 229
Touriga Nacional, 69, 229, 240–241, 244, 430
Tours, France, 170
Tradouw ward, 430
Training staff, 472–475
Traminer, 212
Transdanubia region, 286
Transfer method, 67
Treaty of Methuen, 223
Treaty of Windsor, 223
Trebbiana Toscano, 209
Trebbiano, 196, 197t, 198, 204, 207t, 209
Trebbiano di Soave, 209
Treixadura, 231
Trellising, 6, 24–25, 28, 235
Trentino, 207t, 212
Trentino-Alto Adige, 207t, 211–213
Tre Venezia, 204–213
  wine regions, 207t
Trichloroanisole (TCA), 464–465
Tri-Cities region, 341
Trier, 251, 267
Trimbach, 166
Trincadeira Preta, 245
Trinity Lakes AVA, 522
Trockenbeerenauslese (TBA), 68, *254*, 258–259, 281
  label, *259*
Trocken wines, 260
Truffle, 139
TTB (Alcohol and Tobacco Tax and Trade Bureau), 303, 514, 516
Tucumán region, 411t, 413
Tufo, Italy, 214
Tulbagh, 435–436
Tumbarumba Region, 378t
Tupungato subregion, 412
Turks, 182, 285, 289
Tuscany, 183, 195–204, 197t
  map, *196*
Tutuvén Valley, 403t
Twenty-First Amendment, 301, 513, 514
Twist top closures, *464*, 465
Two-curtain trellis, *25*
Tygerberg District, 437
Tyrrhenian Sea, 215

U

U.C. Davis, 301, 346
  Department of Viticulture and Enology, 531
Uco Valley, *410*
Ugni Blanc
  Bordeaux, 113
  Entre-Deux-Mers, 125
  Southern Rhône, 151t, 157
  South of France, 173, 174
Ukiah Valley, 322
Ukraine, 287
Ullage, 61, 503, *504*
Umami, 79
Umbria, 195, 197t
Umpqua Valley AvA, 348t, 350
Undercropping, 32
Unique selling points (USP), 474–475
United States
  Department of Agriculture, 338
  wine labels, 307–309
  wine laws, 513–517
University of Adelaide, 383, *384*
University of California at Davis, 301, 346
  Department of Viticulture and Enology, 531
University of Stellenbosch, 421, 433
University of Texas, 360
Upper Goulburn Region, 379t
Upper Loire, 169, 169–170, 170t
Upper Mosel, 267
Urzig, 268
USP (Unique selling points), 474–475
U.S.S.R., 287
U-system, *25*

V

Vacqueryas, 151t, 156
Vacuum systems, 488
Valais, Switzerland, 283
Val d'Orbieu, 174

Valée de la Marne, 159
Valence, 149
Valladolid, 231, 232
Valle Central (Central Valley region, Chile), 403t
Valle de, Valle del. *See* specific valley names
Valle del Sur (Southern Valley region, Chile), 403t
Vallejo, Mariano G., 298–299, 319
Valley of the Moon Winery, *64*, *330*
Valley View Vineyard, 346
Valmur, 135t
Valpolicella, 183, 207t, 208–209, 210
Valpolicella Classico, 210
Valpolicella Classico Superiore, 210
Valtellina subzone, 190
Valtellina Superiore, 190, 191t
Value, perceived, 455, *456*, 490
Vancouver Island, 365
Van der Stel, Simon, 422, 437
Van Riebeeck, Jan, 421, 422
Varietal
  cool-climate, 264–265
  wine lists, *494–495*
  as wine name, 12
Vaudésir, 135t
Vaud (Switzerland), 283
VDP (Verbands Deutscher Pradikäts-und-Qualitätsweinguter), *254*, 262, 271
VDQS (Vin delimite de qualite superieure), 108, 109
Vega Sicilia, 231
Vegetative growth, 21, 25
Venantius Fortunatus, 251
Vendage Tardive, 165, 167
Veneto, 207t, 208–211
Venice, 204–205
Véraison, *29*–30, 117
Verbands Deutscher Pradikäts-und-Qualitätsweinguter (VDP), 254, 262, 271
Verdejo, 229, 232
Verdelho, 241, 246
Verdello, 214t
Verdicchio, 197t, 214, 214t
Verduzzi, 207t
Verduzzo di Ramandola, 208
Vermentino, 214t, 215, 216
Vermentino di Gallura, 214t, 215, 216
Vermentino di Sardegna, 214t
Vernaccia, 197t
Vernaccia di San Gimignano, 184, 187, 197t, 203
Vernaccia di Serrapetrona, 195, 197t
Veuve Cliquot, 162, 163
Victoria, 379t, 381–383
  map, *381*
Vidal Blanc, 358, 359, 366
Vielles Vignes, *132*
Vienna, Austria, 282
Vieux Château Certan, 125
Vignes, Jean-Luis, 298
Vilafonté, 435
Viña Caliterra Winery, *399*
Vin de Constance, 437
Vin delimite de qualite superieure (VDQS), 108, 109
Vin de pays, 108, 109, 174, 175–176, 177, 289, 290
Vin de Pays d'Oc, 174
Vin de pays du jardin de France, 167
Vin de table, 108, 109–110
Vin doux naturel, 156, 175
Vinegar, 61
Vineyard & Winery Management Magazine, 531
Vineyard designation, 309
Vinho de mesa, 228
Vinho regional, 228
Vinho Verde, 229, 239
Vino comarcal, 226
Vino da tavola, 185, 187, 201
Vino de Cosecha, 227
Vino de la tierra, 227
Vino de mesa, 226
Vino Nobile de Montepulciano, 196, 197t, 202
Vin ordinaire, 109–110, 174, 175, 177
Vin Santo, 197t, 203–204
Vintage and ageability, 503–504
Vintage Champagne, 162
Vintage Keeper, 508
Vintage Port, 242–243
  aging, 504
Vintage year, 30, 126, 185, 391, 407, 426
Vintners Quality Alliance (VQA), 362, 363, 366
Viognier, 41, *42*, 434
  France, 150, 151t, 153
  South Africa, 430, 435, 436
  Texas, 360
Virginia, 359, 519–520
Viscosity, 79, *80*
Visigoths, 221
Viticulture, 6–8, 9, 24–26
  Austria, 281–282
  Champagne, 159–160
  monks, 129
  organic, 33, 172
  sustainable, 33
Vitis aestivalis, 355
Vitis labrusca, 19, 355, 362, 364–365
Vitis riparia, 355
Vitis rotundifolia, 19
Vitis vinifera, 336, 337, 346, 355–356, 364, 399, 400
  clones, 19
  Virginia, 359
Viura, 229, 234, 23, 2365. *See also* Macabeo
Voerzio, Roberto, 192
Volatile compounds, 78–79, 455
Volatility, 77
Volcanic rock, 164, 212, 213, 214, 286
Volnay, 131t, 133, 142
Volstead Act (Eighteenth Amendment), 11, 300, 513
Volume, 126, 185, 309, 407
Von Othegraven label, *256*, *257*
Von Pessls, Edward and John, 346
Von Simmern, Langwerth, 272
Voor Paardeberg Ward, 435
Vosges mountains, 163
Vosne-Romanee, 131t, 133, 140
Vougeot, 131t, 133, 139
Vouvray, 170t, 171
  aging, 504
Vouvray Moelleux, 171
Vouvray Mousseux, 171
Voyatzis, Yannis, 290
VQA (Vintners Quality Alliance), 362, 363, 366

W

Wachau, 282
Waikato/Bay of Plenty region, 390t, 394
Waipara region, 390t, 394
Wairarapa/Wellington region, 390t, 394
Walk-around tastings, 480, *481*
Walker Bay District, 437–438
Walla Walla Valley AVA, 334, 336, 340t, 343, 348t, 351
War of Independence (Israel), 290
Warre, 243
Washington
  appellations, 340t, 341–345, 520
  climate, 339–341
  history, 336–338
  latitude relative to France, 334
  map, *339*
  Prohibition, 337
  Tri-Cities region, 341
Washington State Liquor Control Board, 337
Washington State University, 337
  Research Station, 342
Washington Wine Commission, 338, 531
Washington Wine Institute, 338
Washington Wine Quality Alliance (WWQA), 338
Water tables, 116
Wax coating, 461, *462*
Weed control, 28
Weil, Robert, 272
Weingut Dr. Loosen, 270
Weis, Nik, 262, 270, 271
Weissburgunder (Pinot Blanc), 38, 265, 276
Wellington Ward, 435
West Australian South East Central Zone, 379t
West Bank, 291
Western Australia, 378t–379t, 386–387
  map, *386*
Western Cape, 428
  map, *425*
Western Plains Zone, 378t
Western Victoria Zone, 379t, 382

Westport Rivers Winery, 358, *359*
West Virginia AVAs, 520
White and red wine, differences, 55–56
White Port, 243
White Riesling, 341, 342. *See also* Riesling
White wine, effects of age, 77, 505
White winemaking, 55–58, *57*
Whole-cluster pressing, 56
Whole cluster pressing, *347*
Wholesalers, 478, 479
Wild Horse Valley AVA, 311t, 314, 325t, 522
Willamette Valley AVA, *335*, 348–350, 348t
William Fèvre, 136
Willow Creek AVA, 522
Winch, *153*, *264*
Wind machines, *27*
Wine & Spirits Wholesalers of America, 531
Wine and food pairing, 94–95, 97t, *449*–450
  experiment in, 96–97
Wine Aroma Wheel, *92*
Wine availability, between states, 478
Wine Brats, 531
Wine Business Monthly, 531
Wine chemist, 76
Wine coaster, 460
Wine composition, *72*
Wine descriptors, 91, 93–94t
Wine Enthusiast, 531
Wineglasses, *83*, 455–458
  breakage, 456
  carrying, 458
  crystal, 83–84
  durability, 456
  polishing, *457*–458
  refilling, 468–469
  replacing, 466
  setting, 458
  size, *456*
  styles, 456
  table setting, 448
  for tastings, 82–84
  varietal specific, 455
Wine Institute, 301, 516, 531
Wine keys, *459*–463
Wine labels
  Australia, 391
  Canada, 363
  Chile, 407
  France, 126
  German, 250–251, 259–60
  Italy, 185
  New Zealand, 391–392
  South Africa, 426–428
  Spanish, *225*
  United States, 307–309
Wine laws
  France, 108–110
  Germany, 255–262
  Greece, 289–290
  Portugal, 227–229
  Port wine, 227
  Spain, 226–227
  United States, 513–517
Wine lists
  blends, 495
  choosing wines, 479–480, 482–485
  country of origin, *493*–494
  descriptions, 496
  displays, 445
  format, 497
  by the glass wines, 482–485
  organization, 493–497
  price point, 496
  pricing, 491–492
  region of origin, *493*–494
  staff favorites, 496–497
  staff training, 473
  style of wine, *495*
  and USPs, 475
  varietal, *494*–495
  vintage, 459
Wine Lovers Page, 531
Winemaking, organic, 33
Wine notes, 308, 363, 392, 407, 426–427
Wine of Origin Scheme, 425, 426, 428, 438
Wine organizations, 531
Wine presses, *139*, *252*
Wine rooms, 450–*451*
Wines & Vines, 531
Wine sales, 478
Wines of Chile, 531
Wine Spectator, 531
Wine steward. *See* sommelier
Wine storage, 450–451, 507–508
Wine tanks, *63*, *261*
Winterkill, 21, 340, 350
Winter vines, *26*
Wire cage, 469, *470*
Wisconsin, 520
Wölffer Estate, 358
Woltner, Frederic, 121
Woltner, Henri, 121
Women's health, 98–99
Woodward Canyon Winery, 343
Worcester district, 430
World's Fair, 114
World War I, 107, 300
World War II, 112, 163, 184–189, 287, 337, 376
World wars, 253
Worm of wine key, 460, 461–*462*, *463*
Württemberg, 275–276
Würzgarten, 268
WWQA (Washington Wine Quality Alliance), 338
Wynberg, 422

X

Xarel-lo, 236

Y

Yakima Valley AVA, *335*, 336, 340t, 342
Yamhill-Carlton AVA, 348t, 349
Yamhill County, Oregon, 349
Yarra Valley Region, 379t, *382*
Yeast, 47, 48
Yields, 26
York Mountain AVA, 325t, 328, 522
Yorkville Highlands AVA, 322, 322t, 522
Yount, George C., 299
Yountville AVA, 311, 311t, 522

Z

Zell, 270
Zinfandel, *24*, *25*, *42*, 434
  California, 312, 313, 314, 317, 318, 319, 325, 328, 330
  forbear, 214
  harvest, *30*
  Oregon, 346
  Texas, 360
Zinfandel Advocates and Producers (ZAP), 531
Zinfandel Port, 69–70
Zitsa, 289, 290
Zonin, *194*, *210*, 211, 216, 359